1992

Biology of
Wastewater Treatment

'To you it's just crap, to me it's bread and butter.'
Spike Milligan
Recollections of the latrine orderly

Biology of Wastewater Treatment

N. F. GRAY
Trinity College
University of Dublin

OXFORD   NEW YORK   TOKYO
OXFORD UNIVERSITY PRESS

Oxford University Press, Walton Street, Oxford OX2 6DP
Oxford New York Toronto
Delhi Bombay Calcutta Madras Karachi
Petaling Jaya Singapore Hong Kong Tokyo
Nairobi Dar es Salaam Cape Town
Melbourne Auckland
and associated companies in
Berlin Ibadan

Oxford is a trade mark of Oxford University Press

Published in the United States
by Oxford University Press, New York

First published 1989
First published in paperback (with corrections) 1992

British Library cataloguing in Publication Data
Gray, N. F.
Biology of wastewater treatment.
1. Waste water. Biological treatment
I. Title
628.3'51
ISBN 0–19–859014–8
ISBN 0–19–856370–1 (Pbk)

Library of Congress Cataloguing in Publication Data
Gray, N. F.
Biology of wastewater treatment/N. F. Gray.
p.   cm.   Bibliography: p.   Includes index.
1. Sewage—Purification—Biological treatment.   2. Sewage
disposal. 3. Biotechnology. p.   I. Title.
TD755.G73   1989   628.3'51—dc19   88–37713
ISBN 0–19–859014–8
ISBN 0–19–856370–1 (Pbk)

Typeset by
Cotswold Typesetting Ltd, Gloucester
Printed in Great Britain by
Bookcraft (Bath) Ltd, Midsomer Norton,
Avon

CONTENTS

PREFACE

Traditionally, the areas of water pollution and pollution control have been dominated by chemists and engineers, in particular civil and chemical engineers. However, in recent years there has been an enormous increase in the number of biologists, especially microbiologists, entering the water industry, so much so that the water industry has become a major employer of graduates and technicians in the biological and environmental sciences.

The overall aim of this text is to provide the reader with a unified and coherent course in the biological aspects of wastewater treatment. It deals with the operational management of processes rather than their design and construction, and examines the role of the biologist in the water industry. The book is intended to provide a clear and comprehensible account of wastewater treatment for non-engineers, and also to explain the theoretical basis of the biological processes in wastewater treatment to engineers. While dealing primarily with biological wastewater treatment, other major areas of interest to biologists are also discussed, for example, sludge disposal, public health, and the reuse or transformation of waste. Special attention is paid to biotechnology.

This is an introductory course, in that every attempt has been made to make it self-explanatory. All the terms used are explained in the text. If any difficulty with them is encountered, then the excellent technical dictionary *Glossary of terms used in water pollution control*, published by the Institute of Water Pollution Control (now the Institute of Water and Environmental Management), can be highly recommended.

In a book of this size, it has not been possible to cover all the relevant subjects in the depth that one should have wished. Therefore at the end of each major section or chapter, guidance for further reading is given. Notation in equations is that used by the original authors, who have been cited in the bibliography, and so some repetition of symbols occurs. For example, authors are extremely fond of using the letter k for all constants, differentiating between them by the subscripts 1, 2, 3, etc., so the reader is recommended to check the text before using any of the equations in order to identify the notation used.

The text is specifically written for final year undergraduate civil engineers and environmental scientists specializing in environmental engineering and water technology respectively. It is the course text for the water pollution control section in the M.Sc. Environmental Engineering and M.Sc. Environmental Science courses here at the University of Dublin.

In the annual report published in March 1989, Her Majesty's Inspectorate of Pollution stated that a quarter of Britain's sewage treatment plants were discharging final effluents outside their consent conditions. In Ireland a similar situation exists, but with many of the smaller towns and villages having only partial treatment, and in some cases with no treatment at all. The rate of improvement in river quality recorded in the 1970's has slowed up drastically in the 1980's, and as we approach the last decade of the century water quality in Britain and Ireland is under severe pressure and is once more declining. In the 1990's the water industry will be faced with the enormous tasks of reversing this trend, of implementing more complex and stringent national and community environmental legislation, and of harnessing the new technological and biotechnological advances for the control of pollution. Wastewater treatment is the front line in the on-going battle for a cleaner and safer environment, and has been for over a century. I hope that readers will play their part.

My intention in writing this book has been to demonstrate to the engineer the important role the biologist has in wastewater technology and at the same time make the engineering aspects intelligible to the biologist; and vice versa.

Nick Gray
TCD
December 1989.

PREFACE TO PAPERBACK EDITION

On March 19th 1991, EEC Environment Ministers in Brussels adopted the *Urban Wastewater Directive*. This legislation is wide scoping and will have a major impact on the development of wastewater treatment within the Community well into the next century. For example, it sets minimum standards for effluents and treatment efficiency; requires the provision of secondary treatment for all discharges from population equivalents of 2000 or more to both freshwaters and estuaries; it establishes sensitive areas where stricter treatment than secondary treatment will be required and lays down standards. All this, and more, must be achieved over the next 10–14 years. Also, new regulations are included on sludge disposal with all sludge dumping at sea to cease by 31st December 1998. This will pose significant problems to the UK Water Companies (formerly the Regional Water Authorities) currently treating the wastewater from coastal towns, and will result in a massive increase in the volume of sludge to be disposed onto land. So far, nearly all these Water Companies have indicated that incineration is the preferred alternative to sludge dumping at sea, but with intense public opposition to incineration, the strict Directives covering new and existing incinerators, and the problem of disposing of the hazardous ash from such operations, then the Water Companies may be forced to look for new alternatives.

With even more legislation in the pipeline, we are going to expect even more from our wastewater treatment plants in the future. More research into new designs and better operational management are going to be needed. The next decade is going to be both very exciting and rewarding for all of us in the water industry.

Nick Gray   September 1991
TCD

ACKNOWLEDGEMENTS

I should like to thank my many colleagues in the water industry who have given me permission to reproduce material in this text. I am particularly grateful to Drs L. D. Benefield, E. G. Carrington, B. Chambers, C. R. Curds, D. H. Eikelboom, H. A. Hawkes, D. Jenkins, M. A. Learner, H. A. Painter, E. B. Pike, F. D. Pooley, C. W. Randall, T. Stones, E. J. Tomlinson, A. D. Wheatley and M. Winkler, not only for permission to draw heavily on their own work, but also for the help and inspiration I have received from them and their own excellent published works.

Acknowledgement is gratefully made to authors and publishers of material which has been redrawn, reset in tables, reproduced directly or reproduced with minor modification. The exact locations can be derived from the references. For permission to reproduce copyright material thanks are due to the following publishers:

Academic Press Ltd.: Table 1.4; Fig. 3.21; Table 4.8, 4.34; Fig. 4.11, 4.69, 4.70, 4.71, 4.72, 4.87; Fig. 7.2.

American Society for Microbiology: Table 6.11, 6.23, 6.26; Table 7.19.

Blackwell Scientific Publications Ltd.: Fig. 3.25, 3.26; Fig. 4.8; Table 6.11.

British Standards Institution: Table 4.5.

Butterworth and Co (Publishers) Ltd.: Table 1.22.

Cambridge University Press: Table 2.5; Fig. 7.3.

E. and F.N. Spon (Publishers) Ltd.: Fig. 2.22.

Edward Arnold (Publishers) Ltd.: Fig. 2.10, 2.12.

Ellis Horwood Ltd.: Fig. 4.29, 4.30; Table 7.7.

Elsevier Science Publishers: Table 7.17.

Environmental and Sanitation Information Centre, Bangkok: Table 6.14.

Institution of Water and Environmental Management: Tables 2.3, 2.4; Fig. 2.11, 2.16, 2.20, 2.21; Fig. 4.35, 4.36, 4.46, 4.47, 4.91; Table 5.9; Fig. 5.5, 5.9, 5.11, 5.12.

John Wiley and Sons Ltd.: Table 1.7; Fig. 1.10; Table 3.7; Fig. 4.34; Fig. 5.4; Table 6.24.

Kluwer Academic Publishers: Table 6.25.

Springer-Verlag, Vienna: Table 4.44.

The Controller of Her Britannic Majesty's Stationary Office: Table 1.2; Table 5.7; Table 6.8, 6.13.

McGraw-Hill Book Company: Fig. 3.2, 3.17, 3.19.

Oklahoma State University: Fig. 4.86.

Open University Press: Table 4.9.

Oslo and Paris Commissions: Table 5.23, 5.24, 5.29; Fig. 5.13.

Pergamon Press PLC: Table 7.13.

Prentice Hall: Fig. 3.3, 3.5, 3.16, 3.18, 3.20, 3.28.

Surveyor Magazine: Fig. 7.6.

University of Pennsylvania Press: Table 4.43; Fig. 4.97.

Water Pollution Control Federation: Table 3.2; Fig. 3.24; Fig. 4.32, 4.62, 4.63; Table 4.45, 4.48; Fig. 5.58; Table 6.6.

Water Research Centre: Table 5.28; Table 6.2, 6.10, 6.18, 6.19, 6.20.

Water Research Commission of South Africa: Table 4.23, 4.25, 4.26; Fig. 4.61, 4.64.

World Health Organization: Table 4.32.

1

HOW NATURE DEALS WITH WASTE

1.1 INTRODUCTION

Each day, approximately $1 \times 10^6 m^3$ of domestic and $7 \times 10^6 m^3$ of industrial wastewater is produced in the UK. This, along with surface runoff from paved areas and roads, and infiltration water, produces over $20 \times 10^6 m^3$ of wastewater requiring treatment each day. To cope with this immense volume of wastewater there were, in 1978 some 7795 sewage treatment plants serving about 80% of the population. The size of these plants varies from those serving small communities of less than 100, to plants like the Crossness sewage treatment works operated by Thames Water which treats the wastewater from over 1.7 million people living in a 240 square kilometre area of London. Over 1000 plants serve populations greater than 10 000, while 17 serve populations of 300 000 or more (Water Data Unit 1979). In terms of volume or weight, the quantity of wastewater treated annually in the UK far exceeds any other product (Table 1.1) including milk, steel or even beer (Wheatley 1985), with vast quantities of wastewater generated in the manufacture of most industrial products (Fig. 1.1). The cost of wastewater treatment and pollution control is high, reaching £1100m in 1978 in the UK alone and rising annually, due not only to inflation but also to the continuous increase in environmental quality that is expected. In England and Wales, operating expenditure on sewage

TABLE 1.1 The quantity of sewage treated in the UK annually far exceeds the quantity of other industrial products processed (Wheatley 1985)

Product	Tonnes/annum ($\times 10^6$)	Price (£/tonne)
Water as sewage	6500	0.10
Milk	16	25
Steel	12	300
Beer	6.6	280
Inorganic fertilizer	3.3	200
Sugar	1.0	350
Cheese	0.2	1300
Baker's yeast	0.1	460
Citric acid	0.015	700
Penicillin	0.003	45 000

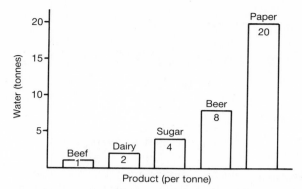
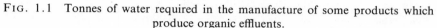

FIG. 1.1   Tonnes of water required in the manufacture of some products which produce organic effluents.

collection, treatment and disposal increased by 187% over the decade 1975/6–1985/6. Whereas gross capital expenditure on sewage treatment and sewerage, which is the expenditure on the construction and provision of new and replacement systems, only increased by 40% over the same period (Table 1.2) (DoE 1987). In the USA, spending on pollution control in all sectors has been estimated at a staggering US$400 000m during the decade preceding 1986 (Congressional Office of Technology 1981).

There are two fundamental reasons for treating wastewater: to prevent pollution and thereby protect the environment; and, perhaps more importantly, protecting public health by safeguarding water supplies and preventing the spread of water-borne diseases.

TABLE 1.2   Comparison of expenditure on sewage collection, treatment, and disposal in England and Wales during 1975/6 and 1985/6 in pounds sterling $\times 10^6$ (DoE 1987)

Water authority area	1975/6	1985/6
Anglian	18.9	54.3
Northumbrian	8.0	18.8
North West	21.6	73.1
Severn–Trent	33.2	91.3
Southern	15.0	38.6
South West	4.9	15.2
Thames	47.1	134.7
Welsh	12.4	38.1
Wessex	8.9	26.6
Yorkshire	18.3	50.3
Total	188.3	541.0

The safe disposal of human excreta is a prerequisite for the supply of safe drinking water, as water supplies can only become contaminated where disposal is inadequate. There are many infectious diseases transmitted in excreta, the most important being the diarrhoeal diseases cholera, typhoid, and schistosomiasis. The faeces are the major source of such diseases with few infections, apart from schistosomiasis, associated with urine. Among the most common infectious water-borne diseases are bacterial infections such as typhoid, cholera, bacillary dysentery, and gastro-enteritis; viral infections such as infectious hepatitis, poliomyelitis, and various diarrhoeal infections; the protozoal infection amoebic dysentery and the various helminth infections such as asariasis, hookworm, and schistosomiasis (bilharzia). Although the provision of clean water supplies will reduce the levels of infection in the short term, in the long term it is vital that the environment is protected from faecal pollution (Feachem and Cairncross 1978). Adequate wastewater treatment and the disinfection of water supplies has effectively eliminated these water-borne diseases from developed countries, but they remain endemic in many parts of the world, especially those regions where sanitation is poor or non-existent. In developed countries where there are high population densities, such as the major European cities, vast quantities of treated water are required for a wide variety of purposes. All the water supplied needs to be of the highest quality possible, although only a small proportion is actually consumed. To meet this demand it has become necessary to utilize lowland rivers and groundwaters to supplement the more traditional sources of potable water such as upland reservoirs. Where the water is reused on numerous occassions, as is the case in the River Severn and the River Thames, adequate wastewater treatment is vital to ensure that the outbreaks of waterborne disease that were so prevalent in the eighteenth and nineteenth centuries do not reoccur. In terms of environmental protection, rivers are receiving large quantities of treated effluent while estuaries and coastal water have vast quantities of partially or completely untreated effluents discharged into them. Apart from organic enrichment endangering the flora and fauna due to deoxygenation, treated effluents rich in oxidized nitrogen and phosphorus can result in eutrophication problems. Where this is a particular problem, tertiary wastewater treatment is required to remove these inorganic nutrients to protect rivers and lakes. Environmental protection of surface waters is therefore a major function of wastewater treatment. In 1985, about 33% of all major rivers in the UK (13 930 km) were classified as being of doubtful quality, or worse, and of little value for recreation or wildlife (DoE 1987). While in Ireland, 74% of rivers surveyed were unpolluted with 24% slightly or moderately polluted and just 2% seriously polluted in 1985/6 (Toner *et al.* 1986). The cost of rehabilitating rivers, as was seen with the River Thames in the period 1960–80, is immense. The River Mersey for example, now Britain's most polluted river, will cost an estimated £3700m over the next quarter of a century to raise to a standard suitable for recreation (DoE 1984). Industrial effluents have, in the past, been a major cause of pollution. The discharge of industrial effluents is generally governed by two objectives: (1) the protection of environmental water quality, and (2) the need to protect sewers and wastewater treatment plants (Table 1.3). To meet these objectives, discharge standards are required that are a compromise between what is needed to protect and improve the environment and the demands of industrial development. Most industrialists accept that the application of the *best practical* technology (i.e. effluent treatment using the best of current technology to meet local environmental requirements at the lowest financial cost) is a reasonable way to comply with the effluent discharge standards set.

TABLE 1.3   Typical effluent standards for discharges to sewers
(Gledhill 1986)

Parameter	Standards	Reasons
pH	6 to 10	Protection of sewer and sewage works fabric from corrosion.
Suspended solids	200–400 mg l^{-1}	Protection from sewer blockages and extra load on sludge disposal system.
BOD$_5$	No general limit	Local authorities would be concerned with large loads on small sewage works and balancing of flows may be required in order not to overload treatment units.
Oils/fats/grease	100 mg l^{-1}	Prevention of fouling of working equipment and safety of men. Soluble fats, etc. can be allowed at ambient temperature.
Inflammables, hydrocarbons, etc.	Prohibited	Prevention of hazards from vapours in sewers.
Temperature	43°C	Various reasons – promotes corrosion, increases solubility of other pollutants, etc.
Toxic metals	10 mg l^{-1}	Prevention of treatment inhibition. The soluble metal is more toxic and different metals can be troublesome. Total loads with a limit on soluble metals more realistic.
Sulphate	500–1000 mg l^{-1}	Protection of sewer from sulphate corrosion.
Cyanides	0–1 mg l^{-1}	Prevention of treatment inhibition. Much higher levels can also cause hazardous working conditions due to HCN gas accumulation in sewer.

However, where discharges contain dangerous or toxic pollutants which need to be minimized, then the application of the *best available* technology is required (i.e. effluent treatment using the best of current technology to minimize local environmental change, especially the accumulation of toxic materials, where financial implications are secondary considerations). Where severe effluent standards are set, which are unobtainable using even the best available technology, then of course industries can no longer continue at that location. With an EEC directive on water quality now listing standards for 60 determinants, there will be a general move towards more stringent effluent discharge standards for industry (Gledhill 1986). The introduction of the 'the polluter pays' charging system throughout Europe and the USA is an attempt to achieve such environmental objectives, at least cost to the community, by reinforcing the philosophy that the polluter is responsible for all aspects of pollution control in relation to its own effluent (Deering and Gray 1986, 1987). Two distinct types of charges exist: *effluent charges* are levied by local authorities for discharges directly to surface waters, whereas *user charges* are levied for the use of the authority's collective treatment system. By charging industry for treating their effluents in terms of strength and volume it encourages them to optimize production efficiency by reducing the volume and strength of their effluent. Most important of all, such charging systems ensure that effluent disposal and treatment costs are taken into account by manufacturers in the overall production costs, so that the cost of the final product reflects the true cost of production (Deering and Gray 1986).

Wastewater treatment is not solely a physical phenomenon controlled by engineers, it also involves a complex series of biochemical reactions involving a wide range of micro-organisms. The same micro-organisms that occur naturally in rivers and streams are utilized, under controlled conditions, to rapidly oxidize the organic matter in wastewater to innocuous end-products that can be safely discharged to surface waters. Compared with other industries which also use micro-organisms, such as brewing or baking, wastewater treatment is by far the largest industrial use of micro-organisms using specially constructed reactors. As treatment plants that were constructed during the early expansion of wastewater treatment in the late-nineteenth and early-twentieth centuries now near the end of their useful lives, it is clear that the opportunities for the biotechnologist to apply new technologies such as genetic manipulation combined with new reactor designs to pollution control are enormous. In the future, cheaper, more efficient, and more compact processes will be developed, with the traditional aims of removing organic matter and pathogens to prevent water pollution and protect public health replaced with a philosophy of environmental protection linked with conservation of resources and by-product recovery. For example, in 1978 the crude fats, proteins, and metals in wastewater disposed to British sewers had an estimated combined value of £150m.

Natural scientists, whether they are trained as microbiologists, biochemists, biologists, biotechnologists, environmental scientists or any other allied discipline, have an important role in all aspects of public health engineering. They already have a significant function in the operation and monitoring of treatment plants, but their expertise is also needed in the optimization of existing plants and in the design of the next generation of wastewater treatment systems.

Further reading

General: DoE (1976); Forster (1985).

1.2   THE NATURE OF WASTEWATER

Although there has been a steady increase in the discharge of toxic materials during the past 20 years, it is still the biodegradable organic wastes that are the major cause of pollution of receiving waters in Britain and Ireland (Gray and Hunter 1985). Organic waste originates from domestic and commercial premises as sewage, from urban runoff, various industrial processes, and agricultural wastes. Not all industrial wastes have a high organic content that is amenable to biological treatment, and those with a low organic content, insufficient nutrients, and that contain toxic compounds require specific chemical treatment, such as neutralization, chemical precipitation, chemical coagulation, reverse osmosis, ion-exchange, or adsorption onto activated carbon.

This book concentrates on non-toxic wastewaters. It is these that are of particular interest to the biologist and biotechnologist in terms of reuse, conversion, and recovery of useful constituent materials. Primarily, sewage containing pathogenic micro-organisms is considered, although other waste-waters, such as agricultural wastes from intensive animal rearing and silage production, food-processing wastes, and dairy industry wastes are also reviewed.

1.2.1   Sources and variation in sewage flow

1.05 gallons

The absolute minimum quantity of wastewater produced per person (per capita), without any excess water, is 4 litres per day. At this concentration the wastewater has a dry solids content in excess of 10%. However, in most communities that have an adequate water supply this minimum quantity is greatly increased. In those countries where technology and an almost unlimited water supply has led to the widescale adoption of water-consuming devices, many of which are now considered to be standard, if not basic, human requirements the volume of wastewater produced has increased by a factor of 100 or more. Flush toilets, baths, showers, automatic washing machines, dishwashers, and waste disposal units all produce vast quantities of diluted dirty water with a very low solids content and all requiring treatment before being discharged to surface waters. For example, a flush toilet dilutes small volumes of waste matter (<1 litre) to between 10 or 30 litres each time it is used. Domestic sewage is diluted so much that it is essentially 99.9% water with a dry solids content of less than 0.1%. Conventional sewage treatment aims to convert the solids into a manageable sludge (2% dry solids) while leaving only a small proportion in the final effluent (0.003% dry solids).

The total volume of wastewater produced per capita depends on the water usage, the type of sewerage system used and the level of infiltration. The volume of wastewater varies from country to country depending on its standard of living and the availability of water supplies (Table 1.4). Generally, the volume and strength of the sewage discharged in a particular country can

TABLE 1.4 Average daily flow of sewage per capita (litres) in selected countries (Gloyna 1971; Pike 1978)

Country	Sewage per capita ($l\,d^{-1}$)	Country	Sewage per capita ($l\,d^{-1}$)
Belgium	100–150	Italy	150–350
Brazil	150–300	Japan	250
Costa Rica	263–379	Luxembourg	150–200
Cuba	190–225	Mauritius	63–144
Denmark	200	The Netherlands	100
Ecuador	100	Nigeria	180
France	100–500	Peru	140
Germany,	180	Saudi Arabia	158
Federal		Sweden	400
Republic of		UK	300
Ghana	90–145	USA	491 *129 gallons*
India	100–250		
Eire	230		
Israel	80–415		

be predicted fairly accurately. For example, the mean daily volume of wastewater, excluding industrial waste but including infiltration, produced per capita in England is $180\,l\,d^{-1}$ compared with $230\,l\,d^{-1}$ in Ireland and $250\,l\,d^{-1}$ in Scotland. The variation in volume depends on a number of variables including the amount of infiltration water entering the sewer. The higher volume of wastewater produced in Scotland is due primarily to the widescale use of a larger flushing cistern, 13.6 l compared with 9.0 l in England and Wales, although other factors also contribute to this variation. New UK guidelines from the Department of the Environment stipulate that all new cisterns manufactured after 1993 will have to have a maximum flushing volume of 7.5 l. However, the reliance of water closets which function on a siphon rather than a valve to release water restricts the minimum volume to between 4–5 l (Pearse 1987). A recent study by the Building Research Establishment (1987) highlights the potential water saving from the adoption of new cistern designs and suggests the need for new British Standards. Comparative studies were carried out using a 'standard turd', which is a 43 mm diameter ball of non-absorbent material with a relative density of 1.08, and with a cohesive shear strength, coefficient of friction, and adhesive properties very close to the real thing.

In rural areas, where water is from bore holes or from small community water schemes, water may be at a premium so the necessary conservation of supplies results in reduced volumes of stronger sewage. Occasionally, the water pressure from such rural supplies is too low to operate automatic washing machines or dishwashers and results in an overall reduction in water usage and subsequent wastewater discharge. The equivalent volume of sewage produced in the USA is, on average, 300 l per capita per day ($= 100$ US gallons d^{-1}).

In the home, wastewater comes from three main sources. Approximately a third of the volume comes from the toilet, a third from personal washing via the wash basin, bath, and shower, and a third from other sources such as washing up, laundry, and food and drink preparation (Table 1.5). Outside the home, the strength and volume of wastewater produced per capita per day will fluctuate according to source, and this variation must be taken into account when designing a new treatment plant. For example, the flow per capita can vary from 50 l d^{-1} at a camping site to 300 l d^{-1} at a luxury hotel (Table 1.6). More detailed tables of the volume of wastewater produced from non-industrial sources, including the strength of such wastewater, are given by Hammer (1977) and also by Metcalf and Eddy (1984).

The diluted nature of wastewater has led to the development of the present system of treatment found in nearly all the technically developed countries, which is based on treating large volumes of weak wastewater. In less developed communities, the high solids concentration of the waste makes it difficult to

TABLE 1.5 Comparison of the percentage consumption of water for various purposes in a home with an office; indicating the source and make-up of wastewater from these types of premises (Mann 1979)

Home (sources)	Total water consumed (%)	Office (sources)	Total water consumed (%)
WC flushing	35	WC flushing	43 ⎫ 63
Washing/bathing	25	Urinal flushing	20 ⎭
Food preparation/ drinking	15	Washing	27
		Canteen use	9
Laundry	10	Cleaning	1
Car washing/garden use	5[a]		

[a] May not be disposed to sewer.

TABLE 1.6 Daily volume of wastewater produced per capita from various non-industrial sources (Mann 1979)

Source category	Volume of sewage (litres/person/day)
Small domestic housing	120
Luxury domestic housing	200
Hotels with private baths	150
Restaurants (toilet and kitchen wastes per customer)	30–40
Camping site with limited sanitary facilities	80–120
Day schools with meals service	50–60
Boarding schools: term time	150–200
Offices: day work	40–50
Factories: per 8 hour shift	40–80

move to central collection and treatment sites while the more diluted wastewater flows easily through pipes and can be transported easily and efficiently via a network of sewers to a central treatment works. In isolated areas or underdeveloped countries human waste is normally treated on-site due to its smaller volume and less fluid properties (Feachem and Cairncross 1978).

The collection and transport of sewage to the treatment plant is via a network of sewers. Two main types of sewerage systems are used, combined and separate. Combined sewerage systems are common in most towns in Britain. Surface drainage from roads, paved areas, and roofs are collected in the same sewer as the foul wastewater and piped to the treatment works. This leads to fluctuations in both the volume and the strength of sewage due to rainfall and although the treatment works is designed to treat up to three times the dry weather flow of wastewater (DWF) problems arise if the rainfall is either heavy or continuous. During such periods the wastewater becomes relatively diluted and the volume too great to be dealt with by the treatment works. Excess flow is, therefore, either directly discharged to a watercourse as stormwater or stored at the treatment works in stormwater tanks. The stored wastewater can be circulated back to the start of the treatment works once capacity is available. However, once the tanks become full then the settled wastewater passes into the river without further treatment where the watercourse, already swollen with rainwater, can easily assimilate the diluted wastewater because of the extra dilution now available.

A separate sewerage system overcomes the problem of fluctuations in sewage strength and volume due to rain, by collecting and transporting only the foul wastewater to the treatment works, and surface drainage is discharged to the nearest water course. Such systems are common in new towns in Britain and are mandatory in Canada and the USA. This type of sewerage system allows more efficient and economic treatment works to be designed as the variation in the volume and strength of the wastewater is much smaller and can be more accurately predicted. A major drawback with separate systems is that the surface drainage water often becomes polluted. All stormwater is contaminated to some degree because of contact during the drainage cycle: it passes over paved areas along roadside gullies to enter the sewer via a drain with a gully pot, which catches and removes solids that might otherwise cause a blockage in the sewer pipe (Bartlett 1981). The quality of urban runoff is extremely variable and biochemical oxygen demand (BOD) values have been recorded in excess of 7500 mg l^{-1} (Mason 1981). It is the first flush of stormwater that is particularly polluting as it displaces the anaerobic wastewater, rich in bacteria, that has been standing in the gully pots of the roadside drains since the last storm. The runoff from roads is rich in grit, suspended solids, hydrocarbons, heavy metals and, during the winter, chloride from road-salting operations. Surprisingly, it also contains organic matter, not only in the form of plant debris such as leaves and twigs, but as dog faeces (Table 1.7). It has been estimated that up to 17 g $m^{-2}y^{-1}$ of dog faeces

TABLE 1.7 Comparison of the concentration of various compounds reported in urban runoff with precipitation, strictly surface runoff from roads and with combined sewer overflow (Pope 1980). All units are in mg l^{-1} except those marked † which are in mg kg^{-1} and ‡ in kg Curb km^{-1}

Parameter	Reported concentration range (mg l^{-1})			
	Precipitation	Road/street runoff	Urban runoff	Combined sewer overflow
COD	2.5–322	300	5–3100	93–2636
BOD	1.1	25–165	1–700	15–685
Total solids	18–24	474–1070	400–15 322	150–2300
Volatile total solids	—	37–86	12–1600	—
Suspended solids	2–13	11–5500	2–11 300	20–1700
Volatile suspended solids	6–16	100–1500	12–1268	113
Settleable solids	—	—	0.5–5400	—
Total dissolved solids	—	66–33050	9–574	—
Volatile dissolved solids	—	1630	160	—
Conductance (μmho cm^{-1})	8–395	10 000	5.5–20 000	—
Turbidity (JTU)	4–7	—	3–70	—
Colour (Pt-Co units)	5–10	—	5–160	—
Total organic carbon	1–18	5.3–49	14–120	—
Total inorganic carbon	0–2.8	—	1–17	—
Oils/hydrocarbons	—	28–400	0–110	—
Phenols	—	—	0–10	—
Total nitrogen N	0.5–9.9	0.18–4.0	1.1–6.2	4.0–63.3
Organic N	0.1–0.32	0.18–3.23	0.1–16	1.5–33.1
Inorganic N	0.69	—	1.0	—
Ammonia N	0.01–0.4	1–2	0.1–14.0	0.1–12.5
Nitrate N	0.02–5.0	0.31–2.62	0.1–2.5	—
Nitrite N	0–0.1	—	0–1.5	—
Total phosphorus	0.001–0.35	0.3–0.7	0.09–4.4	1.0–26.5

TABLE 1.7 (*cont.*)

Parameter	Reported concentration range (mg l^{-1})			
	Precipitation	Road/street runoff	Urban runoff	Combined sewer overflow
Hydrolysable phosphorus	0.8–0.24	—	0.1–10	—
Aldrin	—	—	'trace'	—
Dieldrin	0.003†	6.8×10^{-6}‡	'trace'	—
p, p'-DDD	—	18.9×10^{-6}‡	—	—
p, p'-DDT	—	17.2×10^{-6}‡	—	—
Heptachlor	0.04†	—	'trace'	—
Lindane	—	—	'trace'	—
PCB	—	311×10^{-6}‡	—	—
Bromide	—	—	5	—
Chloride	0.1–1.1	4–70 000	2–25 000	—
Cadmium	0.013–0.056	0.002–0.01	0.006–0.045	—
Chromium	0.023–0.08	0.018–1.0	0.01–27.0	—
Copper	0.06–0.48	0.007–2.55	0.041–0.45	—
Iron	0–3.05	5–440	0–5.3	—
Lead	0.024–10.4	1–113	0.01–14.5	—
Mercury	—	0,029	—	—
Nickel	—	0.02–1.5	—	—
Zinc	0.02–4.9	1–15	0.01–5.23	—
Total coliform (ml^{-1})	—	—	240–99 100	—
Total coliform (organisms km^{-1})	—	15.9×10^{10}	—	—
Faecal coliform (ml^{-1})	—	—	5500–11 200	—
Faecal coliform (organisms km^{-1})	—	0.9×10^{10}	—	—
Faecal streptococcus (ml^{-1})	—	—	120–20 000	—

are deposited on to urban paved areas and that the dog population of a city the size of Manchester will produce an organic load equivalent to the human population of a small town of 25–30 000 people. In New York, the dog population deposits over 68 000 kg of faeces and 405 000 l of urine on to the streets each day, much of which is washed by stormwater into local streams and rivers (Feldman 1974). The degree of contamination of urban runoff during a specific storm depends on: (i) the intensity and duration of the rainfall; (ii) the length of the preceding dry period, which controls the build up of pollutants on roads and in the quality of water stored in gully pots and gutters; (iii) seasonal variations that occur in the rainfall pattern and temperature which affects the degradation of organic matter; including leaf fall and the use of grit and salt during the winter, and (iv) the effectiveness of local authorities to clean roadside gullies and gully pots (Helliwell 1979). Unlike drainage from land, runoff from roads and paved areas is very rapid due to the short length of surface water sewers. The contaminated wastewater, therefore, reaches the receiving watercourse very quickly and before the dry weather flow has increased, so that any pollutants entering will receive minimum dilution. Where there is an accidental or deliberate spillage of chemicals or noxious wastes on roads, or in private yards, serious pollution of receiving waters is bound to occur. However, with combined sewerage systems such spillages can be confined at the treatment works and recovered or treated before reaching the watercourse.

It is common, in both separate and combined sewers, for water not discharged as wastewater to enter the sewer via joints and cracks in the pipework. Infiltration water is normally from groundwater sources and can be especially high during periods of rainfall. Few estimates of the extent of the problem are available although some studies have found infiltration to be as high as 80% of the total volume in badly deteriorated sewers. In the USA it is estimated that a mean value is 70 m^3d^{-1} per km of sewer (30 000 US gallons per day per miles of sewer) (Clarke *et al.* 1977), although Grace (1978) recorded mean values some 50% less. As groundwater is generally very clean, infiltration has the effect of diluting the strength of wastewater and at the same time increasing the volume requiring treatment.

The flow rate of wastewater to treatment works is extremely variable, and although such flows follow a basic diurnal pattern, each treatment works tends to have a characteristic flow pattern. This pattern is controlled by such factors as: the time taken for sewage to travel from households to the treatment works, which is itself a function of sewer length, the degree of infiltration; and the presence of stormwater and the variability in the water consumption practices of communities (Gower 1980). Industrial inputs obviously have a profound effect on flow rates, and industrial practices such as discharging wastes after 8-hour shifts can completely alter the expected normal flow pattern to a treatment plant. The basic flow pattern for a domestic wastewater treatment plant is shown in Fig. 1.2 with the minimum flow normally occurring in the early hours of the morning when water consumption is lowest and the flow consists largely of infiltration water. Flow rate rapidly increases during the morning when peak morning water consumption reaches the plant, followed by a second peak in the early evening. When infiltration, stormwater, and the water used for non-sewered purposes such as garden use, are removed from a basic model of consumption and discharge, then the water supplied is essentially equivalent to the wastewater discharged to the sewer. Thus, the wastewater discharge curve, as measured at the sewage treatment works, will closely parallel the water supply curve, as measured at the waterworks, with a lag of several hours.

Infiltration and stormwater tend to distort the basic shape of the hydrograph of diurnal flow. Infiltration, while increasing the total daily volume, does not alter its characteristic shape. Stormwater, however, can alter the shape of the hydrograph by hiding peaks and troughs or adding new peaks as the rainfall causes rapid increases in the flow. Hourly fluctuations are less clear in large catchments due to the diversity of activities taking place during

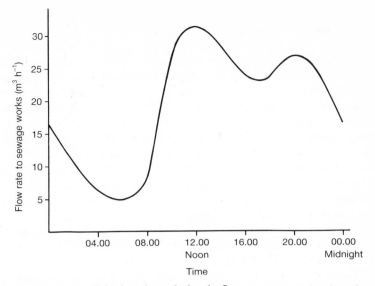

FIG. 1.2   Example of the hourly variation in flow to a sewage treatment works.

the 24-hour period and the presence of industry. The variable distance of households from the treatment works normally results in the hydrograph of the diurnal pattern becoming flattened and extended so that only one trough and one peak is seen daily (Clark *et al.* 1977; Escritt and Haworth 1984). Many problems at small to moderate sized treatment works are associated with the diurnal variation in flow, which is especially serious at the smallest works where often there is no flow at all during the night. Smaller variations of the average daily flow rate are recorded at treatment works serving large catchments (50–200%) compared with smaller communities (20–300%), (Painter 1958; Water Pollution Control Federation 1961). Many works overcome the problem of flow variation by using flow balancing, where the wastewater is stored at times of high flow and allowed to enter the works at a constant rate or by recirculating treated final effluent during periods of low flow.

Variation between weekday flows are negligible, except in those areas where the household laundry is done on specific days. However, with the advent of automatic washing machines this practice has become largely extinct. Although summer discharges normally exceed winter flows by 10–20%, up to 20–30% in the USA, seasonal variations in flow are due mainly to variation in population, as is the case at holiday resorts, schools, universities, and military camps. Other seasonal variations in flow are due to infiltration, which is linked to rainfall pattern and groundwater levels, and seasonal industrial activities such as food processing.

1.2.2   Composition of sewage

Wastewater is defined as domestic (sanitary) or industrial (trade). Domestic wastewater comes exclusively from residences, commercial buildings, and institutions such as schools and hospitals, while industrial wastewater comes from manufacturing plants. Inevitably, large towns and cities have a mixture of domestic and industrial wastewaters which is commonly referred to as municipal wastewater, and normally includes effluents from the service industries such as dairies, laundries, and bakeries, as well as a variety of small factories. It is unusual for modern municipal treatment plants to accept wastewater from major industrial complexes, such as chemical manufacturing, brewing, meat processing, metal processing, or paper mills, unless the treatment plant is specifically designed to do so. The practice in all European countries is now for water authorities to charge industry for the treatment and disposal of their wastewater. Thus, the current trend is for industry to treat its own waste in specifically designed treatment plants. In many cases, it is not cost-effective for an industry to provide and operate its own treatment plant, although most industries partially treat their waste to reduce the pollution load before discharge to the public sewer in order to reduce excessive treatment charges.

It is of prime importance for the designer and operator of a treatment plant to have as much knowledge of the composition of the wastewater to be treated as possible. This is particularly important when new or additional wastes are discharged to existing plants. A full analysis of the wastewater will, for example: (i) determine whether pretreatment is required; (ii) determine whether an industrial waste should be treated alone or with sewage and, if so, in what proportions; (iii) determine whether an industrial waste would attack the sewer; (iv) permit a better selection of the most appropriate treatment process; (v) allow an assessment of the toxicity or disease hazards; (vi) provide an indication of the resultant degree of eutrophication or organic enrichment in the form of sewage fungus in the receiving water; and (vii) an assessment of the recoverable or reusable fractions of the wastewater.

Although there is considerable similarity in the basic content of sewage, the precise volume and characteristics will vary not only from country to country because of climatic conditions and social customs, but also within individual countries due to supply water characteristics, water availability, population size, and the presence of industrial wastes. Data on wastewaters is normally limited to BOD_5 (the five-day biochemical oxygen demand test), COD (chemical oxygen demand), suspended solids and ammonia, while a fuller characterization of the wastewater being treated is rare (Table 1.8). Analysis of wastewater composition can be done directly by laboratory examination of the sewage itself or indirectly by predicting the composition by examination of the gross components. Details of the composition of human faeces and urine

TABLE 1.8 Chemical characteristics of treated effluents from three UK sewage treatment plants (Water Research Centre, Stevenage)

Constituent[a]	Source		
	Stevenage	Letchworth	Redbridge
Total solids	728	640	931
Suspended solids	15		51
Permanganate value	13	8.6	16
BOD (biochemical oxygen demand)	9	2	21
COD (chemical oxygen demand)	63	31	78
Organic carbon	20	13	
Surface-active matter			
Anionic (as Manoxol OT)	2.5	0.75	1.4
Non-ionic (as Lissapol NX)			0.4
Ammonia (as N)	4.1	1.9	7.1
Nitrate (as N)	38	21	26
Nitrite (as N)	1.8	0.2	0.4
Chloride	69	69	98
Sulphate	85	61	212
Total phosphate (as P)	9.6	6.2	8.2
Total phenol			
Sodium	144	124	
Potassium	26	21	
Total hardness	249	295	468
pH value	7.6	7.2	7.4
Turbidity (ATU)[b]			66
Colour (Hazen units)	50	43	36
Coliform bacteria (no./ml)	1300		3500

[a] Results are given in mg l^{-1}, unless otherwise indicated.
[b] Absorptiometric turbidy units.

are available (Table 1.9) although details of other household wastes which are more variable are less well known, therefore, more direct methods of wastewater characterization are preferred. Surprisingly little is known of the composition of sewage and few specific studies have been carried out. Some individual components of sewage, which causes specific problems have been studied. For example, total phosphorus and nitrogen in eutrophication studies, detergents causing foaming, and indole in the control of odours.

Sewage is a complex mixture of natural inorganic and organic materials with a small proportion of man-made substances. The main source of pollution in sewage is human excreta with smaller contributions from food preparation, personal washing, laundry and surface drainage. The chemical and physical nature of wastewaters can be further complicated by the inclusion of industrial wastes which are composed of strong spent liquors from main industrial processes and comparatively weak wastewaters from rinsing, washing, and condensing.

TABLE 1.9   Volume and composition of human faeces and urine
(Gloyna 1971)

	Faeces	Urine
Moist weight per capita per day	135–270 g	1.0–1.3 l
Dry weight per capita per day	35–70 g	50–70 g
Moisture content	66–80%	93–96%
Organic matter content (dry basis)	88–97%	65–85%
Nitrogen (dry basis)	5.0–7.0%	15–19%
Phosphorus (as P_2O_5) (dry basis)	3.0–5.4%	2.5–5.0%
Potassium (as K_2O) (dry basis)	1.0–2.5%	3.0–4.5%
Carbon (as dry basis)	40–55%	11–17%
Calcium (CaO) (dry basis)	4–5%	4.5–6.0%

The reason why sewage composition is normally measured in terms of BOD_5, COD, suspended solids, and ammonia content is because it is from these basic determinants that its polluting strength is assessed. Most charging systems are based on the Mogden formula, which uses these basic determinants (Table 1.10). Other variables are occasionally measured under specific circumstances, such as total phosphorus if the final effluent is discharged to inland lakes, or heavy metals if the sludge is to be subsequently used for agriculture.

The strength of sewage varies widely and depends on such factors as per capita water usage, infiltration, surface and storm water, and local habits. The water usage in the USA is at least three times greater than in Britain, which is why American sewage is usually weaker. Although the per capita production of organic matter is essentially the same in the USA and Britain, the difference in water consumption results in a raw sewage BOD_5 of between 100–700 mg l^{-1} (with a mean BOD_5 of 320 mg l^{-1}) in Britain (Painter 1971). The use of garbage grinders or disposal units, so that household kitchen waste is disposed to the sewer rather than the refuse bin, results in a 30% increase in wastewater BOD_5 and 60% increase in the suspended solids. The concentration of nitrogen in domestic wastewater is directly related to the BOD_5 with about 40% of the total nitrogen in solution as ammonia. Proteins and urea undergo deamination releasing ammonia as the wastewater flows to the treatment plant. The longer sewage is held in the sewer, the greater will be the release of ammonia. The per capita nitrogen production in the UK is 5.9 g N per day (Painter 1958), which is essentially the same as the American figure (Babbitt 1947). The per capita production of phosphorus is about a third of the weight of nitrogen produced – about 1.4 kg per capita per year. Of this up to 70% comes from polyphosphate builders used in synthetic detergents (Table 1.11).

The amount of organic matter produced per capita each day expressed in terms of BOD_5 is also known as the population or person equivalent.

TABLE 1.10 The Mogden formula is used by the water authorities in England and Wales to calculate charges for the disposal and treatment of trade wastes:

$$C = R + V + \left(\frac{O_t}{O_s} \times B\right) + \left(\frac{S_c}{S_s} \times S\right),$$

where C is the total charge in pence per 1000 litres of trade effluent, R is the reception and conveyance cost per 1000 litres, V the volumetric and primary treatment cost per 1000 litres, O_t the COD of trade effluent after one hour quiescent settlement (mg l^{-1}), O_s the COD of settled sewage (mg l^{-1}), B is the cost per 1000 litres of settled sewage, S_c is the total suspended solids of trade effluent (mg l^{-1}), S_s is the total suspended solids of settled sewage (mg l^{-1}), and S the treatment and disposal cost of primary sludge per 1000 litres of sewage.

The charges levied by the regional water authorities of England and Wales in 1986 are given below in pence (sterling) per cubic metre of wastewater (p m^{-3})

Region	Average regional strength		Apportionment of charge (p m^{-3})				Total minimum charge (p m^{-3})
	O_s (mg l^{-1})	S_s (mg l^{-1})	R	V	B	S	
Anglian	680	400	4.64	5.79	8.47	3.19	22.09
Northumbrian	353	178	7.52	4.06	7.15	3.95	22.68
North West	363	258	3.22	2.95	3.74	2.29	12.20
Severn–Trent	331	344	4.33	5.02	5.42	3.13	17.90
Southern	452	512	4.06	8.51	11.12	6.53	30.22
South West	406	343	6.13	5.45	8.73	7.41	27.72
Thames	442	331	3.24	3.79	5.91	7.46	20.40
Welsh	500	350	3.35	3.16	8.26	4.53	19.30
Wessex	351	323	1.90	4.76	5.30	4.69	16.65
Yorkshire	927	316	0.00[a]	7.40	8.08	4.46	19.94

[a] Cost for R and V combined.

TABLE 1.11   Typical chemical composition of raw
wastewaters from the USA and UK

Constituent	USA ($mg\,l^{-1}$)	UK ($mg\,l^{-1}$)
pH	7.0	
BOD	250	326
COD	500	650
Total organic carbon	250	173
Total solids	700	
Suspended solids	220	127
Total nitrogen	40	66
Organic nitrogen	25	19
Ammonia nitrogen	25	47
Nitrite	0	0
Total phosphorus	12	
Organic phosphorus	2	
Inorganic phosphorus	10	

Population equivalent (PE), expressed in kg BOD_5 per capita per day, is
determined as:

$$PE = \frac{\text{Mean flow (l)} \times \text{mean } BOD_5 \ (mg\,l^{-1})}{10^6}$$

Population equivalent is often used in the design of treatment plants, and
the volume and strength of industrial wastewaters are normally expressed in
terms of equivalent population. In the UK, the PE of domestic sewage is
equivalent to 0.055 kg BOD_5 per capita d^{-1}. This ranges from 0.045 kg for an
entirely residential area to 0.077 kg for a large industrial city. American figures
are similar for domestic sewage being 0.052 kg for separate sewers and
0.063 kg for combined sewers. However, the recognized design figures in the
USA and Canada are 0.077 kg BOD_5 and 0.10 kg suspended solids. Apart
from BOD_5 and suspended solids, it is also common to quote total nitrogen or
total phosphorus in terms of PE.

The PE of an industrial wastewater in Britain is calculated using the
relationship:

$$PE = \frac{\text{Mean flow } (m^3 d^{-1}) \times \text{mean } BOD_5 \ (mg\,l^{-1})}{0.055 \times 10^3}$$

Similar to the flow, the strength and composition of sewage changes on an
hourly, daily, and seasonal basis. However, it is the diurnal variation that is
usually the greatest. A similar diurnal pattern of sewage strength is formed, in
terms of BOD_5 and suspended solids, as occurs with flow (Fig. 1.2). The peak
in BOD occurs in the mid-morning but, as with the flow, the actual time
depends on the length of the sewers and the nature of the area served. The
strength of sewage in large cities, with very long and complex sewerage

systems, does not fluctuate as widely as it does in smaller catchments, with maximum values occurring between 10 p.m. and 6 a.m. (Painter 1971). There is a wide diurnal variation in BOD_5 strength. This variation is also reflected by similar fluctuations in the concentrations of the various carbohydrates and fatty acids that largely make up the biodegradable carbon fraction (Painter 1958). Peaks also occur in the concentration of ammonia and urea, occurring in the morning and late at night, reflecting the habits of the local population served. However, only the morning peak is generally discernible.

Strength varies from day to day due to the dilution effect of surface and stormwater. The daily fluctuation in flow and strength is much less at treatment plants fed by a separate sewerage system. Wastewater volumes are greater on Mondays than any other day in the USA (Heukelekian and Balmat 1959), while the concentration of detergents in the UK has been reported as being higher on Mondays than the rest of the week (Eden and Truesdale 1961). However, these data were collected when automatic washing machines were not widely available and the household laundry for the week was done by necessity on a specific day, usually Monday, so it is unlikely that this trend is still discernible today. Seasonally, sewage strength does not vary significantly, although periods of drought and excessive rainfall can effect the dilution ratio in combined sewerage systems due to reduced infiltration and storm water. In the summer bacterial concentrations in the wastewater reach a maximum as do virus concentrations (Painter 1971).

Wastewater can only be treated biologically if sufficient nitrogen and phosphorus is present. Normally, there is a surplus of these nutrients in sewage for biological needs but it is necessary to assess the treatability of wastewaters by checking the ratio of carbon to nitrogen and phosphorus (C:N:P). The optimum C:N:P weight ratio for biological treatment is 100:5:1 (100 mg l^{-1} BOD_5, 5 mg l^{-1} N; 1 mg l^{-1} P). The C:N:P ratio for raw sewage is approximately 100:17:5 and 100:19:6 for settled sewage, both containing abundant nutrients for microbial growth. The nitrogen requirement of the micro-organisms in the biological unit of a treatment plant is satisfied if the ratio of carbon, measured as BOD, to nitrogen equals or is less than 18:1. Even at C:N ratios >22:1 removal still occurs, but much less efficiently.

Micro-organisms required much lower levels of phosphorus compared with nitrogen so the phosphorus requirement will be met if the C:P ratio is less than 90–150:1. Above 150:1 there is an increasing loss of efficiency (Porges 1960; Hattingh 1963a; Komolrit et al. 1967). The exact C:N:P: ratio for optimum biological growth depends on the biological process and the form in which the nutrients are available in the wastewater. Where wastes fail to meet the C:N:P criteria it becomes necessary to add nutrients in order to ensure that biological oxidation occurs. This is often carried out by mixing nutrient deficient wastes with sewage in the correct proportions. Wastewaters from the brewing and canning industries are particularly deficient in nitrogen and phosphorus, therefore nutrient addition is required if optimum carbonaceous oxidation is to be achieved. An example of nutrient deficient wastewater is cited by Jackson and Lines (1972) who found that the BOD removal during the treatment of a cider effluent using low-rate percolating filtration was increased from 92% to 99% when the nitrogen and phosphorus balance was corrected by the addition of an inorganic supplement.

Physical properties

Those who have not actually come into contact with raw sewage often harbour rather strange ideas as to what it looks and smells like. By the time it reaches the sewage treatment plant the vast majority of the large solids, such as faeces and paper, have broken up into very small particles. Thus, apart from a small quantity of floatable material, raw sewage is a rather turbid liquid with visible particles of organic material that readily settle out of suspension. The colour is normally grey to yellow-brown, according to the time of day. However, if all the oxygen has been used up during transit in the sewer then the watewater becomes anaerobic or septic and takes on a much darker colour and in extreme cases turns black. Municipal wastewaters receiving industrial wastes containing dyes will take on the colour of the dye present, and at treatment plants receiving effluents from the textile industry, the raw wastewater undergoes spectacular and frequent colour changes. Under ultraviolet light, domestic sewage has a characteristic coloured fluorescence which is due to a variety of minor constituents present in household detergents.

Generally, domestic wastewater has a musty smell which is not at all offensive, although pungent odours can be produced if the wastewater becomes anaerobic. Certain industrial wastes and contaminants in surface runoff do have distinctive odours. Like a wine connoisseur, the ability of the operator to develop a discerning nose can be extremely helpful in identifying potential problems within the treatment plant and to identify changes in the composition of the sewage entering the plant. The odours produced are usually caused by gases produced by the decomposition of various fractions of the organic matter present in the wastewater. The commonest odour encountered is the smell of rotten eggs caused by hydrogen sulphide, which is produced by anaerobic bacteria reducing sulphate to sulphide. Table 1.12 lists the major categories of odours encountered at sewage treatment plants, although the quantification of such odours for use as a control variable has proved extremely difficult (American Public Health Association *et al*. 1983; Metcalf and Eddy 1984). Some food processing wastewaters produce extremely strong odours, especially during treatment and storage (Gerick 1984; Gray 1988). For example, sugar beet wastewater undergoes partial anaerobic breakdown within the process and subsequently on storage in lagoons, with the production of a variety of odours. The major odours come from the volatile fatty acids that comprise most of the organic fraction of the effluent. The odour threshold concentrations for the volatile acids produced during treatment of sugar beet are 24.3 ppm for acetic acid, 20.0 ppm for

TABLE 1.12 Some characteristic odours produced by compounds present in wastewaters. These degradation products can be categorized into two main groups, either degradation products of nitrogenous or sulphurous compounds. There are other odourous compounds such as those associated with chlorine and phenolic wastes

Compounds	General formulae	Odour produced
Nitrogenous		
Amines	CH_3NH_2, $(CH_3)_3N$	Fishy
Ammonia	NH_3	Ammoniacal, pungent
Diamines	$NH_2(CH_2)_4NH_2$, $NH_2(CH_2)_5NH_2$	Rotten flesh
Skatole	$C_8H_5NHCH_3$	Faecal, repulsive
Sulphurous		
Hydrogen sulphide	H_2S	Rotten eggs
Mercaptans	CH_3SH, $CH_3(CH_2)_3SH$	Strong decayed cabbage
Organic sulphides	$(CH_3)_2S$, CH_3SSCH_3	Rotten cabbage
Sulphur dioxide	SO_2	Pungent, acidic
Other		
Chlorine	Cl_2	Chlorine
Chlorophenol	$Cl.C_6H_4OH$	Medicinal, phenolic

propionic acid, 0.05 ppm for iso-butyric acid, 0.24 ppm for butyric acid, 0.7 ppm for iso-valeric acid and 3.0 ppm for valeric acid. Therefore, by measuring the volatile acid concentration and dividing it by the appropriate odour threshold concentration, a measure of the odour production or concentration known as the odour number can be calculated (Gray 1988). Other odours associated with sugar beet processing include trimethylamine which has a fishy odour, organic sulphides which produce a strong odour of rotting cabbage as do the thiol compounds methyl mercaptan (CH_3SH) and ethyl mercaptan (C_2H_5SH) both of which have very low odour thresholds of 0.0011 ppm and 0.00019 ppm respectively (Shore *et al.* 1979). A detailed list of odour threshold values has been compiled by Fazzalari (1978).

The temperature of sewage is normally several degrees warmer than the air temperature, except during the warmest months, because the specific heat of water is much greater than that of air. Sewage temperatures are normally several degrees warmer than the water supply and because of its high conductivity rarely freezes in temperate climates. In the UK, the temperature of raw sewage ranges from 8–12°C in winter to 17–20°C in summer (Painter 1971). Comparative studies have shown that the variability in the temperature of settled sewage, as it entered the percolating filtration stage at a treatment works in South Yorkshire, was 30–50% less than that recorded for the air temperature, a total range of 11.5°C and 30.4°C respectively. Both the sewage and air temperatures followed similar seasonal patterns, reaching maximum and minimum temperatures during the same periods. However, while the mean daily air temperature varied annually, the mean daily temperature of the sewage remained constant at 12.4°C (Gray 1980). In hotter climates domestic sewage can be much warmer, and in India sewage temperatures of 28–30°C are not unusual (Kothandaraman et al. 1963).

⏿ The pH of sewage is usually above 7 with the actual value depending largely on the hardness of the supply water. Although extreme values are occasionally encountered, soft water catchments generally have a pH range of 6.7–7.5 (modal pH = 7.2) and hard water catchments a range of 7.6–8.2 (modal pH = 7.8) (Painter 1971).

Total solids (i.e. the weight of matter remaining after a known volume of wastewater has been evaporated at 105°C) is a commonly used wastewater variable in the USA. It provides a simple characterization of the wastewater with which the theoretical performance of unit treatment processes can be predicted. Total solids can be classified as either suspended or filterable depending on whether solids will pass through a standard filter (DoE 1972).

The standard filter used in Britain is a Whatman GF/C filter paper, the GF refers to its glass fibre structure, which has a pore size of 1.0 μm. Therefore, all the filterable solids have a particle size of < 1.0 μm. The suspended solids fraction ranges from colloidal particles > 1.0 μm up to recognizable gross matter. A portion of the suspended solids fraction is settleable, which is measured by measuring the volume of solids that settle out of suspension over a 60 min period under quiescent conditions. An Imhoff cone, which is an inverted 1 litre conical flask, is used for this purpose and provides a useful estimate of the solids removal and sludge production during primary sedimentation. The filterable fraction contains colloidal and dissolved material. The colloidal solids are particulate ranging from 1 nm to 1 μm in size, which is too small to be removed by gravity settlement. An assessment of colloidal solids can be made by measuring the light-transmitting properties of the wastewater, the turbidity, as colloidal matter scatters and absorbs light. The dissolved fraction is made up of both organic and inorganic molecules' that are in solution.

Each of these major categories of solids is comprised of both organic and inorganic material, and the ratio of each can be measured by burning off the organic fraction in a muffle furnace at 600°C (Allen 1974). Certain salts are also destroyed by heating, although only magnesium carbonate is decomposed at this temperature and is transformed to magnesium oxide and carbon dioxide at 350°C. The major inorganic salt in domestic sewage is calcium carbonate, but this remains stable up to 825°C. The percentage of each solids fraction in a wastewater depends on the chemical composition of the sewage. However, Painter (1971) separated the particulate solids in domestic sewage as approximately 50% settleable (> 100 μm diameter), 30–70% supra-colloidal (1–100 μm) and the remaining 17–20% as colloidal (1 nm–1 μm). However, a more detailed breakdown of the proportion of particular solids

TABLE 1.13 The organic strength in terms of chemical oxygen demand (COD) of the various solids fractions of sewage (adapted from Rickert and Hunter 1971)

Solids fraction	Total solids		Organic content		COD	
	mg l^{-1}	%	mg l^{-1}	%	mg l^{-1}	%
Settleable	74	15	59	25	120	29
Supracolloidal	57	11	43	18	87	21
Colloidal	31	6	23	9	43	10
Soluble	351	68	116	48	168	40
Total	513	—	241	—	418	—

fractions in domestic wastewater and the organic strength of each fraction is shown in Table 1.13.

Organic properties
Organic matter comprises of carbon, hydrogen, and oxygen with nitrogen frequently present. Sulphur, phosphorus, and iron are only occasionally present. In medium strength sewage 75% of the suspended solids and 40% of the filterable solids fractions are organic. In settled sewage, Painter (1983) estimates that 50% of the organic carbon and between 35–50% of the organic nitrogen is in solution. Three-quarters of the organic carbon can be attributed to the major organic groups carbohydrates, fats, proteins, amino acids, and volatile acids. The remainder comprises other organic molecules such as hormones, vitamins, surfactants, antibiotics, hormonal contraceptives, purines, pesticides, hydrocarbons, and pigments. Many of the synthetic organic molecules are non-biodegradable whereas others are only decomposed biologically at very slow rates. The organic constituents of suspended solids and the filterable fraction of sewage are very different (Table 1.14). Carbohydrates comprise the largest group in British sewage followed by non-volatile and volatile acids, free and bound amino acids, and anionic detergents. Urea is a major component of urine but is hydrolysed so rapidly to ammonia that it is only found in very fresh sewage. The composition of sewage changes rapidly on storage due to anaerobic bacterial action, for example carbohydrates are rapidly converted to volatile acids. Fats are the major organic constituents in the suspended solids fractions, and together with carbohydrates and proteins account for 60–80% of the organic carbon present.

Most of the naturally occurring amino acids, carbohydrates, and organic acids are found in sewage. Of the carbohydrates, glucose, sucrose, and lactose are the major ones, with smaller proportions of galactose, fructose, xylose, and arabinose. Together, they account for 90–95% of all the carbohydrate present which is equivalent to 50–120 mg l^{-1}. A diurnal variation in carbohydrate

TABLE 1.14 Organic constituents of domestic wastewater (Painter 1983)

Constitutent	In solution		In suspension	
	Concentration (mg l^{-1})	Proportion as C of total C in solution (%)	Concentration (mg l^{-1})	Proportion as C of total C in suspension (%)
Fats	—	—	140	50
Carbohydrates	70	31.3	34	6.4
Free and bound amino acids	18	10.7	42	10
Volatile acids	25	11.3	}	
Non-volatile acids	34	15.2	12.5	2.3
Detergents (ABS)	17	11.2	5.9	1.8
Uric acid	1	0.5	—	—
Creatine	6	3.9	—	—
Amino sugars	—	—	1.7	0.3
Amides	—	—	2.7	0.6
Organic carbon				
by direct analysis	90	100	211	100
by addition of above constituents	75.6	84.1	151	71.4

concentration and composition is evident, and although glucose accounts for over 50% of the total carbohydrate content in composite samples, sucrose concentration is greater than glucose in the afternoon. The ratio of hexose to pentose is between 10 and 12. The non-soluble high molecular weight carbohydrates such as starches, cellulose, and wood fibre are restricted to the suspended solids fraction resulting in a low hexose to pentose ratio (2.0–2.6) and a concentration of 30–38 mg l^{-1}.

In wastewater terminology, fats is a general term as is lipids or grease, to describe the whole range of fats, oils and waxes discharged to the sewer. They are among the more stable organic compounds and are not easily degraded biologically. The major source of fats is from food preparation and to a lesser extent excreta, the major sources being butter, lard, margarine, vegetable fats and oil, meat, cereals, nuts, and certain fruit. Fats are only sparingly soluble in water and so are only an important component of the suspended fraction of the wastewater, contributing up to 50% of the total carbon present. Normal concentration ranges for fats in domestic wastewater are between 40–100 mg l^{-1}, although this figure is normally higher than that recorded for American sewages. Fats are broken down by hydrolytic action to yield fatty acids, and a wide variety of free fatty acids have been reported from sewage including all the saturated ones, from C_{8} to C_{14} as well as C_{16}, C_{18}, and C_{20}. The major acids include palmitic, stearic, and oleic acids, which form between 75–90% of those present. Full details of the fat content of domestic sewage is given by Painter (1971).

Acetic acid is the major volatile acid found in sewage, being recorded at concentrations between 6–37 mg l^{-1} and, together with propionic, butyric, and valeric acids, make up 90% of the total volatile acidity in wastewater. The acidity of sewage rapidly increases on storage at the expense of sugars, and if high concentrations are recorded in fresh sewage, anaerobosis should be suspected. Non-volatile soluble acids are present at concentrations between 0.1–1.0 mg l^{-1}, the commonest being glutaric, glycolic, lactic, citric, benzoic, and phenyllactic acids.

Proteins are a comparatively important source of carbon in wastewater although they are less important than soluble carbohydrate or fats in suspension. Protein is the principal constituent of all animal and, to a lesser extent, plant tissue. Thus, waste from food preparation and excreta is rich in protein such as casein from milk, or albumen and gelatine from animal tissue and bone. Apart from containing carbon, hydrogen, and oxygen, proteins also contain a fairly high proportion of nitrogen, which is consistent at about 16%. Proteins, apart from urea, are the chief source of nitrogen in wastewater and supply up to 80% of the total organic nitrogen present. Proteins are made up of long chains of amino acids connected by peptide bonds, and are readily broken down by bacterial action to form free amino acids, fatty acids, nitrogenous compounds, phosphates, and sulphides. In wastewater, the free amino acids generally account for <5 mg N l^{-1}, although this can be occasionally higher, while bonded amino acids in the form of peptides or protein account for between 4–15 mg N l^{-1}.

Apart from amino acids, urea is also a major source of nitrogen in sewage providing between 2–16 mg N l^{-1}. Urea is most abundant in fresh sewage as it is rapidly converted to ammonia under both aerobic and anaerobic conditions. The rate of conversion to ammonia has been estimated at 3 mg N l^{-1} per hour at 12°C in stored samples (Painter 1958).

Not all biodegradable organic matter found in sewage can be classified into one of the major categories. Some natural compounds are in fact combinations of carbohydrates, proteins, and fats such as lipoproteins and nucleoproteins. Of the organic matter in wastewater up to 20–40% will not be degraded within the treatment plant. However, the actual proportion of non-biodegradable material is normally very small. Fractions such as lignins and cellulose are only slightly degraded, which is (a) due to the limited time for decomposition within the treatment plant, and (b) many of the specific microorganisms required are not present in sewage.

Sewage contains a diverse range of organisms which originate not only from faeces, but also from soil and water. They include viruses, bacteria, fungi, protozoans, and a variety of other groups of organisms. Many of these organisms are pathogenic to man and are discussed fully in Chapter 6.

Inorganic properties
There are substantial inorganic components in sewage, especially compounds continuing sodium, calcium, potassium, magnesium, chlorine, sulphur (as sulphates and other forms), phosphates, bicarbonates, and ammonia. Traces of heavy metals are also found. The inorganic content of domestic wastewater depends on the geology of the catchment from which the water supply originated (natural water dissolves minerals from the surrounding rocks and soil of the area), and on the nature of the polluting material itself. This is vividly illustrated by comparison of sewages from a soft and hard water area (Table 1.15). In hardwater catchments, the calcium, sodium, and chloride ions are significantly more concentrated in the supply water, which is reflected in the resulting wastewater. Domestic wastewater contains a very wide range of inorganic salts and trace elements, including all those necessary for biological growth and activity. Among the major ions in wastewater which are worthy of further discussion are chloride, nitrogen, and phosphorus.

Chloride is found naturally in water because of leaching, but it also originates from a wide variety of agricultural, industrial, and domestic sources. Although infiltration by groundwater contaminated with saltwater into the sewer is a major source of chloride and sulphate in some areas, a major seasonal source at treatment plants served by combined sewerage systems is from road runoff during salting operations in icy weather. In hardwater areas, the widespread use of water softeners can result in significant increases in the wastewater chloride concentrations. Without these additional sources, sewage

TABLE 1.15 Concentration of major inorganic constituents of domestic sewage (Painter 1971)

Constituent	Whole sewage USA Soft water area (mg l^{-1})	Settled sewage UK Hard water area (mg l^{-1})
Cl	20.10	68.00
Si	3.90	—
Fe	0.80	0.80
Al	0.13	—
Ca	9.80	109.00
Mg	10.30	6.50
K	5.90	20.00
Na	23.00	100.00
Mn	0.47	0.05
Cu	1.56	0.2
Zn	0.36	0.65
Pb	0.48	0.08
S	10.30	22.00
P	6.60	22.00

normally contains between 30–100 mg l^{-1} of chloride: human excreta contains 6 g Cl per capita per day and urine contains 1% chloride. As chloride is not removed to any great extent by conventional treatment, the detection of higher concentrations in surface waters may indicate that they are being used for wastewater disposal.

Nitrogen and phosphorus are both essential nutrients for plant growth. Nitrogen is also necessary for the synthesis of protein and biological growth generally. In fresh wastewater, nitrogen is primarily present as proteinaceous matter and urea. This organic nitrogen is rapidly decomposed by bacterial action in the case of proteins, or by hydrolysis in the case of urea to ammonia, the concentration of which in wastewater is indicative to some extent of its age. Ammonia N exists in aqueous solution as either the ammonium (NH_4^+) or as ammonia (NH_3) depending on the pH of the wastewater. At pH values of > 7 the equilibrium of the reaction

$$NH_3 + H_2O \leftrightarrow NH_4^+ + OH^-$$

is displaced to the left so that ammonia predominates, and at pH values < 7 equilibrium moves to the right and ammonium predominates. Organic nitrogen is normally measured separate from ammonia, although occasionally they are expressed together as the kjeldhal nitrogen. The normal concentration range of nitrogen in settled sewage in Britain is 41–53 mg N l^{-1} as ammonia, 16–23 mg N l^{-1} as organic nitrogen, and 57–76 mg N l^{-1} as kjeldhal nitrogen (Painter 1971). The oxidized forms of ammonia, nitrite, and nitrate are normally absent from fresh sewage as they are products of the biological oxidation processes within the treatment plant. Therefore, as the total nitrogen includes all chemical forms of nitrogen, the kjeldhal nitrogen can be assumed to be equivalent to the total nitrogen in raw and settled sewage. Phosphorus is present in sewage in three distinct forms, as orthophosphate, polyphosphate, and organic phosphate. Organic phosphate is a minor constituent of sewage and like the polyphosphates requires further decomposition to the more assimilable orthophosphate form, which is normally fairly slow. About 25% of the total phosphorus in settled sewage is present as orthophosphates, such as PO_4^{3+}, HPO_4^{2-}, $H_2PO_4^-$, H_3PO_4, which are available for immediate biological metabolism. Therefore, in terms of utilization, both in the treatment plant and subsequently in receiving waters, it is the inorganic phosphate concentration that is important rather than the total phosphorus concentration. After secondary treatment about 80% of the total phosphorus in a final effluent is in the orthophosphate form. Average phosphorus concentrations in sewage range from 5–20 mg P l^{-1} as total phosphorus, of which 1–5 mg P l^{-1} is the organic fraction and the rest inorganic. Between 1960 and 1980 the world-wide per capital consumption of detergents rose from 3.6 to 6.3 kg per annum, doubling in the USA over this period from 12.8 to 30.1 kg per capita per annum. Since 1965, legislation in the USA and a voluntary ban in Britain has seen a steady reduction in the use of 'hard' or non-biodegradable alkyl-benzene-sulphonate (ABS) detergents in favour of 'soft' biodegradable linear-alkyl-benzene-sulphonate (LAS) detergents. Detergents are made up of a number of compounds each with a specific function during the washing process. All detergents vary in their formulation, although all generally contain the basic functional groups of compounds. Main chemical substances in typical detergent formulations include: surfactant (e.g. linear alkyl benzene sulphonate) 3–15%; builder (e.g. sodium tripolyphosphate) 0–30%; ion-exchanger (e.g. zeolite type A) 0–25%; antiredeposition agent (e.g. polycarboxylic acids) 0–4%; bleaching agent (e.g. sodium perborate) 15–35%; bleach stabilizer (e.g. phosphonate) 0.2–1.0% foam booster (e.g. ethanolamide) 1–5%; enzyme (e.g. protease) 0.3–1.0%; optical brightener (e.g. pyrazolan derivatives) 0.1–1.0%; corrosion inhibitor (e.g. sodium silicate) 2–7%; and a fragrance 0.05–0.3%. However, to increase the washing ability of LAS detergents, 'builders' in the form of polyphosphates, such as sodium tripolyphosphate (STPP), are included. These particular polyphosphates are highly unstable and readily break down to super-phosphates within the treatment plant, which has resulted in the growing problems of high phosphate concentrations in the final effluents being discharged to surface waters. In order to reduce the level of phosphates in the environment, certain countries have introduced legislation to control detergent compounds. Canada and the Federal Republic of Germany both operate limits on phosphorus in detergents used for general laundry; while in the USA the states of Connecticut, Florida, Indiana, Maine, Maryland, Michigan, Minnesota, Montana, New York, Vermont, and Wisconsin have all imposed a ban on detergents using phosphate builders. Although a number of other compounds have been used to replace STPP they are less efficient. A number of inorganic compounds have been used including sodium carbonate, sodium silicate, sodium borate, sodium sulphate, and zeolite, an aluminosilicate, although they are generally of limited value. In contrast there are a number of organic compounds which can be used as detergent builders, the most useful being nitrilotriacetic acid (NTA), polyacrylic acid (PAA), and polycarboxylic acids (PCAs). In Canada, the Netherlands, Switzerland, and West Germany NTA is most widely used, although there is growing concern regarding its environmental and health effects that has led to its ban in parts of New York State. Similar concern has been expressed about both PAA and PCAs. All the organic builders are essentially non-biodegradable with removal in wastewater treatment plants mainly by adsorption on to solids, with subsequent precipitation after dilution. The use of organic detergent builders, and of PCAs in particular, has been reviewed by Hunter et al. (1988). The increasing popularity of biological washing powders containing enzymes is also having an important effect on the nature of domestic sewage.

Sulphur is another essential element in the metabolism of all organisms. However, most micro-organisms only require small amounts of the element which is used in the synthesis of the amino acids cysteine and methionine, which are found in protein. Nearly all the requirement for sulphur comes from sulphate. Trace quantities of all the metals necessary for biological growth are present in sewage. Those metals that are particularly toxic in excessive concentrations and are common in sewage are nickel, lead, chromium, cadmium, zinc, and copper. Vacker et al. (1967), Heukelekian and Balmat (1959), and Painter (1958) have provided mean values of the most important metals in sewage and mean values using this data, which are collected from very different catchments, have been given by Gould (1976). These data, and metal concentrations collected from a domestic sewage treatment works in South Yorkshire, are compared in Table 1.16. The data from South Yorkshire

TABLE 1.16 Concentrations of heavy metals ($mg\,l^{-1}$) found in domestic sewage based on US softwater and UK hard water data (Gould 1976), and settled sewage from a small town in South Yorkshire (Gray 1980)

Heavy metal	Gould (1976) $\bar{x}$	Gray (1980)
Cr	0.4	(0.0–0.1)
Mn	0.2	—
Fe	0.8	(0.15–1.30)
Co	<0.02	—
Ni	<0.02	(0.00–0.33)
Cu	0.88	(0.00–0.16)
Zn	0.50	(0.05–0.84)
Cd	0.02	(0.00–0.05)
Pb	0.25	(0.01–1.78)

show much lower concentrations of chromium and copper as it is a purely domestic sewage. However, the high iron and lead concentrations are due to mine water infiltration and runoff from the nearby M1 motorway. There has been a recent increase in the use of antiseptic creams and hair shampoos that are rich in zinc, and this has led to a significant increase in the zinc concentration of sewage.

Sewage remains aerobic provided that it is not permitted to stand. Normal dissolved oxygen concentrations in flowing sewage is usually in the order of $1-2$ mg l^{-1}. However, long retention times in sewers should be avoided and where this is not possible then reaeration, normally by oxygen injection, should be included in sewer design. Hydrogen sulphide should not be present in sewers if they are properly vented. However, if blockages occur or the flow falls to below 0.52 ms^{-1} then anaerobic conditions can develop and hydrogen sulphide given off.

1.2.3   Other wastewaters

The other wastewaters rich in organic materials and thus readily degraded biologically are the agricultural and food-processing wastes. Of particular importance, and this varies according to agricultural practice and manufacturing processes, are wastes from intensive animal rearing, silage production, food processing, and the dairy industry.

Animal wastes
Specialization and the development of new methods in agriculture have led to the intensification of animal rearing and a departure from the traditional farming practice of returning all wastes back to the land as fertilizer, which avoided pollution and treatment costs. The adoption of intensive farm practices can lead to enormous numbers of animals being kept in comparatively small areas. For example, Lynch and Poole (1979) cite an American example where 35 000 cattle were kept in a feed lot of less than one square mile, whereas some farm animals such as pigs and poultry are now raised almost exclusively indoors in specially constructed units. Sewage is comparatively weak compared to most animal wastes which have very high BOD_5 concentrations (Table 1.17). The characteristics of animal wastes has been extensively reviewed by Evans *et al.* (1978, 1980). Most farm animals produce large quantities of waste each day compared to man, and in terms of mean population equivalents based on the BOD_5 where a man$=1.0$, then a cow$=16.4$, a horse$=11.2$, a chicken$=0.014$, a sheep$=2.45$, and a pig$=3.0$ (Gloyna 1971). Another problem is that the waste has a high solids content and thus, unlike sewage, is not a liquid but either a semi-liquid or semi-solid (Fig. 1.3). It is, therefore, difficult to handle or pump unless dewatered or diluted. For many farmers the limiting factor in the development of intensive

TABLE 1.17 Average volume, strength, and nutrient content of animal wastes

Animal	Volume of waste per adult animal ($m^3 d^{-1}$)	COD ($mg l^{-1}$)	BOD_5 ($mg l^{-1}$)	N ($kg tonne^{-1}$)	P ($kg tonne^{-1}$)	K ($kg tonne^{-1}$)	Moisture content (%)
Cow	0.0500	150 000	16 100	11.1	4.5	13.4	87
Pig	0.0045	70 000	30 000	8.9	4.5	4.5	85
Poultry	0.0001	170 000	24 000	38.0	31.3	15.6	32–75[a]

[a] Depending on housing.

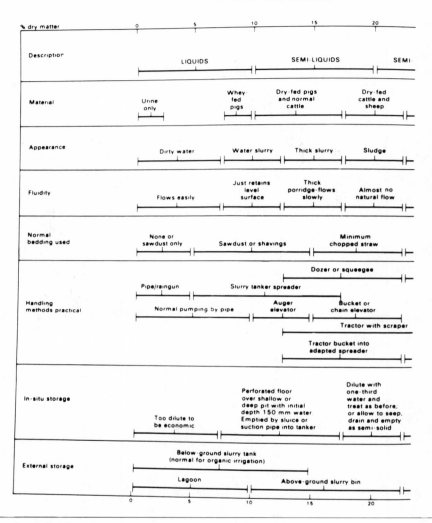

animal rearing is the disposal of the increased amounts of animal waste. The population of farm animals in Britain in 1978 was approximately 3 million cows, 9 million cattle, 7 million pigs, and 130 million poultry. Their waste in terms of population equivalents was of the order of 30 million, 50 million, 17 million, and 13 million respectively, which is almost twice the organic load produced by the present human population in Britain (Weller and Willetts 1977). It is neither feasible nor desirable to discharge this quantity or type of waste to the public sewer. First, because of the vast volume of dilution water required to reduce the BOD to treatable levels by conventional wastewater treatment methods, and secondly because of the cost of increasing the treatment capacity of existing works by 200%. The strength of animal wastes

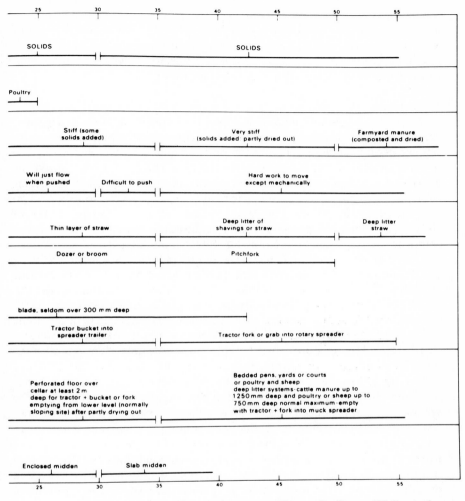

FIG. 1.3 Definition, handling, and storage of livestock effluent (Weller and Willetts 1977).

compared to sewage and other agricultural wastewaters (Table 1.17) must be seen in relation to the dilution of the effluent. The daily volume of effluent produced by the major categories of animals is: dairy cow 0.0445 m^3; beef cattle 0.0198 m^3; sow 0.0117 m^3; fat pig 0.0049 m^3; and poultry 0.0001 m^3 (Gowan 1972). A particular problem with animal wastes is the enhanced metal concentrations that are often present. Concentrated feeds used for fattening pigs in intensive rearing units contain high concentrations of metals, especially copper and zinc. Research on a range of commercial pig foods has shown that these metals are present at concentrations ranging from 116–233 ppm dry weight for copper and 194–300 ppm for zinc. However, little of the metals is retained by the animals with between 70–80% of the copper and 92–96% of the zinc being excreted (Table 1.18). Effluents from intensively reared stock which are fed concentrates, and in particular pigs, will inevitably contain high concentrations of metals (Priem and Maton 1980).

Silage liquor
Ensiling is a large-scale microbial process in which cut grass is degraded anaerobically so that the complex cellulose component is broken down into simpler organic acids, preserving the grass as food for cattle by raising the pH. The effluent (silage liquor) from the clamp in which the grass is stored has a pH of 4.5 or less and is composed mainly of organic acids, in particular lactic, acetic, propionic, and butyric acids, which are very readily broken down biologically (Patterson 1981). Thus, with an average BOD of 50 000–60 000 mg l^{-1}, discharges of the acidic liquor to watercourses leads to complete destruction of all aerobic life. Silage liquor is also rich in nitrogen and can contain up to 2.5 g of nitrogen per litre of liquor (Weller and Willetts 1977). The dry matter content of silage liquor ranges from 4–10% (mean value of 6.5%). The average composition of the liquor as a percentage of the dry matter is crude protein 25.0%, lysine 1.0%, ash 22.0%, calcium 2.2%, phosphorus 1.0%, nitrogen-free extract 53.0%, lactic acid 25.0, and volatile fatty acids 5.5% (Patterson 1980). The amount of liquor produced by silage is closely related to moisture content of the grass at the time of ensiling with the quantity of effluent per tonne of silage being 360–450 l^{-1} at 10–15% dry matter, 90–225 l at 16–20%, and less than 90 l at dry matter of 25% and over (Gibbons 1968). Acidic additives, used to preserve the silage, can increase the volume of liquor produced by up to 25%, although finely chopping silage does not have an effect on effluent production. Although the volume of liquor can be greatly reduced by wilting the grass in the field before ensiling, sometimes wilting is not possible because of time or weather conditions and the disposal of liquor can become a serious problem. It is not generally advisable to dispose of these effluents to the public sewer as small sewage works can be put out of action as a result of toxic shock. Such wastes are usually stored with the animal wastes in slurry tanks or stored in specially constructed acid-resistant tanks and returned to the land when possible.

Dairy industry
Milk production has steadily grown over the past 10 to 20 years and the dairy industry has now become the major agricultural processing industry in Europe.

Wastewater originates from two major processes, from the fluid milk itself at reception and bottling plants but, more importantly, at the processing plants that produce butter, cheese, evaporated and condensed milk, milk powder, and other milk products. Milk itself has a BOD_5 of 100 000 mg l^{-1}

TABLE 1.18 Average composition of the liquid manure from intensively reared pigs fed on commercial food concentrates rich in metals. Feeds A, B, and C contained 116, 233, and 189 ppm dry weight of copper and 194, 300, and 260 ppm dry weight of zinc respectively (adapted from Priem and Maton 1980).

	A		B		C	
	$\bar{x}$	S.D.	$\bar{x}$	S.D.	$\bar{x}$	S.D.
Dry matter (%)	16.2	1.19	15.9	0.75	14.6	2.14
Ash in dry matter (%)	23.9	2.20	24.6	1.79	25.6	2.48
BOD (mg l^{-1})	41 807	1293	41 967	2504	35 546	5253
COD (mg l^{-1})	163 539	26 051	154 026	15 867	143 178	19 997
P (mg l^{-1})	3548	546	3491	360	3294	838
Ammonia nitrogen (mg l^{-1})	6606	1343	6351	1097	5774	1209
Kjeldahl nitrogen (mg l^{-1})	10 345	1272	9998	1000	9083	1367
Cu (ppm in dry matter)	416	44.1	859	68.2	754	66.3
Zn (ppm in dry matter)	851	74.5	1385	107.4	1180	31.2
As (ppm in dry matter)	8.50	0.83	10.93	1.76	11.13	0.87
Se (ppm in dry matter)	0.63	0.05	1.57	0.17	1.08	0.08

and washings from plants producing butter and cheese can have a BOD_5 ranging between 1500 and 3000 mg l^{-1}. Dairy wastes are dilutions of whole milk, separated milk, butter milk, and whey. They are high in dissolved organic matter mainly in the form of the proteins (3.8%) casein and albumin, fat (3.6%), and lactose (4.5%) but low in suspended solids except for the fine curd found in cheese wastes. Apart from whey, derived from the manufacture of cheese which is acidic, most dairy wastes are neutral or slightly alkaline but have a tendency to become acidic quite rapidly because of fermentation of lactose to lactic acid. The average composition and organic strength of milk, milk by-products, and cheese wastes are given in Table 1.19. Details of the various processes used in the dairy industry, with specific reference to wastewater production, are given by Nemerow (1978).

Food-processing industries
Wastes from food processing is similar in nature to the food itself. Some processes give rise to large volumes of weakly polluted effluents such as vegetable washing water, which only contains soil and small amounts of organic matter. More concentrated wastewaters come from processes that either prepare the food or transform it in some way such as blanching of vegetables or pickling meat. Generally, these wastes are rich in organic matter and normally contain sufficient nitrogen, phosphorus, and trace elements for biological growth. The volume and strength of wastewater from food processing greatly depends on the type of process, the size and age of the plant, as well as the season.

Cannery wastewaters are essentially the same as domestic kitchen waste. The waste originates from trimming, culling, juicing, and blanching fruit and vegetables. The wastewaters are high in suspended solids, colloidal and dissolved organic matter, the main components being starch and fruit sugars. For example, 85–90% of the organic waste from a pineapple cannery is sugar in the form of sucrose (Painter 1971). Details of these wastes are summarized by Nemerow (1978). Sugar beet waste is also comprised of sugars, 95% of which is sucrose with raffinose making up most of the remainder, although the waste is particularly low in nitrogen and phosphorus. The sugars are leached from cut and damaged surfaces into the transport and washwater circuits, so that the BOD of these wastes can be as high as 8000–10 000 mg l^{-1} (Table 1.20). The accumulated sugars are rapidly catabolized in the circuits to short chain aliphatic carboxylic acids, principally acetic, propionic, and butyric acids, so that the wastewater requiring treatment is comprised almost exclusively of these acids. However, at low pH concentrations offensive odours from volatile fatty acids and sulphides can be generated, and sufficient lime (CaO) must be added to maintain circulating water at a neutral pH (Shore *et al.* 1984; Gray 1988). Brewery and distillery wastewaters are high in dissolved solids which contain nitrogen and fermented starches and their products. Fermentation wastes and in particular spent yeast is extremely

TABLE 1.19 Composition and organic strength of milk products and associated waste products (Nemerow 1978)

Characteristics	Whole milk (mg l^{-1})	Skim milk (mg l^{-1})	Buttermilk (mg l^{-1})	Whey (mg l^{-1})	Process wastes (mg l^{-1})	Separated whey (mg l^{-1})
Total solids	125 000	82 300	77 500	72 000	4516	54 772
Organic solids	117 000	74 500	68 800	64 000	2698	49 612
Ash solids	8000	7800	8700	8000	1818	5160
Fat	36 000	1000	5000	4000		
Soluble solids				44 000	3956	54 656
Suspended solids				8000	560	116
Milk sugar	34 000	46 000	43 000			
Protein (casein)	38 000	39 000	36 000			
Total organic nitrogen					73.2	1300
Free ammonia					6.0	31
Na					807	648
Ca					112.5	350
Mg					25	78
K					116	1000
P					59	450
BOD$_5$	102 500	73 000	64 000	32 000	1890	30 100
Oxygen consumed	36 750	32 200	28 600	25 900		

TABLE 1.20   Comparative strengths of wastewaters from food-processing industries

	BOD (mg l^{-1})	COD (mg l^{-1})	PV (mg l^{-1})	Suspended solids (mg l^{-1})	pH	Population equivalent per m^3 of waste
Brewery	850	17 000	—	90	4–6	14.2
Cannery citrus	2000	—	—	7000	Acid	33.3
pea	570	—	—	130	Acid	9.5
Dairy	600–1000	—	150–250	200–400	Acid	10.0–16.7
Distillery	7000	10 000	—	Low	—	116.7
Farm	1000–2000	—	500–1000	1500–3000	7.5–8.5	16.7–33.3
Silage	50 000	—	12 500	Low	Acid	833.3
Potato processing	2000	3500	—	2500	11–13	33.3
Poultry	500–800	600–1050	—	450–800	6.5–9.0	8.3–13.3
Slaughterhouse	1500–2500	—	200–400	800	7	25.0
Sugar beet	450–2000	600–3000	—	800–1500	7–8	7.5–33.3

concentrated with the BOD (2000–$15\,000\ \text{mg}\,l^{-1}$), total nitrogen ($800$–$900\ \text{mg}\,l^{-1}$) and phosphate ($20$–$140\ \text{mg}\,l^{-1}$), almost entirely present in the dissolved or colloidal fractions with the suspended solids content rarely in excees of $200\ \text{mg}\,l^{-1}$. Slaughterhouse and meat-packing wastewaters are strong and unpleasant, being comprised of faeces and urine, blood washings from carcasses, floors and utensils, and the undigested food from the paunches of slaughtered animals. These wastewaters are high in dissolved and suspended organic matter, in particular, proteins and fats, which are high in organic nitrogen and grease.

The strengths and volumes of wastewaters from the main food-processing industries are summarized in Table 1.20.

Further reading

Sewage composition: Hunter and Heukelekian 1965; Ligman *et al.* 1974; Painter 1971; Rickert and Hunter 1971.
Sewerage: Bartlett 1981.
Infiltration and urban runoff: Bartlett 1981; Helliwell 1979; Torno *et al.* 1986; Whipple *et al.* 1983.
Agricultural wastewaters: Hobson and Robertson 1977; Taiganides 1977; Gasser 1980.
Food processing wastewaters: Dickinson 1974; Nemerow 1978.
Volume and flowrate: Geyer and Lentz 1964; Hubbell 1962; Metcalf and Eddy 1984.

1.3 MICRO-ORGANISMS AND POLLUTION CONTROL

Micro-organisms have a number of vital functions in pollution control. It is the microbial component of aquatic ecosystems that provides the self-purification capacity of natural waters in which micro-organisms respond to organic pollution by increased growth and metabolism (Section 1.4.1). It is essentially the same processes which occur in natural waters that are utilized in biological treatment systems to treat wastewater (Chapter 4). Apart from containing food and growth nutrients, wastewater also contains the micro-organisms themselves, and by providing a controlled environment for optimum microbial activity in a treatment unit or reactor, nearly all the organic matter present can be degraded (Chapter 3). Micro-organisms utilize the organic matter for the production of energy by cellular respiration and for the synthesis of protein and other cellular components in the manufacture of new cells. This overall reaction of wastewater treatment can be summarized as:

$$\text{Organic matter} + O_2 + NH_4^{2+} + P \rightarrow \text{New cells} + CO_2 + H_2O$$

Similar mixed cultures of micro-organisms are used in the assessment of wastewater and effluent strength by the biochemical oxygen demand test

(BOD$_5$), in which the oxygen demand exerted by an inoculum of micro-organisms growing in the liquid sample is measured over five days to give an estimate of the oxidizible fraction in the wastewater (Section 1.4.2). Many diseases are caused by waterborne micro-organisms, a number of which are pathogenic to man. The danger of these diseases being transmitted via wastewater is a constant threat to public health (Chapter 6). Therefore, the use of micro-organisms, such as *Escherichia coli*, as indicator organisms to assess the microbial quality of water for drinking, recreation, and industrial purposes, as well as in the assessment of wastewater treatment efficiency is an essential tool in pollution control (Section 6.2).

1.3.1 Nutritional classification

In wastewater treatment, it is the bacteria that are primarily responsible for the oxidation of organic matter. However, fungi, algae, protozoans (collectively known as the Protista), and higher organisms all have important secondary roles in the transformation of soluble and colloidal organic matter into biomass, which can be subsequently removed from the liquid by settlement prior to discharge to a natural watercourse. In order to function properly the micro-organisms involved in wastewater treatment require a source of energy and carbon for the synthesis of new cells as well as other nutrients and trace elements. The micro-organisms are classified as either heterotrophic or autotrophic according to their source of nutrients. Heterotrophs require organic matter both for energy and as a carbon source for the synthesis of new micro-organisms, whereas autotrophs do not utilize organic matter but oxidize inorganic compounds for energy and use carbon dioxide as a carbon source.

Heterotrophic bacteria, which are also referred to as saprophytes in older literature, utilize organic matter as a source of energy and carbon for the synthesis of new cells, respiration, and mobility. A small amount of energy is also lost as heat during energy transfer reactions. The heterotrophs are subdivided into two groups according to their dependence on free dissolved oxygen.

Aerobes require free dissolved oxygen in order to decompose organic material:

$$\text{Organics} + O_2 \xrightarrow{\text{aerobic micro-organisms}} \text{Aerobic micro-organisms} + CO_2 + H_2O + \text{energy}$$

Like all microbial reactions it is autocatalytic, i.e. the micro-organisms that are required to carry out the reaction are also produced. Aerobic bacteria predominate in natural watercourses and are largely responsible for the self-purification process. They are also dominant in the major biological wastewater treatment processes such as activated sludge and percolating filtration. Aerobic processes are biochemically efficient and rapid in comparison with other types of reactions, producing by products that are usually chemically simple and highly oxidized such as carbon dioxide and water.

Anaerobes oxidize organic matter in the complete absence of dissolved oxygen by utilizing the oxygen bound in other compounds, such as nitrate:

$$\text{Organics} + NO_3 \xrightarrow{\text{anaerobic micro-organisms}} \text{Anaerobic micro-organisms} + CO_2 +$$

$$N_2 + \text{energy}$$

or sulphate:

$$\text{Organics} + SO_4{}^{2-} \xrightarrow{\text{anaerobic micro-organisms}} \text{Anaerobic micro-organisms} + CO_2$$

$$+ H_2S + \text{energy}$$

Anaerobic bacterial activity is found in fresh water and estuarine muds rich in organic matter, and in the treatment works in the digestion of sludge. Anaerobic processes are normally biochemically inefficient and generally slow, giving rise to chemically complex by-products that are frequently foul-smelling. The end-products of proteins, carbohydrates, and fats that have undergone microbial breakdown under anaerobic and aerobic conditions are summarized in Table 1.21. Facultative bacteria use free dissolved oxygen when available but in the absence of oxygen are able to gain energy anaerobically and are known as facultative aerobes or, equally accurately, as facultative anaerobes. An example of a facultative bacterium is *Escherichia coli*, a common and important coliform, this and other such bacteria are common in both aerobic and anaerobic environments, and treatment systems. Often, the term obligate is used as a prefix to these categories of heterotrophic bacteria to indicate that they can only grow in the presence (obligate aerobe) or absence of oxygen (obligate anaerobe).

Using these basic reactions as guides, it is possible to write balanced equations for the utilization of the organic substrate and the synthesis of new micro-organisms. For example, using glucose as the organic substrate and the formulae $C_5H_7O_2N$ to represent the composition of the organisms, equations for respiration and the production of energy for cell maintenance and synthesis can be expressed as:

$$\text{Respiration: } C_6H_{12}O_6 + 6O_2 \rightarrow 6CO_2 + 6H_2O$$

and the equation for synthesis of new micro-organisms is:

$$\text{Synthesis: } (C_6H_{12}O_6) + NH_4{}^+ \rightarrow C_5H_7O_2N + 3H_2O + H^+$$

The above two 'half-reactions' can be combined to give the basic organic

TABLE 1.21　End-products of the aerobic and anaerobic microbial breakdown of the major organic substrates found in sewage (Berthouex and Rudd 1977)

Substrates	+	Enzymes of micro-organisms	→	Representative end-products	
				(Anaerobic conditions)	(Aerobic conditions)
Proteins and other organic nitrogen compounds		Enzymes of micro-organisms		Amino acids Ammonia Hydrogen sulphide Methane Carbon dioxide Hydrogen Alcohols Organic acids Phenols Indol	Amino acids Ammonia→nitrites→nitrates Hydrogen sulphide→sulphuric acid Alcohols Organic acids }→$CO_2 + H_2O$
Carbohydrates		Enzymes of micro-organisms		Carbon dioxide Hydrogen Alcohols Fatty acids Neutral compounds	Alcohols Fatty acids }→$CO_2 + H_2O$
Fats and related substances		Enzymes of micro-organisms		Fatty acids + glycerol Carbon dioxide Hydrogen Alcohols Lower fatty acids	Fatty acids + glycerol Alcohols Lower fatty acids }→$CO_2 + H_2O$

transformation reaction brought about by aerobic micro-organisms in biological wastewater treatment plants. This is discussed in more detail in Section 3.1.

$$C_6H_{12}O_6 + 0.5\ NH_4^+ \rightarrow C_5H_7O_2N + 3.5\ CO_2 + 5H_2O + 0.5H^+$$

Bacteria are comprised of 80% water and 20% dry matter. Of the dry matter, 90% is organic and the remainder is inorganic. Hoover and Porges (1952) used the equation $C_5H_7O_2N$ to describe the organic fraction of bacteria in wastewater with 53% of the weight of the organism assumed to be organic carbon. More comprehensive equations have been formulated to describe the chemical composition of bacteria, for example, the one used by Mara (1974), which takes account of the phosphorus content of bacterial cells, $C_{60}H_{87}O_{25}N_{12}P$. The remaining 10% of the cells are comprised of phosphorus (50%), sulphur (15%), sodium (11%), calcium (9%), magnesium (8%), potassium (6%), and iron (1%). As all these inorganic elements are required for microbial growth, any deficiency will result in growth limitation or inhibition.

The amount of energy biologically available per unit of organic matter broken down by heterotrophs depends on the oxygen source used. The greatest yield of energy comes from the use of dissolved oxygen in oxidation, and least energy is from strict anaerobic metabolism. With a mixed culture of micro-organisms, as is found in wastewater treatment, the micro-organisms seek the greatest energy yield in order to achieve maximum synthesis. This is illustrated by the microbial activity that occurs when organically enriched water is put into a closed container. At first, aerobic and facultative bacteria will decompose the organic matter, gradually depleting the dissolved oxygen. After all the dissolved oxygen is exhausted, the facultative bacteria continue to use oxygen bound as nitrate and sulphate. At this stage, other facultative and anaerobic bacteria begin to break down the organic matter to organic acids and alcohols that produce least energy:

$$\text{Organics} \rightarrow \text{Organic acids} + CO_2 + H_2O + \text{energy}$$

If methane forming bacteria are present then the anaerobic digestion process is completed by converting the organic acids to methane and carbon dioxide:

$$\text{Organic acids} \rightarrow CH_4 + CO_2 + \text{energy}$$

Autotrophic bacteria cannot utilize organic matter, instead they oxidize inorganic compounds for energy and use carbon dioxide or carbonate as a carbon source. There are a number of autotrophs in the aquatic ecosystem, however, only the nitrifying, sulphur, and iron bacteria are particularly important in wastewater oxidation. The nitrifying bacteria oxidize ammonia nitrogen in a two-step reaction, initially to nitrite which is unstable and finally to nitrate.

$$NH_4^+ + \text{oxygen} \xrightarrow{\textit{Nitrosomonas}} NO_2^- + \text{energy}$$

$$NO_2^- + \text{oxygen} \xrightarrow{\textit{Nitrobacter}} NO_3 + \text{energy}$$

The reaction occurs in secondary treatment units although it is very sensitive to environmental conditions, occurring most efficiently at low organic loadings and warm temperatures.

In sewers, hydrogen sulphide is given off by sulphate reducing bacteria if the wastewater becomes anaerobic. The slightly acidic gas is absorbed into condensation water which collects on the top or crown of the sewer or on the side walls. Here, sulphur bacteria, which are able to tolerate pH levels of 1.0 oxidize the hydrogen sulphide to strong sulphuric acid using atmospheric oxygen:

$$H_2S + \text{oxygen} \xrightarrow{\textit{Thiobacillus}} H_2SO_4 + energy$$

The sulphuric acid reacts with the lime in the concrete to form calcium sulphate, which lacks structural strength. Gradually, the concrete pipe can be weakened so much that it eventually collapses. Crown corrosion is particularly a problem in sewers that are constructed on flat gradients, in warm climates, in sewers receiving heated effluents, with wastewaters with a high sulphur content or in sewers which are inadequately vented. Corrosion-resistant pipe material such as vitrified clay or PVC plastic, prevents corrosion in medium-sized sewers, but in larger diameter sewers where concrete is the only possible material, corrosion is reduced by ventilation which expels the hydrogen sulphide and reduces condensation. In exceptional circumstances the wastewater is chlorinated to prevent sulphate-reducing bacteria forming hydrogen sulphide or the sewer is lined with a synthetic corrosion-resistant coating.

Not all species of iron bacteria are strictly autotrophic, however, those that are can oxidize inorganic ferrous iron to the ferric form as a source of energy:

$$Fe^{2+} \text{ (ferrous)} + \text{oxygen} \xrightarrow{\textit{Leptothrix}} Fe^{3+} \text{ (ferric)} + \text{energy}$$

The bacteria are filamentous and deposit oxidized iron $(Fe(OH)_3)$ in their sheath. They mainly occur in iron-rich mine wastewaters but can also occur in biological wastewater treatment units. For example, they are common in percolating filters that treat domestic effluents receiving infiltration water rich in iron from coal mining areas (Gray 1980). If the domestic water supply contains dissolved iron, the bacteria can become established in water pipes, forming yellow or reddish-brown slimes and tainting the water as the mature bacteria die.

Free dissolved oxygen is essential for the aerobic processes of heterotrophic and autotrophic bacteria. When aerobic organisms utilize organic nutrients they consume dissolved oxygen at the same time. Each molecule of glucose, which is the basic building block of all carbohydrates, requires six molecules of oxygen for complete conversion to carbon dioxide and water by aerobic bacteria:

$$C_6H_{12}O_6 + 6O_2 \xrightarrow{\text{bacteria}} 6CO_2 + 6H_2O$$

There is also a considerable oxygen demand during the nitrification of nitrogenous compounds by autotrophic nitrifying bacteria:

$$2NH_4^+ + 3O_2 \xrightarrow{\text{bacteria}} 2NO_2^- + 4H^+ + H_2O$$

$$2NO_2^- + O_2 + 2H^+ \xrightarrow{\text{bacteria}} 2NO_3^- + 2H^+$$

If the dissolved oxygen is not replaced then aerobic growth will eventually stop when the oxygen is exhausted, allowing only the slow anaerobic processes to continue. Microbial activity is not only oxygen-limited in the case of aerobic micro-organisms, it is also restricted by the availability of adequate supplies of carbon, nutrients such as nitrogen and phosphorus, trace elements, and growth factors. It is the actual composition of micro-organisms that controls the nutrient requirements of organisms, but as proteins are composed mainly of carbon, nitrogen, and smaller amounts of phosphorus, it is these three elements which are essential for microbial growth. The requirements of carbon, nitrogen, and phosphorus by microbial cultures in wastewater treatment processes is expressed as a ratio (C: N: P) and if the waste is deficient in any one of these basic components, complete utilization of the wastewater cannot be achieved.

Autotrophs derive energy from either sunlight (photosynthetic) or from inorganic oxidation–reduction reactions (chemosynthetic). Chemo-autotrophs do not require external sources of energy but utilize the energy from chemical oxidation, whereas phototrophs require sunlight as an external energy source:

$$CO_2 + H_2O \xrightarrow{\text{photosynthetic micro-organisms}} \text{Phototrophs} + O_2$$

Many inorganic ions, mainly metals, are required to ensure that bacterial enzymatic reactions can occur therefore, trace amounts of calcium, magnesium, sodium, potassium, iron, manganese, cobalt, copper, molybedenum, and many other elements are required. These are found in adequate amounts in sewage, as are growth factors, such as vitamins. However, if any of these materials are deficient or absent, then microbial activity will be restricted or may even stop.

The mixed microbial cultures found in biological wastewater treatment units degrade and subsequently remove colloidal and dissolved organic substances from solution by enzymatic reactions. The enzymes are highly specific, catalysing only a particular reaction and are sensitive to environmental factors such as temperature, pH, and metallic ions. The major types of enzyme-catalysed reactions in wastewater biochemistry are:

Oxidation   the addition of oxygen or the removal of hydrogen;
Reduction   the addition of hydrogen or the removal of oxygen;
Hydrolysis   the addition of water to large molecules that results in their breakdown into smaller molecules;
Deamination   the removal of an NH_2 group from an amino acid or amine; and
Decarboxylation   the removal of carbon dioxide.

Microbial energetics, metabolism, population, and community dynamics are fully explored in Chapter 3.

Further reading

General: Mara 1974; Lynch and Poole 1979.

1.4   MICROBIAL OXYGEN DEMAND

It is important to know how much oxygen will be required by microorganisms, as they degrade organic matter present in wastewater, for two reasons: (a) to ensure that sufficient oxygen is supplied during wastewater treatment so that oxidation is complete; and (b) to ensure receiving waters do not become deoxygenated due to the oxygen demand of these microorganisms, which results in the death of the natural fauna and flora. The amount of organic matter that a stream can assimilate is limited by the availability of dissolved oxygen. This is largely determined by the rate oxygen is utilized by microbial oxidation and the rate at which it can be replaced by reaeration and other processes.

1.4.1   *Self-purification*

The term self-purification is defined as the restoration, by natural processes, of a river's natural clean state following the introduction of a discharge of polluting matter. In natural river systems, organic matter is assimilated by a number of processes including sedimentation which is enhanced by mechanical and biological flocculation, chemical oxidation, and the death of enteric and pathogenic micro-organisms by exposure to sunlight. Of course, the assimilative capacity of rivers, i.e. the extent to which the river can receive waste without significant deterioration of some quality criteria, usually the dissolved oxygen concentration, varies according to each river because of available dilution, existing quality, and self-purification capability (Benoit 1971). The most important process in self-purification is biochemical oxidation, i.e. the aerobic breakdown of organic material by micro-organisms. Biodegradable organic matter is gradually eliminated in rivers, which is primarily caused by bacterial action, by methods very similar to those occurring in wastewater treatment. Complex organic molecules are broken down to simple inorganic molecules in a process requiring oxygen. In this process of self-purification it is the attached micro-organisms, collectively known as periphyton, that are normally responsible for the greatest removal. The suspended micro-organisms that are mainly supplied with the discharge are less important in the removal of organic material. However, although the decomposition of organic waste by micro-organisms is advantageous, the process removes oxygen from solution, and in order to prevent the destruction of the natural fauna and flora, aerobic conditions must be maintained.

Water at normal river temperatures holds very little oxygen compared to air. In the atmosphere, gas molecules diffuse or move from an area of high concentration to an area of low concentration. In the same way, oxygen molecules diffuse through the air–water interface into the water where they become dissolved. At the same time, oxygen diffuses in the opposite direction, but when the volume of oxygen diffusing in either direction per unit time is equal then the water is said to be in equilibrium and therefore saturated with oxygen (100%). The solubility of oxygen depends on three major factors: pressure, temperature, and the concentration of dissolved minerals (salinity). A decrease in atmospheric pressure causes a decrease in oxygen; therefore, streams at high altitudes have less oxygen at saturation concentration at a standard temperature than a lowland stream, but in Ireland and the UK this is of little practical significance. It is standard practice to express the solubility of oxygen at 1 atmosphere of pressure. Freshwater at 1 atmosphere of pressure at 20°C contains 9.08 g of oxygen per m^3 ($gm^{-3} \equiv mgl^{-1}$) and as the temperature increases so the saturation concentration (the maximum amount of oxygen that can dissolve into water) decreases (Fig. 1.4). The concentration of dissolved salts lessens the saturation concentration of oxygen and this is why seawater has a lower saturation concentration than freshwater at various temperatures and pressures.

While the dissolved oxygen concentration is affected by factors such as temperature, BOD_5, and salinity, oxygen depletion is prevented primarily by reaeration, although other sources of oxygen, such as photosynthesis may also be important under certain conditions. It is important to know how quickly oxygen dissolves into water, and this depends to a large extent on the concentration of oxygen already in solution in relation to the saturation concentration, i.e. the oxygen deficit. For example, water containing 7 g m^{-3} of oxygen but with a saturation concentration of 9 g m^{-3} has an oxygen deficit of 2 g m^{-3}. The oxygen concentration can become supersaturated, up to

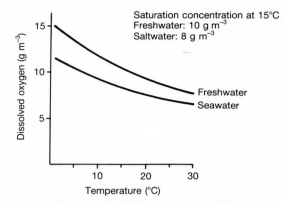

FIG. 1.4   Variation of saturation concentration of dissolved oxygen with temperature at a pressure of 1 atmosphere.

200%, under conditions of agitation at waterfalls and weirs. Supersaturation can also occur on bright sunny days because of photosynthesis when algal growth is abundant (Fig. 1.5). In both cases, the oxygen concentration will quickly return to equilibrium by the excess oxygen being lost to the atmosphere by diffusion. In general terms, the greater the organic load to the river the greater the response in terms of microbial activity, resulting in a larger demand for the available dissolved oxygen.

Reaeration
Oxygen diffuses continuously over the air–water interface in both directions. In the water, the concentration of oxygen will eventually become uniform due to mixing or in the absence of mixing by molecular diffusion. The rate of diffusion is proportional to the concentration gradient which has been described by Flick's law as:

$$\frac{dM}{dt} = K_d A \frac{dC}{dx}$$

where M is the mass transfer in time t (mass–transfer rate), K_d the diffusion coefficient, A the cross-sectional area across which transfer occurs, C the concentration, and x the distance of transfer (concentration gradient).

If a uniform concentration gradient is assumed then:

$$\frac{dM}{dt} = K_d A \frac{(C_s - C_t)}{x}$$

where C is the concentration: at saturation (C_s) and after time t (C_t). The equation can be solved as:

$$C_t = C_s - 0.811 (C_s - C_o)(e^{-K_d} + (1/9)e^{-9K_d} + (1/25)e^{-25K_d} + \dots)$$

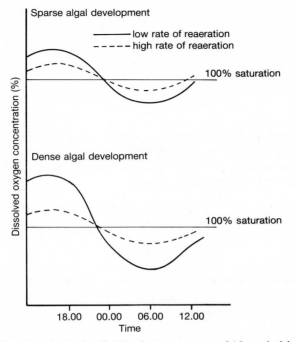

F<small>IG</small>. 1.5 Diurnal variation in dissolved oxygen over a 24 h period in a river with macrophytes and algae present.

where C_o is the concentration after time o and

$$K_d = \frac{K_d \pi^2 t}{4x^2}$$

The diffusion coefficient K_d can be expressed in mm^2s^{-1} or cm^2s^{-1}. The K_d for oxygen in water at 20°C is $1.86 \times 10^{-3} mm^2s^{-1}$.

Aeration in time or distance can be expressed:

$$\frac{dC_t}{dt} = K_2 (C_s - C_t)$$

integrating with limit $C_t = C_o$ at $t = 0$

$$\int_{C_o}^{C_t} \frac{dC_t}{C_s - C_t} = k_2 \int_0^t dt$$

i.e.

$$\log_e \frac{(C_s - C_t)}{(C_s - C_o)} = -K_2 t.$$

If D_t, D_o are the dissolved oxygen deficit at times t and 0 respectively, and K_2 is the reaeration constant:

$$\log_e \frac{D_t}{D_o} = -K_2 t$$

thus,

$$D_t = D_o e^{-K_2 t}.$$

A more useful parameter than the reaction constant (K_2) is the exchange coefficient (f). The exchange coefficient, also known as the entry or exit coefficient, is the mass of oxygen transferred across unit area of interface in unit time per unit concentration deficit.

$$f = \frac{K_d}{x}$$

$$\frac{dM}{dt} = fA \, (C_s - C_t)$$

and if a finite volume of water (V) is assumed, then:

$$\frac{dC_t}{dt} = \frac{dM}{dt} \frac{1}{V} = \frac{fA}{V} \, (C_s - C_t)$$

i.e.

$$\frac{dC_t}{dt} = K_2 \, (C_s - C_t)$$

where

$$K_2 = f \frac{A}{V}$$

$$f = K_2 \frac{V}{A} = K_2 \hbar$$

where V is the volume of water below interface, A is the area of the air–water interface, and $\hbar$ is the mean water depth.

The exchange coefficient f, is expressed in units of velocity (mm h^{-1}) and at 20°C in British rivers it can be estimated by the formula:

$$f = 7.82 \times 10^4 U^{0.67} H^{-0.85}$$

where U is the water velocity (m s^{-1}) and H the mean depth (mm). Typical values for f range from 20 for a sluggish polluted lowland river to over 1000 for a turbulent unpolluted upland stream. Values for the exchange coefficient for various aeration systems expressed in cm h^{-1} have been collated by Klein (1972b) and summarized in Table 1.22.

A rise in temperature can increase the rate of reaeration and vice versa. The

TABLE 1.22 Typical values found for the magnitude of the exchange coefficient f (Klein 1972b)

Aeration system	f(cm h^{-1})
Stagnant water	0.4–0.6
Water flowing at 0.4 m per min in a small channel	
water polluted by sewage	0.4
clean water	0.5
Water flowing at 0.6 m per min in a channel	1
Polluted water in dock and tidal basin	1–3
Sluggish polluted river (Sincil Dike)	2
Sluggish clean water about 51 mm deep	4
Thames Estuary under average conditions	5.5
Water flowing at 10.06 m per min in a small channel	7.5
The open sea	13
Water flowing at 14.94 m per min in a channel	30
Turbulent Lakeland beck	30–200
Water flowing down a 30° slope	70–300

reaeration rate constant (K_2) can be related to temperature (T) by:

$$K_{2(T)} = K_{2(20)} 1.047^{(T-20)}.$$

In general terms, an increase in temperature of 1 degree will result in an increase in the exchange coefficient f, by about 2%.

A number of physical factors affect reaeration. The transfer of oxygen at the air–water interface results in the surface layer of water becoming saturated with oxygen. If the water is turbulent as is the case in upland streams, the saturated surface layer will be broken up and mixing will ensure that reaeration is rapid. When no mixing occurs, as in a small pond to take an extreme example, then oxygen has to diffuse throughout the body of the water. In some cases the diffusion rate may be too slow to satisfy the microbial oxygen demand and anaerobic conditions may occur at depth. In rivers, velocity, depth, slope, channel irregularity, and temperature will all affect the rate of reaeration. To increase the rate of aeration and speed up the self-purification process, weirs can be built below discharges. In recent years, floating aerators have been employed on rivers during periods of high temperature when the critical point has fallen dangerously low. More sensitive rivers, containing salmonid fish, have been protected by pumping pure oxygen into the river at times of particular stress. This technique has been developed specifically to control deoxygenation caused by accidental discharges of pollutants (Anon. 1979a). The use of a compressor with a perforated rubber hose has also been successfully employed. In emergencies, for example where a deoxygenated plug of water is moving downstream, the local fire brigade has been able to prevent total deoxygenation by using the powerful pumps on their tenders to recirculate as much water as possible, with the water returned to the river via high-pressure hoses. The main advantage of this method is that the fire crews can make their way slowly downstream, keeping abreast of the toxic plug.

Thames Water Authority have acquired a barge which is capable of injecting 10 tonnes of pure oxygen per day into the River Thames. The barge is primarily for use in the estuary where the discharge of storm sewage from the combined sewers of London during periods of heavy rainfall reduces the dissolved oxygen concentration of the water to dangerously low levels. The oxygen is processed on board by a pressure swing adsorption plant, although liquid oxygen stored in special tanks could also have been used, and injected into the water via a 'Vitox' injection system (Sections 4.2.3.2). The barge is ideal for use in large rivers where it can quickly move to threatened areas (Griffiths and Lloyd 1985).

The oxygen sag curve
When an organic effluent is discharged into a stream it exerts a biochemical oxygen demand with the processes of oxygen consumption and atmospheric reaeration proceeding simultaneously. Although other processes, such as photosynthesis, sedimentation, and oxidation of the bottom deposits can also affect oxygen concentration, oxygen consumption and reaeration are the primary processes affecting oxygen status.

In many cases, the oxygen demand will initially exceed the reaeration rate, and the dissolved oxygen concentration will fall downstream of the outfall (discharge point). The rate of diffusion across the air–water interface is directly proportional to the oxygen deficit and if the rate of consumption lowers the oxygen concentration, the oxygen mass transfer rate will increase. At some point downstream the rate of reaeration and consumption become equal and the oxygen concentration stops declining. This is the critical point of the curve where the oxygen deficit is greatest (D_c) and the dissolved oxygen concentration is lowest (Fig. 1.6.). Thereafter reaeration predominates and the dissolved oxygen concentration rises to approach saturation. The characteristic curve which results from plotting dissolved oxygen against time or distance downstream is known as the oxygen sag curve. The long tail associated with the recovery phase of the curve is due to the rate of mass transfer of oxygen. As the river's dissolved oxygen concentration recovers, the oxygen deficit is reduced and as the rate of mass transfer is proportional to the oxygen deficit, the rate of reaeration slows, thus extending the curve. For example, water containing 10 g m^{-3} of oxygen but with a saturation concentration of 12 g m^{-3}, has an oxygen deficit of 2 g m^{-3} of oxygen. Because the rate of diffusion is directly proportional to the oxygen deficit, if the same water now contained only 4 g m^{-3} and thus has an oxygen deficit of 8 g m^{-3}, the oxygen would diffuse four times faster.

The shape of the curve remains more or less the same except that the critical point will vary according to the strength of the organic input. It is possible for the dissolved oxygen concentration to be reduced to zero and an anaerobic or

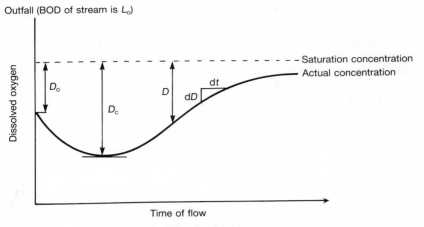

FIG. 1.6   Dissolved oxygen sag curve.

septic zone to be formed (Fig. 1.7). De-oxygenation is usually a slow process and the critical point may occur some considerable distance downstream of the outfall. The degree of de-oxygenation not only depends on the strength of the discharge, but also on dilution, BOD of the receiving water, nature of the organic material in terms of availability and biodegradability, temperature, reaeration rate, dissolved oxygen concentration of the receiving water, and the nature of the microbial community of the river.

The oxygen sag curve can be expressed mathematically for ideal conditions in terms of the initial oxygen demand, the initial dissolved oxygen concentration in the river, and the rate constants for oxygen consumption (K_1) and reaeration (K_2). These mathematical formulations were derived by Streeter and Phelps (1925) when working on the Ohio River. This large river

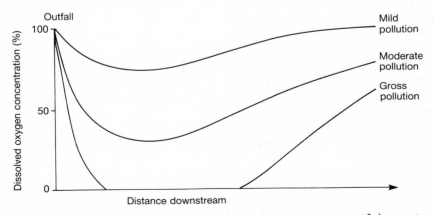

FIG. 1.7   The effect of an organic discharge on the oxygen content of river water.

had long uniform stretches between pollution discharges, also relatively little photosynthesis, and the only major factors affecting the oxygen status were oxygen consumption and reaeration. They considered that the rate of biochemical oxidation of the organic matter was proportional to the remaining concentration of unoxidized organic material, typified by the first-order reaction curve (see Fig. 3.7).

Assuming first-order kinetics, the oxygen demand with no aeration can be shown as:

$$\frac{dL_t}{dt} = -K_1 L_t$$

$$L_t = L_o - D_t \cdot d(L_o - D_t) = -dD_t$$

thus,

$$\frac{dD_t}{dt} = K_1 L_t.$$

Reaeration with no oxygen demand:

$$\frac{dC_t}{dt} = K_2 (C_s - C_t)$$

therefore,

$$\frac{d(C_s - C_t)}{dt} = K_2 D_t$$

thus,

$$\frac{dD_t}{dt} = -K_2 D_t.$$

It is possible to express both demand and aeration in terms of change in the oxygen deficit (dD_t/dt). Thus, for simultaneous oxygen demand and reaeration:

$$\frac{dD_t}{dt} = K_1 L_t - K_2 D_t$$

where D is the dissolved oxygen deficit at time $t(D_t)$, L is the ultimate BOD at time t (L_t) or initially (L_o), K_1 the BOD reaction rate constant, and K_2 the reaeration rate constant.

Providing oxygen is not a limiting factor, the oxygen demand is not dependent on the oxygen deficit. Thus, by substituting L_t according to the equation:

$$L_t = L_o e^{-K_1 t}$$

$$\frac{dD_t}{dt} = K_1 L e^{-K_1 t} - K_2 D_t.$$

When this equation is integrated with limit $D_t = D_o$ when $t = 0$:

$$D_t = \frac{K_1 L_o}{(K_2 - K_1)} \{e^{-K_1 t} - e^{-K_2 t}\} + D_o e^{-K_2 t}.$$

This is the well-known Streeter and Phelps equation.

By changing to base 10 ($K = 0.4343\ k$):

$$D_t = \frac{K_1 L_0}{(K_2 - K_1)} \{10^{-K_1 t} - 10^{-K_2 t}\} + D_0 10^{-K_2 t}.$$

The minimum dissolved oxygen concentration (the critical point) that occurs at maximum oxygen deficit D_t when

$$\frac{dD_t}{dt} = 0,$$

is given by:

$$\frac{dD_t}{dt} = 0 = L_1 L_o - K_2 D_c$$

thus,

$$D_c = \frac{K_1}{K_2} \cdot L_e{}^{-K_1 t_c}$$

therefore,

$$t_c = \frac{1}{K_2 - K_1} \cdot \log_e \left\{ \frac{K_2}{K_1} \left[1 - \frac{(K_2 - K_1))D_o}{K_1 L_o} \right] \right\}$$

where the critical (maximum) deficit (D_c) occurs at time t_c.

Both K_1 and K_2 in the model are assumed to be constant. However, although K_1 is measured by running a BOD determination in the laboratory, it may vary with time. The K_2 value will vary from reach to reach within the river and must be measured in the field. Both these constants are temperature functions and so temperature effects must be taken into consideration. For domestic sewage, K_1 approximates to 0.1 at 20°C, whereas K_2, which is mainly a function of turbulence, can be assessed using the equation developed by O'Connor and Dobbins (1958):

$$K_2 = \frac{(K_d U)^{1/2}}{H^{3/2}}$$

where K_2 is the reaeration coefficient (base e) per hour, K_d the diffusion coefficient of oxygen into water, U is the velocity of flow, and H the depth. Approximate values can be obtained. Low values represent deep, slow-moving rivers and high values shallow fast-flowing upland streams. In reality, K_2 is at best a crude estimate, and often an assumed value which can have severe effects on the predictive estimate. The measurement of K_1 is even more critical.

Boyle (1984) has examined the effects of biological films (including sewage fungus growths) on BOD decay rates (K_1) in rivers. He cites several examples where the dissolved oxygen sag below outfalls could only be modelled if K_1 was assumed to be an order of magnitude greater than expected. Clearly, the decay rate is enhanced by the presence of a biological film on submerged surfaces, although the decay rate was dependent on both nutrient concentration and water velocity. For example, Boyle and Scott (1984) quote K_1 values for a small English river receiving papermill waste and supporting sewage fungus of between 3.9–4.2 d, although other workers, mainly in New Zealand, have recorded BOD rate coefficients of up to 10.56 d.

The oxygen sag curve can be more accurately assessed by providing a third point on the curve. This is provided by the point of inflexion, that is where the net rate of aeration is at a maximum, when $(d^2D_t/dt^2)=0$ then:

$$t_i = \frac{1}{(K_2-K_1)} \log_e\left\{\left(\frac{K_2}{K_1}\right)^2\left[1-\frac{(K_2-K_1)\,D_o}{K_1L}\right]\right\}$$

where the inflexion deficit (D_i) occurs at time t_i.

It is now possible to plot the oxygen sag curve and to predict the minimum oxygen concentration downstream of a point discharge of organic waste such as sewage (Fig. 1.6).

Although the Streeter–Phelps model provides an extremely useful basis for the study of the sequence of events that occur in an organically polluted river, it must be applied with care, particularly to rivers where conditions change frequently, where there is appreciable photosynthesis, deposition of debris, and sediment, or discharges of inhibitory or toxic substances. The model is only valid for a single pollution discharge and where there is no dilution from tributaries. Where these occur the river must be split into discrete sections according to changes in flow or discharge, so that each section can be treated as an individual case and the model applied. The output data from one section provides the input data for the next, and in this way the entire river system can be covered to provide an overall calculation. This type of model is the basis of many other predictive water quality models where many other variables, such as benthic and nitrogenous oxygen demand, salinity and temperature are also able to be included (Forster *et al.* 1985).

In Ireland, An Foras Forbartha (AFF) has developed a deterministic, steady-state model based on a series (approximately 20) of integrated differential equations (McGarrigle 1984). The model has been developed for use on an Apple II microcomputer and can be used in an online mode. Like the Streeter–Phelps model the principle outputs from the AFF model are dissolved oxygen concentration and BOD. However, like most other river models it makes the following assumptions: (i) no longitudinal dispersion; (ii) effluent is mixed immediately and uniformly across the section of river; (iii) coefficients are uniform and constant throughout a reach; and (iv) that flow is steady and one-dimensional. Although none of these assumptions are completely justified they are not likely to lead to significant errors in the predictions made by the model. The factors influencing dissolved oxygen and BOD, are summarized in Table 1.23 and how they interrelate in Fig. 1.8. Although this model has been widely applied to clean sections of river with some success, problems arise in more organically polluted sections, with the BOD decay rate cited as a particular problem in the prediction of downstream BOD and dissolved oxygen (AFF 1984). AFF observed that, in the field, the BOD value drops much more quickly than can be predicted using a theoretical laboratory constant. The same problem as observed by Boyle (1984). For example, in the River Suir the BOD dropped from 80 mg l^{-1} to 20 mg l^{-1} over a distance of just 1700 m in under 80 minutes. The measured laboratory

TABLE 1.23   Principle factors affecting DO and BOD which are accounted for in the River Suir Ecological Model (AFF 1984)

Factors influencing dissolved oxygen directly
1. BOD decay
2. Atmospheric interchange
3. Plant respiration
4. Plant photosynthesis
5. Mud respiration

Factors influencing BOD
1. Temperature
2. Dissolved oxygen availability
3. Settlement
4. Resuspension

Factors influencing atmospheric interchange of dissolved oxygen
1. Temperature
2. Depth
3. Velocity
4. Nature of water surface
5. Weirs, waterfalls
6. Dissolved oxygen

Factors influencing extent of plant respiration
1. Plant biomass
2. Dissolved oxygen
3. Temperature

Factors influencing effect of mud respiration
1. Current velocity
2. Temperature
3. Dissolved oxygen
4. Water depth

Factors influencing plant photosynthesis
1. Plant biomass
2. Light intensity
3. Temperature
4. Depth

Factor influencing nitrification
1. Ammoniacal nitrogen concentration

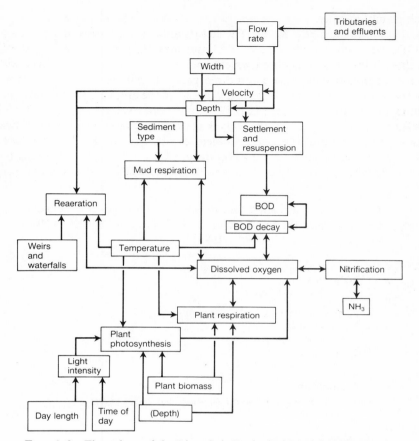

FIG. 1.8   Flow chart of the River Suir Ecological Model (AFF 1984).

decay constant of 0.000139 d, which appears very low, only accounted for a fraction of this drop in BOD. Clearly, the rapid assimilation by dense microbial growths such as sewage fungus must also be incorporated into model equations.

A final word of caution on the use of the Streeter–Phelps model. The model assumes that the flow does not vary over time, that the organic matter is distributed uniformly across the stream's cross-section, and that there is no longitudinal mixing. The effects of algae and bottom sediments are not considered in the equation. In reality, however, the dissolved oxygen sag curve can be affected by other factors apart from microbial oxygen demand and reaeration rate. Among those worthy of further consideration are photosynthesis – the addition of oxygen during the day and the uptake of oxygen by plant respiration at night, benthic oxygen demand, the removal of oxygen by gases released from the sediments and the release of soluble organic material from the sediments which has an oxygen demand, and finally, the input of oxidizable material from surface water. These inputs and the dissolved oxygen are constantly being redistributed within the water column by longitudinal mixing. Some of these factors can be easily predicted and built into the existing model, whereas other factors are less quantifiable.

Microbial interactions
The microbial response to organic enrichment in streams and rivers is essentially the same as those which occur in the biological unit processes at wastewater treatment plants. In natural waters these responses occur longitudinally, often occurring over many miles, whereas in treatment units these changes are accelerated, occurring over a much shorter distance and time basis within a single or series of reactors.

The rate of biological oxidation of organic waste is a time–temperature function, with the concentration of available oxygen decreasing as the temperature rises. It is possible to describe the changes that occur during the self-purification process over a time basis. A river polluted with organic matter responds in a characteristic way. This has been categorized into a number of stages or zones: a polluted zone, a recovery zone (which is normally split into two, depending on the rate of oxidation), and a recovered or clean zone (Fig. 1.9). The best known descriptive classification of the degree of pollution is the Saprobrien system, which uses four terms to describe these zones: *polysaprobic* (p), the heavily polluted or septic zone; the *mesosaprobic* zone the zone of recovery and active oxidation, which is split up into the *α-mesosaprobic* (α) zone of heavy pollution and the *β-mesosaprobic* (β) zone of moderate pollution. The final area is the *oligosaprobic* (o), the zone of very slight pollution and of complete oxidation.

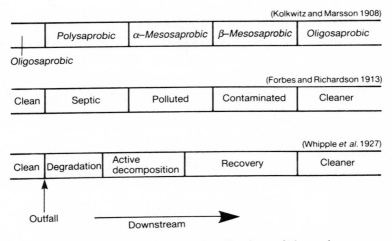

FIG. 1.9 Summary of early descriptive classifications of the various zones of a polluted river.

Immediately below an outfall the density of bacteria rapidly increases in response to the increase in available organic substrate with direct counts of up to 36×10^6 ml^{-1} common (Fig. 1.10). Most of these bacteria are suspended with <10% attached to surfaces (Edwards and Owens 1965), but as the effluent proceeds downstream attached growths of bacteria and fungi dominate, utilizing the breakdown products of polysaccharides and producing a thick growth which covers the entire surface of the bottom substrate. Sewage fungus takes a number of macroscopic growth forms and can form fronds, cotton wool-like growths or gelatinous growths in rivers depending on the flow rate and the form of carbon substrate available. This high level of

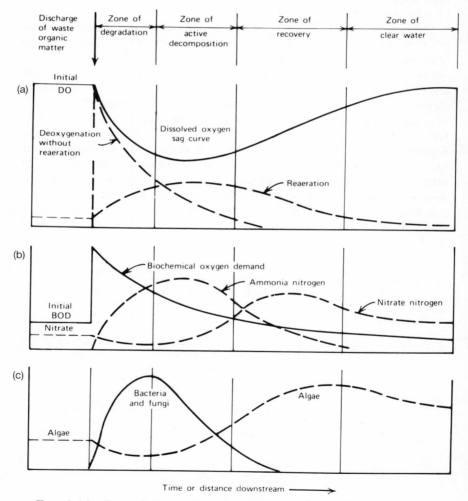

FIG. 1.10 General effects of organic pollution in streams (Hammer 1977).

heterotrophic activity exerts a huge oxygen demand which may exceed the available oxygen, causing de-oxygenation, but also rapidly utilizes the available BOD. As bacterial breakdown of proteinaceous compounds continues, ammonia and phosphorus are released, which increases in concentration downstream. As degradation proceeds, less suitable carbon substrate is left for heterotrophic activity and this subsequently declines and is eventually replaced by algae. Stimulated by the high nitrogen and phosphorus concentration, filamentous algae are first to colonize the stream and complete the oxidation of organic nitrogen to nitrate. There is a discernible change in the microbial population as the organic matter is oxidized (Fig. 1.11). Bacteria

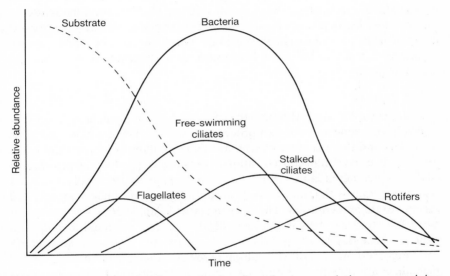

FIG. 1.11 Relative abundance of micro-faunal groups relative to remaining substrate. Bacteria thrive and finally become prey of the ciliates, which in turn are food for the rotifers and crustaceans.

decrease downstream of the outfall due to natural death and predation by ciliate protozoans. Further downstream, as the protozoan population becomes food limited because of low bacterial numbers, the rotifers and crustaceans increase. The rotifers and crustaceans not only feed on the ciliates, but are able to utilize the remaining bacteria due to the reduced competition from the ciliates. Within the stream the pathogenic bacteria naturally die off rapidly although predation by protozoans is also an important factor.

The conditions that characterize each zone are summarized below.

Polysaprobic zone. This zone occurs directly below the discharge point and rapid breakdown occurs so that anaerobic conditions can occur depending on the organic load and the assimilative capacity of the receiving water. The higher degradation stages of the proteins are present, partly as peptones, polypeptides, oligopeptides, and peptides, but degradation can extend to amino acids. Chemically, this zone is characterized by the presence of albumens, polypeptide, and carbohydrates, with hydrogen sulphite, ammonia, and carbon dioxide being produced as end-products of anaerobic digestion. Physically, the water has a dirty-grey colour, a faecal or mouldy smell and is turbid due to the presence of enormous quantities of bacteria and colloids present. The bottom of the watercourse can be covered with a black digesting sludge and the reverse sides of stones will be coloured black by a coat of iron sulphide. Most autotrophic organisms are missing, although bacteria will be abundant reaching densities of $> 10^6$ ml^{-1}. If the organic matter is sewage, then *Escherichia coli* will be abundant although absent if the organic waste is from an industrial or vegetable processing source. Other microorganisms are scarce but some blue-green algae, flagellate protozoans and amoebae are present.

α-Mesosaprobic zone. Although the level of pollution is still heavy, recovery begins in this zone as oxidation processes speed up. There are no anaerobic sediments and there is more oxygen available to allow aerobic oxidation to proceed. There is a high concentration of breakdown products such as amino acids and their degradation products, mainly fatty acids. Physically, the water is grey in colour with mouldy smells produced due to residues of protein and carbohydrate fermentation. The oxygen status is still $< 50\%$ saturation but never falls to zero. This is the zone of most active microbial activity and although bacterial density has fallen and is $< 10^5$ ml^{-1}, filamentous bacteria and fungi are common, often resulting in sewage fungus growths developing. Few algae are present and both flagellate (*Bodo* spp.) and ciliate protozoa (*Paramecium* spp. and *Colpidium* spp.) are common.

β-Mesosaprobic zone. This is still a zone of active oxidation, although the level of pollution has been reduced significantly. Degradation products such as amino acids, fatty acids, and ammonia are found in low concentrations, although ammonical compounds are abundant. The water has plenty of available oxygen which never falls below 50% saturation, although diurnal variations in dissolved oxygen are possibly due to photosynthesis. Degradation no longer affects the oxygen status of the water as much. The water is physically cleaner, being only slightly turbid, and is free from odour and any discolouration. The bacterial concentration is always $< 10^5$ ml^{-1} and bacteria are no longer the dominant organisms, with filamentous algae, such as *Cladophera* (blanket weed), dominating. The protozoans are dominated by the ciliates, with stalked species (*Peritrichia*) being abundant.

Oligosaprobic zone. All the waste products have now been broken down to stable organics and inorganic salts. The dissolved oxygen concentration is normally 100%, although if algae is still present there may be some diurnal fall. The water is clear, odourless and colourless, with bacterial density < 100 ml^{-1}. Filamentous algae are largely replaced by macrophytes and mosses, although diatoms and a few green or blue-green algae may be present (Hawkes 1972).

The changes occurring during self-purification by the microbial component of the river are summarized in Fig. 1.10. Not all the zones may be present below an outfall, for example, if there is sufficient assimilative capacity available in the river system, then only the β-mesosaprobic conditions occur before the river returns to its natural oligosaprobic state.

Dispersed bacteria, present either as individual cells or as small suspended flocs, and free-living protozoans in rivers and streams are essentially the same micro-organisms responsible for biological wastewater treatment in mixed reactors such as activated sludge. Attached micro-organisms, such as bacterial and fungal slimes, stalked protozoans, and algae are similar to those growing on fixed-film reactors, e.g. percolating filters.

Sewage fungus

Sewage fungus growths are predominatly heterotropic communities that lie between the autotrophic–heterotrophic and the heterotrophic–phototrophic continuum. The term 'sewage fungus' was devised by Butcher (1932) who felt that such growths were generally associated with sewage and formed fungus-like growths. This term is rather misleading as sewage fungus is not only associated with sewage (Table 1.24) but with all organic effluents. Also, as fungi are rarely a major component of such growths, the term heterotrophic slime would seem to be more descriptive and appropriate. Slime-forming organisms are probably part of the normal flora of all rivers, and these attach

TABLE 1.24 Percentage of outbreaks of heterotrophic slime caused by specific effluent sources or mixtures of effluents in Irish rivers ($n = 148$)

%	Effluent source
16.2	Farms
14.9	Agricultural industries
14.2	Domestic sewage
8.8	Domestic sewage and agricultural industries
7.4	Waste-tips
6.8	Domestic sewage and industrial
6.1	Industrial
2.7	Domestic sewage and farms
2.0	Industrial and agricultural industries
0.7	Other

themselves to any suitable stable material to form visible macroscopic slimes only when there are significant amounts of readily assimilable organic nutrients to serve as growth substrate. Slimes are complex assemblages of micro-organisms – filamentous bacteria, fungi, zoogloeal bacteria, proto-zoans, and occasionally algae (Table 1.25). A grazing population of protozoans, rotifers, and macro-invertebrates are supported by feeding off the slime. In Ireland, three slime-forming organisms are frequently found forming these growths in rivers, two bacteria *Sphaerotilus natans* and *Zoogloea* spp., and a fungus, *Leptomitus lacteus* (Gray 1982a, 1987) (Table 1.25).

TABLE 1.25 The occurrence of the commonest slime-forming organisms expressed as a percentage of the total sites examined in the UK (Curtis and Harrington 1971), and Ireland

Organism	UK			Ireland		
	Dominant	Secondary	Total	Dominant	Secondary	Total
Sphaerotilus natans	52.1	37.1	89.2	52.8	23.3	76.1
Zoogloea spp.	58.5	34.0	92.5	11.1	43.3	54.4
Beggiatoa alba	6.3	21.4	27.7	5.5	23.3	28.8
Carchesium polypinum	6.3	10.1	16.4	2.8	0	2.8
Geotrichium candidum	4.4	3.1	7.5	0	0	0
Flavobacterium sp.	3.1	37.1	40.2	0	0	0
Leptomitus lacteus	3.1	0.6	3.7	22.2	3.3	25.5
Fusarium aquaeductuum	1.9	0	0	5.5	0	5.5
Stigeoclonium tenue	3.1	7.6	10.7	0	3.3	3.3

Sphaerotilus natans is a filamentous bacterium made up of Gram-negative non-sporing rod-shaped cells with rounded ends, each $1–4 \ \mu m \times 4–10 \ \mu m$ in size, enclosed within a sheath of varying thickness. It uses simple sugars such as glucose, maltose, sucrose, fructose, and mannose, although amino acids, glycerol, and even organic acids can be utilized if sugars are not available. Its nitrogen requirement can be satisfied, using organic nitrogen sources in the form of amino acids or short-chain peptides, whereas inorganic nitrogen can only be used if vitamin B_{12} is also available. Calcium is required to develop the sheath which is a protein–polysaccharide–lipid complex. The capsule is a simple polysaccharide, the composition of which varies with the nutrient regime. The cells contain globular inclusions that are food reserves of poly-β-hydroxybutyrate, which can make up to 40% of the dry weight of the cell. A high C:N ratio normally results in an increased formation of food reserves (Fig. 1.12). The bacterium is a strict aerobe and is restricted by dissolved oxygen concentrations of $< 1 \ mg \ l^{-1}$. It can exert an enormous demand for oxygen in the river, requiring 10–20 times more oxygen than the equivalent biomass of macrophytes. It has a wide temperature tolerance, being found in waters between 4–40°C, although its optimum growth rate occurs between

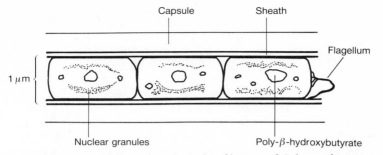

FIG. 1.12 Diagramatic section through a filament of *Sphaerotilus natans*.

25–30°C. *Sphaerotilus natans* requires a minimum water velocity of 0.05 m s^{-1} to ensure oxygen and nutrient transfer, but at velocities $>0.6 \text{ m s}^{-1}$ the growth is scoured away. The pH for growth varies between 6.8–9.0, but in more acidic waters slimes are dominated by fungi. *Sphaerotilus natans* tends to form slimey fronds in rivers, whereas zoogloeal bacteria are restricted to slow-flowing waters and form more gelatinous growths which are easily broken up and washed away. *Zoogloea* is not well defined taxonomically but includes *Zoogloea ramigera*, *Pseudomonas* spp., and zoogloeal forms of *S. natans*. Zooloeal bacteria comprise Gram-negative non-sporing cells $(0.5–1.0 \ \mu\text{m} \times 1.0–3.0 \ \mu\text{m})$ not arranged in filaments but embedded in a gelatinous matrix, forming lobed and unlobed spherical masses (Curtis 1969; Gray 1982a). They have the same nutritional requirements to *S. natans*. In Ireland, the fungus *Leptomitus lacteus* is a major slime-forming organism. It has a macroscopic growth form similar to that of *S. natans* except that it is less slimey in texture as no external mucilage is produced. Unlike *S. natans* it does not form fronds but long characteristic streamers composed of overlapping spherical cotton wool-like growths. This *Phycomycete* is non-septate (no cell walls) but has constrictions at irregular intervals along the hyphae. Spherical plugs, made out of cellulin, move along the length of the hyphae between constrictions, blocking the constricted gaps and preventing movement of cellular material along the hyphae. The plugs are composed of a polysaccharide which remain in the slime matrix, even after the fungal mycelium has degraded, leaving a high density of the spherical plugs mixed with the remaining slime. The fungus requires a high dissolved oxygen concentration and grows preferentially at acid pH values. Sugars do not support growth and it proliferates in wastes rich in acetate, most low-molecular weight fatty acids, and is especially associated with waste-tip leachates, dairy wastes, and paper pulp wastes (Gray 1985a). *Leptomitus lacteus* is found in moderate- to fast-flowing water and normally dominates growths upstream of *S. natans* (Fig. 1.13).

It is difficult to quantify the effects of outbreaks of heterotrophic slime in rivers, although a survey (Gray and Hunter 1985) found that the major effects

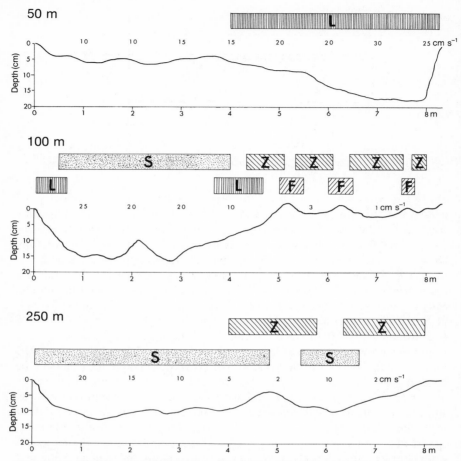

FIG. 1.13 Transects across the River Big in Co. Louth at 50 m, 100 m, and 250 m below the effluent outfall. Complete mixing of the effluent with the receiving water occurred at 70 m. The diagram shows the depth profile across the river (cm), the water velocity at 1 m intervals along each transect (cm s^{-1}) and the dominant slime-forming organism present. L, *Leptomitus lacteus*; S, *Sphaerotilus natans*; Z, zoogloeal bacteria; F, *Fusarium aquaeductuum*.

was damage to the amenity value of the river, while and actual damage to fish was restricted to 40% of the outbreaks. Severity of the problem is related to the length of outbreaks with damage to fish stocks and the problem of sloughed flocs increasing with length. The percentage of sites where slimes caused no adverse effects decreased with increasing length (Table 1.26). The length of river affected by slime growth can be extensive and in Ireland outbreaks are generally greater than in England and Wales, with 55.1% of outbreaks <0.8 km in length and 31.4% >1.6 km in Ireland, compared with 73.5% and

TABLE 1.26 The influence of effluent source, total length, and duration of heterotrophic slimes on the severity of effects measured in Irish rivers

	n	Effects of slime outbreaks						
		Appearance and amenity (%)	Smell and de-oxygenation (%)	Smell only (%)	De-oxygenation only (%)	Sloughed flocs (%)	Damage to fish (%)	None (%)
Effluent source								
Farm	26	92.3	46.2	3.9	7.8	11.5	34.6	11.5
Agricultural industry	27	92.6	48.1	7.4	7.4	55.6	66.7	7.4
Industrial	14	100.0	57.1	28.6	42.9	35.7	42.9	7.1
Domestic sewage	53	84.9	50.9	11.3	5.7	22.6	22.6	17.0
Waste-tip	4	50.0	25.0	25.0	0	0	25.0	25.0
Length of outbreak								
0–20 m	11	72.7	27.3	9.0	0	0	9.0	36.4
20–100 m	14	78.6	21.4	0	0	14.3	21.4	28.6
100–500 m	21	85.7	57.1	9.5	19.1	14.3	19.1	14.2
0.5–1 km	13	100.0	53.9	0	15.4	30.8	46.2	0
1–5 km	18	77.8	50.0	38.9	22.2	38.9	55.6	5.6
5 km+	12	100.0	50.0	16.7	0	50.0	75.0	0
Duration								
Permanent	21	85.5	66.7	23.8	19.1	38.1	47.6	9.5
Spring	7	100.0	57.1	0	0	14.3	28.6	0
Summer	27	88.9	48.2	7.4	11.1	22.2	40.7	18.5
Autumn	12	91.7	41.7	16.7	0	41.7	50.0	16.7
Winter	11	100.0	54.6	18.2	0	54.6	72.7	0

15.5% respectively in England and Wales (Curtis 1972; Gray and Hunter 1985). Between 1982 and 1983, 34% of outbreaks in Ireland were longer than 1 km and 13.5% in excess of 5 km, reaching a maximum length of 29 km in the River Barrow below a sugar beet factory. Large outbreaks have been recorded elsewhere in the world, for example, a 64 km outbreak on the River Altamaha in Georgia, USA, caused by a Kraft process waste (Phaup and Gannon 1967). Heterotrophic slimes can have direct or indirect effects on the river ecosystem and are summarized in Fig. 1.14. Such slimes can remove soluble carbohydrate extremely rapidly from solution. Starch and intermediate polymers are not removed by the slime, but short-chain carbohydrates (mono- to pentasaccharides) are readily utilized. Under ideal conditions, the biomass of slime produced is proportional to the concentration of organic carbon in the effluent, when this is in the form of glucose or acetate (Curtis 1972). The rate of removal of organic carbon is also directly proportional to the slime biomass at a rate of 0.3 g C g^{-1} dry weight of slime d^{-1}, although less than 20% of the organic carbon is incorporated into the slime (Curtis *et al.* 1972). Although slimes are effective in removing nutrients from solution, this does not involve a complete removal from the river as nutrients can be released back into solution by the slime. The economic coefficient (i.e. the ratio of pollutant incorporated into slime to the total amount removed from solution) of river slimes varies from 60% at the time of colonization, falling to 11% as heavy slimes build up (Curtis 1969).

It is difficult to estimate the severity of a sewage fungus outbreak and normally only total length or presence/absence of outbreaks are used. A knowledge of the position of the recovery zone below a sewage fungus outbreak allows the exact length of river affected to be determined. Also maximum oxygen demand exerted by the slime will occur upstream of the recovery zone and so it is important to know where the recovery zone occurs. If the degree of pollution increases then the recovery zone will be pushed further downstream and vice versa. The resultant eutrophication zone may be as extensive as the sewage fungus outbreak and causes many problems. Rapid and accurate plotting of the development of the various zones is therefore very useful. Gray (1987) has used three indices to evaluate sewage fungus growth in rivers and streams.

The first index allows the degree of recovery to be determined at any particular site along the watercourse (Table 1.27). This identifies exactly where the site is in relation to the recovery curve (Fig. 1.10) and whether the situation is improving or getting worse.

The whole width of the river should be examined for 2–5 m upstream. Always work upstream so that the area under examination is not damaged. Remember that the fronds of sewage fungus will be broken off by your feet and could be mistaken for sloughed material downstream. The index is calculated by estimating the degree of heterotrophic growth and subtracting the estimate for phototrophic growth. This gives a rating for the river, which, if positive,

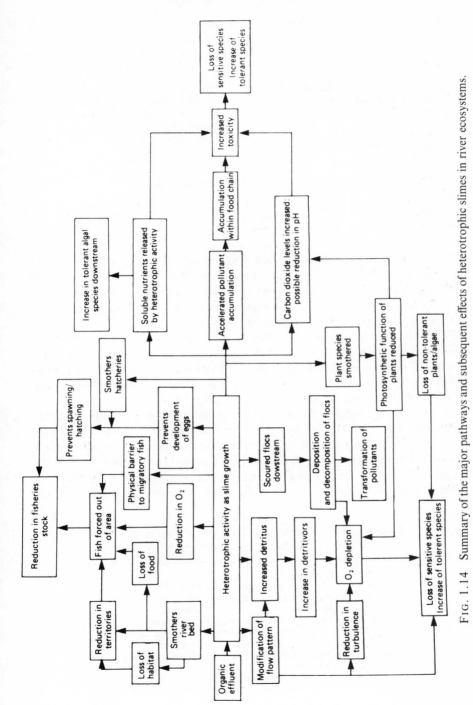

FIG. 1.14 Summary of the major pathways and subsequent effects of heterotrophic slimes in river ecosystems.

TABLE 1.27　Index to determine the degree of recovery in rivers containing sewage fungus

Heterotrophs	Score	Phototrophs
Not visible on hand-held boulders	1.	Not visible on hand-held boulders
Visible on hand-held boulders	2.	Visible on hand-held boulders
Present as clearly visible colonies on river bed	3.	Present as clearly visible colonies on river bed
Covering many surfaces	4.	Covering many surfaces
Covering most surfaces	5.	Covering most surfaces
Covering all surfaces	6.	Covering all surfaces

Degree of recovery = Heterotrophic rating − Phototrophic rating
Where:

$$+5 \longleftarrow \qquad\qquad 0 \qquad\qquad \longrightarrow -5$$

Heterotrophic　　　　　　　Recovery zone　　　　　　　Phototrophic

indicates heterotrophic activity predominating (i.e. there is an ample supply of organic matter to support sewage fungus growth), whereas a negative rating indicates a high level of algal growth (i.e. reduced organic matter but abundant organic and inorganic forms of nitrogen). The closer to zero the rating is then the closer you are to the recovery zone.

The second index is purely descriptive and allows a rapid estimation of the degree of recovery (Table 1.28). It is less sensitive than the first index but is useful for comparing the situation at individual sites over a long time period, as the length of the various zones will vary according to factors such as organic load, river flow, temperature, and pulsing frequency.

Once a sewage fungus outbreak has been identified then the severity of the growth can be estimated by using the third index (Table 1.29). Where deep rivers are studied then examination will be restricted to the banks only and the 'degree of cover' section of the index omitted and the index scored out of a total of nine. When very small or shallow streams are examined it may be necessary to modify the sloughed floc section of the index to present (1) or absent (0) and

TABLE 1.28　Descriptive index of the degree of hetero-trophic and phototrophic growth in rivers

1. Heterotrophs only
2. Heterotrophs dominant: some algal growth visible
3. Heterotrophs dominant: algal growth common
4. Heterotrophs dominant: algal growth abundant
5. Heterotrophs and phototrophs equally abundant
6. Phototrophs dominant: sewage fungus abundant
7. Phototrophs dominant: sewage fungus common
8. Phototrophs dominant: small colonies of sewage fungus
9. Phototrophs only
10. Discrete colonies of phototrophs only

TABLE 1.29 Index of sewage fungus development in rivers and streams (0, lowest; 12, highest)

Score	0	1	2	3
Cover	Occasional/rare 0–1%	Common 1–20%	Frequent 20–40%	Abundant >40%
Frond length	Thin film only no fronds visible	Visible fronds formed	Short–medium fronds > 50 mm	Medium–long fronds > 100 mm
Sloughed flocs	None	Occasional (small particles)	Common (small–medium particles)	Heavy (medium–large particles)
Surface mats	No	Small area	Large area	—
Algae	Present	Absent	—	—

the index scored out of 10. Initial experiments have shown that the standing
crop of sewage fungus is directly related to the rank obtained using this index.
It is not possible to give exact biomass values for particular ranks as this differs
for each river and effluent source. However, by plotting this association, the
oxygen demand on the river exerted by the sewage fungus can be predicted
when required, or an estimation made of the total biomass of sloughed flocs
being released downstream. A review on heterotrophic slimes has been
prepared by Gray (1985a).

1.4.2   Biochemical oxygen demand

1.4.2.1   The test

The most important effect that organic wastes can cause in receiving waters is
a reduction in the dissolved oxygen concentration, which is due to the
microbial breakdown of the organic matter present. It is possible to determine
the theoretical oxygen demand of a specific compound in wastewaters from
the stoichiometry of its oxidative breakdown, although it is impossible to
calculate the oxygen demand in this way for complex wastes such as domestic
sewage. In order to determine the gross oxygen demand that will be exerted in
a river or a wastewater treatment plant, a test is required that will estimate the
amount of oxygen needed to oxidize all the compounds present, both the
major and minor components of the waste.

Although the total organic carbon (TOC) content of the waste could be
measured, using a carbon analyser, it is more useful, in terms of predicting
effects in watercourses, to measure the oxygen demand that will be exerted by
these wastes on the watercourse. There are two widely used measures of
oxygen demand: chemical oxygen demand (COD), which measures the
organic content in terms of biodegradable and non-biodegradable compounds, and the biochemical oxygen demand (BOD) test, which measures the
biodegradable fraction of the wastewaters by monitoring the assimilation of
organic material by aerobic micro-organisms over a set period of time under
strictly controlled conditions. The COD test employs a potassium dichromate
reflux with concentrated sulphuric acid using silver sulphate (Ag_2SO_4) as
catalyst and mercuric sulphate ($HgSO_4$) to complex any chlorides present
which could interfere with the reaction. The sample is refluxed for 2 h in an
acidified potassium dichromate solution of known strength so that the
amount of oxidizable organic matter in the sample is proportional to the
potassium dichromate consumed in the oxidation reaction. The excess
dichromate is titrated in the ferrous ammonium sulphide to calculate the
amount of dichromate consumed. Although nearly all organic compounds are
oxidized by this procedure, some aromatic compounds such as benzene,
pyridine, and toluene are either unaffected or only partially oxidized during

the test. The COD will always be higher than the BOD as the former includes substances that are chemically oxidized as well as biologically oxidized. The ratio of COD:BOD provides a useful guide to the proportion of organic material present in wastewaters that is biodegradable. However, some polysaccharides such as cellulose can only be degraded anaerobically and will not be included in the BOD estimation. The COD:BOD relationship varies from 1.25 to 2.50 depending on the waste being analysed. The ratio increases with each stage of biological treatment as biodegradable matter is consumed but non-biodegradable organics remain and are oxidized in the COD test. The relationship remains fairly constant for specific wastes, although the correlation is much poorer when the COD values are < 100 mg $O_2 l^{-1}$ (Aziz and Tebbutt 1980). This correlation can be expressed by the simple linear regression equation:

$$COD = a \times BOD_5 + b$$

where a and b are constants, the value of which depends on the wastewater. For domestic wastewater:

$$COD = 1.64 \times BOD + 11.36 \text{ (Ademoroti 1986)}.$$

The biochemical oxygen demand test (BOD), often incorrectly referred to as the biological oxygen demand test, is a laboratory simulation of the microbial self-purification process occurring in rivers. The test measures the amount of oxygen consumed in 5 days at a temperature of 20°C by the biological oxidation of any biodegradable organic material present. The oxygen is consumed by the micro-organisms, mainly bacteria, via respiration and metabolism. The organic matter is broken down to carbon dioxide, although some of it is incorporated into cellular material or oxidized for energy. If the sample contains large amounts of organic matter the micro-organisms will require proportionately larger volumes of oxygen in order to degrade it. The amount of dissolved oxygen consumed does, however, depend on temperature and the duration of the test. Originally, the test was carried out at 18.3°C (65°F) for 5 days, the reason being that British rivers do not have a flow time to the sea in excess of 5 days and have a mean summer temperature of 18.3°C. So the use of these values ensured that the maximum oxygen demand which could occur under British conditions would be measured for each sample.

Not all the substrate within the BOD bottle will be oxidized to CO_2, some will be converted to new cells. Thus, if a simple organic source like glucose is oxidized both chemically and biologically, there will be a discrepancy. For example, the COD test will predict an oxygen consumption of 192 g $O_2 mol^{-1}$ of glucose compared with only 150 g $O_2 mol^{-1}$ using the BOD test. The BOD test, therefore, does not give a measure of the total oxidizable matter present in wastewaters, because of the presence of considerable quantities of carbonaceous matter resistant to biological oxidation. However, it does indicate the potential possessed by a wastewater for de-oxygenating a river or stream. The test also provides a useful theoretical example of the oxygen balance in aquatic ecosystems, thus allowing a clearer understanding of the role of micro-organisms in oxygen–food limited environments (Stones 1981).

Complete breakdown of even the most biodegradable wastes can take several weeks, therefore during the 5-day test only a proportion of the organic material will be broken down. Some organic materials, such as cellulose, can remain virtually unaffected by aerobic micro-organisms, only being broken down anaerobically. When the organic fraction has been aerobically broken down as completely as possible, the oxygen consumed is termed the ultimate BOD or ultimate oxygen demand. The test can incorporate two discrete oxygen demands forming the characteristic BOD curve (Fig. 1.15). The basic

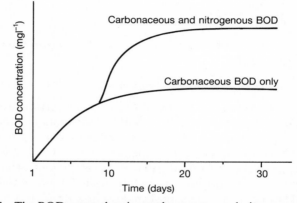

FIG. 1.15   The BOD curve showing carbonaceous and nitrogenous oxidation.

curve represents the carbonaceous material, which can take up to 3 weeks to be fully degraded at 20°C. The second source of oxygen demand comes from the oxidation of the nitrogenous material present (nitrification). The oxygen, demand from nitrification is only important in wastewaters. In raw wastewaters nitrification only becomes a significant source of oxygen demand after 8–10 days, whereas in partially treated effluents nitrification can dominate the oxygen demand after just a few days (Fig. 1.16).

The standard 5-day BOD test (BOD_5) measures only the readily assimilable fraction of organic material present in a wastewater, with the low-molecular weight carbohydrates in particular being readily utilized. However, the BOD_5 gives a far more reliable estimation of the possible oxygen demand that a waste will exert on a receiving river than the COD test, as the latter also measures the more refractory (non-biodegradable) compounds. Because of the similarity between the self-purification process and wastewater treatment process, the BOD test has been widely used as a measure of organic strength of river water

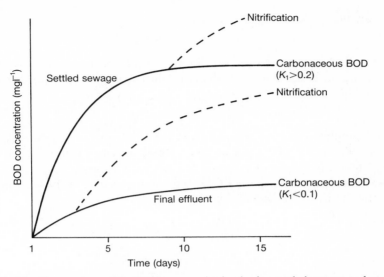

FIG. 1.16 Comparison of the BOD curves obtained using settled sewage and treated wastewater.

and effluents. The low capital cost, unlike TOC analysers, and low running costs of the test have ensured that it remains popular even today, some 70 years after its introduction by the Royal Commission on Sewage Disposal in 1913, although a similar test was being used as early as 1868 (Phelps 1944). However, there are many problems associated with the BOD test, most of them are associated with the way it is carried out (Section 1.4.2.2). The test is used for numerous purposes, including assessing the quality of river water, the strength of wastewaters, the assimilative capacity of receiving waters, and the effect of effluent discharges on receiving waters, as well as being used in the design and operation of treatment processes.

Stoichiometry
With domestic wastewaters only 60–70% of the total carbonaceous BOD is measured within 5 days at 20°C (BOD_5), and only the most biodegradable fraction utilized. For most materials, an incubation period of about 20 days (BOD_{20}) is required for complete breakdown, even though some more recalcitrant organic compounds, such as some polysaccharides, will not have been degraded even then. The test is essentially the oxidation of carbonaceous matter:

$$C_xH_yO_z \rightarrow CO_2 + H_2O$$

However, whereas this first stage may be the only component of the BOD curve, often a second stage is present – nitrification. The oxidation of nitrogenous matter proceeds as:

$$NH_3 \rightarrow NO_2^- \rightarrow NO_3^-$$

Glucose is used as a reference for the BOD test and is also useful for examining the stoichiometry of the test. Glucose is completely oxidized as:

$$C_6H_{12}O_6 + 6O_2 \rightarrow 6CO_2 + 6H_2O$$
(glucose)

For complete oxidation, a glucose solution of 300 mg l^{-1} concentration will require 320 mg l^{-1} of oxygen at 20°C. However, using the standard 5-day BOD_5 test only 224 mg of oxygen is utilized, with complete oxidation taking longer than 5 days. Thus, the BOD_5 only measures part of the total oxygen demand of any waste, and in this case:

$$\frac{BOD_5}{BOD_{20}} = \frac{224}{320} = 70\%$$

Kinetics

The BOD has been traditionally modelled as a continuous first-order reaction (Section 3.1.2), so that the rate of breakdown of carbonaceous material is proportional to the amount of material remaining. In this type of reaction the rate of breakdown is at first rapid when the organic content is high, but gets progressively slower as the organic material is utilized. This can be expressed as:

$$\frac{dL}{dt} = -K_1L$$

where K_1 is the BOD reaction rate constant and L the ultimate BOD (carbonaceous only). This integrates to:

$$L_t = L_o e^{-K_1 t}$$

where the initial BOD (L_o) is L_t after time t. The amount of oxygen consumed during the BOD test period (Y) is:

$$Y = L_o - L_t.$$

Thus

$$Y = L_o(1 - e^{-K_1 t})$$

or using base 10:

$$Y = L_o(1 - 10^{-K_1 t}).$$

Thus, for a test where 65% of the carbonaceous material is broken down within the 5 days, K_1 will equal 0.223 d^{-1}, and the removal rate is approximately 20% per day. Therefore, 95% removal will take 13 days and

99%removal 21 days, although adherence to the relationship between K_1 at base e and base 10 is:

$$K_e = 2.303\ K_{10}.$$

It is convention to quote K_1 to the base 10. The rate constant K_1 varies according to the quantity and nature of the organic matter present, the temperature and the type of micro-organisms in the wastewater. This can be best illustrated by considering the way in which micro-organisms utilize the available organic material present. Essentially, two reactions take place within a BOD bottle, a rapid synthesis reaction in which there is a rapid consumption of oxygen due to the high concentration of available organics, which is characteristic of raw wastewaters or effluents high in low molecular weight carbohydrates, followed by a slower endogenous metabolism (Fig. 1.17). In treated effluents, most of the organics originally present in the wastewater have been removed and oxygen is consumed at the lower endogenous rate. Therefore, the greater the rate of reaction due to the concentration of assimilable organic material, the larger the K_1 value. The average BOD rate constant at 20°C ranges from 0.04–0.08 for rivers with low pollution, 0.06–0.10 for biologically treated effluents, 0.12–0.22 for partially treated effluents, and those using high-rate systems, to 0.15–0.28 for untreated wastewaters. It is possible for samples with different reaction rates to have the same BOD_5 (Fig. 1.18).

The rate constant K_1 and the ultimate BOD (L) are traditionally calculated using graphical methods. Three techniques are widely used, the method of moments developed by Moore *et al.* (1950), the log-difference method, and the graphical method of Thomas (1950). Sheehy (1960) subsequently developed a

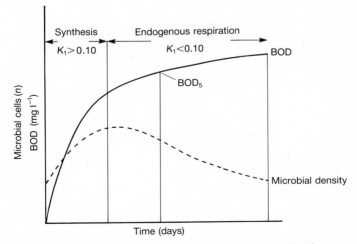

FIG. 1.17   Reactions which occur in the BOD bottle.

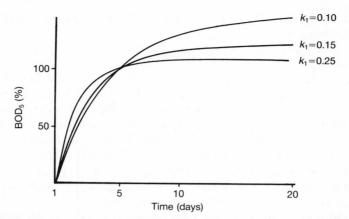

FIG. 1.18   The effect of various rate constants in the calculation of the same BOD.
The result is expressed as a percentage (Tebbutt 1983).

specially developed slide rule and today a range of computer models are
available to calculate these values.

Of all the available methods of calculating the BOD constants, the Thomas
method is perhaps the simplest. The procedure is based on the function:

$$\left(\frac{t}{y}\right)^{1/3} = (2.3 \; K_1 L)^{-1/3} + \frac{K_1^{2/3}}{3.43 \; L^{1/3}} \cdot t$$

where y is the BOD exerted in time t, K_1 the reaction rate constant (base 10)
and L the ultimate BOD.

This equation forms a straight line with $(t/y)^{1/3}$ plotted as a function of time
t. The slope $K_1^{2/3}/(3.43 \; L)^{1/3}$ and the intercept $(2.3 \; K_1 L)^{-1/3}$ of the line of
best fit of the data is used to calculate K_1 and L.

Using the form $Z = a + bt$ for the straight line where $Z = (t/y)^{1/3}$,
$a = (2.3 \; K_1 L)^{-1/3}$ and $b = K_1^{2/3}/(3.43 \; L)^{1/3}$:

$$K_1 = 2.61 \; (b/a)$$

$$L = \frac{1}{2.3 \; K_1 a^3}.$$

For example, over a 10-day period the BOD was measured every second day.
From this data $(t/y)^{1/3}$ can be calculated:

$t(d^{-1})$	2	4	6	8	10
$y(mg \; l^{-1})$	14	22	27	30	32
$\left(\dfrac{t}{y}\right)^{1/3}$	0.523	0.567	0.606	0.644	0.679

The graph of $(t/y)^{1/3}$ is plotted against t (Fig. 1.19) and from this the

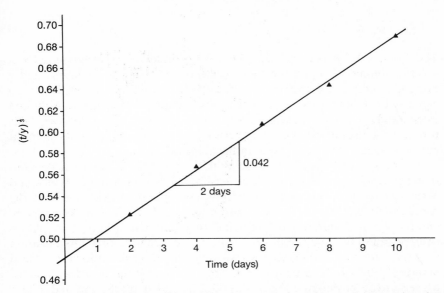

FIG. 1.19   Determination of the BOD constant K_1 (reaction rate constant) and L (the ultimate BOD) from BOD data using the Thomas method.

intercept a can be measured ($a = 0.481$) and slope b calculated:

$$\text{slope } b = \frac{0.042}{2} = 0.021.$$

From these values the rate reaction rate K_1 and the ultimate BOD (L) can be estimated:

$$K_1 = 2.61 \ (b/a)$$

$$= 2.61 \ \frac{(0.021)}{0.481} = 0.114$$

$$L = 1/2.3 \ K_1 a^3$$

$$= \frac{1}{2.3 \ (0.114) \ (0.481)^3} = 34.3 \text{ mg } l^{-1}$$

Although the kinetics of the BOD test have been modelled as a first-order reaction it has been argued that the BOD process is so complex that it cannot be adequately described solely by the first-order reaction equation (Young and Clark 1965; Landine 1971; Stones 1981, 1982). This is fully discussed in Section 3.1.

1.4.2.2 Methodology

The BOD_5 is defined universally as the mass of dissolved oxygen required by a specific volume of liquid for the process of biochemical oxidation under prescribed conditions over 5 days at 20°C in the dark. The result is then expressed as milligrams of oxygen per litre of sample (HMSO 1983). Although three standard methods have been published for the test, each follows a similar procedure with only minor discrepancies regarding the strength and volume of reagents added at various stages in the analysis.

The most widely used standard is that published as 'Oxygen demand (biochemical)' in the Standard Methods for the Examination of Waste and Wastewater. This comprehensive reference work is published jointly by the American Public Health Association, American Water Works Association, and the Water Pollution Control Federation, and is now in its 16th edition (American Public Health Association 1985). The International Organization for Standardization (ISO) is a world-wide federation of national standard bodies of which Ireland and the UK are members. It published a standard procedure for the determination of BOD (ISO: 5815) in 1983, entitled 'Water quality determination of biochemical oxygen demand after "n" days (BOD_n), dilution and seeding method'. Finally, a new British standard method, 'Biochemical Oxygen Demand (1981)' has been published by the Department of the Environment as part of their series 'Methods for the examination of wastes and associated materials' (HMSO 1983). It has superseded the forerunner to this series, 'The analysis of raw, potable and waste-waters (HMSO 1972) known commonly as the 'green book'. The new standard methods are published individually in booklet form, which slot into a loose-leaf folder, the idea being that any changes can be made rapidly without the necessity of reprinting the entire series. The preparation of these standard methods is the responsibility of the Standing Committee of Analysts, which is one of the joint technical committees of the Department of the Environment and the former National Water Council.

The methodology described below is the method recommended by this committee, and a fuller description of the method and further discussion of the test can be found in the standard method (HMSO 1983).

Sample preparation
It is only in a small number of cases that dilution and seeding will not be necessary. Dilution can only be omitted when the BOD of the sample is expected to be <4 mg l^{-1}; and seeding is not necessary if the sample already contains adequate numbers and a suitable diversity of acclimatized micro-organisms. Although these conditions exist in some treated effluents and most river water, if doubt exists then a series of dilutions and a seed should be used. In practice, it is difficult to know whether a sample does contain suitable

micro-organisms, and there is a growing tendency to seed all samples, regardless. Samples need to be analysed as quickly as possible, preferably within 2 h. If this is not possible the organic decomposition must be inhibited, as a significant proportion of the available organic substrate could be oxidized giving a low BOD value. For example, samples stored at 20°C for 4 h and 22 h resulted in decreases in the BOD value of 14% and 22% respectively (American Public Health Association 1985). Chemical inhibition will obviously interfere with the test as well, and samples should be stored at between 2–4°C and be analysed ideally within 6 h but never more than 24 h after collection. Influences of the methods and period of storage on the BOD have been reviewed by Ranchet *et al.* (1981). They found that freezing samples, as recommended in the ISO standard, depressed the BOD results. If samples are frozen, seeding with bacteria acclimatized to 20°C, must be carried out to replace those destroyed by the low temperature. The storage period is also critical and if it exceeds 24 h then samples must be discarded. In the case of composite samples collected using a 24 h sampler, the container must be kept as near to 4°C as possible during collection, and all the samples must be analysed within 24 h of the last aliquot being collected (Water Pollution Research Laboratory, 1967). There is some evidence to suggest that diluted samples can be successfully stored at 4°C for up to 4 days without any effect on the BOD. It appears that the low bacterial density in the sample after dilution remains so low at the reduced temperature as to have little impact on the substrate in solution (Tyers 1988).

As chlorine and chloramines severely inhibit microbial activity, tapwater is unsuitable for use as dilution water. Until fairly recently, clear natural waters, especially groundwaters, were used for diluting samples in the BOD test. However, the variability in nutrient content had a significant effect on the microbial activity of the micro-organisms, with nutrients often limiting full microbial oxidation. This has been overcome by the introduction of a standard synthetic mineral nutrient dilution water. Freshly prepared distilled or de-ionized water is used, although distilled water from a copper still should not be used as residual copper concentrations in excess of 0.01 mg l^{-1} can inhibit bacterial activity. By adding small amounts of chemicals to the distilled water, a dilution water with a standard pH, reasonable buffering capacity and salinity, and sufficient inorganic nutrients to support microbial activity can be produced. The chemicals are added in the form of a phosphate buffer solution to provide the phosphorus requirement and to maintain an optimum pH of 7.2; potassium, sodium, calcium, and magnesium salts, which are essential nutrients for the growth and metabolism of micro-organisms; and, finally, ferric chloride, magnesium sulphate, and ammonium chloride to provide iron, sulphur, and nitrogen. Together, these solutions should be added to the dilution water to give a BOD:N:P ratio of 60:3:1. Four stock solutions are made and 1 ml of each is added to each litre of dilution water prepared, in the following order: ferric chloride (0.0124% m/V), calcium chloride

(2.75% m/V), magnesium sulphate (2.5% m/V), and phosphate buffer solution (pH 7.2). The HMSO standard dilution water differs from the ISO and the US standard in that a 50% lower concentration of ferric chloride is used and a 10% higher magnesium sulphate concentration is used. The former is to reduce the possibility of bacterial inhibition. The dilution water is saturated with oxygen and stabilized before use, with the temperature being maintained at 20°C (Section 1.4.2.3).

When samples are diluted then it is vital that the dilution water used has a very low oxygen demand. Unseeded dilution water can be used so long as it has an oxygen demand of <0.3 mg $O_2 l^{-1}$. High dilution water BODs are usually caused by a combination of factors such as the use of dirty glassware or storage vessels, glassware containing trace amounts of detergents, the presence of volatile organic materials in the distilled water, and the topping up of old dilution water with freshly prepared dilution water. There is much dissatisfaction in this maximum oxygen demand standard for dilution water, especially as it is extremely difficult to measure a BOD of <0.3 mg l^{-1}. Also, as a period of 5 days incubation is required to establish the BOD of the dilution water it is not possible to check the suitability of the dilution water before use, so if the BOD has exceeded the 0.3 mg l^{-1} maximum then that set of BODs will have to be discarded (Fitzmaurice and Gray 1987a). If dilution water constantly exceeds the limit then it can be stored at 20°C in the dark long enough for it to satisfy its own BOD. However, if this is done a small amount of seed is subsequently required to ensure an oxygen uptake of 0.1 mg $O_2 l^{-2}$ and a nitrification inhibitor should be added to preserve the ammonia and prevent the growth of nitrifying bacteria. When seeded, the dilution water should have an oxygen demand of <0.5 mg $O_2 l^{-1}$, although the ISO and US standards allow the BOD to be <1.0 mg $O_2 l^{-1}$. In an inter-laboratory study of 23 state and semi-state water laboratories in Ireland, Fitzmaurice (1986) found that the BOD of dilution water seeded with a dehydrated propriatory seed ranged from 0.2–2.1 mg l^{-1}. Five laboratories exceeded the ISO standard of 1.0 mg l^{-1} and of those recording a BOD of <1.0 mg l^{-1} the mean value was 0.51 mg l^{-1}. The fact that 13 of the 23 laboratories failed to reach the more stringent level of 0.5 mg l^{-1} may indicate that this figure may, in practice, be too low. Stover and McCartney (1984) addressed the problem of high blank dilution values and formulated a seed correction factor. A sample of unseeded dilution water and several dilutions of the seed material are incubated along with the samples under test. After the incubation period, a plot of dissolved oxygen depletion versus ml of seed added is made. This results in a straight line (Fig. 1.20). Simple linear regression is applied to the results with the intercept on the y-axis, at zero seed concentration, corresponding to the unseeded dilution water BOD. The slope of the line corresponds to the dissolved oxygen depletion of 1 ml of seed. The sum of these two correction factors is then substituted in the calculation formulae for $(B_o - B_n)$. This method provides a dilution water correction and a seed correction as separate and independent factors.

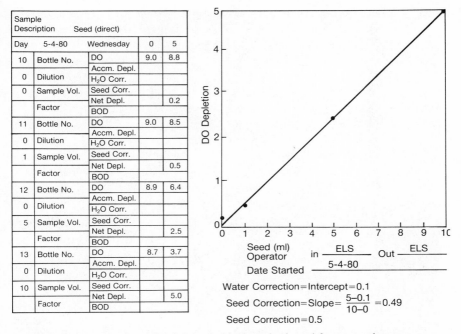

Sample Description	Seed (direct)			
Day	5-4-80	Wednesday	0	5
10	Bottle No.	DO	9.0	8.8
		Accm. Depl.		
0	Dilution	H₂O Corr.		
0	Sample Vol.	Seed Corr.		
		Net Depl.		0.2
	Factor	BOD		
11	Bottle No.	DO	9.0	8.5
		Accm. Depl.		
0	Dilution	H₂O Corr.		
1	Sample Vol.	Seed Corr.		
		Net Depl.		0.5
	Factor	BOD		
12	Bottle No.	DO	8.9	6.4
		Accm. Depl.		
0	Dilution	H₂O Corr.		
5	Sample Vol.	Seed Corr.		
		Net Depl.		2.5
	Factor	BOD		
13	Bottle No.	DO	8.7	3.7
		Accm. Depl.		
0	Dilution	H₂O Corr.		
10	Sample Vol.	Seed Corr.		
		Net Depl.		5.0
	Factor	BOD		

Seed (ml) Operator in $\dfrac{\text{ELS}}{\text{5-4-80}}$ Out $\dfrac{\text{ELS}}{}$

Date Started

Water Correction = Intercept = 0.1

Seed Correction = Slope = $\dfrac{5-0.1}{10-0}$ = 0.49

Seed Correction = 0.5

FIG. 1.20   An example of the graphical method used for correcting oxygen depletion due to dilution water and microbial seed used in the BOD test (Strover and McCartney 1984).

There are a number of methods of diluting samples. The least satisfactory is pipetting the sample directly into the BOD bottle and then filling with dilution water. The problem with this approach is that the volume of the BOD bottles are never exactly 150 cm³ or 250 cm³ and vary enough to significantly affect the final BOD when large dilutions have been employed. Also, as the top is placed into position some of the sample will be lost, and as it is not possible to completely mix the sample before placing the stopper the loss of sample and dilution water will most likely be unproportional, thus affecting the final dilution. The most widely adopted method is the jug technique. Here, the sample and dilution water are mixed in a graduated cylinder using a plunger-type mixing rod, so as not to entrain air in the sample, and then transferred into the BOD bottles. A development of this technique is the automated mixing chamber, which is perhaps the most efficient method of ensuring adequately diluted and mixed samples. The apparatus consists of a glass aspirator with a tap, a separating funnel, a three-way large bore stopcock assembly, and associated tubing (Fig. 1.21). Seeded or unseeded dilution water is stored in the aspirator which acts as a reservoir, and flows under gravity through the three-way stopcock into a graduated separating funnel. The dilution water enters at the base of the funnel thus avoiding the

FIG. 1.21.   BOD mixing chamber: 1, dilution water reservoir (5 or 10 l aspirator bottle); 2, glass stopcock assembly with key; 3, tubing; 4, graduated cylindrical separating funnel (1 l); 5, three-way T-form glass stopcock; 6, BOD bottle (250 ml).

entrainment of air, and is allowed to partially fill the funnel (25%) before the flow is stopped by closing the stopcock. The required amount of sample is added by means of pipetting it through the neck and allowing it to flow down the side of the funnel. The dilution water is then allowed to flow into the funnel up to the required volume, usually 600 cm^3 to fill two 250 cm^3 BOD bottles. The contents of the funnel are mixed using a plunger-type mixing rod to ensure no air is entrained and the flow diverted via the free leg of the three-way stopcock to rinse and fill the BOD bottles.

If the Winkler method is used (see below), two BOD bottles must be prepared for each sample. However, if the electrode method is used for oxygen analysis only one bottle is required as it can be used for both determination of the initial and final dissolved oxygen concentrations (Fitzmaurice and Gray 1987a,b).

Measuring the oxygen concentration

There are two standard methods for determining the dissolved oxygen concentration in the BOD test, a titrimetric procedure and a membrane electrode method. The classical wet chemistry iodometric technique developed by L. W. Winkler in 1888 is still widely used today albeit in modified forms. It is a titrimetric procedure based on the oxidizing property of dissolved oxygen and is usually referred to as the Winkler method. Due to the variability of the chemical composition of natural waters and wastewaters, the iodometric method has always been prone to chemical interference. To overcome the more common interferring substances, a number of modifications of Winkler's original method have been developed. For example: the permanganate modification used in the presence of ferrous iron, the alum flocculation modification used in the presence of suspended solids, the copper sulphate-sulphamic acid flocculation modification developed for use with activated sludge mixtures and the azide modification which is used in the presence of nitrite. It is this last modification that is recommended for the analysis of dissolved oxygen in sewage, effluents and river waters. Complete details of this method are given elsewhere (HMSO 1983) and only a résumé of the azide modification of the Winkler method is given here.

The basis of the method is the production of a white precipitate of manganous hydroxide which reacts with the dissolved oxygen to form a brown hydroxide of manganese in higher valency states. The sample is then acidified with sulphuric acid, which in the presence of iodide liberates free iodine equivalent in amount to the original concentration of dissolved oxygen. The iodine is titrated with a standard solution of thiosulphate. The end point of the titration can be determined electrometrically but is normally detected visually using soluble starch as an indicator. Five reagents are required for these reactions: manganous sulphate solution to produce the manganous hydroxide precipitate; alkali-iodide-azide reagent provides the iodide concentration and the azide counteracts any interference due to the presence of nitrites; sulphuric acid to acidify the manganese hydroxide precipitate; starch solution to detect the end point of the titration, and sodium thiosulphate is the titrant used to measure the free iodine concentration. The analysis begins with the addition of 2.0 ml of manganous sulphate and 2.0 ml of alkali-iodide-azide solutions to the sample within the BOD bottle. The stopper is replaced and the content of the bottle mixed vigorously producing a precipitate which is allowed to settle to the lower half of the bottle with a clear supernatant discernible. Then 4.0 ml of sulphuric acid solution is added to the sample which is repeatedly mixed to ensure the precipitate is fully dissolved. A portion of the acidified solution (100–200 ml) is transferred to a conical flask and titrated with sodium thiosulphate using starch as an indicator. The end point of the titration is reached at the first disappearance of the blue coloration, with any re-coloration being ignored. One per cent starch glycollate has been shown to give a sharper and more reliable end point (Vogel 1978). If 0.0125 M solution thiosulphate titrant is used to titrate 200 ml of the acidified solution, then the amount used is equivalent to the dissolved oxygen concentration of the sample. In cases where the dissolved oxygen measurement is not part of BOD analysis a correction factor is applied to compensate for the displacement of a small portion of the sample caused by the addition of the reagents.

The more recent membrane electrode method is based on the rate of diffusion of molecular oxygen across a permeable membrane. A modified oxygen electrode is used to measure the dissolved oxygen in the BOD test. The electrode is usually of the polarographic type and is manufactured to a size capable of being inserted into the standard BOD bottle with a wide neck. In order to contain the small volume of sample which is displaced as the electrode is inserted into the BOD bottle, a special funnel is supplied which forms a seal at the neck of the BOD bottle and contains the displaced liquid in a chamber above the neck. A fixed magnetic stirring bar is attached to the end of the funnel to provide the necessary flow across the electrode. The advantage of this method over the Winkler method include: its speed of measurement; simplicity; less chance of errors in measurement; can be used for continuous monitoring of uptake; allows K_1 values to be calculated without the need of many replicate bottles; only a single bottle required; not susceptible to interfering substances as is the iodometric method. However, it is subject to interference, particularly from gases that undergo reduction at the same potential as oxygen such as nitrous oxide, chlorine, nitric oxide, hydrochloric acid, and formaldehyde as well as the presence of hydrogen sulphide.

Each day before use, the electrode must be calibrated using single point calibration at both high and low dissolved oxygen concentrations. The calibration should be repeated as frequently as practicable, but especially at the end of the day. Checks can be made by analysing split samples of air saturated and de-oxygenated samples of distilled water using the iodometric method as a reference by which the electrode is calibrated. There are several ways of preparing oxygen-free water (HMSO 1983; American Public Health

Association 1985), but a particularly successful one is to deoxygenate distilled water by boiling and then bubbling with oxygen-free nitrogen overnight before use. Once calibrated, the electrode can be inserted into the BOD bottle. The mixing mechanism is then switched on and a period of at least 60 s allowed to elapse before the oxygen concentration is read. The electrode is carefully removed to allow the displaced liquid retained by the collar to drain back into the bottle, and if the oxygen consumption is to be measured over more than 5 days, the volume of the BOD bottle is made up using fresh dilution water if necessary and the stopper replaced. The electrode is rinsed with distilled water and is then ready to measure the next sample. The linearity of response of the electrode will alter over a long period, so it should be checked at monthly intervals by plotting the dissolved oxygen concentrations, in BOD bottles over as wide a range as possible, measured by the Winkler method against the electrode method. A small fixed bias will have no effect on the BOD calculation (Fig. 1.22) whereas a relative bias, which either increases or decreases with increasing sample concentration will severely affect the calculation of BOD (Fig. 1.23) (Fitzmaurice 1986).

The BOD calculation
Two forms of BOD calculation are used. For *undiluted, unseeded* samples:

$$BOD_n = (D_o - D_n) \text{ mg } O_2 l^{-1}$$

where D_o and D_n are the dissolved oxygen concentrations before and after n days incubation respectively. For *diluted samples, seeded or unseeded*:

$$BOD_n = f[(D_o - D_n) - \frac{(f-1)}{f}(B_o - B_n)] \text{ mg } O_2 l^{-1}$$

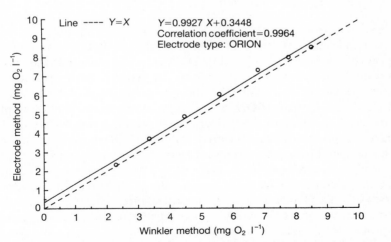

FIG. 1.22 Oxygen electrode calibration test showing a small fixed bias in response (Fitzmaurice 1986).

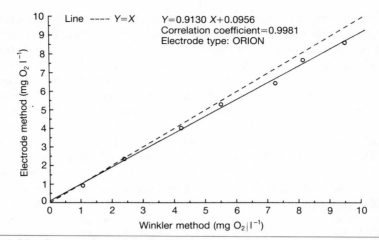

FIG. 1.23   Oxygen electrode calibration test showing a relative bias in response
(Fitzmaurice 1986).

where B_o and B_n are the dissolved oxygen concentrations of the seed control
(blanks) before and after n days respectively and f is the dilution factor. Thus,
$([f-1]/f)(B_o-B_n)$ is the oxygen demand of the seed, but the dilution factor
correction $([f-1]/f)$ becomes insignificant with dilutions in excess of 1:100. If
a chemical inhibitor is added to the dilution water to suppress nitrification,
this should be stated when expressing the result. For example, if allythiourea is
used the result should be expressed as BOD $(ATU)_n$.

1.4.2.3   Factors affecting the test

Temperature
As bacterial activity is a function of temperature, the BOD test is temperature
dependent. Although the ultimate BOD (L_o) is slightly affected, because
oxidizability increases with temperature, in practice the temperature only
affects the rate of oxidation (K_1) and not the amount of waste oxidized,
therefore, the ultimate BOD will always be the same regardless of the
temperature at which the test is performed. Although not widely used, the
breakdown process can be accelerated within the BOD bottle by incubating at
higher temperatures. The time lapse of 5 days between sample preparation and
the result is a severe limitation, and with a 5-day working week this means
that, in practice, BOD analyses cannot be commenced on Mondays or
Tuesdays without incurring overtime attendance. The time restraint also
means that the use of the BOD for process control and effluent monitoring
purposes is meaningless. To overcome these limitations, rapid BOD tests have
been developed, the most popular, based on raising the incubation
temperature resulting in increased bacterial activity. For example, there is

close agreement between the 5-day BOD test at 20°C and a 2.5 day test at 35°C (Fig. 1.24), allowing the BOD test to be completed within a working week, and the Ministry of Health (1936) published tables of BOD_5 at 20°C and BOD_3 at 27°C for a range of effluents with differences rarely exceeding 5%. Good correlations between BOD_2 at 37°C and BOD_5 at 20°C were achieved by Orford and Matusky (1959) and Robbins (1961) suggested a BOD_1 at 37°C for effluent treatment plant control. However, it is advisable to calibrate

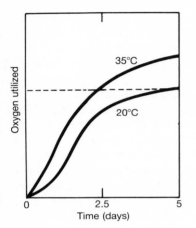

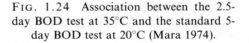

FIG. 1.24 Association between the 2.5-day BOD test at 35°C and the standard 5-day BOD test at 20°C (Mara 1974).

particular effluents at the two different temperatures first. In warmer climates, 30°C is a more appropriate temperature at which to carry out the BOD test, as the naturally occurring bacteria are acclimatized to this higher temperature and a BOD_3 at 25°–30°C is commonly used in tropical countries. However, the adoption of a single 20°C standard over 5 days means that BOD results are comparable internationally.

The reaction rate constant (K_1) in the BOD test increases with temperature according to Vant Hoff's law:

$$K_{1(T)} = K_{1(20)} \cdot \theta_T^{(T-20)}$$

where θ_T is the temperature coefficient with values between 1.047 to 1.135 and the reaction rate constant (K_1) measured at temperatures T°C and 20°C. The value of θ depends on the temperature and its mean value is 1.047 (Streeter and Phelps 1925) which essentially means that the speed of reaction increases by 4.7% for each 1°C rise in temperature. The mean value proposed by Streeter and Phelps is however inaccurate at low temperatures, so two values of θ are generally now used: 1.135 between 4–20°C and 1.056 over the temperature range 20–30°C.

Once the incubation temperature has been decided every attempt must be made to ensure that the exact temperature is maintained over the incubation period, which includes preheating the dilution water to the incubation temperature. A 1°C deviation from the 20°C incubation period can produce an error of up to 5% over the 5 days (Gray 1989a,b). The effect of temperature on BOD stoichiometry and oxygen uptake rate has been reviewed by Flegal and Schroeder (1976).

Dilution

It is vital that some dissolved oxygen remains after incubation in order to ensure that the oxygen assimilation during the test can be calculated. If the waste is too strong then all the dissolved oxygen will be utilized before the 5-day incubation period has elapsed. In contrast, if the waste is too dilute then only a small proportion of the dissolved oxygen will have been used, which leads to any analytical errors in determining the oxygen becoming excessively significant. As a general guide, if the BOD_5 is <7 mg l^{-1} no dilution is required; however, if it exceeds 9 mg l^{-1} all the dissolved oxygen will be depleted after 5 days, resulting in a zero value on completion of the test. This means that the BOD_5 calculation cannot be worked out and the result will be limited to '>9 mg l^{-1}'. Thus, apart from river waters, most samples require dilution. Dilution of samples should be done outside the BOD bottle and the dilution water itself should be capable of sustaining bacterial growth. It should contain a mixture of salts including nitrogen, phosphorus, sulphur, and iron, as well as a range of trace elements and have a neutral pH and contain sufficient ions to give an ionic strength to the water which favours microbial growth. It is important that it contains as little organic material as possible so that it does not exert a significant BOD in its own right, and, of course, it must be aerated for 40 minutes before use to ensure supersaturation and then left to stand for a further 30 minutes to ensure all the excess oxygen is released and that 100% saturation has been achieved. The water is normally preheated to 20°C and constantly stirred.

The dilution required depends on the actual BOD of the original sample. The most accurate BOD estimation will be obtained when between 35–50% of the dissolved oxygen is utilized. If the dissolved oxygen remaining in the bottle falls to below 1 mg l^{-1}, aerobic breakdown is inhibited leading to a misleading result. Choosing the correct dilution is vital to the successful operation of the test. For example, if the original sample has a BOD_5 of 250 mg l^{-1} then a × 50 dilution is required that will give a predicted BOD_5 value for the diluted sample of 5 mg l^{-1}. However, if only a × 20 dilution is used then the diluted sample will have a BOD_5 of 12.5 mg l^{-1} and as this is greater than the available oxygen in the BOD bottle, where the maximum is 9.08 mg l^{-1} at 20°C, all the available dissolved oxygen will be used within the 5 days so that no result can be calculated. Therefore, if the approximate BOD value of a sample is not known, a range of dilutions should be used to cover all the most likely ranges, and the dilution resulting in between 35–50% dissolved oxygen utilization used to calculate the BOD. Typical dilution ranges are given in Table 1.30. For non-river samples the most efficient dilution ranges to use are

TABLE 1.30 Recommended dilution factors for the determination of BOD (ISO 1983)

Expected BOD (mg $O_2 l^{-1}$)	Dilution factor	Report to nearest mg $O_2 l^{-1}$	Applicable to
3–6	1–2	0.5	R
4–12	2	0.5	R, E
10–30	5	0.5	R, E
20–60	10	1	E
40–120	20	2	S
100–300	50	5	S, C
200–600	100	10	S, C
400–1200	200	20	I, C
1000–3000	500	50	I
2000–6000	1000	100	I

R, river water; E, biologically treated domestic effluents; S, clarified domestic effluents; C, raw domestic effluents; I, heavily contaminated industrial effluents.

1:20 (5%), 1:50 (2%), and 1:100 (1%), which will ensure an accurate estimation of the BOD_5 over a range of 40–700 mg l^{-1}. Each dilution gives a certain degree of overlap with the 1:20 dilution covering the 40–180 mg l^{-1} range, 1:50 the 100–350 mg l^{-1} range and 1:100 the 200–700 mg l^{-1} BOD range.

Microbial influences on the test
The BOD test is a microbial growth system and it is important to make certain that a suitable microbial community is present in order to ensure that the test proceeds efficiently. Four factors can have a significant effect on the BOD result: (i) a low initial bacterial density; (ii) the use of unacclimatized bacteria; (iii) the presence of nitrifying bacteria; and (iv) the presence of algae (Mara 1974).

Most effluent samples contain sufficient bacteria to allow biological oxidation to proceed immediately. However, if the density of bacteria is low initially there may be a delay before a sufficient population of bacteria have developed to allow oxidation to proceed at its optimum rate (Fig. 1.25). If samples have less than 10^3 bacteria per ml then seeding will be necessary. In practice, all samples should be seeded. The standard seed is settled sewage acclimatized to the BOD test temperature for 24–36 h, with 1–2 ml added to each litre of dilutant or, in the case of undiluted samples, per litre of sample tested. This provides a mixture of micro-organisms capable of metabolizing a range of substances that may be present. The effluent from the biological treatment unit can also be used and, like settled sewage, it should be settled for at least 1 h at 20°C before use. Another source of seed is river water, especially below effluent outfalls, and where it is impossible to obtain seed from one of these sources, garden soil can be used. About 100 g of soil is added to 1 litre of

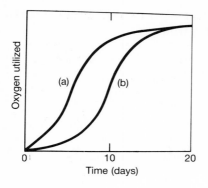

FIG. 1.25   Effect of initial bacterial population on the BOD bottle. The BOD curves obtained using (a) normal initial population of acclimatized bacteria, and (b) low initial bacterial population or an unacclimatized seed (Mara 1974).

water, well mixed and allowed to stand for 10 minutes; 10 ml of the supernatant is then diluted with water to 1 litre and used as seed in the usual way. Propriatory seeds are also available which are made up of a number of bacterial and fungal species. They are supplied in a capsule containing the correct weight of dried organisms that when mixed with 1 litre of water will provide a seed of standard quality (Section 7.4.3). As the microbial quality and diversity of settled sewage is very variable the avantages of a standard seed are obvious. However, the sample must provide a reasonable nutritional balance to allow the micro-organisms to thrive. Some industrial wastewaters may have a limiting range of nutrients which will only support a restricted range of micro-organisms, whereas others may be toxic and completely inhibit microbial activity. For example, phenol wastewaters are toxic to normal micro-organisms and a BOD test will give a zero result as no dissolved oxygen is utilized. Thus, it is important to use specially adapted microbial cultures, which can be purchased or collected from biological treatment units treating phenol wastes in order to obtain a BOD value. It is possible to produce your own acclimatized microbial culture for seeding difficult industrial wastes, although it is time consuming. A 1 litre plastic or glass bottle is three-quarters filled with settled wastewater and aerated using a small aquarium aerator. Starting with small amounts, the industrial effluent is added to the system over a period of several weeks. As the wastewater becomes cloudy this indicates that the culture is acclimatized. Alternatively, instead of settled wastewater, activated sludge can be used with the final settled effluent being used as the acclimatized seed. It is not only industrial wastes that are nutritionally deficient. Many biodegradable wastes from the food processing and drink manufacturing industries are deficient in either nitrogen or phosphorus, which need to be supplemented, normally via the dilution water, otherwise degradation proceeds more slowly producing a low BOD value. By removing protozoans that prey on the bacteria and by filtering the seed, it is possible to increase the oxidation rate by allowing the concentration of bacteria to rapidly increase, thus allowing carbonaceous oxidation to be completed more rapidly and thereby allowing a shorter incubation period to be used. Le Blanc (1974) found that, compared to the standard BOD_5, samples seeded with protozoan-free inocula produced more reproducible BOD results after just 2 days incubation. So far this rapid BOD method has not been widely adopted.

In the oxidation of ammonia by nitrifying bacteria, considerable quantities of oxygen can be utilized which can represent a significant fraction of the total oxygen demand of a wastewater. For example, it is normal for the nitrogenous fraction to account for two or even three times more than the carbonaceous fraction in the BOD test. Considerable oxygen is required to oxidize ammonia, as can be seen from the stoichiometry:

$$2NH_4 + 3O_2 \xrightarrow{\text{Nitrosomonas}} 2NO_2 + 4H^+ + 2H_2O + \text{energy}$$

$$2NO_2 + O_2 \xrightarrow{\text{Nitrobacter}} 2NO_3 + \text{energy}$$

with the overall reaction:

$$NH_4 + 2O_2 \rightarrow NO_3 + 2H + H_2O + \text{energy}.$$

Theoretically, 3.4 g of molecular oxygen is required by *Nitrosomonas* to oxidize 1 g of ammonia to nitrite and a further 1.14 g of molecular oxygen by *Nitrobacter* to oxidize 1 g of nitrite to nitrate. However, small amounts of nitrogen are assimilated as cell material during synthesis and this amount must be subtracted from the theoretical requirement. Montgomery and Bourne (1966) calculated the oxygen equivalent of the assimilated ammonia at 0.02 g and nitrite at 0.02 g. From these values, the following equation can be used to predict the extent of nitrogenous oxygen demand (NOD) in the BOD test:

$$NOD = 3.23 \times \text{increase in nitrite-N} + 4.35 \times \text{increase in nitrate-N}$$

Thus, a partially nitrified effluent containing 20 mg l^{-1} of ammonia would exert a NOD in the order of 80 mg O_2 l^{-1}. The extent of nitrification in the BOD test is more easily measured by incubating a parallel set of samples, one with and one without nitrification suppressed, the difference being the NOD. Nitrification only occurs when ammonia and nitrifying bacteria are present in sufficient concentration and numbers, and nitrification inhibitors are absent. In non-nitrified effluents, only ammonia is present and the density of nitrifying bacteria is extremely low. Nitrifying bacteria multiply very slowly, having a doubling time of between 2–6 days (Downing *et al.* 1964), therefore, nitrification generally occurs towards the end of the carbonaceous oxidation phase in the BOD test. Generally, it will be upwards of 10 days before nitrification begins to exert an oxygen demand. In partially nitrified effluents, both ammonia and nitrifying bacteria will be abundant, therefore, nitrification exerts a high oxygen demand after about 5 days that will be far in excess of the carbonaceous oxygen demand (Figs 1.15 and 1.16). This produces a problem in interpreting the BOD of sewage before and after treatment and raises the question that is so often posed by wastewater treatment plant operators: 'Should nitrification be included in the measurement of BOD when sewage treatment processes are based on the removal of organic material only?' Normally, the nitrogenous oxygen demand that occurs in the BOD test is much greater than will occur in natural water, with greatest nitrification occurring in natural waters during the summer months. Thus, in general, nitrification should be suppressed during the test using an inhibitor, so only the carbonaceous demand is measured. This is now standard practice both in the UK and the USA (National Water Council 1978; Carter 1984). Two inhibitors are widely used, allythiourea (ATU) or 2-chloro-6-(trichloro-methyl) pyridine (TCMP) added to either the dilution water or the sample. A doseage rate of 0.5 mg l^{-1} ATU prevents the onset of nitrification for a period of up to nine days with no effect on carbonaceous oxidation, and unlike thiourea, ATU does not interfere with the azide modification of the Winkler method. This dosage rate of ATU exerts an average oxygen (iodine) demand of 0.06 mg $O_2 l^{-1}$. ATU only inhibits *Nitrosomonas* and does not inhibit nitrification by *Nitrobacter*, but the second stage of nitrification rarely proceeds in the absence of the first. Although ATU is recommended by HMSO (1983), the US standard recommends the use of TCMP for inhibiting nitrification. TCMP was originally developed for the fertilizer industry to prevent the leaching of nitrogen based fertilizers through the soil, and a concentration of 10 mg l^{-1} will effectively inhibit nitrification without affecting carbonaceous oxidation for a much longer period than ATU (Young 1973). Numerous other chemicals can inhibit nitrification and these have been listed by Richardson (1985). Mara (1974) suggests exceptions when nitrifica-tion should be taken into account in the overall determination of the BOD exerted on receiving waters. These are: (i) when the river temperature is greater than about 20°C; (ii) when the effluent is discharged into an estuary; (iii) when the effluent is discharged into a river which has a flow time in excess of five days from the point of discharge to the sea; and (iv) when effluent flow contributes more than 50% of the total river flow. The Environmental Protection Agency in the USA have recommended the use of an approximation to calculate the Ultimate Oxygen Demand (L_o) from the carbonaceous oxygen demand ($BOD_{(ATU)}$):

$$L_o = (1.5 \times BOD_{(ATU)}) + (4.6 \times NH_3\text{-}N).$$

Although it is possible for carbonaceous oxidation and nitrification to occur simultaneously, with the resultant BOD a mere composite of the two reactions, nitrification normally begins some time after carbonaceous oxidation has started, resulting in the characteristic two stage BOD curve (Fig. 1.15).

The second stage reaction (nitrification) can be described mathematically as:

$$Y_2 = L_n(1 - e^{-K_n t})$$

so that the overall two stage BOD curve can be expressed as:

$$Y = L_o(1 - e^{-K_1 t}) + L_n(1 - e^{-K_n t})$$

where L_o is the ultimate oxygen demand and L_n the ultimate nitrogenous demand, K_1 is the rate constant for carbonaceous demand and K_n for nitrogenous demand. The rate constant K_n is usually less than the rate constant for carbonaceous material (K_1) and has been approximated by the Water Research Centre for river water taken from the Thames estuary as:

$$K_n = 0.0317(1.017)^t \, d^{-1}.$$

Thus, at 20°C $K_n = 0.044$.

The presence of algae in samples can cause significant problems. Normally, the production of oxygen by algae is prevented by incubating BOD bottles in the dark. However, like other aerobic micro-organisms, algae respire even in the dark and so exert an oxygen demand. Although their overall contribution is normally slight, samples containing high algal populations, such as highly eutrophic lake waters or samples from oxidation ponds, will have a significant oxygen demand exerted by the algae. This will distort the BOD value and provide an inaccurate measure of the biodegradable fraction in the sample. In these cases it is essential that the algal cells are removed by filtration through a Whatman GF/C filter. The effect of algae on the BOD is shown in Fig. 1.26.

In the dark, algal cells only survive for a short time after which they die and may contribute to the organic content of the sample, thus increasing the BOD. The BOD is unrepresentative of the deoxygenation processes occurring in eutrophic lakes or other systems where algae is abundant, as no estimation is made of the benefits of reaeration via photosynthesis.

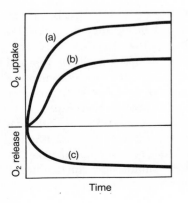

FIG. 1.26 Effect of algae on the BOD curve: (a) containing algae incubated in the dark; (b) filtered sample containing no algae; (c) oxygen is released when sample contains algae and is incubated in the light (Mara 1974).

Suspended solids and turbulence
The suspended solids content of wastewaters, especially those from food-processing industries, are likely to be composed of both a biodegradable and non-biodegradable organic matter. These wastewaters normally require high dilutions for BOD analysis and it is difficult to ensure that the small sub-sample used is representative of the wastewater. Therefore, the presence of suspended solids can lead to erroneous results. Another problem with suspended solids is during incubation when the solids will settle to the bottom of the BOD bottle causing stratification of the dissolved oxygen concentration, this being greater in the top half than that in the lower half. Mixing of the sample during incubation will equalize the dissolved oxygen concentration and is widely employed in respirometric BOD apparatus; however, the resultant turbulence in the BOD bottle may break up the solid particles into a more readily usable substrate with a consequent increase in BOD_5. Ali and Bewtra (1972) found that the average increase in the BOD_5 due to mixing ranged from 7% for a synthetic wastewater to 44% for final effluents. They suggest that turbulence around the bacterial cells increases the rate of material transport into the cell and the rate of removal of by-products accumulating on the cell membrane. Turbulance also increases the contact between the bacterial cells and the substrate, thereby increasing the rate of assimilation (Fitzmaurice 1986). The optimum mixing speed during incubation is in the range of 300–400 rpm (Morrissette and Mavinic 1978), while higher speeds cause the flocs to shear and increases the rate of CO_2 production with a consequent reduction in pH which may cause bacterial inhibition.

Filtering the sample through Whatman GF/C filter paper removes the problem of interference of the BOD test by suspended solids. However, this will nearly always result in a significant reduction in the BOD and a possible change in the K_1 constant (Fig. 1.27). Filtered samples are used as a measure of soluble BOD.

Aeration
Achieving the correct dissolved oxygen concentration in the bottle before the test commences is often difficult. If the sample is being diluted with 100% saturated dilution water then there is no problem. However, there must be at least 7 mg l^{-1} of dissolved oxygen initially available for the BOD_5 test and undiluted samples may need to be aerated prior to commencement of incubation. Saturation of individual samples can be achieved by aerating or shaking, but they must be left to stand for 20 minutes to allow the excess air to be released. Over-aerating, using supersaturated dilution water or shaking a partially filled BOD bottle will result in excess oxygen being released after the test has commenced, causing gross errors in the test. The presence of algae can cause oxygen to be released and all BOD bottles should be incubated in the dark. Furthermore, anaerobic samples have a high instantaneous oxygen

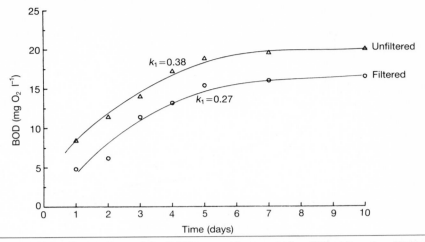

FIG. 1.27 Comparison of the BOD curves for a sample of filtered and unfiltered treated domestic effluent (Fitzmaurice 1986). Where BOD (unfiltered), 18 mg O_2 l^{-1}; BOD (filtered), 15 mg O_2 l^{-1}; suspended solids, 7 mg l^{-1}.

demand and pre-aeration is vital, even when diluted. In such samples, aerobic micro-organisms may take some time to become established so seeding is recommended.

Inhibitory and toxic wastes

Various chemical compounds present in wastewaters are toxic to micro-organisms. At high concentrations, these compounds will kill the micro-organisms and at sub-lethal concentrations, their activity can be significantly reduced. Non-tolerant bacteria may be unable to degrade wastewaters containing toxins to the same extent as they would in the absence of the toxin, resulting in a depressed BOD value. Much work has been done on the effects of heavy metals, with the BOD_5 severely suppressed by even small concentrations $(1–2$ mg $l^{-1})$ of Cu or Cr (Fig. 1.28). Stone (1979) measured the percentage suppression caused by 1 mg l^{-1} of selected heavy metals, dosed in the form of inorganic salts, on the BOD of domestic sewage (Fig. 1.29). The results showed metal toxicity was in the order of Ag > Hg > Cu > Cr > Ni > Pb > Cd > Zn. Although not considered a heavy metal, ferrous iron in concentrations exceeding 1 mg l^{-1} will also interfere with the BOD test. The ferrous iron reacts with the oxygen in the dilution water producing falsely high BOD results. However, the concentration of heavy metals which cause total inhibition of bacterial activity in the BOD test tends to be very high, far in excess of those normally encountered in either domestic wastewaters or river samples. Berkun (1932) found that 6 mg l^{-1} of mercuric chloride, 40 mg l^{-1} of copper sulphate, or 30 mg l^{-1} of potassium dichromate were required to completely inhibit bacterial activity on a sample of glucose.

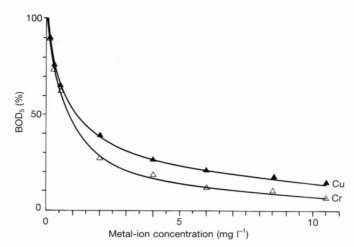

FIG. 1.28   The effect of metal-ion concentration, using copper and chromium, on BOD.

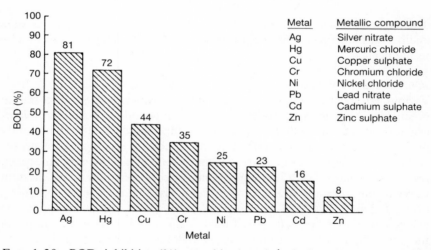

FIG. 1.29   BOD$_5$ inhibition (%) caused by 1 mg l^{-1} of selected metal compounds during the BOD test of a 1:100 dilution of settled domestic sewage (adapted Stones 1979).

Toxicity is usually suspected when the BOD increases with increasing dilution. Also, as chemical (COD or Permanganate Value) and instrumental (TOC) methods are unaffected by the presence of toxins, they can be used to check if the BOD is being depressed. A ratio of COD or TOC:BOD should be established for a specific wastewater and this then used to check for inhibition. For example, the normal COD:BOD ratio for domestic sewage is 2:1, so if the ratio was found to be $> 4:1$ the presence of toxic compounds in the sewage should be suspected as a possible cause. Toxicity in effluents can be overcome by pretreatment. Toxic metals can be complexed with chelating agents such as EDTA or precipitated out of solution. Volatile compounds and residuals of chlorine can be reduced by allowing the sample to stand for several hours or by gentle agitation, while high concentrations of chlorine residuals can be neutralized by sodium sulphite. Reseeding is required after samples have been dechlorinated. Acidic samples should be neutralized to pH 6.5–7.5 using sodium hydroxide and alkaline samples using sulphuric acid. In both cases, neutralization should not dilute the sample, by the addition of reagents, by more than 0.5%. The most effective way to overcome the presence of inhibitory or toxic compounds in wastewaters is to use an acclimatized seed.

1.4.2.4 Sources of error

Any analytical determination will inevitably include some errors and the fact that the BOD test is biological in action, depending on active aerobic micro-organisms, is another possible source of error. However, with practice duplicate results can be within 5%, and certainly should not exceed 8–10%.

The residual BOD in the dilution water is measured by carrying out blank tests using duplicate bottles containing only dilution water. The dilution water blanks are treated in the same way as samples and this residual BOD_5 value is subtracted from the overall BOD value. The BOD of the dilution water will be significantly increased by seeding and a separate determination of seeded dilution water is necessary. In order to ensure that this residual BOD does not affect the overall reaction within the BOD bottle, it should not exceed 0.2 mg l^{-1} for unseeded and 0.5 mg l^{-1} for seeded dilution water.

Probably the most common source of error in the test is the measurement of the oxygen concentration. It should be remembered that the determination of the BOD of a single unknown sample will involve a minimum of eight oxygen determinations, which includes three dilutions and one blank. Therefore, the chances of making an error that is carried through to the final calculation are large. There are two widely used methods of determining the dissolved oxygen concentration in the BOD test. The Winkler method is a chemical titration method whereas electrodes, which incorporate stirrers, are also now widely used. In the chemical method, high concentrations of suspended solids can adsorb iodine and give low oxygen values, and in these instances settlement, filtration or flocculation may be required. The starch–iodine titration requires skill and experience, with the recognition of the correct end-point particularly important. Whereas the acid-titration ensures that the BOD bottles are kept clean, the use of an electrode requires the bottles to be acid-rinsed in between use. Apart from the obvious problems of calibrating the electrode, it is necessary to establish standardized measuring techniques, for example, employing fixed stirring speeds and taking readings after a specific period of stirring.

Among the most frequently cited sources of error are poor analytical and laboratory technique; inadequate preparation of dilution water; the use of contaminated glassware and sampling bottles; incorrect dilution and poor mixing of samples; failure to use seeds and to pretreat samples when necessary; utilization of more than 50% of the dissolved oxygen in the second bottle over the incubation period; poor titration technique and in particular the end-point determination; infrequent and poor calibration of dissolved oxygen electrodes and meters; inefficient incubation both in allowing exactly 120 hours (5 days) and ensuring the temperature is 20°C. When incubation periods are less than 5 days it is possible to apply a correction factor which will allow a rough approximation of what the BOD_5 would have been. The BOD_n value is multiplied by the correction factor:

$$BOD_5 = K^1(BOD_n)$$

where K^1 is 1.58 for BOD_2, 1.243 for BOD_3, and 1.10 for BOD_4. These correction factors were computed by determining the BOD_2, BOD_3, BOD_4, and BOD_5 of different strengths of domestic and industrial wastewater. Plots were made of the BOD results versus wastewater strength. The slope of each line was computed by linear regression and the ratio of BOD_2, BOD_3, and BOD_4 slopes to the BOD_5 slope is the K^1 value (Ademoroti 1984). This technique can be successfully applied to wastewaters for which specific K^1 values have been determined as a rapid BOD technique.

Basic analytical technique should be checked periodically using inter-laboratory harmonization studies and more frequently by using standard samples of known BOD strength (Committee for Analytical Quality Control 1984; Fitzmaurice 1986). A useful test is a mixture of 150 mg l^{-1} glucose and 150 mg l^{-1} glutamic acid, seeded with fresh settled sewage. This should give a BOD_5 of 218 ± 11 mg l^{-1}. An error of up to 5% is acceptable, even using this standard solution, however, the greater the error the poorer the analytical technique.

Fitzmaurice and Gray (1987a) carried out an inter-laboratory precision test between 23 Irish water pollution laboratories. They measured the repeatability (within-laboratory precision) and reproducibility (between laboratory precision) (BSI 1979; 1987) of the BOD test at three test levels using sterile synthetic solutions of glucose and glutamic acid representing expected concentrations of 40, 200, and 400 mg O_2 l^{-1} (Table 1.31).

The results showed that the membrane electrode method is more precise than the Winkler method at each of the test levels. They suggested that the poor performance recorded for the Winkler method was probably due to the influence of random errors caused by poor quantitative techniques.

The azide modification of the Winkler titrametric procedure involves 10 steps before a dissolved oxygen result can be calculated. Apart from the

TABLE 1.31 Interlaboratory BOD precision test results using the membrane electrode (E) and Winkler methods (W) expressed in $mg\ O_2\ l^{-1}$ (Fitzmaurice and Gray 1987a)

Test level	1		2		3	
Method	E	W	E	W	E	W
Expected BOD_5	40	40	100	100	400	400
Mean BOD	41.2	35.6	196.1	168.4	356.6	336.4
Maximum result	51.0	52.0	225.0	239.5	429.0	515.0
Minimum result	33.5	18.0	181.5	102.0	302.5	173.5
Range	17.5	34.0	43.5	137.5	122.5	341.5
Standard deviation	6.0	10.8	15.0	48.0	39.3	109.6
Repeatability	4.9	9.3	19.6	23.8	34.5	46.3
Reproducibility	17.2	24.1	45.1	148.7	113.9	306.3

preparation of the reagents and the titrant, these steps involve: the addition of reagents to the BOD bottle; the transfer of the sample from the BOD bottle to a titration flask via a graduated cylinder; the filling of the burette with the titrant; the titration of the sample; the addition of the indicator; the visual detection of the end-point in the titration; the reading of the burette; and the calculation of the dissolved oxygen concentration. The transfer of reagents and samples using various types and sizes of volumetric glassware and the filling/reading of burettes are all common sources of random errors. The volume of sample titrated is very important as the loss of iodine during transfer from the BOD bottle to the titration flask may result in negative bias of up to 2% (DoE 1980, 1983). For this reason it is recommended that large sample volumes, > 200 ml, should be titrated. The rate at which the titrant is added to the sample may introduce a significant random error as the colour change from blue to colourless is very rapid, a final drop of 0.05 ml is sufficient to affect the end-point in the titration. Therefore, it is very important to add the titrant very slowly after the addition of the starch indicator. The occurrence of small random errors in the Winkler method tend to become significant when the results are used in the calculation of BOD because of the multiplicative factor introduced by high sample dilutions. With so many sources of random error the precision of the Winkler method is dependent upon the skill of the analyst, who should have a firm understanding of the principles and procedures of good quantitative techniques. This skill can only be acquired by frequent analysis but unfortunately the Winkler method tends to be used by the laboratories with a low turnover of BOD determinations.

The precision of the membrane electrode method was comparable to the results of studies conducted by the Environmental Protection Agency (1978) in the USA. In contrast to the Winkler method, the membrane electrode method only involves two steps and is a much faster technique requiring half the sample volumes used in the Winkler method. After setting up the electrode and checking its linearity of response the procedure only involves the insertion of the electrode into the same BOD bottle both before and after incubation. A direct reading in mg $O_2 l^{-1}$ is obtained after a response time of around 60 seconds. Provided the calibration check is carried out regularly and the membrane is maintained there is little chance of significant errors occurring in the analysis. Even if the electrode has a fixed positive or negative bias the result of the BOD test remains unaffected because the BOD calculation is based on the depletion of dissolved oxygen over a given period rather than the actual concentration of dissolved oxygen in the sample at any instant.

The criterion for judging the acceptability of analytical methods as developed by McFarren (1970) was applied to the results for both analytical methods used in the BOD precision test. The percentage total error for both methods at each test level is calculated, and based on the result, the analytical methods are divided into the following categories:

1. Excellent: Total Error < 25%.
2. Acceptable: Total Error > 25% < 50%.
3. Unacceptable: Total Error > 50%.

The analytical acceptability of the membrane electrode and Winkler methods as measured in the inter-laboratory BOD precision test are tabulated in Table 1.32, which shows that the membrane electrode method is rated as excellent for test level 2 and acceptable for test levels 1 and 3 whereas the Winkler method is rated as being unacceptable for each of the test levels.

The BOD test is widely used in Ireland as a parameter in the measurement and control of water pollution. BOD tests are frequently used to assess the degree of pollution in prosecutions; to check compliance with effluent discharge licences; to determine the deoxygenating effects of effluents discharged to rivers and streams; to determine (in conjunction with other parameters) charges for effluent treatment and to classify the quality of rivers. BOD results that do not achieve an acceptable level of precision are meaningless in court cases, cause disputes between industrialists and local authorities, and risk misclassifying the quality of rivers. Fitzmaurice and Gray concluded that there is no justification for poor precision due to bad quantitative techniques. Where the constituents of the sample cause precision problems other parameters such as TOC or COD should be used, but in no case should the BOD test be performed in isolation to other parameters.

TABLE 1.32   The analytical acceptability of the membrane electrode (E) and Winkler (W) methods for determining BOD as measured in the inter-laboratory BOD precision test (Fitzmaurice and Gray 1987a)

Test level	1		2		3	
Analytical method	E	W	E	W	E	W
Total error (%)	33	65	17	64	31	71

Specific actions to avoid the commonest sources of error in the BOD test identified by Fitzmaurice and Gray are summarized below:

Glassware
- All glassware should be cleaned with an acidic iodide/iodine wash solution irrespective of the method used to determine dissolved oxygen.
- Volumetric flasks should be used in preference to graduated cylinders for preparing sample dilutions.

Sample dilution
- Freshly prepared distilled water from an all-glass water still should be used.
- Nutrient solutions should be prepared monthly and stored in the dark at all times.
- Dilution factors should be chosen by reference to Table 1.30.
- Where the expected BOD concentration of a sample is unknown, COD analysis should be carried out to determine the optimum dilution factor.

Seeding
- A dehydrated microbial seed should be used in preference to seed from a biological effluent treatment process.
- The composition of seed from effluent treatment plants is rarely if ever determined before use and tends to be very variable both within and between plants. On the other hand manufactured seeds contain homogeneous microbial cultures, is easy to prepare and produce more repeatable BOD results than seed from biological effluent treatment plants. The universal use of such seeds should, in theory, eliminate a significant variable from interlaboratory BOD analysis. However, there is a need for more research and development on such products to ensure standard species composition and density of micro-organisms between batches.

The Winkler method
- Where the Winkler method is used on an infrequent basis or by inexperienced personnel, factory prepared volumetric solutions of sodium thiosulphate titrant (e.g. the 'Convol' range from BDH Chemicals) should be used.
- Automatic dispensers should be used to add the other reagents to the BOD bottles. This reduces the hazards associated with strong alkaline and acidic solutions and avoids cross contamination of reagents.
- Sodium starch glycollate (0.5 m/V) is more stable than soluble starch powder and should be used as the indicator solution (DoE 1980).
- Automatic zero burettes with a reservoir are both more convenient and faster than ordinary burettes. Their use also considerably reduces the problems of contamination of standard solutions.

The membrane electrode method
- Prior to the recording of any BOD measurements the electrode should be calibrated at both high and low dissolved oxygen concentrations by reference to the Winkler method. This type of calibration check is preferable to air calibration checks which are recommended by some manufacturers.
- The linearity of response of the electrode should be checked at monthly intervals.

Sample incubation
- Only standard BOD bottles should be used for incubation. The well in the bottle neck should be filled with dilution water and 'Parafilm' should be wrapped tightly around the neck of the bottle totally enclosing the glass stopper and the water seal.
- Reagent bottles with polypropylene stoppers should never be used for incubating BOD samples.
- The temperature in the incubator should be checked by placing a water filled BOD bottle coupled with a thermometer into the centre shelf of the incubator.

BOD calculation
- Both the initial and final dissolved oxygen concentrations should be recorded to two significant figures and the standard BOD formula, which incorporates the seed correction factors, should be used to calculate the BOD.

Further reading

Self-purification: Benoit 1971; Klein 1972*a,b*; Hynes 1971; Welch 1980.
Biochemical oxygen demand: Young *et al.* 1981; HMSO 1983; Carter 1984; American Public Health Association *et al.* 1985; Fitzmaurice and Gray 1987*a,b*.

2

HOW MAN DEALS WITH WASTE

2.1. BASIC TREATMENT PROCESSES

The aims of wastewater treatment are: to convert the waste materials present in wastewaters into stable oxidized end products, which can be safely discharged to inland or coastal waters without any adverse ecological effects; to protect public health; to ensure wastewater is effectively disposed of on a regular and reliable basis without nuisance or offence; to provide an economical method of disposal; and, more recently, to recycle and recover the valuable components of wastewater. A wastewater treatment plant is a combination of separate treatment processes or units designed to produce an effluent of specified quality from wastewater (influent) of known composition and flow rate. The treatment plant is also usually required to process the separated solids to a suitable condition for disposal. The amount of treatment required depends largely on the water quality objectives for the receiving water and also the dilution available. It has been the general practice in the past, and still remains the norm in Ireland, to base treatment plant design on the production of a Royal Commission Standard effluent which has a BOD_5 of 20 and a suspended solids concentration of 30 mg l^{-1}. This effluent quality is acceptable so long as there is at least an eight-fold dilution of clean water, with a BOD_5 <2 mg l^{-1}, available in the receiving watercourse. The 20:30 standard has led to the final effluent quality of treatment plants to be normally quoted in this manner, such as 10:10 or 5:10, with the first value generally referring to the BOD_5 and the second to the suspended solids concentration in mg l^{-1}. More relaxed standards apply to effluents discharged to estuarine, coastal or marine waters as the available dilution is vast. However, inland the final effluent quality will be 'tailor-made' for the specific receiving water course depending on available dilution and general assimilative capacity of the water which may be more or less stringent than the Royal Commission Standard (Warn and Brew 1980; Warn 1982a,b). Lakes, for example, may have eutrophication problems so that nutrient limits must also be set. For example, sewage from Mullingar (Ireland) must have the phosphorus removed by precipitation prior to discharge in order to prevent algal blooms in Lough Ennell.

Consent conditions for discharges are set according to river users downstream and the existing assimilative capacity. River quality objectives are the recognized uses to be made of each stretch of river which may range from salmonid fisheries requiring a BOD_5 <4 mg l^{-1} and a ammonical nitrogen concentration <0.5 mg l^{-1} to minor watercourses where the objective is merely to prevent a nuisance developing so that a BOD $<$ 20 mg l^{-1} and a ammonical nitrogen concentration of <10 mg l^{-1} are acceptable. Water quality criteria have been set locally, nationally, and internationally in order to achieve specific water quality objectives (Table 2.1).

TABLE. 2.1 River water quality criteria in terms of $BOD_{(ATU)}$ and ammonical nitrogen concentration (Warn 1982a)

	Water quality criteria (95-percentile)	
Type of use	$BOD_{(ATU)}$ (mg l^{-1})	Ammoniacal nitrogen (mg l^{-1})
Fisheries		
F1: trout, dace, perch, roach	4	0.5
F2: perch, roach, bream	6	1.0
F3: roach, bream	9	2.0
Public water supply direct	6	1.0
Public water supply (impounded)	9	2.0
High amenity	9	2.0
General amenity	12	5.0
No recognized river use; minor watercourses where the objective is merely to prevent nuisance developing	20	10.0

Calculation of consent conditions are based on the mass-balance equation:

$$T = \frac{FC + fc}{F + f}$$

where F is the river flow upstream of the discharge; C is the concentration of pollutant in the river upstream of the discharge, f is the flow of the discharge, c is the concentration of pollutant in the discharge, and T is the concentration of pollutant downstream of the discharge.

Thus, the permitted concentration of a pollutant for a discharge can be calculated by rearranging the mass-balance equation:

$$c = \frac{T(F + f) - CF}{f}$$

where T is now the water quality criteria set for that stretch of river, that is the safe concentration required to be achieved in the river downstream of the discharge (Warn 1982a).

Wastewater treatment is essentially a mixture of settlement and biological or chemical unit processes. Little real purification takes place as the action of the treatment is one of separation of the suspended and soluble nutrients from the water by adsorption onto particles large enough to be removed from suspension by settlement. The concentrated particles form a sludge which then requires to be processed and disposed which can itself be a major problem, especially at smaller plants. Treatment plants are assembled from combinations of unit processes, and as the range of available unit processes is large (Table 2.2) by using a suitable combination of the available processes it is possible to produce a final effluent of a specified quality from almost any type of influent wastewater (Fig. 2.1). Unit treatment processes can be classified into five stages:

1. *Preliminary treatment*: the removal and disintegration of gross solids, the removal of grit, and the separation of stormwater. Oil and grease are also removed at this stage if present in large amounts.

2. *Primary* (*sedimentation*) *treatment*: the first major stage of treatment following preliminary treatment, which usually involves the removal of settleable solids which are separated as sludge.

3. *Secondary* (*biological*) *treatment*: the dissolved and colloidal organics are oxidized in the presence of micro-organisms.

4. *Tertiary treatment*: further treatment of a biologically treated effluent to

TABLE. 2.2 Summary of major unit processes available in wastewater treatment

Group	Specific unit processes
Solids conversion	Biochemical precipitation including activated sludge and biofiltration
	Chemical precipitation
	Chemical oxidation/reduction
	Biochemical oxidation/reduction
	Combustion
Solids conditioning	Chemical coagulation
	Mechanical flocculation
Screening	Macro-screening
	Fine-screening
	Ultrafiltration
	Reverse osmosis
	Electrodialysis
Sedimentation	Plain sedimentation (gravity settling)
	Fluidized bed sedimentation
	Dissolved gas flotation
	Centrifugation
Filtration	Sand filtration
	Grass plot irrigation
	Membrane filtration
Ion-exchange	Softening
	Demineralization
	Metal recovery
	Phosphate, nitrate, ammonia removal
Absorption	Activated carbon removal of surfactants, trace organics, etc.
Gas transfer	Aeration to bring oxygen into solution or strip supersaturated gases such as carbon dioxide, ammonia, etc.

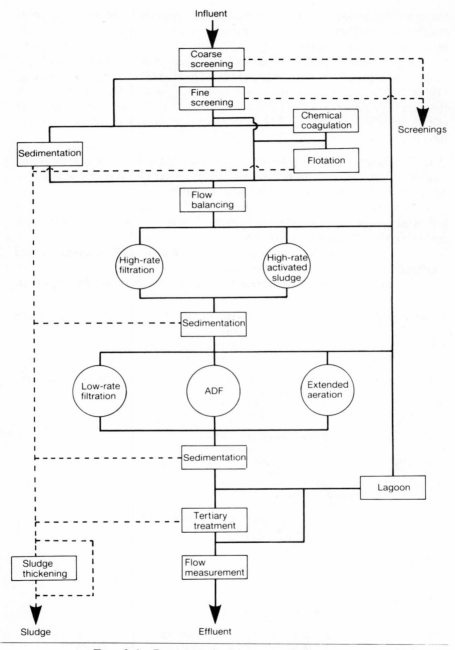

FIG. 2.1   Process options in wastewater treatment.

remove BOD_5, bacteria, suspended solids, specific toxic compounds or nutrients to enable the final effluent to comply with a standard more stringent than 20:30 before discharge.

5. *Sludge treatment*: the dewatering, stabilization, and disposal of sludge (Institute of Water Pollution Control 1975).

Depending on the quality of the final effluent required, not all the stages may be utilized. Preliminary treatment only may be given to effluents which are discharged to coastal or marine waters to prevent floating debris and gross faecal solids being washed ashore later, whereas primary treatment is generally given to wastewaters discharged to estuaries with the sludge often dumped at sea from special vessels. Most effluents discharged to rivers receive secondary treatment while tertiary treatment is often required if water is abstracted for potable supply downstream of a discharge. The degree of treatment is dictated by the dilution factor. However, there are those that feel that the maxim *the solution to pollution is dilution* is no longer appropriate and that all wastewaters should receive full treatment regardless of the dilution available in the receiving water.

All wastewater treatment plants are based on the generalized layout shown in Fig. 2.2.

2.1.1 Preliminary treatment

Screens
Raw sewage enters the plant containing a variety of coarse solids that need to be removed prior to treatment so that pumps are not damaged and pipework is not blocked. Screens, which are vertical rows of steel bars which can be curved, vertical straight or inclined straight in design, are used to physically remove such solids from the influent wastewater. Screens are in three sizes, coarse, medium, or fine. In domestic wastewater plants it is normal only to use a medium size screen with apertures of between 19 to 25 mm. Coarse screens (apertures > 50 mm) are also used at large plants to remove larger debris such as pieces of wood, while fine screens are normally constructed out of mesh and are installed after grit removal. The use of fine screens is generally limited to certain industrial and food processing treatment plants. Floating solids, rags, paper, wood, and plastic are caught on the screens, which are raked manually in smaller treatment works and the screen angled at 60° to the flow to facilitate cleaning. However, if the flow exceeds approximately 1000 m^3d^{-1} mechanically operated rakes are used. Mechanical raking systems operate either on a time basis or by depth using some form of water level detector because as the solids are retained and block the screen the flow is impeded so that the level of the wastewater downstream of the screen rises. At a small works a rough design figure for manually raked screens is 0.14 m^2 of submerged area per 1000

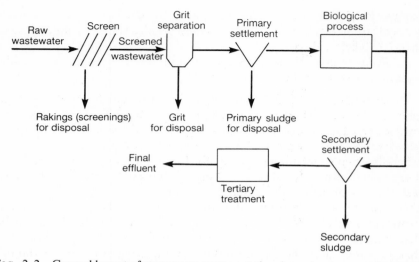

FIG. 2.2   General layout of wastewater treatment showing primary, secondary, and tertiary treatment stages.

population. This figure assumes that the screen will be cleaned frequently and if this is not possible then the area is increased accordingly. Flow velocity is important for the successful operation of screens. Above 0.9 m s^{-1} the trapped solids are scoured from the screen whereas below 0.3 m s^{-1} grit deposition occurs within the screen chamber (Institute of Water Pollution Control 1984).

The materials removed from screens are usually called screenings or rakings, and because they contain material such as gross faecal solids and sanitary towels they are very unpleasant and unhygienic to handle. Between 0.01 and 0.03 m^{3}d^{-1} of screenings are produced per 1000 population although the exact quantity depends on the aperture of the bars or mesh. For example, the volume of screenings produced at a treatment plant serving a population of 100 000 would be 0.9 m^{3}h^{-1} using a screen with an aperture of 10 mm, 0.6 m^{3}h^{-1} at 25 mm, and 0.25 m^{3}h^{-1} at 45 mm. With a moisture content of 69–85%, screenings weigh between 600–1000 kg m^{-3} and thus can be difficult to handle. Careful disposal is essential and the main options are incineration, burying or tipping. Recent research has indicated that it is unwise to store such material uncovered as gulls will readily feed on the waste which could result in contamination of supply reservoirs used by the gulls as roosting sites (Gray 1979) (Section 6.2.4.1). If buried, screenings must be covered with soil immediately at a sufficient depth to prevent birds and rats being attracted. Various dewatering equipment has been recently introduced at larger plants which is much cleaner and reduces the bulk of material to be disposed. These units can be attached to automated bagging units to make this unpleasant task even easier (Institute of Water Pollution Control 1984).

Many plants use macerators, comminutors, or disintegrators to finely chop up the coarse solids so that they can be dealt with by the normal treatment system, mainly by settlement. Whereas some plants macerate the screenings returning them to the inlet, other plants macerate the whole flow. However, if the latter option is employed then macerators must be placed after coarse screens and grit removal. Maceration increases the organic loading to the plant so that extra treatment capacity must be provided.

Grit separation

After screening and/or maceration the wastewater is termed screened or macerated sewage. Although it no longer contains gross solids it does contain grit which is not only mineral aggregate but a mixture of silt, sand, gravel, as well as fragments of metal, glass, and even dense plastic. Grit needs to be removed from the influent wastewater to prevent pump damage and silting.

The density of grit is much greater than wastewater and there is a tendency for it to settle whenever the rate of flow falls below 0.3 m s^{-1}. This can occur in sewers if the gradient is inadequate or if flows are low, such as might happen during the night. Therefore, the quantity of grit arriving at the treatment plant will vary according to the flow conditions in the sewerage system. Particularly large quantities of grit can be flushed to treatment plants after periods of dry weather when grit accumulates in roadside gully pots and in the sewer which is then displaced by a heavy storm. Grit originates mainly from surface drainage, and thus is a problem of combined sewerage systems only, coming from road construction and especially from newly constructed housing estates where the roads have not been properly surfaced. Sandy material can enter damaged and cracked sewers with the groundwater if the water table is above the level of the sewer. Some industrial processes such as vegetable washing can also produce enormous quantities of silt. The normal quantity of grit in domestic wastewater ranges between 0.005 to 0.05 m^3 per 1000 m^3.

There are several proprietary systems available such as the detritor, pista grit trap, and the vortex grit separator which have been described elsewhere (Clark *et al.* 1977; Institute of Water Pollution Control 1984; Viessman and Hammer 1985).

The most widely used, and perhaps the simplest, is the constant velocity grit channel (detritor) where the flow velocity is reduced enough to allow the heavier grit to settle but still high enough to maintain organic matter in suspension. The channel was developed by Townsend (1937) who was able to maintain a constant velocity within the channel regardless of the rate of the influent wastewater by placing a standing wave flume immediately downstream of the grit channel. In essence, the flume controls the depth of flow in the channel and by having the cross-sectional area of the channel at any level proportional to the rate of flow, it is possible to maintain a constant velocity in it. This is achieved by making the grit channel parabolic in cross-section, but due to constructional problems channels are built in trapezoidal cross-section.

The optimum velocity of wastewater for grit settlement is 0.3 m s^{-1} and if the velocity falls outside the range of 0.2–0.4 m s^{-1} then organic matter is deposited or fine grit fails to settle respectively.

At smaller plants the grit is removed from channels manually while pumps or suction devices are used at larger works which are mounted on a travelling gantry.The grit contains between 15–30% organic matter, which at the larger plants will be reduced by washing using a cyclone-type washer. Even after washing, grit will still contain about 15% organic matter and so must be dumped at landfill sites. Other methods of grit separation are to remove the organic matter either *in situ* as is the case with the Pista grit trap and the vortex grit separator, or as a separate but integral part of the process. For example, grit removed in the shallow square tank of the detritor has the organic matter removed by an inclined cleansing channel where the reciprocating action of the inclined rake pushes the grit slowly up the channel against a counterflow of wastewater, which removes much of organic matter present.

Stormwater

When it rains the normal concentration of the wastewater is reduced according to the volume of surface water entering the sewerage system. This diluted wastewater is known as either stormwater or storm sewage. Wastewater treatment plants are not designed to provide secondary treatment for all stormwater entering during wet weather and only a proportion of the flow arriving at the plant will receive full treatment, the rest will be stored and treated later. The design of treatment plants is based on dry weather flow (DWF), which is 'the average daily flow to the treatment plant during seven consecutive days without rain following seven days during which the rainfall did not exceed 0.25 mm in any one day' (Institute of Water Pollution Control 1975). Plants are normally designed to be able to fully treat three times the dry weather flow (3 DWF) with the excess stored in extra settlement tanks known as stormwater tanks. The maximum flow to the plant is equivalent to twice this flow (6 DWF) and at this loading rate the excess wastewater is discharged directly to the receiving water without any treatment. The separation of flow is achieved by overflow weirs set at the appropriate heights situated downstream of the preliminary treatment units so that the gross solids can be removed or macerated and grit removed before storage in the storm tanks or direct discharge.

The storm tanks provide a buffer between the excess wastewater and the watercourse. When a storm occurs the surface runoff reaches the treatment plant faster than it is able to significantly increase the flow of the watercourse. Thus, it is important that the diluted wastewater, which cannot be treated at the plant, should not be discharged until the flow in the watercourse has increased sufficiently to produce adequate dilution for the stormwater in order to prevent pollution. The first flush of storm water is usually the most polluted as the sediment in the sewers and roadside gully pots is usually flushed out, therefore, the first flush is able to be stored in the storm tanks which have a capacity equivalent to 6 hours at DWF. Stormwater tanks are almost identical to primary settlement tanks in design and have two functions: (1) to store as much storm diluted sewage as possible so that it can be returned for full treatment after the influent flow rate has returned to normal; and (2) to remove as much as possible of the suspended solids and associated BOD_5 from the stormwater that cannot be stored and which overflows to the watercourse. The stormwater tanks can also be used for emergency storage of sewage contaminated by toxic or dangerous substances resulting from spillages.

The 3 and 6 DWF systems were superseded by a more specific formula that takes into account the main factors more efficiently (Ministry of Housing and Local Government 1970). The maximum flow to a plant is estimated as:

$$(PG+I+E)+1.36P+2E \text{ m}^3\text{d}^{-1}$$

where P is the population, G the average domestic waste consumption $(\text{m}^3\text{hd}^{-1}\text{ d}^{-1})$, which is generally considered to be 180 l $\text{hd}^{-1}\text{d}^{-1}$, although recent research has put it as low as 120 l $\text{hd}^{-1}\text{d}^{-1}$ (National Water Council 1982), I the infiltration $(\text{m}^3\text{d}^{-1})$, and E the industrial effluent $(\text{m}^3\text{d}^{-1})$. The DWF is represented by $(PG+I+E)$ in this formula, and before the stormwater flow operates the domestic sewage must be diluted by 1.36 m^3 per person and the industrial effluent by a factor of 2. The maximum flow to receive full secondary treatment is expressed as:

$$3PG+I+3E$$

the infiltration component remaining constant.

2.1.2 Primary treatment

Primary treatment or sedimentation is the process by which settleable solids are removed from the screened sewage by passing it through a specially constructed tank at such a velocity that the solids settle out of suspension by gravity. The settleable solids collect at the base of the tank where they are removed as primary sludge. The retention time in sedimentation tanks is usually a minimum of 2 h at 3 DWF and when operated correctly can significantly reduce the organic loading to the secondary treatment stage. Comparing the unit processes, sedimentation is far cheaper than biological treatment in terms of unit removal of pollution and for that reason the majority of plants have incorporated primary sedimentation into their design. Removal of suspended solids is due to a number of processes occurring simultaneously within the sedimentation tank such as flocculation, adsorption, and sedimentation, and maximum removal efficiencies can be as high as 70% for suspended solids and 40% for BOD_5. The overflow from primary sedimentation tanks is known as settled sewage and contains fine suspended solids and dissolved material which passes onto the next treatment stage (Fig. 2.2).

There are a variety of tank designs which are fully described in Section 2.2.

2.1.3 Secondary treatment

Secondary treatment is a biological process where the settled sewage enters a specially designed reactor where under aerobic conditions organic matter in solution is oxidized or incorporated into cells. The reactor provides a suitable environment for the microbial population to develop and as long as oxygen and food, in the form of settled sewage, are supplied then the biological oxidation process will continue. Biological treatment is primarily due to bacteria, which form the basic trophic level in the reactor food chain. The biological conversion of soluble material into dense microbial biomass has essentially purified the wastewater and all that is subsequently required is to separate the micro-organisms from the water by settlement. Secondary sedimentation is essentially the same as primary sedimentation except that the sludge is comprised of biological cells rather than gross solids (Section 2.2).

There are two main types of reactors, those where the micro-organisms are attached to a fixed surface (e.g. percolating filter) and those where the micro-organisms mix freely within the wastewater (e.g. activated sludge). In the latter, the biological cells which are separated after secondary settlement are returned to the reactor to maintain a high microbial density in order to maintain maximum microbial breakdown of the wastewater (Figs 2.3 and 2.4). The secondary treatment phase may comprise of other biological systems, both aerobic or anaerobic, or incorporate a mixture of several systems. Biological treatment is further explored in Section 2.3, and the individual processes reviewed in Chapter 4.

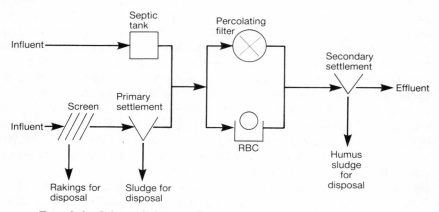

FIG. 2.3   Schematic layout of sewage treatment by fixed film reactor.

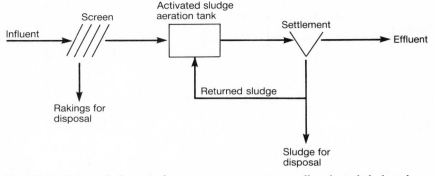

FIG. 2.4 Schematic layout of sewage treatment at a small activated sludge plant.

2.1.4 Tertiary treatment

Primary and secondary treatment will normally be able to produce a 20:30 effluent, often even better, but when the dilution available in the receiving water is inadequate and the water quality is already under stress then further treatment of the secondary effluent will be necessary.

Tertiary treatment processes, or polishing processes as they are often referred to, are normally filtration or straining devices, although the various processes now available are extremely varied. They can be used to reduce the BOD_5 or suspended solids concentration, eliminate bacteria or pathogens, remove nitrates and phosphates, and meet the more stringent conditions that are unobtainable by using biological treatment only (Section 2.4).

2.1.5 Examples of treatment plants

Nearly all textbooks on wastewater treatment give specific examples of treatment plants. For example, Mason (1981) describes the Crossness treatment plant which serves 1.7 million Londoners and treats an average of 580 000 m^3 of sewage daily. However, the majority of treatment plants are much smaller than this, serving populations of less than 25 000 PE (population equivalent). In Ireland, most treatment plants built in the past 20 years have been based on extended aeration activated sludge systems, mainly of the oxidation ditch type and to a lesser extent on surface-aerated activated sludge using vertical shaft aeration (Section 4.2). In such plants, the primary sedimentation tank is often omitted with the extra organic loading taken into account when designing the capacity of the secondary treatment unit (Fig. 2.5).

The Leixlip wastewater treatment plant in Co. Kildare serves a population of 20 000 and has a DWF of 4500 $m^3 d^{-1}$. The wastewater enters the plant via

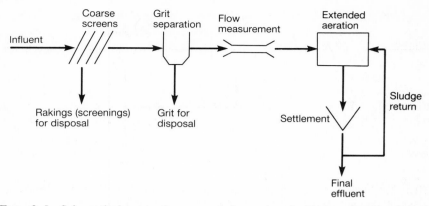

FIG. 2.5 Schematic layout of an extended aeration (oxidation ditch) activated sludge plant for populations of less than 25 000 where primary settlement is omitted.

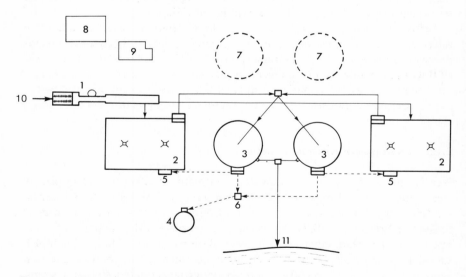

FIG. 2.6 Plan of the Leixlip Wastewater Treatment Works in Co. Kildare. Solid lines are wastewater or effluent pipes while the broken lines are sludge pipes.

1. Inlet works
2. Aeration tanks
3. Sedimentation tanks
4. Sludge holding tank
5. Return sludge pumping station
6. Excess sludge pumping station
7. Site for future sedimentation (primary) tanks
8. Generator house
9. Administration house
10. Inlet sewer
11. Final effluent

the main trunk sewer and has to be raised using two Archimedean screw pumps to provide enough head so that the wastewater can pass through the plant by gravity (Fig. 2.6). The wastewater passes through a preliminary treatment stage where it is mechanically screened and grit is removed. There is no primary treatment and the sedimentation tanks are omitted so that the screened sewage passes directly into two activated sludge aeration tanks, each 40×20 m, with cone aeration and providing a total reactor capacity of 5900 m^3. The activated sludge (mixed liquor) is displaced, as the screened wastewater enters the aeration tanks, into two 24 m diameter radial flow final sedimentation tanks. These are mechanically scraped, with the settled activated sludge returned to the aeration tanks by two airlift pumps. Excess activated sludge is pumped to a sludge holding tank with a capacity of 77 m^3 for disposal on to land. Like all well defined plants this also has available space for future development to an expected population of 50 000, a DWF of 11 250 m^3d^{-1}. The extra treatment capacity in this case will be achieved by incorporating primary treatment by constructing two 28 m diameter radial flow sedimentation tanks to remove the bulk of the organic loading. The organic loading to the aeration tanks will be significantly reduced, as will be the retention time in the aeration tanks which is reduced by more than half. No other expansion is required except an extra screw pump at the inlet and two extra sludge-holding tanks to bring up the capacity to 375 m^3. The final effluent produced by the existing and enlarged plant should be below the 20:30 standard and is discharged to the River Liffey, which eventually flows through Dublin.

Further reading

General: BSI 1983.
Preliminary processes: Institute of Water Pollution Control 1984.

2.2 SEDIMENTATION

Sedimentation is the most widely used unit operation in wastewater treatment. The terms sedimentation and settling are the same and can be used interchangeably. Sedimentation is the process by which the suspended solids, which are heavier than water and commonly referred to as the settleable fraction, are removed from wastewater by allowing the particles to gravitate to the floor of a tank to form a sludge under near quiescent conditions. The process is used in primary treatment to remove settleable organic and inorganic material to reduce the organic load to the secondary treatment processes. It is also used for secondary sedimentation, to remove material converted to settleable solids during the biological phase of treatment, such as the removal of humus from filter effluents and the recovery of activated sludge.

2.2.1 The settlement process

The way in which particles settle out of suspension is a vital consideration in the design and operation of sedimentation tanks as well as other unit processes. There are four types of settling: type I or discrete, type II or flocculant, type III or hindered which is occasionally referred to as zonal settling, and type IV or compression settlement. Where particles are dispersed or suspended at a low solids concentration then types I or II settlement occurs. Types III or IV settlement only occurs when the solids concentration has increased to such an extent that particle forces or particle contact affects normal settlement processes. During settlement, it is common to have more than one type of settling occurring at the same time and it is possible for all four to occur in a settlement tank simultaneously.

Type I or discrete particle settling
Discrete particles settle out of a dilute suspension as individual entities. Each particle retains its individual characteristics and there is little tendency for such particles to flocculate, thus settlement remains solely a function of fluid properties and particle characteristics. Settling of heavy inert particles such as grit and sand particles is an example of type I settlement.

Type I settlement is analysed by means of the classic laws of sedimentation formed by Newton and Stokes. Newton's law yields the terminal particle velocity by equating the gravitational force of the particle to the frictional resistance or drag. The rate of settlement of discrete particles varies with the diameter of the particle, the difference in density between the particles and fluid in which it is suspended and the viscosity, as shown by Stokes' law. The theory of settlement is fully explained elsewhere (Clark *et al.* 1977; White 1978*b*; Metcalf and Eddy 1984).

Under quiescent conditions, discrete particles settle out at their terminal velocity, which will remain constant provided the fluid temperature does not alter. A knowledge of this velocity is fundamental in the design of sedimentation tanks and in the evaluation of their performance. A sedimentation tank is designed to remove all particles which have a terminal velocity equal to or greater than V_c. The selection of V_c will depend on the specific function of the tank and on the physical characteristics of the particles to be removed. The rate at which clarified water is produced in a sedimentation tank is given by

$$Q = V_c \cdot A \ (\mathrm{m^3 d^{-1}})$$

where V_c is the terminal velocity of the particle and A is the surface area of the settling chamber. Therefore, in theory, the rate of clarification by type I settlement is independent of depth (Dick 1976). The overflow rate or surface loading, which is the average daily flow in cubic metres per day divided by the total surface area of the sedimentation tank in square metres and equivalent to the terminal settling velocity, is the basis of sedimentation tank design:

$$V_c = \frac{Q}{A} = \text{Overflow rate } (\text{m}^3\text{m}^{-2}\text{d}^{-1}).$$

Effluent weir loading is also a widely used design criterion and is equal to the average daily overflow divided by the total weir length, being expressed as cubic metres of effluent flowing over per metre of weir per day ($\text{m}^3\text{m}^{-1}\text{d}^{-1}$).

Modern sedimentation tanks are operated on a continuous-flow basis. Therefore the retention time of the tank should be long enough to ensure that all particles with the desired velocity V_c will settle to the bottom of the tank. Retention time is calculated by dividing the tank volume by the influent flow:

$$T = 24 \frac{V}{Q} \text{ (h)}$$

where T is the retention time in hours, V the tank volume (m^3), and Q the mean daily flow (m^3d^{-1}). The depth of a sedimentation tank is taken as the water depth at the side wall, measuring from the tank bottom to the top of the overflow weir. This excludes the additional depth resulting from the slightly sloping bottom in both rectangular and circular sedimentation tanks. The terminal velocity, tank depth and retention time are related as:

$$V_c = \frac{D}{T} \text{ (m s}^{-1})$$

where D is the depth (m).

Type II or flocculant settling
Type II particles in relatively dilute solutions do not act as discrete particles but coalesce or flocculate with other particles during gravitational settlement. Subsiding particles coalesce with small particles falling at lower velocities to form larger particles, which then settle faster than the parent particle. Settlement of solids in the upper layer of both primary and secondary sedimentation tanks are typical of type II settlement, and flocculation is particularly important in the separation of the sludge in the activated sludge process where rapid separation is required, so that the sludge can be returned to the aeration basin. The degree of flocculation is dependent on the opportunity for particle contact, which increases as the depth of the settling tank increases. Thus, the removal of suspended material depends not only on the clarification rate (Q) but also on depth, which is a major difference between type I and type II settlement. Surface overflow rate, the velocity gradients in the system, the concentration of particles and the range of particle sizes are also important considerations in flocculant settling. As there is no adequate mathematical relationship to determine the effect of flocculation on sedimentation, the effects of all these variables can only be determined by settling column analysis.

   Settling column analysis is carried out in a cylinder of the same depth as the
proposed tank and with a minimum width of 150 mm to minimize wall effects.
Sample ports spaced at 0.5 m intervals allow samples to be taken for analysis
from all depths. The wastewater is poured into the column and mixed to
ensure that there is a uniform distribution of particles throughout the column
at the beginning of the analysis. Care must also be taken to ensure that
uniform temperature is maintained throughout the column to avoid
convection currents affecting the settlement rate. Water samples are taken
from the ports at regular intervals and are analysed for suspended solids. The
percentage removal from the original suspended solids concentration of the
wastewater is plotted against time and depth which allows curves of equal
percentage removal, iso-concentration curves, to be plotted (Fig. 2.7). These
curves allow depth and retention time to be compared so providing the
optimum design and operating conditions for particular wastes. The degree of
flocculation is indicated by the slope of the iso-concentration curve, increasing
as the slope increases. Overall removal of solids can be estimated by reading
values directly from the plotted curves. For example, in Fig. 2.7 the percentage
removal after 25 minutes retention is 40% at 1 m and 36% at 2 m. The curves
also represent the settling velocities of the particles for specific concentrations
in a flocculant suspension where V_c = depth/retention time. Thus, after
25 minutes, 40% of the particles in suspension at 1 m will have average
velocities of $\geq 6.7 \times 10^{-4}$ m s^{-1} and 36% $\geq 1.3 \times 10^{-3}$ m s^{-1} at 2 m.

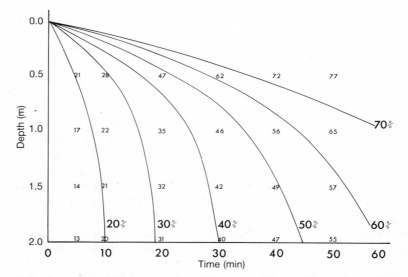

FIG. 2.7   Iso-concentration curves for solids removal from a flocculent suspension.
The numbers plotted represent the percentage removal of solids based on settling
column analysis.

Type III or hindered (zonal) settling
In dilute suspensions, particles settle freely at their terminal velocity until they approach the sludge zone when the particles decelerate and finally become part of the sludge. In concentrated suspensions (>2000 mg l^{-1} of suspended solids) hindered settlement occurs. Due to the high concentration of particles there is a significant displacement of liquid as settlement occurs which moves up through the interstices between particles reducing their settling velocity and forming an even more concentrated suspension (Fig. 2.8). The particles

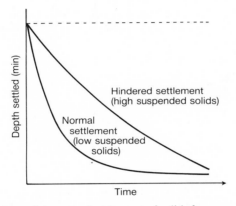

FIG. 2.8   Comparison of the rate of settlement of solids for a wastewater with a high suspended solids concentration showing hindered settlement compared with a normal wastewater containing a low suspended solids concentration.

are close enough for interparticulate forces to hold them in fixed positions relative to each other so that all the particles settle as a unit or blanket. A distinct solids–liquid interface develops at the top of the blanket with the upper liquid zone relatively clear and free from solids. This type of settling is normally associated with secondary settlement after a biological unit, in particular, the activated sludge process. As settlement continues a compressed layer of particles begin to form at the base of the tank, where the particles are in physical contact with one another. The solids concentration increases with depth and this gradation of solids is even apparent throughout the hindered settling zone from the solids–liquid interface to the settling–compression zone.

Hindered settling and compression can be clearly demonstrated using a graduated cylinder, by measuring the height of the interface between the settling particles and the clarified liquid at regular intervals (Fig. 2.9*i*). This is known as a batch settling test and can be used to plot a settlement curve (Fig. 2.9*ii*). On the curve, A–B shows hindered settling of the solid–liquid interface, B–C deceleration as there is a transition between hindered and compressive settling, and C–D, which represents compression. Further

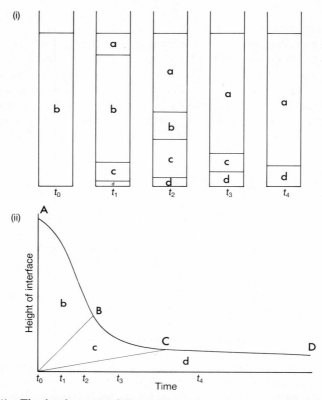

FIG. 2.9(*i*)   The development of discernible settlement zones at various times during the settling test on a flocculant suspension. Where (a) is the clarified effluent, (b) the wastewater with solids at its initial concentration and distributed evenly throughout its volume, (c) the zone where particle concentration is increasing and compaction is prevented by constant water movement up through the sludge blanket due to water being displaced in the lower zone, (d) where the particles have compacted.

   (*ii*)   The data derived from the settling test is used to plot a settlement curve. Where A–B shows hindered settling of the solid–liquid interface, B–C deceleration as there is a transition between hindered and compressive settling, and C–D, which is compression.

settlement in the compression zone is due to physical compression of the particles. The surface area required in a continuous-flow system designed to handle concentrated suspensions depends upon the clarification and thickening capabilities of the system. Batch settling tests are used to estimate these factors using the method developed by Talmadge and Fitch (1955).

Type IV or compressive settling
Consolidation of particles in the compressive zone is very slow as can be seen in Fig. 2.9*ii*. With time, the rate of settlement decreases as the interstices

between the particles become progressively smaller until eventually the release of liquid from the zone is inhibited. The particles are so concentrated that they are in physical contact, forming a thick sludge. Further settling is only possible by the weight of the new particles physically compressing the sludge layer. Gentle agitation enhances further compaction of the sludge by breaking up the flocs and releasing more liquid. Similarly, stirring has also been shown to increase the rate of settlement in the hindered zone (Dick 1967; Dick and Ewing 1967).

Compression normally occurs in sludge thickening tanks where specifically designed stirrers are used to encourage consolidation, or in the bottom of deep sedimentation tanks.

2.2.2 Design of sedimentation tanks

Sedimentation tanks have two functions, the removal of settleable solids to produce a clarified effluent and the concentration of solids to produce a handleable sludge. The design of a sedimentation tank takes both of these functions into consideration and the eventual size will depend on whichever function is limiting. For example, in the design of tanks for activated sludge settling, the limiting factor is usually sludge thickening. The criteria for sizing sedimentation tanks are the overflow rate, tank depth, and retention time. If wastewater was only composed of discrete particles then primary sedimentation tanks would be constructed as shallow as possible to optimize removal efficiency, as the rate of type I settlement is independent of depth. However, wastewater also contains some flocculant particles and the depth of primary sedimentation tanks are roughly the same depth as secondary tanks, which normally have a minimum depth of 2.1 m. Although increasing the depth does not increase the surface loading rate, it does, however, increase the retention time, which allows a greater degree of flocculation and enhances settlement. Recent research has indicated that retention time should be used as the primary design criterion instead of surface loading or overflow rate. However, whether either retention time or surface loading rate is used, one is normally kept at an accepted value. Furthermore, because the depth varies only between narrow limits, the choice of criterion has had little influence, in practice, on the dimensions of the tanks constructed (White 1978b).

Primary sedimentation
Primary sedimentation tanks clarify sewage by flotation as well as by settlement. Any suspended matter which is less dense than water such as oil, grease or fat will rise to the surface and form a scum. Skimmer blades attached to the scraper bridge push the scum across the surface of the tank to where it is discharged into a special sump or a mechanically operated skimming device which decants the scum from the surface. Although primary sedimentation

tanks are designed to remove both floating and settleable material, special flotation tanks may be required before the primary settlement stage if large amounts of floatable material are present in the wastewater. There is little difference between the efficiency of horizontal or radial-flow units for primary settlement and selection is normally based on constructional costs or plant configuration. For example, a circular tank of the same volume and surface area as a rectangular tank usually costs less to build (White 1978b). However, if a group of tanks is required, then rectangular tanks are normally used with common dividing walls, which are not only cheaper to construct than individual circular tanks but also take up considerably less space. Except at very small rural plants, there is always more than one primary sedimentation tank to allow for maintenance, with the flow divided equally between the tanks. Primary sedimentation tanks are designed to remove between 50–70% of the suspended solids and 25–40% of the BOD of the raw wastewater. Effectiveness depends on the nature of the wastewater and, in particular, the proportion of soluble organic material present, which can severely affect BOD removal efficiency. Some industrial wastes may contain higher proportions of settleable organic matter resulting in BOD removals at the primary sedimentation stage in excess of 60–70%. Overflow rates vary between $16–32 \text{ m}^3\text{m}^{-2}\text{d}^{-1}$, with $24 \text{ m}^3\text{m}^{-2}\text{d}^{-1}$ the commonly used value for design. The accumulated sludge density is influenced by the overflow rate forming a dense sludge at loadings of $< 24 \text{ m}^3\text{m}^{-2}\text{d}^{-1}$, with the scrapers enhancing consolidation, or a dilute sludge at $> 32 \text{ m}^3\text{m}^{-2}\text{d}^{-1}$ as the extra hydraulic movement inhibits consolidation. At higher overflow rates the turbidity of the effluent will increase, especially during periods of maximum flow. Turbid effluents from primary tanks are also caused by excessive weir loadings, which should not exceed $120 \text{ m}^3\text{m}^{-1}\text{d}^{-1}$ for plants receiving $< 4000 \text{ m}^3\text{d}^{-1}$ or $180 \text{ m}^3\text{m}^{-1}\text{d}^{-1}$ at flows of $> 4000 \text{ m}^3\text{d}^{-1}$ if solids are not to be carried over the final effluent weir. Over-storage of sludge, especially if wasted activated sludge has been discharged to the primary sedimentation tank for settlement, can result in anaerobic bacterial activity in the sludge layer. Gas is produced and becomes trapped in the solids making them more buoyant, expanding the sludge blanket, reducing consolidation and reducing the solids concentration of the sludge. In severe cases, gas production can become extremely active with unpleasant odours being produced, solids rising and floating to the tank surface, and the effluent turning darker and eventually grey-black in colour.

Secondary sedimentation
Secondary sedimentation tanks separate the microbial biomass produced in the biological treatment unit from the clarified effluent. The biological growth or humus which is flushed from filter medium comprises of well-oxidized particles, mainly in the form of dense microbial film, and also living invertebrates and invertebrate debris, which settle readily. The volume of sludge produced is much less compared with the activated sludge process. The

sludge will form a thin layer at the base of the sedimentation tank, rarely exceeding 0.3 m in depth, if the tank is regularly desludged twice a day. The overflow rate can be up to 32 $m^3m^{-2}d^{-1}$ with similar weir loadings and depth as a primary sedimentation tank.

Settlement of activated sludge is more variable as the flocs are light and more buoyant than solids from percolating filters, which reduces their settling velocities. This is made worse by denitrification at the centre of the tiny sludge flocs, which results in small bubbles being produced which buoy up the clusters of particles. This results in a thick sludge blanket which can often take up half of the tank volume at peak flows and is normally between 0.8 and 1.0 m in depth. Therefore, a rapid and continuous method of sludge removal is necessary. This is achieved by not having a central sump but instead, a floor slope of 30° to the horizontal, which allows the sludge to be quickly and continuously withdrawn. Another common method of removal is to have numerous uptake tubes attached to the scraper, which forces the sludge up and out by the hydrostatic pressure. Rapid withdrawal is achieved across the entire bottom of the tank, which ensures that the retention time of solids settling at the periphery of the tank is no greater than at the centre, thus reducing the age of the floc and the chance of gas production that results in reducing settleability. Sludge removed by this method encourages a vertical, rather than horizontal movement in the tank, which encourages flocculation and increases sludge density. Other problems causing reduced settleability of activated sludge effluents are discussed in Section 4.2.

The volume of effluent from a primary sedimentation tank is equivalent to the influent, as the volume of sludge withdrawn from the tank is negligible compared with the total volume of wastewater. However, in a secondary settlement tank receiving activated sludge the influent is equivalent to the effluent plus the returned sludge. In this case, the overflow of effluent is used for design purposes and not the influent, which includes recirculation. Secondary and primary sedimentation tanks are similar in design except that special attention is paid to the large volume of flocculant particles in the liquid. The capacity of tanks receiving activated sludge effluents are usually based on a retention time of 1.5–2.0 h at maximum flow rate. The maximum rate of flow will be 3 DWF plus 1 DWF for returned sludge which will give a DWF retention time of 6–8 h which is the same capacity as a primary sedimentation tank. A retention time of 2–3 h at maximum flow rate for primary sedimentation is adequate to allow almost all settleable solids to settle to the bottom of the tank, but is not long enough for septic conditions to result. However, at flows below DWF, septicity of the sludge can occur due to the extended retention time.

Horizontal- and radial-flow tanks

Two designs of sedimentation tank are widely used at wastewater treatment works, these are the rectangular horizontal-flow and circular radial-flow units.

Rectangular horizontal-flow tanks are commonly used for primary sedimentation and are particularly favoured at works at coastal sites where the
wastewater receives primary treatment only, before discharge to the sea via a
long outfall pipe. The length of rectangular tanks are usually 2.5–4.0 times the
width, which is limited by the availability of mechanical scrapers which rarely
exceed 24 m. Depths vary from 2.5 m but is normally 3.0–3.5 m. The size,
location, and choice of rectangular tanks depends on such considerations as
site layout and structural economy. The design of such a tank is shown in
Fig. 2.10. Baffles deflect the incoming sewage producing an even flow in the
tank and preventing high velocity streams within the body of liquid which
could cause short-circuiting of the liquid and hydraulic disturbances
(Clements 1966). The flow through the tanks is rarely laminar, and although
the unit has been designed to provide an optimum velocity of particles of
5 mm s^{-1} eddies and reversals in flow result in greater local velocities
(Fig. 2.11). The majority of solids settle at the inlet end of the tank from where
sludge is removed. The tanks gently slope towards the inlet end where the
sludge is scraped mechanically and collects in a sump and allowed to
consolidate. The sump also provides a degree of storage for sludge so that
sludge withdrawal can be balanced. Sludge is withdrawn from the sump under
hydrostatic head by opening the valve in the sludge well. The floor of the tank

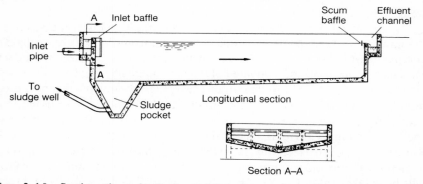

FIG. 2.10   Sections through a horizontal-flow rectangular sedimentation tank. The
scrapping machinery is omitted (White 1978*b*).

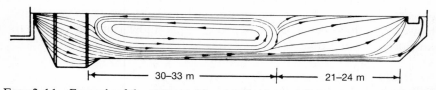

FIG. 2.11   Example of the pattern of flow in a horizontal-flow tank measured using
the radioisotope ^{82}Br (Institute of Water Pollution Control 1980).

is swept at regular intervals by scraper blades that push the sludge towards the sump. In the UK, blades are normally attached to a moving bridge which scans the width of the tank, whereas in the USA, flight scrapers, which are multiple blades attached to a continuous belt or chain, are commonly used. Scum baffles at each end of the tank prevent floating material passing over the weir into the effluent channel. Skimmer bars attached to the moving bridge or the exposed returning blades on the flight scrapers, push the floating material to the end of the tank where it is collected in a channel which scans the width of the tank. The major advantage of rectangular tank design is its compact nature, however, such units have a restricted weir length.

Radial-flow sedimentation tanks are circular in plan and a typical design is shown in Fig. 2.12. The normal diameter for such tanks ranges between 15 and 30 m and are the same depth as rectangular tanks, although circular tanks used as secondary clarifiers tend to be smaller in diameter than those used for primary sedimentation. The wastewater is piped under the floor of the tank

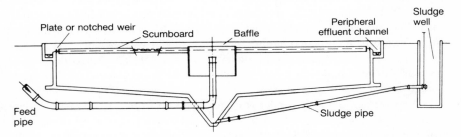

FIG. 2.12   Cross-section of a radial-flow circular sedimentation tank. The scraping machinery and centre supports are omitted (White 1978*b*).

and up a centrally placed vertical pipe which ends in a bell mouth just below the liquid surface. A deep cylindrical baffle plate prevents the flow from streaming across the surface and induces a gentle radial flow. Although the flow is nearly horizontal there is a slight upward rise and due to the radial design the velocity decreases outwards. As with rectangular tanks, complex flow patterns can develop within the tanks which can adversely effect settlement. The clarified effluent passes over a peripheral weir that is often protected by a plate or notched weir. The sludge is swept into a central sump for collection and consolidation by blades supported from a rotating radial bridge. As with rectangular tanks, sludge withdrawal is by hydrostatic pressure. Floating solids migrate slowly to the edge of the tank and are prevented from passing above the peripheral weir with the clarified effluent by a scumboard set in front of the weir. The skimmer attached to the rotating bridge concentrates the floating scum which is discharged into a mechanically operated scum box which drains into the sludge well. Advantages of circular tank design include long weir length and simpler scraping mechanisms compared with rectangular designs, but they are not so compact. Also, the installation and maintenance costs of circular tanks are lower for small- to medium-sized treatment plants.

2.2.3 Performance evaluation

The efficiency of sedimentation tanks is normally evaluated in terms of suspended solids and BOD removal. Examples of typical performance data for horizontal- and radial-flow primary sedimentation tanks are given in Tables 2.3 and 2.4. The suspended solids of primary settled sewage is normally in excess of 100 mg l^{-1} and the percentage removal of suspended solids is between 40–70% for horizontal-flow and 50–75% for radial-flow tanks (Institute of Water Pollution Control 1980). Although the dry solids content of the sludge depends on the skill of the operator, it is normally within the range of 3.0–6.5%. The efficiency of secondary sedimentation tanks is linked to the efficiency of the biological unit and depends on the settleability of the microbial biomass formed and, as stated above, this is particularly critical in the separation of activated sludge from the treated wastewater. The sludge volume index (SVI) and sludge density index (SDI) are long-established measures of activated sludge settling ability. Both are obtained by allowing a sludge sample to settle under standardized conditions. The SVI is measured by filling a 1 litre graduated cylinder with mixed liquor from the activated sludge aeration basin and allowing it to stand undisturbed for 30 minutes. The volume of settled sludge (V) is then read in millilitres. The suspended solids concentration is determined using a sample of the mixed liquor from the aerating basin, which is known as the mixed liquor suspended solids (MLSS), and the SVI expressed as the volume in millilitres occupied by 1 gram of settled suspended solids:

$$\text{SVI} = \frac{V \times 1000}{\text{MLSS}} \text{ (ml g}^{-1})$$

where V is the volume of settled sludge (ml l^{-1}) and the MLSS is the mixed liquor suspended solids concentration (mg l^{-1}). The SDI $= 100/\text{SVI}$. A high SVI indicates a poor settleability and in general a sludge with an SVI > 120 has poor settling properties. A good sludge would have an SVI < 80 and a very good one around 50. These indices are easily and rapidly performed on site and are routinely used by operators to check the condition of activated sludge. They are not used for design purposes. A more accurate measure of settleability, which is used for research and design studies, is the stirred specific volume index (SSVI). This is measured in a special settling column 0.5 m deep and about 0.1 m in diameter with settlement impeded by a wire stirrer at one revolution per minute. This test reproduces the non-ideal situation in settling

TABLE. 2.3 The performance of horizontal flow primary sedimentation tanks (Institute of Water Pollution Control 1980)

Sewage treatment works	Average daily flow (tcm)	Primary tanks								Suspended solids			BOD			Sludge production		Sludge dry solids	
		No.	Length (m)	Width (m)	Average depth (m)	Ratio of length to width	Weir over-flow rate* ($m^3\,m^{-1}\,d^{-1}$)	Surface loading* ($m^3\,m^{-2}\,d^{-1}$)	Retention period* (h)	Crude sewage ($mg\,l^{-1}$)	Settled sewage ($mg\,l^{-1}$)	Removal (%)	Crude sewage ($mg\,l^{-1}$)	Settled sewage ($mg\,l^{-1}$)	Removal (%)	Average ($m^3\,d^{-1}$)	$l\,hd^{-1}\,d^{-1}$	%	Volatile content (%)
Cambridge	40.2	2 / 5	33.9 / 38.3	16.3 / 19.7	1.8 / 1.7	2.1 / 2.0	305	8.4	5.0	260	135	49	210	150	29	202 (B)	1.7	3.8	72
Darlington	13.4	2[a]	47.0	12.8	2.8	3.7	530	11.2	6.2	295	195	34	280	215	24	137[d](B)	1.5	5.9	64
Kew	32.0	11	31.1	9.1	1.8	3.3	320	10.3	5.0	310	150	51	210	160	24	264 (AL)	3.2	3.7	75
High Wycombe	25.2	11	41.3[b]	11.3[b]	1.3[b]	3.7[b]	205	17.6	6.4	300	115	62	270	150	44	54 (BL)	0.9	6.6	76
Nottingham (Stoke Bardolph)	154.1	6	91.4	34.1	1.9	2.7	750	8.2	5.6	240	140	42	310	220	30	1714 (AL)	3.7	2.2	72
Oldham	49.2	8	39.0	11.0	2.0	3.6	545	14.1	3.4	305	130	57	290	160	44	110[d] (ABL)	0.7	6.8	63
Oxford	12.5	3	45.7	15.2	3.3	3.0	275	6.0	13.1	390	105	73	350	180	49	103 c	2.2	5.3	73
Oxford	16.3	2	45.7	15.2	3.3	3.0	535	11.7	6.7	415	125	70	315	150	52	108 c	2.3	4.0	79
Rotherham (Aldwarke)	27.5	2	93.0	12.0	1.8	7.8	1150	12.3	3.7	200	80	61	215	145	32	224 (AL)	2.3	3.8	73
Scunthorpe	11.3	4[e]	16.0	8.0	2.7	2.0	370	22.5	2.9	310	150	51	325	207	36	61 (ABL)	1.1	6.6	—

* Based on average daily flow; a two tanks (out of three) in operation; b average of all tanks; c same tanks but under different operating conditions; d sludge from horizontal flow tanks and radial flow tanks is combined; e consists of 4 hoppers and 4 tanks constructed as one unit.

A, includes surplus activated sludge; B, includes sludge from biological filters; L, includes works' liquors.

TABLE. 2.4   The performance of radial flow primary sedimentation tanks (Institute of Water Pollution Control 1980)

Sewage treatment works	Average daily flow (tcm)	No.	Primary tanks						Suspended solids			BOD			Sludge production		Sludge dry solids	
			Diameter (m)	Side wall depth (m)	Ratio of radius to side wall depth	Weir over-flow rate* $(m^3 m^{-1} d^{-1})$	Surface loading* $(m^3 m^{-2} d^{-1})$	Retention period* (h)	Crude sewage $(mg\,l^{-1})$	Settled sewage $(mg\,l^{-1})$	Removal (%)	Crude sewage $(mg\,l^{-1})$	Settled sewage $(mg\,l^{-1})$	Removal (%)	Average $(m^3 d^{-1})$	$l\,hd^{-1}\,d^{-1}$	%	Volatile content (%)
Barnsley Lundwood	15.6	6	16.7	3.0	2.4	52	11.8	7.0	370	150	60	400	250	38	114 (BL)	1.5	4.6	78
Bournemouth (Holdenhurst)	8.6	2	24.4	2.7	4.5	56	9.2	7.9	420	125	71	305	150	50	59 (AL)	1.3	4.1	78
(Kinson)	8.9	2	18.3	2.1	4.4	77	16.9	3.5	275	135	51	265	155	41	45 (AL)	1.1	5.5	78
Darlington	14.3	2	27.5	2.1	6.6	83	12.0	6.0	295	140	53	280	180	36	137† (B)	1.5	5.9	64
Hogsmill Valley	9.8	4	30.5	3.2	4.8	108	15.0	6.7	305	100	67	275	150	45	327 (AL)	1.75	4.4	75
Malling	4.5	2	19.8	2.4	4.1	38	7.3	9.7	490	135	72	405	170	58	46	1.5	5.6	75
Oldham	17.0	1	33.5	3.3	5.1	161	19.4	5.2	165	95	43	195	115	41	110 (ABL)	0.7	6.9	63
Oxford	23.2	8	19.5	4.2	2.3	47	9.7	7.8	405	100	75	370	175	53	86	1.0	5.3	73

* Based on average daily flow; † sludge from radial flow tanks and horizontal flow tanks is combined.
A, includes surplus activated sludge; B, includes sludge from biological filters; L, includes works' liquors.

tanks, whereas the SVI is measured under complete quiescence (White 1976). The use of the SSVI test is becoming more widely used for normal operational management as the SVI has been shown to have serious limitations. For example, marked changes in settleability that have been reflected by the SSVI have not been detected by the SVI (Johnstone *et al.* 1979, 1980; Rachwal *et al.* 1982), also, the SVI can vary in an inconsistent way with the initial concentration of suspended solids (White 1982). Rachwal *et al.* (1982) have also demonstrated that SVI has a much more limited working range than SSVI due to the independence of SSVI on solids concentration. At high solids concentration, the settled volume of the sludge, as measured in an unstirred one litre cylinder, increases at approximately the same rate as the increase in MLSS, thus suppressing the SVI. They found that by comparing sludges with the same settleability as measured by the SSVI (SSVI = 80–85 ml g^{-1}) the equivalent SVIs ranged from 80–290 ml g^{-1}. By plotting SVI against solids concentration, a steady rise in SVI is discernable up to a solids concentration of 4 g l^{-1}, above this concentration there is virtually no settlement, indicating that 4 g l^{-1} is the maximum limit of the test (Fig. 2.13). The wider working range of SSVI compared with SVI is clearly illustrated by plotting the percentage volume, occupied by sludges with the same settleability (SSVI = 80–85 ml g^{-1}) after 30 minutes settlement in the respective cylinders, against solids concentration. The SVI curve ends at just below 4 g l^{-1} with virtually no settlement above 4 g l^{-1}. The relationship between settled volume in the SSVI cylinder and solids concentration is very close to linear (Fig. 2.14).

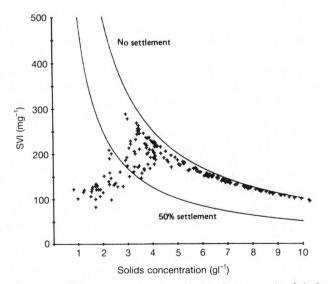

FIG. 2.13   SVI plotted against solids concentration for a sample of sludges with the same settleability as measured by the SSVI (Rachwal *et al.* 1982).

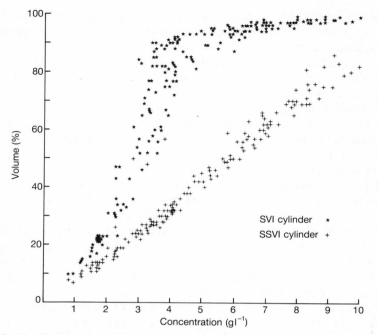

FIG. 2.14 Comparison of the working range of SVI with SSVI. The percentage volume of sludges with the same settleability (SSVI 80–85 ml g^{-1}), after settlement for 30 min in the respective cylinders, is plotted against concentration (Rachwal *et al.* 1982).

Sludge settleability in relation to biological processes is discussed further in Chapter 4.

The level of the sludge blanket in an activated sludge sedimentation tank can vary quite dramatically over 24 hours (Fig. 2.15), especially if the flow is not well balanced. The level of the sludge blanket can be very important, especially if the settleability of the sludge particles is low and there is a risk that sludge particles could be discharged with the clarified effluent. Sludge level detectors are available, based on optical or ultrasonic attenuation (Fig. 2.16). Such instruments not only detect the level or follow the rise and fall of the sludge–water interface but also measure the concentration or density of the sludge (Institute of Water Pollution Control 1980). Equipment is available to reduce the variation in the level of the sludge blanket by linking the rate of return to the rate of incoming sewage flow. However, an automatic sludge blanket level control, which would improve final settlement tank performance, has yet to be fully developed.

The pattern of flow and retention period in sedimentation tanks can be measured using chemical and radioactive tracers. The range of available tracers has been reviewed by the Institute of Water Pollution Control (1980).

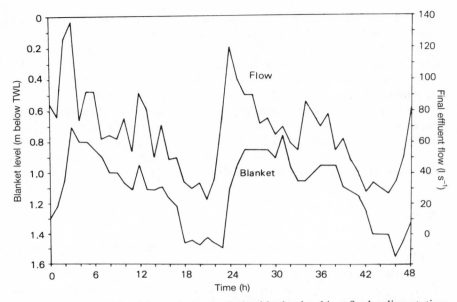

FIG. 2.15 Example of the variation in sludge blanket level in a final sedimentation tank with flow. The tank is receiving the effluent from a carousel-type activated sludge unit (Rachwal *et al.* 1982).

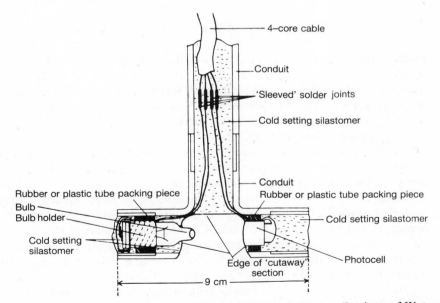

FIG. 2.16 Basic design of a photoelectric sludge-level detector (Institute of Water Pollution Control 1980).

However, the tracer commonly employed is lithium chloride, although any element which is easily detected and is not already present in large or irregular concentrations in sewage can be used. Sodium chloride has been employed using conductivity to measure the concentration of the salt (Gray 1981). Where the quality of a chemical tracer required is excessively large, then radioactive tracers such as ^{82}Br are used. The advantage of radioactive tracers is that they can be easily and continuously monitored without the necessity of taking water samples for laboratory analysis. Dyes are only used for a quick, semi-quantitative assessment of flow pattern as they can be followed visually. The flow pattern can also be studied using float techniques (Clements and Price 1982) or temperature differentials (Cox 1971).

Further reading

General: White 1978*b*; Institute of Water Pollution Control 1980; Metcalf and Eddy 1984.
Design: Dick 1976; White 1976.
Theory: Dick 1976; White and Allos 1976; Scottish Development Department 1980; Metcalf and Eddy 1984; Lumley 1985.
Settleability tests: White 1975; Rachwal *et al.* 1982.
Operation and maintenance: Institute of Water Pollution Control 1980.

2.3 SECONDARY (BIOLOGICAL) TREATMENT

After preliminary and primary treatment the wastewater will still contain significant amounts of colloidal and dissolved material that needs to be removed before discharge to a watercourse. Secondary treatment biologically converts this unsettleable material into biological cells which can then be removed by settlement using sedimentation tanks similar to those used for primary sedimentation (Section 2.2).

Methods of purification in secondary treatment units are similar to the '*self-purification process*' that occurs naturally in rivers and streams, and involves many of the same organisms. Removal of organic matter from settled wastewaters is carried out by heterotrophic micro-organisms, predominately bacteria but also occasionally fungi. The micro-organisms break down the organic matter by two distinct processes, biological oxidation and biosynthesis, both of which result in the removal of organic matter from solution. Oxidation or respiration results in formation of mineralized end-products that remain in solution and are discharged in the final effluent, while biosynthesis converts the colloidal and soluble organic matter into particulate biomass (new cells) which can then be subsequently removed by settlement. If the food supply, in the form of organic matter, becomes limiting then the microbial cell tissue will be endogenously respired (auto-oxidation) by the micro-organisms to obtain energy for maintenance. All three processes occur simultaneously in the reactor and can be expressed stoichiometrically as:

Oxidation:

$$COHNS + O_2 + bacteria \rightarrow CO_2 + NH_3 + \text{other end-products} + \text{energy}$$
(organic matter)

Biosynthesis: $COHNS + O_2 + bacteria \rightarrow C_5H_7NO_2$
                          (organic matter)   (new cells)

Auto-oxidation: $C_5H_7NO_2 + 5O_2 \rightarrow 5CO_2 + NH_3 + 2H_2O + \text{energy}$
                          (bacteria)

In natural waters, soluble organic matter is principally removed by oxidation and biosynthesis, but in the intensified microbial ecosystem of the biological treatment plant, adsorption is perhaps the major removal mechanism, with material adsorbed and agglomerated onto the dense microbial mass. The adsorptive property of the microbial biomass is particularly useful as it is also able to remove from solution non-biodegradable pollutants present in the wastewater such as synthetic organics, metallic salts, and even radioactive substances. The degree to which each removal mechanism contributes to overall purification depends on the treatment system used, its mode of operation, and the materials present in the wastewater.

In nature, heterotrophic micro-organisms occur as thin films (Periphyton) growing over rocks and plants, or, in fact, over any stable surface, or as individual or groups of organisms suspended in the water. These natural habitats of aquatic heterotrophs have been utilized in wastewater treatment to produce two very different types of biological units, one using attached growths and the other suspended microbial growths. The design criteria for secondary treatment units are selected to create ideal habitats to support the appropriate community of organisms responsible for the purification of wastewater. Therefore, attached and suspended microbial growth systems will require fundamentally different types of reactors. Both treatment systems depend on a mixed culture of micro-organisms, but grazing organisms are also involved so that a complete ecosystem is formed within the reactor, each with distinct trophic levels. In its simplest form the reactor food chain comprises:

Heterotrophic bacteria and fungi→Holozoic protozoans→
rotifers and nematodes→insects and worms→birds.

Because of the nature of the reactor, suspended growth processes have fewer trophic levels than attached growth systems (Fig. 2.17). These man-made ecosystems are completely controlled by operational practice, and are limited by food (organic loading) and oxygen (ventilation\aeration) availability.

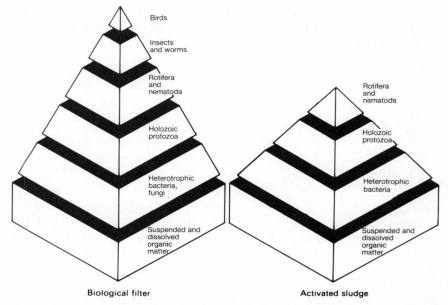

FIG. 2.17   Comparison of the food pyramids for percolating filter and activated
sludge treatment systems (Wheatley 1985).

Chemical engineers have been able to manipulate the natural process of self-purification, and by supplying ideal conditions and unlimited opportunities for metabolism and growth, have intensified and accelerated this biological process to provide a range of secondary wastewater treatment systems. However, a number of basic criteria must be satisfied by the design. In order to achieve a rate of oxidation well above that found in nature, a much denser biomass in terms of cells per unit volume must be maintained within the reactor. This will result in an increased oxygen demand which must be met in full, so as not to limit the rate of microbial oxidation. Essentially, this is done by increasing the air–water interface. The wastewater containing the polluting matter must be brought into contact for a sufficient time with a dense population of suitable micro-organisms and with excess oxygen, to allow oxidation and removal of unwanted material to the desired degree. Finally, inhibitory and toxic substances must not be allowed to reach harmful concentrations in the reactor.

The main methods of biological treatment rely on aerobic oxidation. To ensure that oxidation proceeds rapidly, it is important that as much oxygen as possible comes into contact with the wastewater so that the aerobic micro-organisms can break down the organic matter at maximum efficiency. Secondary treatment units of wastewater treatment plants are designed to bring this about. Oxidation is achieved by three main methods:

(1) by spreading the sewage into a thin film of liquid with a large surface area so that all the required oxygen can be supplied by gaseous diffusion (e.g. percolating filter);

(2) by aerating the sewage by pumping in bubbles of air or stirring vigorously (e.g. activated sludge); and

(3) by relying on algae present to produce oxygen by photosynthesis (e.g. stabilization pond).

In systems where the micro-organisms are attached, a stable surface must be available. Suitable surfaces are provided by a range of media such as graded aggregate or moulded plastic and even wooden slats, retained in a special reactor, on which a dense microbial biomass layer or film develops. These reactors are generally categorized as fixed-film reactors, the most widely used type being the percolating or trickling filter. Organic matter is removed by the wastewater as it flows in a thin layer over the biological film covering the static medium. Oxygen is provided by natural ventilation which moves through the bed of medium via the interstices supplying oxygen to all parts of the bed. The oxygen diffuses into the thin layer of wastewater to the aerobic micro-organisms below. The final effluent not only contains the waste products of this biological activity, mainly mineralized compounds such as nitrates and phosphates, but also particles of displaced film and grazing organisms flushed from the medium. These are separated from the clarified effluent by settlement, and the separated biomass disposed as secondary sludge (see Chapter 5), (Fig. 2.18). In suspended growth processes the micro-organisms are either free-living or flocculated to form small active particles or flocs which contain a variety of micro-organisms including bacteria, fungi, and protozoans. These flocs are mixed with wastewater in a simple tank reactor, called an aeration basin or tank, by aerators that not only supply oxygen but also maintain the microbial biomass in suspension to ensure maximum contact between the micro-organisms and the nutrients in the wastewater. The organic matter in the wastewater is taken out of solution by contact with the active suspended

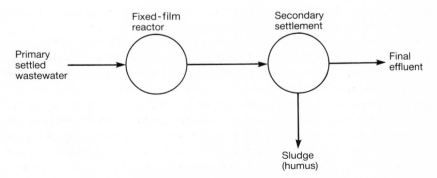

FIG. 2.18   Schematic layout of a fixed-film reactor.

biomass. The purified wastewater is displaced from the reactor by the incoming flow of settled wastewater and contains a large quantity of micro-organisms and flocs. The active biomass is separated from the clarified effluent by secondary settlement, but if the biomass was disposed of as sludge the concentration of active biomass in the aeration basin would rapidly fall to such a low density that little purification would occur. Therefore the biomass, called activated sludge, is returned to the reactor to maintain a high density of active biological solids ensuring a maximum rate of biological oxidation (Fig. 2.19). Excess biomass, not required to maintain the optimum microbial

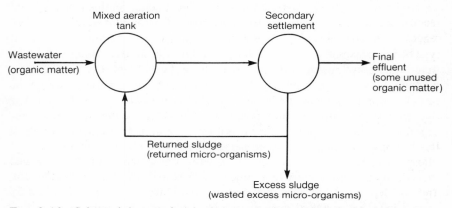

FIG. 2.19   Schematic layout of a mixed reactor (activated sludge) with sludge return.

density in the reactor, is disposed of as surplus sludge (Chapter 5). Suspended growth processes are more intensive than attached growth systems and are able to treat up to 10 times more effluent per unit volume of reactor making them much cheaper in terms of capital costs. However, suspended growth systems are more difficult to operate and maintain. They have much higher operating costs due to mixing, aeration, and pumping sludge from the settlement tank back to the reactor, and also produce comparatively large quantities of surplus biomass as sludge.

Biological wastewater treatment differs from more traditional fermentation processes, such as the production of bakers yeast, in a number of important ways. Primarily, wastewater treatment is aimed at removing unwanted material, whereas the commercial fermentation processes are all production systems. These production fermentations use highly developed, specialized strains of particular micro-organisms to synthesize the required end-product, whereas in wastewater treatment a broad mixture of micro-organisms are used. These are largely self-selecting with nearly all the organisms that can contribute to substrate removal welcome. Unlike commercial fermenters, wastewater reactors are not aseptic and because production fermenters require highly controlled conditions they are more complex and comparatively more expensive than those used in wastewater treatment.

Secondary treatment processes are extensively reviewed in Chapter 4.

2.4 TERTIARY AND ADVANCED TREATMENT

Most biological treatment plants are designed to produce a 20:30 effluent when loaded correctly, with extended aeration processes normally producing an even better effluent. However, treatment plants are all affected by a wide range of environmental variables and it is not possible to rely on a specific level of treatment all of the time. Furthermore, as the demand for water increases it is necessary to produce effluents to even higher standards in order to protect the quality of receiving water. This is particularly important if receiving water is extracted for domestic or industrial supply downstream of the effluent outfall, or if there is inadequate dilution. In order to meet standards more stringent than 20:30 on a reliable basis some form of tertiary treatment is required. Most tertiary treatment methods remove suspended solids and also the associated BOD_5 and pathogens. A removal of 10 mg l^{-1} of suspended solids from a normal sewage effluent is likely to reduce the BOD_5 by up to 3 mg l^{-1}. Methods that remove nutrients such as ammonia, nitrates, phosphates or other soluble material are more strictly termed as advanced treatment methods. Such techniques are used in eutrophication control, water reuse for potable or industrial supply, and perhaps most widely in desalination. There is confusion between tertiary and advanced treatment: the former removes fine suspended matter, whereas the latter removes soluble components from secondary treated wastewaters. In essence, both can be classed as the third stage in the treatment process (Institute of Water Pollution Control 1987a).

2.4.1 Tertiary treatment

Tertiary treatment is a method of improving good quality effluents, not for converting poor effluents to an acceptable standard. The processes are expensive and so cannot be considered as a method of uprating substandard treatment plants. It is often more cost-effective to use other methods rather than incorporating tertiary treatment to protect water quality, such as augmenting low river flow, the direct oxygenation of the river when necessary or the adoption of a more efficient treatment plant design. Tertiary treatment methods are predominantly physical in action relying largely on flocculation, settlement, and filtration. There are four basic methods: prolonged settlement; irrigation on to grassland; straining through a fine mesh; or filtration through media, such as sand or gravel.

Lagoons

If sufficient land is available shallow lagoons are the most efficient tertiary treatment process available (Potten 1972). The lagoons allow further settlement, and as the retention time is much longer than secondary sedimentation tanks, flocculation and settlement of some of the remaining suspended material will occur. Lagoons, also known as clarification lakes or maturation ponds, provide a combination of settlement and biological oxidation depending on the retention time. At short retention times (<60 h) purification is mainly by flocculation and settlement with suspended solids removal of 30–40% normal. There is a marked improvement in performance at longer retention times (14–21 days) with 75–90% removal of suspended solids, 50–60% BOD_5 and 99% coliform removal possible (Table 2.5). Fish

TABLE. 2.5   Comparison of the performance and relative cost of commonly used tertiary treatment units (Forster 1985)

| Method | Percentage removal | | | Approximate cost (£ per 1000 m³) |
	Suspended solids	BOD	Coliform bacteria	
Grass plots	70	50	90	0.38
Shallow lagoons (detention 3–4 d)	40	40	70	0.29
Deep lagoons (detention 17 d)	80	65	99	0.54
Slow sand filters	60	40	50	1.92
Rapid gravity sand filters	80	60	30	0.67
Upward flow sand filters	70	55	25	0.46
Microstrainers	70	40	15	0.75
Upward flow gravel bed clarifier	50	30	25	0.63

(1966) reported BOD_5 removal in a single lagoon of 29% at 52 h and 70% at 84 h retention. At longer retention times algae develop which can be discharged with the final effluent resulting in a raised suspended solids and BOD_5, and so offsetting the potential advantage gained by the increased settlement. Phosphate removal is linked to algal biomass and at the Rye Meads Treatment Works phosphate removal varied seasonally from a maximum removal of 73% in May to a minimum of 2% in January. The removal of nitrate is unaffected by algal biomass (Fig. 2.20) (DoE 1973). Research has shown that in eutrophic lagoons, fish can be reared very rapidly (White and Williams 1978), and that by harvesting fish commercially the cost of effluent treatment can be largely offset (Wert and Henderson 1978). Of all the tertiary treatment processes, lagoons, under ideal conditions, provide the most efficient bacteriological removal, except for grass plots. They are particularly useful for protecting receiving waters used as a raw water supply,

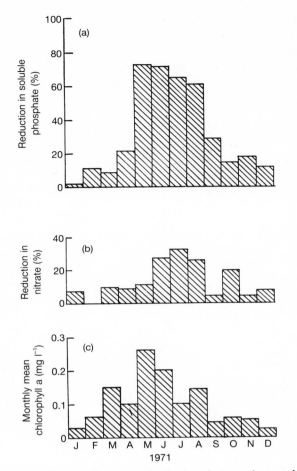

FIG. 2.20 Seasonal cycles in removal of phosphate and growth of algae in maturation lagoons after 10 days retention. (Institute of Water Pollution Control 1974).

for water recreation, and for the protection of shell-fisheries. An optimum depth for lagoons is about 1 m in the UK, which will restrict growth of macrophytes, and at a retention time of <60 h will provide maximum treatment efficiency without giving rise to excessive algal growths. The lagoon is a buffer between the treatment plant and the river, with retention times optimized by using baffles to prevent short-circuiting. Scumboards will also be required to prevent the loss of sludge which may rise from the bottom of the lagoon as anaerobic breakdown progresses. Loading rates depend on the efficiency of the secondary treatment processes but if a 20:30 effluent is produced then a maximum loading of between 3500–5000 $m^3ha^{-1}d^{-1}$ can be used. If the effluent quality entering the lagoon is worse than 20:30 then the loading rate must be reduced accordingly down to 1500 $m^3ha^{-1}d^{-1}$. If lagoons are small it is advisable to have at least two in series rather than one of the same capacity. A greater buffering effect, as well as improved nitrification, is obtained with several well-mixed lagoons (Fig. 2.21). The use of maturation ponds is fully examined in Section 4.4.2.

Grass plot irrigation
Land irrigation on grass plots is probably the most economical method of tertiary treatment and is especially suitable for treatment plants serving small

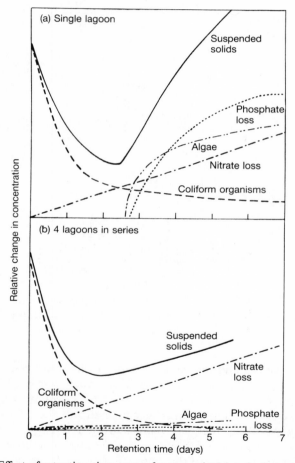

FIG. 2.21   Effect of retention time on performance in (*a*) a single lagoon 1 m deep and (*b*) four lagoons in series also 1 m deep when operated in a temperate climate. Phosphate removal and algal growth were measured from April to September (Institute of Water Pollution Control 1974).

communities. Effluent is distributed evenly over grassland via channels or spray guns. The plot is graded to a slope of between 1:60 to 1:100 to ensure that the effluent percolates downhill where it is collected in a channel. Except in very porous soil, sites should be under-drained to prevent waterlogging and to encourage growth of grass. The suspended solids are largely retained in the soil and some of the nutrients present are used for plant growth. Maximum removals of 80% suspended solids and up to 70% of BOD_5 is possible, with entrobacteria greatly reduced (Table 2.5). At least two plots are needed, preferably more, as the ground needs to be used intermittently to allow full drainage and subsequent mowing. There is no minimum or maximum size of plots but a recorded range of 0.1 to 3.0 ha is reported by the Institute of Water Pollution Control (1987a). Loadings to grass plots are expressed in terms of volume of wastewater applied per unit area of land per day. This figure should be based on the total land area available including those plots not currently in use. Loadings range from 2000 to 5000 $m^3ha^{-1}d^{-1}$ depending on climatic conditions (rainfall) and soil structure. If the secondary effluent fails to reach the 20:30 standard then the loading rate to the grassland must be reduced.

Short grasses are preferable, although the sowing of special grass mixtures does not appear to be worthwhile. The growth of grass irrigated with wastewater is prolific because of the nutrients present and the grass plot needs to be mown regularly to a height of about 5 cm in order to discourage weeds. It is usually best to remove the cuttings from the plot, and there is some indication that they can be safely used as silage. However, the area involved is usually too small to be of interest to farmers. Although most of the suspended solids removed by the plot are degraded in the soil the plot will require periodic renovation. This is done by allowing the plot to fully drain and dry off, and then harrowing, rotovating or even ploughing the land before regrading and reseeding. As in lagoons, mosquitoes can breed on grass plots if the humidity is high thus allowing a serious fly nuisance to develop, and care must be taken not to overload plots. Regrading is also important because if the slope becomes too steep channelling occurs, which causes erosion and usually an increase in the suspended solids concentration of the final effluent.

Microstrainers
Suspended solids in secondary wastewater can be efficiently removed by passing it through a fine mesh screen made from stainless steel. Such screens have been widely adopted for tertiary treatment of wastewaters since their introduction in 1948. Because of the small apertures that make up the mesh the screen rapidly becomes blocked when used with treated effluents and the screens require almost constant backwashing. This can only be achieved by fixing the mesh around the periphery of a rotating drum which is open at one end. Drums range from 1–3 m in diameter and 0.3–4.5 m in length. The drum is fixed horizontally allowing the wastewater to enter the drum to about two-thirds of its depth so that the effluent passes laterally through the sides of the

drum under a small hydraulic head ($=150$ mm) out into a collecting chamber. As the drum rotates, a small jet of filtered effluent is used to continuously backwash the screen as it passes the highest point of rotation, with the filtered sediment being flushed into a small collecting trough inside the drum, which is then piped away to the inlet of the plant.

Microstrainers are only used for 20:30 effluents, with the loading rate depending on the grading of mesh used. For example, the range of mesh sizes normally used ranges from 90 to 390 apertures per mm^2, with the finest grades only used for potable water treatment, allowing loadings of secondary wastewaters of 300 to 700 $m^3m^{-2}d^{-1}$. As a rough guide, a mesh aperture of 35 μm would give a 15:15 effluent, whereas a smaller mesh with a 23 μm diameter aperture would give a 10:10 effluent. Therefore, 30–80% removal of suspended solids and a 25–70% reduction in BOD_5 are possible depending on the wastewater quality and mesh grade used (Table 2.5). However, there is a very poor removal of enterobacteria by this method compared to either lagooning or grass plot irrigation. Between 2–5% of the filtrate is pumped from the collecting chamber and used for backwashing. Apart from blocking (blinding) due to the suspended solids in the wastewater, the mesh can also become clogged due to bacterial growth, which can result in an excessive head loss. Such growths are difficult to remove manually or with chemicals and so are prevented by using a UV lamp. Microstrainers are self-contained and relatively small units that are normally kept under cover in specially constructed buildings. Although the meshes can last for up to 10 years they require a high degree of maintenance and at least two full-scale units are required. This restricts their use to medium- and large-scale treatment plants (Institute of Water Pollution Control 1987a).

Sand filters

Gravity filtration through beds of sand have been used for many years in potable water treatment. In their simplest form, the filters are downflow units, although developments on the original design have been made such as mixed media and upflow units. There are three types of sand filter used for the tertiary treatment of domestic wastewater, the slow sand filter, the rapid downward flow sand filter, and the upward flow sand filter.

Slow sand filters are comprised of a layer of fine sand approximately 450 mm in depth overlying a 200–300 mm layer of pea gravel with a nominal diameter between 20–30 mm. Settled secondary effluent can be filtered at a maximum rate of 2.5–3.5 $m^3m^{-2}d^{-1}$ ($=0.15$ $m^3m^{-2}h^{-1}$). Because of the relatively low loading, slow sand filters require a large area and so often prove too expensive except for small treatment plants, although operating and maintenance costs are low. They are also particularly suitable for small treatment plants as they only require periodic cleaning. This is done by removing the top few millimetres of accumulated sludge and sand, which must be replaced occasionally with clean sand to maintain the depth of the filter. The contaminated sand is then washed and settled ready for reuse and the removed solids returned to the inlet of the plant. The surface layer of sand retains some of the suspended solids and quickly becomes covered with zoogloeal bacteria, which can significantly reduce the BOD_5 of the effluent. However, if the bacterial slime becomes too thick then the flow through the filter is impeded. Lower down in the filter nitrifying bacteria develop providing extra nitrification of the final effluent. Up to 60–80% of suspended solids are removed when a 20:30 effluent is passed through such a filter, and 30–50% reduction in BOD_5 and coliform bacteria are possible (Table 2.5) (BSI 1972).

Rapid downward flow sand filters have a greater hydraulic head than slow sand filters (3–4 m), and when operated under pressure give a hydraulic loading 40–50 times greater than slow sand filters at 200–250 $m^3m^{-2}d^{-1}$. Loaded at this high rate the surface layer of sand rapidly blinds under the pressure and backwashing is required every 24–48 h to clean the sand medium and to reduce compaction of the bed. Most rapid sand filters have an automatic backwashing facility which uses 2–3% of the total throughput for this purpose at a rate of 10 l $m^{-2}s^{-1}$. The washwater can contain significant quantities of sand and must be returned to the inlet so that the sand can be removed during grit extraction. The beds comprise of sand with a nominal particle size between 0.8 to 1.7 mm at a depth of 1.5 m. When operated at a hydraulic loading of 200 $m^3m^{-2}d^{-1}$ a rapid downward flow sand filter should remove 65–85% of the suspended solids and 20–35% of the BOD_5 from a 20:30 effluent (Table 2.5). Performance is not significantly affected by hydraulic loading or by the use of finer sand particles. Because of the regular backwashing little biological activity occurs and the BOD_5 removal is lower than in slow sand filters and there is little nitrification. Bacterial reductions are also poor and this type of tertiary treatment is not suitable if bacterial quality is important. However, rapid downward flow sand filters are widely used at large plants although they do require a high level of maintenance and operational control (Institute of Water Pollution Control 1987*a*).

Upward flow sand filters achieve similar removal rates of suspended solids and BOD_5 to downward flow sand filters (Table 2.5), but at much higher hydraulic loadings (400 $m^3m^{-2}d^{-1}$), which makes them especially suitable for large treatment plants. The effluent is forced upwards through a bed of graded media under pressure or hydraulic head. The bed is made out of layers of graded media with the coarser particles at the base so that the effluent has to pass through layers of increasing finer grade media as it progresses upwards. The entire bed of the filter is used to remove suspended solids making it more efficient than other filter systems, and does not require washing so frequently. A typical arrangement of media would be a 0.15 m layer of 40–50 mm gravel at the base with 0.25 m of 8–12 mm gravel followed by 0.25 m of 2–3 mm coarse sand, and, finally, a 1.5 m layer of 1–2 mm sand at the top. Such filters are washed by increasing the flow to 900 $m^3m^{-2}d^{-1}$ so that the interstices in the media are opened up and the accumulated solids flushed out in the washwater which is returned to the inlet.

Upward flow clarifiers are either separate or more usually incorporated as part of the secondary sedimentation tank. Clarifiers act as a sedimentation tank with a coarse medium upward flow filter at the outlet. They were originally developed for use with small percolating filtration units to obtain better effluents than from just secondary settlement only. Upward flow clarifiers consist of a 150 mm layer of pea gravel (5–20 mm diameter) supported on a perforated metal platform near the top of a horizontal flow sedimentation tank with a surface overflow rate of between 15–25 $m^3m^{-2}d^{-1}$. Other porous materials, apart from gravel, can also be used. Suspended solids are not physically strained from the effluent by the medium as the interstices are too large (1000–300 μm diameter) compared to the solids (< 100 μm). However, the passage through the gravel promotes flocculation of the suspended particles, which then settle on top of the gravel. The depth of water above the top of the gravel is therefore critical if flocculated solids are not to be carried over with the final effluent, thus the surface of the medium must be at least 150 mm, preferably 300 mm, below the surface.

Backwashing is required periodically to remove entrained solids which is done by lowering the level of the water in the clarifier to below the level of the gravel bed and hosing it down. The exact frequency of backwashing will depend on the size of the gravel used and the suspended solids content of the effluent. No washwater is produced, because all the solids washed from the medium settle to the bottom of the tank. Desludging must be done frequently and at least once a week, and the flocculated solids should be drained off as necessary and returned to the inlet. A 30–50% removal of suspended solids is possible, although this depends largely on the size of the gravel, with up to 35% removal of BOD_5 (Table 2.5). Loading rates of up to 3 DWF are permitted, although the suspended solids from percolating filters flocculate far more readily than those from activated sludge processes, and higher loadings to clarifiers are permitted from percolating filters (42 $m^3m^{-2}d^{-1}$) than from activated sludge (30 $m^3m^{-2}d^{-1}$). The unit is capable of dealing with a wide fluctuation in suspended solids concentrations and hydraulic flows. Only a small hydraulic head is required (25 mm) and clarifiers are much smaller than other tertiary treatment units that are suitable for small treatment plants. For example, a Satec clarifier capable of treating a PE (population equivalent) of 880 is only 2.74 m × 2.74 m in area (Barnes and Wilson 1976) (Fig. 2.22). This clarifier is now widely used at smaller extended aeration activated sludge plants.

Mann (1979) has compared the tertiary treatment methods most commonly adopted for small treatment plants and has listed their potential advantages and disadvantages (Table 2.6). These methods are illustrated in Fig. 2.23.

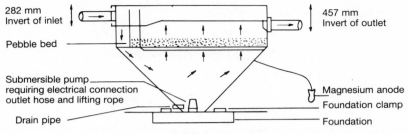

282 mm
Invert of inlet

457 mm
Invert of outlet

Pebble bed

Submersible pump
requiring electrical connection
outlet hose and lifting rope

Magnesium anode

Foundation clamp

Drain pipe

Foundation

FIG. 2.22   The Satec pebble bed clarifier (Mann 1979).

TABLE. 2.6   Comparison of the performance and operation of tertiary treatment methods shown in Fig. 2.23 (Mann 1979)

System	Hydraulic loading rate $(m^3 m^{-2} d^{-1})$	Removal (%) BOD	Suspended solids	Specific advantages	Potential disadvantages
Upward flow gravel clarifier	24	30	50	Can be fitted in the top of the humus tank	Requires regular backwashing; not less than weekly
Sand filter	3.0	40	60	Positive system. Little possibility of short-circuiting	Regular cleaning necessary. Two filters are needed in alternate use. Highest cost system
Grass plots	0.85	50	70	Very low cost, low maintenance, high efficiency	Can be unsightly, if maintenance neglected. Can encourage breeding of flies. A spare plot is needed to permit resting and maintenance
Lagoons	0.5	40	40	Efficiency can be increased by use of lower application rates. Very low maintenance requirements	Lagoons must be watertight. Can be unsightly if overloaded owing to formation of scum. May encourage breeding of insects

2.4.2   Advanced wastewater treatment

The term tertiary treatment is often used as a synonym for advanced water treatment (AWT) and although similarities exist the two are not precisely the same. Whereas tertiary treatment is an additional step applied after secondary

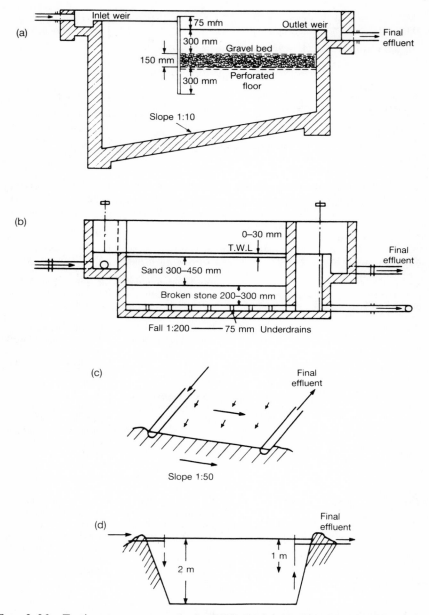

FIG. 2.23 Tertiary treatment methods widely used at small-scale treatment works. (a) Upward flow gravel bed clarifier which is installed in the final sedimentation tank, (b) slow sand filter of which at least two will be required for continuous operation, (c) grass plot or land treatment, (d) maturation or tertiary treatment lagoon (Mann 1979).

treatment to reduce the suspended solids and, to some extent, the BOD_5, AWT is any process or system used after conventional treatment, or to modify or replace one or more steps. AWT systems remove refractory and mainly soluble pollutants which are not readily removed by conventional biological treatment. Several of these pollutants can affect aquatic life. For example, ammonia is highly toxic to fish and can cause de-oxygenation as it is oxidized; nitrogen and phosphorus promote eutrophication in rivers and lakes respectively; while nitrogen compounds, trace organics, and viruses can hinder the re-use of surface water for supply. The main treatment methods include: the removal of ammonia by nitrification; air-stripping or breakpoint chlorination; the removal of inorganic nitrogen by denitrification; phosphate removal by algal synthesis, chemical precipitation using iron or aluminium coagulents and lime, or modifications such as the 'Bardenpho' and 'Phostrip' processes; reduction of dissolved organics (residual organic matter) using activated carbon, chemical or ozone treatment; removal of inorganic salts, especially sulphates and chlorides, by ultrafiltration, reverse osmosis or electrodialysis; and disinfection of effluents to control pathogens, especially viruses. Some of these techniques are dealt with elsewhere in the text, where appropriate.

Further reading

General: Hammer 1977; Metcalf and Eddy 1984; Viessman and Hammer 1985; Eilbeck and Mattock 1987.
Tertiary treatment: Institute of Water Pollution Control 1987a.
Advanced wastewater treatment: Culp and Culp 1978; McCarty and Reinhard 1980.

3

THE ROLE OF ORGANISMS

3.1. STOICHIOMETRY AND KINETICS

In biological wastewater treatment, the most widely occurring and abundant group of micro-organisms are the bacteria, and it is this group which are most important in terms of utilizing the organic matter present in wastewater. The following section deals with the kinetics of biological treatment systems and concentrates mainly on the bacteria, although the basic principles are applicable to the other groups of micro-organisms.

The organic matter in wastewater is utilized by micro-organisms in a series of enzymatic reactions. Enzymes are proteins, or proteins combined with either an inorganic or low molecular weight organic molecule. They act as catalysts forming complexes with the organic substrate which they convert to a specific product, releasing the original enzyme to repeat the same reaction. The enzymes have a high degree of substrate specificity and a bacterial cell must produce different enzymes for each substrate utilized. Two types of enzymes are produced: *extracellular* enzymes, which convert substrate extracellularly into a form that can be taken into the cell for further breakdown by the *intracellular* enzymes, which are involved in synthesis and energy reactions within the cell. Normally, the product of an enzyme-catalysed reaction immediately combines with another enzyme until the final end-product required by the cell is reached after a sequence of enzyme-substrate reactions.

A portion of the absorbed material in the bacterial cell is oxidized to provide energy while the remainder is used as 'building blocks' in cellular synthesis. In the oxidation of organic matter, for example the carbohydrate glucose, oxygen only becomes involved at the end of a series of meticulously integrated enzyme chemical transformations:

$$C_6H_{12}O_6 \rightarrow 6CO_2 + 6H_2O + 678 \text{ cal energy}$$

The oxidation reaction that has occurred can be described in a number of ways such as, the use of oxygen, the loss of hydrogen from the substrate (normally the food source), which in this case is glucose, or the loss of electrons from the glucose substrate. These reductions are described in terms of hydrogen or electron donors (substrate) and hydrogen or electron acceptors (in this case oxygen).

The electron donor gives up electrons that are transported via complicated biochemical pathways to the ultimate or terminal electron acceptor, which is oxygen for an aerobic reaction. Organic electron donors are utilized in heterotrophic metabolism whereas autotrophic metabolism uses inorganic electron donors. It is the terminal electron acceptor that determines the amount of energy which is available from the substrate. The general relationship between energy yields of aerobic and anaerobic metabolism is summarized in Fig. 3.1. Energy stored in organic matter (AH_2) is released in the process of biological oxidation by dehydrogenation of substrate followed by transfer of an electron or electrons to an ultimate acceptor. The higher the ultimate electron acceptor is on the energy or electromotive scale, the greater the energy yield from the oxidation of 1 mole of substrate, aerobic metabolism using oxygen as the ultimate electron acceptor yields the greatest amount of energy. Aerobic respiration can be traced in Fig. 3.1 from reduced organic matter (AH_2) at the base through the hydrogen and electron carriers to oxygen. Facultative respiration, using oxygen bound in nitrate or sulphate, yields less energy than aerobic metabolism. The least energy results from strict anaerobic respiration, where the oxidation of AH_2 is coupled with reduction of B (an oxidized organic compound) to BH_2 (a reduced organic compound).

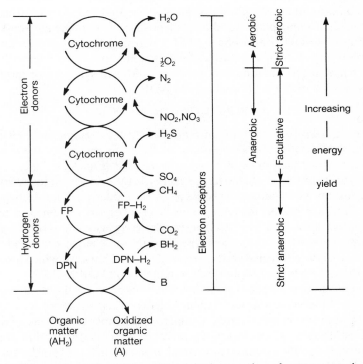

FIG. 3.1 Schematic representation of the dehydrogenation of wastewater, showing the relative energy yields for various ultimate electron acceptors (adapted Clark *et al.* 1977).

The preferential use of electron acceptors based on energy yields in a mixed bacterial culture is shown in Table 3.1. Electron acceptors are used in the general descending order of dissolved oxygen, nitrate, sulphate, and oxidized organic compounds. Therefore, hydrogen sulphide formation follows nitrate reduction but precedes methane production (Clark *et al.* 1977).

TABLE 3.1   The preferential selection of electron acceptors during microbial oxidation of organic matter

Aerobic	$AH_2 + O_2 \rightarrow CO_2 + H_2O + energy$	
|	$AH_2 + NO_3^- \rightarrow N_2 + H_2O + energy$	Decreasing
Facultative	$AH_2 + SO_4^{2-} \rightarrow H_2S + H_2O + energy$	energy
↓	$AH_2 + CO_2 \rightarrow CH_4 + H_2O + energy$	yield
Anaerobic	$AH_2 + B \rightarrow BH_2 + A + energy$	↓

During the oxidative reaction, energy is conserved by the cell in two ways. The micro-organisms function by capturing energy liberated by exergonic (energy-releasing) reactions, using this energy to drive endergonic (energy-requiring) reactions. The energy that is released in the cell by oxidizing organic or inorganic matter or by photosynthesis is captured and stored within the cell in the form of a phosphate bond on a adenosine diphosphate molecule (ADP) forming adenosine triphosphate (ATP), which is known as substrate-level phosphorylation. When required, the stored energy can be released to drive other reactions, forming needed metabolites for cell synthesis or mobility by the hydrolysis of the ATP molecule to ADP:

$$ATP + H_2O \rightarrow ADP + H_3PO_4$$

Under standard conditions of concentration (0.1 M), temperature (30°C) and pH (7), 8.4 kcals of energy becomes available to the micro-organisms for biosynthesis for each mole of phosphate released (Fig. 3.2).

The second method of recovering energy is by oxidative phosphorylation. Electrons, normally pairs, are produced by the oxidation of an electron donor (DH_2) and are passed through an electron-transport system to a terminal electron acceptor (A). The electron-transport system is a series of electron carriers arranged so that the large amount of energy produced by the oxidation of the electron donor is released in small packets, which are used to drive the endergonic phosphorylation reaction of ADP to ATP.

In Fig. 3.1, the oxidation-reduction reactions are carried out exclusively by enzymes. While some enzymes depend solely on their structure as proteins for activity, some require co-factors. Co-factors are either metal ions or complex organic molecules, generally known as co-enzymes, which function as carriers of electrons, specific atoms or functional groups which are transferred during the enzymatic reaction. The co-enzyme component of the enzyme determines what chemical reaction will occur. One of the most important co-enzymes is nicotinamide adenine dinucleotide, known as NAD in its oxidized form and

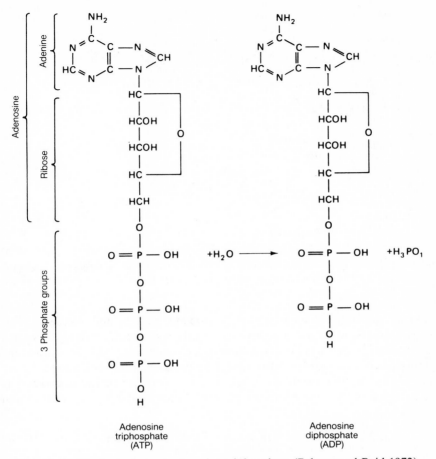

FIG. 3.2   The hydrolysis of adenosine triphosphate (Pelczar and Reid 1972).

NADH + H in its reduced form (Fig. 3.3). Flavoproteins (FP) which are responsible for hydrogen transfer are also important co-enzymes. Cytochromes are respiratory pigments that can undergo oxidation reduction and serve as hydrogen carriers. The various pathways of energy metabolism of organisms are considered in more detail in the following sections of this chapter.

3.1.1   Stoichiometry

It is essential in the design of biological treatment systems to be able to calculate the necessary inputs into the system such as oxygen and nutrients as

NAD NADH

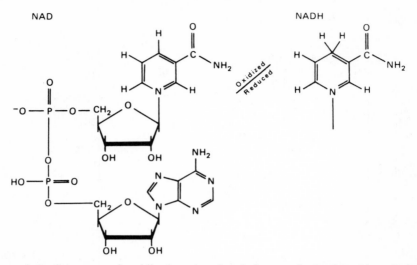

Fig. 3.3 Structure of oxidized and reduced forms of nicotinamide adenine dinucleotide (Brock 1971a).

well as the outputs such as carbon dioxide, sludge, and methane. This stoichiometric approach to energy production and cell synthesis has been developed by McCarty (1971). A table of half-reactions (Table 3.2) allows the organic substrate that has been utilized by the bacteria to be divided into two portions; first, that which has been oxidized to produce energy, and secondly that which is used for synthesis of new cellular material. The energy and

TABLE 3.2 Oxidation half reactions (Christensen and McCarty 1975)

Equation no.	Half-reaction		$\Delta G°(W)^*$ kcal per electron equivalent
	Reactions for bacterial cell synthesis (R_c)		
	Ammonia as nitrogen source:		
1	$1/5CO_2 + 1/20HCO_3^- + 1/20NH_4^+ + H^+ + e^-$	$= 1/20C_5H_7O_2N + 9/20H_2O$	
	Nitrate as nitrogen source:		
2	$1/28NO_3^- + 5/28CO_2 + 29/28H^+ + e^-$	$= 1/28C_5H_7O_2N + 11/28H_2O$	
	Reactions for electron acceptors (R_a)		
	Oxygen:		
3	$1/4O_2 + H^+ + e^-$	$= 1/2H_2O$	-18.675
	Nitrate:		
4	$1/5NO_3^- + 6/5H^+ + e^-$	$= 1/10N_2 + 3/5H_2O$	-17.128

TABLE 3.2—*Continued*

Equation no.	Half-reaction		$\Delta G°(W)$* kcal per electron equivalent
		Reactions for bacterial cell synthesis (R_c)	
	Sulphate:		
5	$1/8SO_4^{2-}+19/16H^++e^-$	$=1/16H_2S+1/16HS^-+1/2H_2O$	5.085
	Carbon dioxide (methane fermentation):		
6	$1/8CO_2+H^++e^-$	$=1/8CH_4+1/4H_2O$	5.763
		Reactions for electron donors (R_d)	
	Organic donors (heterotrophic reactions)		
	Domestic wastewater:		
7	$9/50CO_2+1/50NH_4^+ +$ $1/50HCO_3^-+H^++c^-$	$=1/50C_{10}H_{19}O_3N+9/25H_2O$	7.6
	Protein (amino acids, proteins, nitrogenous organics):		
8	$8/33CO_2+2/33NH_4^+ +$ $31/33H^++e^-$	$=1/66C_{16}H_{24}C_5N_4+27/66H_2O$	7.7
	Carbohydrates (cellulose, starch, sugars):		
9	$1/4CO_2+H^++e^-$	$=1/4CH_2O+1/4H_2O$	10.0
	Grease (fats and oils):		
10	$4/23CO_2+H^++e^-$	$=1/46C_8H_{16}O+15/46H_2O$	6.6
	Acetate:		
11	$1/8CO_2+1/8HCO_3^-+H^++e^-$	$=1/8CH_3COO^-+3/8H_2O$	6.609
	Propionate:		
12	$1/7CO_2+1/14HCO_3^-+H^++e^-$	$=1/14CH_3CH_2COO^-+5/14H_2O$	6.664
	Benzoate:		
13	$1/5CO_2+1/30HCO_3^-+H^++e^-$	$=1/30C_6H_5COO^-+13/20H_2O$	6.892
	Ethanol:		
14	$1/6CO_2+H^++e^-$	$=1/12CH_3CH_2OH+1/4H_2O$	7.592
	Lactate:		
15	$1/6CO_2+1/12HCO_3^-+H^++e^-$	$=1/12CH_3CHOHCOO^- +$ $1/3H_2O$	7.873
	Pyruvate:		
16	$1/5CO_2+1/10HCO_3^-+H^++e^-$	$=1/10CH_3COCCO^-+2/5H_2O$	8.545
	Methanol:		
17	$1/6CO_2+H^++e^-$	$=1/6CH_3OH+1/6H_2O$	8.965
	Inorganic donors (autotrophic reactions)		
18	$Fe^{3+}+e^-$	$=Fe^{2+}$	−17.780
19	$1/2NO_3^-+H^++e^-$	$=1/2NO_2^-+1/2H_2O$	−9.430
20	$1/8NO_3^-+5/4H^++e^-$	$=1/8NH_4^++3/8H_2O$	−8.245
21	$1/6NO_2^-+4/3H^++e^-$	$=1/6NH_4^++1/3H_2O$	−7.852
22	$1/6SO_4^{2-}+4/3H^++e^-$	$=1/6S+2/3H_2O$	4.657
23	$1/8SO_4^{2-}+11/16H^++e^-$	$=1/16H_2S+1/16HS^-+1/2H_2O$	5.085
24	$1/4SO_4^{2-}+5/4H^++e^-$	$=1/8S_2O_3^{2-}+5/8H_2O$	5.091
25	H^++e^-	$=1/2H_2$	9.670
26	$1/2SO_4^{2-}+H^++e^-$	$=1/2SO_3^{2-}+1/2H_2O$	10.595

* Reactants and products at unit activity except $(H^+)=10^{-7}$.

synthesis reactions add together to become the overall reaction or total metabolism (Christensen and McCarty 1975).

Using Table 3.2, it is possible to calculate the energy yield for the metabolism of all the commonly encountered substrates under aerobic, anoxic or anaerobic conditions. For example, if we assume the composition of domestic sewage to be $C_{10}H_{19}O_3N$, the energy yields for different electron acceptors can easily be calculated by using equation 7 for electron donor and equation 3 for aerobic conditions, equation 4 for anoxic conditions or equation 6 for anaerobic conditions.

Thus, the energy yield from the metabolism of domestic sewage under aerobic conditions is calculated by writing the half reactions:

Donor: $$1/50C_{10}H_{19}O_3N + 9/50H_2O$$

$$= 9/50CO_2 + 1/50NH_4^+ + 1/50HCO_3^- + H^+ + e^-$$

$$\Delta G° = +7.6 \text{ kcal/e}^-$$

Acceptor: $$1/4O_2 + H^+ + e^- = 1/2H_2O \quad \Delta G° = -18.675 \text{ kcal/e}^-$$

Adding the half reactions yields:

$$1/50C_{10}H_{19}O_3N + 9/25H_2O + 1/4O_2 + H^+ + e^-$$
$$= 9/50CO_2 + 1/50NH_4^+ + 1/50HCO_3^- + H^+ + e^- + 1/2H_2O$$

If the above equation is balanced by multiplying by 50, then:

$$C_{10}H_{19}O_3N + 18H_2O + 12.5O_2 = 9CO_2 + HN_4^+ + HCO_3^-$$

The energy yield by the reaction is:

$$\Delta G° = \Delta G° \text{ of products } - \Delta G° \text{ of reactions}$$

As each mole of wastewater utilized results in the transfer of 50 electrons (Table 3.2). The energy resulting from the aerobic metabolism of wastewater is:

$$\Delta G° = \frac{50 \text{ electrons}}{\text{mole wastewater}} \times [-18.675 \text{ kcal/e}^- - 7.6 \text{ kcal/e}^-]$$

$$= -1313.75 \text{ kcal per mole wastewater}$$

In a similar way, the energy yield from wastewater under anoxic conditions using NO_3^- as the electron acceptor or anaerobic conditions using CO_2 as the electron acceptor can be calculated as -1236.4 and -91.85 kcal per mole of wastewater respectively (Table 3.2). This demonstrates that the energy yields for specific substrate are essentially the same under aerobic and anoxic conditions but significantly less from anaerobic metabolism.

Synthesis is the biochemical process of substrate utilization to form new protoplasm for growth and reproduction. The new cell material formed is a complex of organic compounds including proteins, carbohydrates, and lipids. Therefore, protoplasm is mainly composed of carbon, hydrogen, and oxygen, although on a dry weight basis it also contains 10–20% nitrogen, about 2.5% phosphorus, together with other essential elements present in trace amounts. The general formula used to characterize bacterial cells is $C_5H_7O_2N$, which is used in all stoichiometric calculations. The primary product of metabolism is energy and the major use of energy is for synthesis, and as energy release and synthesis are coupled then the maximum rate of synthesis occurs simultaneously with maximum rate of energy yield or metabolism. In wastewater treatment terms, the maximum rate of oxidation of organic matter for a given population of heterotrophic micro-organisms, occurs during maximum biological growth and the lowest oxidation rate occurs when growth ceases.

Calculating the proportion of substrate utilized as either energy or synthesis can be useful to the wastewater engineer, as both the amount of electron acceptor (oxygen) required and the end-products produced can be calculated by knowing what portion of the substrate is synthesized into new cellular material.

As part of the substrate goes into energy formation and the remainder goes to cell synthesis, all the reacting material can be expressed as:

$$f_e + f_s = 1$$

where f_e is the fraction of electron donor used for energy and f_s the fraction of electron donor used for synthesis. McCarty (1975) calculates the amount of substrate metabolized by bacteria to form energy and new cells by using a balanced half equation. The overall reaction is constructed from three half reactions (Table 3.2), one for the synthesis of bacterial cells, which are assumed to be $C_5H_7O_2N$ (R_c), one for the electron acceptor (R_e) and one for the electron donor (R_d), which combine to give the relationship where R is the overall reaction:

$$R = f_s R_c + f_e R_e - R_d$$

The ratio of f_e to f_s depends on the age of the cell culture as well as the substrate electron donor. An increase in the age of the cells in the system will reduce the net amount of substrate material converted to new cell mass: The age of cells is commonly referred to as sludge age or mean cell residence time (θ_c) and is defined as the average time in days an organism remains within the treatment system. Maximum values of f_s are given in Table 3.3 for various substrates. However, these values are for young, rapidly growing cultures and represent a maximum value of f_s. In older cultures the f_s value is lower and in oxidation ditches that have very long cell residence times f_s can be as low as 20% of $(f_s)_{max}$. It is convenient to estimate the fraction f_s by using the relationship:

$$f_s = (f_s)_{max}\, 1 - \frac{0.8b\theta_c}{(1 + b\theta_c)}$$

where θ_c is the sludge age in days and the coefficient b represents the rate of cell death and decay which is assumed to be 0.03.

TABLE 3.3 Maximum cell yield $(f_s)_{max}$ (i.e. the maximum fraction of the electron donor used for synthesis) for various electron donors and electron acceptors

Electron donor	Electron acceptor	$(f_s)_{max}$
Heterotrophic reactions		
Carbohydrate	O_2	0.72
Carbohydrate	NO_3	0.60
Carbohydrate	SO_4	0.30
Carbohydrate	CO_2	0.28
Protein	O_2	0.64
Protein	CO_2	0.08
Fatty acid	O_2	0.59
Fatty acid	SO_4	0.06
Fatty acid	CO_2	0.05
Glucose	O_2	0.79
Lactose	O_2	0.74
Sucrose	O_2	0.75
Glycine	O_2	0.52
Alanine	O_2	0.52
Propionate	O_2	0.58
Acetate	O_2	0.58
Methanol	NO_3	0.36
Methanol	CO_2	0.15
Propionate	CO_2	0.07
Acetate	CO_2	0.06
Glucose	CO_2	0.27
Sewage sludge	CO_2	0.11
Autotrophic reactions		
S	O_2	0.22
S_2O_3	O_2	0.11
S_2O_3	NO_3	0.20
NH_4	O_2	0.10
H_2	O_2	0.24
H_2	CO_2	0.04
Fe	O_2	0.07

As an example, the oxygen required by the aerobic biological treatment of domestic sewage can be calculated by first estimating $(f_s)_{max}$. No value of $(f_s)_{max}$ is given for wastewater in Table 3.3, therefore it must be calculated from basic constituents. Domestic sewage consists of approximately 50% protein, 40% carbohydrate, and 10% fat (Section 1.2). Thus, by reference to Table 3.3, $(f_s)_{max}$ is calculated as:

$$(f_s)_{max, sewage} = (\% \text{ protein} \times (f_s)_{max, protein})$$
$$+ (\% \text{ carbohydrate} \times (f_s)_{max, carbohydrate}) + (\% \text{ fat} \times (f_s)_{max, fat})$$
$$(f_s)_{max, sewage} = (0.50 \times 0.64) + (0.40 \times 0.72) + (0.10 \times 0.59)$$
$$(f_s)_{max, sewage} = 0.67$$

If the value of f_s is 20% of $(f_s)_{max}$ for an oxidation ditch in which the cells are undergoing endogenous respiration, with a long mean cell residence time, then the value of f_s is 0.13, under these conditions.

The approximate half-reactions are selected from Table 3.2.

R_c: $1/5CO_2 + 1/20HCO_3^- + 1/20NH_4^+ + H^+ + e^- = 1/20C_5H_7O_2N + 9/20H_2O$

R_e: $1/4O_2 + H^+ + e^- = 1/2H_2O$

R_d: $9/50CO_2 + 1.50NH_4^+ + 1/50HCO_3^- + H^+ + e^- = 1/50C_{10}H_{19}O_3N + 9/25H_2O$

The overall reaction is constructed using $f_s = 0.13$. In accordance with $R = f_s R_c + f_e R_e - R_d$, the donor reaction (R_d) is not changed, the cell reaction (R_e) is multiplied through by $f_e = 1 - f_s = 0.87$ to obtain the following three equations:

$R_c \times f_s$: $0.026CO_2 + 0.0065HCO_3^- + 0.0065NH_4^+ + 0.130H^+ + 0.130e^-$

$= 0.0065C_5H_7O_2N + 0.059H_2O$

$R_e \times f_e$: $0.218O_2 + 0.870H^+ + 0.870e^- = 0.435H_2O$

R_d: $0.18CO_2 + 0.02HN_4^+ + 0.02HCO_3^- + H^+ + e^-$

$= 0.02C_{10}H_{19}O_3N + 0.36H_2O$

When added, the overall reaction R is:

$0.02C_{10}H_{19}O_3N + 0.218O_2 = 0.0065C_5H_7O_2N + 0.0135HCO_3^- + 0.0135NH_4^+ + 0.154CO_2 + 0.134H_2O$

Multiplying by $1/0.02 = 50$

$C_{10}H_{19}O_3N + 10.9O_2 = 0.325C_5H_7O_2N + 0.675HCO_3^- + 0.675NH_4^+ + 7.70CO_2 + 6.7H_2O$

Thus, in the oxidation ditch system 1 mol of domestic sewage $(C_{10}H_{19}O_3N)$ will yield 0.325 mol of micro-organisms $(C_5H_7O_2N)$ and will require 10.90 moles of oxygen. It is also possible to express these values in terms of mg oxygen required per mg COD or mg volatile suspended solids per mg COD (Mandt and Bell 1982).

Another example of energy and synthesis calculations is given in Section 3.4.3, where the methane production of sewage sludge by heterotrophic fermentation is estimated.

In anaerobic decomposition, the low energy yield per unit of substrate due to a lack of electron acceptors is a limiting factor, which results in the incomplete breakdown of the substrate. Metabolism and synthesis cease when the supply of biologically available energy is exhausted. With aerobic metabolism, the biologically available carbon is the limiting factor, with no shortage of electron acceptors because of the abundance of oxygen.

However, the supply of substrate carbon is rapidly exhausted because of respiration of carbon dioxide and synthesis (Clark *et al.* 1977). The energy conversion efficiencies of aerobic and anaerobic metabolism are shown in Fig. 3.4. Incomplete metabolism, a small amount of biological growth, and the production of high-energy products, such as acetic acid and methane, characterizes the anaerobic reaction. In contrast, the complete metabolism and synthesis of substrate, ending up with large amounts of biological growth, is characteristic of aerobic metabolism. The end-products of these metabolisms are compared in Table 1.21.

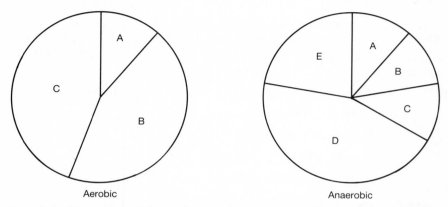

FIG. 3.4   Proportion of the total energy in wastewater utilized under aerobic and anaerobic metabolism as: (A) lost heat energy; (B) respiration; (C) synthesis; (D) energy bound in end-products; and (E) unused energy due to a lack of hydrogen acceptors.

3.1.2   Bacterial kinetics

In terms of substrate removal, the rate of carbonaceous oxidation, nitrification and denitrification depends on the rate of microbial growth, and in particular the rate of bacterial growth. This is best illustrated by observing the development of a microbial population in batch culture. When a small inoculum of viable bacterial cells are placed in a closed vessel with excess food and ideal environmental conditions, unrestricted growth occurs. Monod (1949) plotted the resultant microbial growth curve from which six discrete phases of bacterial development can be defined (Fig. 3.5). The microbial concentration is usually expressed as either the number of cells per unit volume or mass of cells per unit volume of reactor. However, it is not possible to directly translate one set of units from the other as the size of the cells and the creation of storage products by cells dramatically affects the mass but not the number of cells (Fig. 3.6).

The lag phase represents the acclimatization of the organisms to the substrate with the bacterial cells having long generation times and zero growth

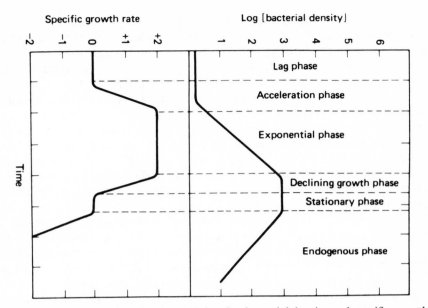

FIG. 3.5   The microbial growth curve showing bacterial density and specific growth rate at various microbial growth phases (Benefield and Randall 1980).

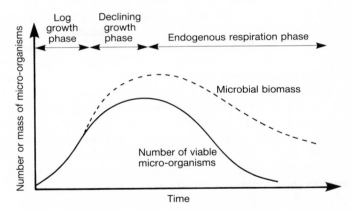

FIG. 3.6   Microbial growth curves comparing total biomass and viable biomass (Clark *et al.* 1977).

rates. Nutrients are taken into the cells and both the size and the mass of bacteria increase as the amount of enzymes and nucleic acid increases. However, depending on the size and degree of adaptation of the inoculum to its new environment, the lag phase may be very short or even absent.

The bacterial inoculum normally comprises of cells in the stationary growth phase, however if log-phase cells are used instead, then the lag phase can be shortened even further. Cells only begin to divide when a sufficient concentration of the appropriate enzymes have built up, but once division has started the population density of bacteria rapidly increases. In the acceleration phase, the generation time decreases and there is a discernable increase in the growth rate leading to the exponential or log phase. In this phase, the generation time is minimal, but constant, with a maximum and constant specific growth rate resulting in a rapid increase in the number and mass of micro-organisms. This is the period when the substrate conversion is at its maximum rate. The steady-state condition of growth is indicated by a near constant ratio of DNA/cell, RNA/cell, and protein/cell as well as constant cell density and minimum cell size. The rate of metabolism and, in particular, the growth rate is limited only by the microbial generation and its ability to process substrate. The exponential phase continues until the substrate becomes limiting. This produces the declining growth phase where the rate of microbial growth rapidly declines as the generation time increases and the specific growth rate decreases as the substrate concentration is gradually diminished. It is at this stage that the total mass of the microbial protoplasm begins to exceed the mass of viable cells as many of the micro-organisms have ceased reproducing due to substrate-limiting conditions (Fig. 3.6). In batch situations the accumulation of toxic metabolites or changes in the concentration of nutrients, or other environmental factors such as oxygen or pH can also be responsible for the onset of the declining growth phase. The microbial growth curve flattens out as the maximum microbial density is reached with the rate of reproduction apparently balanced by the death rate. This is the stationary phase where the substrate and nutrients are exhausted and there is a high concentration of toxic metabolites. It has been suggested that the majority of cells remain viable during this phase but in a state of suspended animation, without the necessary substrate or environmental conditions to continue to reproduce (Williams 1975). The final phase is the endogenous or log death phase. The substrate is now completely exhausted and the toxic metabolites have become unfavourable for cell survival. The microbial density decrease rapidly with a high micro-organism death rate resulting in the rate of metabolism and hence the rate of substrate removal also declining. The total mass of microbial protoplasm decreases as cells utilize their own protoplasm as an energy source (endogenous respiration) and as cells die and lyse they release nutrients back into solution, although there is a continued decrease in both the number and mass of micro-organisms.

The microbial growth curve is not a basic property of bacterial cells but is a response to their environmental conditions within a closed system. Because biological treatment processes are continuous flow systems it is not directly applicable, although it is possible to maintain such systems at a particular growth phase by controlling the ratio of substrate to microbial biomass, commonly referred to as the food to micro-organism ratio (f/m). The f/m ratio maintained in the aeration tank in an activated sludge system controls the rate of biological oxidation as well as the volume of microbial biomass produced by maintaining microbial growth either in the log, declining or endogenous growth phase. The type of activated sludge process can be defined by the f/m ratio as being high-rate, conventional or extended (Fig. 4.34). For example, at a high f/m ratio micro-organisms are in the log-growth phase which is characterized by excess substrate and maximum rate metabolism. Whereas at low f/m ratios the overall metabolic action in the aeration tank is endogenous, with the substrate limiting microbial growth so that cell lysis and resynthesis occurs. The effect of the f/m ratio on the microbial dynamics of wastewater treatment systems are considered fully in Chapter 4.

Rates of reaction

The rate at which components of wastewater, such as organic matter, are removed and the rate at which biomass is produced within a reactor are important criteria in the design and calculation of the size of biological reactors. The most useful method of describing such chemical reactions within a biological reactor is by classifying the reaction on a kinetic basis by reaction order. Reaction orders can differ when there is variation in the micro-organisms, the substrate or environmental conditions and they must be measured experimentally.

The relationship between rate of reaction, concentration of reactant and reaction order (n) is given by the expression:

$$\text{Rate} = (\text{concentration})^n$$

or, more commonly, by taking the log of both sides of the equation:

$$\text{Log rate} = n \log (\text{concentration}).$$

This equation is then used to establish the reaction order and rate of reaction. Thus, if the log of the instantaneous rate of change of the reactant concentration at any time is plotted as a function of the log of the reactant concentration at that instant, then a straight line will result for constant order reactions and the slope of the line will be the order of the reaction (Fig. 3.7). The rate of reaction is independent of the reactant concentration in zero-order reactions, which results in a horizontal line when plotted. The rate of reaction in first-order reactions is directly proportional to the reactant concentration, and with second-order reactions the rate is proportional to the concentration squared.

In practice, it is simplest to plot the concentration of reactant remaining against time in order to calculate the reaction rate. Zero-order reactions are linear when the plot is made on arithmetic paper (Fig. 3.8), following the equation:

$$C - C_0 = -kt$$

where C is the concentration of reactant (mg l^{-1}) at time t, C_0 the constant of

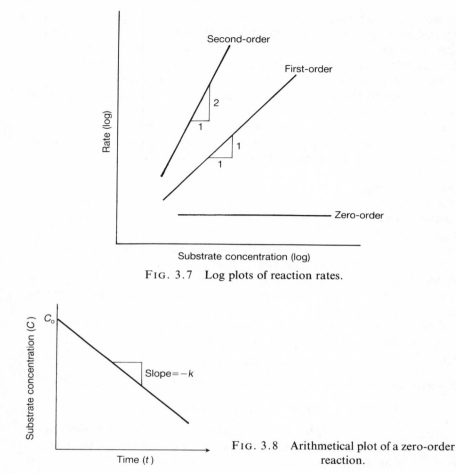

FIG. 3.7   Log plots of reaction rates.

FIG. 3.8   Arithmetical plot of a zero-order reaction.

integration (mg l^{-1}), which is calculated as $C = C_o$ at $t = 0$, and k the reaction rate constant (mg l^{-1}d^{-1}). As first-order reactions proceed at a rate directly proportional to the concentration of one reactant, the rate of reaction depends on the concentration remaining, which is decreasing with time. The plot of variation in concentration with time on arithmetic paper does not give a linear response but a curve. However, first-order reactions follow the equation:

$$\text{Log} \frac{[C_o]}{[C]} = \frac{kt}{2.3}$$

so that a plot of log C (the log of the concentration of the reactant remaining) against time will give a linear trace (Fig. 3.9). Second-order reactions proceed at a rate proportional to the second power of the concentration of a single

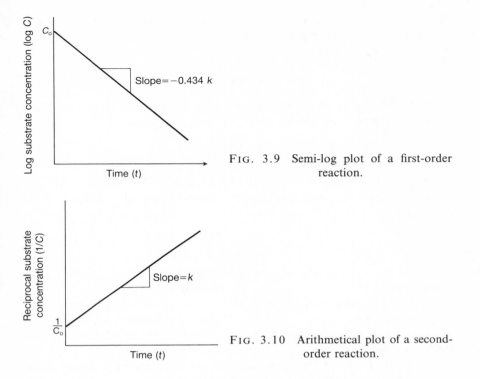

FIG. 3.9 Semi-log plot of a first-order reaction.

FIG. 3.10 Arithmetical plot of a second-order reaction.

reactant and obey the function:

$$\frac{1}{C} - \frac{1}{C_o} = kt$$

with the arithmetic plot of the reciprocal of the reactant concentration remaining ($1/C$) against time giving a linear trace, the slope of which gives the value of k (Fig. 3.10).

Thus, for any set of values C and t, such as the rate of removal of organic matter as measured by BOD at regular intervals, the reaction rate equations can be tested by making the appropriate concentration versus time plots and noting any deviation from linearity. It should be noted that although fractional reaction orders are possible it is normal for an integer value for reaction order to be assumed. A detailed and excellent account of reaction rate kinetics is given by Benefield and Randall (1980).

Enzyme reactions
The overall rate of biological reaction within a reactor is dependent on the catalytic activity of the enzymes in the prominent reaction. If it is assumed that enzyme-catalysed reactions involve the reversible combination of an enzyme

(E) and substrate (S) in the form of a complex (ES) with the irreversible decomposition of the complex to a product (P) and the free enzyme (E), then the overall reaction can be expressed as:

$$E+S \underset{k_2}{\overset{k_1}{\rightleftharpoons}} ES \xrightarrow{K_3} E+P$$

where k_1, k_2, and k_3 represent the rate of the reactions. Under steady-state conditions the various rate constants can be expressed as:

$$\frac{(k_2+k_3)}{k_1} = k_m$$

where k_m is the saturation or Michaelis constant. The Michaelis–Menten equation allows the reaction rate of enzyme-catalysed reactions to be calculated:

$$r = \frac{R_{max}-[S]}{K_m+[S]}$$

where r is the reaction rate, R_{max} the maximum rate at which the product is formed (mg $l^{-1}d^{-1}$), and S is the substrate concentration (mg l^{-1}). If this equation is plotted graphically (Fig. 3.11) it can be seen that the rate of an enzyme-catalysed reaction is proportional to the substrate concentration at low substrate concentrations (first-order). However, as the substrate concentration increases, the rate of reaction declines finally becoming constant and independent of the substrate concentration (zero-order). In practical terms, this is seen when a batch reactor is started and no further substrate is added.

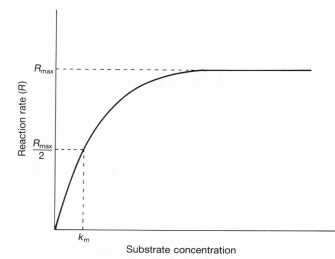

FIG. 3.11   The rate of enzyme-catalysed reactions as presented by the Michaelis–Menten equation.

Initially, the rate of reaction is only restricted by the ability of the enzymes to utilize the substrate which is in excess thus the reaction kinetics are zero-order. However, as the substrate is utilized the reaction begins to become substrate-limited resulting in fractional-order reactions until the substrate concentration is so low that the rate of reaction becomes totally limited by the substrate concentration and thus first-order kinetics result. First-order kinetics are assumed when $[S] \leq k_m$. From Fig. 3.11 it can be seen that the saturation constant k_m is equal to the substrate concentration when the reaction rate is equal to $(R_{max}/2)$.

Most biological wastewater treatment systems are designed to operate with high micro-organism concentrations which results in substrate-limiting conditions. Therefore, the majority of wastewater treatment processes can be described by first-order kinetics.

Environmental factors affecting growth
A number of environmental factors affect the activity of wastewater microbial populations and the rate of biochemical reactions generally. Of particular importance are temperature, pH, dissolved oxygen, nutrient concentration, and inhibition by toxic compounds. It is possible to control all these factors within a biological treatment system, except for temperature, in order to ensure that microbial growth continues under optimum conditions. The majority of biological treatment systems operate in the mesophilic temperature range, growing best in the temperature range 20–40°C. Aeration tanks and percolating filters operate at the temperature of the wastewater, 12–25°C, although in percolating filters the air temperature and the rate of ventilation can have a profound effect on heat loss. The higher temperatures result in increased biological activity that in turn increases the rate of substrate removal. The increased metabolism at the higher temperatures can also lead to problems of oxygen limitations. Generally, activated sludge systems perform better than percolating filters below 5–10°C although heterotrophic growth continues at these temperatures. However, the practice in colder climates of covering filters and controlling the rate of ventilation, thus reducing heat loss within the filter bed, has largely overcome this problem. Van't Hoff's rule states that the rate of biological activity doubles with every 10°C rise in temperature within the range 5–35°C. The variation in reaction rate with temperature is represented by the modified Arrhenius expression:

$$k_T = k_{20}\theta^{T-20}$$

where T is the temperature (°C), k is the reaction rate constant at temperature $T(d^{-1})$, k_{20} is the reaction rate constant at 20°C (d^{-1}), and θ is the temperature coefficient. There is a rapid decrease in growth rate as the temperature increases above 35°C which falls to zero as the temperature approaches 45°C (Benefield and Randall 1980; Barnes *et al.* 1983). Anaerobic digestion tanks are normally heated as near to the optimum temperature of the mesophilic range as possible (35–37°C). The pH and dissolved oxygen concentration in a biological reactor can be controlled to any level by the operator. The optimum pH range for carbonaceous oxidation lies between 6.5–8.5. At pH >9.0 microbial activity is inhibited whereas at pH <6.5 fungi dominate over the bacteria in the competition for the substrate. Fluctuations in the influent pH are minimized by completely mixed aeration reactors that offer maximum buffering capacity. If the buffering capacity is not sufficient to maintain a pH within the acceptable range, then pH adjustment will be required. Anaerobic bacterium have a smaller pH tolerance ranging from pH 6.7–7.4 with optimum growth at pH 7.0–7.1. A dissolved oxygen concentration between 1–2 mg l^{-1} is sufficient for active aerobic heterotrophic microbial activity, although optimum growth is dependent on sufficient essential nutrients and trace elements being present. Apart from organic carbon, nitrogen, and phosphorus, true elements such as sulphur, iron, calcium, magnesium, potassium, manganese, copper, zinc, and molybdenum must also be available for bacterial metabolism. Normally, all these nutrients and elements are present in excess in sewage although they may have to be supplemented in industrial wastewaters where nitrogen and phosphorus are usually only present in low concentrations. It is desirable to maintain a BOD$_5$:N:P ratio of 100:5:1 in order to ensure maximum microbial growth (Section 1.2). Toxic compounds such as phenol, cyanide, ammonia, sulphide, heavy metals, and trace organics can totally inhibit the microbial activity of a treatment plant if the concentration exceeds threshold limits for the micro-organisms. However, constant exposure to low concentrations of these substances results in the microbial community becoming acclimatized and increasing the concentration to which they can tolerate. Completely mixed aeration tanks are able to dilute shock toxic loads reducing the influent concentration to the final effluent concentration, whereas in filters the contact time between the micro-organisms and the toxic material is relatively short. The effect of environmental factors on both aerobic and anaerobic heterotrophic as well as autotrophic micro-organisms is discussed more fully in Sections 3.2 and 3.3.

Kinetic equations of bacterial growth
The various growth phases on the microbial growth curve (Fig. 3.5) can be represented quantitatively. The common autocatalytic equation is used to describe microbial growth during the exponential growth phase:

$$\frac{dX}{dt} = \mu X$$

where X is the concentration of micro-organisms (mg l^{-1}), μ is the specific growth rate (d^{-1}), and t is the time in days. The integrated form of this equation when plotted on semi-log arithmetic graph paper results in a straight line, hence the term logarithmic growth phase. The equation assumes that all micro-organisms are viable and although this may be true for a test-tube culture, it cannot be the case for a wastewater treatment unit with long retention times. It is, however, assumed that a constant fraction of the organisms within the biological treatment unit will remain viable (Weddle and Jenkins 1971).

Exponential growth will continue so long as there is no change in the composition of the biomass and the environmental conditions remain constant (Pirt 1975). In a batch reactor, a change in environmental conditions will inevitably occur, usually substrate limitation, which will cause a derivation from exponential growth into the declining growth phase. The most commonly used model, relating microbial growth to substrate utilization, is that of Monod (1942, 1949, 1950). Monod observed that the growth rate dX/dt was not only a function of organism concentration but also of some limiting substrate or nutrient concentration. He described the relationship between the residual concentration of the growth limiting substrate or nutrient and the specific growth rate of biomass (μ) by the classical function which takes his name:

$$\mu = \mu_m \frac{S}{k_s + S}$$

where μ_m is the maximum specific growth rate at saturation concentration of growth limiting substrate (d^{-1}), S is the substrate concentration (mg l^{-1}), and K_s is the saturation constant (mg l^{-1}), which is the concentration of limiting substrate at which the specific growth rate equals one-half of the maximum specific growth rate ($\mu = \mu_m/2$).

From this relationship it can be seen that specific growth rate can have any value between zero to μ_m provided that the substrate can be held at a given constant value. This is the basis for all continuous flow treatment processes in biological wastewater treatment in which micro-organisms are continuously cultivated but the overall rate of metabolism is controlled by the substrate concentration. When plotted, the Monod relationship between specific growth rate and the growth limiting substrate concentration (Fig. 3.12) has the same form as the Michaelis–Menton equation, which describes the rate of reaction of an enzyme with the substrate concentration (Benefield and Randall 1980). Figure 3.13 shows that the microbial growth rate increases as the availability of substrate increases until the maximum specific growth rate is achieved at which point a factor other than substrate, such as generation rate or a specific nutrient, becomes growth-limiting.

The specific growth rate of the equation of microbial growth under exponential growth conditions $dX/dt = \mu X$ can be replaced by the Monod function so that

$$\frac{dX}{dt} = \mu_m \frac{SX}{k_s + S}$$

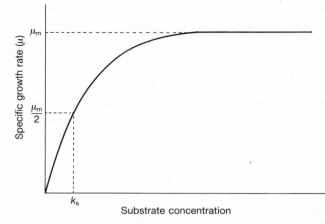

FIG. 3.12   The relationship between specific growth rate and growth-limiting nutrient concentration.

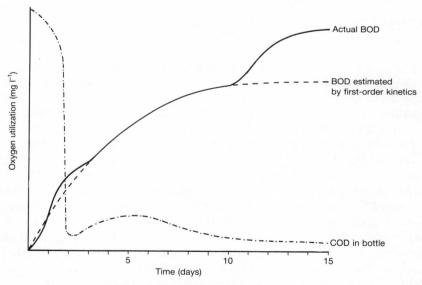

FIG. 3.13   BOD progression curve.

where the growth rate (dX/dt) is directly proportional to substrate concentration (first-order). When the substrate concentration is much larger than k_s the expression for growth rate reduces to

$$\frac{dX}{dt} = \mu_m x$$

where the growth rate is independent of substrate concentration (zero-order).

The organic carbon as measured by BOD or COD is usually considered to be the rate-limiting substrate in aerobic wastewater treatment systems, although the growth rate of the micro-organisms can also be controlled by other substances such as ammonia, phosphate, sulphate, iron, oxygen, carbon dioxide, and light (Andrews 1983). In low substrate concentrations, the rate of mass transfer into the cell may also control the rate of growth. The thickness of microbial flocs or film may be orders of magnitude greater than the size of an individual micro-organism and the rate of mass transfer may limit growth under certain conditions (Powell, 1967; Baillod and Boyle 1970). However the Monod function can still be used if k_s is considered as a variable dependent upon the degree of mixing in the reactor.

In nearly all biological wastewater treatment systems the retention time of micro-organisms is such that the endogenous phase occurs. Some unit processes, such as oxidation ditches, are designed to operate specifically in this phase. The basic equation for microbial growth can be modified to incorporate endogenous decay:

$$\frac{dX}{dt} = (\mu - k_d)X$$

where k_d is the specific endogenous decay rate which includes endogenous respiration, death, and subsequent lysis. The specific endogenous decay rate k_d is of little significance when the retention time is short, being an order of magnitude less than μ. However, when the system is operated in the endogenous growth phase, k_d is important in the calculation of the net amount of micro-organisms produced and the oxygen utilization rate.

The mass of organisms produced is related to the mass of substrate consumed, using the expression:

$$\frac{dX}{dt} = -Y\frac{[ds]}{[st]}$$

where X is the concentration of micro-organisms, s is the concentration of substrate, Y the yield coefficient, and t the time. The yield coefficient is a function of the species of micro-organisms present, the type of substrate, and environmental conditions. However, it is usually assumed to be constant for a given biological treatment process treating a specific wastewater. The yield coefficient is determined experimentally and such factors as formation of storage products (i.e. glycogen and poly-β-hydroxybutyrate), temperature, pH, and the variation in the fraction of viable cells, all of which can significantly affect the coefficient, must be taken into account. This is done by ensuring that the experimental conditions under which the coefficient is measured are the same as those encountered in practice, or by taking into account the factors used in the model of yield. BOD or COD are used usually as a measure of substrate concentration and volatile suspended solids as an index of organism concentration. Yield coefficient is expressed as mass (or mole) of organism produced per mass (or mole) of substrate consumed. Similar relationships can be established for the utilization of other substances, such as oxygen or light energy, or the formation of products, such as methane or carbon dioxide.

The expression of yield can be combined with the constant growth rate equation $(dX/dt = \mu X)$ to give the rate of substrate utilization:

$$\frac{ds}{dt} = -\frac{\mu}{Y} X$$

When the amount of substrate utilization for cell maintenance is expected to be significant, i.e. at high temperatures, long residence times and high cell concentrations, the above equation should be modified:

$$\frac{ds}{dt} = -\frac{\mu}{Y} X - k_m X$$

where k_m is the specific maintenance coefficient (Benefield and Randall 1980).

Kinetic constants should be determined experimentally using bench or pilot scale methods (Giona *et al.* 1979). However, because of the biological and chemical nature of such experiments the wastewater engineer and designer often prefer to rely on data presented in the literature. The variation in the kinetic constants quoted in the literature, due to the variation in wastewater quality and environmental conditions, is large. For example, $Y = 0.35 - 0.45$ mg VSS mg COD^{-1}, $k_d = 0.05 - 0.10$ d^{-1}, $k_s = 25 - 100$ mg l^{-1}, and $\mu_m = 3.5 - 10.0$ d (Mandt and Bell 1982), thus, clearly, kinetic analysis can only be performed with any degree of confidence on experimentally collected data.

3.1.3   The BOD test

The biochemical oxygen demand (BOD_5) test was introduced in Section 1.4.2 and is a quantitative measure of the oxygen uptake by aerobic micro-organisms, mainly bacteria, as they oxidize organic matter. The test initiates the heterotrophic microbial activity that occurs both during the self-purification process in rivers and in the breakdown of organic watewaters in biological treatment processes. The test provides a useful laboratory model describing what happens during the biodegradation of organic matter.

Microbial growth
Unlike the wastewater treatment plant or river, the BOD bottle is a closed microbial ecosystem which is devoid of external influences such as light, turbulence, nutrient imbalance, and factors that affect the oxygen balance. The test is done under controlled conditions, in darkness at 20°C and by plotting the oxygen uptake against time the characteristic BOD curve is obtained (Fig. 1.15). Similar to the microbial growth curve (Fig. 3.5), it can be

divided into the various growth phases: lag, log (acceleration), stationary, and endogenous respiration phases. There is a fifth phase in the BOD curve, that of nitrification (Fig. 1.16). Of course, the exact nature and extent of each growth phase is dependent on the composition of the sample being tested, thus in certain cases one or more phases may not be in evidence.

The lag phase is dependent upon the nature of the micro-organisms present in the sample. Its duration is a function of the number of bacteria present and whether or not they are acclimatized to the substrate. Normally, only small numbers are present and these have to acclimatize to their new environment in the BOD bottle, resulting in a delay before log growth and maximum oxidation occurs. Initially, the soluble substrate is rapidly assimilated and while cell numbers remain constant, cell size and mass increase. There is relatively little oxygen utilized during this phase as synthesis reactions require much less oxygen per unit of substrate utilized than does oxidation to carbon dioxide. Therefore, in order to avoid underestimating the BOD, the lag phase should be as short as possible, which is achieved by using laboratory grown micro-organisms to seed samples that are acclimatized both to the test temperature and to the substrate. Organisms transferred from a log growth phase will have a shorter lag phase than if transferred from any other growth stage (Young and Clark 1965). The effect on the BOD of prolonged lag phases because of insufficient bacteria when a sample has not been seeded, using unacclimatized seed or in the presence of inhibitory or toxic substances, is shown in Fig. 3.14 (HMSO 1983a).

During the log-growth phase the organic matter present in the sample is utilized by the heterotrophs for energy and growth. Once the mass of micro-organisms is large enough and is acclimatized to the substrate, the synthesis reaction rate declines and the oxidation reaction rate increases. During the log

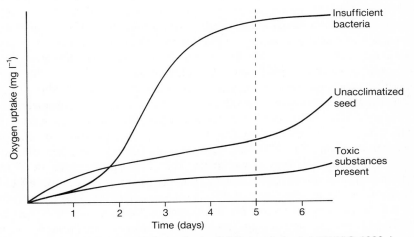

FIG. 3.14   Lag growth phases in the BOD test (adapted HMSO 1983a).

growth phase the available food supply (substrate) is in excess, resulting in a high food to micro-organism ratio (f/m), so that microbial growth is unrestricted. However, the micro-organisms are at their most vulnerable during this period because their generation time is reduced to a minimum, cell size is decreasing, and their resistance to toxic and inhibitory substances is low (Young and Clark 1965). This phase generally lasts between 24–36 hours during which time about 50% of the organic matter will be oxidized. Eventually, with the micro-organisms having reached maximum concentration, the substrate becomes limiting (a low f/m ratio) and more difficult to assimilate. With a consequent decrease in growth rate, there is a decrease in the rate of oxygen uptake. The rate continues to decrease and levels out at near zero uptake rate over a period of up to 30 hours, resulting in a plateau called the stationary growth phase.

In the stationary phase all the available soluble substrate has been utilized. Continued oxygen uptake rate after this phase is due to endogenous respiration and the presence of predators, such as protozoans feeding on the bacterial biomass, produced during the log-growth phase (Busch 1958; Schroeder 1968). Bhatla and Gaudy (1965) found that the occurrence and duration of the plateau is usually linked to the population of predators, especially the protozoans, present at the end of the log growth phase. For example, if protozoan growth lagged significantly behind bacterial growth, then a long plateau could be expected.

With a low f/m ratio, the declining bacterial population is forced into endogenous respiration. This normally occurs 3–5 days after incubation commences. Endogenous respiration is the oxidation of the portion of organic matter that was converted into cellular tissue during the log-growth phase. The first compounds to be oxidized are the expendable storage products such as glycogen and PHB (poly-β-hydroxybutyrate), leaving the more essential cellular constituents such as amino acids, proteins, and nucleic acids to degrade only when these have been utilized (Painter 1983). Oxidation occurs at a much slower rate than during the log growth phase. By the fifth or sixth day, the reduced bacterial population begins to limit the growth of protozoans, which are also forced into endogenous respiration. All the micro-organisms continue their endogenous respiration for about a further 15 days after which time biochemical oxidation should be over 90% complete.

The final stage in the BOD curve is only observed in samples containing a high concentration of ammonia and a significant population of nitrifying bacteria (Section 3.5). With the exception of partially nitrified effluents, nitrification proceeds very slowly during the initial stages of the BOD test, with no appreciable demand for oxygen. Thus, in practice, the effect of nitrification on the 5-day test (BOD_5) is usually negligible. This is due to the very slow reproduction rate of the nitrifying bacteria, which is between 2–6 days for most genera and results in a noticeable nitrification oxygen demand only after 10 days incubation. Nitrification can exert an appreciable oxygen demand in the BOD_5 when partially nitrified sewage effluents are tested. Such samples contain a high concentration of ammonia as well as an established population of nitrifying bacteria.

If the oxidation of organic matter is allowed to proceed in the BOD bottle, complete oxidation is achieved between 20–100 days depending on substrate. The total oxidation of all organic carbon, nitrogen, and hydrogen is referred to as the ultimate BOD (L). In practice, L is approximated mathematically from the rate of oxygen uptake during the initial incubation period, rather than actually incubating samples for such a lengthy period. (Fig. 1.19).

BOD kinetics
The kinetics of the BOD test have been fully explained in Section 1.4.2.1 and are based on first-order reaction principles. However, the concept of the oxygen uptake rate in the BOD test conforming to a first-order reaction has been widely disputed (Rivera *et al.* 1965; Young and Clark 1965; Landine 1971; Stones 1981, 1982). Stones (1982) applied both first- and second-order equations to the results of an experiment in which a 0.01 dilution of settled domestic sewage was incubated at 20°C and the residual dissolved oxygen concentrations determined at daily intervals over 15 days, from which k_1 values were determined (Table 3.4). His results clearly show that the values of

TABLE 3.4 Values of the velocity coefficients $k_{1.1}$, $k_{1.2}$, and $k_{1.3}$ computed using first- and second-order reaction equations (Stones 1981, 1982, 1985)

Days	$k_{1.1}$	$k_{1.2}$	$k_{1.3}$
1	0.350	0.200	0.0397
2	0.318	0.170	0.0390
3	0.295	0.149	0.0386
4	0.277	0.132	0.0383
5	0.263	0.119	0.0381
6	0.252	0.109	0.0381
7	0.244	0.100	0.0381
8	0.236	0.092	0.0381
9	0.230	0.085	0.0382
10	0.224	0.080	0.0382
12	0.218	0.070	0.0391
15	0.212	0.059	0.0404

$k_{1.1}$, velocity coefficient calculated using a *first-order* reaction equation in which it is assumed that the rate of biochemical oxidation varies with the *unsatisfied BOD*.

$k_{1.2}$, velocity coefficient calculated using a *first-order* reaction equation in which it is assumed that the rate of biochemical oxidation varies with the *residual DO concentration*.

$k_{1.3}$, velocity coefficient calculated using a *second-order* reaction equation in which it is assumed that the rate of biochemical oxidation varies with *both* the *unsatisfied BOD* and the *residual DO concentration*.

$k_{1.1}$ and $k_{1.2}$, computed using first-order reaction kinetics, decrease rapidly as oxidation proceeds. This is in contrast to the $k_{1.3}$ values, computed using second-order reaction kinetics, which are virtually constant from the second to the tenth day of incubation, with the observed decline after 10 days attributable to the onset of nitrification. There is little doubt that the second-order reaction equation gives a better mathematical description of the log growth phase, However, the overall BOD curve is made up of a minimum of three growth phases each having an entirely different growth rate. Therefore, the use of a first-order reaction equation is to be preferred as it more accurately describes the summation of all the individual phases that make up the BOD curve.

Further reading

General: Benefield and Randall 1980; Andrews 1983; Jank and Bridle 1983; Atkinson and Mavituna 1983; Schugerl 1987; Bu'lock and Kristiansen 1987.
Stoichometry: Lawrence and McCarty 1970; McCarty 1972, 1975; Christensen and McCarty 1975; Schugerl 1987.
Kinetics: Giona *et al.* 1979; Schugerl 1987; Sinclair 1987.

3.2 ENERGY METABOLISM

There are three general methods by which heterotrophs obtain energy: by fermentation; and by either aerobic or anaerobic respiration. Fermentation and respiration are the two fundamental types of metabolism, and differ in that respiration requires an external electron acceptor; oxygen in aerobic respiration and carbon dioxide, sulphate or nitrate in anaerobic repiration. Fermentation does not require an external electron acceptor as the substrate is the electron donor and the other participating organic molecule acts as the electron acceptor.

Heterotrophic micro-organisms, including bacteria, fungi, and, to a lesser extent, protozoans, are capable of utilizing a wide range of organic compounds as a carbon source. These vary from simple sugars, alcohols or amino acids to complex carbohydrates, lipid polymers and proteins. However, these carbon sources are converted to a comparatively small number of intermediates such as pyruvate and acetyl-Co enzyme A (acetyl-CoA), which are common to all living tissue and are the 'building blocks' for cell synthesis. The anabolic pathways of cell synthesis are essentially the same in all micro-organisms whether heterotrophs or autotrophs. As carbohydrates are broken down to glucose before entering more complicated biochemical routes and as the majority of bacteria can metabolize carbohydrates, by taking glucose as the basic substrate then it is possible to compare all the major steps in heterotrophic metabolism (Fig. 3.15). The steps from carbohydrate to

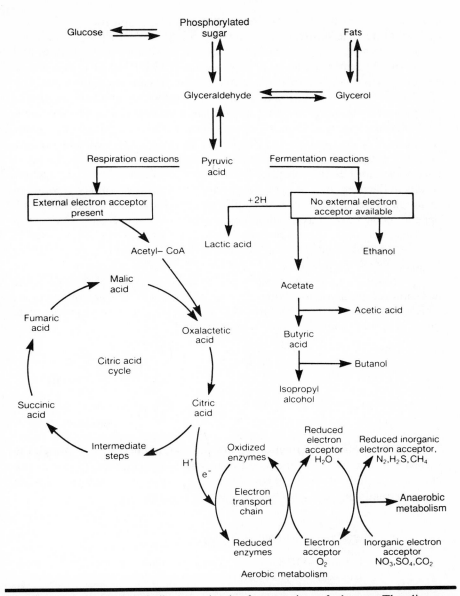

FIG. 3.15 Major metabolic steps in the fermentation of glucose. The diverse pathways after a pyruvic acid intermediate stage can be exploited to manufacture a variety of organic chemicals. Aerobic metabolism occurs by way of the citric acid cycle. Oxygen enters the reaction only at the last step as an electron acceptor.

glucose and then to pyruvic acid are common to all organisms metabolizing carbohydrate. The substrate is broken down in a series of enzyme-catalysed reactions with small amounts of energy released at each oxidation stage and recovered by substrate-level phosphorylation (ADP→ATP). As this is achieved without the use of an electron acceptor, it is a fermentation reaction. The term fermentation describes any anaerobic metabolism of energy production that does not involve an electron transport chain. The glucose is broken down to pyruvate in three stages (Fig. 3.16). Stage 1 is an endergonic reaction using energy from ATP, whereas in stage 2 the 6-carbon sugar is cleaved to form two interconvertible 3-carbon compounds. The final stage is energy-yielding, with substrate-level phosphorylation occurring, which results in the formation of pyruvate. Pyruvic acid is the pivotal compound in metabolism, and from it many different products can be metabolized, depending on the organism and the environmental conditions (Fig. 3.17). If an external electron acceptor is present then the pyruvic acid can be converted to acetyl-CoA which enters the citric acid cycle (also known as the Krebs or tricarboxylic acid cycle) (Fig. 3.15). However, if there is no external electron acceptor available then the pyruvic acid may undergo a further stage in the fermentation pathway involving any one of a series of alternative reactions, which not only serve to regenerate NAD and NADH, but also produce a wide range of fermentation products such as ethanol.

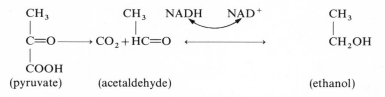

$$
\begin{array}{ccc}
\underset{\text{(pyruvate)}}{\overset{\displaystyle \text{CH}_3}{\underset{\displaystyle \text{COOH}}{\overset{\displaystyle |}{\underset{\displaystyle |}{\text{C=O}}}}}} \longrightarrow & \underset{\text{(acetaldehyde)}}{\text{CO}_2 + \overset{\displaystyle \text{CH}_3}{\overset{\displaystyle |}{\text{HC=O}}}} \xrightarrow{\ \ \text{NADH}\quad\text{NAD}^+\ \ } & \underset{\text{(ethanol)}}{\overset{\displaystyle \text{CH}_3}{\overset{\displaystyle |}{\text{CH}_2\text{OH}}}}
\end{array}
$$

The end-products of fermentation are still rich in energy even though the fermentation pathway is complete. For example, although the breakdown of glucose to lactic and acetic acids yields energy, the complex end-products are still potentially energy-rich.

$C_6H_{12}O_6 \rightarrow 2CH_3CHOHCOOH + $ energy (22.5 kcal)
(glucose)   (lactic acid)

$C_6H_{12}O_6 \rightarrow 3CH_3COOH + $ energy (15 kcal)
(glucose)   (acetic acid)

When external electron acceptors are present, pyruvic acid is converted to acetyl-CoA, which enters the citric acid cycle (Fig. 3.18). Metabolism now becomes respiration and not fermentation. However, in the cycle a hydrogen ion and electrons are released which enter the electron transport system where they react with either oxygen (aerobic respiration) or an inorganic electron acceptor (anaerobic respiration).

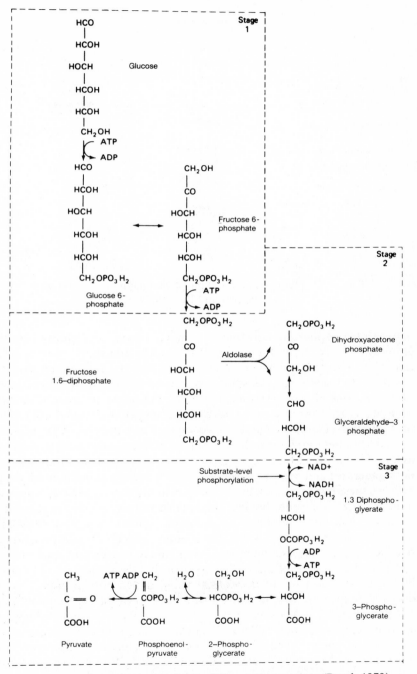

FIG. 3.16   Fermentation pathway of glucose to pyruvate (Brock 1970).

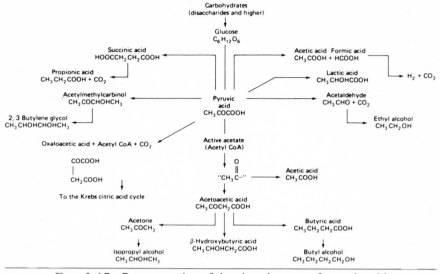

FIG. 3.17　　Representation of the pivotal nature of pyruvic acid
(Pelczar and Reid 1972).

In aerobic respiration, most energy comes from oxidative phosphorylation as electrons are carried through the electron transport system, which reoxidizes reduced co-enzymes that are formed as a result of oxidative reactions. In the electron transport system used by most wastewater bacteria there are three phosphorylation sites along the transport chain (Fig. 3.19). Thus, if the electrons enter at the NAD level each pair of electrons will result in the formation of three ATP molecules compared with only two ATP molecules if electrons enter the chain at the FAD level. Whereas the metabolic pathways followed in the breakdown of the carbon and the energy source is the same for both anaerobic and aerobic respiration. There are two important differences between these two processes. In anaerobic respiration, inorganic acceptors are utilized, such as carbon dioxide, sulphate or nitrate, which are reduced to methane, hydrogen sulphide, or ammonia, nitrous oxide or molecular nitrogen respectively (Fig. 3.20), The amount of ATP formed by oxidative phosphorylation is also different, being much less during anaerobic as opposed to aerobic respiration. The reason is that the amount of ATP formed by the passage of pairs of electrons through the electron transport system depends on the difference in redox potential between the electron donor and acceptor (Fig. 3.21). As oxygen has a lower redox potential than the inorganic electron acceptors, more ATP is normally released during aerobic respiration. In terms of energy yield, the fermentation of glucose realizes two ATP molecules per hexose unit compared with 38–39 ATP molecules per hexose unit during aerobic respiration. The greater energy utilization during aerobic

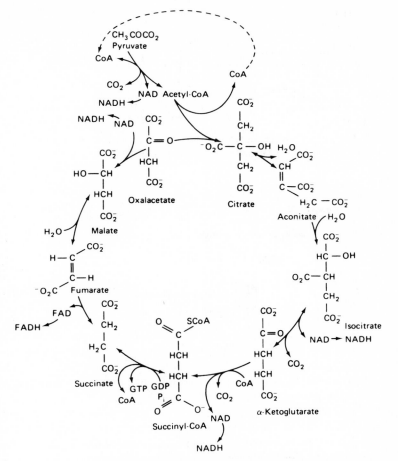

FIG. 3.18 The Krebs cycle (Brock 1970).

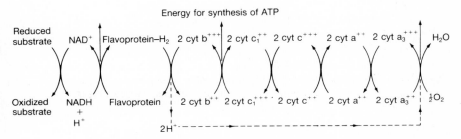

FIG. 3.19 Electron transport system common to most bacteria
(Palczar and Reid 1972).

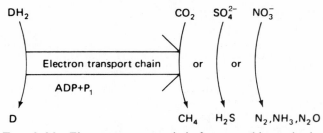

FIG. 3.20   Electron transport chain for anaerobic respiration
(Benefield and Randall 1980).

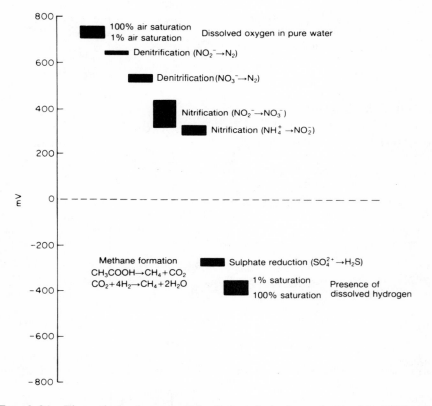

FIG. 3.21   Theoretical redox potentials of selected reactions calculated for 25°C and
for normal wastewater treatment conditions (Mosey 1983).

respiration means that a much greater proportion of the glucose will be assimilated as cell material compared with anaerobic respiration (Table 3.2) (Benefield and Randall 1980). The energy-generating mechanisms have been excellently reviewed by Anderson (1980).

Further reading

General: Mandelstam and McQuillen 1973; Haddock and Hamilton 1977; Lynch and Poole 1979; Anderson 1980.

3.3   AEROBIC HETEROTROPHIC MICRO-ORGANISMS

3.3.1   The organisms

Bacteria
Bacteria are the most versatile of all the organisms associated with wastewater treatment in terms of the conditions under which they can thrive and the substrates they can metabolize. Undoubtedly, it is the bacteria that are the most important group of organisms in biological treatment systems, although fungi, protozoans, and a range of other invertebrate organisms also play important but comparatively minor roles.

The bacteria form the basic trophic level in all the biological treatment systems and so form the major proportion of the biomass, with the exception of waste stabilization ponds in which algae may predominate (Section 4.3). The dominant bacteria are the aerobic heterotrophs that degrade and eventually mineralize organic compounds present in wastewater to carbon dioxide and water. It is the small size of bacteria and their resultant large surface area to volume ratio which makes them so efficient, in terms of nutrient and catabolic exchange, with the liquid medium in which they are either suspended or are in contact. Their short doubling times, which can be as little as 20 minutes in pure culture, enable bacteria to rapidly take advantage of increased substrate availability compared with other organisms. Protozoans, for example, have an average doubling time in the order of 10 hours, although some species do have doubling times of less than 2 hours, such as *Tetrahymena pyriformis* (Curds and Cockburn 1971).

Estimation of the numbers of bacteria in sewage and treatment processes can be made by direct microscopic observation using a Hebler counting chamber (Harris and Sommers, 1968) or by culturing bacteria on various nutrient media at 22°C or 30°C (Pike 1975). Serious discrepancies exist between total counts of bacteria in treatment processes made by microscopy, which cannot distinguish between viable and non-viable cells, and by various culturing techniques that measure only the viable bacteria (Table 3.5). The

THE ROLE OF ORGANISMS

TABLE 3.5 Numbers of total and viable bacteria (geometric means) at different stages of treatment (Pike and Carrington 1972)

| Source (and number) of samples[b] | Bacterial counts[a] | | | | Viability (%) | Total suspended solids ($mg\,l^{-1}$) |
| | In samples (No./ml) | | In suspended solids (No./g) | | | |
	Total	Viable	Total	Viable		
Settled sewage (46)	5.6×10^8	6.3×10^6	3.0×10^{12}	3.4×10^{10}	1.1	190
Activated sludge mixed liquor, conventional rate (18)	5.9×10^9	4.9×10^7	1.3×10^{12}	1.1×10^{10}	0.83	4600
Activated sludge mixed liquor, high rate (24)	1.4×10^{10}	2.4×10^8	3.0×10^{12}	5.0×10^{10}	1.7	4800
Filter slimes (18)	6.2×10^{10}	1.5×10^9	1.3×10^{12}	3.2×10^{10}	2.5	54 000
Secondary effluents (16)	5.4×10^7	1.1×10^6	1.9×10^{12}	4.1×10^{10}	2.1	28
Effluents, high-rate activated-sludge plants (24)	4.8×10^7	1.4×10^6	3.3×10^{12}	1.0×10^{11}	3.0	14
Tertiary effluents (11)	2.9×10^7	6.6×10^4	3.0×10^{12}	6.8×10^9	0.23	9.7

[a] Samples from sewage works and laboratory pilot plants; high-rate plants are pilot scale, working at loadings of 0.46–2.5 kg BOD removed/kg MLSS day, secondary effluents from nine filters and seven activated-sludge plants, tertiary effluents from ten lagoons, one grass plot.

[b] Total counts obtained with Helber counting chamber, viable counts by plate-dilution frequency method on CGY agar, incubation for 6 days at 22°C; viability expressed as percentage ratio of viable count to total count.

latter technique gives results that are often two orders of magnitude less due to the proportion of the microbial biomass, which is either dead or non-viable, because of most processes being operated in the stationary or declining phase of the microbial growth curve (Fig. 3.5). Culturing techniques of aerobic heterotrophic bacteria are reviewed by Pike (1975) who concludes that cultural methods cannot be used as estimates of biomass or microbial activity, as the enzymes of the non-viable bacteria remain active within the treatment system. However, this does not prevent culturing techniques being used for measuring the rates of increase or decline in the number of viable bacteria, for which they are selective. A major problem in culturing is that none of normal growth media supports the growth of all nutritional types of aerobic heterotrophic bacteria present in wastewater and treatment systems, although optimal counts are obtained using a casitone–glycerol–yeast extract agar (CGY) which is inoculated by either the spread plating or plate dilution methods and incubated for 6 days at 22°C (Pike et al. 1972). Bacteria from treatment plants are usually aggregated as either sludge flocs or filter film, which must be completely broken up into individual cells randomly dispersed, before plating on to medium and incubating. Ultrasonics are used in preference to standard homogenization procedures, although many cells are inevitably damaged or destroyed. Thick filter film can only be broken down to individual cells by extreme homogenization methods which result in massive cell mortalities and produce counts which are misleadingly low.

Metabolic methods are used to estimate biomass and microbial activity. Substrate utilization can be measured by determining the uptake of substrate itself, by oxygen consumption or dehydrogenase activity (Brock 1971b). As adenosine triphosphate (ATP) is only found in living cells and its concentration is linked to the organic content of cells, attempts have been made to use it to estimate biomass as well as the viability of sludges and films. However, the presence of large organisms, especially protozoans and rotifers, make it less useful as a specific measure of bacterial biomass or bacterial viability (Pike 1975).

Bacterial identification is essentially a task for a specialist microbiologist, primarily because few bacterial species have been adequately described. Also, the biochemical tests used for positive identification are complex and time consuming. Bergey's manual (Buchanan and Gibbons 1975) is the definitive work on identification because it provides an extremely comprehensive list of bacteria, which are fully described and characterized in the literature. However, it is possible to identify the more common bacterial components of treatment systems and those species producing gelatinous matrices or filaments of cells, in particular, by using simple microscopic and histological tests (Farquhar and Boyle 1971a,b; Eikelboom 1975). The recent key by Eikelboom and Van Buijsen (1981) contains many excellent photomicrographs of all the major filamentous bacteria found in wastewater. The major isolation methods for filamentous micro-organisms have been compared by

Strom and Jenkins (1984) and are discussed further in Section 4.2, when bulking of activated sludge is examined.

A healthy Western adult excretes approximately 135 g dry weight of faeces and 1400 ml of urine each day, with 30% of biomass being bacteria. As these bacteria form part of the gut flora they are predominantly obligate anaerobes, out-numbering aerobes by a factor of 40 to 1. Urine contains few bacteria and only exceed $> 10^5 ml^{-1}$ when the urinary tract is infected. The major aerobic species in urine are Gram-negative, including *Escherichia coli*; *Proteus* spp., and *Pseudomonas aeruginosa*. The bacterial flora of excreta is dealt with in more detail in the section on anaerobic heterotrophs (Section 3.4).

The bacterial count of sewage shows a marked diurnal variation which is linked to the water usage and excretion pattern of the population, following a similar curve to the BOD of domestic sewage (Fig. 1.2), but with a more differentiated peak in the evenings. Seasonal patterns have also been identified (Tomlinson *et al.* 1962; Gameson *et al.* 1967; Harkness 1966), with maximum numbers found during August and September. However, it is likely that this has been due to other factors, such as the temporary removal of toxic and inhibitory substances in the waste, or an increase in population size because of tourists. Pathogenic bacteria are also present in sewage, and these are discussed fully in Chapter 6.

The bacterial communities found in wastewater treatment plants are complex, with a variety of genera present, in addition to the dominant genera that are usually aerobic, heterotrophic, Gram-negative, rod-shaped bacteria with polar flagella. The bacterial flora of all aerobic treatment systems are basically the same, with *Zoogloea*, *Pseudomonas*, *Chromobacter*, *Achromobacter*, *Alcaligens*, and *Flavobacterium* the major genera present. All are able to oxidize organic compounds to carbon dioxide and water. *Escherichia coli* and the other faecal indicator bacteria are universally present in treatment systems but are not indigenous members of the microbial community (Table 3.6).

The bacterial communities of activated sludge systems are far more specialized than those associated with fixed-film reactors and also have a lower diversity. The bacteria in activated sludge comprise of two major types. Free-swimming species are constantly removed by protozoan grazing and by being discharged with the effluent. The free-swimming species grow faster than the second major type, the flocculating bacteria, which form flocs and, due to the design of the process, are retained in the system. Although the activated sludge process selects the flocculating bacteria in preference to the free-swimming species, the latter play a major part in substrate utilization. Fixed-film reactors have a greater range of micro-habitats compared with activated sludge and so support a less specialized flora of bacteria comprising a much greater diversity. There is a noticeable variation in the numbers and abundance of bacteria with filter depth, whereas the community structure in both treatment processes are affected by seasonal and operational factors. The individual bacterial communities associated with each treatment process are discussed in Chapter 4. However, two bacteria, *Zoogloea* and *Sphaerotilus natans*, which are important components of wastewater treatment systems, can be considered as typical examples of aerobic heterotrophic bacteria and are considered in detail here. Much research has been done on these species as they are easily isolated and subsequently cultured. They are extremely common in wastewater treatment systems and are also found in natural watercourses receiving organic enrichment. Another important feature is that they are among the easiest species of bacteria to identify.

Sphaerotilus natans is a Gram-negative non-sporing bacterium made up of individual rod-shaped cells with rounded ends, each $1-4 \times 4-10$ μm in size, enclosed within a sheath of varying thickness (Stokes 1954; Mulder and Van Veen 1963; Phaup 1968). It is a variable species, the appearance of the cells and the amount of sheath material varying with nutritional regime. The sheath is a protein–polysaccharide–lipid complex which is itself enclosed in a capsule composed of a simple polysaccharide material whose composition varies with the nutrient regime (Fig. 1.12). The cells often contain globular inclusions of the food reserve material poly-β-hydroxybutyrate. These inclusions are visible even in the early phases of growth and has been shown to be utilized under nutrient-limited conditions. A high C:N ratio favours the formation and storage of poly-β-hydroxybutyrate which can comprise up to 40% of the dry weight of the cells. However, as the filaments age so the individual bacterial rods become smaller in size and the inclusions disappear (Mulder and Van Veen 1963; Sanders 1982). In young cultures, the filaments sometimes appear to be non-septate and as they age the cells become more distinct. Occasionally, a single sheath contains two or even three rows of cells with false branching occurring by single rods slipping sideways and growing into a new filament (Phaup 1968). Two varieties of *S. natans* are found in treatment plants and polluted waters (Gray 1982). When the organic carbon concentration is low the form *S. natans* var. *diochotoma* is common, the filaments appearing highly branched and enclosed in a thin sheath. Whereas, if the organic carbon concentration is high, especially in the form of low polymer carbohydrate and organic acids, then *S. natans* var. *typica* is found. This latter variety is normally associated with treatment plants and branching is weakly developed and often absent. The sheath is also thicker and in polluted rivers has a more slimey consistency (Gray 1985a). Reproduction occurs by fragmentation of the filament or by the production of motile flagellated cells. The flagellated cells, commonly referred to as swarm cells, become attached to new suitable substrates, by a gelatinous holdfast formed at the non-flagellated end, which then lose their flagella and grow into new filaments (Phaup 1968). Isolation of *S. natans* into pure culture is relatively simple, colonies on agar medium being dull white and of variable form (Gaudy and Wolfe 1962; Marcus *et al.* 1978). Identification should be made using the characters described by Gaudy and Wolfe (1962) or Buchanan and Gibbons (1975). *Sphaerotilus natans* was

TABLE 3.6 Outline of useful organisms and reactions (Berthoux and Rudd 1977)

Organism	Reaction	Oxygen requirement	pH	Temp. (°C)	Comments
1. Organic degraders					
Escherichia coli	Organics to CO_2 and H_2O	Facultative	6–8	10–30	Indicator organism
Zooglea ramigera	"	"	"	"	Active in sewage treatment–utilizes
Sphaerotilus natans	"	"	"	"	polysacharides and carbo-
Pseudomonas sp.	"	"	"	"	hydrates
Chromobacteria sp.	"	"	"	"	
Achromobacter sp.	"	"	"	"	
Corynebacteria	"	"	"	"	
Clostridia sp.	"	Anaerobic	7–8	20+	Proteolytic
Bacillus sp.	"	Facultative			
2. Nitrogen users					
Nitrosomonas	$NH_3 \rightarrow NO_2^-$	Aerobic	7–8	10–25	Nitrification–common in soils and streams
Nitrobacter	$NO_2^- \rightarrow NO_3^=$	Aerobic	"	"	
Rhizobium	$N_2 \rightarrow NO_3^=$				Symbiotic with legumes
Azobacter sp.	"				Non-symbiotic
Clostridium pasteurianum					Non-symbiotic
Thiobacillus denitrificans	$NO_3^{2-} + 2S + H_2O \rightarrow$ $HSO_4^- + SO_4^{2-} + N_2$	Anaerobic			Autotrophic
Micrococcus denitrificans	$NO_3^{2-} \rightarrow N_2$	Anaerobic			Heterotrophic
Pseudomonas	"	"			Heterotrophic

		Aerobic/Anaerobic	pH	Temp.	Remarks
3. Sulphur bacteria	*oxidation of S*				
Beggiatoa	$2H_2S+O_2\rightarrow 2S+H_2O$	Aerobic	Low		Autotrophic
Thiobaccillus thiooxidans	$2S+2H_2O+3O_2\rightarrow 2SO_4^{2-}+4H^+$	"	"		"
	reduction of S				
Desulfovibrio	$CH_3COOH+SO_4^{2-}\rightarrow 2CO_2+H_2S+2OH^-$	Anaerobic	Low		Organic consumed in reaction
	$4H_2+SO_4^{2-}\rightarrow H_2S+2H_2O+2OH^-$				Autotrophic reduction
4. Methane bacteria					
Methanobacillus omelianskii	$Alcohols\rightarrow CH_4+CO_2$	Anaerobic	6.4-7.2	35	Methane fermentation exploited in sludge
Methanococcus vanieli	$Formate+H_2\rightarrow CH_4$	"	"	"	Digestion and other anaerobic sewage treatment processes
Methanobacterium formicicum	$CO_2+H_2\rightarrow CH_4$				Carbon dioxide reduction
Methanobacterium sohngeni	$Acetate\ butyrate\rightarrow CH_4$				
Methanomonas	$CH_4+2O_2\rightarrow CO_2+H_2O$	Aerobic			Methane oxidation
5. Iron bacteria					
Leptothrix sp.	$Fe^{2+}\rightarrow Fe^{3+}$	Aerobic			Autotrophic – causes rust and corrosion
Gallionella sp.	"	"			
6. Others					
Pseudomonad	$CO+1/2O_2\rightarrow CO_2$	Aerobic	7	35	
Thiocyanoxidans	oxidize cyanide				
Cellulomonas sp.	cellulose-protein				

originally described by Kutzing in 1833, who considered it to be an alga, and since then much confusion has developed over the taxonomy of this genus which is now firmly placed in the family *Chlamydobacteriaceae*. Many synonymous genera and species have been described in the past including *Cladothrix*, *Chlamydothrix*, *Clonothrix*, and *Streptothrix*. However, *S. natans* and *S. discophorous* are now considered to be the only valid species (Pringsheim 1949). The genus *Leptothrix* is distinguished from *Sphaerotilus* on the basis of its differing ability to oxidize manganese and iron (Van Veen *et al.* 1978).

Studies on wastewater microbiology have generally not tried to characterize zoogloeal-forming organisms as they can be difficult to isolate and are not well defined taxonomically. The category includes *Zoogloea ramigera*, *Pseudomonas* spp., and zoogloeal forms of *Sphaerotilus*. The organism comprises small Gram-negative non-sporing cells ($0.5–10 \times 1.0–3.0 \ \mu m$) not arranged in filaments but embedded in a gelatinous matrix, forming lobed and unlobed spherical masses (Curtis 1969; Buchanan and Gibbons 1975; Gray 1982*a*). Since the proposal of *Z. ramigera* in 1867, there has been much confusion regarding both the genus and its species as all the original type cultures have been lost, although a new strain has been proposed as neotype (Crabtree and McCoy 1967; Zvirbulis and Hatt 1967). This new strain is non-proteolytic, oxidizes carbohydrates but is unable to ferment them. Ammonia is utilized as a sole source of nitrogen in the presence of vitamin B_{12}. However, zoogloeal forms are so widespread it would appear that they have an ability to utilize other nitrogen sources. Friedman and Dugan (1968) have isolated several *Zoogloea* strains, which although morphologically distinct from the neotype proposed by Crabtree and McCoy (1967), are biochemically similar. Unz and Dondero (1967*a,b*) isolated 65 *Zoogloea* strains using single cell isolations taken from naturally branching wastewater zoogloeal masses, all of which were able to reduce nitrate to nitrogen, possessed urease, catalase and an oxidative reaction, but could not produce acid from carbohydrates. It would appear, therefore, that zoogloeal bacteria in terms of wastewater treatment describes all non-fermentative bacteria forming zoogloeal masses or flocs (Gray 1985*a*).

Fungi
Heterotrophic fungi can be as efficient as heterotrophic bacteria in the removal of organic matter from wastewater (Tomlinson and Williams 1975). But fungi produce a greater biomass per unit weight of substrate utilized than many of the important bacteria in wastewater treatment units, which results in more film and sludge production per kg BOD removed. Although this is disadvantageous in conventional treatment systems, this higher biomass production from fungi is potentially extremely useful in the recovery and reuse of energy-rich substrates discharged from the food-processing industries.

A range of recovery and isolation techniques have been used which are fully reviewed by Cooke (1963) and Tomlinson and Williams (1975). Fungi are allowed to colonize a new substrate placed in the treatment units, such as glass slides (Tomlinson 1941) or a bait, such as hemp seeds. They are subsequently isolated from this substrate using special agar medium such as rose-bengal agar which contains aureomycin to reduce bacterial contamination on plates. However, a number of fungi, in particular *Subbaromyces splendens* and *Sepedonium* spp., are very sensitive to antibiotics and are not recovered by this method. A more recent technique is to plate out the actual film or sludge floc directly on to the medium using the ring-plate method. The material is collected from the treatment plant and scanned under the microscope so that any macro-invertebrates can be removed. It is then placed within the glass ring which stands on three glass bead feet which is set into modified Czapek-Dox agar, which does not contain any antibiotics. The bacteria are physically retained within the ring, whereas the fungi are able to grow through the medium between the feet of the glass ring and colonize the outer areas of the plate (Tomlinson and Williams 1975). Estimation of fungal abundance is difficult. Direct counts have been used but problems in the interpretation as to what constitutes a viable growth unit is difficult (Gray 1983b). As each intact cell could potentially give rise to a new colony, direct counts are affected by the degree that film or sludge samples have been broken down during collection and microscopic preparation. A more accurate method is to measure the concentration of muramic acid which will give an accurate estimation of fungal activity.

Identification of fungi is difficult as most species produce extensive vegetative growth in the presence of wastewater but rarely produce spores. Some common wastewater fungi rarely produce spores even in pure culture. Identification is made mainly on morphological characters using the laboratory guide to fungi in polluted water by Cooke (1963) and the excellent taxonomic summaries and photographs produced by Tomlinson and Williams (1975). Examination of species in pure culture may also be useful, especially to distinguish between superficially similar species, such as *Subbaromyces splendens* and *Ascoidea rubescens*. Some species are morphologically very distinct, such as *Leptomitus*, which is aseptate and has intermittent constrictions in the hyphae and contains spherical cellulin plugs, whereas other species, such as *Fusarium aquaeductuum* and *S. splendens*, often produce conidia within the treatment plant so aiding identification. As a general guide, *Geotrichium candidum*, *Sepedonium* sp., and *F. aquaeductuum* all form visable mats over the surface of the medium in percolating filters. *Geotrichium candidum* forms a soft textured mat, grey to brown in colour, which loosely follows the contours of the medium, whereas *Sepedonum* sp. forms much thicker and tougher mats which often have a leathery appearance and are olive-green to yellow-grey in colour. *Fusarium aquaeductuum* also form mats which have a distinctive orange to pink coloration, but can also form discrete lumps of growth on the sides of the medium or as fluffy growths in surface pools when the interstices are blocked. This discrete growth pattern is the normal growth form for *S. splendens* and *A. rubescens*. *Leptomitus lacteus* and *Trichosporon cutaneum* have never been recorded as forming visible growths in percolating filters and thus, can only be seen under the microscope. A simple key to these common fungi which regularly occur in percolating filters is given in Table 3.7. Well over 100 fungi have been isolated from wastewater treatment plants (Cooke 1958), and although the majority of these species will have entered the plant with the influent, few can survive or flourish within the treatment units. For example, of the 112 species Cooke isolated from

TABLE 3.7   A simple guide to the major fungi occurring in percolating filters

1.	Hyphae:.. Non-septate 2	
	Septate 3	
2.	Non-septate hyphae with constrictions .... *Leptomitus lacteus*	
3.	Septate hyphae with conidia or arthrospores: ....Absent 4	
	Present 5	
4.	No conidia or arthrospores formed but chlamydospores present .... *Sepedonium* sp.	
5.	Conidia formed: .... Conidia boat-shaped and septate (25–35 × 4–5 μm) .... *Fusarium aquaeductuum*	
	.... Conidia formed singly, oval shaped (35–40 × 10–15 μm), non-sepatate, produced on sparingly branched conidiophores .... *Subbaromyces splendens*	
	.... Conidia formed successively at conidiophore apex to form tight clusters, conidiophore unbranched, conidia less oval than *S. splendens* but more spherical (20–25 × 12–15 μm), endoconidia formed .... *Ascoidea rubsescens*	
	Arthrospores formed: .... Blastospores present .... *Trichosporon cutaneum*	
	Blastospores absent .... *Geotrichium candidum*	

percolating filters, only 29 were isolated frequently enough or in sufficient numbers to be considered as permanent members of the microbial community. Further, few appeared to be important components of the film. Only two species were considered to be primarily associated with wastewater treatment plants, *Ascodesmis microscopia*, which is only known from mammalian dung and *Subbaromyces splendens*, which has only been isolated from percolating filters and its natural habitat remaining unknown (Hesseltine 1953), the other species are equally abundant in natural habitats.

The origin of sewage and other wastewaters is such that viable units, either as hyphal fragments or spores, of a wide range of fungal species will be present. The conditions in the aerobic treatment units of excess substrate and nutrients plus sufficient oxygen ensures that some of these species will colonize and successfully develop. Fungi are commonly associated with all wastewater units, including sludge tanks and anaerobic digesters, although most of the records relate to their occurrence in the film of percolating filters. Large amounts of fungi are considered to be undesirable in all types of fixed-film reactors and percolating filters because they cause blockage of the interstices between the stone medium, which impedes drainage and aeration and eventually leads to ponding (Section 4.1). Fungal films are tougher than predominantly bacterial films and are less readily sloughed off. All the fungi have seasonal patterns of growth reaching a peak in winter or early spring and disappearing or becoming scarce by mid-summer. Most of the fungi which flourish in the fixed-film reactors show largely or entirely vegetative growth with spores being rare. At the filters in Minworth (Birmingham), six fungi accounted for the majority of the mycelium in filter film, these were *Ascoidea rubescens*, *Fusarium aquaeductuum*, *Geotrichium candidum*, *Sepedonium sp.* *Subbaromyces splendens*, and *Trichosporon cutaneum*. Other fungi which were also present, but much less common, were *Phoma* and the phycomycetes *Leptomitus lacteus* and *Pythium gracile*. Fungal filaments are often observed in activated sludge flocs, although fungi are rarely dominant. However, fungi are especially associated with industrial effluent and will grow profusely if the bacterial component is inhibited, especially by low pH conditions. When fungi dominate, the reduced density of flocs reduces their settling velocity and may eventually lead to bulking. In a survey of activated sludge fungi, Cooke and Pipes (1968) found that 90% of the species isolated belonged to four common genera. *Geotrichium candidum* was the most abundant fungal species recorded being present in nearly all the activated sludge units examined together with a species of *Trichosporon* (Table 3.8). Predacious fungi are common in treatment units. They either capture prey by modified hyphal traps or parasitize hosts via a range of conidial adaptations, attacking a wide range of mesofauna in particular, nematodes, protozoans, and rotifers (Gray 1984a).

Fungal growth is similar to that of bacteria, with logarithmic growth occurring in the early stages of colonization following the Michaelis–Menten equation as expressed by Monod, where the maximum rate of growth occurs

TABLE 3.8 Occurrence of the major wastewater fungal species in the activated sludge and percolating filter processes (+, presence; −, absence)

Fungi	Activated sludge	Percolating filtration
Ascoidea rubescens	−	+
Fusarium aquaeductuum	−	+
Geotrichium candidum	+	+
Leptomitus lacteus	+	+
Sepedonium sp.	−	+
Subbaromyces splendens	−	+
Trichosporon cutaneum	+	+

when the substrate concentration is not limiting. Maximum theoretical growth rates (μ_{max}), doubling times (t_d), and saturation constants (K_s) for the major fungi are given by Tomlinson and Williams (1975) who were able to express K_s in terms of BOD_5 (mg l^{-1}), (Table 3.9). They estimated expected growth rates of fungi, expressed as doubling times (t_d) using the expression:

$$t_d = t_{dmin}\ 1 + \frac{K_s}{s}\ (\text{h})$$

TABLE 3.9   Maximum theoretical growth rates (μ_{max}), doubling times (t_d), and saturation constants (k_s) of six filter fungi (Tomlinson and Williams 1975)

Fungus	Medium	μ_{max} (h^{-1})	t_d (h)	k_s (gl^{-1})	k_s (BOD mg l^{-1})
Sepedonium sp.	CYG	0.124	5.6	2.5	1850
	CYGS	0.167	4.2	1.25	900
Subbaromyces splendens	CYG	0.0628	11.0	3.84	2460
	CYG	0.077	9.0	1.54	990
Ascoidea rubescens	CYG	0.0887	7.8	1.16	1000
	CYG	0.053	13.1	1.54	1375
Fusarium aquaeductuum	CYG	0.188	3.7	0.435	278
Geotrichum candidum	CYG	0.30	2.3	0.875	560
Trichosporon cutaneum	CYG	0.22	3.2	1.0	640
	'Complan'	0.178	3.9	0.182	100

Note: All fungi grown in batch culture in liquid media at 25°C, the flasks continuously shaken.

where t_{dmin} is the minimum doubling time shown in Table 3.9 and s is 250 mg l^{-1}, where *Sepedonium* 33 h (79 h), *Subbaromyces* 80 h (216 or 560 h), *Ascoidea* 58 h (135 h), *Fusarium* 8 h (13h), *Geotrichium* 7.5 h (9 h), and *Trichosponon* 7.8 h (22 h) at the summer temperature of 25°C (and winter temperature of 12°C). If these values of t_d are compared with similar values for zoogloeal bacterial grown at 25°C which ranged from 10–25 h, then clearly only *Fusarium*, *Geotrichium*, and *Trichosporon* are able to compete with bacteria at summer temperatures. Tomlinson (1941) measured the maintenance energy requirements of two fungi by measuring the increase in biomass at different substrate loadings and estimating the minimum loading at which zero growth occurred (Table 3.10). As bacteria have lower K_s values than fungi, the growth of bacteria is not significantly affected by reducing the substrate loading below that required for maintenance of fungi. Therefore, in percolating filters weak sewages (BOD < 280 mg l^{-1}) limit the growth of fungi especially at summer temperatures, whereas stronger sewages and lower winter temperatures favour fungi. The fact that excessive fungal growths are rarely reported in American filters is due to the weaker wastewater treated (Section 1.2).

TABLE 3.10   Change in weight of *Sepedonium* and *Subbaromyces* after 3 days at 25°C in sewage diluted with water (Tomlinson and Williams 1975)

BOD mg l^{-1}	*Sepedonium*	*Subbaromyces*	BOD mg l^{-1}	*Sepedonium*	*Subbaromyces*
	Initial wt. (mg)			Initial wt. (mg)	
	3.54	2.46		0.62	0.67
200	4.40	2.46	220	2.93	2.60
100	3.85	2.10	110	1.71	2.46
50	3.15	2.24	55	1.02	1.35
25	3.30	1.88	27.5	0.45	1.00
12.5	2.60	1.80	14	0.32	0.66
			0	0.39	0.55

3.3.2   Nutrition

Domestic sewage, and to a lesser extent wastewater from slaughter houses, breweries, and food-processing industries, contain a rich variety of organic and inorganic compounds including important trace elements and organic growth factors (Section 1.2). All the nutritional requirements for bacterial and fungal growth are present in the wastewater. Thus, providing the environmental factors are favourable, a wide and diverse range of heterotrophic micro-organisms will develop within the biological reactor of a wastewater treatment plant. The micro-organisms present are not only those species which can metabolize the raw constituents of the wastewater, but also those that utilize the breakdown products of other micro-organisms and species which prey upon the micro-organisms. As already discussed in Section 1.3, heterotrophs need two categories of nutrients: those specifically required to produce energy for growth and metabolism; and those chemical elements required for biosynthesis. Although phototrophs utilize light energy in wastewater treatment processes, the chemotrophs are most important and they require a chemical source of energy.

Plant and animal tissue is primarily composed of carbon, hydrogen, oxygen, nitrogen, phosphorus, and sulphur and bacterial and fungal metabolism will require these specific elements in the greatest quantities. Other chemical constituents are also necessary for successful heterotrophic growth including potassium, magnesium, manganese, and calcium (which are used for the production of enzyme co-factors and in protein construction), trace amounts of iron, cobalt, copper, zinc, and molybdenum (which are important in the production of co-factors for specific enzymes), and a range of other growth factors. It is interesting that sodium has rarely been shown as an essential element in microbial metabolism, although it is always present in excess in wastewaters. The specific chemical nature of the wastewater will determine the range of heterotrophic species present. For example, the nature of nitrogen will limit the species present as will the C:N:P ratio. Certain industrial wastes are deficient in one or more important elements, e.g. citrus wastes lack nitrogen, whereas coke oven liquors lack phosphorus. Thus, these elements will have to be added as suitable soluble salts if biological treatment of wastewater is to be successful (Painter 1983).

Carbon

Most aerobic heterotrophs utilize the same compound or compounds for both the energy and carbon source. A wide variety of organic compounds can be utilized, although the majority of species can readily utilize the soluble compounds, such as sugars, organic acids, and amino acids. Most of the naturally occurring sugars, and in particular glucose, are assimilated by bacteria and fungi. However, fats and proteins and their breakdown products are less acceptable, with a varying ability in utilization between species. Although insoluble and relatively insoluble substances, such as cellulose, lignin, and higher fatty acids, are utilized by specialist micro-organisms, many of these species only utilize them as secondary substrates.

Work on the nutritional requirements of wastewater micro-organisms is limited (Painter 1983), and has been largely restricted to a number of key species that are also important in aquatic ecosystems (Gray 1985a). However, these key species do provide a useful indication of the requirements of aerobic heterotrophic micro-organisms found in wastewater treatment plants. Considerable work done on the utilization of carbon sources by *Sphaerotilus natans* has been done in laboratory pure culture studies (Mulder and van Veen 1963; Phaup 1968). The carbon sources providing favourable growth are glucose, mannose, galactose, fructose, maltose, lactose, sucrose, mannitol, succinate, lactate, fumarate, pyruvate, acetate, and glycerol. Other compounds, such as amino acids, alcohols, and organic acids, are able to substitute for the carbon requirements if glucose is absent (Scheuring and Hohnl 1956). When *S. natans* and *S. discophorous* are grown in media containing glucose, they only have a limited capacity to oxidize sugars, amino acids, and organic acids (Stokes and Powers 1967), but when grown in the absence of glucose these compounds are readily utilized. These results suggest that represssion of enzyme synthesis is involved. Starch and peptone were used to culture *Sphaerotilus* by Bisset and Brown (1969), and Roberts (1977) observed that both of his strains of *S. natans* could utilize soluble oxidized starch from a paper mill effluent. Roberts also found that high molecular weight polysaccharides (molecular weight > 10 000) present in the effluent were also utilized by *S. natans* in pure culture. This is clearly due to the mechanism by which the bacterium utilized these large polymeric molecules. Extracellular enzymes are produced to degrade the polymer to a molecular size that is able to penetrate the bacterial cell wall membrane. In batch culture, carried out at comparatively high concentrations, the activity of extracellular amylase is able to build up at a level at which starch can be degraded at an acceptable rate. This is because amylase production is a function of substrate concentration and in batch culture the enzyme remains undiluted. However, at the lower concentrations present in the biological reactor of a wastewater treatment plant, extracellular enzyme production is reduced and may be unable to build up to effective concentrations of amylase, resulting in a very slow starch degradation (Roberts 1978; Gray 1985a). The fungi *Leptomitus lacteus* and *Fusarium aquaeductuum* both utilize fatty acids up to C_8. Glucose and maltose both produce excellent growths of both fungi in pure culture, and xylose and arabinose to a lesser degree. Ethanol supports sparse growth of *L. lacteus*, whereas glycerol is utilized by *F. aquaeductuum*. Pyruvic acid is utilized by both organisms, whereas only *F. aquaeductuum* uses succinic and citric acids. Malic and fumaric acids both inhibit the growth of these fungi. *Ascoidea rubescens*, *Geotrichum candidum*, and *Trichosporon cutaneum*, like *Fusarium*, are able to utilize glucose, sucrose, arabinose (except *G. candidum*), and glycerol. Growth of *Subbaromyces splendens* is not stimulated by glucose but, together with *Sepedonium*, grows extremely well on complex organic compounds, with high molecular weights, such as casein. *Sepedonium* grows well on starch as it has levels of strong amylase activity (Tomlinson and Williams 1975). Full details of the carbon substrates utilized by wastewater fungi are given by Tomlinson and Williams (1975).

Nitrogen, basal salts, and other growth factors
Species vary in their ability to utilize forms of nitrogen. Although most bacteria and fungi are able to utilize ammonia as the sole nitrogen source, better growth is usually obtained when an organic nitrogen source is available for use. Growth of *S. natans* is certainly less luxuriant on inorganic nitrogen. Most amino acids support growth (Wilson 1960; Phaup 1968), and in pure culture the nitrogen requirements of *S. natans* can be met using organic nitrogen in the form of amino acids or short-chain peptides (Cawley 1958). However, those that are not suitable or toxic, are serine, histidine, lysine, ornithine, threonine, tryptophane, tyrosine, and valine (Phaup 1968) and, although necessary, most of the amino acids can be toxic in high concentrations. Asparagine, glutamine, aspartic, and glutamic acids are used as both carbon and nitrogen sources, whereas methionine and glycine are only used as a nitrogen source (Mulder and van Veen 1963). The growth of *S. natans* is also promoted when inorganic forms of nitrogen are available, and ammonia, nitrite, and nitrate ions can all be utilized, although vitamin B_{12} is also required under these conditions (Okrend and Dondero 1964; Dias and Huekelekian 1967). The fungi *Geotrichium candidum* and *Trichosporon cutaneum* are both able to utilize ammonia and organic nitrogen, although *Sependonium* sp. and *L. lacteus* are supported only by organic sources of nitrogen (Schade 1940; Painter 1954; Cantino 1966). *Ascoidea rubescens* and *Subbaromyces splendens* both require an organic source of nitrogen for growth, whereas *F. aquaeductuum* was the most versatile of the filter fungi

being able to utilize nitrate, ammonia, and organic nitrogen (Tomlinson and Williams 1975). Comparing the nutritional requirements of *L. lacteus* and *F. aquaeductuum*, Williams (1983) tested a wide variety of possible nitrogen sources. *Leptomitus lacteus* could not utilize methylamine, acetamide or urea, although *F. aquaeductuum* did use them as sole nitrogen source. Leucine, asparagine, and glutamine were utilized as nitrogen sources by *L. lacteus*, whereas neither glutamine nor glycine could be used. *Fusarium aquaeductuum* appears to utilize glutamine and leucine as both nitrogen and carbon sources. The ratio of BOD_5 to N (C:N) should be equal to, or less than, 18 if the nitrogen requirements of micro-organisms are to be fully satisfied and optimum microbial growth is to occur. As the ratio increases above 22 then BOD_5 removal efficiency will begin to fall off as microbial activity becomes nitrogen-limited. Under high C:N conditions, there will be a tendency for increased filamentous growth in activated sludge tanks, possibly resulting in bulking (Hattingh 1963a,b).

A large number of other elements are also required, although the concentrations required for optimum growth depends on species and other factors, such as growth rate. Phosphorus is required in comparatively large quantities, although at much lower levels than nitrogen. Phosphates are common constituents of sewage and satisfy all the requirements of the micro-organisms provided that the BOD:P (C:P) ratio is less than 90–150. If the C:P ratio rises above 165–170 then microbial treatment efficiency in terms of BOD_5 removal falls, resulting in enhanced filamentous growth and possible bulking in an activated sludge plant (Greenberg *et al.* 1955; Hattingh 1963a,b). Laboratory growth studies of *S. natans* have shown that the bacterium requires a basal salt supplement of sodium, potassium, calcium, magnesium, phosphate, sulphate, and chloride (Gray 1985a). Iron is required in trace amounts but is toxic to *S. natans* in low concentrations (Waitz and Lackey 1959), and Razumor (1961) used this relationship with iron to separate *Sphaerotilus* from *Leptothrix* and *Cladothrix*. Dias and Dondero (1967) found that both *S. natans* and *S. discophorous* required calcium. Using a continuous-flow apparatus, they found that calcium is required for the development of the sheath (Dias *et al.* 1968). The percolating filter fungus *Sepedonium* sp. requires calcium $(5.0–12.5 \text{ mg l}^{-1})$, zinc $(0.5–1.0 \text{ mg l}^{-1})$ and trace amounts of iron, copper, and manganese for optimum growth (Painter 1954). Imbalances of trace metals, in particular, that often results when metal wastes are present, can reduce the diversity of micro-organisms present in sludges and filter film and can result in excessive growths of tolerant species that can cause bulking or ponding (Pfeffer *et al.* 1965).

Some micro-organisms require organic growth factors such as B-group vitamins, which act as co-enzymes or enzyme precursors, although many species are able to synthesize their own growth factors. Agricultural wastes, domestic sewage, and some industrial wastes contain adequate concentrations of required growth factors. Sewage, in particular, is rich in B-group vitamins, and riboflavin, nicotinamide, pantothenic acid, thiamine, and biotin have all been isolated (Painter 1983). Painter (1954) demonstrated that sewage also contained other growth factors by showing that enhanced growth by *Sepedonium* sp. was achieved by the addition of small amounts of sewage to growth medium containing all the B vitamins and a number of other known growth factors. *Zoogloea* spp. and *Comamonas* spp. isolated from activated sludge, required either amino acids or vitamins, or both, and that biotin, thiamine, and cobalamine were the most frequently required growth factors (Painter 1983). Laboratory growth studies indicate that *S. natans* does not require a supply of accessory growth factors apart from trace amounts of vitamin B_{12} (cyanocobalamin) or methionine, from which B_{12} can be synthesized (Okrend and Dondero 1964). Vitamin B_{12} was shown to be necessary for the growth of 34 *Sphaerotilus* and *Leptothrix* strains, although it could be replaced by much higher concentrations of methionine. The presence of vitamin B_{12} was shown to be necessary for the utilization of inorganic nitrogen by *S. natans*, whereas *S. discophorous* had a growth requirement for adenine, guanine, thiamine, and biotin as well as vitamin B_{12}. *Sphaerotilus natans* does not require any growth factor except vitamin B_{12}. Among those tested and found not to be required were thiamine, riboflavin, nicotinic acid, pryridomine, pantothenic acid, biotin, folic acid, *p*.-amino benzoic acid, adenine, xanthine, guanine, and urea (Mulder and van Veen 1963). *Fusarium aquaeductuum* does not require thiamine and, like *Geotrichium candidum*, is independent of an external supply of B-group vitamins, although both *T. cutaneum* and *Sepedonium* sp. require thiamine, with the latter also requiring biotin (Painter 1954). Although a number of the more common wastewater micro-organisms have a requirement for external sources of vitamins and other growth factors, the majority do not. Therefore, any deficiency would only result in a slight reduction in species diversity, and treatment efficiency would remain unimpaired (Painter 1983).

3.3.3 Environmental factors

Dissolved oxygen
Dissolved oxygen serves only as an electron acceptor for heterotrophic aerobes in wastewater. In terms of substrate utilization and energy production, aerobic heterotrophs are far more efficient than anaerobic micro-organisms. The facultative micro-organisms also grow better and more efficiently in the presence of oxygen. The growth rate of an aerobe increases with the concentration of oxygen until a critical dissolved oxygen concentration is reached, at which time maximum growth rate occurs under prevailing conditions and no further increase is possible. The critical concentration is normally below 1 mg l^{-1} for dispersed bacteria, but as the critical concentration is governed by the rate of diffusion and the diameter of the organism, the

critical concentration in the activated sludge process is somewhat higher due to the aggregation of bacteria into flocs. Filamentous fungi also have a critical concentration of > 1 mg l^{-1}.

Whereas percolating filters obtain oxygen from natural ventilation, dissolved oxygen has to be supplied by mechanical means in the activated sludge process. The higher the concentration at which the dissolved oxygen is maintained in the aeration tank, the greater the energy input required. Also, as the oxygen deficit of the wastewater is reduced then the aeration efficiency also declines. The cost of aerating wastewater is extremely expensive and it is important to maintain the wastewater in the activated sludge tank as near as possible to the critical dissolved oxygen concentration in order that maximum micro-organism efficiency can be achieved at minimum cost (Fig. 3.22). Turbulence is an important factor in the activated sludge process as it reduces localized dissolved oxygen and substrate deficiencies. Efficient mixing ensures that the dissolved oxygen and substrate concentration are maintained at maximum concentration, which maintains a maximal diffusion rate through the cell–liquid interface. However, diffusion through flocs of activated sludge require higher critical concentrations of dissolved oxygen to give maximum oxygen uptake, compared with dispersed bacterial cultures where diffusion is simply across the cell wall. For example, in zoogloeal flocs, maximum uptake occurs at dissolved oxygen concentrations of between 0.6 to 2.5 mg l^{-1}, depending on floc size, compared with < 0.1 mg l^{-1} for dispersed zoogloeal cells. However, the normal floc size in activated sludge is such that maximum oxygen uptake occurs at concentrations of 2 mg l^{-1} or less, although a survey of activated sludge plants showed that 75% contained flocs with nominal

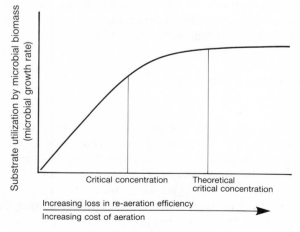

FIG. 3.22   The effect of oxygen concentration on activated sludge activity showing critical and theoretical critical oxygen concentration.

diameters of $<43\mu m$, indicating maximum uptake rates to be <0.6 mg l^{-1} (Muller et al. 1968). The dissolved oxygen concentration in aeration tanks is controlled by oxygen electrodes placed in the tank and linked to the aeration equipment. The desired oxygen concentration in aeration tanks is normally between 0.6–2.0 mg l^{-1}, with 1.0 and 1.2 mg l^{-1} being generally used. Selection is dependent on such factors as tank design, wastewater characteristics, and microbial characteristics, and a precise critical concentration for a specific treatment unit can only be derived from practical operational experience. It has been suggested that enhanced agitation encourages the growth of dispersed bacteria and inhibits filamentous forms such as S. natans.

In fixed-film reactors oxygen only diffuses to a maximum depth of 0.1–0.2 mm, below which the film is either anoxic aerobic. When filamentous fungi are present, the critical depth increases to 2 mm as the diffusion of oxygen is assisted by protoplasmic streaming (Tomlinson and Snaddon 1966). Where the film is well developed and the interstices are small, thus restricting the rate of ventilation, the oxygen concentration can become a limiting factor (Section 4.1).

Temperature
Temperature has a profound effect on wastewater micro-organisms, not only in governing the rate of reaction (Section 3.1.2), but changes in temperature also give rise to significant alterations in the community structure. In terms of removal efficiency, higher temperatures give rise to higher rates of BOD removal up to 35–40°C, after which bacterial cells become reduced in size and number, thus growing in a dispersed phase and resulting in turbid effluents (Rogovskaya et al. 1969). In pure culture, S. natans has a wide temperature tolerance ranging from 4–40°C with optimal growth between 25–30°C (Sanders 1982). The rate of metabolism and growth is very much reduced at lower temperatures which is illustrated when the filamentous bacteria form microscopic slime growths (sewage fungus) in polluted rivers. For example, the slime growth in River Altamaha (Georgia, USA) in summer (30°C) was limited to 150–250 m below the effluent source, whereas in winter (10°C) it extended for 24–64 km (Phaup and Gannon 1967). *Sphaerotilus natans* has the ability to out-compete other bacteria for available nutrients at lower temperatures, although fungal components are known to have more rapid growth rates than the bacterial component at reduced temperatures providing suitable nutrients are available. This is clearly demonstrated by experimental channel experiments where the standing crop of zoogloeal/*Fusarium* slime is closely related to temperature. Growth was initially more rapid at 20°C than at 10°C. However, a much more luxuriant and heavier *Fusarium* growth was eventually established at the lower temperature with the biomass six times greater at 10°C compared with 20°C after 30 days (Ministry of Technology 1970).

The rates of growth of all the major wastewater fungi increase linearly with temperature. At 25°C, which is the maximum summer temperature of wastewater, the fungal growth rate is approximately double the growth rate at 12°C, the average winter temperature. *Fusarium aquaeductuum* and *L. lacteus* both grew well over the range 10–20°C with optimum growth occurring between 15–25°C and 20–25°C respectively (Williams 1983). *Leptomitus lacteus* has been recorded growing in the heated effluent from an alcohol factory at 27°C (Gray 1985*a*), and Zehlender and Boek (1964) recorded optimum growth at 8°C at pH 6. *Sepedonium* sp. and *G. candidum* have similar optimum temperatures at 20°C and 20–25°C respectively (Tomlinson 1942; Pipes and Jones 1963). Clearly, the fungi in percolating filters do not all have the same temperature range. For example, in contrast to *Sepedonium* sp., *S. splendens* predominates in filters during the summer and autumn because of its high optimum temperature of 25–27°C. Conidia fail to germinate below 5°C and are restricted up to 15°C, limiting the dominance of this fungus to the warmer months, although it is present in filters throughout the year (Williams 1971; Gray 1983*b*).

Activated sludge liquors are reported as being less filamentous at higher temperatures (Hunter *et al.* 1966) and have been shown to take longer to acclimatize to changes in temperature than to toxic compounds. Complete acclimatization takes at least 2–3 weeks and takes up to five times longer at 5°C compared with 30°C (Rogovskaya *et al.* 1969). In percolating filters, maximum microbial activity occurs between 20–35°C, although some yeasts are active up to 40°C. However, the activity of most of the important micro-organisms is inhibited at temperatures < 7°C and > 36°C, although viability of most of the organisms are not lost if the temperature only exceeds these limits for a few days.

pH

The importance of pH on microbial growth has been demonstrated by studies on polluted rivers in Norway. The community structure of heterotrophic slimes was different in rivers with different pH values. For example, the River Dams supported a *S. natans*-dominated slime (pH 7.0), whereas the River Otta (pH 5.8) supported a *Fusarium*-dominated slime (Ormerod *et al.* 1966; Baalsrad 1967). The same affect is found in fixed and mixed biological reactors, with fungi growing over a wider pH range than bacteria dominating at acidic pH values (pH < 6.5), with the bacteria preferring neutral pH values and generally dominating at pH values > 7.0. Domestic wastewaters have a pH range of between 6–8, and are amenable to biological oxidation. However, some industrial wastewaters can be quite acidic or alkaline and need to be neutralized before biological treatment. The optimum pH range for carbona-ceous oxidation lies between 6.5 and 8.5.

In pure culture studies, *S. natans* can tolerate a wide pH range (pH 5–10) although most rapid growth occurs between 6.5–8.1, with optimum growth at pH 7.5 and growth inhibited at pH <6.2. *Zoogloea* spp. are found over a similar range although the pH has shown to affect the growth form, being dispersed at pH 7 but forming aggregates at pH 6 (Angelbeck and Kirsch 1969). All the fungi can grow over a wide range of pH values with *F. aquaeductuum* (pH 4–9), *G. candidum* (pH 3–9), *L. lacteus* (pH 2.5–7.5), and *T. cutaneum* (pH 4–9) all having wide pH tolerances (Painter 1954; Pipes and Jones 1964; Zehlender and Boek 1964). Although *Sepedonium* sp., which is found over a wider pH range (pH 4–10), has a restricted optimum growth range 7.0–8.5 (Tomlinson and Williams 1975). Tomlinson and Williams (1975) suggest that the effect of pH on *F. aquaeductuum*, *G. candidum*, and *T. cutaneum* depends largely on the composition of the growth medium. They concluded that the common filter fungi can be classified as 'colonial' species (*F. aquaeductuum*, *G. candidum*, *T. cutaneum*), which can tolerate low pH values but are susceptable to inhibition by undissociated organic acids and the 'spreading' species (*Sepedonium* sp., *S. splendens*, *A. rubescens*), which have a pH optimum in the neutral range.

Light

Heterotrophic micro-organisms are independent of light, often growing in complete darkness within the biological reactor. However, in facultative waste stabilization ponds (Section 4.4) light is vital in order for the algae to produce enough oxygen to maintain the aerobic heterotrophic demand. The organic loading controls the ratio of phototrophic to chemotrophic organisms. Reducing the loading will increase the ratio of sunlight energy to chemical energy into the system, increasing the ratio of phototrophic to heterotrophic organisms. Plants play a very minor role in other biological treatment processes and are rarely recorded except from the surface of percolating filters. Algae and diatoms are the commonest plants recorded, although excessive growths of filamentous algae and mosses can result in a reduction in performance by causing filters to pond (Benson-Evans and Williams 1975; Hussey 1982; Gray 1984b).

Fungi are generally independent of light with the exception of *F. aquaeductuum*. Both conidia formation and pigmentation in this species are light-dependent, the pigment being a photo-induced carotenoid (Rau 1967), which is why this species is restricted to the surface of percolating filters.

3.3.4 Inhibition

Many heavy metals and organic compounds are toxic to aerobic heterotrophs, both in pure culture and in the treatment plant. However, in the mixed culture of micro-organisms that makes up activated sludge, these toxic compounds need to be at far greater concentrations compared with pure cultures before any inhibitory effect is noticed. This is due to physical adsorption of the compounds onto organic matter and flocs, also to chemical reactions such as precipitation or chelation with other constituents of wastewater including other toxic substances. However, as the concentration of the toxin increases, the inhibition of life processes becomes increasing severe until the cells eventually die and lyse. When treating wastewater containing inhibitory substances, some degree of acclimatization and selection of tolerant species will occur. However, the performance of such systems will rarely be as good as toxic-free systems (Moulton and Shumate 1963). Metals such as copper and mercury are particularly toxic, complexing with enzymes and other metabolic agents connected with respiration and rendering them inactive. The common filter fungus *Sepedonium* sp. is surprisingly sensitive to zinc, being inhibited at concentrations of between 4–10 mg l^{-1}, when other trace metals are also deficient in the medium. Other substances, especially organic complexes containing nitrogen, and occasionally sulphur, compete with enzymes for essential metals which act as co-enzymes and catalysts, whereas phenols and detergents act by transforming the cell or causing it to disintegrate (Painter 1983). Wastewaters with high concentrations of salts or nitrogen compounds also inhibit biological treatment processes and affect the aerobic heterotrophs, in particular. Salt concentrations of 30 000 mg l^{-1} severely inhibit the activated sludge processes due to osmotic stress (Kincannon and Gaudy 1966), although shock loads of sodium chloride have little effect on the micro-organisms forming the film in percolating filters (Gray 1981). *Fusarium aquaeductuum*, *G. candidum*, and *T. cutaneum*, which are able to use ammonia as a sole nitrogen source, are not inhibited by ammonia at concentrations up to 2000 mg N l^{-1}. However, *A. rubescens*, *Sepedonium* sp., and *S. splendens*, which all require an organic nitrogen source, are inhibited by ammonia. Many bacterial species cannot tolerate high ammonia concentrations and the activated sludge process is severely inhibited if ammonia is present in excess. Anionic detergents, which are normally present in concentrations of 35 mg l^{-1} as Manoxol OT in domestic sewage, inhibit a wide range of heterotrophs. At concentrations of 5 mg l^{-1} there is little inhibitory effect on fungi, but at concentrations in excess of 20 mg l^{-1} many fungal species are inhibited or eliminated. *Sepedonium* sp. is very susceptible to concentrations of detergents above 10 mg l^{-1} and *S. splendens* is affected at concentrations in excess of 30 mg l^{-1}, when conidia formation is also inhibited, which is probably why conidia are so rarely seen in filters treating domestic sewage. *Ascoidea rubescens* is unaffected by anionic detergents up to 50 mg l^{-1} and *T. cutaneum* is also very tolerant (Tomlinson and Williams 1975). *Geotrichum candidum* and *F. aquaeductuum* are able to inhibit the growth of certain other fungi by the production of antibiotics. *Geotrichum candidum* inhibits the growth of *F. aquaeductuum* and *T. cutaneum*, and *F. aquaeductuum* inhibits *Sepedonium* sp., *S. splendens*, and *T. cutaneum* (Table 3.11).

TABLE 3.11 Interactions between five species of fungi on nutrient agar plates (Tomlinson and Williams 1975)

	Se	Su	G	F	T
Sepedonium (Se)		=	=	>	=
Subbaromyces (Su)	=		=	>	=
Geotrichum (G)	=	=		<	<
Fusarium (F)	<	<	>		<
Trichosporon (T)	=	=	>	>	

=, indicates no interaction; >, indicates that organism in vertical column was inhibited by organism in horizontal column; <, indicates that organism in horizontal column was inhibited by organism in vertical column.

Further reading

General: Haddock and Hamilton 1977; Painter 1983.
Bacteria: Phaup 1968; Pike 1975; Painter 1983.
Fungi: Tomlinson and Williams 1975.
Nutrition and environmental factors: Tomlinson and Williams 1975; Painter 1983.

3.4 ANAEROBIC HETEROTROPHIC MICRO-ORGANISMS

3.4.1 Introduction

Anaerobic heterotrophic bacteria are either obligate and unable to grow in the presence of oxygen, or facultative and can adapt to environments either with or without oxygen. The latter form a bridge between obligate aerobes and obligate anaerobic species. Growth and metabolism of obligate anaerobes are inhibited by oxygen, with oxygen toxicity depending on the redox potential (E_h), partial pressure, composition of the substrate, growth rate, and cell density (Hughes 1980).

The major role of anaerobic heterotrophs is in sludge digestion converting unstable sewage into a more stabilized form. Digestion normally takes place in specially constructed reactors, although anaerobic digestion also occurs in the sludge blanket of waste stabilization ponds. The pressure of anaerobic activity in aerobic treatment processes is generally undesirable with the commonest effects of anaerobosis being the production of foul odours and interference of floc settling due to denitrification occurring in the sedimentation tank after the aeration basin in the activated sludge process.

Anaerobic treatment is an effective method for the complete treatment of many organic wastes, especially animal wastes and organic effluents from the food processing industries (Stafford *et al.* 1980). The organic substrate is degraded in the absence of oxygen to carbon dioxide and methane with only a small amount of bacterial growth. Approximately 90% of the available chemical energy, in the form of organic material, is retained as methane production (McInnery *et al.* 1980). Apart from the economic value of the methane gas produced, anaerobic treatment has many advantages over aerobic treatment processes, such as less biomass produced per unit of substrate utilized (the lower biomass production means a lower requirement for nitrogen, phosphorus, and other nutrient and growth factors), higher organic loadings are possible as anaerobic processes are not limited by oxygen transfer rates, and the lower constructional and operational costs compared with aerobic processes. However, the major disadvantage in temperate and colder climates is the elevated temperatures required to maintain microbial activity at a reasonble level.

3.4.2   Presence in the treatment plant

Wastewater
Anaerobic bacteria are common in sewage and other wastewaters. They originate from the human intestine, although other sources include land drainage, stormwater, and biological processes in industry. In sewage, the faecal bacteria outnumber the other micro-organisms present, although no really satisfactory methods of counting anaerobic bacteria have been developed. But as coliform counts are in excess of 10^6–10^7ml^{-1} in sewage, and coliforms only represent a small portion of total anaerobic bacteria, the total count must be very high.

Treatment plant operators try to prevent incoming wastewater from becoming anaerobic. Not only because this makes it more difficult to treat but to prevent anaerobic activity within the sewers. Anaerobic activity is usually indicated by offensive smells such as amines, skatole, and, more commonly, hydrogen sulphide (Table 1.12). These gases have to be vented from the sewer as they are extremely toxic to men working in the sewer environment and can also cause corrosion of concrete pipework.

Primary sedimentation tanks
The wastewater in primary sedimentation tanks is normally aerobic and inhibits the anaerobic bacteria present. The retention times in the tanks are normally <8 h, which is not long enough for the sulphide bacteria to build up. Therefore, odours are rarely produced unless the wastewater is already anaerobic when entering the treatment works. When desludging is irregular or the radial scrapers are not effective, anaerobic fermentation can occur in the compacted sludge layer causing the sludge to rise.

Aerobic treatment units
Although present, obligate anaerobes are not active in aerobic treatment systems and only dominate when anaerobic conditions exist. In percolating

filters, anaerobic conditions only occur as a result of gross overloading when the interstices of the medium become blocked by excessive film accumulation; or after a toxic shock which kills and subsequently inhibits the aerobic micro-organisms. In the activated sludge process, anaerobic conditions occur because of factors such as poor design, insufficient aeration, and gross overloading. It is less likely for a percolating filter to become completely anaerobic, therefore some treatment capacity will normally always remain. However, once an activated sludge aeration unit becomes anaerobic the entire treatment capacity of the system is lost, which results in foul odours and sludges with extremely poor settleabilities. The role of facultative bacteria in aerobic systems is unclear, and although they are capable of utilizing oxygen they have not been recorded as dominant in the activated sludge processes (Lighthart and Oglesby 1969).

Percolating filters and, to a lesser extent sludge aeration basins, provide a wide diversity of micro-habitats for micro-organisms and therefore there are microniches where anaerobic bacteria can flourish, albeit in small numbers. Oxygen can only diffuse to a maximum depth of 0.2 mm into filter film and the inner layers are normally anaerobic (Section 4.1). The centre of sludge flocs can also be anaerobic because of the problem of oxygen diffusion, which is reduced considerably as the oxygen concentration in the aeration basin falls below $1 \, \text{mg} \, l^{-1}$.

Final settlement tanks
It is in the final settlement process that anaerobic bacteria can most frequently cause problems. In the final settlement tank of the activated sludge process, solids are retained for only a short time in order to maintain the aerobic character of the sludge, which is returned to maintain the microbial density in the aeration basin. Two problems may arise. If the sludge is not removed continuously or very frequently, but instead allowed to remain in the tank, then facultative anaerobes begin to multiply and can become incorporated into the activated sludge thus reducing efficiency. If anaerobic conditions are allowed to develop, gas will be produced and floc particles will rise to the surface because of bubbles becoming entrapped or attached to particles. This is nearly always due to the conversion of nitrate, formed by nitrification in the aeration tank, to nitrogen gas, a process known as denitrification (Section 3.4.5).

Anaerobic processes
It is in the digestion of sludge, in specially designed anaerobic reactors (fermentors or digesters), that the activities of anaerobic micro-organisms are used to advantage in wastewater treatment. The settled sludge, from both primary and final sedimentation tanks, is converted from its highly putrid state to a stable and disposable product that neither smells or undergoes further decomposition on storage. The main advantages of treating sludges

anaerobically include: the high degree of stabilization; the low production of extra biological sludge; the low nutrient requirements; no oxygen requirement; and the production of useful end-products, in the form of methane and a useful biomass.

The main constituents of raw sludge are protein, fats, and polysaccharides. Typical chemical analysis of a sewage sludge is protein (25%), cellulose and lignin (25%), and fats (20%) (Heukelekian 1957). If anaerobic breakdown is complete then the end-products will be methane, carbon dioxide, water, and new bacterial cells. However, incomplete breakdown due to environmental conditions or inhibitors could result in intermediate products being formed, such as volatile fatty acids, alcohols, and ammonia.

3.4.3   Anaerobic digestion

The process of sludge digestion is generally considered as a two-phase process, the non-methanogenic followed by a methanogenic phase. It is more accurate to describe the process of anaerobic digestion as being comprised of three discrete stages; hydrolysis and acid formation (the non-methanogenic phase), followed by methane formation (methanogenic phase) (Fig. 3.23).

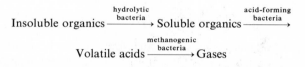

It is convenient to think of these stages as different trophic levels, and although all three stages are normally occurring simultaneously within a

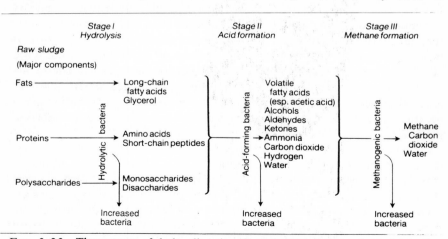

FIG. 3.23   Three stages of sludge digestion. Some hydrolytic bacteria are known to be able to carry out both stages I and II.

digester, the micro-organisms involved at each stage are metabolically dependent on each other for survival. For example, the methanogenic bacteria require the catabolized end-products of the acid-forming bacteria. However, the latter species would eventually become inhibited by their own end-products if these were not degraded by the methanogenic bacteria. Although bacteria are the major group of micro-organisms involved in anaerobic digestion, fermentative ciliate and flagellate protozoa and some anaerobic fungi also occur (Hobson et al. 1974). The process does not readily occur in the presence of electron acceptors such as oxygen, sulphate, or nitrate. Energy transformation is by the ATP system with energy being stored by the reaction of ADP and inorganic phosphate (P_i) to form ATP. The energy conserved in the pyrophosphate bond is used by splitting ATP into either $ADP + P_i$ or adenosine-5-monophosphate (AMP) and pyrophosphate (PP_i).

In the first stage, the major substrates in the sludge are hydrolysed to basic components; proteins to amino acids, fats to glycerol and long-chain fatty acids, and polysaccharides to mono- or disaccharides (Fig. 3.23).

Proteins are hydrolysed to smaller units such as polypeptides, oligopeptides, or amino acids by extracellular enzymes called proteases, which are produced by only a small proportion of the bacteria. The majority of bacteria are able to utilize these smaller peptides or the amino acids, which pass through the cell wall and are broken down intracellularly. The production of protease is far in excess of that required, even though protease-producing bacteria represent such a small percentage of the total bacteria present. Estimates indicate that this over-production could be in the order of 50 times more than is required (Hattingh et al. 1967). The most active proteolytic bacteria are the spore-forming Clostridium sp. In anaerobic sludge, proteolytic bacteria can reach concentrations of $6.5 \times 10^7 ml^{-1}$, of which 65% of the isolates examined by Toerien (1970) were spore formers, 21% cocci, and the remainder non-sporing rods and bifid-like bacteria. Little is known about the lipolytic bacteria even though they have been shown to be highly effective in anaerobic digesters (Crowther and Harkness 1975). They are present in densities of up to $7 \times 10^4 ml^{-1}$ and the addition of vegetable oil to digesters to enhance gas production is commonly practised in some countries. Anaerobic cellulolytic bacteria are present in anaerobic sludge at concentrations of between 10^4–$10^5 ml^{-1}$ and are predominantly Gram-negative coccobacci, with Bacteroides ruminicola being a particularly common species. Other species, including a Gram-positive species forming curved rods which formed short chains, have been isolated by Hobson and Shaw (1971). In sewage sludge, the ability to hydrolyse starch is the most common activity of these bacteria (Table 3.12).

The heterogeneous group of facultative and anaerobic bacteria, which are responsible for hydrolysis, are also responsible for acid formation. In this second stage, the hydrolysed substrate is converted to organic acids and alcohols, with new cells also being produced. Various biochemical pathways

TABLE 3.12   Enumeration and identification of anaerobic bacterial
populations in sewage sludge digesters (Zeikus 1980)

Group	Numbers (per ml)	Generic identity
Hydrolytic bacteria		Majority unidentified
Total	10^8–10^9	Gram-negative rods
Proteolytic	10^7	*Eubacterium*
Cellulolytic	10^5	*Clostridium*
Hydrogen-producing		Unidentified
acetogenic bacteria	10^6	Gram-negative rods
Homoacetogenic bacteria	10^5–10^6	*Clostridium*
		Acetobacterium
Methanogens	10^6–10^8	*Methanobacterium*
		Methanospirillum
		Methanococcus
		Methanosarcina
		'*Methanothrix*'
Sulphate reducers	10^4	*Desulfovibrio*
		Desulfotomaculum

are utilized, including fermentation and β-oxidation. There is very little stabilization of the substrate in terms of BOD_5 or COD removal, with the products of acid fermentation being large organic molecules. Obligate anaerobic bacteria occur in much larger numbers than the facultative or aerobic bacteria, with ratios of 1:100 and numbers in the order of 10^7–10^8ml^{-1} are not uncommon (Kirsch 1968). The major acid-forming organisms are *Bacillus* sp., *Micrococcus* sp., and *Pseudomonas* sp., although little taxonomic work has been done (Crowther and Harkness 1975). Mono- and disaccharides, long-chain fatty acids, glycerol, amino acids, and short-chain peptides provide the main carbon source for growth, with saturated fatty acids, carbon dioxide, and ammonia being the main end-products. Alcohols, aldehydes, and ketones are also produced but only in minute quantities (Fig. 3.23). The concept of pyruvate as the pivotal compound in metabolism was discussed earlier. When no external electron acceptor is present, as is the case in acid fermentation, pyruvate can undergo several alternative reactions which regenerate NAD from NADH (Fig. 3.17). Acetic, propionic, butyric, and lactic acids are the most frequently produced end-products during stage II and Toerien (1970) found that these acids were produced by 87, 67, 10, and 70% respectively of the 92 acid-forming bacterial isolates he examined. Propionic and longer chain fatty acids are degraded by an intermediate microbial group called the obligate hydrogen-producing acetogenic bacteria, whereas other acid-producing bacteria, referred to as the homacetogenic bacteria, produce acetic and sometimes other acids (McInerney *et al.* 1980).

The third and final stage in anaerobic digestion is methane fermentation where the end-products of acid fermentation are converted to gases, mainly methane and carbon dioxide, by several different species of obligate anaerobic bacteria. In this stage, complete stabilization of the substrate occurs and as the end-products are only gases it is more efficient than complete aerobic stabilization. Methane is an ideal end-product as it is non-toxic, easily escapes from the site of production with the use of a separation process, is not very soluble, inert under anaerobic conditions, and can be readily collected and used as an energy source. In the overall anaerobic fermentation of carbohydrates to carbon dioxide and methane, equal volumes of each gas are produced.

$$(C_6H_{12}O_6)_x \times xH_2O \rightarrow xC_6H_{12}O_6 \rightarrow 3xCH_4 + 3xCO_2$$

The carbon dioxide evolved only partially escapes as gas, because, unlike methane, it is relatively soluble in water. It also reacts with any hydroxide ions (OH^-) in the system to produce bicarbonate ions (HCO_3^-). The evolution of carbon dioxide gas is therefore a function of factors such as pH, bicarbonate concentration, temperature, and substrate composition. The biodegradable protein is deaminated to produce ammonia which reacts with water:

$$NH_3 + HOH \rightleftharpoons NH_4^+ + OH^-$$

This is the major source of hydroxide ions which react with the carbon dioxide evolved during methanogenesis to form bicarbonate ions:

$$CO_2 + HOH \rightleftharpoons H_2CO_3 \rightleftharpoons H^+ + HCO_3^-$$

$$H_2CO_3 + OH^- \rightleftharpoons HCO_3^- + HOH$$

Therefore, the protein content of the wastewater substrate will significantly affect the quantity of carbon dioxide actually released from solution as well as the buffering capacity of the system in terms of bicarbonate. The portion of carbon dioxide incorporated in the bicarbonate ion is eventually removed from the reactor in the liquid rather than in the gas phase (Pfeffer 1980).

Anaerobic digestion and methane production are not unique to anaerobic digesters, they occur in natural environments including the digestive tract of most animals, in the sediments of lakes and rivers, and in estuaries, swamps, marshes, and peat bogs. The bacteria responsible for methanogenesis in digesters are similar to those found in other environments, although little is known about them because of the problems of isolating and maintaining cultures of bacteria under anaerobic conditions. Most methanonegenic bacteria belong to the genera *Methanobacterium*, *Methanosarcina*, *Methanospirillum*, and *Methanococcus*. The methanogenic genera isolated so far, are limited to the catabolism of either one carbon (e.g. H_2/CO_2, CH_3OH, CO, HCOOH, CH_3NH_2) or two carbon compounds (e.g. CH_3COOH). Methanogens are unusual in that they are composed of many species with very different cell morphology. They require a strict anaerobic environment for growth with a redox potential below -300 mV. They have simple nutritional requirements; carbon dioxide, ammonia, and sulphide. Ammonia is the essential nitrogen source for growth and no methanogenic species are known to utilize amino acids or peptides, sulphide is the most common sulphur source, although some species can use cysteine instead. Methanogens contain a number of unique co-enzymes, for example, co-enzyme 420 which is involved in electron transfer instead of the usual ferredoxin, co-enzyme M which is used in methyl transfer reactions, and factor B which is required for the enzyme formation of methane from methyl co-enzyme M. The synthesis of ATP appears to be via electron transport linked to phosphorylation. The ultrastructure and cellular composition of methanogens make them very different from typical bacteria and some workers have suggested that they are in fact a primitive group of bacteria. The group does not contain muramic acid in their cell walls and this, linked with their unique co-enzymes and the unique oligo-nucleolide sequences of the 16S ribosomal RNA molecule, has been used to re-classify the micro-organisms of the group (Table 3.13). Further taxonomic and physiological details are given by Wolfe (1971), Crowther and Harkness (1975), Morris (1975), Thauer *et al.* (1977), and Zeikus (1977, 1980).

There is a close relationship between gas production and the numbers of methanogenic bacteria present in a digester, generally being present in numbers between 10^6–10^8 ml^{-1} (Table 3.12). It is from acetic and, to a lesser extent, propionic acid that the greatest percentage of methane is derived during methane fermentation, whereas formic acid fermentation and methane fermentation associated with β-oxidation of long-chain fatty acids probably accounts for most of the small percentage of methane not derived from acetic and propionic acids (Fig. 3.24).

Methane formation from acetic acid is a single-step process carried out by one group of methanogenic bacteria:

$$CH_3COOH \rightarrow CH_4 + CO_2$$
(acetic acid) (methane)

Methane fermentation of propionic acid is a two-step process involving two groups of methanogenic bacteria, with acetic acid as the intermediate step.

Step 1: $CH_3CH_2COOH + 0.5H_2O \rightarrow CH_3COOH + 0.25CO_2 + 0.75CH_4$
            (propionic acid)   (acetic acid)   (methane)

Step 2: $CH_3COOH \rightarrow CO_2 + CH_4$
            (acetic acid)   (methane)

Overall: $CH_3CH_2COOH + 0.5H_2O \rightarrow 1.25CO_2 + 1.75CH_4$
            (propionic acid)   (methane)

The acid-fermenting bacteria are tolerant to changes in pH and temperature and have a much higher rate of growth than the methane fermenting bacteria. This difference in growth rate, linked with a greater sensitivity to environmental factors, results in the methanogenic bacteria becoming the major factor controlling the overall rate of anaerobic digestion (Benefield and Randall

TABLE 3.13 Proposed taxonomic scheme based on comparative cataloguing of the 16S ribosomal RNA and substrates used for growth and methanogenesis (McInerney et al. 1980)

	Type strain	Former designation	Substrates for growth and CH_4 production
Order I. *Methanobacteriales* (type order)			
Family I. *Methanobacteriaceae*			
Genus I. *Methanobacterium* (type genus)			
1. *Methanobacterium formicicum* (neotype species)	MF	*Methanobacterium formicicum*	H_2, formate
	M.o.H.	*Methanobacterium* sp. strain M.o.H.	H_2
2. *Methanobacterium bryantii*			
Methanobacterium bryantii strain M.o.H.G.	ΔH	*Methanobacterium* sp. strain M.o.H.G.	H_2
3. *Methanobacterium thermoautotrophicum*	ΔH	*Methanobacterium thermoautotrophicum*	H_2
Genus II. *Methanobrevibacter*			
1. *Methanobrevibacter ruminantium* (type species)	MI	*Methanobacterium ruminantium* strain MI	H_2, formate
2. *Methanobrevibacter arboriphilus*	DHI	*Methanobacterium arboriphilicum*	H_2
Methanobrevibacter arboriphilus strain AZ		*Methanobacterium* sp. strain AZ	H_2
Methanobrevibacter arboriphilus strain DC		*Methanobacterium* sp. strain DC	H_2
3. *Methanobrevibacter smithii*	PS	*Methanobacterium ruminantium* strain PS	H_2, formate
Order II. *Methanococcales*			
Family I. *Methanococcaceae*			
Genus I. *Methanococcus*			
1. *Methanococcus rannielii* (neotype species)	SB	*Methanococcus rannielii*	H_2, formate
2. *Methanococcus roltae*	PS	*Methanococcus* sp. strain PS	H_2, formate
Order III. *Methanomicrobiales*			
Family I. *Methanomicrobiaceae* (type family)			
Genus I. *Mathanomicrobium* (type genus)			
1. *Methanomicrobium mobile* (type species)	BP	*Methanobacterium mobile*	H_2, formate
Genus II. *Methanogenium*			
1. *Methanogenium carioci* (type species)	JR1	Cariaco isolate JR1	H_2, formate
2. *Methanogenium marisnigri*	JR1	Black Sea isolate JR1	H_2, formate
Genus III. *Methanospirillum*			
1. *Methanospirillum hungatii*	JF1	*Methanospirillum hungatii*	H_2, formate
Family II. *Methanosarcinaceae*			
Genus II. *Methanosarcina* (type genus)			
1. *Methanosarcina barkeri* (type species)	MS	*Methanosarcina barkeri*	H_2, CH_3OH, CH_3NH_2, acetate
Methanosarcina barkeri strain 227		*Methanosarcina barkeri* strain 227	H_2, CH_3OH, CH_3NH_2, acetate
Methanosarcina barkeri strain W		*Methanosarcina barkeri* strain W	H_2, CH_3OH, CH_3NH_2, acetate

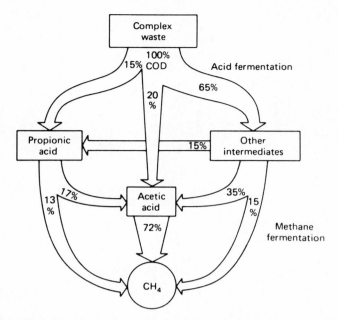

FIG. 3.24 Pathways in methane fermentation of complex wastes. Percentages represent conversion of waste COD by various routes (McCarty 1964).

1980). However, in the digestion of other wastes, the first stage of anaerobic digestion, the enzymic hydrolysis of polymers to monomers and cellulose hydrolysis in particular, has been reported as the rate-limiting step in the conversion of cellulose to methane (Chan and Pearson 1970) and in the digestion of household refuse (Pfeffer 1974). The effect of environmental factors and the inhibition of toxic compounds in the anaerobic process is considered in Section 4.4.

The utilization of the organic fraction of wastewater is directly related to methane production and vice versa. Where the exact chemical nature of the sludge is known the quantity of methane produced can be estimated by the following equation (Buswell and Mueller 1952):

$$C_nH_aO_b+\left(n-\frac{a}{4}-\frac{b}{2}\right)H_2O \rightarrow \left(\frac{n}{2}-\frac{a}{8}+\frac{b}{4}\right)CO_2+\left(\frac{n}{2}+\frac{a}{8}-\frac{b}{4}\right)CH_4$$

However, McCarty (1964) estimated the theoretical methane production from the complete stabilization of 1 kg of COD as 0.348 m^3 at standard temperature and pressure.

The volume of bacteria biomass and methane produced in an anaerobic sewage sludge digester can be calculated using stoichiometry (Section 3.1). Using the tables devised by McCarty (1975) (Tables 3.2 and 3.3), the appropriate half reactions can be selected with carbon dioxide as the electron acceptor:

Rc: $1/5CO_2 + 1/20HCO_3^- + 1/20NH_4^+ + H^+ + e^- = 1/20C_5H_7O_2N + 9/20H_2O$

Re: $1/8CH_4 + 1/4H_2O = 1/8CO_2 + H^+ + e^-$

Rd: $1/50C_{10}H_{19}O_3N + 9/25H_2O = 9/50CO_2 + 1/50NH_4^+ + 1/50HCO_3^- + H^+ + e^-$

Table 3.3 gives a $(f_s)_{max}$ value of 0.11 for sewage sludge and as the digester has a mean cell residence time of 20 days, f_s is calculated as:

$$f_s = 0.11\left(1 - \frac{0.8(0.03)(20)}{1 + 0.03(20)}\right) = 0.077$$

$$\text{As } f_s = 1 - f_e, \quad f_e = 0.923$$

Therefore, the revised reactions using the relationship $R = f_sRc + f_eRe - Rd$ are

Rc $\times f_s$: $0.0154CO_2 + 0.0039HCO_3^- + 0.0039NH_4^+ + 0.007H^+ + 0.007e^- = $
$0.0039C_5H_7O_2N + 0.035H_2O$

Re $\times f_e$: $0.115CO_2 + 0.923H^+ + 0.923e^- = 0.115CH_4 + 0.231H_2O$

Rd: $0.02C_{10}H_{19}O_3N + 0.36H_2O = 0.18CO_2 + 0.02NH_4^+ + 0.02HCO_3^- + H^+ + e^-$

When added and normalized by dividing by 0.02, the overall reaction R is:

$C_{10}H_{19}O_3N + 4.7H_2O = 0.195C_5H_7O_2N + 5.75CH_4 + 2.48CO_2 + 0.81NH_4^+ + $
$0.81HCO_3^-$

Therefore, for each mole of sewage sludge 0.195 mole of new cells are produced and 5.75 moles of methane is released.

3.4.4 Sulphide production

The production of hydrogen sulphide is the most common and well-known manifestation of anaerobiosis. Sulphide can be produced by anaerobic micro-organisms in two ways.

Protein is broken down to amino acids, and those containing sulphur, e.g. cysteine, cystine, and methionine, are degraded further with sulphide being produced. Most anaerobic bacteria are able to produce sulphide from protein, e.g. *Proteus*, *Bacteroides* spp., and some *Clostridium* spp. Although all can grow anaerobically only *Bacteroides* spp., which can be present in faeces at concentrations up to $10^{10}g^{-1}$, are obligate species (Houte and Gibbons 1966). In wastewater systems, most sulphide is produced from sulphate reduction by the anaerobic sulphate-splitting bacterium *Desulfovibrio desulfuricans*, although species of the genus *Desulfotomaculum* are also routinely isolated from digesters (Zeikus 1980). The former species is present in low numbers in sewage $(60–600 \text{ ml}^{-1})$ but rapidly increases on storage up to concentrations in excess of $100\,000 \text{ ml}^{-1}$ over 14 days.

The sulphate-reducing bacteria only utilize a restricted range of carbon compounds, such as lactate and malate, and rely on the metabolic products of other anaerobic bacteria that are able to utilize more complex organic compounds. Sulphate-reducing bacteria are found in a wide range of anaerobic environments where there is a supply of sulphate, which they utilize instead of oxygen for respiration, organic matter, and a suitable bacterial population able to utilize the organic matter to produce compounds such as lactate (Lynch and Poole 1979). Sulphate is a major ion in seawater and bacterial sulphate reduction is an important reaction in anoxic estuarine and marine environments. When heavy metals are present in the sediment where sulphate-reducing bacteria are active then the sulphide produced reacts to form insoluble salts; the black discoloration so often associated with anaerobic sediments is due largely to the formation of ferrous sulphides. If no metals are present, the sulphide escapes into the water column or the atmosphere as hydrogen sulphide. The metabolism of the sulphate-reducing bacteria has been extensively studied and reviewed (Postgate 1965; Dart and Stretton 1977). The release of hydrogen sulphide depends on the pressure, temperature, and pH. Hydrogen sulphide is the compound responsible for the unpleasant smell associated with anoxic estuaries, such as Dublin Bay and many of the estuaries receiving partially treated wastewaters. When produced in sewers, sulphide can cause severe corrosion of the concrete pipework (Section 1.3.1) and is extremely toxic, especially in enclosed environments where it is fatal to air breathing organisms including man. The sulphide can be used as a source of energy by sulphur oxidizing bacteria or as an electron donor by some of the photosynthetic bacteria. More commonly, it is chemically oxidized, thus maintaining anoxic conditions (Fig. 3.25).

3.4.5 Denitrification

Nitrate can be converted via nitrite to gaseous nitrogen under low dissolved oxygen conditions by the process known as denitrification. The process occurs in any nitrified effluent when deprived of oxygen. Therefore, unless a specific denitrifying reactor is constructed (Section 4.4), nitrification is only likely to occur in the sludge layer of the sedimentation tank after a biological reactor where nitrification occurs. The process can only proceed under anoxic conditions (when the dissolved oxygen conditions are very low but not necessarily zero), with the dissolved oxygen concentration <2% saturation (Kiff 1972), and when a suitable carbon source is available. There is enough organic carbon remaining in the treated effluent and sludge within secondary settlement units to allow denitrification to proceed, although in specifically constructed denitrifying reactors the carbon is supplied as either methanol or

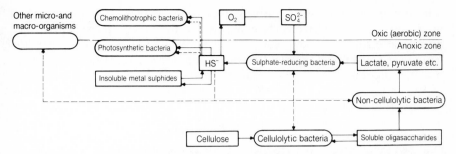

FIG. 3.25 How sulphate-reducing bacteria interact with the anoxic sediment zone (Lynch and Poole 1979).

settled sewage. The process is carried out by a wide range of facultative anaerobes, the most common being *Pseudomonas* sp., with *Achromobacterium* sp., *Denitrobacillus* sp., *Spirillum* sp., *Micrococcus* sp., and *Xanthomonas* sp. frequently present (Painter 1970). The overall stoichiometric equations for denitrification using methanol as the carbon source has been calculated by McCarty *et al.* (1969) as:

Overall nitrate removal:
$$NO_3^- + 1.08CH_3OH + H^+ \rightarrow 0.065C_5H_7O_2N + 0.47N_2 + 0.76CO_2 + 2.44H_2O$$

Overall nitrite removal:
$$NO_2^- + 0.67CH_3OH + H^+ \rightarrow 0.04C_5H_7O_2N + 0.48N_2 + 0.47CO_2 + 1.7H_2O$$

Overall de-oxygenation:
$$O_2 + 0.93CH_3OH + 0.056NO_3^- + 0.056H^+ \rightarrow 0.056C_5H_7O_2N + 0.65CO_2 + 1.69H_2O$$

3.4.6 Redox potential

Oxidation-reduction (or redox) reactions in biological reactions are normally defined in terms of loss and gain of hydrogen or electrons. Each oxidation is accompanied by a reduction and can be summarized as:

$$AH_2 \quad B$$
$$A \quad BH_2$$

where AH_2 is the hydrogen donor and B the hydrogen acceptor. Each couple $(AH_2/A$ or $B/BH_2)$ has a tendency to either donate reducing equivalents and be oxidized $(AH_2 \rightarrow A)$ or accept them and be reduced $(B \rightarrow BH_2)$. When both couples combine in a complete redox reaction, the net flow of the reaction can be determined by the relative tendency of each couple to donate or accept reducing equivalents, which is the redox potential. This can be measured using a galvanic cell consisting of two electrodes connected by a conducting solution: oxidation occurs at the negative electrode (anode) and electrons are produced, whereas electrons are consumed and reduction takes place at the positive electrode (cathode). The redox potential is quantified by comparison with a standard redox couple. By convention, the standard redox couple is that present at a hydrogen electrode consisting of a platinum electrode, with hydrogen ions in solution. In the presence of platinum as the catalyst, the reaction is:

$$H_2 \rightleftharpoons 2H^+ + 2e^-$$

and the tendency to donate reducing equivalents, as electrons in this case, is measured as the voltage (potential) of the electrical current generated, when the electrode is coupled in series with another redox couple electrode. Under standard conditions, 25°C, 1 atm of H_2 and at pH 0, the redox potential of the hydrogen electrode is zero. At pH 7, the potential of the redox couple $H_2/2H^+ + 2e^-$ is -420 mV. The symbol E_h is used for the redox potential under standard conditions. The theory of redox measurement is fully discussed by Kokholm (1981).

A couple of lower redox potential will always donate reducing equivalents to a couple of higher potential and, during the oxidation of a substrate, reducing equivalents are transferred in the direction of increasing potential. This transfer is accompanied by the release of free energy, the magnitude of which is given by the standard free energy change,

$$\Delta G^{o\prime} = -nF\Delta E_h$$

where n is the number of electrons transferred in the reaction, F the charge on one mole of electrons, which is Faraday's constant (96.649 kJV^{-1} mol^{-1}), and ΔE_h the standard electrode potential. Biological systems have evolved to conserve this energy and to convert it into biologically useful forms; i.e. oxidative phosphorylation.

In practical terms, the redox potential can be used to indicate which oxidative-reduction reactions will occur within a wastewater system and is particularly useful in the management of anaerobic systems (Fig. 3.21) (Dirasian 1968a,b). The redox potential gives a measure of the general condition of the liquid. Anaerobic processes similar to those found in sludge lagoons and digesters will have low values of $E_h (< -200$ mV$)$, whereas aerobic processes will have higher values $(> +50$ mV$)$. More precisely, values of $E_h -150$ mV to -420 mV are found in anaerobic environments, whereas aerobic environments vary between -200 mV and $+420$ mV. Facultative environments change from aerobic to anaerobic systems at about $+100$ mV (Hughes 1980). Redox potential is a more reliable measure of aerobic conditions than a dissolved oxygen concentration measurement as is often used in the control of environmental conditions in activated sludge tanks. Because odours are produced from anaerobic rather than aerobic processes, redox potential appears to be a convenient measure to determine the onset of odour production during the treatment of high strength wastes, such as animal wastes. Odours at a redox potential that is far more negative than that for zero-dissolved oxygen. Thus, the control of an aeration system to prevent odour production will be far more effective if based on redox potential rather than on dissolved oxygen concentration. This is because of the closer correlation between redox potential and odour breakthrough, compared with dissolved oxygen concentration and odour breakthrough. In practice, a minimum redox potential of $E_h = +40$ mV has been shown to be most efficient, although minimum thresholds of -100 mV have been used. Therefore, by automatically aerating the surface wastewater when E_h falls below $+40$ mV, no odours are produced and, by using redox potential instead of dissolved oxygen concentration, savings in aeration costs of up to 75% have been achieved (Barnes *et al.* 1985).

Most aquatic sediments are prone to low oxygen conditions due to the problems of oxygen diffusion from the water column into the sediment. This depends on the oxygen concentration of the water, the physical and chemical nature of the sediment, and the burrowing activities of bottom-dwelling fauna. Once the sediment becomes anoxic the fauna population will become progressively excluded. Under these conditions, the redox potential is a useful indicator of which electron acceptor is being used by the anaerobic bacteria present in the sediment (Fig. 3.26).

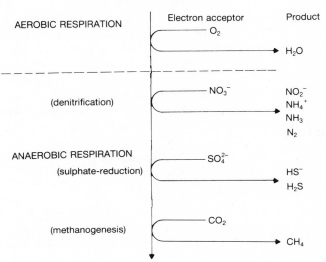

FIG. 3.26 The redox potential of water and sediment indicates which inorganic electron acceptor is being used by the microbial community (Lynch and Poole 1979).

Further reading

General: Lynch and Poole 1979; Benefield and Randall 1980; Stafford *et al.* 1980; Schugerl 1987.
Organisms and physiology: Wolfe 1971; Morris 1975; Zeikus 1977, 1980.
Biochemistry: Thauer *et al.* 1977.
Denitrification: Barnes and Bliss 1983; Winkler 1984.
Redox: Dirasian 1968*a,b*; Kokholm, 1981; Barnes *et al.* 1985.

3.5   AUTOTROPHIC MICRO-ORGANISMS

3.5.1   Introduction

Unlike heterotrophic micro-organisms, autotrophs are unable to utilize organic matter. They use carbon dioxide or carbonate as their sole carbon source for the synthesis of cellular material, and obtain energy for metabolism by the oxidation of reduced inorganic compounds (chemo-autotrophs) or by photosynthesis (photo-autotrophs).

Chemo-autotrophic bacteria are able to oxidize a range of elements that are present in a reduced form (i.e. low oxidation state) (Table 3.14). However, the conditions required for the active growth of these bacteria are critical, and in some ways conflicting. The group requires a plentiful supply of the reduced inorganic compound (e.g. NH_3, NO_2^-, H_2S), oxidizing agent (e.g. O_2 or for some species NO_3^-), which must have a more positive redox potential than the reducing agent, and carbon dioxide as the carbon source. Chemo-autotrophs are aerobic organisms, although they can grow at low partial pressures of oxygen and some species are able to utilize other oxidizing agents. Unlike the photo-autotrophs, they do not require light and do not contain pigments. However, they are restricted in wastewater treatment units to the areas where there are plentiful supplies of reduced inorganic compounds, carbon dioxide, and some oxygen. As a group, they can oxidize a wide range of reduced inorganic compounds, although individual species are highly specific

TABLE 3.14   Primary energy yielding reactions of chemo-autotrophic bacteria

	Electron donor	Product	ΔG° $(kJ\,mol^{-1})$	Major genus
Sulphur bacteria	$S^2 + 2O_4^{2-} \rightarrow$	SO_4^{2-}	-715	*Thiobacillus*
	$S^0 + 1.5O_2 + H_2O \rightarrow$	$SO_4^{2-} + 2H^+$	-502	*Thiobacillus*
Nitrifying bacteria	$NH_3 + 1.5O_2 \rightarrow$	$NO_2^- + H_2O + H^+$	-274	*Nitrosomonas*
	$NO_2^- + 0.5O_2 \rightarrow$	NO_3^-	-73	*Nitrobacter*
Hydrogen bacteria	$H_2 + 0.5O_2 \rightarrow$	H_2O	-234	*Hydrogenomonas*

in the reactions they catalyse and in the environmental conditions they require. The energy metabolism of autotrophs is essentially the same as heterotrophs, with ATP formation by oxidative phosphorylation. The electrons produced from the oxidation of reduced inorganic molecules enter the electron transport system at the level of cytochrome-c (Fig. 3.19). In all chemo-autotrophic reactions, oxygen is the terminal electron acceptor, and the subsequent electron flow through the electron transport chain, via a series of biological redox carriers in exergonic reactions to oxygen, will produce a negative free energy change which is used for the synthesis of ATP (Anderson 1980). For example, in the case of the oxidation of NO_2^- to NO_3^- only one ATP molecule per NO_2^- unit oxidized is produced:

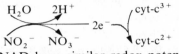

Because ADP and NAD have similar redox potentials, $NADH_2$ can be synthesized instead of ATP as cytochrome-c is reduced (Fig. 3.27).

Photo-autotrophs use free carbon dioxide as a carbon source and derive energy from sunlight. There are many photo-autotrophic bacteria present in small numbers in wastewater treatment systems, but the most important group is the eukaryotic photo-autotrophs which include the true algae, bryophytes, and the higher plants. With the exception of bryophytes, algae are the only eukaryotic phototrophs found in conventional wastewater treatment systems, although macrophytes are used in warmer climates (Section 4.5).

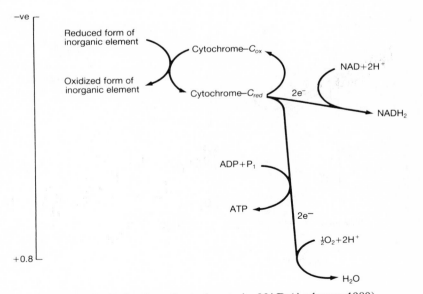

FIG. 3.27 Reduction of cytochrome by NAD (Anderson 1980).

Although algae are found in small numbers in most aerobic treatment systems they are particularly important in waste stabilization ponds where they provide oxygen for heterotrophic and chemo-autotrophic bacteria (Section 4.3). Like other plants, algae use free carbon dioxide as a carbon source and derive all their energy requirements from sunlight by the process of photosynthesis (photophosphorylation). Essentially, water is oxidized by releasing electrons which are then used to reduce carbon dioxide to a carbohydrate. Two basic reactions occur, a light and a dark reaction, both of which occur during the hours of daylight, although the dark reaction utilizes the product of the light reaction which is dependent on light energy. Short-wave light energy is absorbed in the light reaction by chlorophyll, which supplies the energy to oxidize water, releasing oxygen and electrons. The energy level of the electrons is reduced by passing through an electron transport system, which releases energy in the form of ATP. The electrons are re-energized by the absorption of long-wave light energy by pigment P700 in the plant tissue. This is used to reduce nicotinamide adenine dinucleotide phosphate (NADP) to NADPH, which provides the energy to drive the dark reaction (Fig. 3.28). The ATP and NADPH produced by the light reaction are utilized to reduce six molecules of carbon dioxide to a hexose molecule in the dark reaction with the enzyme 6-ribulose-1,5-diphosphate required to catalyse the reaction that is regenerated at the end of the cycle:

$$6\text{-Ribulose-1,5-diphosphate} + CO_2 + 18ATP + 12NADPH + 12H^+ \rightarrow$$
$$6\text{-Ribulose-1,5-diphosphate} + hexose + 18P_i + 18ADP + 12NADP$$

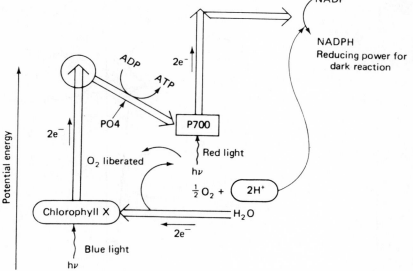

FIG. 3.28   Electron flow during photosynthesis (Benefield and Randall 1980).

There are important differences in the pigments between various groups of eukaryotic photo-autotrophs, which reflect their adaptation to the environment in which each grows. For example, green algae are found close to the surface of the water in waste stabilization ponds where they absorb red light, whereas red algae grow at lower depths in ponds where there is poor penetration of red light and they use the pigment phycoerythrin to absorb light of shorter wavelengths.

Two major groups of photo-autotrophic bacteria are associated with wastewater treatment. The green photosynthetic bacteria, the most important family being the *Chlorobacteriaceae*, are the simplest of the photo-autotrophs. They contain the pigment bacteriochlorophyll and require carbon dioxide, light, anaerobic conditions, and either hydrogen or a reduced form of sulphur (e.g. H_2S) for growth. They are commonly isolated from anaerobic environments rich in hydrogen sulphide, such as sludge digesters or the sludge layer in waste stabilization ponds, there they are referred to as green photosynthetic sulphur bacteria. In the oxidation of hydrogen sulphide by *Chlorobium* sp., elemental sulphur (S°) is produced, which is released into the environment:

$$2H_2S + CO_2 \xrightarrow{\text{light }(h\upsilon)} [CH_2O] + 2S^\circ + H_2O$$
$$\text{(new cells)}$$

Hydrogen sulphide is not a sufficiently strong reducing agent (i.e. the redox potential of hydrogen sulphide is insufficiently negative) to reduce carbon dioxide. Oxidation is, in fact, light-dependent, with visible light absorbed by the bacteriochlorophyll, which oxidizes the hydrogen sulphide, releasing electrons that are subsequently trapped by a compound with a more negative redox potential (Anderson 1980).

Purple photosynthetic bacteria also contain the green pigment bacteriochlorophyll but appear red or purple due to the presence of carotenoid pigments. There are two main groups. The purple sulphur bacteria or the *Thiorhodaceae* (e.g. *Chromatium* sp.) are motile and require light, anaerobic conditions, and either hydrogen sulphide or hydrogen for the assimilation of carbon dioxide. Species of this group occur in the sludge layers of waste stabilization ponds and produce elemental sulphur from hydrogen sulphide, which is deposited in bacterial cells where it accumulates. The purple non-sulphur bacteria, the *Athiorhodaceae* (e.g. *Rhodopseudomonas* sp. and *Rhodospirillum* sp.) are unable to use reduced forms of sulphur but utilize hydrogen instead. They are also restricted to anaerobic environments and require light and carbon dioxide. This group is also capable of utilizing simple organic compounds such as acetate instead of carbon dioxide, although the assimilation of the organic carbon is still enhanced by light. This ability shows that all photosynthetic organisms are autotrophic and that this group is more accurately described as being photoheterotrophic.

The blue-green algae, *Cyanophyceae*, are not an important group in wastewater treatment, although they are occasionally isolated from aerobic and anaerobic systems. Under anaerobic conditions, they use light to reduce gaseous nitrogen to ammonia, a process known as nitrogen fixation, whereas under aerobic conditions they carry out light-dependent carbon dioxide assimilation without the use of any strong reducing agent, such as hydrogen sulphide or hydrogen. They produce oxygen at the same rate as carbon dioxide is assimilated.

3.5.2   Nitrification

The principal sources of organic nitrogen are domestic wastes, animal slurry from intensive farming, and high-protein wastes from certain processing industries, especially the meat trade. In the sewerage system, organic nitrogen is rapidly de-aminated and urea is hydrolysed by the enzyme urease to release ammonia:

$$\begin{array}{l} NH_2 \\ | \\ C=O+2H_2O \xrightarrow{\text{urease}} (NH_4{}^+):CO_3{}^{2-} \\ | \qquad\qquad\qquad\qquad\text{(ammonium carbonate)} \\ NH_2 \\ \text{(urea)} \end{array}$$

By the time raw domestic sewage enters the treatment plant, 90% of the nitrogen is either present as ammonia or unstable organic compounds, which are readily transformed to ammonia (Culp *et al.* 1978). Domestic sewage, with an ammonia–nitrogen concentration of 35 g m^{-3}, is extremely weak in terms of ammonia–nitrogen concentration compared with other nitrogen-rich wastewaters, due to the large dilution it receives.

Domestic sewage has nitrogen in excess of the microbial requirement to oxidize the amount of carbon present (Section 1.2), therefore only part of the nitrogen is removed by conventional heterotrophic activity, being incorporated into the microbial biomass. This is also true for phosphorus. The residual nitrogen stimulates autotrophic activity, which, if discharged into a watercourse, will be in the form of photo-autotrophic activity, i.e. eutrophication. The utilization of nitrogen by photo-autotrophs produces a large quantity of biomass in the form of algae, because the proportion of nitrogen in the biomass is small. The problem is that the nutrient is assimilated into the cells, for the synthesis of amino acids, enzymes, and nucleic acids, thus a direct relationship between nutrient removal and biomass production exists. Only a small proportion of the ammonia–nitrogen is assimilated into the heterotrophic biomass during wastewater treatment and under low loading conditions in most fixed-film and mixed biological treatment units the remainder is oxidized by chemo-autotrophic bacteria. Autotrophic bacteria are able to use the nitrogen in a non-assimilative way, as an energy source, so only small amounts of biomass are produced. Ammonia, the reduced form of nitrogen, is oxidized by autotrophic nitrifying bacteria to nitrate, via nitrite, a process known as nitrification.

The microbial oxidation of ammonia and ammonium ions occurs in two distinct stages, each involving different species of chemo-autotrophic nitrifying bacteria. The chemo-autotrophs utilize ammonia or nitrite as an energy source, oxygen as the terminal electron acceptor, ammonia as the nitrogen source, and carbon dioxide or carbonate as a carbon source.

The first stage is the oxidation of ammonium ions to nitrite (nitrosification):

$$NH_4^+ + 1.5O_2 \rightarrow NO_2 + 2H^+ + H_2O \quad -58 \text{ to } 84 \text{ kcal}$$

This reaction is generally considered to be catalysed by the genus *Nitrosomonas*, and indeed two species of this genus *N. europa* and *N. monocella* are frequently isolated. However, other genera have also been identified as being able to carry out the first stage of nitrification, these include *Nitrosococcus*, *Nitrosospira*, *Nitrosocystis*, and *Nitrosogloea* (Belser 1979). The hydrogen ions released in the oxidation of ammonia to nitrite lowers the pH, and during nitrification a discernible fall in the pH of the wastewater can be recorded. This can be a problem in enclosed systems, or systems with a long retention time, where there is sufficient ammonia present for nitrification to occur as the pH will be reduced until nitrification is inhibited and eventually stopped.

In the second stage, nitrite is oxidized to nitrate:

$$NO_2^- + 0.5O_2 \rightarrow NO_3^- \quad -15.4 \text{ to } 20.9 \text{ kcal}$$

The genus *Nitrobacter* is considered to be responsible for the second nitrifying reaction, with *N.winogradskyi* being the only well-defined species. A second species, *N. agilis* has been isolated more recently but whether it is distinct enough from *N. winogradskyi* to be considered a separate species is still under debate. *Nitrocystis*, *Nitrococcus*, and *Nitrospina* have also been cited as being able to oxidize nitrite to nitrate (Belser 1979; Winkler 1981).

The overall nitrification reaction shows that the oxidation of ammonia to nitrate requires a high input of oxygen, about 4.5 kg for each kg of ammonia–nitrogen (NH_4–N) oxidized:

$$NH_4^+ + 2O_2 \rightarrow NO_3^- + 2H^+ + H_2O \quad -73.4 \text{ to } 104.9 \text{ kcal}$$

There is some evidence that the nitrifying bacteria may not be obligate autotrophs, but have the ability to utilize certain organic compounds, as do the heterotrophic nitrifying bacteria found in soil (e.g. *Arthrobacter* sp., *Aspergillus* sp.).

Kinetics and environmental factors

Growth of nitrifying bacteria is represented by Monod kinetics (Section 3.1.2):

$$\mu = \frac{\mu_m \cdot S}{K_s \cdot S}$$

where μ is the specific growth rate of the nitrifying bacteria, μ_m the maximum specific growth rate, K_s the saturation constant, and S the residual concentration of the growth-limiting nutrient.

As nitrite is not accumulated under steady-state conditions the rate-limiting step in nitrification is assumed to be the oxidation of ammonia to nitrite. Therefore, it is more convenient to model nitrification on the specific growth rate of *Nitrosomonas* (μ_{NS}):

$$\mu_{NS} = \mu_{m \cdot NS} \frac{[NH_4-N]}{K_N + [NH_4-N]}$$

where $\mu_{m \cdot NS}$ is the maximum specific growth rate of *Nitrosomonas* (d^{-1}), NH_4-N the ammonia–nitrogen concentration of the wastewater in the reactor ($g\ m^{-3}$), and K_N the saturation constant ($g\ m^{-3}$) (Table 3.15).

However, the specific growth rate, thus the rate of nitrification, is affected by a number of environmental factors. In particular, nitrification is inhibited by short retention times, low dissolved oxygen concentrations, low temperatures, a wide range of inorganic and organic compounds, extreme pH, and deficiencies of key nutrients.

Nitrifying bacteria, when compared with the heterotrophic organisms, are very much slower growing. Nitrification, therefore, proceeds at a much slower rate, 3–4 times slower in fact, than carbonaceous oxidation (Section 1.4). In order to maintain an effective population of nitrifying bacteria within a biological reactor, a long retention time is required. This provides sufficient contact between the wastewater and the bacteria to ensure maximum nitrification. Also, a long retention time or sludge residence time (SRT) prevents the rate of loss of nitrifying organisms exceeding the rate of production of new organisms, thus maintaining the nitrifying population.

TABLE 3.15 Normal values for the various bio-kinetic constants applicable to the nitrification process (Hultman 1973)

Constant	Value
$(\mu_{max})_{NS[20°C,\ (pH)_{opt}]}$	0.3–0.5 d^{-1}
K_N	0.5–2.0 mg/l
Y_N	≈ 0.05 mg VSS/mg $[NH_4^+-N]$
$(pH)_{opt}$	8.0–8.4

Nitrification is generally only associated with low-rate loadings, with the degree of nitrification being progressively lost as loading increases. In the activated sludge process the shortest possible SRT to avoid wash-out of the nitrifiers and to maintain nitrification is determined by the specific growth rate of the organism, which is dependent on the other environmental or operating factors, especially temperature and pH. A generalized expression of the estimation of the minimum SRT $(t_{s(min)})$ at temperature $(T^{\circ}C)$ has been devised by Marais (Jones and Sabra 1980):

$$t_{s(min)} = 3.05 \times (1.127)^{(T-20)} \, d^{-1}$$

Dissolved oxygen is used as the terminal electron acceptor by nitrifying organisms and, in general terms, nitrification is inhibited at low dissolved oxygen concentrations. In the activated sludge process, it is generally accepted that nitrification does not occur below 0.2–$0.5 \, mg \, l^{-1}$. However, no inhibition is found at oxygen concentrations $>1.0 \, mg \, l^{-1}$ (Wild et al. 1971), and as long as the aeration system is adjusted to maintain a minimum dissolved oxygen concentration of $2 \, mg \, l^{-1}$, then the effects of dissolved oxygen on nitrification can be neglected. In high purity oxygen systems, high dissolved oxygen concentrations can be achieved. Nitrification does not appear inhibited up to concentrations of $20 \, mg \, l^{-1}$ (Winkler 1981), although there is some evidence to suggest that above this oxygen concentration some inhibition may occur. Other workers, however, have reported nitrification at even higher dissolved oxygen concentrations, up to a maximum of $60 \, mg \, l^{-1}$ (Haug and McCarty 1972). Little or no loss in nitrifying ability occurs if activated sludge is stored anaerobically for short periods $<4 \, h$, although over longer periods $>24 \, h$ the nitrifying bacteria are killed (Department of Scientific and Industrial Research 1963a). Clearly, oxygen is a growth-limiting substrate in terms of nitrifying bacteria, and this can be expressed using a Monod-type relationship:

$$\mu_{NS} = \mu_{m \cdot NS} \frac{[DO]}{K_{O_2} + [DO]}$$

where DO is the dissolved oxygen concentration and K_{O_2} the oxygen saturation coefficient. Reported values of K_{O_2} are in the range of 0.15–$2.0 \, g \, m^{-3}$ for nitrification in the activated sludge process. A mid-value of $1.3 \, g \, m^{-3}$ is usually used, although Eckenfelder and Argaman (1978) have reported a K_{O_2} value of approx. $1.0 \, g \, m^{-3}$. The effect of dissolved oxygen on nitrification has been excellently discussed elsewhere (de Renzo 1978; Benefield and Randall 1980; Stenstrom and Poduska 1980; Winkler 1981).

The overall rate of nitrification has been shown to decrease with a decrease in temperature (Downing and Hopwood 1964). *Nitrosomonas* isolated from activated sludge has an optimum growth rate at $30^{\circ}C$ (Loveless and Painter 1969), although a slightly higher range has also been reported 30–$35^{\circ}C$ (Buswell et al. 1954). There is little growth below $5^{\circ}C$ and no growth of

Nitrobacter below 4°C. *Nitrobacter* has a slightly higher optimum temperature for growth at 35°C, although maximum growth has been reported up to
42°C (Deppe and Engel 1960; Laudelout and Tichelen 1960). Thus, the
specific growth rate of *Nitrosomonas* ($\mu_{M \cdot NS}$) and the saturation constant K_N
are affected by the temperature according to the relationship:

$$\mu_{m \cdot NS} = \mu_{m \cdot NS \ (15°C)} e^{0.95(T-15)}$$

and $$K_N = 10^{0.051(T)-1.158}$$

where $\mu_{m \cdot NS}$ is the maximum specific growth rate of *Nitrosomonas* at operating
temperature $T°C$ (d^{-1}) and $\mu_{m \cdot NS \ (15°C)}$ is the maximum specific growth rate of
Nitrosomonas at 15°C (d^{-1}), and K_N is the saturation constant in g m^{-3} as N.
Therefore, the specific growth rate of *Nitrosomonas* should always be adjusted
for temperature before use in the calculation of nitrification rates or yields. The
Q_{10} value (20–30°C) for *Nitrosomonas* in pure culture is ~ 1.8 (Buswel *et al.*
1954) but 3.3 in activated sludge (Downing and Hopwood 1964).

Organic matter is known to inhibit nitrification and increases in organic
loadings result in rapid decreases in the rate of nitrification. This is probably
due to the increased activity of heterotrophs, which because of their more
rapid growth rates, successfully compete with the nitrifying bacteria for
dissolved oxygen and nutrients. This direct competition from heterotrophs, or
photo-autotrophs if light is available (Winkler and Cox 1980), is a major cause
of nitrification failure in biological treatment systems. Most organic and
inorganic compounds, especially metals, inhibit nitrification (Painter 1977;
Stankewich and Gyer 1978).

Nitrification is favoured by mildly alkaline conditions, pH 7.2–9.0, with an
optimum pH of between 8.0 and 8.4. Below pH 8 the rate of nitrification
decreases becoming completely inhibited at pH < 5 (Wild *et al.* 1971), even
though nitrifying bacteria can be acclimatized to slightly more acidic pH
values. Acclimatization to a different pH may take several weeks. For
example, a pH shift from 7 to 6 required 10 days of acclimatization before
nitrification returned to its former rate (Haug and McCarty 1972).
Nitrosification produces hydrogen ions and if there is insufficient alkalinity
present to buffer the wastewater, as would be the case in a soft water catchment
area and when there was a high concentration of ammonium salts in the
wastewater, then the pH will gradually decrease. However, this phenomenum
is not restricted to soft waters. Painter (1983), cites an example of a domestic
sewage in a hard water area with a hardness of 300–400 mg $CaCO_3$ l^{-1} in
which 60 mg NH_4-N l^{-1} was being oxidized, resulting in pH values as low as
5.0–5.5. This decrease in pH will not only inhibit nitrification but also
heterotrophic activity, reducing BOD_5 removal. Thus, when this occurs a
characteristic cycle in effluent quality can be identified, with high nitrate and
low BOD_5 concentrations alternating with low nitrate and high BOD_5
concentrations. There is very little carbon dioxide in the atmosphere, about

0.05% by weight, and instead the carbon dioxide produced from heterotrophic activity is utilized by the nitrifying organisms. However, due to the acidity of nitrification, the carbon dioxide from heterotrophic activity is often in the form of carbonate or bicarbonate ions. In purely nitrifying systems (e.g. nitrifying filters) where there is no associated heterotrophic activity, a supplementary source of carbon is required, normally in the form of a carbonate or bicarbonate supplement. Therefore, if nitrifying bacteria consume carbonate ions, then the reduction in alkalinity may seriously affect the buffering capacity of the system. The maximum specific growth rate of *Nitrosomonas* ($\mu_{m \cdot NS}$) can be adjusted for pH variation by using the equation devised by Hultman (1971):

$$\mu_{m \cdot NS} = \frac{(\mu_{m \cdot NS} \text{ at optimum pH})}{1 + 0.04(10(pH_x - pH) - 1)}$$

where pH_x is the optimum pH for growth of *Nitrosomonas* (usually taken as 8.2) and pH is the operating pH.

The process of nitrification does not remove nitrogen from the wastewater, it transforms it from ammonia to nitrate, with the latter being discharged in soluble form in the final effluent. During the process of nitrification the ammonia and nitrite are used as an energy source and little biomass is produced in terms of either NH_4–N or NO_2–N removed (Water Research Centre 1971):

Stage 1:
$$55NH_4^+ + 5CO_2 + 76O_2 \rightarrow C_5H_7O_2N + 54NO_2^- + 52H_2O + 109H^+$$
$$\text{(bacterial biomass)}$$

Stage 2:
$$400NO_2^+ + 5CO_2 + NH_4^+ + 195O_2 + 2H_2O \rightarrow C_5H_7O_2N + 400NO_3^- + H^+$$
$$\text{(bacterial biomass)}$$

Thus, in the oxidation of 1 kg of NH_4–N only 150 g (dry weight) of bacterial biomass in the form of *Nitrosomonas* is produced in the first stage and 20 g (dry weight) of *Nitrobacter* in the second stage. Only approximately 2% of the original NH_4–N is assimilated into the biomass, this being used to form new cellular material, therefore all the nitrification reactions can be considered as stoichiometric.

The biomass of *Nitrosomonas* produced (Y_N) and the rate of ammonium oxidation can be expressed as:

$$\mu_{NS} = Y_N(q)_{NS}$$

where μ_{NS} is the specific growth rate of *Nitrosomonas* (g m^{-3}), Y_N the yield coefficient, or the biomass of *Nitrosomonas* produced per unit of ammonium oxidized, and $(q)_{NS}$ the specific ammonium oxidation rate (d^{-1}).

Typical kinetic coefficients for stage 1 (*Nitrosomonas*), stage 2 (*Nitrobacter*), and overall nitrification reactions in the activated sludge process are given by Metcalf and Eddy (1984) (Table 3.16). The kinetic models in this section can be used for any stage in the nitrification process.

TABLE 3.16   Typical kinetic coefficients for nitrification in the activated sludge process obtained from pure culture experiments[a] (Metcalf and Eddy 1984)

Coefficient	Basis	Value	
		Range	Typical[b]
Nitrosomonas			
μ_m	d^{-1}	0.3–2.0	0.7
k_s	NH_4^+–N, mg l^{-1}	0.2–2.0	0.6
Nitrobacter			
μ_m	d^{-1}	0.4–3.0	1.0
k_s	NO_2^-–N, mg l^{-1}	0.2–5.0	1.4
Overall			
μ_m	d^{-1}	0.3–3.0	1.0
k_s	NH_4^+–N, mg l^{-1}	0.2–0.5	1.4
Y	NH_4^+–N, mg VSS[c] mg^{-1} d^{-1}	0.1–0.3	0.2
k_d		0.03–0.06	0.05

[a] Values for nitrifying organisms in activated sludge will be considerably lower than the values reported in this table.
[b] Values reported are for 20°C.
[c] VSS = volatile suspended solids.

Similar to other micro-organisms, the nitrifying bacteria require trace elements; *Nitrosomonas* require trace quantities of calcium, magnesium, and copper (Loveless and Painter 1968); and *Nitrobacter* trace quantities of molybdenum (Finstein and Delwiche 1965). *Nitrosomonas* and *Nitrobacter* are more susceptible to inhibition than heterotrophic micro-organisms and nitrification in percolating filters is less susceptible to inhibition compared with the activated sludge process (Department of Scientific and Industrial Research 1963a). However, details on inhibition are difficult to obtain because in mixed cultures (i.e. within the biological treatment unit) the concentration of a toxic compound causing inhibition is much higher than is found in pure culture studies. Also, nitrifying bacteria are able to acclimatize to much higher concentrations of the inhibitory compound if they are allowed to become accustomed slowly to increasing concentrations. This is discussed fully by Painter (1983) who cites the example of thiourea toxicity to *Nitrosomonas*. Thiourea is thought to utilize any available copper, which is an essential requirement for an enzyme system in *Nitrosomonas*. The inhibition of *Nitrosomonas* will also reduce the *Nitrobacter* population, which grows on the product of the former genus, thus thiourea (or allylthiourea) is used as an inhibitor of nitrification in the BOD_5 test to differentiate between carbonaceous oxidation and ammonia oxidation. In pure culture, the genus is completely inhibited at 0.5 mg l^{-1} thiourea, but in activated sludge 10 times this concentration can be tolerated. In time, *Nitrosomonas* can build up a tolerance to thiourea up to 92 mg l^{-1}. Although pure culture studies have shown nitrifying bacteria to be slightly sensitive to a wide range of inorganic and organic compounds, their ability to acclimatize to relatively high concentrations means that, in practice, they are rarely inhibited by the range of ions found under normal wastewater conditions. A comprehensive list of chemicals that cause nitrification inhibition during sewage treatment has been prepared by Richardson (1985).

Further reading

General: Anderson 1980; Painter 1983.
Nitrification: Wild *et al.* 1971; Hultman 1973; Poduska and Andrews 1975; Sharma and Ahler 1977; Hall and Murphy 1980.

4

BIOLOGICAL PROCESSES

4.1. FIXED-FILM REACTORS

It is clear, when examining stones or other submerged objects in an enriched river or the wall of a sewer, that micro-organisms will readily colonize any suitable surface providing that sufficient nutrients are present. This principle has been utilized in fixed-film or attached growth systems where the microbial biomass is present as a film which grows on the surface of an inert and solid medium. Purification is achieved when the wastewater is brought into contact with this microbial film. Because the active biomass is largely retained within the reactor there is no need to recirculate any displaced biomass back to the reactor in order to maintain a sufficient density of micro-organisms, as is the case in the activated sludge process (Section 4.2). The required contact between the film and the wastewater is achieved in most fixed-film reactors by allowing the wastewater to pass over the stationary medium in which the film has developed. However, it is not essential for the medium to be stationary, and in recently developed reactors the medium itself moves through the wastewater. Fixed-film reactors are designed as secondary treatment processes to partially treat (high-rate filter) or fully treat settled wastewater (percolating filter, rotating biological contractor, anaerobic filter), and for tertiary treatment to provide nitrification or denitrification. The most widely used fixed-film reactor is the percolating filter, which is considered in detail below.

4.1.1 Percolating filters

The design and function of biological filters has been described by numerous workers (Bruce 1969; Warren 1971; Pike 1978; Bruce and Hawkes 1983). In its simplest form, the filter consists of a bed of graded hard material, 'filter medium', about 2 m deep. The medium has interstices or voids that allow air and applied wastewater to reach all parts of the bed. The filter has a ventilating system to ensure free access of air to the bed and a distributor to regulate the volume and frequency of application of the sewage (influent) over the surface. The medium provides the necessary base for the attachment of non-motile micro-organisms, principally bacteria and fungi, which form a film. Motile organisms, both micro- and macroscopic, live in the shelter of the interstices, feeding on and controlling the accumulated film. The action of this grazing fauna prevents heavy film growths blocking the interstices (Hawkes 1963), which would cause ponding and anaerobic conditions within the filter bed. The accumulation of the film follows a seasonal pattern, becoming thicker during the winter months. The action of the grazing fauna loosens and breaks down the film, resulting in a large removal of film each spring which is known as sloughing. The nutrients in the wastewater promote the growth of the micro-organisms that the film is comprised of, thus, as the wastewater percolates downwards over the surface of the film-covered medium, biological oxidation and conversion takes place.

Percolating filters are the most widely used secondary treatment process in Europe being equally common in other temperate regions including the USA, where they are employed at over 3700 separate municipal sewage treatment plants. Although biofiltration was historically the first process used, it still has certain advantages over the activated sludge process (Section 4.2). Filters require virtually no skilled maintenance or close control. In energy terms, percolating filters are more economical than the activated sludge process, and are more versatile in responding to changes in flow and character of wastewater (Hawkes 1963). Filters are more tolerant of continual or shock discharges of certain pollutants compared with activated sludge (Cook and Herning 1978), including toxic industrial wastes containing heavy metal ions, phenols, cyanides, sulphides, and formaldehyde. They are widely used for both total and partial treatment of a wide range of industrial wastewaters (Bruce 1969; Calley et al. 1987; Pike 1978). Their major disadvantage is capital cost, being normally uneconomic in serving populations greater than 50 000. This is due not only to higher capital costs compared to activated sludge systems but also to the larger area of land they occupy, which is often at a premium in urban areas (Jeger 1970) (Table 4.1). For this reason, the activated sludge process predominates at very large sewage treatment works. Although the proportions of the population of England and Wales served by these two bio-oxidation processes are about the same, many more of the 5000 or so sewage treatment plants use percolating filters rather than the activated sludge process (Institute of Water Pollution Control 1972). Bruce (1969) concluded, in his review on percolating filtration, that there was no indication that the use of percolating filtration was likely to be outmoded and this remains true even with the introduction of new processes and the development of packaged plants.

The term 'percolating filter' in Ireland and the UK is still used, although there are many derivations of the name such as 'biological filter', 'bacteria bed', 'percolater', and, in the USA, 'trickling filter'. However, the term 'filter' often causes confusion as the process is essentially biological, even though there is some physical removal of fine solids, but when the process was first developed it was thought that purification was brought about by physical filtration, as in a sand filter. Percolating filtration was developed from such physical filter processes. Experiments were carried out at the Lawrence Experimental

TABLE 4.1 Comparison of the activated sludge and percolating filter processes

	Activated sludge	Percolating filtration
Capital cost	Low	High
Area of land	Low advantageous where land availability is restricted or expensive	Large: 10 times more area required
Operating cost	High	Low
Influence of weather	Works well in wet weather, slightly worse in dry weather, less affected by low winter temperatures	Works well in summer but possible ponding in winter
Technical control	High: the microbial activity can be closely controlled; requires skilled and continuous operation	Little possible except process modifications. Does not require continuous or skilled operation
Nature of wastewater	Sensitive to toxic shocks, changes in loading, and trade wastewaters; leads to bulking problems	Strong wastewaters satisfactory, able to withstand changes in loading and toxic discharges
Hydrostatic head	Small: low pumping requirement, suitable for site where available hydraulic head is limited	Large: site must provide natural hydraulic head otherwise pumping is required
Nuisance	Low odour and no fly problems. Noise may be a problem both in urban and rural areas	Moderate odour and severe fly problem in summer possible. Quiet in operation
Final effluent quality	Poor nitrification but low in suspended solids except when bulking	Highly nitrified, relatively high suspended solids
Secondary sludge	Large volume, high water content, difficult to dewater, less stabilized	Small volume, less water, highly stabilized
Energy requirement	High: required for aeration, mixing, and maintaining sludge flocs in suspension and for recycling sludge	Low: natural ventilation, gravitational flow
Synthetic detergents	Possible foaming, especially with diffusers	Little foam
Robustness	Not very robust, high degree of maintenance on motors, not possible to operate without power supply	Very sturdy, low degree of maintenance, possible to operate without power

Station of the Massachusetts State Board of Health (USA) using small filter beds to test the efficiency of various soils and aggregates, e.g. different grades of sand and gravel, to physically remove solids from wastewater. They found that effective purification of settled sewage could be achieved using quite coarse gravel (19–25 mm grading) and at very high rates of application ($0.2–0.5 \ m^3m^{-2}d^{-1}$), so that it was not always necessary to use sand filtration or land treatment, both of which required much lower hydraulic loading rates to be as effective. Although it was quickly realized that this new type of treatment must involve a degree of biological oxidation, it was not until the process was well established that the physical role of filtration was found to be minor in comparison to the biological function, by which time the terms, 'trickling' and 'percolating filtration' had become established. The impact on sewage treatment practice was immediate, because this was a process that required only one-tenth of the area that existing land treatment required and, as the beds were specially constructed, treatment became independent of the suitability of the soil for the first time. The first percolating filters were constructed in Salford (England) in 1892. This first installation was designed by Corbett and it was here that many of the practical problems of design and construction were solved. The first municipal trickling filter installed in the USA was in Atlanta, Georgia, and was commissioned in 1903 (Stanbridge 1976; Bruce and Hawkes 1983).

4.1.1.1 Design and operation

Design

The design of percolating filters has changed very little since they were first introduced and, with a working life of 80 years, many of the original filters are only now reaching the end of their useful lives. However, replacement is more often due to an increase in the loading in excess of their original design, rather than structural or mechanical failure (Fig. 4.1). The major factors that need to

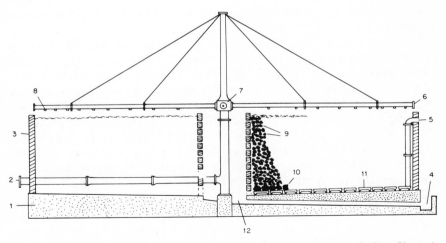

FIG. 4.1 Basic constructional features of a conventional percolating filter. 1, foundation floor; 2, feed pipe; 3, retaining wall; 4, effluent channel; 5, ventilation pipe; 6, distributor arm; 7, rotary seal; 8, jets; 9, main bed of medium; 10, base layer of larger medium; 11, drainage tiles; 12, central well for effluent collection (Bruce and Hawkes 1983).

be taken into account in the design and operation of percolating filters are: medium type, and, in particular, specific surface area and voidage, depth, area of the bed, organic, and hydraulic loading rates.

The depth of filters is arbitrary, and when all types of percolating filtration are taken into consideration there is a total range of between 0.9 and 15.0 m. The standard (low-rate) percolating filter in the UK is usually about 1.8 m deep and rarely less than 1.5 m or in excess of 2.5 m. Filter beds in the rest of Europe tend to be deeper than this and in Germany most filters are between 3–4 m in depth. Low-rate filters in the USA are a similar depth to those in the UK, at between 1.5–2.0 m, although shallow beds (< 1 m) are used for high-rate filtration. High-rate filters using modular plastic medium are used to treat industrial wastewaters. The medium is very light and can be stocked, like building blocks, into a tall tower which is essentially free-standing. These filter towers are housed in lightweight prefabricated material or are built out of breeze blocks and can be constructed as high as 12–15 m. In terms of treatment efficiency, the majority of BOD and suspended solids removal occurs within the top 750 mm of the filter bed, which is also about the minimum depth when using conventional mineral medium in order to avoid short-circuiting of wastewater through the bed. Apart from constructional costs, the major limiting factor in the selection of the depth of filters is the loss of hydraulic head, with the greater the depth the greater the loss of head occurring. Unless the plant is built on a slope, pumping will be required. From experience, it would appear that a depth of 1.8 m, when using conventional medium, is a good compromise between treatment efficiency, loss of hydraulic head, and cost. The influence of depth on performance is considered in Section 4.1.1.4.

There appears to be a wide variety in the configuration, or plan, of percolating filters (Learner 1975a), although it is dependent to a large extent on the site and the total loading. They are usually either rectangular or circular in plan. Rectangular filters are normally used for large populations because they are more compact and are cheaper to construct because they share retaining walls. Maximum size of such filters rarely exceeds 10 m in width and 100 m in length. They can also be used for very small populations (< 40 people) when the rectangular shape is more compatible with the mechanical distribution system used. This comprises of a tipping trough that discharges the wastewater intermittently into a fixed set of distribution channels that are laid across the filter. Circular plan filters are generally preferred as they allow greater control over the frequency of wastewater application. The maximum size of these filters is limited by the maximum diameter of rotating distributor available, which is 40 m.

Percolating filters are built on a concrete base covered with drainage tiles that provides a raised floor on which the medium rests. This provides an unrestricted area of underdrainage, which takes the final effluent away. It is important to ensure that the underdrainage area is deep enough to allow the passage of air from ventilation pipes lining the base of the filter bed. The ventilation pipes are connected to the atmosphere so that there is a free flow of air through the whole filter, providing the necessary aeration. Natural convection causes aeration because of the difference between the air temperature and the temperature within the bed. The concrete base is sloped to at least $1:100$ to ensure a minimum flow of 0.6 m s^{-1}, which prevents settlement and the accumulation of solids washed from the medium. Drainage tiles can be replaced by a series of half-round field drains laid out in a herringbone pattern in rectangular filters, or arranged radiating out from the centre of circular shaped filters to a central drainage channel or well. As is the case with the tiles, the drains must be of sufficient size to allow for adequate ventilation.

Filter bed walls have to be able to support a considerable weight of medium, attached film, and water (Table 4.2). The retained material exerts lateral pressure on the walls, which increases towards the base and it is thus cheaper and safer to construct filter beds partially or completely buried in the ground. This allows breeze blocks or a single layer of bricks to be used for the retaining wall and also ensures maximum thermal insulation provided by the surrounding earth. When constructed entirely above ground the retaining walls must be able to take the maximum bulk density of the filter medium contained within the bed. In this case, the walls are constructed out of reinforced concrete, thick brickwork or prefabricated concrete sections bolted together. In the early years, some small filter beds were built free-standing. This was done by inclining the sides sufficiently so that the media formed a stabilized bank which was then sometimes covered using a larger grade of aggregate. Drystone walling was also popular, although both systems were liable to the walls collapsing and surface water overflowing down the sides. Modular plastic medium exerts a negligible amount of lateral pressure on the retaining walls and therefore lightweight cladding or single thickness breeze block housing are adequate. This is not the case with random plastic filter medium, as is seen when the maximum bulk density exerted on the retaining

TABLE 4.2 Comparison of the bulk density of a mineral medium and a random plastic medium of the same nominal grade (50 mm)

	Dry weight (kg m^{-3})	Total bulk density due to film accumulation (kg^3 m^{-3})		Weight when totally saturated with water (kg^3 m^{-3})
		minimum film	maximum film	
Flocor R. C.	97.84	133.73[a] 140.68[b]	290.59[a] 354.77[b]	1004.84
50 mm slag	886.31	1014.55[a] 1050.01[b]	1277.32[a] 1180.33[b]	1401.31

[a] Low-rate loading; [b] high-rate loading.

walls can be almost equal to that of mineral medium if ponding occurs (Table 4.2) (Gray 1983*b*).

The main function of the distribution system is to apply the wastewater to the top of the filter bed as evenly as possible. Distribution systems can be categorized as either static or moving. Static distributions are a network of pipes or open troughs which are permanently fixed in position above the surface of the medium. The wastewater is discharged through fixed outlets which are usually spray nozzles (where pipes are used) or via V-notches from open channels. However, the uniformity of dosing and the completeness of wetting the surface of the medium is not as good as obtained when using moving distributors, and static distribution systems are restricted to very small percolating filters. High-rate filters using modular media are often square or rectangular in plan (Section 4.1.1.2) and because it is important to have even distribution in order to utilize all the potential surface area of the medium, static spray nozzles are frequently used. There are two types of moving distributors, rotary distributors which are used for circular filters and reciprocating distributors that are employed on rectangular beds. Rotary distributors are either 2 or 4 radial sparge pipes which are pivoted in the centre of the filter on a rotating column (Fig. 4.1). The outlet holes are spaced along the arm to ensure a uniform loading, with the holes becoming more closely spaced further away from the centre. The outlets must be large enough to ensure that blockages from solids carried over from the settlement tank or accumulated grease rarely occur. Once blocked, that area of the bed fed by that outlet hole will be without any influent until the blockage is cleared. For this reason, simple holes, with splash plates situated below to ensure maximum dispersion of the wastewater, are preferred to jets or spray nozzles. The distributor arms can be driven by water wheels, electrically powered wheels which drive the distributor arm around or, more commonly, by jet reaction as the water leaves the outlet. Whereas jet reaction does not require a power supply and is mechanically simple, it does require a liquid head of 0.5 m as well as an extremely low friction bearing to support very well-balanced distributor arms. The speed of rotation varies according to the hydraulic loading and such systems are affected by changes in wind direction, which can result in the distributor stopping altogether and causes local overloading of the bed and a subsequent reduction in effluent quality.

Reciprocating distributors are associated only with filters that are rectangular in plan and are only common in the UK. The walls of the bed serve as a track to convey as well as support the distributor as it moves back and forth along the bed. Wastewater is fed into the moving sparge pipe from a longitudinal supply channel by syphonic action. The moving distributor is propelled either by a water wheel mechanism or an electric motor using cables. Because the distributor cannot travel in excess of 10 m min^{-1} it takes a considerable time to travel from one end to the other. This can result in long intervals between successive dosing of wastewater, especially at the far ends of the bed, thus particularly long beds may employ several distributors each operating over a particular area (Stanbridge 1972).

Loading
For conventional single-pass filtration, where settled sewage passes through a single percolating filter bed before having the solids (humus) removed in a secondary sedimentation (humus) tank, loading can be categorized as either low- or high-rate. Hydraulic loading is expressed in cubic meters of settled wastewater per cubic metre of filter medium per day ($m^3m^{-3}d^{-1}$) so that loadings $<3\,m^3m^{-3}d^{-1}$ are classed as low-rate and $>3\,m^3m^{-3}d^{-1}$ as high-rate filtration. Hydraulic loading is not always expressed as a volumetric loading. It is also expressed as a surface loading, which is often referred to as the irrigation rate, in cubic metres of wastewater per square metre of superficial bed area per day ($m^3m^{-2}d^{-1}$). The hydraulic loading rate depends solely on the physical characteristics of the medium and the degree of film accumulation within the bed. The upper hydraulic limit is determined by the onset of flooding because the wastewater is being applied at a faster rate than it can percolate through the bed. Where this occurs then the interstices of the medium become flooded and the access of air to the interior of the bed is prevented. The lower limit is known as the minimum wetting rate, which is the loading rate at which the medium is kept sufficiently moist to prevent drying and the microbial film to remain active (Winkler 1981). Organic loading is measured as the BOD per cubic metre of filter medium per day ($kg\,BOD\,m^{-3}d^{-1}$) with low-rate filtration classed as filters receiving loads of $<0.6\,kg\,BOD\,m^{-3}d^{-1}$ and high rate $>0.6\,kg\,BOD\,m^{-3}d^{-1}$. In practice, in order to produce a Royal Commission standard effluent (20 mg l^{-1} BOD, 30 mg l^{-1} suspended solids) after settlement with a high degree of nitrification, filters treating domestic sewage should receive an organic loading of between $0.07–0.10\,kg\,BOD\,m^{-3}\,d^{-1}$ and a hydraulic loading between $0.12–0.60\,m^3m^{-3}d^{-1}$. Generally, increases in organic loading in excess of $0.10\,kg\,BOD\,m^{-3}\,d^{-1}$ will result in heavier film growths, which may result in ponding. In an attempt to produce more efficient percolating filters which would operate at much higher loadings a number of modifications of the basic process have been adopted. By using larger mineral medium, for example, greater loads can be applied to filters without the risk of ponding. Such high-rate filtration will produce a 20:30 effluent with an increased hydraulic loading of up to $1.8\,m^3m^{-3}d^{-1}$, but with little or no nitrification (Institution of Public Health Engineers 1978). If a less stabilized effluent is required, such as roughing treatment for strong industrial wastes, then loadings of up to $12\,m^3m^{-3}d^{-1}$, with organic loads up to $1.8\,kg\,BOD\,m^{-3}d^{-1}$ will give 60–70% removal. Treatment at such high rates is facilitated by using modular or random plastic medium (90% voids) in tall towers, in place of the usual stone medium (40% voids), thus reducing the risk of ponding. Modifications of the percolating filter process are dealt with in Section 4.1.1.2.

If the amount of biological film growing per unit area of medium remains more or less constant then the organic loading corresponds to the sludge loading or the food to micro-organism ratio (f/m ratio) as used in the activated sludge process (Section 4.2). This is difficult to measure directly as film accumulation is rarely even throughout the filter bed and is certainly more abundant in the upper areas of the bed where food is a non-limiting factor. Also, a considerable proportion of the biomass will be non-active being comprised of inert macro-invertebrate fragments, and the proportion of non-active material will increase towards the base of the filter. Thus, in theory, using sludge loading in a percolating filter offers a degree of control on par with that of the activated sludge process.

However, in practice, accurate estimations of film accumlation can only be made for high-rate filters, especially those using modular plastic medium, where seasonal variation in film development and the problem of inert solids being retained in small intertices have been largely eliminated. Another important factor is that none of the medium in a high-rate filter should be either food or oxygen-limited, thus ensuring an even growth of film throughout the bed. Estimates of film accumulation have been made. For example, the active depth of film is known to be about 0.2 mm, which gives an average weight of active film per m^2 of medium of around 0.2 kg. If it is assumed that the water content of film is about 97%, then each square metre of medium is supporting 6 g (DW) of active microbial biomass. The exact amount in the filter will depend on the surface area of the medium, but if this is assumed to be 100 $m^2 m^{-3}$ then the active microbial biomass will be 0.6 kg m^{-3}. Thus, at an organic loading of 0.1 kg BOD m^{-3} d^{-1} the sludge loading will be about 0.2 kg BOD kg^{-1} d^{-1}. For high-rate systems, assuming the same surface area for the medium, the sludge loading will be approximately 1 kg BOD kg^{-1} d^{-1} at an organic loading of 0.6 kg BOD $m^{-3} d^{-1}$. Therefore, there would appear to be agreement between the term low- and high-rate as used in both the percolating filtration and the activated sludge processes (Winkler 1981).

Mineral media

The function of the medium in a percolating filter is to provide an extensive surface to support the biological film and the associated fauna necessary for the purification of settled wastewaters. At the same time it must allow sufficient ventilation for the process to operate aerobically and ensure maximum contact between the active film and the waste liquid.

The important features of mineral media were summarized in the original British Standard in 1948. It stated that a mineral filter medium should be selected for 'high surface area consistent with adequate voidage, and for satisfactory grading (within specified size limits), durability, roughness of texture, satisfactory shape characteristics and low cost'. Thompson (1925) had shown that the material selected had to be mechanically and chemically stable if the medium was not to degrade filling the voids with smaller pieces of aggregate broken off the parent medium. Whereas Levine *et al.* (1936), comparing a number of different filter media, found that the performance was directly related to the surface area of the medium. After the publication of the 1948 British Standard, it was standard practice for engineers to choose a medium which was predominantly cubic in shape. This view was based on the supposition that a filter medium containing a high proportion of flattish pieces would have an undesirably low voidage, and in many textbooks this still remains the view. The more irregular a material is for a given nominal size, the greater the surface area. For example, although more costly, blast furnace slag or clinker is to be preferred to gravel or pebbles. For material of a given shape or uniform grade, the voidage is independent of size. However, the important factor is the actual dimensions of the void spaces (the interstices), as these will determine whether or not a given medium will clog during operation or will allow adequate ventilation. Excessive accumulation of film also reduces the effectiveness of the surface area. The rejection of flaky material (i.e. particles with one excessively thin dimension) by British Standard 1438:1948 was proved to be unjustified by Schroepfer (1951). When examining the effect of particle shape on voidage and surface area, he found that the more cubical material possessed a lower voidage than the flaky material and concluded that an increase in angularity of particles, of which flaky particles are an extreme example, resulted in an increase in both voidage and surface area. Compaction will reduce the available voidage by between 5–7% (Moncrieff 1953). In a review of the relationship of particle shape and voidage, Bruce (1968) supported the earlier findings of Schropfer, that flaky media possess a higher voidage and surface area than regular media of the same sieve size. However, the average volume of flaky particles is smaller and this controls the actual size of the voids, although the use of flaky material of a large grading would compensate for this. These findings led to a relaxation of the 'index of flakiness' and the withdrawal of the 'index of elongation' in the revised British Standard on Percolating Filter Medium, 1971.

Numerous comparative investigations into the ideal characteristics of filter media followed the publication of the British Standard in 1948, in particular, those carried out at Minworth near Birmingham (Hawkes and Jenkins 1955, 1958) and at Stevenage (Wilkinson 1958; Truesdale *et al.* 1962). These investigations showed that the smaller media consistently produced better quality effluents, and that medium with a rough surface gave marginally better performance than the smooth surface materials such as gravel. Truesdale *et al.* (1962) found that although small grade, rough-textured medium was extremely efficient in treating large organic loads of settled sewage during the summer (0.18 kg BOD $m^{-3}d^{-1}$), it suffered from excessive film accumulation during the colder winter months. This resulted in ponding and the eventual clogging of the filter. Experience has shown that in order to achieve maximum efficiency throughout the year, a 50 mm mineral medium with a rough surface provides the best compromise between large surface area and the provision of large voidage, and that this will produce a high quality effluent in a conventionally operated British plant (Hawkes 1963).

Learner (1975a), in a survey of percolating filters, found that granite, clinker, and blast furnace slag were the most frequently used media (Table 4.3). While the pitted nature of the clinker and blast-furnace slag makes

TABLE 4.3   Types of media most commonly used in percolating filters in the UK (Learner 1975a)

Type of medium	%
Granite	26
Clinker	24
Blast-furnace slag	23
Rounded gravel	6
Limestone and clinker	6
Limestone	4
Coke	4
Clinker and gravel	3
Slag and coke	1
Saggar chippings	1

them superior to other mineral media, the choice of filter medium does not only depend on its suitability but also on availability and cost. As Learner points out, the majority of filters using clinker as medium were constructed prior to 1956, and since then clinker has become more expensive and extremely difficult to obtain. Another interesting aspect of the survey was that the majority of the filters sampled in Scotland used granite as a filter medium; although this may not be the most suitable medium it was the most readily available in the area. A similar situation exists in Ireland where neither blast furnace slag or clinker are available (Earle and Gray 1987). For this reason, the majority of percolating filters use the locally available stone which is normally basalt or granite. Learner (1975a) found that the commonest range of media grades used was 13 to 51 mm (Table 4.4), based on particle size analysis. Bruce (1969) states, in his review of percolating filters, that the commonest grades of media used in the UK are the nominal sizes of 38–51 mm. As previously stated, 50 mm mineral media provides a good

TABLE 4.4   Percentage occurrence of the grades of filter media used in UK percolating filters (Learner 1975a)

Grades of medium (mm)	0–13	13–25	25–38	38–51	51–64	64–76	> 76
%	2	21	39	29	6	3	0

compromise between surface area and voidage and has been shown to produce good quality effluents under low-rate conditions (Hawkes and Jenkins 1955, 1958).

Physical nature of media
The shape of media is defined by using a descriptive classification based on visual inspection. They can be classed as being flat, angular, irregular, rounded, or spherical. The different shapes are beautifully compared in a series of photographs in British Standard 1438 (1971). As already mentioned, the voidage increases with an increase in the proportion of flaky media, whereas angular shaped media have a higher voidage than rounded media. Typically, basalt and granite media are either flat or angular (50–54% voidage), blast-furnace slag is irregular (50%), and gravel is rounded or spherical (38–40%). The voidage can be determined directly using a cylindrical metal vessel with spouted outlets at two levels and having a volume of at least 25 litres between the two outlets. The voidage is calculated by filling the vessel full of media and then filling with water. It may be necessary to leave the medium submerged in water for 24 hours in order to ensure that absorption by the aggregate is complete before topping up the vessel with water. The highest outlet is opened and allowed to drain fully and the volume of water filling the interstices of the known volume of medium between the two spouts is determined by measuring the volume of liquid which drains from the lower outlet. The results are expressed as a percentage. The volume and overall diameter of the cylinder must be as large as practicable in order to minimize the wall effects (Bruce 1968).

Grading is normally done by the supplier of the medium who will issue a certificate showing the nominal grade. It is important to obtain a medium which has a fairly uniform particle size, otherwise if there is a wide discrepancy between the sizes, the voidage and more importantly the size of the voids will be seriously reduced. When the aggregate is crushed, considerable dust and small fines are produced which must be removed by washing with high pressure hoses before the medium is placed in the filter bed. If this is not done, localized blockages of the voids deep within the filter will eventually occur. It is impossible, unless the media is manufactured, e.g. plastic media, for the medium to be comprised of particles of a single size. For this reason, the medium is graded by determining the percentage by weight passing each sieve and the nominal size of the medium is then calculated using the limits set by British Standard 1438 (Table 4.5). Small quantities of over- or under-sized medium do not affect the voidage significantly, thus for a nominal size of 50 mm, for example, up to 15% of the medium by weight can be > 50 mm and up to 30% < 37.5 mm. Particle size analysis is done using square hole sieves, according to British Standard 410, using a minimum sample of 100 kg of medium. In practice, grading proves more difficult as often the percentage passing each sieve does not conform exactly to the limit set in Table 4.5. For

TABLE 4.5   Grading limits for media in British Standard 1438 (1971)

BS410 square hole perforated plate test sieves (mm)	Nominal sizes (mm)					
	63	50	40	38	20	14
	% by weight passing					
75	100	—	—	—	—	—
63	85–100	100	—	—	—	—
50	0–35	85–100	100	—	—	—
37.5	0–5	0–30	85–100	100	—	—
28	—	0–5	0–40	85–100	100	—
20	—	—	0–5	0–40	85–100	100
14	—	—	—	7	0–40	85–100
10	—	—	—	—	0–7	0–40
6.3	—	—	—	—	—	0–7

example, slag medium crushed to a nominal size of 50 mm was found on sieve analysis to have 100% (by weight) passing the 63.0 mm sieve; 70.5%, 50 mm; 15.1%, 37.5 mm, and 0%, the 28.0 mm sieve. This medium does not conform to the limits set for 50 mm as 29.5% was retained in the 50 mm sieve, so crushing was not done adequately. However, it would not appear serious enough to abandon the media, especially after it has been delivered, so what nominal grade is it? It clearly does not fit the limits set for 63 mm either, so the nominal grade of the medium lies somewhere between these two values. Table 4.5 does make it clear that the bulk of the medium of the 50 mm nominal size should be between 37.5–50.0 mm in size, and in this case 55.4% was within this particular range. As up to 30% of the medium is allowed to pass the 37.5 mm sieve then Table 4.5 indicates that 55% within the 50–37.5 mm is legitimate for a medium of 50 mm nominal grade even though it specifies limits of 85–100. Therefore, the sample tested appears to fit best within this category rather than the 63 mm nominal size grade. Interestingly, as filter media ages its particle-size distribution expands as pieces of medium are weathered and fractured (Table 4.6). For this reason, the use of an average size or median size for filter media has been proposed (Earle 1986).

In selecting the medium, a compromise must be reached between the conflicting requirements of a high specific area, which requires a fine grade and large voids, which are obtained by using a coarse grade. Surface area ($m^2 m^{-3}$) is affected by surface texture and particle shape, although the nominal size of the medium is the major influence (Table 4.7). A rough porous medium, such as clinker or blast-furnace slag can have a surface area up to 17% greater than smooth-surface media, such as gravel of the same nominal size. However, as the film develops many of the small holes in rough-textured medium become filled with organic debris, fauna, or film itself, thus, in practice, the surface area will tend to approach that for smooth-medium types. The numerous holes of various sizes and the irregular areas on rough-textured medium do provide a

TABLE 4.6 Problems of calculating nominal size of mineral filter medium using Table 4.5 when the medium has undergone weathering within the filter. Examples are given for Irish percolating filters and the nominal size compared with the mean size (mm) (Earle 1986)

Site	Bulk percentages passing the sieve grading (mm)					
	16	20	28	37.5	50	63
Ballygar	5	28	26	8	25	8
Enniskerry	0	0	1	13	55	31
Gort (i)	5	10	40	40	5	0
Gort (ii)	10	10	34	30	3	14
Kildare (i)	9	19	59	13	0	0
Kildare (ii)	9	18	59	14	0	0
Killincarrig	41	36	23	0	0	0
Leighlinbridge (i)	2	4	54	38	3	0
Leighlinbridge (ii)	8	9	45	34	3	0
Loughrea	11	14	24	29	22	0
Mountbellew	0	0	2	41	56	0
Newcastle	1	2	11	41	46	0
Rathmore	8	19	45	23	6	0
Roundwood	1	2	8	24	46	19
Saggart (i)	3	3	9	17	45	22
Saggart (ii)	0	2	2	38	45	13

Site	Voidage (%)	Nominal size (mm)	Mean size (mm)
Ballygar	42	37.5	25
Enniskerry	49	50	44
Gort	44–45	37.5	27
Kildare	42–43	28	23
Killincarrig	44	20	16
Leighlinbridge	46–47	37.5	28
Loughrea	44	37.5	29
Mountbellew	45	37.5	40
Newcastle	49	37.5	37
Rathmore	44	37.5	24
Roundwood	43	50	42
Saggart	45–48	50	40

much greater variety of niches for both the micro-organisms and the larger grazers than on smooth-medium types. The rougher texture also prevents the film being so readily sloughed off and allows increased weights of biomass to accumulate within the filter.

It is not normally necessary to measure the surface area of medium directly, as the type and the grade being used will most likely conform to one of the three media in Table 4.7. Where necessary, surface area can be determined by the paint dipping technique. Using a known bulk volume of medium, each piece is coated with a free-flowing paint using a standard dipping procedure

TABLE 4.7 Values for specific surface area (m^2m^{-3}) of three different types of filter medium in relation to nominal size and grading. The range for graded material represents maximum oversize or undersize material permitted by British Standard 2438 (1971)

Nominal maximum size of medium (mm)	Granite		Type of medium Blast-furnace slag		Crushed gravel	
	Single-size	Range for graded material	Single-size	Range for graded material	Single-size	Range for graded material
25	194	185–237	208	200–246	176	169–208
37.5	135	129–149	146	140–163	125	120–140
50	97	94–111	104	101–118	89	86–101
63	75.5	73–85	81	79–90	69	67–77

(Schroepfer 1951). The weight of paint adhering to the medium is calculated by weighing the paint container before and after dipping. The increase in weight per unit area is calculated by dipping test blocks of known surface area into the same paint. These test blocks can be of wood or aggregate provided that the paint adheres to them and the surface area can be accurately calculated. Problems are encountered in obtaining an even distribution of paint over all the pieces of medium even after several dippings. Truesdale *et al.* (1962) noted that filter media has different absorptive capacities and, in practice, absorption of paint varies between individual pieces of the same medium. The problem of uneven coating, and of paint collecting and thickening at the base is overcome by painting individual pieces by hand and then calculating the increase of weight for each piece of medium. Using red lead paint (British Standard 2523, type B), which is thick enough to give good coverage and heavy to ensure a good increase in weight after application, the medium is coated individually using a 10 mm thick paint brush. This procedure is more time consuming than paint dipping, but does ensure that the coat of paint is even and that all the pores are covered. Surface area is then calculated by comparing the increase in weight of the media with the mean increase in weight obtained with the test blocks of directly measurable surface area. To overcome the varying absorptive capacity of the media the pieces of medium will have to be painted a number of times. Experiments have shown that the paint retained on test blocks varied considerably on the first painting but this variability became less with each subsequent coating of paint. Three coats of paint appear sufficient for the most porous media.

Media must be strong enough to take the weight of the medium and accumulated film above it, be chemically non-reactive, and be durable so that the media does not need to be replaced for many years. The media will undergo weathering within the bed because of constant irrigation with wastewater, and surface microbial activity. On the surface of the bed the media will also be subject to normal weathering processes, such as freezing and heating. Once media begins to disintegrate the voids will become gradually clogged with fragmented particles. Some media are totally unsuitable for use and with time they will be broken down until the bed is full with a 'mush', e.g. peat bricketts. In general, most sedimentary rocks are unsuitable for use as percolating filter media. The media should remain completely stable for at least 50 years. It suitability can be assessed by the sodium sulphate 'soundness' test. Test pieces of medium are immersed in saturated solutions of sodium sulphate for a number of days before being dried at 105°C. This cycle is repeated 20 times and the material is then examined for signs of possible defects or loss of weight. Any medium that passes such a severe test will certainly survive 50 years in a percolating filter. However, some media which fail this test may also be suitable. Media should be chemically non-reactive with all types of wastewater that are normally treated in percolating filters. Problems of leaching from media do occur, especially from metals, and it has been suggested that this may interfere with some sensitive species, successfully colonizing the media, such as the nitrifying bacteria. Sulphur compounds can be leached from certain slag media and may lead to damage of the concrete base of the filter under certain circumstances (Lavender 1970). Although, under normal operating conditions, excess sulphate will be diluted by the wastewater. Limestone is widely used both in Ireland and the UK, and can be used successfully, provided that the wastewater is either neutral or alkaline. Soft limestones and chalk are eventually broken down even in neutral wastewater, and where the wastewater includes the drinking water supplied from a catchment with acidic geology, dissolution will occur and the limestone medium will literally be dissolved away. Limestone should not be used to treat industrial wastewaters for this reason.

Plastic media
This medium has been freely available in Europe since the end of the 1950s and is now usually made out of PVC. It has a very high voidage (normally $>90\%$), wide interstitial spaces, and is only about 10% of the weight of the mineral medium when in operation. This allows tall filters to be built using prefabricated and light-weight structures, which is significantly cheaper than building conventional mineral filters. The plastic filter media must be robust and strong enough to withstand the weight of 3–12 m of media and the accumulated film. This strength is supplied not by the material but in the way it has been prefabricated. The other major advantage over mineral medium is that blockages because of accumulated film are very unlikely due to the high voidage and so high organic loads can be applied. The wide interstices allow sufficient air to reach all areas of the filter, whereas oxygen availability can be limiting in mineral filters. The specific surface area of plastic medium varies between 83–330 m^2m^{-3}, although it appears that there is very little practical difference in surface area between plastic and mineral media (Tables 4.7 and 4.8). Apart from the requirements already mentioned, the main requirements for plastic media are similar although not exactly the same as for mineral media. The advantage of manufacturing a filter medium is that no compromise is required. Generally, the main features of the ideal plastic media are: an ability to treat high hydraulic and organic loads; an open structure in order to avoid restricting film development or subsequent blockages due to excessive film growth; wide interstices to permit thorough aeration; chemical inertness; durability; physical strength to support sufficient media and associated film to ensure adequate treatment; light bulk density so that structural costs can be reduced; configuration of the media which ensures uniform distribution and minimizes channelling; and, finally, it must be cost-effective (Chipperfield 1968; Hemmings 1979).

There are two main categories of plastic media, modular (ordered) and random, with a wide variety of commercially available designs within each category. These have been described and their performances compared by

TABLE 4.8 Comparison of the physical properties of widely used random and modular plastic filter medium with the mineral medium 50 mm blast furnace slag

Medium	Composition	Specific surface area $(m^2 m^{-3})$	Voidage (%)
Random			
Norton			
Actifil 90E	Polypropylene	101	95
Actifil 50E	Polypropylene	124	92
Actifil 75	Polypropylene	160	92
MT			
Filterpak 1127	Polypropylene	120	93
Filterpak 1130	Polypropylene	190	93
Mini ring (2)	Polypropylene	118	93
Mini ring (3)	Polypropylene	79	94
ICI			
Floor RC	PVC	330	91
Modular			
Cloisonyle	PVC	220	94
ICI			
Flocor E	PVC	90	95
Flocor M	PVC	135	95
Munsters			
Plasdek B2760	PVC	100	95
Plasdek B19060	PVC	140	95
Plasdek B12060	PVC	230	95
Mineral			
Blast furnace slag (50 mm)		143	51

various workers (Bruce and Merkens 1973; Forster 1977; Besselievre and Schwartz 1976; Winkler and Thomas 1978; Porter and Smith 1979; Rowlands 1979).

Two main designs of modular medium are generally used, vertical sheet and vertical tube media. The former is the most widely used design of modular media and is made up of sheets of corrugated ribbed plastic assembled to form a rectangular module with a large number of open zig-zag channels (Fig. 4.2). The modules can be prefabricated on site or delivered already made up into blocks. The media is simply stacked into the filter bed and the modules can be cut to size to exactly fit any configuration of bed. The individual sheets of plastic which comprise the modules are orientated vertically so that both surfaces of the channel are exposed to the wastewater. The design of the zig-zag channel is such that wastewater is unable to free fall through the medium without coming into contact with the film. The media is available in a range of spacings which allow for extra accumulation of film. The smooth texture and vertical placement of the media discourages excessive build-up of film, and the film regularly sloughs because of its own weight (Pearson 1965; Eden *et al.*

FIG. 4.2　Configuration of a modular plastic filter medium (Flocor E). (a) The top surface over which wastewater is applied via spray nozzles and (b) the high surface area of the medium on which the biological film develops.

1966). The most widely used media of this type is Flocor E, which has a dry bulk density of 40 kg m^{-3} and increases to up to 300 kg m^{-3} as the film develops. Because of such a large increase in weight when in operation, beds in excess of 3 m depth will need intermediate support (Hemmings 1979; Porter and Smith 1979). It has a specific surface area of 90 m^{2}m^{-3}, comparable to blast-furnace slag of a nominal size of 63 mm, which is also used for high-rate treatment (Tables 4.7 and 4.8). Cloisonyle is the most frequently used vertical tube medium. It consists of a block of closely assembled vertical tubes between 4 to 6 m in length, which extends the whole depth of the filter bed. Each 80 mm diameter tube is subdivided into 14 smaller concentric tubules each about 15 mm in diameter. The medium has a high voidage (94%) and specific surface area (220 m^{2}m^{-3}) (Table 4.8). However, successful operation is due entirely to adequate distribution of wastewater, because as the tubes are not interconnected, any tubule not receiving a wastewater supply will remain dry throughout. Also, as the surfaces are entirely vertical there is a chance that wastewater can fall for some distance within a tubule without coming into contact with the attached film. It was quickly realized that modular media were of limited value in the complete treatment of domestic wastewater. However, modular media are extremely effective as roughing filters, removing large weights of BOD per unit volume of media at relatively low levels of efficiency in terms of BOD concentration, i.e. 50–80% (Ministry of Technology 1968).

Even with careful design the short contact time between the influent wastewater and the film, and the possibility of the free fall of wastewater, prevents the modular forms of plastic media from producing well-purified effluents. In order to increase the contact time but retain the high voidage and high surface area per unit volume of modular media, various new random plastic media were designed. These were shaped in such a way that free fall of wastewater experienced in some modular designs was prevented. Its random nature increased contact time with the film while maintaining the high voidage, thus reducing the problem of film accumulation and lack of ventilation experienced with mineral media (Ramsden 1972). Random packed media are made up of individual pieces of PVC or polypropylene moulded into specific shapes in the same nominal sizes used for mineral media such as 37.5, 50, and 65 mm, although some designs can be obtained in sizes up to 90 mm in diameter. The media are generally rings or short tubes with the surface area increased by external or internal fins and ribs (Fig. 4.3). The media are packed randomly into the filter bed in the same way as mineral media. The advantages of random plastic over mineral media include large voidage, higher specific surface areas (Table 4.8), and extremely low bulk density (Table 4.2), which permits easy handling. Its lightweight nature allows it to be housed in prefabricated beds such as PVC sheeting attached to a mild steel framework, glass reinforced plastic (GRP), or brick. The higher voidage allows the influent to be continuously loaded using fixed sprays, although conventional

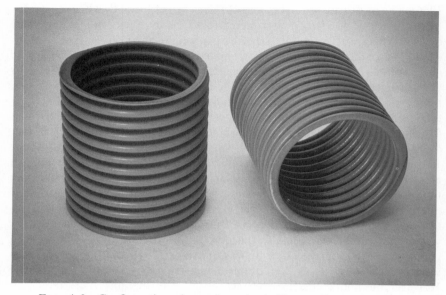

FIG. 4.3   Configuration of a random plastic filter medium (Flocor RC).

distribution systems can also be used. The types of random plastic media available have been reviewed by Porter and Smith (1979).

Plastic media have been successfully used to treat effluent wastewater from a wide range of industries (Besselievre and Schwartz 1976). However, whereas modular media are used for high rate filtration, random plastic media are essentially used for low-rate applications. On the other hand, random plastic does not appear any more efficient than comparable mineral media at conventional low-rate loadings (<0.1 kg BOD $m^{-3}d^{-1}$) and may even be less efficient in terms of nitrification (Eden *et al.* 1966). Also, a comparable plastic low-rate filter will be more expensive than a mineral system. Random plastic media are generally used for the primary biological treatment of strong wastewaters prior to their treatment by established methods (Anon 1979*b*; Hemming 1979). The performance of different media is compared in Section 4.1.1.4.

Purification and film development
The microbial film takes 3–4 weeks to become established during the summer and up to 2 months in the winter, and only then has the filter reached its maximum purification capacity, including nitrification. Unlike many industrial wastewaters, it is not necessary to seed domestic wastewaters treated by percolating filters because all the necessary micro-organisms are present in the sewage itself, with the dipteran grazers flying on to the filter and colonizing it. The micro-organisms quickly form a film over the available medium.

However, the film only develops on the surfaces that receive a constant supply of nutrients. Therefore, the effectiveness of media to redistribute the wastewater within the filter, in order to prevent channelling and to promote maximum wetting of the medium is an important factor affecting performance. The film is a complex community of bacteria, fungi, Protozoa, and other mesofauna, plus a wide diversity of macro-invertebrates , such as enchytraeids and lumbricid worms, dipteran fly larvae, and a host of other groups which all actively graze the film (Curds and Hawkes 1975).

The organic matter in the wastewater is degraded aerobically by the heterotrophic micro-organisms that dominate the film. The film has a spongy structure, rather similar to activated sludge flocs, which is made more porous by the feeding activities of the grazing fauna which are continually burrowing through the film. The wastewater passes over the surface of the film and, to some extent, through it, although this depends on film thickness and the hydraulic loading (Fig. 4.4). In low-rate filters a large proportion of the wastewater may be flowing through the film matrix at any one time, and it is the physical straining action of this matrix that allows such systems to produce extremely clear effluents. Another advantage is that the greater the proportion of wastewater that flows through the film the greater the micro-organism–wastewater contact time, which is known as the retention time or hydraulic

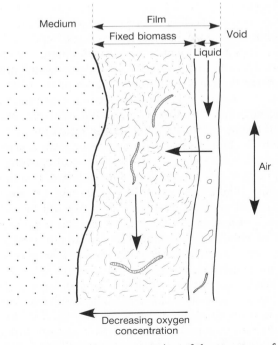

FIG. 4.4 Diagramatic representation of the structure of film.

retention time (HRT). The higher the hydraulic loading the greater the proportion of the wastewater passing over the surface of the film, which results in a lower HRT and a slightly inferior final effluent. Whether the liquid forms a separate layer passing over the surface of the film or is actually making its way slowly through the film matrix, the first stage of purification is the adsorption of organic nutrients onto the film. Fine particles are flocculated by extracellular polymers secreted by the micro-organisms and adsorbed on to the surface of the film, where together with organic nutrients, which have been physically trapped, they are broken down by extracellular enzymes secreted by heterotrophic bacteria and fungi. The soluble nutrients in the wastewater, and those produced from this extracellular enzymatic activity, are directly absorbed by the micro-organisms comprising the film and synthesized. Oxygen diffuses from the air in the interstices, first into the liquid and then into the film. Conversely, carbon dioxide and the end-products of aerobic metabolism diffuse in the other direction. The thickness of the film is critical, as the oxygen can only diffuse for a certain distance through the film before being utilized, leaving the deeper areas of the film either anoxic or anaerobic. The depth to which oxygen will penetrate depends on a number of factors, such as composition of the film and its density and the rate of respiration within the film itself. It has been estimated as being between 0.06–2.00 mm thick (Schulze 1957; Ekenfelder 1961). Tomlinson and Snaddon (1966) calculated this critical depth experimentally to be 0.2 mm for a predominately bacterial film. However, where fungal mycelium is present, oxygen diffusion is enhanced by movement within the hyphae, increasing the critical depth to 4 mm. Only the surface layer of the film is efficient in terms of oxidation, and only a thin layer of film is required for efficient purification. Tomlinson and Snaddon estimated that the optimum thickness in terms of performance efficiency was only 0.15 mm. This means that it is the total surface area of active film that is important and not the total biomass of the film. The film accumulates and subsequently thickens within the filter during operation. This is due to an increase in microbial biomass from synthesis of the waste and to accumulation of particulate material by flocculation and physical entrainment where the accumulation rate exceeds the rate of solubilization and assimilation by the micro-organisms. Where the wastewater is largely soluble, such as food processing wastewaters, most of the film increase will be due to microbial growth. Whereas with domestic wastewater, that has been poorly settled, the accumulation of solids may account for the major portion of the film accumulation. Once the film exceeds the critical thickness an anoxic, and subsequently an anaerobic, environment is established below the aerobic zone. As the thickness continues to increase, most of the soluble nutrients will have been utilized before they can reach the lower micro-organisms thus forcing them into an endogenous phase of growth. This has the effect of destabilizing the film because the lower micro-organisms lose their ability to hold on to the surface of the medium, when results in portions of the film becoming detached and washed away in the wastewater flow, a process known as sloughing. Although thick film growths do not reduce the efficiency of the filter, excessive growths can reduce the volume of the interstices, reducing ventilation and even blocking them completely, preventing the movement of wastewater. Severe clogging of the interstices is known as ponding and is normally associated with the surface of the filter when whole areas of the surface may become flooded.

The accumulation of film within a filter bed follows a seasonal pattern; being low in summer due to high metabolic rates and grazing rates but high in the winter when the growth rate of the micro-organisms is reduced, as is the activity of the grazing fauna. As the temperature increases in the spring there is a discernible sloughing of the film that has accumulated over the winter months (Fig. 4.5). Wheatley and Williams (1976) found that no single factor directly influenced film accumulation within their experimental filters, but suggested that it was controlled by the interaction of a number of factors, namely ambient temperature, organic load, the distribution system, the

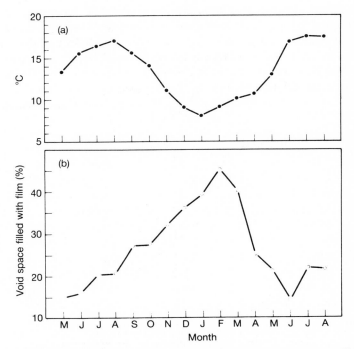

FIG. 4.5 Seasonal variation in the temperature of the sewage applied to a percolating filter and in the percentage of void space in the filter filled with film and water. (a) Monthly average temperature of applied sewage; (b) monthly average proportion of voids filled with film. Quantity of film and water measured by neutron scattering technique (Bruce *et al.* 1967).

microbial characteristics of the film, and the activity of grazers. Temperature has been shown to be an important factor in film accumulation by Hawkes and Shephard (1971), who demonstrated that below 10°C the rate of film accumulation increased rapidly. In a comparative experiment using laboratory filters with and without macro-invertebrate grazing fauna at 5 and 20°C, they examined the effects of grazers and temperature in film accumulation (Shephard and Hawkes 1976). They found that at higher temperatures a greater proportion of the BOD removed by adsorption was oxidized, and therefore fewer solids accumulated. The rate of oxidation decreased as the temperature fell, although the rate of adsorption remained unaltered. Therefore, at the lower temperatures there was a gradual increase in solids accumulation which eventually resulted in the filters becoming clogged. This relationship is shown in Fig. 4.6, where it can be seen that in the warmest months the high microbial activity may exceed the rate of adsorption, thus reducing the overall film biomass. Seasonal variation in film accumulation in the USA in percolating filters is thought to be due to this differential microbial activity at different temperatures (Holtje 1943; Heukelekian 1945; Cooke and Hirsch 1958). However, such variations in the UK have been primarily attributed to the activity of the grazing fauna, with the film accumulating during the winter months when both population densities and the activity of grazers were suppressed (Reynoldson 1939; Lloyd 1945; Tomlinson 1946b; Hawkes 1957). It appears that the grazing fauna do play a significant role in reducing the overall film biomass by directly feeding on the film, converting it into biomass which is flushed from the filter in the effluent (Hawkes and

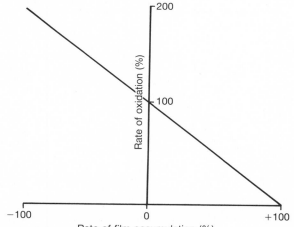

FIG. 4.6 Rate of film accumulation in percolating filters showing that as the rate of adsorption is constant, the rate of oxidation is proportional to the temperature of the film.

Shephard 1972). There can be little doubt that the spring sloughing of the accumulated film is pre-empted by the increased activity of the grazers, which loosen the thick film from the medium. Once the film has become detached it strips accumulated film from other regions of the bed as it is washed out. Although grazers suppress maximum film accumulation and maintain minimum film accumulation for a long period after sloughing, it is temperature which primarily controls film accumulation. Whereas hydraulic loading in low-rate filters is of little significance compared to the action of macro-invertebrate grazers in controlling film as the hydraulic loading increases, as is the case after modifications to the process such as recirculation of double filtration, physical scouring of the film, by the wastewater becomes increasingly important. In high-rate filters, especially those employing modular plastic media, the high hydraulic loading controls the film development by scouring the film from the smooth surfaced media as it reaches critical thickness. In such filters sloughing tends to occur on a more regular basis rather than seasonally.

The organic matter in the settled sewage applied to a percolating filter is present as fine or colloidal matter, or in solution. The film transforms this material by biophysical processes (flocculation and adsorption) and bio-chemical processes (bio-oxidation and synthesis) to produce a variety of end-products. Some of these are soluble or gaseous end-products, such as nutrient salts or carbon dioxide from oxidation processes, or soluble products from the lysis of the micro-organisms comprising the film. The remainder are present as solids that require separation from the final effluent. These solids are of three types: flocculated solids; detached fragments of the accumulated film and the grazing fauna; and fragments of their bodies and faeces. These solids are collectively known as humus and like all secondary solids they require further treatment after settlement (Section 5.1). The mode of operation will influence the nature of the humus sludge. High-rate operation will produce a humus sludge mainly comprised of flocculated solids and detached fragments of film, whereas a low-rate filter sludge contains a large proportion of grazing fauna and animal fragments. Sludges containing animal fragments and grazing fauna will be more stable than sludges from high-rate systems where the grazing fauna is absent or reduced. The production of humus varies seasonally with mean production rates for low-rate filters between 0.20–0.25 kg humus per kg BOD removed. It varies from 0.1 kg kg^{-1} in the summer to a maximum of 0.5 kg kg^{-1} during the spring sloughing period. In high-rate systems a greater volume of sludge is produced due to the shorter HRT, which results in less mineralization. The humus production does not vary seasonally to the same extent having a mean production rate of 0.35 kg kg^{-1}, although this is dependent on the nature of the influent wastewater. Sludge production is much less compared with activated sludge, being more stabilized and containing less water (Table 4.9), although as loading increases the mode of purification in percolating filters approaches that of the activated sludge

TABLE 4.9   Typical sludge volumes and characteristics from biological unit processes (Open University 1975)

Source	Volume (l per head d^{-1})	Dry solids (kg per head d^{-1})	Moisture content (%)
Primary sedimentation	1.1	0.05	95.5
Low-rate percolating filtration	0.23	0.014	93.9
High-rate percolating filtration	0.30	0.018	94.0
Activated sludge (wasted)	2.4	0.036	98.5

process and the sludge alters accordingly both in quality and quantity. Separation takes place in secondary settlement tanks traditionally called humus tanks. They are similar in principle to primary settlement tanks (Section 2.2.2), although often of different design. For example, deep square tanks of small cross-sectional area with an inverted pyramid bottom with a maximum upflow rate of <1 m h^{-1} are common. Percolating filters usually achieve high levels of nitrification, and a long sludge retention time within the humus tank may give rise to anoxic conditions and problems of denitrification, which results in the carry over of sludge in the final effluent. High-rate sludges are more susceptible to denitrification than low-rate sludges that are more stabilized and so exert less of an oxygen demand within the settlement tank. Temperature is also important with gas production being much heavier during the summer. Whereas the bulk of the solids settle easily, there is a fraction of fine solids that do not, and are carried out of the humus tank in the final effluent. These fine solids are responsible for a significant portion of the residual BOD in the final effluent, and if a high quality effluent is required some form of tertiary treatment may be necessary. The residual BOD can be successfully removed by any tertiary treatment process, such as microstraining, sand filtration, upflow clarification, or land treatment (Section 2.4.1).

4.1.1.2   Process modifications

Percolating filters originally were designed to treat domestic wastewater on a single pass through one filter bed containing mineral media of nominal size between 25–50 mm. At a hydraulic loading of <0.4 m^{3}m^{-3}d^{-1} and an organic loading of 0.12 kg BOD m^{-3}d^{-1} a final effluent with a BOD and suspended solids concentration not exceeding 20 and 30 mg l^{-1} respectively can be obtained. Assuming a third of the organic load is removed during primary sedimentation, the per capita loading rate of a single-pass low-rate filter is equivalent to approximately 2.6 people per cubic metre of filter medium. Over time, the general trend of increasing loads to treatment plants plus more stringent disposal standards has resulted in the necessity of expanding low-rate filters. However, because of cost and of land availability restrictions, ways of uprating the efficiency of this original system have been examined. Currently, there are a number of modifications in use that have effectively increased the per capita loading per unit volume of medium.

Filtration can be separated into single- and multi-stage processes. Wastewater treated by a single filter or by filters operated in parallel is known as single-stage filtration. Examples include low- and high-rate single filtration, and recirculation. In multi-stage processes the wastewater is treated progressively by filters operated in series and includes double filtration, alternating double filtration, and two-stage systems (Fig. 4.7).

Single-stage systems

Conventional single-pass low-rate filtration (Fig. 4.7) aims to produce a final effluent of a very high standard. With good operational conditions, almost complete purification including nitrification, is possible. To achieve this, both the hydraulic and organic loadings must be kept low in order to obtain a low growth rate of film. Humus production is low and the sludge highly mineralized and easy to dewater. Using medium of a nominal size of 37.5–50 mm, the maximum loading rate, in the UK and Ireland, does not generally exceed 0.10–0.12 kg BOD m^{-3}d^{-1}, if a Royal Commission standard final effluent is required. In practice, the BOD and suspended solids concentration fluctuates seasonally and is highest in the spring. Such systems

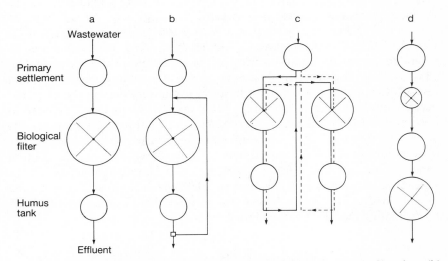

FIG. 4.7 Main systems of operation of percolating filters. (a) Single filtration; (b) recirculation; (c) alternating double filtration (ADF); (d) two-stage filtration with high-rate primary filter.

are employed at small sewage treatment plants, and where an appropriate hydraulic head is available they can be operated without a power supply. Where the flow is $> 1000 \ m^3 d^{-1}$ modifications to conventional single-pass filters are usually used.

The earliest modification to the percolating filter process was recirculation and is now built into most single filtration systems (Fig. 4.7). Recirculation has generally been used to dilute strong wastewaters or wastewater containing a large proportion of industrial waste, particularly to reduce toxic components to below inhibitory concentrations, or occasionally to 'even-out' the diurnal fluctuations in flow to the filters, which ensures a uniform hydraulic loading. This modification results in an increase in the hydraulic loading to the filter, which has a number of advantages. The stronger flushing action prevents surface accumulation of film and reduces the film accumulation within the bed encouraging a thinner, faster-growing film which is more efficient in removing organic matter. The greater hydraulic loading increases the area of the media utilized. The diluted influent also restricts film accumulation at the surface of the filter because of a reduction in the available nutrients in the wastewater, whereas the recirculated fraction contains dissolved oxygen, nitrate, and micro-organisms that may be beneficial. Thus, by diluting the influent with recirculated treated effluent, higher organic loadings can be applied than to single-pass low-rate filters because excessive film accumulation is controlled. These factors allow the loading to be safely increased to between 0.15 and 0.20 kg BOD $m^{-3} d^{-1}$ with little effect on the final effluent quality (90–95% BOD removal), although nitrification will be slightly suppressed. When the nutrients in wastewater are soluble and readily oxidized, such as in food-processing wastes, the organic loading may be as high as 0.5–0.6 kg BOD $m^{-3} d^{-1}$ (Oliver and Walker 1961; Jackson and Lines 1972; Peacock 1977). At higher hydraulic loadings, more of the filter depth is used by heterotrophic micro-organisms because the food is more evenly distributed. This results in the zone of carbonaceous oxidation extending much deeper and restricts the area available for nitrification to the very base of the filter and occasionally excludes it altogether. The ratio of recirculated effluent to influent wastewater is the recirculation ratio and for domestic wastewater it is generally either 1 : 1 or 2 : 1. The recirculation ratio can rise to as high as 25 : 1 for very strong industrial wastewaters. The ideal loading rate after dilution should be between 0.1 and 0.15 kg BOD $m^{-3} d^{-1}$, although the actual effluent BOD concentration is normally used to determine the ratio and therefore the organic loading is normally higher. The recirculation ratio can be calculated by estimating the BOD of the diluted influent fed to the filter (S_o):

$$S_o = \frac{Q_i S_i + Q_r S_r}{Q_i + Q_r}$$

where the flow rate (Q) and the BOD (S) are known for the influent (i) and the recycled or treated effluent (r). The recycle ratio R_r is $Q_r : Q_i$, and the equation can be rearranged as:

$$S_o = \frac{S_i + R_r S_r}{1 + R_r}.$$

Because the effluent BOD in low-rate systems is usually very low, S_o can be estimated as:

$$S_o = \frac{S_i}{1 + R_r}$$

Recirculation can be considered as medium-rate filtration, whereas recirculation in high-rate systems has been found to have little effect on BOD removal efficiency. This is probably because the flushing action of the wastewater in plastic media filters is of less importance than in mineral beds. Recirculation has been studied in detail by Lumb and Eastwood (1958).

Conventional single-pass filtration is also used for partial treatment of wastewater (Fig. 4.7). Settled or raw wastewater that has been finely screened can be applied to filters containing mineral media with a large nominal size > 63 mm, random plastic medium or more commonly modular plastic medium at high organic and hydraulic loadings. High-rate filtration is used for partial treatment before discharge to a sewer or to surface water, especially estuaries and coastal waters, where the effluent discharge standards permit. The high voidage prevents ponding and heavy accumulations of film, so that film growth is rapid with a constant and high removal of film from the filter. The humus from such filters is poorly oxidized making it less stable than conventional low-rate humus and more difficult to dewater (Bruce and Merkens 1970). Almost any degree of partial treatment can be obtained (50–70% BOD reduction), depending on the loading rate. High-rate filters are operated at hydraulic loading $> 3 \ m^3 m^{-3} \ d^{-1}$ and at organic loadings $> 0.6 \ kg \ BOD \ m^{-3} d^{-1}$. In general a loading of $1 \ kg \ BOD \ m^{-3} d^{-1}$ will give a BOD removal rate of between 80–90%, whereas this falls to 50% at loadings of between 3–$6 \ kg \ BOD \ m^{-3} d^{-1}$. The specific removal rate depends on the nature of the wastewater and generalizations are difficult. Performance data for a range of wastewaters is given elsewhere (Chipperfield 1967; Jackson and Lines 1972; Anderson 1977). Coarse mineral media (63–150 mm) must be loaded at significantly lower organic loading rates than modular media. Modular and random plastic media have been successfully used to treat macerated and finely screened raw sewage. This eliminates the use of primary sedimentation which results in a considerable saving in primary treatment and subsequent sludge handling (Hemming 1979; Hoyland and Roland 1984). The efficiency of high-rate filters is generally measured in terms of weight of BOD removed per unit volume of medium per day, rather than percentage removal as is the case with low-rate filters. Low-rate filters remove a high percentage of the influent organic matter but the rate of removal is low in terms of mass of nutrients removed per unit volume. Conversely, high-rate systems remove a smaller proportion of the influent organic matter but at a much higher rate. This has given rise to the general terms of 'polishing' and 'roughing' filters for low- and high-rate systems respectively. The percentage removal of BOD decreases as the loading increases, whereas the mass of BOD removed increases with loading up to a critical limiting value above which the amount removed remains constant (Winkler 1981).

Multi-stage systems

The most widely used multi-stage system is double filtration (DF). In this modification two sets of similar beds are used in series. The media is essentially of the same grade (63–75 mm) and normally sedimentation is employed between each stage to minimize the risk of the humus solids clogging the interstices of the medium in the second filter. Intermediate sedimentation is not absolutely necessary, but if not used it will reduce the performance slightly and increase the amount of solids discharged from the second filter (Tomlinson and Hall 1953). The major problem with using identical filters was that the primary filter often became ponded during the winter (Tomlinson and Hall 1951; Barraclough 1954), which severely limited the overall loading rate. This has been overcome by a number of modifications. Some degree of control of film accumulation in the primary stage was achieved by using low frequency dosing, although the system was vastly improved by using high-rate medium in the first stage (two-stage filtration) and by alternating double filtration.

Two-stage filtration is a direct development of DF, although some textbooks fail to distinguish between the two. Here, different media are used in each filter with the first stage operated as a roughing filter and the second stage for polishing the partially treated wastewater. The first stage is operated as a high-rate filter at high loading rates using either a coarse mineral medium (75–130 mm) or plastic media with a high voidage. This first stage reduces the organic loading by 70% allowing the second stage, after intermediate sedimentation, to be operated as a conventional low-rate filter. At loading rates of between 1.6–2.3 kg BOD $m^{-3}d^{-1}$ for the first filter, and 0.04–0.12 kg BOD $m^{-3}d^{-1}$ for the second stage, a 20:30 effluent with a high degree of nitrification is possible (Forster 1977). The system can be further enhanced by using recirculation. Where very strong wastewaters are being treated it may be necessary to have several roughing filters in series to reduce the organic loading sufficiently to allow complete treatment by the final low-rate stage. This multiple unit system is known as multi-stage filtration (Fig. 4.7).

A further development of the DF system is alternating double filtration (ADF). Identical filters are used, each containing the same medium, with the film accumulation in the first stage controlled by periodically reversing the sequence the filters are used, so that each filter is subject to successive periods of feeding and starvation (Fig. 4.7). The basic research was carried out at Minworth in Birmingham, where so much pioneering work on percolating filtration has been done (Wishart *et al.* 1941; Mills 1945; Tomlinson 1946*b*). The filters are filled with a coarser medium (63–75 mm) than normally used for single filtration, which allows a relatively high loading to the filter. The film is controlled to a certain extent by the high flow, distributing the growth of film more evenly through the filter and providing some restriction of film accumulation by washing out film. The rate of film growth at the high organic loadings used is very rapid in the first filter of the sequence. Loading in this order continues for 1–2 weeks until the filter has accumulated a heavy growth of film and is close to becoming clogged. At this point, the flow is reversed so that the second filter is being fed with dilute partially purified wastewater. This puts the film in the second filter under nutrient-limited conditions and forces the micro-organisms to enter an endogenous growth phase, which reduces the overall film biomass but at the same time retains a healthy and active film. Humus is removed after every stage and the ADF process allows the organic loading to be increased by at least a factor of two. For domestic wastewater, a 20:30 effluent can be produced at a loading of 0.24 kg BOD m^{-3} d^{-1} (1.5 m^{3}m^{-3}d^{-1}) (Calley *et al.* 1977), although loadings of 0.30–0.47 kg BOD m^{-3}d^{-1} are normal (Forster 1977). The starvation effect on the second filter is important to ensure film reduction, and with stronger wastes, such as meat-processing wastewaters, recirculation may also be necessary to achieve a wastewater dilute enough for the second stage. A major disadvantage with the ADF process is that nitrification is suppressed. Nitrifying bacteria are normally found at the base of conventional filters where all the carbonaceous material has been utilized. The nitrifying bacteria are slower growing than heterotrophs and are unable to compete for available space. Thus, by alternating the filters in this way they never become established, which results in poor nitrification (Bruce *et al.* 1975).

Modern percolating filter plants are designed with enough flexibility to allow the filters to be operated at single or double, and even ADF filtration, depending on the strength of the incoming wastewater and the degree of treatment required.

4.1.1.3   The organisms and their ecology

Whereas the activated sludge process is a truly aquatic habitat, the percolating filter provides a variety of habitats, ranging from aquatic to terrestrial. In this respect the filter bed is similar to the sea-shore, with the aquatic micro-organisms forming the film that supports a wide range of air breathing macro-invertebrate grazers. The natural habitat of these micro-organisms is the surface of stones and other submerged objects in rivers and streams, especially those receiving large nutrient inputs. The micro-organisms are also associated with similar aquatic habitats such as lakes and ponds, wetlands, and other damp environments.

The food-chain in a percolating filter contains several extra-trophic levels compared with the activated sludge process. The meiofauna provides the top trophic level in activated sludge, whereas in fixed-film reactors there is also present a diverse range of macro-invertebrates (Fig. 4.8). The basic unit of purification is the heterotrophic film, which converts the soluble and suspended organic matter in the influent wastewater into the bacterial and fungal biomass on which the rest of the community depends. Protozoa graze mainly on the film or on free-living bacteria, although a few are predators of other protozoans or saprophytes. The remaining meiofauna, mainly nematodes and rotifers, also feed on the film, although several are predatory. The grazing fauna, which are dominated by dipteran larvae and oligochaete worms, feed mainly on the film. The main energy transformations are given in Fig. 4.8. A typical list of species from a percolating filter is given in Table 4.10 and contains species of: Bacteria, Fungi, Algae, Protozoa (Sarcomastigophora and Ciliophora), Nematoda, Rotifera, Annelida, Insecta, Crustacea,

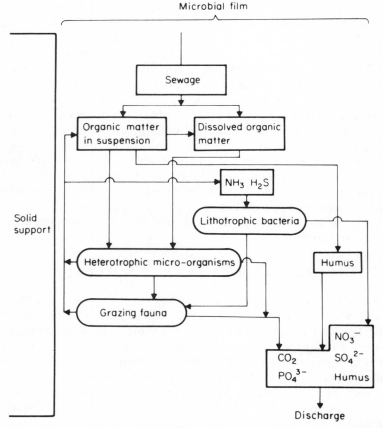

FIG. 4.8. Food web in the microbial film of percolating filters (Lynch and Poole 1979).

TABLE 4.10   Typical list of species recorded from a single percolating filter

Bacteria
Zoogloeal forms, mainly *Zoogloeal ramigera*
Sphaerotilus sp.
Leptothrix sp.
Beggiatoa sp.

Fungi
Subbaromyces splendens Hesseltine
Sepedonium sp.
Fusarium aquaeductuum (Radimacher and Rabenhorst) Saccardo
Geotrichum candidum Link

Algae
Chlorella sp.
Scenedesmus sp.
Stigeoclonium sp.

Protozoa: Sarcomastigophora
Bodo sp.
Amoeba sp. mainly *Amoeba radiosa* Ehrenberg
Euglena sp.

Protozoa: Ciliophora
(Holotrichia)
Trachelophyllum pusillum Perty-Claparède and Lachmann
Hemiophrys fusidens Kahl
Hemiophrys pleurosigma Stokes
Chilodonella cucullulus (Müller)
Chilodonella uncinata Ehrenberg
Colpoda cucullus Müller
Uronema nigricans (Müller) Florentin
Glaucoma scintillans Ehrenberg
Colpidium colpoda Stein
Colpidium campylum (Stokes)
Paramecium aurelia Ehrenberg
Paramecium caudatum Ehrenberg

(Peritrichia)
Vorticella microstoma Ehrenberg
Vorticella convallaria Linnaeus
Vorticella vernalis Stokes
Opercularia minima Kahl
Opercularia microdiscum Faure-Fremiet
Opercularia coarctata Claparède and Lachmann
Epistylis rotans Svec

(Spirotrichia)
Stentor roeseli Ehrenberg
Aspidisca costata (Dujardin) = *cicada*
Tachysoma pellionella (Müller-Stein)
Oxytricha ludibunda Stokes

(Suctoria)
Acineta cuspidata Stokes
Acineta foetida Maupas
Podophrya maupasi Bütschli
Podophrya carchesii Claparède and Lachmann
Podophrya mollis Bütschli
Sphaerophrya magna Maupas

TABLE 4.10—*Continued*

Nematoda

Rotifera
(Bdelloidea)
Philodina roseola Ehrb.

(Monogonota)
Lecane sp.
Dicranophorus sp.

Annelida
(Oligochaeta)
Enchytraeidae
  Enchytraeus buchholzi Vejdovsky
  Lumbricillus rivalis Levinsen

Lumbricidae
  Dendrobaena rubida (Sav.) f. Subrubicunda (Eisen)
  Eiseniella tetraedra (Savigny)

Insecta
(Collembola)
Isotomidae
  Isotoma olivacea-violacea gp.

(Coleoptera)
Hydrophilidae
  Cercyon ustulatus (Prey.)

Staphylinidae
  Unidentified sp.

(Diptera)
Anisopodidae
  Sylvicola fenestralis (Scop.)

Psychodidae
  Psychoda alternata Say.
  Psychoda severini Tonn.

Chironomidae
  Hydrobaenus minimus Mg.
  Hydrobaenus perennis Mg.
  Metriocnemus hygropetricus Kieff.

Ephydridae
  Scatella silacae Lw

Sphaeroceridae
  Leptocera sp.

Cordyluridae
  Spathiophora hydromyzina Fall.

(Chilopoda)
  Lithobius forficatus Linn.

Crustacea
(Cyclopoida)
Cyclopidae
  Paracyclops fimbriatus-chiltoni (Thomson)

Table 4.10—*Continued*

Arachnida
(Acari)
Acaridae
  Histiostoma carpio (Kramer)
  Rhizoglyphus echinopus (Fumouze and Robin)
Anoelidae
  Histiostoma feroniarum (Dufour)
Ascidae
  Platyseius italicus (Bertese)
(Araneida)
Linyphiidae
  Unidentified sp.
Mollusca
(Gastropoda)
Limacidae
  Agriolimax reticulatus (Müll.)

Arachnida (Acari and Araneida), and the Mollusca. This is by no means a definitive list and a vast diversity of species from all phyla have been recorded (Clay 1964; Learner 1975a; Curds and Hawkes 1975). Larger species also regularly visit the surface of the filter beds to feed, especially insectivorous birds such as Wagtails and various small mammals such as rodents and hedgehogs, although access to the mammals is often restricted by the filter bed wall. The ecology of percolating filters has been fully reviewed by Hawkes (1959, 1963, 1983b), with each major group examined in detail by Curds and Hawkes (1975).

Heterotrophs and phototrophs
Bacteria form the basic trophic level in filter beds and are predominant both in terms of numbers and biomass, with aerobic, anaerobic, and facultative species all present. The bacterial flora in percolating filters are similar to those found in activated sludge. These species are able to grow over an inert medium dominating filamentous forms. The dominant aerobic genera are the Gram-negative rods *Zoogloea*, *Pseudomonas*, *Achromobacter*, *Alcaligens* and *Flavobacterium* (James 1964; Harkness 1966; Pike and Carrington 1972; Pike 1975). Filamentous bacteria such as *Sphaerotilus natans* and *Beggiatoa* sp. are frequently associated with percolating filters (Cooke 1959; Bruce *et al.* 1970; Wheatley 1976), but are rarely dominant. Their abundance is directly associated with organic loading in filters, although they do not cause operational problems as they do in activated sludge (Section 4.2). In a conventional low-rate single-pass filter, the heterotrophic bacteria and fungi are responsible for the primary oxidation of the organic matter in the wastewater. The actual microbial composition of the film varies seasonally

with filter depth and the composition of the wastewater, and it is difficult to characterize the microbial community structure. An example of seasonal and depth variation in a pilot filter containing 50 mm blast-furnace slag medium treating domestic wastewater is given in Fig. 4.9. As the wastewater passes through the filter bed it is being purified by the film, and changes occur as it percolates over the medium resulting in the wastewater having a different composition at various depths. As a consequence, the microbial community structure varies in composition at different depths, having a greater number of species nearer the surface than at the base. This is why there is a greater diversity of species in percolating filters than in completely mixed systems such as activated sludge. The stratified distribution of species within the filter bed allows it to withstand shock loadings.

Fungi are more common in percolating filters than in activated sludge, as they are among the first species to colonize the media. Although many fungi can be isolated from percolating filters only a few manage to take advantage of the habitat and flourish (Cooke 1959). Among those that do flourish are *Geotrichium candidum*, *Fusarium aquaeductuum*, *Sepedonium* sp., *Subbaromyces splendens*, *Ascoidea rubescens*, and *Trichosporon cutaneum*. Fungi dominate over heterotrophic bacteria under specific conditions. They outgrow bacteria at colder temperatures, under conditions of low pH such as wastewater from fruit and vegetable processing or effluents containing mineral acids, and also dominate filters treating strong wastewaters with a high organic content. For this reason, fungi are associated mainly with industrial effluents (Hesseltine 1953; Watson *et al.* 1955; Hawkes 1957, 1965; Tomlinson 1946b; Sladka and Ottova 1968). Fungi have a lower surface area to volume ratio compared to bacteria and thus have a lower affinity for available substrate than heterotrophic bacteria. Therefore, bacteria will predominate when the substrate concentration is low, i.e. when the wastewater is weak, or at lower depths in the filter bed after a proportion of the substrate has been utilized and removed. Conversely, fungi dominate when the substrate concentration is high and usually at the surface of the filter bed before the strength of the influent has been reduced. Fungi have the ability to attach strongly to medium and are not easily dislodged by the wastewater. The mycelium physically entrains solids and fungal dominant films grow rapidly compared to bacterial films that rely on adsorption only. This results in very thick layers of fungal film developing which, due to the reduced grazing activity during the winter, is why the interstices of the medium become clogged and ponding often results. The hyphae protrude into the wastewater passing over the surface of the film, and fungal-dominated films are less likely to be nutrient-limited as the nutrients can diffuse throughout the film, resulting in continuous growth. Fungal-dominated films are so thick that if the film was bacterial in nature, oxygen would become limiting in the deeper regions. However, the nature of the hyphae allows oxygen transfer deep into the film layer by a process of protoplasmic streaming within the hyphae, preventing

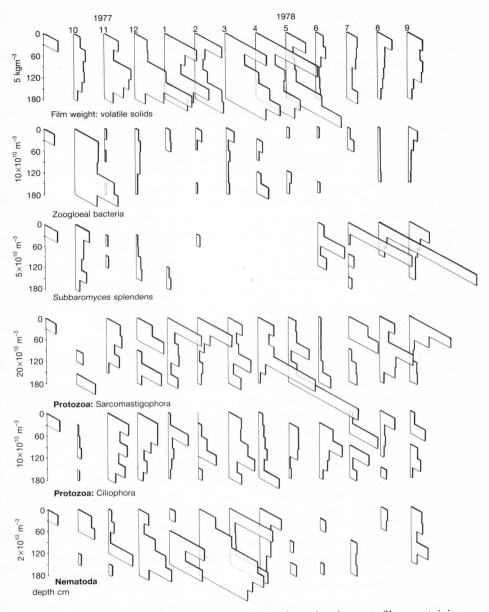

FIG. 4.9 Vertical distribution of film and microfauna in a low-rate filter containing 50 mm blast-furnace slag medium.

anaerobic conditions associated with thick bacterial-dominated films (Tomlinson and Snaddon 1966). Fungi are generally considered undesirable as dominant members of the film community, as they cause solids accumulation and eventual ponding (Hawkes 1963). Many authors have also associated heavy fungal films in filters with very large populations of fly larvae, which result in fly problems later on. However, the fungi have the same removal efficiencies as the bacteria, although the fungi produce a greater biomass per unit BOD removed resulting in faster film accumulation and, eventually, a greater sludge production (Water Pollution Research Laboratory 1955).

Fungi often dominate the film that develops on plastic media, and is able easily to colonize the smooth surface. One particular fungus, *Subbaromyces splendens*, grows particularly well on random plastic filter medium. The fungus is unusual because it has only ever been isolated from percolating filters, and its natural habitat is unknown (Hesseltine 1953). It is also unusual because it is associated with weak domestic wastewaters (Hawkes 1965; Hawkes and Shephard 1972; Wheatley 1976; Gray 1983b). Fungi are generally found in greatest abundance on the surface and in the top 150 mm of the filter bed (Hawkes 1963). They follow a seasonal pattern of abundance, which slowly increases in autumn and reaches a peak in winter or spring, becoming scarce by mid-summer. A comparative study by Gray (1983b) showed that *S. splendens* produced more extensive growth on random plastic medium than on conventional mineral medium. The fungus was not restricted to the upper 150 mm but produced extensive vegetative growth throughout the depth of the filter causing localized ponding both at the surface and inside the bed. An increase in the hydraulic loading normally leads to a reduction in the fungal growth at the surface of a filter, resulting in a more even distribution of film throughout the bed (Hawkes 1957). This was also the case in the comparative study, when the loading was increased from $1.68 \text{ m}^3\text{m}^{-3}\text{d}^{-1}$ ($0.28 \text{ kg BOD m}^{-3}\text{d}^{-1}$) to $3.37 \text{ m}^3\text{m}^{-3}\text{d}^{-1}$ ($0.63 \text{ kg BOD m}^{-3}\text{d}^{-1}$). However, at the higher loading the total biomass of the fungus in the filter containing the plastic medium was increased nine-fold, with ponding occurring in the lower half of the filter. Fungal films are structurally stronger than bacterial ones and less readily sloughed off (Tomlinson and Williams 1975). It is this aspect that may help fungi, and, in particular *S. splendens*, to dominate in filters containing the smoother plastic medium. Other factors include: a higher medium suface area resulting in less competition for space with bacteria; the lower diversity of niches within plastic media compared with mineral media for supporting grazing fauna; and the large voidage permitting a high ventilation rate resulting in the temperature of the bed being significantly cooler than a mineral filter bed and permitting a large diurnal fluctuation in temperature, which will be similar to the variation in the air temperature thus inhibiting bacterial growth. The ability of *S. splendens* to cause ponding at depth within the filter has a practical significance, apart from problems in operational control. It has generally been accepted that random plastic filter media can be housed in prefabricated or weaker structures than normally used for mineral media, due to the difference in bulk densities (Table 4.2). However, if ponding was to occur at depth, then the bulk density would increase proportionally to the extra volume of sewage retained. It may, therefore, be possible that under extreme conditions, ponding by the fungus might increase the bulk density of a filter, housing random plastic medium, in excess of the designed strength of the unit. This could result in the structure becoming damaged, weakened or even failing and collapsing. Three control options are suggested:

1. *Recirculation or alternating double filtration:* Hawkes (1965) suggests that fungal growths in percolating filters can be controlled by these modifications. However, the study by Gray did not show that the fungus was restricted at increased hydraulic loadings or that the growth of the fungus was related to the sewage strength at a particular depth.

2. *Periodic dosing:* Continuous dosing has been shown to favour the fungus *S. splendens*, and Hawkes and Shephard (1972) demonstrated that the fungus could be controlled by reducing the frequency of dosing. Although surface growths would be controlled, careful consideration would have to be given to the retention time as this may even out the flow at lower depths in the filter resulting in enhanced growth of the fungus deeper in the filter bed.

3. *Design:* The safest precaution, but the most expensive, is to ensure that the structure housing the medium is able to withstand maximum possible bulk density, i.e. total saturation with water (Table 4.2), making the filter completely safe. Such a design, however, would make random plastic filter media less attractive economically, because one of the major advantages of using such media is the saving of capital cost of engineering work (Gray 1983*b*).

The fungal ecology of percolating filters has been reviewed by Cooke (1954, 1963), Becker and Shaw (1955), and Tomlinson and Williams (1975).

The surface of the filter is the only part of the medium exposed to sunlight and photosynthetic algae and bacteria are restricted to the top 50 mm of the bed. Phytosynthetic species, and algae in particular, play a very minor role in the purification process, but can interfere with the efficient operation of the filter. Algae may be present as thin or dispersed incrustations of unicellar algae and diatoms, or filamentous species can form thick luxuriant surface mats covering the surface of the medium. The algal sheets impair distribution, decrease ventilation, and may even cause ponding. The major species causing the dark-green, thick surface mats so often seen in percolating filters is *Phormidium* sp., which forms a characteristic leathery mucilaginous sheet over the medium. Thin cellular and filamentous algae are equally abundant in percolating filters with the commonest species being *Chlorella*, *Chlorococcum*, *Phormidium*, *Oscillatoria*, *Stigeoclonium*, and *Ulothrix* (Benson-Evans and

Williams 1975). Mosses and liverworts are also occasionally found. Moss growth can become so extensive that the entire surface of the bed can become affected. The moss traps solids forming a thick layer of debris in the interstices of the medium which interferes with percolation. The most frequently occurring species is *Leptodictyum* (= *Amblystegium*) *riparium*, although Hussey (1982) recorded 12 species of moss in a survey of 64 filters. Factors that encourage moss growth include: a rough, pitted medium; wide spacing of the distribution nozzles providing interjet zones; a low organic loading; and the absence of strong industrial or inhibitory wastewaters. The ecology of the filter is altered, with significant changes in the grazing fauna occurring due to the presence of the moss (Table 4.11) (Gray 1984b). Raking may not effectively

TABLE 4.11   Comparison of macro-invertebrate densities (no. l^{-1} medium) in filter medium with and without moss growth. The 95% confidence limits and the level of significance (*P*) of the Mann–Whitney test statistic *U* are also given

Macro-invertebrate group	Medium		*P*
	Moss present $\bar{x}$ CL	Moss absent $\bar{x}$ CL	
Lumbricidae	50 ± 8	4 ± 6	<0.001
Enchytraeidae	1040 ± 420	356 ± 205	<0.05
Psychodidae (larvae)	1514 ± 370	809 ± 317	<0.05
Tipulidae (larvae)	11 ± 3	0 ± 0	<0.001

control the moss, and may even help to spread it over the rest of the surface of the bed. Herbicides are effective but can also destabilize the community structure of the filter bed by killing other species. Success has been obtained by excluding the light from the surface of the bed or, alternatively, by using a flame gun to burn off the moss. Increasing the organic strength of the influent by reducing the number of filters in operation eliminates moss growth over a period of six months, although there is also normally a deterioration of the final effluent quality. Control can be obtained by increasing the frequency of dosing, reducing the interjet spacing on the distributors, or by using splash plates to eliminate the injet zones on the surface of the medium. Replacing or covering the surface layer of medium with a smooth surfaced random plastic filter medium will also discourage growth. Prevention is better than trying to eradicate extensive moss growth. Thus, when the surface of the filter is checked periodically for accumulated inorganic debris (plastic strips, etc.) any moss can be weeded out before it becomes established.

Nitrification
Nitrification is a two-stage process with ammonia oxidized to nitrite by bacteria of the genus *Nitrosomonas*, and nitrite to nitrate by *Nitrobacter* spp.

(Section 3.5.2). In low-rate single pass filters containing 50 mm stone medium, virtually full nitrification can be obtained throughout the year with a specific ammonia removal rate of between 120–180 mg m^{-2}d^{-1} when loaded at 0.1 kg BOD m^{-3}d^{-1}m resulting in a final effluent low in ammonia but rich in nitrate. In a percolating filter, nitrifying bacteria tend to become established later than heterotrophs, with *Nitrosomonas* becoming established before *Nitrobacter*, as ammonia is abundant. Therefore, the first sign of nitrification in a filter is the production of nitrite rather than nitrate. The number of nitrifying bacteria and the level of nitrifying activity increase with depth (Tomlinson 1942; Harkness 1966; Painter 1970). This results in the upper level of single pass filters being dominated by heterotrophs, and the lower section containing a proportionately higher number of nitrifying bacteria. The reason for this apparent stratification is due to a number of factors. The autotrophic bacteria responsible for nitrification are slow growing compared to heterotrophs and have an even more reduced growth rate in competitive situations. Therefore, in the upper layers of the bed where there is abundant organic matter, the heterotrophs will dominate. Nitrifiers are extremely sensitive to toxic compounds in the wastewater, especially heavy metals, and so the presence of such compounds will limit the growth of the bacteria until the compounds have been removed from the wastewater by adsorption by the heterotrophic film as it passes through the filter bed. Research has shown that the process is also inhibited when the oxygen concentration in the influent wastewater is limited (Heukelekian 1947; Hawkes 1963; Tomlinson and Snaddon 1966). Nitrifying bacteria are strict aerobes and will be inhibited by reduced aerobic conditions caused by high heterotrophic activity, and as nitrification is a high oxygen consuming process, adequate supplies of air are required. Although Painter (1970) showed that organic matter did not *directly* inhibit nitrification, he indicated that the nitrifying bacteria needed to be attached to a stable surface, suggesting inhibition may be due to competition for space. When loadings are increased, extending the depth of heterotrophic activity, nitrifying bacteria are overgrown and killed by the more rapidly growing heterotrophic bacteria. Nitrification is virtually eliminated by hydraulic loadings of domestic wastewater in excess of 2.5 m^{3}m^{-3}d^{-1}. This enhances heterotrophic growth which can extend throughout the depth of the bed and exclude the nitrifying organisms (Bruce *et al.* 1970; Joslin *et al.* 1971; Bruce *et al.* 1975). Temperature also has a marked influence on nitrification (Painter 1970), and the large fluctuations in temperature seen in random plastic filters account for the low degree of nitrification that occurs with such media (Gray 1980). Although the threshold temperature for the process is 10°C, with domestic wastewater rarely falling to below 12°C (normal annual range 12–18°C), a few degrees reduction in the temperature below 10°C is likely to have a disproportionate reduction on nitrification (Bruce *et al.* 1975). This is clearly seen in comparative studies between mineral and plastic medium filters where the high voidage of the plastics medium allows a greater degree of ventilation and large diurnal changes in temperature similar to the air temperature.

Many workers have noticed discrepancies in the total nitrogen balance of their filters, with the decrease in ammonia concentration not corresponding to the increase in nitrate concentration (Bruce *et al.* 1975; Hemming and Wheatley 1979; Gray 1980). This is due to the ammonia being supplied not only in the applied wastewater, but also within the filter from the de-animation of organic nitrogen, endogenous respiration, and from cell lysis. Ammonia, on the other hand, is not only removed by nitrification, but also by volatilization of free ammonia and by metabolism into new cellular material. Gray found that at times of low film accumulation, when the grazing fauna population was still large, the ammonia concentration in the final effluent was high. He concluded that the observed increase in the ammonia concentration was due to the excretion products of the grazing fauna. The oxygen profile formed by diffusion through the film results in nitrate ions being lost by denitrification as gaseous nitrogen from the anoxic zone near the biomass–medium interface.

Meiofauna

Protozoa are particularly abundant in percolating filters and over 218 species have been isolated. There are 35 species of Phytomastigophorea, 30 species of Zoomastigophorea, 31 species of Rhizopodea and 7 species of Actinopodea. However, the bulk of the species, some 116, belong to the class Ciliatea. The abundance of ciliates, amoebae, and flagellated species in filters has resulted in most studies concentrating on these groups alone.

The role of protozoans in percolating filters is similar to that of the activated sludge process, with species feeding mainly on free-living bacteria and clarifying the effluent as well as stimulating bacterial growth by reducing the population density. Some protozoans are able to feed on non-living particulate, or soluble organic material, whereas a few species are predators of other protozoans. It is generally thought that the ciliates are numerically dominant over the flagellates (Frye and Becker 1929; Brink 1967), although Barker (1942, 1946) and Gray (1980) both found flagellates to be dominant in percolating filters. Gray found the ratio of Sarcomastigophora to Ciliophora to range from 3.7 to 4.3 in low- and high-rate mineral media filters and 2.0 to 4.4 in low- and high-rate random plastic media filters respectively. However, flagellates are significantly smaller than ciliates both in overall dimensions and in volume, and, in terms of biomass, ciliates are always dominant. The population density of protozoans increase with organic loading, although the flagellated and free-swimming ciliate protozoans are more susceptible to being flushed out of the plastic medium filter at higher loading. It has been reported that the number of ciliate species in filters increases with depth (Baker 1949; Baines *et al.* 1953; Curds 1975; Wheatley 1976), and Hussey (1975) found that ciliate diversity was negatively associated with the film weight. Many of these workers suggested that ciliates are unable to compete effectively with the other organisms normally associated with the film in the low-dissolved oxygen conditions prevalent at the filter surface. Therefore, the greatest diversity and largest abundance of ciliates occur in the lower depths of filters where there is an increasingly smaller concentration of organic matter in the partly treated sewage. Vertical distribution of various ciliate species within filter beds have been reported (Lackey 1924, 1925; Frye and Becker 1929; Cutler *et al.* 1932; Holtje 1943; Barker 1946; Ingram and Edwards 1960). Generally, they found that particular species tended to predominate at certain depths, stratification being dependent on any one factor such as nutrition, or a mixture of environmental and biological interactions. Liebmann (1951) showed that whereas bacterial feeding species were present throughout the bed, being especially abundant in the upper areas, predatory species of protozoans were restricted to the lower half of the filter. Liebmann (1949) also put forward the saprobity theory, that as the wastewater passed through the filter it was gradually purified so that the saprobic nature of the liquid environment changed with depth. This was polysaprobic at the top, followed by α- and β-mesosaprobic zones as the depth increased. Thus, for example, *Vorticella microstoma*, *Glaucoma scintillans*, and *Colpidium colpoda* are restricted to the surface of the filter (polysaprobic), whereas *Paramecium caudatum*, *Chilodonella uncinata*, *Uronema nigricans*, *Opercularia coarctata*, and *Podophyra fixa* are all typical of the middle regions of the filter (α-mesosaprobic). *Aspidisca costata* is usually recorded in the lower regions of filters and is associated with the lowest levels of organic matter (β-mesosaprobic). Gray found that at the lower loading, ciliates were widely distributed throughout the depth of his pilot filters, and although certain species were limited to specific regions, that, generally, there was no increase in either diversity of species or abundance with depth. At the higher loading, maximum species diversity and the abundance of protozoans occurred in the lower half of the filters. The increased abundance of certain species in the lower depths of the filters at this loading, such as *Opercularia microdiscum*, was not due to the lower organic content of the influent sewage at that depth as suggested previously, because this species is tolerant of both organic load and film accumulation. It is more probable that the increased occurrence of protozoans at the lower depth was a consequence of the increased loading. This is confirmed by a greater number of individuals being washed out with the final effluent at the higher loading. In conventional filters, there would be a greater tendency for the protozoan fauna to be forced deeper into the filters by using the normal instantaneous and heavy system of sewage application, compared to the finer distribution system used by Gray on his pilot filters (Gray and Learner 1983).

The class ciliatea is divided into four subclasses:

1. *Suctoria* are predatory and do not possess cilia or locomotary organs, they catch other protozoans with tentacles. Two types of tentacles are seen, sharp tentacles that capture prey by piercing its body, or tentacles that hold

the prey by suction via an adapted terminal knob. Typical genera include *Tokophyra*, *Acineta*, *Podophyra*, and *Sphaerophrya*.

2. *Peritrichia* are sesscile species with no cilia on the body but conspicuous oral ciliatures. The body is barrel or inverted bell-shaped and is borne on a stalk which may be contractile. In some species, the body can break free from the stalk to form a free-swimming form (a telotroch). By closing the mouth the telotroch can use the ring of aboral cilia to propel the body through the water. Typical genera are *Vorticella*, *Carchesium*, *Opercularia*, and *Epistylis*.

3. *Holotrichia* are free-swimming protozoans, having a body covered in uniform ciliature. Typical *Holotrichia* include the genera *Chilodonella*, *Colpoda*, *Colpidium*, *Paramecium*, and *Uronema*.

4. *Spirotrichia*, in contrast, do not have cilia uniformly distributed over their bodies, but the cilia are bound together to form thick tufts known as cirri. Among the common genera are *Stentor*, *Aspidisca*, *Euplotes*, *Stylonychia*, and *Oxytrichia*. The taxonomy and identification of the protozoans is dealt with by Curds (1969) (Fig. 4.10).

In a survey of 52 percolating filters in the UK, Curds and Cockburn (1970*a*) found that all the filters including those treating industrial wastewater contained protozoans, with ciliates generally dominant. They found that the protozoan fauna of percolating filters closely resembled those of activated sludge but that certain species were found in either one or the other of the processes (Table 4.12). The ciliates identified in the survey consisted of 19 holotrichs, 20 peritrichs, 11 spirotrichs, and 3 suctorians. The most common ciliates found are listed in Table 4.13, with *Chilodonella uncinata* present in 90% of the samples, *Vorticella convallaria* in 83%, *Opercularia microdiscum* in 81%, and *Carchesium polypinum* in 62%. However, although species are widely distributed they are not always abundant or the dominant protozoan in

TABLE 4.12   Characteristic ciliate protozoans identified in percolating filters and activated sludge plants (Curds and Cockburn 1970*a*)

Percolating filters	Activated sludge plants
Opercularia microdiscum	*Aspidisca costata*
Carchesium polypinum	*Vorticella convallaria*
Vorticella convallaria	*Vorticella microstoma*
Chilodonella uncinata	*Trachelophyllum pusillum*
Opercularia coarctata	*Opercularia coarctata*
Opercularia phryganeae	*Vorticella alba*
Vorticella striata var. *octava*	*Carchesium polypinum*
Aspidisca costata	*Euplotes moebiusi*
Cinetochilum margaritaceum	*Vorticella fromenteli*

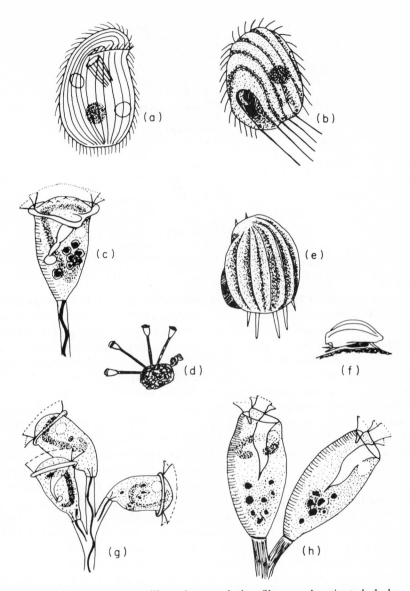

FIG. 4.10 Some common ciliates in percolating filters and activated sludge: (a) *Chilodonella unicata*; (b) *Cinetochilum margaritaceum*; (c) *Vorticella convallaria*; (d) *V. convallaria* growing on a sludge floc; (e) *Aspidisca costata*; (f) *A. costata* lateral view showing crawling habit; (g) *Carchesium polypinum*; (h) *Opercularia microdiscum* (Curds 1975).

TABLE 4.13 Most frequently observed species of protozoans recorded in percolating filters (Curds and Cockburn 1970*a*)

Class	Species
Phytomastigophorea	*Peranema trichophorum*
Zoomastigophorea	*Bodo caudatus*
	Trepomonas agilis
Rhizopodea	Small amoebae
	Arcella vulgaris
Ciliatea	*Aspidisca costata*
	Carchesium polypinum
	Chilodonella uncinata
	Cinetochilum margaritaceum
	Opercularia coarctata
	Opercularia microdiscum
	Trachelophyllum pusillum
	Vorticella convallaria
	Vorticella striata var. *octava*

a filter. Curds and Cockburn only recorded eight ciliates in large numbers, *Chilodonella uncinata* (4% of sites), *C. polypinum* (15%), *Vorticella alba* (2%), *Vorticella convallaria* (10%), *Vorticella striata* var. *octava* (2%), *Opercularia coarctata* (2%), *Opercularia microdiscum* (44%), and *Opercularia phryganeae* (4%). In his pilot-scale study, Gray (1980) observed that species were restricted by a variety of factors including saprobity, flow rate, film accumulation, season, and medium type. He found that five species dominated his filters, four holotrichs, *Paramecium aurelia*, *Uronema nigricans*, *Glaucoma scintillans*, and *Colpidium colpoda*, and one peritrich *Opercularila microdiscum*, all of which feed on bacteria (Table 4.14). Under low-rate conditions *Opercularia microdiscum*, *Uronema nigricans*, and *Paramecium aurelia* were most frequently recorded, whereas under high-rate conditions, only *O. microdiscum* was regularly recorded as abundant (Fig. 4.11). *Paramecium aurelia* is inhibited by high hydraulic loadings, most likely due to its relatively large size (120–150 μm in length). It was present in large numbers frequently comprising 70–80% of the total ciliate population (Table 4.15). The abundance of this species is linked to the film accumulation, being more abundant in the winter months and scarce when the film was at its minimum thickness during May to August (Fig. 4.12). At the low-rate loading, *P. aurelia* was recorded throughout the bed, whereas at the higher rate it was restricted to the lower half of the filter until eventually washed out. *Uronema nigricans* is also unable to cope with high hydraulic loadings but is found in greatest abundance in the summer, with numbers declining as the film accumulates. The density of *Uronema* is negatively correlated with film accumulation, which is in contrast to *P. aurelia*. *Uronema*'s preference for light film development results in *P. aurelia* and *U. nigricans* being rarely found together (Fig. 4.12).

TABLE 4.14 Number of months when specific ciliate species were dominant in experimental filters loaded at 1.68 $m^3m^{-3}d^{-1}$ (low loading) and 3.37 $m^3m^{-3}d^{-1}$ (high loading) (Gray 1980)

Species	Low loading			High loading			Total		Total Dominance
	Slag	Filter Mixed	Plastic	Slag	Filter Mixed	Plastic	Low loading	High loading	
Uronema nigricans	2	5	5	1	1	2	12	4	16
Paramecium aurelia	5	1	4	1	0	1	10	2	12
Opercularia microdiscum	5	5	2	8	7	5	12	20	32
Colpidium colpoda	0	0	1	1	2	3	1	6	7
Glaucoma scintillans	0	1	0	0	1	0	1	1	2

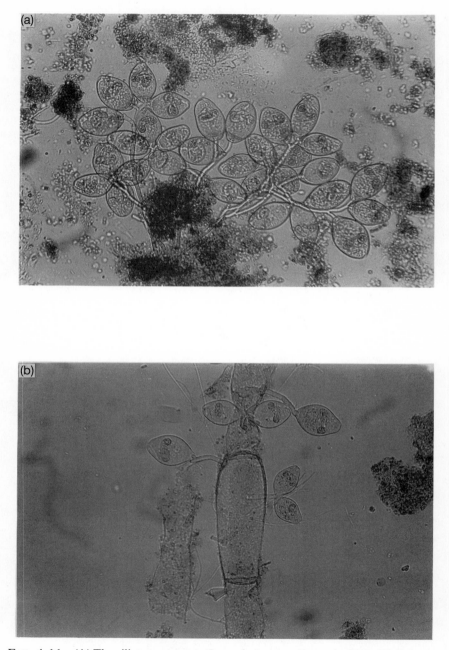

FIG. 4.11   (A) The ciliate protozoan *Opercularia microdiscum* (× 500). (B) Colonies of *Opercularia microdiscum* attached to the hyphae of the fungus *Subbaromyces splendens* (× 500).

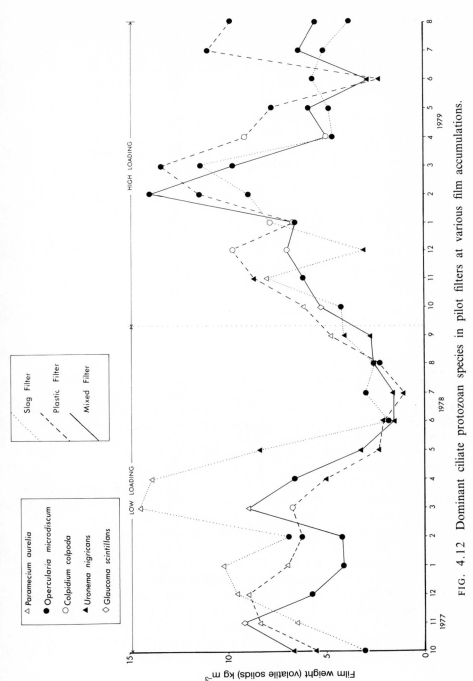

FIG. 4.12 Dominant ciliate protozoan species in pilot filters at various film accumulations.

TABLE 4.15   Maximum percentage of the total ciliate population comprised by individual component species in each filter during both low- and high-rate loadings in the experimental filters operated by Gray (1980)

Percentage of total population	Slag filter	Mixed filter	Plastic filter
>90	Opercularia microdiscum	Opercularia microdiscum	Opercularia microdiscum
80–89		Uronema nigricans	Uronema nigricans
70–79	Paramecium aurelia		Paramecium aurelia
60–69			Colpidium colpoda
50–59	Uronema nigricans Colpidium colpoda	Colpidium colpoda	
40–49		Paramecium aurelia Glaucoma scintillans	
30–49			
20–29	Glaucoma scintillans		Glaucoma scintillans Chilodonella uncinata
10–19			
0–9			

Opercularia microdiscum is a sessile organism attached to the medium or other suitable substrate by a non-contractile stalk and feeding passively on free-swimming bacteria. It is unable to actively search for food and, unlike the holotrichs, it is unable to move away from any adverse environmental changes, predators or the activities of the macro-invertebrate grazing fauna, except in its telotroch phase. The population density of the opercularian remained relatively small under low-rate conditions, but when the loading was increased to high-rate, the mean population density increased five-fold, occasionally making up between 90–100% of the total ciliate population in all three pilot filters (Table 4.15). However, at neither loading was *O. microdiscum* able to compete successfully with *U. nigricans* during periods of light film accumulation, but was found in greatest abundance during January/February and July/August when film accumulation was heavy. The sedentary ciliates attach themselves to a variety of substrates including zoogloeal bacteria, fungal hyphae, insect debris, and the larger filamentous bacteria. *Opercularia microdiscum* was positively correlated with both zoogloeal bacteria and the fungus *Subbaromyces splendens*, which it used as a substrate for attachment. In the pilot filters, the peritrich was restricted by competition for nutrients and space under low-rate conditions, and limited only by the lack of suitable surfaces for attachment at the higher loading. In all the pilot filters, *O. microdiscum* was found throughout the depth of the beds during the low loading, except at depths with heavy film accumulation or high abundance of *P. aurelia*. With the increase in loading rate, however, the population increased mainly in the lower half of the beds, although still avoiding those areas of heaviest film accumulation. The clear association of the ciliate with *S. splendens*, the tough hyphae of which grow out into the liquid layer flowing over the film, provides an ideal niche for the sessile protozoan so that it is away from any danger of being overgrown by the rapidly developing film (Fig. 4.11). Where the film is bacterial in nature then no suitable substrates for attachment may exist when holotrichs dominate. Although *O. microdiscum* has been previously associated with low concentrations of organic matter (Barritt 1940; Barker 1946; Tomlinson and Snaddon 1966), more recent work has suggested that the species may have a more general distribution (Curds 1969; Curds and Cockburn 1970a; Hussey 1975). Learner (1975a) also noted the importance of this species in wastewater treatment and found it to be the dominant organism in the majority of filters examined. He recorded a positive correlation of the organic and hydraulic loading with *O. microdiscum*, noting that maximum abundance was recorded in filters receiving loads in excess of 0.25 kg BOD $m^{-3}d^{-1}$. Some years earlier, Bruce and Merkens (1970) had found large numbers of the species, but no other ciliate, in experimental filters receiving an organic loading of 2.0 kg BOD $m^{-3}d^{-1}$ at a hydraulic loading of 6 $m^3m^{-3}d^{-1}$. *Colpidium colpoda* occupies the same niche as *P. aurelia* and is present at the same times. The species is associated with the lower range of film accumulations at which *P. aurelia* dominates, and is found reaching maximum numbers at film weights of between 4 and 9 kg m^{-3}. At the higher loading, the density of *C. colpoda* increased due to the exclusion of *P. aurelia*, where it dominated at periods of moderate film accumulation. At the higher loading, the sequence of species from heavy to light accumulation can be seen quite clearly, from *O. microdiscum* to *C. colpoda* and then to *U. nigricans* at the lighter weights of film, with each peak in the population density clearly separated from the next. *Colpidum colpoda* and *O. microdiscum* are both high-rate species and compete directly; when one species is found in large numbers the other is present only in small numbers. It does seem, however, that *O. microdiscum* is more successful than *C. colpoda* when there are enough suitable surfaces for attachment. The succession of these species, which are either controlled by film accumulation or temperature, at low- and high-rate loadings, is summarized in Fig. 4.13. The final dominant species, *Glaucoma scintillans*, competes directly with *U. nigricans* and the former is more successful at moderate film weights (5–9 kg m^{-3}), whereas *U. nigricans* is dominant at low film weights (< 5 kg m^{-3}).

Curds (1973c, 1975) divided the wastewater protozoan fauna into three groups according to their habitat: those that swim freely in the liquid phase and are prone to being washed out; those that crawl over the surface of the film

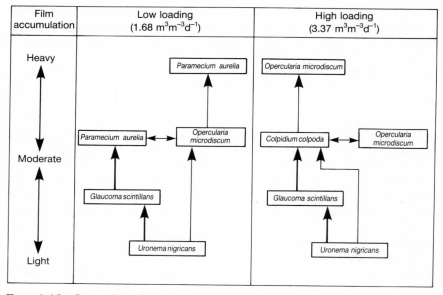

Film accumulation	Low loading (1.68 m³m⁻³d⁻¹)	High loading (3.37 m³m⁻³d⁻¹)

F IG. 4.13   Succession of dominant protozoan species with film accumulation at loadings of 1.68 and 3.37 $m^3m^{-3}d^{-1}$.

and are occasionally washed out; and those that are attached directly to the film, or some other material, and are only removed during sloughing. Obviously, habitat preference of those species found in percolating filters is important in survival terms and dictate which species are to be successful. Bungay and Bungay (1968) found that a peritrich, such as *Opercularlia microdiscum*, is always present even after sloughing in quite large numbers, and, therefore, potentially able to build up the population rapidly. However, in any given situation, in the competition between species for food, the organism that is fastest both to grow and reproduce under the prevailing conditions will become dominant (Moser 1958).

Opercularia microdiscum and *P. aurelia* were found at all depths by Gray in his low-rate filters, whereas *C. colpoda* was found in the upper regions of the bed where the organic concentration of the wastewater was greatest. *Aspidisca costata* was limited to the lower half of the pilot filters and was continuously washed out of the filters in large numbers. The dominant species at the higher loading were restricted to particular depths in line with the saprobic index proposed by Liebmann (1949). *Colpidium colpoda* was found in the top and middle regions of the pilot filters, *O. microdiscum* in the middle and lower regions, whereas *A. costata* and *U. nigricans* were restricted to the lower portion of the filters. Suctoria, which are mainly predators on other protozoan species, were found in the middle and lower areas of the filters where they would have maximum opportunity to come into contact with suitable prey.

However, Gray concludes that no individual reason can account for the stratification of the various protozoan species, but that it is the result of a number of environmental (e.g. nature of wastewater, organic load, temperature, hydraulic flow, food availability, surface area) and biological factors (e.g. competition, predation, type of film), which change continuously and alters the distribution of the protozoans within the film.

The role of rotifers in wastewater treatment processes has been reviewed by Doohan (1975). Unfortunately, most of the research carried out on this group has been in connection with the activated sludge process (Curds and Vandyke 1966; McKinney 1957; Calaway 1968; Sydenham 1968, 1971), thus, relatively little is known concerning the rotifers found in the percolating filter environment (Donner 1966). In the completely mixed environment of the activated sludge process, rotifers have two distinct functions: they break up floc particles providing nuclei for new floc formation; and they clarify the effluent by removing non-flocculated bacteria that are in suspension. They are powerful feeders and the strong ciliary currents produced by rotifers allows them to feed effectively even when the concentration of bacteria on which they feed is very low. This is in contrast to protozoans such as Vorticellids, which have weaker ciliary action and therefore are unable to survive at such low bacterial concentrations. Thus, in the activated sludge process, Vorticellids are normally succeeded by rotifers, and for the same reason rotifers are generally found in the lower sections of percolating filters. From field experiments Doohan (1975) supposed that it was food availability which exerted the greatest influence on the reproduction rate of the Rotifera, rather than other environmental factors, such as temperature. Therefore, their distribution within the filter environment would appear related to food availability, and ability to compete with other bacterial feeders. Rotifers have greater mobility compared to protozoans, for example, Lecane spp. have a specialized foot that enables them to crawl over the film and medium. This specialization allows rotifers to move within the filter bed and extends their distribution area when conditions permit. When protozoans are under stress because of food limitation, rotifers usually move into that area and compete for the available bacteria. Rotifers also help in the production of discrete faecal pellets that consist of undigested material bound together by mucus and which rapidly settle. A comprehensive list of species found in wastewater treatment processes is given by Doohan (1975) (Fig. 4.14).

Little is known regarding the role of Nematoda in wastewater treatment processes, but as they are present in such large numbers in percolating filters they must be important members of the filter community. Large populations of nematodes have been recorded in filters (Peters 1930; Lloyd 1945; Calaway 1968). The population of nematodes is dominated by bacterial feeders, although predator species feeding on other nematodes and rotifers are also present, but are far less abundant (Schiemer 1975). Weninger (1964) recorded maximum population densities of 180 individuals per millilitre, whereas

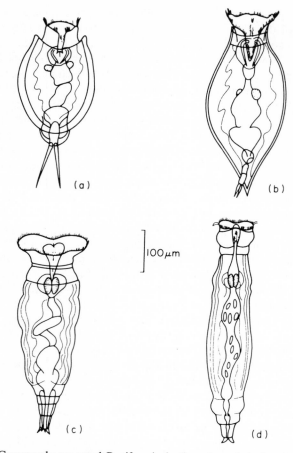

Fig. 4.14 Commonly reported Rotifera in both activated sludge and percolating filters; (a) *Lecane* sp.; (b) *Notommata* sp.; (c) *Philodina* sp.; (d) *Habrotrocha* sp. (Doohan 1975).

Scherb (1968) found maximum densities of up to 280 ml^{-1} in a bench-scale-activated sludge plant. Schiemer (1975) suggested that experiments carried out by Pillai and Taylor (1968) showed that the maximum population in conventional low-rate filters could be in the order of 1000 ml^{-1}, which corresponds closely with the results obtained by Gray (1980), where a maximum population density of 940 ml^{-1} was recorded in his low-rate pilot filter containing slag medium. The seasonal variation in abundance coincides with film accumulation, with maximum population densities of nematodes recorded during the early spring and with minimum population densities occurring immediately after sloughing. Weninger (1971), and Murad and Bazer (1970) observed that population density was inversely related to temperature, resulting in maximum populations between 7–10°C, although Chaudhuri *et al.* (1965) recorded maximum nematode densities between 17–18°C. Hawkes and Shephard (1972) regarded the Nematoda as being important grazers that were closely associated with film accumulation. Nematodes live in the film rather than on the surface or in the liquid phase and thus are directly related to film accumulation and film thickness. They were appreciably less abundant on smoother plastic media where film thickness is less, compared with rough mineral media. Another advantage of nematodes is their high reproductive potential (Schiemer 1975), which is similar to that of the astigmatid mites and the enchytraeid worms, all three groups being able to respond rapidly to increases in the available food. Schiemer considered the role of nematodes in the filter environment and estimated that the nematode population is responsible for 0.001% of the community's respiration. He identified that they had three functions: grazing of bacteria, which affects bacterial density and growth; decomposition of organic matter; and recycling of energy-rich substances inside the filter because of excretion products, faeces, and dead body tissues, although these may be lost to the system via the final effluent. Thus, although nematodes may have important effects on bacterial activity, they have probably only a minor role in the decomposition of organic matter in wastewater.

Grazing fauna
Percolating filters are dominated by two phyla of grazers, the Annelida and Insecta. Other grazers from the Arachnida, Crustacea, and Mollusca are also present but are generally of less significance.

Only two families of the phylum Annelida, the Enchytraeidae and the Lumbricidae, are common in wastewater treatment. They are almost exclusively found in percolating filters because, except for certain naids and lumbriculids, they are not active swimmers. The most frequently observed enchytraeid is *Lumbricillus rivalis* which was found in 91% of the filters surveyed by Learner (1975a). This species can make up 95–100% of the total enchytraeid population in filters with maximum population densities reaching 10 000 per litre of mineral medium (Gray 1980). *Enchytraeus buchholzi* is the next most frequently observed species, being recorded in low numbers in 57% of the filters examined by Learner. Gray observed that *L. rivalis* was distributed throughout the depth of his pilot filters at the low-rate loading, reaching maximum abundance in the central and lower areas of the filters, with the enchytraeid restricted to the lower half of the filters at the high-rate loading. Previous workers had found *L. rivalis* principally in the upper portions of the filters near the surface (Reynoldson 1947; Solbe *et al.* 1967; Williams *et al.* 1969). Maximum numbers of enchytraeids are found during February with minimum densities occurring in August, indicating the population density is directly linked to film accumulation. However, Enchytraeidae are restricted by the presence of Psychodid larvae. They reach maximum population densities the month preceding the occurrence of the maximum number of Psychodid larvae, but both reach maximum abundance in response to the heavy film accumulation. The *Psychoda* spp. and enchytraeids are in direct competition both for food and space in the filter, with the larvae being more successful because of their size and rapid growth rate. The increased competition from the larvae caused a dramatic decline in the numbers of enchytraeids present (Gray 1980). Enchytraeidae have also been found to dominate the grazing fauna under reduced competition from the fly larvae caused by higher hydraulic loadings or lower temperatures ($< 10°C$) (Tomlinson and Hall 1950; Hawkes 1955; Solbe *et al.* 1974). Hawkes (1955) demonstrated in his experiments on dosing frequencies that *L. rivalis* could withstand certain hydraulic flows because it possessed strong, curved setae and that as the cocoons were firmly attached to the medium they could withstand periods of even higher flow-rates. Enchytraeidae are restricted at higher flow rates ($> 2 \text{ m}^3\text{m}^{-3}\text{d}^{-1}$) with large numbers lost in the effluent especially during sloughing (Reynoldson 1941, 1948), and are generally absent from high-rate filters using either plastic or mineral media at loading rates $> 6 \text{ m}^3\text{m}^{-3}\text{d}^{-1}$ (Bruce and Merkens 1970).

Lumbricillus rivalis is capable of increasing its population density rapidly because of the large number of cocoons present in the filter at specific periods, and its high population growth rate (Learner 1972). This allows the Enchytraeidae to respond quickly to changes in the film accumulation and also to changes in community structure. The number of cocoons in the filter bed is controlled by the film accumulation. More are retained as the film increases, either by adhesion to the film or some other suitable substrate, and by being mechanically filtered out of the sewage by the film and humus. At times of low film accumulation there is a corresponding low abundance of cocoons which are mainly adhering to the actual surface of the media. Reynoldson (1947) and Solbe *et al.* (1967) both found that the abundance of cocoons followed a seasonal pattern reaching maximum numbers in the spring and autumn. In his pilot filters, Gray found that the seasonal abundance of cocoons followed a similar pattern to that of the adult enchytraeids, reaching maximum numbers during February to April and minimum numbers during the summer months after sloughing. Both the adult enchytraeids and the cocoons are eaten by a number of predatory dipteran larvae including *Hydrobaenus minimus*, *Metriocnemus hygropetricus*, and *Psychoda severini* (Lloyd 1945).

Only three lumbricids are frequently recorded in percolating filters, *Eisenia foetida* (31% of the filters surveyed by Learner), *Dendrobaena subrubicunda* (43%), and *Eiseniella tetraedra* (52%). Considerable information regarding these species in filters has been gathered (Tomlinson 1946a; Hawkes 1963; Solbe *et al.* 1967; Solbe 1971) and this has been reviewed in detail by Solbe (1975). Of the two commonest species, *E. tetraedra* is an amphibious species, whereas *D. subrubicunda* is terrestrial in nature being commonly found in compost heaps (Gerard 1964). Therefore, it is not surprising that *E. tetraedra* is more successful at higher hydraulic loading than the other species. Lumbricids tend to be more numerous in smaller media (Terry 1951), with worms rarely found near the surface in filters containing larger media or media with a high voidage, due to the flushing action of the applied wastewater. Solbe (1971) examining the depth distribution of lubricids reported that the population density of *D. subrubicunda* increased towards the base of the filters, whereas *E. tetraedra* is found in the middle regions of the bed. Therefore, the distribution of the lumbricids is apparently related to the hydraulic flow and the size of the interstices.

The presence of lumbricids is generally accepted as an indication that a filter has matured and that the community structure has become stable. However, during high rainfall large numbers of worms are flushed out of the soil on to paved areas and are washed away in the surface runoff. Thus, where combined sewerage systems are employed these worms eventually arrive at the treatment plant and those not removed by primary sedimentation will become established in the filter, with large population densities becoming established quite quickly. The function of annelids in percolating filters is mainly one of film control. *Dendrobaena subrubicunda*, for example, can ingest film at a rate of 133 mg $g^{-1}d^{-1}$ at 15°C, with the entire lumbricid population of a mature filter capable of ingesting an amount of film equivalent to 55% of the daily input of carbonaceous material to the filter (Solbe 1975). Annelids therefore prevent the accumulation of excess solids and may also be the prime cause of the sloughing of film in some filters during spring. Annelids are able to ingest a wide range of substances including those not readily degraded by the filter bacteria. Thus, Annelida are able to reduce the organic matter directly by absorption and indirectly by the formation of excreted aggregates (faeces) which readily settle in the humus tank. Although some heterotrophic bacteria flourish in the gut of certain annelids and are passed out with the aggregates, they are also responsible for the consumption of many pathogenic micro-organisms. The respiration of the phylum accounts for a small, but useful, proportion of the carbon dissipated by the filter, which in some filters may be as high as 8.5%. The group causes little nuisance in the filter, although the accumulation of large numbers on the surface can cause alarm to operators even though they are quite harmless. If large numbers are washed out into the humus tank they rapidly die and putrefy causing a reduction in the dissolved oxygen concentration that can cause problems. They do, however, compete directly with fly larvae and so reduce the potential nuisance caused when adult flies emerge from filters.

The members of the various orders of Insecta are principally associated with percolating filters, rarely being found in other kinds, or at other stages, of wastewater treatment. Many species lists have been compiled (Lloyd 1945; Tomlinson 1946*a*; Terry 1951; Hawkes 1963; Solbe *et al.* 1967). However, the first comprehensive survey into the fauna of percolating filters was undertaken by Learner (1975*a*) and a comprehensive species list prepared, containing 186 species of insects belonging to 38 families (Learner 1975*b*).

Sixteen species of Collembola (springtails) have been recorded from percolating filters, the most important species being *Hypogastrura viatica* (= *Achorutes subviaticus*), which was recorded in 57% of the filters surveyed by Learner (1975*b*), with *Tomocerus minor* (28%) and *Proisotoma minuta* (21%) also frequently observed. However, only *H. viatica* is found in large numbers, reaching maximum densities of 3800 individuals per litre of medium. They are active grazers found throughout the depth of the filter and feed on a wide variety of organic material including fungal hyphae and spores, bacteria, dead and decaying plant material, and algae (Learner 1975*b*). *Hypogastrura viatica* is extremely sensitive to increased rates of filtration (Hawkes and Jenkins 1955, 1958; Wheatley 1976). For example, Tomlinson and Hall (1950) recorded maximum abundance of the springtail at $1.5 \ m^3 m^{-3} d^{-1}$, with none found at loads $> 3.0 \ m^3 m^{-3} d^{-1}$. *Isotoma olivaceaviolacea* is more abundant in random plastic than mineral media (Gray 1980), because of its preference for drier areas, such as interjet zones in conventional filters (Hawkes 1959). Each module of random plastic media, such as Flocor RC or RS has a dry area suitable for such organisms as *Isotoma* sp.

Beetles (Coleoptera) are not common in percolating filters. Only members of the Hydrophilidae and Staphylinidae are frequently encountered, with the most widely occurring species being *Cercyon ustulatus* and *Platystethus arenarius*, which were observed by Learner during his survey. Coleoptera are rarely found as larvae in the filter, only as adults, and presumably do not breed within the filter environment. The most important insects found in filters are the Diptera, with 28 species from 11 dipteran families regularly occurring (Learner's survey). The principal filter species belong to the genera Psychodidae, Chironomidae, and Anisopodidae, and can all breed successfully in the filter bed, with the larvae being abundant and feeding on the film. Many more species of Diptera are associated with filters and are captured in very large numbers on insect traps set on, or near to, filters. However, it is these species that are able to breed in the filter bed environment and which have an important role in the actual purification process.

The dipteran larvae and Collembola are very active grazers, preventing the filters from becoming blocked with excessive film accumulation (Williams and Taylor 1968). They are most active during the warmer months, although large populations of some species, such as *Sylvicola*, are still active even during winter. The function of the Insecta in the purification process is the same as that of the Annelida; they digest the film and absorb some of the organics, and form dense faecal pellets that rapidly settle. Williams and Taylor demonstrated that solids from filters containing *Psychoda* and/or *Lumbricillus* spp. settled far more rapidly, with 65–70% of the total settleable solids settling within one hour compared with 34% in the control filter without macro-invertebrates. The sludge from the filters containing macro-invertebrates contained a high density of animal fragments and faecal pellets.

There are three species of Psychodidae that occur in filters. *Psychoda alternata* is the most frequently observed species, occurring in 89% of the filters sampled by Learner and reaching maximum densities of $44\,700\,1^{-1}$ of medium. Therefore, this species, together with *Lumbricillus rivalis*, must be considered the major grazer in percolating filters. *Psychoda severini* is also widely distributed (72%) but reaches lower densities ($1240\,1^{-1}$) than *P. alternata*, whereas the third species, *Psychoda cinera*, is rare being found in only 4% of filters and in low numbers (Fig. 4.15). *Psychoda severini* is parthenogenetic and, like *P. alternata*, is able to carry out its entire life-cycle without leaving the filter bed. It is therefore quite common to find adult flies deep within the medium. This is in contrast to many other dipterans where the adult flies need to leave the filter after pupation to swarm and mate before returning to lay eggs. These two species do not actively compete as *P. severini* is able to reproduce at temperatures below 10°C and is more abundant than *P. alternata* during the winter and spring, whereas the latter is dominant in summer. The abundance of *P. alternata* is positively correlated with temperature. Learner (1975a) clearly illustrated that the reproductive potential (life cycle) of *P. alternata* is controlled by temperature and that it had the most rapid development rate at temperatures in excess of 10°C of any of the insects found in the percolating filter environment. Obviously, there is a 'lag phase' between maximum food availability and the resultant increase in the number of grazers. In the case of *P. alternata*, this phase was of 1–2 months duration, depending on the temperature within the filter (Solbe and Tozer 1971). This explains why no direct correlation between film weight and density of Psychodid larvae was observed by Gray (1980). The restriction in the surface accumulation of *Psychoda* spp. is due partly to its sensitivity to the hydraulic flow (Tomlinson and Hall 1950; Hawkes 1955; Lumb and Eastwood 1958). The effect of various distribution systems was studied by Hawkes (1959), who found that splash plates produced an even distribution of sewage resulting in a large accumulation of film and a high density of *Psychoda* in the top 600 mm. By increasing the velocity of application, the *Psychoda* populations were reduced in the surface layers and were forced below 600 mm. This resulted in more film in the top layer. Tomlinson and Hall (1950) found that the abundance of the *Psychoda* larvae decreased if the hydraulic load exceeded $3.6\ m^3m^{-3}d^{-1}$, even though the film was thick. In their experimental filters, Bruce and Merkens (1970) found that large numbers of *Psychoda* larvae were present in filters loaded at $6.0\ m^3m^{-3}d^{-1}$. This was probably due to the higher voidage as it has been shown that smaller media restricted the natural life-cycle of the insects (Tomlinson and Stride 1945; Hawkes and Jenkins 1951, 1955).

Three chironomid species were frequently isolated by Learner (1975a), *Hydrobaenus perennis* (24%), *Metriocnemus hygropectricus* (54%), and *Hydrobaenus minimus* (61%), all being able to breed in filters with maximum larval densities reaching 1400, 1892, and $3460\,1^{-1}$ medium respectively. The larvae of *Hydrobaenus perennis*, currently named *Chaetocladius perennis*

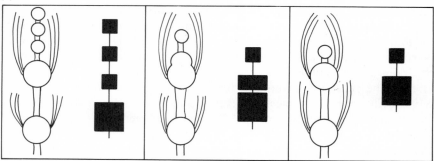

FIG. 4.15 The dipteran most associated with filters are small moth-like flies less than 5 mm long of the genera *Psychoda*. (a) *Psychoda alternata* is extremely common and is known as the 'sewage filter fly'. It is readily distinguished from the other two species of *Psychoda* found in filters by dark tufts of hairs at the tips of some wing veins. *Psychoda cinerea* and *P. severini* have hair-less wings. (b) Flies can be positively identified by low-power microscopic examination of their antennae. *Psychoda cinerea* terminates in three small segments, *P. alternata* has two large segments broadly joined and a smaller terminal segment, while *P. severini* in a single small segment.

(Pinder 1978), migrate in a general downward direction during their development. This results in the maximum abundance of larvae being found at the base of filters, high numbers of larvae being washed out in the final effluent and a scarcity of pupae in comparison with other species. The final larval instar finally burrows deep into the film to pupate (Lloyd *et al.* 1940). Like other chironomids, the species is common in all types of filter media although it is more readily flushed out of the smoother faced media, and is abundant from February to August. Chironomid larvae are generally only found in large numbers in lightly loaded filters (Tomlinson and Stride 1945), when the film accumulation is thin (Terry 1956; Hawkes and Shepherd 1972). *Metriocnemus hygropectricus* is found in greatest numbers during the autumn, and due to the upward migration of the species prior to pupation (Dyson and Lloyd 1936), the larvae are found in large quantities in the upper section of filters. *Hydrobaenus minimus*, now renamed *Limnophytes minimus* (Pinder 1978), is restricted to low-rate filters. It is found from May to October throughout the filter, and is most successful when the film is thin. Chironomid larvae are only found in relatively small numbers in filters and have only a minor role in the purification process. However, the larvae are capable of successfully competing with the other macro-invertebrates of the filter in favourable conditions. Lloyd *et al.* (1940) reported that chironomid larvae were able to compete with psychodid larvae for the available food, reducing the population densities of *Psychoda* spp. at the surface of the filter and causing the extension of the species distribution deeper into the filter. In heavy film conditions, the *Psychoda* and *Sylvicola* larvae are more able to cope than the larvae of the chironomid species. They utilize their respiratory siphons when buried in the thick layer of film. Both *H. minimus* and *M. hygropetrcus*, like *P. severini*, have shorter life cycles than *P. alternata* at the lower range of temperatures recorded in filters. This may account for their relative success in the areas of the bed most affected by exposure to the air and which is frequently cold in comparison to the other areas of the filter. Examples of vertical and seasonal variation of some grazing organisms, including *Psychoda* spp. and *Hydrobaenus perennis* larvae, are shown in Fig. 4.16. *Sylvicola* (= *Anisopus*) *fenestralis* is the only member of the Anisopodidae recovered from filters as it is widely distributed (49% of filters) and found in large numbers (1680 1^{-1} medium). The adult fly has extremely large wings in comparison with other filter flies and is a powerful insect with large larvae. Like *Psychoda*, it can complete its life cycle within the filter bed with maximum numbers occurring between April to June. The larvae are generally found in largest numbers in the top 600 mm of filters irrespective of the organic or hydraulic loading (Tomlinson and Hall 1950; Gray 1980) (Table 4.16), while the pupa reach maximum density between 300–900 mm. Hawkes (1963) found that the vertical distribution of the species could be altered by high instantaneous rates of domestic wastewater. *Sylvicola fenestralis* abundance has been shown to be closely and directly related to film distribution (Hawkes 1952a). However, recent studies

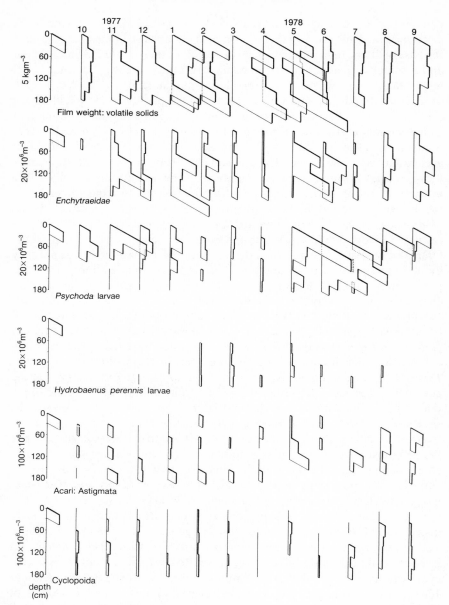

FIG. 4.16   Vertical distribution of film and macrofauna in a low-rate filter containing 50 mm blast-furnace slag medium.

TABLE 4.16 Vertical distribution of insects at Minworth (filter block B) during August 1968 (Learner 1975b)

Depth (m)	No. per litre of medium						
	Anurida granaria	Hypogastrura viatica and H. purpurescens	Cercyon ustulatus	Sylvicola fenestralis	Psychoda alternata	P. cinerea	P. severini
0.0–0.15	18	18	—	—	89	89	—
0.15–0.30	71	18	36	107	36	214	53
0.30–0.46	142	89	36	18	18	231	18
0.46–0.61	1175	53	53	71	—	303	18
0.61–0.76	552	125	18	—	36	71	—
0.76–0.91	445	71	—	—	—	89	—
0.91–1.07	677	142	18	—	—	36	—
1.07–1.22	944	71	—	—	18	36	18
1.22–1.37	427	53	—	—	—	53	18
1.37–1.52	374	—	—	—	—	18	—
1.52–1.68	214	71	—	—	—	—	—
1.68–1.83	641	89	—	—	—	—	—

suggest that the amount of film is not always an important factor determining the vertical distribution of this species. This species was reported as being more successful in random plastic medium rather than slag medium by Gray (1980), with the greatest number of larvae and pupae being recorded in the plastic filter. Each module of random plastic media has some of its surface area free from the film which is often quite dry, and this may be the reason why the pupae in particular were found in comparatively large numbers in this filter. *Sylvicola fenestralis* requires a drier environment for successful pupation than that tolerated by the larvae (Hawkes 1952a), and this is thought to be the reason why the larvae are reported to migrate to drier areas in conventional plants (Learner 1975b). *Sylvicola fenestralis* appears unaffected by competition with either chironomid or psychodid larvae.

Many of the dipteran flies form dense swarms as they emerge from the filter bed, and for some chironomids swarming is a prelude to mating. The numbers of flies emerging can be so large that from a distance it has the appearance of smoke coming from the beds. Emergence is affected by temperature, light intensity, and wind velocity. Insects will not fly unless the temperature is above a threshold value, for example 10°C for *P. alternata*, 7–8°C for *H. minimus*, and 4°C for both *Metriocnemus* spp. and *Sylvicola fenestralis* (Learner 1975b). Hawkes (1961b) found that for every 1.2°C rise in temperature above the threshold value up to 24°C the number of *Sylvicola fenestralis* flies in flight doubled. Diel periodicity is observed for all the species with the peak of emergence for *P. alternata* and *P. severini* in the early afternoon, *S. fenestralis* at dusk, and a smaller peak at dawn. Wind velocity is an important factor with swarms quickly broken up and individuals unable to fly if the wind is too strong. In essence, the stronger the fly the greater the wind velocity it can withstand with *S. fenestralis* able to withstand velocities up to 6.7 m s^{-1} (Hawkes 1952b, 1961b). At the treatment plant, the density of flies can be problematic making working conditions unpleasant as flies are drawn into the mouth and nostrils, and are caught in the eyes. None of the commonly occurring flies, including *S. fenestralis*, bite. However, although harmless, the large size and intimidating appearance of *S. fenestralis* can alarm people, which results in frequent complaints from nearby residents. Psychodid flies are found up to 1.6 km away from the treatment plant, although this does depend on the direction of the prevailing wind. *Sylvicola fenestralis* flies are found in large numbers up to 1.2 km, although few reach farther than 2.4 km from the plant. However, flies are well known to cause both aesthetic and public health problems to those living or working close to treatment plants, with *P. alternata*, *P. severini*, *S. fenestralis*, *H. minimus*, and *M. hygropetricus* being the main nuisance species. However, Learner (1975a) found that *P. alternata* comprised 80% or more of the total annual emergence of flies from 10 out of 17 filters where emergence traps were located.

Three control options are available for remedying fly nuisance. Physical methods have been least successful and are not recommended. In the USA filters were flooded to eliminate fly species. This approach damaged the ecology of the system and affected performance, providing at best just a temporary reduction in the number of flies emerging (Otter 1966). Covering the surface of the medium with a layer of finer media (13–19 mm) to a depth of 250 mm reduces the numbers of adult *Psychoda* and *Sylvicola* emerging, but results in severe ponding during the winter. Enclosing filters would appear effective but rather expensive. However, Painter (1980) suggests that this remedy may not always work and quotes an example when after covering a filter the extremes in temperature were reduced and led to an overall increase in fly production, with flies escaping via ventilation ports. The most widely adopted control method is the use of chemical insecticides. DDT (dichlorodiphenyl trichloroethane) and HCH (1,2,3,4,5,6-hexachlorocyclohexane, also known as Gammexane) are effective in killing the larger larvae without affecting the other grazers. However, such non-biodegradable pesticides are no longer permitted in Europe. Also, some flies, such as *S. fenestralis*, become immune to HCH after prolonged use (Watson and Fishburn 1964). Currently, two insecticides are widely used in the UK for controlling filter flies, Actellic (pirimiphos methyl) and Dimilin (diflubenzuran), which are *O*-2-diethylamine-6-methylpyrimidin-4-yl-*O*-*O*-dimethyl phosphorothioate and 1-(4-chlorophenyl)-3-(2,6-difluorobenzoyl) urea respectively. However, neither insecticide appears particularly effective against *S. fenestralis*. Limited control can be obtained by changing the operational practice of the plant. For example, by limiting the amount of film accumulation, especially at the surface. This can be effected by reducing the f/m ratio or by using one of the modifications such as recirculation, double filtration, or ADF. Increasing the hydraulic loading may also make it more difficult for dipterans to complete their life-cycles in the filter. Continuous dosing using nozzles prevents emergence and severely reduces the available surface area for the flies to lay eggs. The problem of fly nuisance associated with percolating filters and the various possible control measures are fully reviewed by Painter (1980). Houston *et al.* (1989*a,b*) have successfully used the entomopathogenic bacterium *Bacillus thuringiensis* var. *israelensis* to control filter flies. The bacterium, which is available as a commercial preparation known as *Teknor*, is effective against fly larvae only and does not affect other filter fauna or performance.

Astigmata mites (Acari) are extremely abundant in percolating filters and are associated with drier areas of the medium. They are found in the dry interjet zone of filters and are abundant in random plastic medium where each module supports a dry niche used by for such mites. Maximum abundance occurs in spring, with maximum densities in slag medium reaching 3900 l^{-1}, although this exceeds 32 000 l^{-1} in random plastic media (Gray 1980). They are more abundant in low-rate systems but are found in reduced numbers in high-rate filters provided that a suitable niche is available. Four species are particularly common, *Histogaster carpio*, *Histiostoma feroniarum*, *Rhizoglyphus echinopus*, and the predatory mesostigmata *Platyseius italicus* (Baker 1961). The Astigmata are able to respond quickly to increases in film accumulation in comparison to the other macrograzers. The reproductive potential of the Astigmata and, in particular *H. feroniarum*, is far greater than that of the dipteran larvae (Learner 1975*b*) or the Enchytraeidae (Learner 1972). Hughes (1961) found that *H. feroniarum* completed its life cycle in only 2 to 4 days at 20–25°C, whereas *R. echinopus* takes 9 to 13 days over a similar temperature range (Zachvatkin 1941). Initially, the mites take advantage of the large accumulation of film due to their fast reproductive rate, but the enormous numbers of Psychodid larvae that will develop subsequently force the mites into other areas of the filter bed or cause a reduction in the total population density. *Platyseius italicus* has been shown to feed on a wide variety of invertebrates, but mainly on *Lumbricillus rivalis*, although it does not eat the cocoons (Baker 1961). It is closely associated with the Enchytraeidae density, often being found together on the surface in large numbers (Tomlinson 1946*a*; Gray 1980).

Not all the macrofauna found in filters are grazing on the microbial film. The spiders and mesostigmatid mites are predators as are some of the chironomid larvae when the film becomes scarce. Where humus (dead organic matter) accumulates within the filter saprophytic species, such as *Cercyon ustulatas* and the larvae of *Spaziphora hydromyzina*, are found. A number of the dipteran larvae are predated on or parasitized by other flies.

The ecology of percolating filters has been excellently reviewed by Hawkes (1983*b*).

Biological analysis
Regular biological analysis of the film can often provide answers to operational problems as well as being an ideal habitat for ecological research (Gray 1982*b*). Learner (1975*a*) used a single sample of medium collected from the surface of each filter plus a litre sample of the final effluent for biological analysis during his survey of percolating filters. However, only limited information can be obtained by taking surface samples. In practice, it is not possible to collect medium from depths in excess of 500 mm by digging a hole in the surface of the bed. Thus, little information can be obtained regarding the vertical distribution of the film or of the organisms restricted to the lower half of the filter by this method. Also, where high hydraulic loadings are used or simple flow-on distributors, then little film and a restricted fauna will be present in the top layer of the medium anyway. It is important, therefore, to be able to sample the medium throughout the depth of the filter in such a way as to cause as little disturbance and damage as possible. This is facilitated by the provision of sampling tubes which allows the vertical distribution of film and fauna to be measured both quantitatively and qualitatively.

Sampling tubes are perforated pipes into which closely fitting perforated baskets, containing medium, fit. The baskets can then be lifted out allowing the medium and associated film to be sampled at any depth. Gray and Learner (1983) used perforated thick-walled (25 mm) ABS plastic pipes with an internal diameter of 150 mm. The pipes were exactly the same depth as the bed with the top of the plastic pipe just below the surface of the medium. Each pipe was perforated with 38 mm diameter holes some 6 mm apart. Hawkes and Shephard (1972) used smaller holes, less closely spaced, in their sampling columns but examination of the voids within full-scale percolating filters indicated that larger holes would simulate actual operating conditions more closely. The larger holes also allow greater redistribution of wastewater and movement of loose solids and filter fauna between the sampling column and the surrounding medium. However, the diameter of the perforations was not so large that the blast-furnace slag or random plastic media (50 mm nominal size) used by Gray and Learner would protrude through them sufficiently during settlement to interfere with the removal of the sampling baskets. Each sampling column contained six rigid plastic-coated wire-mesh baskets, each 300 mm long and 146 mm in diameter. Two plastic-coated wires were welded on to each sampling basket to aid their removal. The wires were just thin enough to pass between the baskets and the wall of the sampling column. Film accumulated on these wires and on pieces of medium that protruded slightly through the holes in the wall of the pipe, thus restricting the free vertical passage of sewage through the gap between the baskets and the pipe. Removable baskets have been used by many other workers. The baskets enable the medium to be sampled throughout the depth of a filter, although a wide variety of designs and materials have been used (Williams *et al.* 1969; Hawkes and Shephard 1972; Wheatley 1976; Rowlands 1979; HMSO 1988). Horizontal distribution of the biota can be investigated by using extra sampling baskets housed in shorter sections of perforated pipes positioned in the surface of each filter, thus providing details of distribution of film in the top 300 mm. This provides useful information on the development of film and of the possibility of ponding.

Sampling procedure. Removing sampling baskets from their tube obviously causes some disturbance and so they should be left as long as possible (at least a month) between sampling. The sampling procedure should always be the same. The distribution system must be turned off prior to sampling, the baskets removed one at a time, and carefully labelled. The basket should be left for five minutes for excess liquid to drain before a subsample of the medium is taken. Gray and Learner (1983) carefully graded the medium within their sample baskets so that four pieces of medium were equivalent to 250 cm^3 of medium, and this was taken along with its attached film for analysis. The pieces were removed randomly from different points within the basket to give a representative sample and placed in a labelled plastic bag that was sealed to prevent evaporation and loss of material. The remaining medium was carefully replaced in the baskets in the same order in which they had originally been removed, with the medium from the previous sampling taking the place of the sampled pieces. The sampling procedure should not exceed 20 minutes for each filter in order to minimize the damage to the rest of the biota. It is important to carefully replace all the pieces of medium, matching up the disturbed surface film so that only a small quantity of film is dislodged and washed away when the distribution system is switched on again. The sample bag should be checked for leaks, double sealed, and then returned to the laboratory as quickly as possible where it can be weighed to give an estimate of the biomass of film on the medium (Section 4.1.1.4). The addition of preservatives such as formaldehyde or alcohol makes identification of the microfauna, especially the Protozoa, almost impossible, therefore it is advisable to store samples at 4°C until they can be processed further.

Sample processing. The film needs to be removed from the sampled medium within a few days of collection. The procedure used by Gray (1980) is summarized in Fig. 4.17. The sample bag is opened carefully and the pieces of medium removed individually, taking care not to allow any of the animals, including individual flies, to escape. The loose film and associated animals are removed from the medium by gently brushing the surface with a soft artist's paintbrush (red sable, bright) in a shallow dish of water. Any large animals present are removed at this stage, identified, and counted, in order to prevent them from being damaged during the more vigorous washing procedure later on. The piece of medium is then placed into a beaker and stirred vigorously using a magnetic stirrer, and the next piece of medium is removed from the bag and initially brushed clean. The first piece of medium is then removed from the beaker and put into a second dish of water and scrubbed with a short-haired paintbrush (hogs hair, flat), which removes any tenacious film still adhering to the surface. The medium and the sample bag are finally rinsed with clean water to remove any remaining debris and the medium is then resealed in the bag. By standing the bags of freshly cleaned medium on a 40°C heater for about an hour, many animals, mainly enchytraeids and dipteran larvae, not removed from deep within the pores of pitted media such as slag or clinker, are driven to the surface by the heat. Both the bag and the medium are subsequently rewashed and the extra animals collected are identified and counted. The bags of clean medium are finally left to dry completely and then reweighed. The weight of the film is estimated by subtracting the dry weight of the bag and its contents from the wet weight which had been measured immediately after the samples had been returned to the laboratory.

All the debris, solids, and animals removed from the sample of medium collected from each basket are added together and the total volume made up to 500 cm^3 with distilled water. The liquid sample is then mixed thoroughly using a magnetic stirrer and subsamples taken for the various analyses summarized in Fig. 4.17. The subsamples taken for biological analysis are stored in glass bottles, at 4°C. The remaining 150 cm^3, left after all the various subsamples are removed, is poured on to a white examination tray so that an assessment of the relative abundance of the various macro-invertebrates can

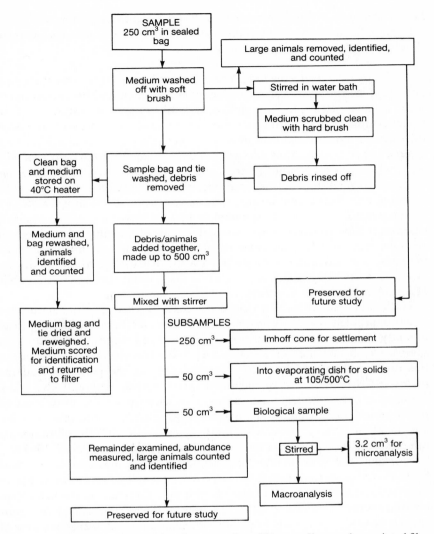

FIG. 4.17   Flow chart illustrating the processing of filter medium and associated film for chemical, physical, and biological analysis.

be made. All the animals are identified and the larger ones are counted. Afterwards, as much of the sample as possible should be retained and preserved for future study and reference.

An assessment of the volume of solids present can be made using the Imhoff cone method (DoE 1972). The 250 cm^3 subsample (Fig. 4.17) is allowed to settle for 45 minutes in the Imhoff cone, which is then gently twisted to remove any debris adhering to the glass sides. The quantity of solids settled after one hour is recorded. Although most of the larger invertebrates should be removed prior to settlement, the samples contain large numbers of organisms and these should be included in this assessment of film accumulation.

Micro- and meiofauna analysis. The 50 cm^3 subsample taken for biological analysis (Fig. 4.17) is shaken to produce complete mixing within the container then, using a sterile pasteur pipette, a small volume is transferred to a haemacytometer type counting chamber of the Mod–Fuchs Rosenthal type. A total area of 36 mm^2 split up into 0.25 mm^2 squares is examined for each sample under a compound microscope. Three magnifications are generally required, × 100 for counting large micro-organisms such as *Paramecium caudatum*, nematodes and also large bacterial and fungal colonies, and × 200 for counting the other micro-organisms that are normally identified at × 400. By measuring the depth of the sample under the cover slip of the counting chamber, using the microscope, the total volume of the sample examined per chamber can be calculated. Details of other counting chambers, staining methods, and separation methods for parasitic ova, cysts, and nematodes are given by Fox, Fitzgerald, and Lue-Hing (1981) and the American Public Health Association *et al.* (1985).

The microfauna can only be identified accurately when alive and this poses a practical problem regarding protozoans, in particular, because of their greater mobility. This is partly overcome by the addition of 1% nickel sulphate to the sample, which has a narcotic effect on the protozoans. However, general use of this method should be avoided where possible as the peritrichs and the suctorians are more easily identified when active. Alternatively, small amounts of 2% xylocaine can be used to slow down the movement of protozoans without significantly affecting their physical form, thus allowing identification. The xylocaine will eventually evaporate and therefore treated samples should be processed reasonably quickly (Norouzian *et al.* 1987).

Problems will arise in the identification of the fungi and filamentous bacteria, and in deciding how many cells or what length of filament constitutes the presence of a countable and reproducible unit. Gray set minimum limits in his study (Gray 1983*b*). For a fungal hypha, this was 10 cells or 6 cells plus either a growing tip or conidium, and for bacteria only filaments in excess of 0.2 mm in length or complete zoogloeal colonies were counted. The main identification keys available for each major group are summarized in Table 4.17. Furthermore, Fox *et al.* (1981) have produced a beautifully illustrated key, which covers many groups including parasitic helminths, parasitic, and free-living protozoa, algae, and many other meiofauna groups. However, this is not a key in the true sense but rather a collection of photomicrographs of micro-organisms isolated at the sewage treatment plants of the Metropolitan Sanitary District of Greater Chicago, USA. Photographs can be a useful aid to identification and prove invaluable for monitoring the changes in the surface film. By using colour transparencies, the photographs of

TABLE 4.17 Identification keys to the micro- and macrofauna found in percolating filters and other wastewaster treatment units

Microfauna group	Key references	Macrofauna group	Key references
Bacteria	Farquhar and Boyle 1971a; Eikelboom 1975; Eikelboom and Van Buijsen 1981; Jenkins et al.1984	Annelida	Brinkhurst 1971; Nielson and Christensen 1959, 1961, 1963; Gerard 1964; Sperber 1950; Tynen 1966
Fungi Algae	Cooke 1963; Tomlinson and Williams 1975 Belcher and Swale 1976; George 1976; Bellinger 1980; Sykes 1981	Insecta	Lawrence 1970; Satchell 1947, 1949; Coe et al. 1950; Bryce 1960; Brindle 1962; Mason 1968; Bryce and Hobart 1972; Unwin 1981
Protozoa	Kudo 1932; Martin 1968; Calaway and Lackey 1962; Page 1976; Curds 1969, 1982; Curds et al. 1983; Bick 1972	Arachnida Crustacea Mollusca	Evans et al. 1961 Harding and Smith 1974 Janus 1965; Cameron et al. 1983
Nematoda Rotifera	Tarjan et al. 1977 Donner 1966; Ruttner-Kolisko 1972; Pontin 1978		

General reference work: Edmondson 1959; Armitage et al. 1979; Martin 1968

General reference work: Macan 1959; Tomlinson 1946a; Armitage et al. 1979.

the surface film growth in the filter can be enlarged so that the extent of film accumulation, dominant species, action of grazers, and effects of surface ponding can be carefully examined and recorded.

Macrofauna analysis. The remainder of the subsample used for the microfaunal analysis is shaken as before and poured into a large Hartley pattern Buchner funnel, and the sample container rinsed out. The sample is gently filtered at low pressure, so as not to damage the annelids, through Whatman 113 (150 mm diameter) filter paper. The filter paper is then cut into four equal sections and examined in a low form plastic dish under a stereomicroscope. The larger invertebrates, such as dipteran larvae and enchytraeid worms can be identified, counted and removed at × 5–10 magnification, whereas the other invertebrates have to be located before identification and counted by a systematic search using fine needles at × 15–20 magnification. Mites have to be identified and counted separately using a 1 cm^2 illuminated background plate that fits under the plastic dish containing the filter paper. This allows only a specific area to be illuminated which can then be carefully searched at × 40 magnification. A total area of 6 cm^2 is searched in this way, the 1 cm^2 areas being chosen at random on the four sections of the filter paper. Apart from the general key by Tomlinson (1946a), which covers only a few of the most common percolating filter grazing organisms and is now taxonomically very out of date, there is no general key available. Therefore, identification must be made using specialized keys (Table 4.17). The number of grazers is expressed as total number per litre or per cubic metre of medium.

Curds and Cockburn (1970a) found that a greater variety of protozoan species were to be found in the effluent from filters than in the film collected just from the surface of the filter. Therefore, in order to obtain a comprehensive list of species present in the filter and to discover which micro- and macro-organisms are being washed out of the filters, the effluent should be regularly examined. The same methods are used for the effluent as are used for the media. The effluent can be collected as spot 1 litre samples or by using a plankton net.

In order to assess the rate at which flies emerge from a filter a special emergence trap is required (Solbe *et al.* 1967) (Fig. 4.18). Styles (1979) has developed a much simpler trap of smaller dimensions which is ideal for trapping Chironomidae. It is a rectangular perspex box 280 × 140 mm and 70 mm high, with a removable sliding lid that is covered with fly adhesive on the under surface and so acts as the trapping area. Each end of the box is made of nylon mesh to allow ventilation. It was designed specially to be placed in the interjet zone between the jets from the distributor and does not incorporate a system of redistributing the wastewater as seen in the larger traps. An accurate assessment of the aerial density of flies over or near the filter can be more easily obtained. A variety of traps have been designed and successfully used (Tomlinson and Stride 1945; Hawkes 1951; Taylor 1951, 1955; Hawkes 1983b). Although traps are by far the most accurate and efficient method of

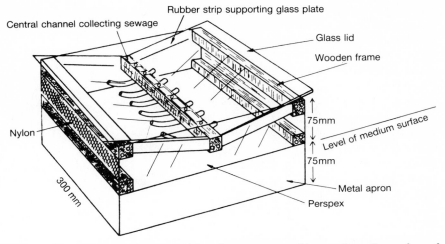

FIG. 4.18 Insect trap for determining the emergence of insects from the surface of percolating filters. The inclined glass or plastic sheets are covered with an adhesive and placed sticky side downwards so that the emerging insects become trapped (Solbe *et al.* 1967).

measuring aerial density they are expensive to construct. However, large sticky paper sheets stretched out between supports is simple and fairly accurate, although identification of the smaller species can be difficult because as they struggle the flies become covered in the adhesive used on the paper. Fly abundance can also be assessed by using a large entomological aspirator to catch as many flies as possible over a unit period of time.

4.1.1.4 Factors affecting performance

A number of factors which affect the performance of percolating filters have already been discussed in some detail, especially the effect of hydraulic and organic loadings. Of course, the wastewater must be amenable to treatment by filtration and have a satisfactory $C:N:P$ ratio for aerobic oxidation (Section 1.2). Other important factors include the media, temperature, retention time, depth, film, oxygen, and frequency of dosing, all of which are dealt with below.

Media
Performance is generally related to the specific surface area of the medium provided film accumulation does not result in blocking of the interstices. Therefore, smaller grades of media will produce a better quality final effluent than a larger grade of the same medium, when loaded at the same rate (Table 4.18) (Truesdale *et al.* 1962). The potential for film accumulation is an important selection criterion for media, which is dependent on the size of the interstices, the specific surface area, surface texture of the medium, and the

TABLE 4.18 Mean results of analysis over a 12-month period, of influent settled sewage and settled effluents from filters containing different types and sizes of media. Mean rate of application of sewage 0.59 $m^3 m^{-3} d^{-1}$. Figures in brackets, specific surface area ($m^2 m^{-3}$) of bulk medium as determined by paint dipping test (Truesdale et al. 1962)

| | Influent | Settled effluent from filter containing | | | | | | | |
| | | 25 mm nominal size | | | | 63 mm nominal size | | | |
		Clinker (202)	Slag (196)	Rock (142)	Rounded gravel (146)	Clinker (123)	Slag (108)	Rock (91)	Rounded gravel (65)
BOD ($mg\,l^{-1}$)	308	12	10	18	21	18	21	26	28
Ammonia (as N) ($mg\,l^{-1}$)	61	4.0	1.8	13.1	13.5	13.3	18.6	32.5	40.6
Oxidized nitrogen (as N) ($mg\,l^{-1}$)	—	52.7	50.7	43.3	44.0	40.1	34.2	24.7	18.3

organic loading. Smoother media are slightly less effective than rough-surfaced media, although the latter is more susceptible to excessive accumulation of solids. Performance varies seasonally with the larger grades of media producing a better effluent in the winter, whereas the smaller grades perform best in the summer (Hawkes and Jenkins 1955). Therefore, selection of media must be a compromise between having sufficient specific surface area and large enough interstices to prevent clogging in the winter. The effects of media type on performance has already been considered in an earlier section (Section 4.1.1.1).

Temperature
Temperature is an important operational parameter in all biological wastewater treatment systems. In percolating filters, the metabolic rate of micro-organisms increases with temperature, with the rate of treatment doubling for every 10°C increase within the range 5–30°C. Whereas the diffusivity of nutrients and oxygen increases with temperature and the solubility of oxygen decreases. As the BOD removal efficiency (E) falls with declining temperature (T) the relationship can be expressed as:

$$E_T = E_{20} \cdot A_T^{(T-20)}$$

where E_T and E_{20} are the BOD removal efficiencies at T°C and 20°C respectively, and A_T is a temperature coefficient with a value between 1.035–1.047 (Roberts 1973; Shriver and Bowers 1975). Film accumulation is affected by temperature by reducing the rate of oxidation (Fig. 4.6) and the activity of grazers, which is severely restricted at temperatures below 10°C (Hawkes 1957; Bayley and Downing 1963; Solbe *et al.* 1974; Shephard and Hawkes 1976). Nitrification is also temperature-dependent with little activity below 10°C (Bruce *et al.* 1967) (Fig. 4.19).

The normal range of domestic wastewater is between 10–20°C, although Gray (1980) recorded a range of 6.5–18.0°C for settled influent wastewater loaded onto his pilot filters (Table 4.19). So the design loading for a percolating filter is restricted by the lowest winter temperature. The temperature of the influent wastewater and of the film varies seasonally (Fig. 4.20). Gray (1980) examined the effect of temperature on percolating filter performance in his pilot filters which were built above ground. The concrete walls, although thick enough to offer some insulation, allowed heat transfer between the filter medium and the air outside. Changes in the air temperature, wind velocity affecting heat loss from the walls, and the absorption of heat by solar radiation all caused temperature changes to occur in the medium immediately adjacent to the walls within the filters, which were closely recorded by the thermo-couples inside the pilot filters (Gray and Learner 1983). There was a clear diurnal variation in the temperature in the medium immediately adjacent to the wall, getting warmer during the day and cooler at night. The temperature of the central core of the filters remained far

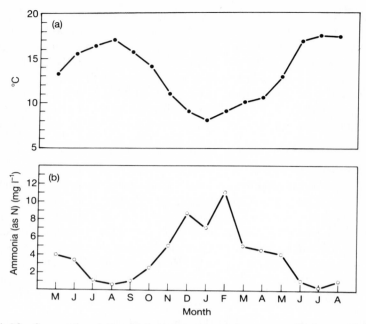

FIG. 4.19   Seasonal variation in the temperature of sewage applied to a percolating filter and the concentration of ammonia remaining in the effluent. Average concentration of ammonia over the whole period was 5.1 mg N l^{-1}. (a) Monthly average temperature of applied sewage; (b) monthly average concentration of ammonia in filter effluent (Bruce *et al.* 1967).

more constant. Gray compared 50 mm blast-furnace slag with a random plastic filter medium of the same nominal size (Flocor RC). He found that in fact the core temperature in the slag medium, as measured by the central thermo-couple, remained extremely constant, with small changes in temperature occurring over long periods, i.e. in excess of 6 h, whereas the temperature in the plastic medium changed more rapidly, often by 1°C within 30 minutes. During the winter, the temperature gradually increased with depth due to heat produced from microbial activity, although the medium adjacent to the walls of the filters remained cooler. In the summer, the influent retained more of its original heat and the greater microbial activity meant that again the temperature of the influent sewage increased with depth but to a greater degree than before. The metabolic rate increases exponentially with temperature and, therefore, greater temperature productions are to be expected during the warmer months. The temperature of the slag filter was closely related at both low- and high-rate operation to the temperature of the influent. The temperature of the influent wastewater was affected by the air temperature (Fig. 4.20), and during extremely cold conditions the influent was also cooled considerably in the fall from the distributor to the surface of the medium.

TABLE 4.19 Summary of the mean performance of all three pilot filters at the various loadings

Loading (duration) Filter	1.68 m³m⁻³d⁻¹ (13 months)			3.37 m³m⁻³d⁻¹ (13 months)			5.72 m³m⁻³d⁻¹ (3 months*)		
	Slag	Mixed	Plastic	Slag	Mixed	Plastic	Slag	Mixed	Plastic
Organic load (kg BOD m⁻³d⁻¹)	0.280	0.280	0.280	0.628	0.628	0.628	0.854	0.854	0.854
Effluent BOD (mg l⁻¹)	20.20	20.53	22.54	33.09	25.03	27.12	33.70	26.25	32.00
Percentage removal BOD	87.58	87.73	86.26	82.36	87.18	85.50	68.00	74.50	68.60
Suspended solids load (kg m⁻³d⁻¹)	0.201	0.201	0.201	0.417	0.417	0.417	0.743		
Effluent suspended solids (mg l⁻¹)	24.73	24.72	27.53	31.08	26.32	28.79	71.00	39.50	27.00
Percentage removal suspended solids	77.82	77.67	75.66	72.20	77.00	78.30	41.60	60.10	66.40
Ammonia load (kg m⁻³d⁻¹)	0.054	0.054	0.054	0.075	0.075	0.075	0.114	0.114	0.114
Effluent ammonia (mg l⁻¹)	14.71	17.80	20.73	17.98	15.75	19.09	16.90	16.60	16.70
Percentage removal NH₃	52.81	44.97	34.86	21.08	32.42	17.75	7.75	8.45	8.2
Total oxidized nitrogen (mg l⁻¹)	14.32	12.09	9.03	6.19	6.95	4.30	1.05	1.15	0.75
Effluent temperature (°C)	8.97	8.78	8.94	10.86	10.64	11.09	—	—	—

* These results were collected over a shorter period during maturation of the filters, and therefore are not directly comparable to the results collected during the other loadings.

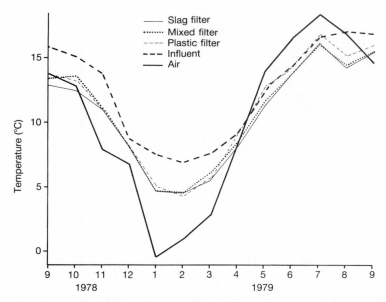

F I G. 4.20   Seasonal changes in influent, final effluents, and air temperatures during
high-rate loading of experimental filters.

Although the temperature of the surface layer of the medium was clearly
related to the influent temperature, the core temperature of the filter remained
more or less constant due to heat production from biological oxidation. The
slag filter comprised of 48.5% solid material that was able to retain the heat
compared with 8.7% in the plastic filter. Therefore, with more heat-retaining
material, less voidage and so lower potential ventilation than the plastic filter,
the slag filter was a far more effective buffer against changes in the air
temperature; the central core of the filter being maintained at a constant
temperature by the heat produced by biological oxidation. The temperature of
the plastic medium filter was influenced far more by the air temperature than
that of the influent because of the greater voidage giving rise to excessive
ventilation and therefore more heat exchange. The plastic filter was seen to
respond quickly to changes in air temperature. This close association resulted
in variations in the diurnal temperature of the film that was not reflected in the
seasonal data (Table 4.19). Bayley and Downing (1963) suggested that with
synthetic media with a high voidage, the air temperature and rate of flow of air
influenced the temperature in the voids, but that the temperature of the
influent was still the main factor in maintaining the temperature, and to a
lesser extent the rate of reaction within the microbial film. The temperature of
the influent always prevented extremes of temperature within the plastic filter
and this was seen clearly at the higher loading. In tall towers containing
modular plastic media it is necessary to control the rate of ventilation through

the filter, by adjusting the air vents at the base, in order to maintain the temperature, especially on cold windy days, to prevent excessive cooling.

Warm wastewaters are advantageous in reducing seasonal fluctuations in temperature, preventing low temperatures in the winter. In colder climates, filters are covered to conserve heat and to prevent freezing, although covers are expensive to construct.

Retention time
The longer intimate contact is maintained between wastewater and the active film supported on the medium of percolating filters, then the better the final effluent quality. Clearly, the duration of liquid retention within the filter is potentially an extremely useful and important parameter (Eckenfelder 1961). However, prediction of the performance of filters using the retention time has proved to be largely unsuccessful. Although retention time is considered a major factor in filter efficiency (Tariq 1975), the exact meaning of such data has remained unclear, resulting in its infrequent use in the assessment of filter efficiency.

The importance of retention time, also referred to as residence or contact time, in the assessment of the efficiency of filters has been stressed from the earliest times (Royal Commission on Sewage Disposal 1908). The theory that increased retention allows more time for the wastewater to be in intimate contact with the film, therefore allowing greater adsorption of particulate matter and maximum exchange of nutrients, has been widely studied (Eden *et al.* 1964; Craft *et al.* 1972; Craft and Ingols 1973; Cook and Katzberger 1977). Many of these workers found that the retention time was associated with, although not directly related to, performance. Eden *et al.* (1964), when examining the measurement and significance of retention time in filters, found that there was three controlling factors, the hydraulic flow, film accumulation, and the size and shape of the filter medium. The physical characteristics of the medium remain constant, and any variation in retention time of a particular filter will be due to diurnal or daily changes in the hydraulic flow or seasonal changes in the film accumulation. Increases in hydraulic flows normally result in a decrease in retention time, except in certain random plastic filter medium where an increase in hydraulic loading results in better redistribution of the wastewater within the bed, thus maintaining the retention time (Bruce *et al.* 1975; Porter and Smith 1979). Escritt (1965) found a strong inverse relationship between the hydraulic loading and retention time in a conventional full-scale filter, 1.8 m deep. He was able to calculate the retention time (t_r) in minutes as:

$$t_r = 40 \ln(9.4/L_v)$$

where L_v is the hydraulic retention time $(m^3 m^{-3} d^{-1})$. Generally, the fluctuations in hydraulic flow to filters are small and will only have slight effects on the retention time. The effects of film accumulation on retention time are less clear, and as no direct relationship has been established it would appear that the film modifies the flow pattern of the wastewater at particular accumulations. This was demonstrated by Gray (1983c), who showed that film and humus had a pronounced effect on retention time, but that retention time was not directly related to removal efficiency. Large variations in the retention time had little effect on the organic removal efficiency of low-rate filters, whereas small variations in retention time had large effects on the removal efficiency of high-rate filters. The results of retention tests are commonly expressed as the time (in minutes) required for the recovery of 16% (t_{16}) and 50% (t_{50}) of the added tracer, taken from the plot of percentage tracer recovered against time on logarithmic paper on which a log normal distribution is a straight line. The 16 percentile and the 50 percentile (or median) values are chosen as $(t_{50}/t_{16}) = \sigma$, the standard deviation of the log normal distribution (Eden et al. 1964). Gray used t_{50}/t_{16} to predict the flow pattern of wastewater within the filter. He found that, when channels are formed in filters because of excessive film accumulation, some of the tracer passed through the filter extremely quickly, causing the rate of discharge to decrease after a short period producing a ratio (t_{50}/t_{16}) of > 3.125. Under this film condition, and during periods of normal film accumulation, the median retention time (MRT) increased as the film accumulation increased, with a ratio at 3.125 or less. It was during this period that the MRT was directly associated with the performance. When the MRT increased rapidly to a very high value and the ratio became more than 3.125, this indicated that the influent was being retained in reservoirs formed by excessive film conditions. Although in this instance the MRT is longer, it is not associated with increased performance. Because the influent is stored in reservoirs only a very small proportion of the liquid is in intimate contact with the film, thus nutrient transfer does not take place so efficiently. Also, wastewater stored in this way will be low in dissolved oxygen and effectively eliminates oxygen transfer into the film, severely reducing activity. An increase in film accumulation either leads to the production of channels or if the excess film is not sloughed or removed by the grazing fauna, the filter becomes completely blocked. These processses are summarized in Fig. 4.21.

The size, shape, and surface area of the media are also important, with the MRT increasing with the specific surface area of the media (Fig. 4.22). However, as the medium within a filter remains constant, it can be ignored in terms of operation. Work on the periodicity of dosing has shown that retention time can be increased by large intermittent doses of influent (Shephard 1967) due to storage of liquid in horizontal chambers within the filters. These reservoirs are found in filters at times of high accumulations of film where the tracer enters quickly but is slowly discharged as more influent wastewater replaces and dilutes it. This explains why the large increase in MRT recorded before the rapid decline in MRT is not directly associated with performance. In contrast, a number of workers consider that retention is

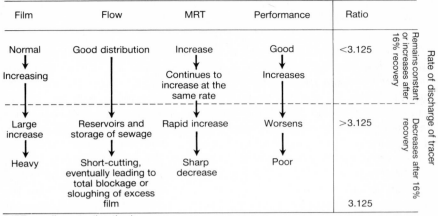

Film	Flow	MRT	Performance	Ratio	Rate of discharge of tracer
Normal ↓ Increasing ↓ Large increase ↓ Heavy	Good distribution Reservoirs and storage of sewage ↓ Short-cutting, eventually leading to total blockage or sloughing of excess film	Increase ↓ Continues to increase at the same rate Rapid increase ↓ Sharp decrease	Good ↓ Increases Worsens ↓ Poor	<3.125 >3.125 3.125	Remains constant or increases after 16% recovery Decreases after 16% recovery

(MRT=median retention time)

FIG. 4.21   Summary of observed flow characteristics in relation to the ratio of the 16 percentile to the 50 percentile (t_{50}/t_{16}) of the retention time in pilot percolating filters.

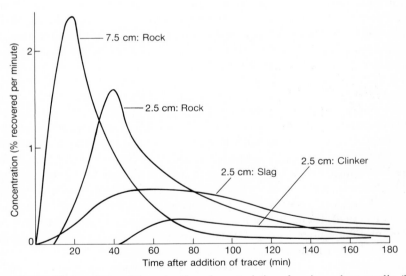

FIG. 4.22   Comparison of the retention characteristics of various clean media (i.e. without any film development) in pilot filters 1.8 m deep and loaded at 0.6 $m^3 m^{-3} d^{-1}$ (Eden *et al.* 1964).

probably irrelevent to percolating filter performance (Atkinson *et al.* 1963; Mehta *et al.* 1972).

Tracers can only measure their own retention and not necessarily the retention characteristics of waste liquids (Eden *et al.* 1964). Originally, dyes, salt solutions or ammonium salts were used (Tomlinson and Hall 1950), but these often suffered from prolonged retention due to adsorption on to the biological film. In the late 1950s, however, experiments with radioactive tracers showed them to be ideal for retention analysis. This was because: (a) only small quantities of such tracers are required; (b) of the ease and sensitivity of detection; and (c) of the negligible tendency of some of the radioactive substances to be adsorbed on to the film (Eden and Melbourne 1960). Since that time most of the research on retention analysis has involved the use of radioactive tracers, although Tariq (1975) published a method to calculate the mean time of retention by measuring the influent and drainage rates. A major disadvantage of Tariq's method is that the filter has to be shut down in order to measure the drainage characteristics, thus taking it out of operation for a number of hours, which in a smaller works would prove very difficult. Due to the technical and financial problems of using radioactive isotopes, and also the possible environmental and health aspects of using them as tracers for the measurement of retention time, Gray (1981) used a simple technique using the traditional tracer sodium chloride and continuously monitored the conductivity of the final effluents to determine its recovery. He found the method of negligible cost, extremely quick, and produced reproducible results, making it possible to include the regular measurement of retention time in the routine assessment of filter efficiency.

Depth
Most filters are built between 1.5–2.5 m in depth, with the majority approximately 1.8 m (6′) deep. It is traditional in some countries, such as Germany, to build them slightly deeper (3–5 m) but there seems little justification for constructing conventional filters deeper than 2.5 m. Depth does not appear to be a major factor in determining performance, although total available surface area of the media is. Therefore, a shallow filter will give the same performance as a deeper filter of the same volume. Bruce and Merkens (1970) found no significant difference in performance between two filters of different depths containing plastic media, one 7.1 m deep the other 2.1 m, treating the same high volumetric loading.

Shallow filters (< 1 m) do have significant disadvantages. The risk of ponding and short circuiting is far more serious in shallow filters, without the extra depth to compensate, and as the depth increases it becomes easier to obtain uniform distribution within the medium. Although shallow filters are structurally cheaper to construct they require more land area and the distribution cost will be increased. In contrast, tall filters take up less area and there is a significant increase in retention time with the greater depth.

However, increased pumping costs and increased structural costs make depth selection critical and design engineers should avoid adding an extra 0.5 m or so for good luck.

Gray (1980) studied the performance of low- and high-rate filters at various depths. In single-pass low-rate filters three-quarters of all the available BOD is removed in the top 900 mm of the bed, with between 45–50% removal occurring in the top 300 mm. The removal rate is affected by the film accumulation, with large removals of BOD associated with relatively thin active films. This association continues until the film becomes so thick that it begins to restrict the flow of wastewater through the filter, resulting in a short retention time due to channelling and a subsequent decrease in removal efficiency. At depths >900 mm, little BOD removal occurs (Table 4.20). More depth is utilized at high-rate loadings, illustrating the need for a greater surface area to oxidize the increased organic load. This is why most high-rate filters using plastic media are > 5 m in height although they rarely exceed 7 m. Higher loading rates will increase the recirculation within filters using random plastic filter media, due to the high specific area of the medium, which reduces the possible increase in depth required. Suspended solids removal occurs mainly in the top 900 mm at both low- and high-rate loadings, and is clearly linked with the adsorption capacity of the film. In the same filters, nitrification was restricted to depths >900 mm which increased to >1500 mm, or even excluded altogether from the filter in the high-rate filters. Therefore, a minimum depth of 1 m is required to ensure adequate BOD and suspended solids removal in low-rate filters, although if nitrification is also required a minimum depth of 1.5 m is required. At higher loadings a greater depth is required if enough surface area is to be made available for maximum BOD removal. If random plastic medium is used sufficient depth can also result in nitrification. Examples of depths of percolating filters in various countries are given by Pike (1978).

Film
Removal of organic matter occurs in four stages: (a) adsorption to the surface of the film; (b) extracellular breakdown of adsorbed material; (c) absorption by heterotrophic micro-organisms; and (d) mineralization. As the first step in purification is adsorption, any limitation on the availability of adsorption sites will reduce purification. Removal efficiency of both BOD and suspended solids in wastewater is linked to thin actively growing film. The grazing fauna are very important in helping to maintain an actively growing film thus ensuring maximum rates of adsorption. The rate of adsorption has already been shown to be independent of temperature, whereas the rate of oxidation is temperature-dependent (Fig. 4.6).

The factor that most influences retention time, assuming the hydraulic flow remains constant, is the film accumulation. Eden *et al.* (1964) demonstrated that retention time increased directly with film accumulation up to an

TABLE 4.20 The percentage removal of the total BOD, in relation to depth, at the low loading ($1.68 \ m^3 m^{-3} d^{-1}$) in experimental filters (Gray 1980)

Depth (mm)	Slag filter				Mixed filter				Plastic filter			
	0–300	300–900	900–1500	1500–1800	0–300	300–900	900–1500	1500–1800	0–300	300–900	900–1500	1500–1800
Nov. 77	69.6	16.4	5.7	5.9	58.9	32.5	2.8	0.0	63.1	26.0	0.0	2.8
Dec. 77	30.0	17.1	27.5	4.3	26.9	15.4	29.7	0.0	30.0	27.1	16.3	0.0
Jan. 78	20.6	61.6	12.2	0.0	40.0	46.1	0.0	3.4	47.5	45.4	3.7	1.5
Feb. 78	32.7	21.6	45.6	0.0	43.7	46.1	0.0	4.5	66.9	31.9	0.0	0.8
Mar. 78	40.8	33.8	13.3	6.5	53.2	19.4	22.5	0.3	58.7	22.5	3.4	6.6
Apr. 78	—	—	—	—	—	—	—	—	—	—	—	—
May. 78	59.0	27.9	0.0	2.7	62.0	19.3	14.4	0.0	56.9	24.6	6.9	2.8
Jun. 78	38.8	37.3	0.0	12.4	34.4	50.3	0.6	4.5	68.3	0.4	10.1	3.9
Jul. 78	58.6	25.7	3.7	7.5	57.2	26.8	3.0	6.8	45.5	27.1	19.0	0.0
Aug. 78	73.1	15.9	0.0	1.9	48.1	39.5	3.6	1.9	45.7	37.9	9.9	0.0
Sep. 78	44.2	20.3	17.9	7.9	68.4	6.8	12.3	0.0	42.2	27.3	10.2	8.1
Mean	46.74	27.76	12.59	4.91	49.28	30.22	8.89	2.14	52.48	27.02	7.95	2.65
S.d.	17.56	13.96	14.69	3.91	13.08	14.96	10.47	2.50	12.29	11.62	6.43	2.84

optimum weight, after which retention characteristics changed due to excessive film. Heavy weights of film alter the flow pattern resulting in the storage of effluent as localized reservoirs within filters, channelling of the wastewater, and eventual ponding, which results in a loss of performance (Gray 1983c). The film accumulation also follows a seasonal pattern, becoming thicker during the winter months due to reduced bioxidation and reduced grazing at the lower temperatures (Gray 1983b) (Fig. 4.5). Therefore, the most important influence on filter efficiency is the volume and growth rate of film, which not only controls the retention time, but also the rate of adsorption of suspended and dissolved nutrients, the rate of bioxidation, and the sludge production. In operational terms, a knowledge of the film accumulation within a filter is required if maximum performance efficiency is to be maintained (Honda and Matsumoto 1983). Ideally, the quantity and quality of active biomass within the filter should be controlled as in the activated sludge process, which can only be achieved in percolating filters by having a greater fundamental understanding of the kinetics of film growth and accurate methods of determining film accumulation. A number of modifications of the percolating filter process have been developed such as random and modular plastics media, recirculation, alternating double filtration (ADF), control of the frequency of dosing, and modifications to the distribution system. One major advantage of these modifications is they attempt to achieve optimal film growth. However, even with such modifications it is desirable to have some indication of the film accumulation within the filter, especially as depth studies have shown that the surface film accumulation rarely resembles the film accumulated within the filter (Gray 1983b). With modifications such as ADF and recirculation, the stage of film development is an important factor in deciding whether the filter series should be changed or the ratio of returned effluent modified.

Although a number of possible methods of estimating the degree of film accumulation are available, for example, the adsorption of dyes (Smith and Coackley 1983), or comparative retention time analysis, the relationship between these parameters and film weight is far from clear. The information obtained in this way is also rather imprecise, with one value obtained for the condition of the entire filter but no estimation of film accumulation at specific depths or areas of the filter being possible. Two approaches to film measurement are generally used at present. Direct measurement using sampling tubes containing baskets of medium which can be removed and film estimated by a number of gravimetric methods, or an indirect assessment method where the film is measured within the filter without disturbing the medium using a neutron modulation technique.

The film is comprised mainly of water between 92–97% in plastic media filters and 95–98% in mineral media filters (Gray 1980). This allows film accumulation to be estimated in filters by measuring the abundance of hydrogen atoms present in the water molecules using neutron modulation.

Fast neutrons emitted from a radioactive isotope (beryllium–americium) form a cloud of slow neutrons on collision with hydrogen atoms. By using a slow neutron detector the hydrogen atoms in any water present in the filter can be measured (Bell 1973) (Fig. 4.23). There are several commercial neutron probes available, although the most widely used probe in wastewater studies is the Wallingford Soil moisture probe (manufactured by D. A. Pitman Ltd, Weybridge, Surrey). The probe is lowered into the filter via a 50 mm diameter aluminium access tube, which is sealed when not in use to keep it perfectly dry. The probe has to be calibrated for each medium tested using a 210 l drum containing the medium over a range of moisture levels. Specific and significantly different calibration curves are obtained for each medium used, with the difference between the slope of the calibration curves for different media because of absorption of thermal neutrons by the media (Harvey *et al.* 1963). Therefore, the appropriate calibration curve must be used when converting the probe readings into percentage saturation of the voids. The use of neutron modulation in the estimation of film accumulation has been fully described by Harvey *et al.* (1963) and Gray (1984c).

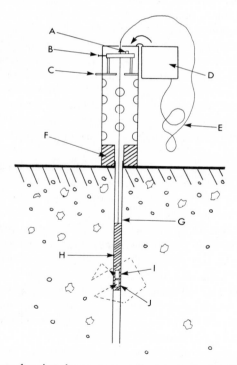

FIG. 4.23   Diagram showing the neutron probe in use. A, depth indicator; B, cable clamp; C, lock; D, ratemeter; E, cable; F, plastic shield containing isotope; G, access tube; H, probe; I, slow neutron detector; J, neutron source.

Gray and Learner (1984*b*) compared direct gravimetric measurements of film weight with the neutron scattering technique. Four gravimetric methods of film accumulation were used, total film weight, total dry solids, volatile solids, and percentage settlement of solids, which incorporated both weight and volume measurements. Total film was the weight increase of the sampling basket and the medium due to accumulated film. The wet weight of the clean medium in each sampling basket was measured at the start of the investigation. Once a month, each basket was removed from the sampling tube, excess liquid drained off for 5 minutes and then weighed, the increase in weight being due to film growth. Total dry solids were measured by evaporating a 50 cm^3 subsample of film, removed from a 1 litre sample of filter medium, in a weighed evaporating dish to dryness at 105°C for 24 hours (DoE 1972). The dish was reweighed after cooling to room temperature. The quantity of volatile solids was determined by removing the organic fraction of the dried sample by burning in a muffle furnace at 500°C for 1 hour and then reweighing the dish after cooling to room temperature (Allen 1974). An assessment of the volume of solids present was made using the Imhoff cone method (DoE 1972). A 250 cm^3 subsample of the removed film was allowed to settle for 45 minutes in the Imhoff cone, which was then gently twisted to remove any debris adhering to the glass sides. The quantity of solids which settled after 1 hour was recorded as the percentage settlement. Although any large lumbricid worms were removed prior to settlement, all the samples contained large numbers of invertebrates and these were included in this assessment of film accumulation. All the gravimetric estimations were made at monthly intervals at six depths within each filter corresponding with the depths at which the film was estimated by the neutron scattering technique. Before neutron scattering can be used, all the excess liquid has to be drained out of the filters, once the distribution system has been shut off. Readings were taken at 150 mm intervals throughout the depth of the filters, and each value taken was a mean of five readings each of which was a mean count per second integrated over a 16 second sample period. Replicate access tubes were used to allow the film accumulation to be studied from either side of each filter and the mean of the two average readings at each depth was taken to represent the mean percentage saturation of the voids. Neutron probe readings were always taken within 24 hours of the gravimetric sampling. The results of the comparative study by Gray and Learner (Table 4.21) showed good correlations ($P < 0.001$) between all the methods tested in low-rate conditions (1.68 m^{3}m^{-3}d^{-1} and 0.28 kg BOD m^{-3}d^{-1}), although the neutron scattering results were not significantly correlated ($P > 0.05$) with any of the gravimetric methods at the higher loading (3.37 m^{3}m^{-3}d^{-1} and 0.63 kg BOD m^{-3}d^{-1}). They concluded that although neutron scattering provided a rapid and sensitive measure of hydrogen atoms in the filter, the results expressed as percentage saturation of the voids are not directly transferable to film weights and should be treated separately, not as a true measure of film accumulation.

TABLE 4.21 Summary of the range of film accumulations measured by the neutron scattering technique (NP), and also as total film (TF), volatile solids (VS), total solids (TS), and percentage settlement (PS) in the mineral and plastics media at each loading (Gray and Learner 1984b)

	Mineral medium					Plastics medium				
	NP (% sat)	TF	VS (kg m^{-3})	TS	PS (%)	NP (% sat)	TF	VS (kg m^{-3})	TS	PS (%)
Low loading rate (1.68 m^{3}m^{-3}d^{-1})										
Mean	23.0	257	7.25	10.08	16.0	3.5	104	5.01	6.60	11.0
Minimum	11.3	96	1.04	1.48	2.4	1.1	28	0.36	0.44	0.6
Maximum	39.0	616	23.08	31.44	60.0	15.6	315	16.44	21.28	37.2
Range	27.7	520	22.04	29.96	57.6	14.5	287	16.08	20.84	36.6
Skewness	0.09	0.75	0.99	0.96	1.37	2.22	0.72	0.69	0.60	0.69
n	66	54	66	66	66	66	54	66	66	66
High loading rate (3.37 m^{3}m^{-3}d^{-1})										
Mean	26.5	216	5.91	7.68	11.1	4.5	173	8.34	10.91	18.8
Minimum	16.9	83	1.32	1.88	2.2	1.3	34	1.20	1.36	1.4
Maximum	40.2	494	23.68	32.56	48.0	12.2	474	25.08	33.92	78.4
Range	23.3	411	22.36	30.68	45.8	10.9	440	23.88	32.56	77.0
Skewness	0.25	1.31	2.22	2.37	1.85	1.06	0.91	1.20	1.28	1.62
n	12	72	72	72	72	12	72	72	72	72

Neither could the technique provide qualitative information about the film that was obtained by visually inspecting the exposed cores of film covered medium. It does, however, clearly indicate to the operator areas of high or low water saturation. It therefore appears to be ideal for use in the routine operational control of filter modifications, such as ADF and recirculation, and also in predicting the onset of ponding, especially below the surface of the medium. However, accurate measurement of film weight and a detailed analysis of the condition and composition of the film is only possible by using gravimetric analyses.

Film accumulation data is therefore useful in indicating where biomass is building up within the filter and can inform the operator if potential ponding problems exist, if the filter is being over or underloaded, or if film development is inhibited. There are a number of ways of visually interpreting such data. Traditionally, kite diagrams have been used to illustrate film accumulation at various depths, with a separate kite plot for each month or sampling period (Fig. 4.24). A development of this is the horizontal histogram, which, when turned on its side, gives a three-dimensional, view of the changes over the sampling period of vertical film accumulation within the filter bed (Fig. 4.16). More recently, contour mapping of film data allows film concentration contours to be plotted, with depth and time being the only axes. This approach allows interpolation between the available data points, normally by simple linear regression. By selecting the number of concentration contour levels to

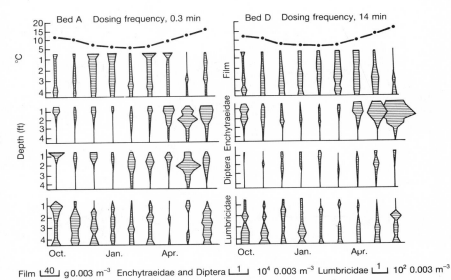

Film ⌊40⌋ g 0.003 m⁻³ Enchytraeidae and Diptera ⌊1⌋ 10⁴ 0.003 m⁻³ Lumbricidae ⌊1⌋ 10² 0.003 m⁻³

FIG. 4.24 Kite diagram showing seasonal changes in the abundance and vertical distribution of film and grazing fauna in high-frequency dosed filter A and low-frequency dosed filter D (Hawkes and Shephard 1972).

be plotted, either simple or very detailed views of seasonal film accumulation within the filter can be obtained (Fig. 4.25).

Oxygen

Oxygen is supplied to the aerobic micro-organisms comprising the film almost exclusively by absorption from air as it passes through the filter by natural ventilation. When ventilation ports are blocked or partially restricted, the performance of the filter is seriously affected; therefore good ventilation is vital for the successful operation of percolating filters. Ventilation currents are caused by the temperature difference between the air outside and the liquid and film inside the filter, and the resultant flow of air may be in either direction. In the winter, the air entering the filter will be warmed by the wastewater passing through the bed causing a net upward flow while in the summer the opposite may occur with the air possibly being cooled in the filter resulting in a downward flow of air. The rate of flow of air through the filter is directly proportional to the difference in the temperature between the outside air and the wastewater, with a difference of $10°C$ inducing a rate of flow of air of about $20 \, m^3 m^{-2} d^{-1}$. Other factors also affect ventilation, such as wind currents

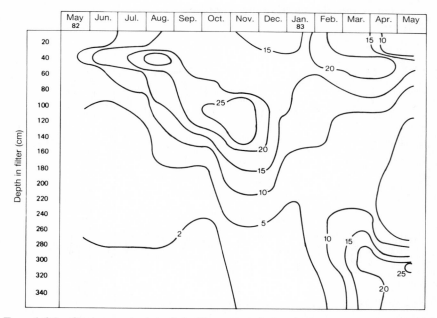

FIG. 4.25 Contour mapping of the film accumulation (as percentage saturation of the voidage) in a random plastic filter medium (Flocor RC) receiving finely screened sewage (Hoyland and Roland 1984).

(Mitchell and Eden 1963) and humidity. The humidity of the air inside the filter will increase as it passes through, and as an increase in humidity reduces the density of air, this encourages an upward flow. While the oxygen concentration of air in the interstices within the filter remains very close to that of normal air (20.9% v/v), reduction of the oxygen concentration is possible during periods of high carbonaceous oxidation. It is unlikely that normal heterotrophic activity will be affected by low oxygen concentrations, with no reduction in BOD removal at oxygen concentrations of 2% v/v, although nitrification is far more sensitive to oxygen limitation (Department of Scientific and Industrial Research 1956).

The rate of oxygen uptake in percolating filters varies from $0.273 \text{ kg O}_2\text{m}^{-3}\text{d}^{-1}$ in low-rate filters, where $0.187 \text{ kg m}^{-3}\text{d}^{-1}$ is required for heterotrophic oxidation and $0.086 \text{ kg m}^{-3}\text{d}^{-1}$ for nitrification (Montgomery and Borne 1966), to $1.490 \text{ kg m}^{-3}\text{d}^{-1}$ for high-rate filters (Bruce and Merkens 1970). The rate of oxygen utilization depends on the specific surface area of the media on which the film develops, thus, for a medium with a specific surface area of $200 \text{ m}^2\text{m}^{-3}$ the potential uptake rate of oxygen could be as high as $1.900 \text{ kg m}^{-3}\text{d}^{-1}$. The rate of oxygen utilization is fairly uniform throughout low-rate filters with demand from heterotrophs in the upper layer declining with depth, and the demand from nitrifiers increasing with depth. The rates at which oxygen and organic nutrients diffuse into the film depends on their relative concentrations in the liquid phase covering the film. For example, when the nutrient concentration is high then the concentration gradient is large causing more rapid diffusion to a much greater depth. In fact, the depth of penetration into the film is approximately proportional to the substrate concentration (Atkinson and Fowler 1974). The depth of penetration of organic nutrients into the film doubled from 0.06 to 0.15 mm when the substrate concentration was increased from 10 to 500 g COD m^{-3}. In practice the maximum substrate concentration is found in the upper 900 mm of filters, where the maximum film accumulation is also recorded. The rate of diffusion in this area of the filter may exceed the rate at which oxygen can diffuse into the film for aerobic metabolism, resulting in incomplete oxidation. Therefore, in the top section of the filter the oxidation rate will be oxygen-limited, whereas with depth, the rate of oxidation will decrease due to a reduction in the substrate concentration. At lower substrate concentrations the organic nutrients can only penetrate a short distance into the film before being consumed so here the rate of oxidation is substrate limited (Fig. 4.26). The remaining thickness of film is important as it acts as a buffer against environmental and operational changes. For example, if the temperature fell, the nutrients would penetrate further into the film before being consumed thereby using more of the microbial population. This has been modelled by Harris and Hansford (1976) who suggested that oxygen limitation occurs when the substrate concentration exceeded 400 g COD m^{-3}. Williamson and McCarty (1976) have also modelled this phenomenon and suggested that

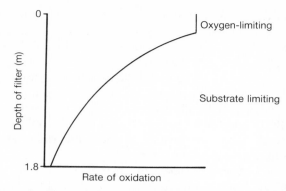

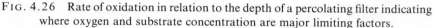

FIG. 4.26   Rate of oxidation in relation to the depth of a percolating filter indicating
where oxygen and substrate concentration are major limiting factors.

oxygen becomes the limiting factor in film when:

$$C \cdot K_c < S \cdot K_s$$

where C and S are the concentrations of oxygen and BOD in the film
respectively, and K_c and K_s their respective saturation coefficients.

Numerous methods have been employed to improve ventilation including
leaving gaps between stonework, using special air bricks, providing ventilation tubes through the medium, and even using fans and pumps to force air
through the bed. However, it appears that such systems are unnecessary as
normal ventilation systems using air vents attached to the under drains will
provide adequate amounts of oxygen for biological activity unless the voids
are excessively clogged with excessive film accumulation.

Frequency of dosing
Influent wastewater is distributed on to the surface of percolating filters via
sprays or nozzles mounted on either a moving distributor or a system of static
distributor pipes. Although the same loading rate is applied to both types of
filters the former distributor arrangement doses each specific area intermittently whereas the latter doses the entire surface of the filter continuously. It is
general practice to use moving distributors on all filters, except very small
installations where a fixed system of distributor pipes is used and intermittent
dosing is achieved by using balancing tanks that are emptied at the required
interval, or tipping-trough systems. Filters containing plastic media that have
a high voidage are generally loaded continuously from a fixed distribution
system as this gives better distribution of wastewater within the filter. Under
low-rate conditions the best performance is obtained using large doses at
infrequent intervals, i.e. low speeds of rotation. In this way, each segment of
the filter bed is subjected to a surge of liquid followed by a long period without
any flow before the next dose. The slower the speed of rotation the greater the

surge of wastewater and the longer the intervening period between doses. In theory, the long period between dosing ensures that the micro-organisms comprising the film are in a nutrient-limited condition so that they are in the endogenous respiration growth phase, i.e. a low f/m condition, which limits film accumulation. This has been found to be especially effective in preventing fungal growths in the winter. The surge of wastewater ensures a more uniform distribution of organic nutrients within the bed, extending the depth of heterotrophic activity and encouraging a more even distribution of film. The mechanical scouring effect also limits film development, but this is of minor significance compared with the limitation of nutrients. By lowering the frequency of dosing the retention time is also reduced. Therefore, dosing frequency must be a compromise between maximizing the retention time so that a sufficient wastewater–film contact time is maintained in order that maximum adsorption of organic nutrients can occur, and excessive film accumulation controlled. High frequency dosing approaches the conditions found with continuous dosing with a large accumulation of microbial film at the surface. High frequency dosing is ideal for random plastic medium or media with a high voidage that is able to support a high film accumulation in the top 900 mm. It is also suitable for mineral media filters that have been uprated by replacing the surface layer of medium with a random plastic medium (Gray and Learner 1984b).

The optimum frequency of dosing for low-rate filters is between 15–30 minute intervals (Tomlinson and Hall 1950). This prevents ponding in the winter, reduces the population of the fly Sylvicola fenestralis emerging during the warmer months, and encourages the enchytraeid worms to become the dominant grazer replacing the dipteran larvae (Hawkes 1955). However, the optimum frequency of dosing for a specific filter varies with the operating conditions, the type of medium, and the nature of the wastewater. The optimum frequency of dosing will also change seasonally, being much lower in the winter in order to control excessive film accumulation, compared to the summer when other factors control the film (Bruce and Hawkes 1983).

To control the frequency of dosing, the rotating arms need to be propelled using a motor driven rotor or a peripheral wheel driven by the wastewater flow using a water-wheel arrangement. Although the speed of rotary arms driven by the reaction of the spray jets alone can be controlled to a certain extent by altering the angle of the jets, reaction driven rotation requires a minimum head of 0.5–1.0 m and is severely affected by changes in wind direction. Some form of mechanical drive is required therefore, for accurate periodic dosing of filters.

4.1.2 Other fixed-film reactors

There are a variety of other wastewater treatment processes that require a biological film to develop on the surface of a support media. Among the more commonly used are rotating biological contactors (filters), fluidized beds, and nitrifying filters. These processes are all aerobic. However, anaerobic filters, which involve the development of an anaerobic film on a suitable support medium, are also widely used.

Anaerobic filters are used for treating wastewater with a high concentration of soluble BOD (>500 mg l^{-1}) and which contain little suspended solids. Under anaerobic conditions, the organic matter is converted to gaseous end-products with little sludge being produced. The most widespread use of anaerobic filters is for the removal of oxidized nitrogen from purified effluents by denitrification, with the nitrogen present reduced to elemental nitrogen and released from the filter as a gas. Denitrification requires a carbon source, and in the removal of oxidized nitrogen (nitrite, nitrate) from purified effluents this is usually supplied as methanol (Anderson et al. 1984; Bailey and Thomas 1975). Anaerobic fixed-film reactors are also known as anoxic filters. The use of anaerobic fixed-film reactors is fully reviewed in Section 4.4.3.3.

Submerged or upflow filters can be operated not only as anoxic or anaerobic systems but also as an aerobic process. The design of aerobic submerged filters is the same as for anaerobic systems except that an aeration device is required. The media used is normally 6 mm thick asbestos cement panels spaced 37 mm apart. They are positioned in such a way as to encourage the wastewater to flow up and down each section ensuring maximum contact with the available surface area. Kato and Sekikawa (1967) used net sheets instead of solid sheets, which took up only 4% of the tank volume and, more recently, random plastic media with very high voidage has been used. The process is very similar in action to activated sludge. Submerged biological filters are also used for nitrification (Schlegel 1988).

4.1.2.1   Rotating biological contactors

Rotating biological contactors (RBC) became commercially available in 1965, although the system was first developed in the late 1920s (Doman 1929). They are now widely used throughout the world and although particularly well suited for treating the wastewater from small communities, they are now being used to treat large domestic and industrial loads. RBCs are widely available as packaged plants under a variety of commercial names such as Biosurf, Biodisc, Biodrum, and Biospiral. The basic design consists of a series of flat or corrugated discs 2–3 m in diameter mounted on a horizontal shaft and driven mechanically so that the discs rotate at right angles to the flow of settled wastewater. The discs are usually plastic, corrugated polythene, PVC or expanded polystyrene, although other materials such as asbestos cement and expanded metal are also used. Each disc is 10–20 mm thick and spaced about 20 mm apart (Fig. 4.27). The mounted discs are placed in a contoured tank that fits fairly closely to the rotating medium so that 40% of their area is immersed, and are slowly but continuously rotated. Discs are usually but not

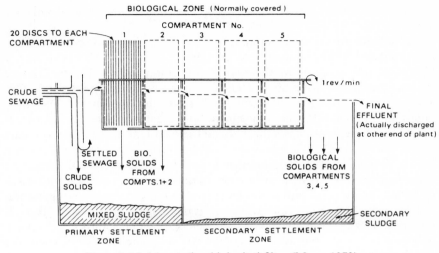

FIG. 4.27 A rotating biological filter (Mann 1979).

exclusively used as support media. Other configurations, such as lattice constructions or wire mesh containers filled with random plastic media have also been used successfully. The flow of wastewater through the tank and the action of the rotating medium produces a high hydraulic shear on the film that ensures efficient mass transfer from the liquid into the film and, at the same time, controls excessive film accumulation. Discs are arranged in groups separated by baffles to minimize short circuiting and the effect of surges of flow. Normally, a minimum of four compartments, separated by baffles, are used to simulate plug flow with a small head loss of between 10–20 mm between each compartment. With large installations, plug-flow conditions are achieved by having several complete RBC units in series (Steels 1974). The system consumes very little energy as the lightweight plastic media is evenly balanced throughout the drive shaft and a small motor is used to drive the rotor, which results in a low power consumption for the amount of BOD removed. For example, in a 300 PE unit, the power consumption of the motor is only 0.3 kW. RBC units are invariably covered, with small units covered with a moulded plastic or GRP cover and larger units housed in buildings. Covering the media is important in order to protect the exposed film on the discs from the weather, especially frost and wind which can damage the film. Wind can also increase the load on the motor driving the rotor. Covering insulates the system and reduces heat loss and increases the rate of oxidation. By insulating the cover such units can operate successfully in arctic conditions. The cover also eliminates fly nuisance and even controls odours to some extent. Using green covers makes such plants inconspicuous, especially as they have the same head loss as a septic tank (< 100 mm), and therefore are low-form structures, making them ideal for use in open areas where landscape quality is important, for example, golf clubs, hotels, and general amenity areas.

As the discs rotate, wastewater enters the spaces between the discs when immersed and is then replaced with air when out of the liquid. In this way, a biological film builds up on the discs in the same way as it develops on the medium in a percolating filter. Therefore, purification occurs with the film alternately adsorbing organic nutrients from the wastewater and then obtaining oxygen from the atmosphere for oxidation. After the biological phase, the wastewater passes through a settlement tank to remove any solids before discharge. There is a gradation of microbial species along the reactor path with the first discs developing a heavy growth of heterotrophic bacteria and fungi. A succession of protozoan species are seen on successive discs following the sapropic rating of species (Liebmann 1951). Depending on the organic loading, nitrification can occur on the first group of discs, although nitrifying activity is usually limited. Like percolating filters, nitrifying bacteria in RBCs are found towards the end of the system on the discs in the third and fourth compartments of a four-compartment unit. Due to the plug flow nature of the system there is a tendency for nitrite to accumulate in the compartment where nitrification first occurs. There is a net loss of nitrogen through the system which approaches 20% because of denitrification at the media–biomass interface (Ellis and Banaga 1976). The nitrogen oxidation rate is similar to that found in filters at approximately $1 \text{ g NH}_3\text{–N m}^{-2}\text{d}^{-1}$ at 20°C (Antonie 1976). Major variables affecting nitrification include total disc area, wastewater temperature, influent nitrogen, and BOD concentrations and flow rate (Barnes and Bliss 1983). The speed of rotation is limited by the shearing of the film at the periphery of the discs, but the normal range is between 0.5–10 rpm. Film is sloughed off continuously and will remain in the liquid phase adding to the overall treatment process. Eventually, sloughed film is carried away to a separate sedimentation tank although in smaller units the secondary settlement tank is built into the chamber. Smaller units also have a digestion chamber below the discs to reduce the load to the secondary settlement tank thereby reducing the quantity and sludge produced and the frequency of desludging, the solids settle and collect in the digestion chamber where they undergo anaerobic digestion (Fig. 4.27).

An electric motor is not always necessary to drive the rotor and it is possible to use air-induced drive systems. For example, air is sprayed into the rotating medium from below and to one side of the horizonal shaft. The liquid movement and asymmetric buoyancy from air collecting in one side of the medium induces it to rotate (Winkler 1981). This method reduces the risk of anaerobiosis in the first compartment or in the first RBC unit of a multiple unit system used for large populations. It is also used to uprate the activated sludge process. The disc system can be placed in an existing activated sludge aeration tank to increase significantly the biomass concentration within the tank without the need to increase the aeration capacity in order to satisfy the oxygen demand of an increased biomass (MLSS) concentration (Guarino *et al.* 1980).

Small plants are designed on the basis of BOD loading per unit area of discs, using both sides of each disc in the calculation. Bruce and Merkens (1975) found that a Royal Commission effluent (20:30) could be produced at organic loading rates of <6 g BOD m^{-2}d^{-1}. A conventional low-rate percolating filter containing 50 mm mineral medium and loaded at 0.10–0.12 kg BOD m^{-3}d^{-1} can be loaded at a surface loading of 1.0–1.2 g BOD m^{-2}d^{-1}, which is five or six times less than an RBC, to produce a 20:30 effluent (Pike 1978). Higher loading figures have been given for 80–90% BOD reductions, ranging from 6–20 g BOD m^{-2}d^{-1}, although the organic loading can be increased by a factor of 10 for efficient partial treatment (Hao and Hendricks 1975*a*, *b*; Forster 1971). Loadings to larger units are based on hydraulic loading per unit area of disc. The calculation of loadings is complex and explained in detail by Steels (1974) and Wilson (1981). For obvious reasons, manufacturers feel that loading to RBCs are too conservative making them less attractive when compared to the activated sludge process. Loadings and their calculation has been reviewed by Lumbers (1983).

The RBC is the intensification of the percolating filtration process with the density of biomass on the discs reaching 200 g (DW) m^2, which is equivalent to a sludge loading of 40–60 kg MLVSS m^{-3} in the activated process. RBCs have many advantages over the percolating filter process, for example: complete wetting of the media; no clogging of the media due to excessive film accumulation reducing the retention time (in practice the retention time of an RBC is the same as a percolating filter); and regular exposure to the air ensures unlimited oxygen and the area of land used is only 10% of that required by the equivalent treatment capacity supplied by low-rate filtration. Other advantages include low sludge production, excellent process control, ease of operation, high degree of BOD removal including substantial nitrification (Barnes and Bliss 1983), good settleability of solids, low power consumption, no distribution problems, and no recirculation required.

There are some potential disadvantages, which include frequent motor drive and bearing maintenance. Also, the process is affected by surges of flow and is sensitive to overloading. Power failure causes total loss of efficiency and where the discs are left standing in the wastewater for any length of time, uneven film distribution results with thicker film developing on the submerged surfaces and the film on the exposed surfaces drying out. This causes imbalance of the discs and severe motor wear, which results in eventual failure. Finally, RBCs are very expensive compared to other biological units, especially for small treatment plants. But their inconspicuousness, lack of odour, low fly nuisance, and quietness make them ideal where treatment must be done close to houses. The problem of cost has been overcome by Wexford County Council, Ireland, who construct their own units, making the discs out of asbestos cement. Also, Imhoff tanks have been uprated into RBC units by incorporating a disc system.

4.1.2.2   Fluidized beds

In order to provide sufficient surface area in a percolating filter to act as support for microbial film development, a medium is used. However, when 50 mm blast-furnace slag is used, 48.7% of the volume of the filter is taken up with the medium itself. This falls to 8.7% for the random plastic medium Flocor RC. The problem of providing sufficient fixed biomass for complete treatment is overcome by using sand as the medium and passing the wastewater through the bed of medium in an upward direction at sufficient velocity to fluidize the sand. The sand provides a large increase in specific surface area compared to other media ($3300 \text{ m}^2\text{m}^{-3}$), which allows considerable biomass to develop, equivalent to a MLSS concentration of up to $40\ 000 \text{ mg l}^{-1}$. In addition, fluidization of the sand particles prevents clogging or excessive film accumulation as the interstices are not stable. This mode of operation allows the density of the biomass to be maintained at high concentrations without having to recycle solids as in the activated sludge process. Also, the quantity of biomass is controllable so there are no wide seasonal fluctuations in biomass concentration as seen in other fixed-film reactors.

The basic layout of a fluidized bed system is shown in Fig. 4.28. The high biomass concentration results in a very high oxygen demand that can only be satisfied by injecting pure oxygen into the influent wastewater stream as it

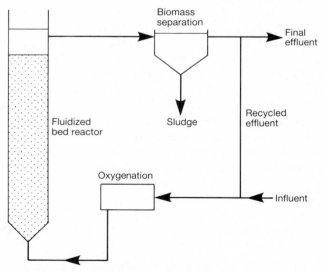

FIG. 4.28   Schematic diagram of a fluidized bed reactor system.

enters the base of the reactor, giving an oxygen concentration approaching 100 mg l^{-1}. Even distribution of the influent wastewater into the medium is achieved by injecting the wastewater downwards into a conical base (Cooper and Wheeldon 1982). The expansion of the bed within the reactor is controlled by the rate of wastewater input, so that the effluent can be removed from the reactor below the level of the expanded layer without loss of media. A secondary sedimentation tank can be supplied but is not necessary as the biomass can be retained within this upflow system by careful operation. The sand-covered biomass is regularly removed and the biomass separated from individual sand particles by passing it through a hydroclone where a high shear strips away any attached microbial growth. The clean sand is returned to the reactor and the sludge can be further treated before disposal.

The above systems are ideal for treating high strength wastewater and are particularly useful for treating industrial wastewaters where the loading is variable (Cooper and Wheeldon 1980, 1982). They operate at short retention times with 95% removal of BOD achieved within 16 minutes. When operated as a nitrifying process, a 20 mg l^{-1} concentration of ammonia–nitrogen can be reduced by 99% in 11 minutes at 24°C, which represents an ammonia–nitrogen loading of 900 mg m^{-2}d^{-1} at a hydraulic loading of 140 m^{3}m^{-3}d^{-1}. This is similar to that of a standard nitrifying filter (Section 4.1.2.3) (Jeris et al. 1977).

Fluidized beds are compact systems compared to activated sludge, therefore land requirement and capital cost is lower. However, operating costs are significantly greater because high purity oxygen and pumping are used. The sludge is more concentrated at 10% solids compared with 2% in gravity-thickened activated sludge. There are no odour or fly nuisances either. The fluidized bed can be operated aerobically, anoxically, or anaerobically. Anaerobic beds tend to expand rather than fluidize and are used for the conversion of carbonaceous wastes into gaseous end-products, therefore these reactors require a gas/liquid separation stage. Operation is in the mesophilic temperature range and heating is required (Section 4.4.3.4). Anoxic beds have been developed for denitrification (Gauntlett and Craft 1979).

Although sand (0.2–2.0 mm diameter) is the most common support media in aerobic systems, anthracite, carbon, and glass are also used. Small particles have a higher specific surface area but a lower settling velocity than large particles. Therefore, larger fabricated media have been developed which are porous allowing the biomass to grow within the media as well as on the outer surface. For example, reticulated polyester foam pads (25 × 25 × 10 mm). There are two full-scale aerobic fluidized beds on the market. Oxitron is manufactured by Dorr–Oliver Inc and uses sand particles 250–550 μm in diameter with a specific surface area > 3000 m^{2}m^{-3}. It is able to support a large biomass of between 10–17 kg MLVSS m^{-3}, with the excess biomass removed by shear. Oxygen is supplied by dissolving high purity oxygen into the influent wastewater stream (Fig. 4.29) (Sutton et al. 1981; Hoyland and

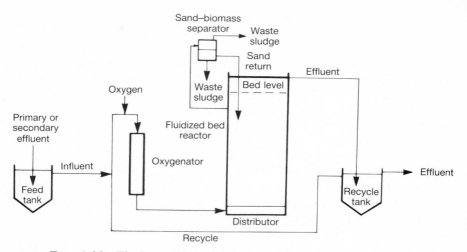

FIG. 4.29   The Dorr–Oliver Oxitron pilot plant (Sutton *et al.* 1981).

Robinson 1983). Capitor is the only commercially available bed using plastic foam pads. Manufactured by Simon–Hartley Ltd, the excess biomass is removed by passing the pads through a compression roller, which releases a concentrated sludge with the pads returned to the reactor with a reduced microbial population. Oxygen is supplied by using a proprietary oxygenator (Fig. 4.30) (Walker and Austin 1981).

4.1.2.3   Nitrifying filters

Many biological treatment processes, because of design or loading, are able to complete the carbonaceous oxidation of wastewater but are unable to successfully oxidize the ammonia–nitrogen (NH_3–N) present. In the activated sludge process, failure to provide full nitrification of wastewater may be due to inadequate retention time or aeration capacity, in percolating filtration, modifications such as ADF, insufficient surface area, or high organic loadings may result in incomplete oxidation of ammonia to nitrate. Ammonia–nitrogen is very toxic and many discharge licences will include an upper limit for ammonia being discharged to fresh waters. Therefore, a separate purification stage just to remove ammonia is often required to meet such standards.

The sole purpose of nitrifying filters is to remove ammonia from effluents that have undergone complete carbonaceous oxidation. It is based on the stratification of microbial species in conventional low-rate filters, where nitrifying bacteria are located towards the base of the bed. In this way, separate filters were developed for carbonaceous oxidation and nitrification, although nitrifying filters can be used to nitrify the effluent from other types of treatment units. Nitrifying filters are very similar to percolating filters in

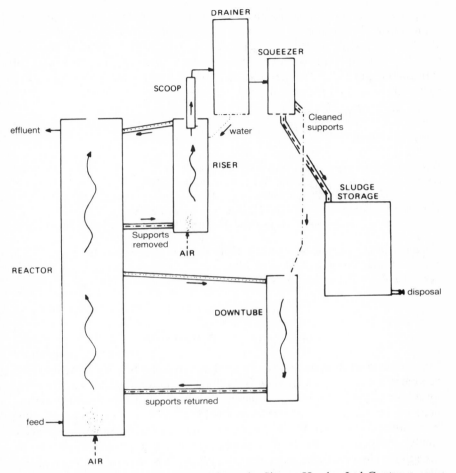

FIG. 4.30 Biomass recovery scheme from the Simon–Hartley Ltd Captor process (Walker and Austin 1981).

design except smaller grades of media are used (<25 mm) in order to maximize the specific surface area. The use of smaller grades of media is made possible as the wastewater contains very little organic matter and only minimal development of heterotrophic film can occur. Therefore, excessive film accumulation that could cause clogging does not occur. The nitrifying bacteria themselves are very slow-growing and produce only a small volume of biomass in comparison with heterotrophs. For example the cell yield of *Nitrosomonas* spp. is 0.053 g per g NH_3–N oxidized. The film accumulation is also restricted by the high hydraulic loading of between 2–6 $m^3 m^{-3} d^{-1}$. Wastewater is distributed on to the filter via a fixed distribution system using fine spray nozzles to ensure maximum utilization of the media and adequate oxygen. Using spray nozzles also allows maximum loss of ammonia by volatization. Removal rates of ammonia–nitrogen can rise to 120 g m^{-3}d^{-1}, although this depends on the specific surface area of the media, so removal rates of ammonia–nitrogen per unit surface area are in the order of 500–1000 mg m^{-2}d^{-1} (Barnes and Bliss 1983). The nitrification process has been fully examined in Section 3.5.1.

Further reading

Percolating filters: Bruce and Hawkes 1983; Pike 1978.
Design and operation: Pike 1978; Allen and Kingsbury 1973
Process modifications: Pike 1978; Bruce and Hawkes 1983; Hawkes 1983*b*.
Organisms and ecology: Curds and Hawkes 1975; Hawkes 1983*b*.
Factors affecting performance: Bruce and Hawkes 1983; Hawkes 1983*b*; Adin *et al.* 1985.
Rotating biological contactors: Antonie 1976; Steels 1974; Lumbers 1983.
Fluidized beds: Cooper and Atkinson 1981.
Nitrifying filters: Barnes and Bliss 1983; Schlegel 1988.

4.2 ACTIVATED SLUDGE

The activated sludge process is the most widely used biological wastewater treatment process today, treating both domestic and industrial wastewaters. Since its conception in the late-nineteenth century and subsequent development into a full-scale unit process in 1913 by Ardern and Lockett at the Davyhulme Treatment Works in Manchester, the basic process has been widely adopted and further developed, giving it a unique flexibility of operation. The process is more than simply a refinement of the percolating filter process, it is conceptually a very different treatment system. Whereas fixed-film reactors can be compared with the periphyton growing over the surface of stones and submerged plants in rivers, the activated sludge process utilizes the dispersed bacterial flocs and free-living micro-organisms that are suspended within the body of the water.

The process relies on a dense microbial population being mixed with the wastewater under aerobic conditions. With unlimited food and oxygen, extremely high rates of microbial growth and respiration can be achieved, resulting in the utilization of the organic matter present either as oxidized end-products (CO_2, NO_3, SO_4, PO_4) or the biosynthesis of new micro-organisms. Purification occurs as a number of successive steps, but as the microbial biomass is mixed with the wastewater within a single reactor (the aeration tank), the individual steps are not discernible but occur simultaneously. In settled sewage, the organic material is present partly as soluble material

(<1 mμm) and partly as finely suspended (1–100 μm) and colloidal (1 mμm–1 μm) particulate matter, with the organic material predominantly associated with the particulate fraction (Heukelekian and Balmat 1959). In sewage, 64% of the total solids are classed as soluble, whereas 34% are particulate, with 80% of the particulate fraction organic, compared with only 20% in the soluble fraction. Therefore, the majority of the organic loading to the activated sludge unit will be in the form of colloidal or larger sized particulate solids with little present as simple low molecular weight nutrients. In the first stage of purification, the particulate and colloidal material is rapidly adsorbed or agglomerated on to the microbial floc, so that initially treatment is primarily physical in nature. Although the soluble material present can be utilized immediately, the colloidal or suspended matter has to be broken down extracellularly before becoming available for microbial oxidation. Rickert and Hunter (1967, 1971, 1972) have demonstrated that particulate solids are removed more efficiently than the soluble fraction by biological treatment, which supports the importance of the physico-chemical processes in wastewater treatment. The varied nature of the organic matter present ensures that it is oxidized at different rates. For example, high respiration rates are measured at the inlet of the tank reaching up to 500 mg $O_2g^{-1}h^{-1}$ with a gradual decline in respiration rate with retention time falling to <50 mg $O_2g^{-1}h^{-1}$ at the outlet. Part of the organic matter is oxidized to simple end-products (mineralization), while the remainder is converted into new cellular material (assimilation).

The activated sludge process consists of two phases, aeration and sludge settlement (Fig. 4.31). In the first phase, wastewater is added to the aeration tank containing the mixed microbial population and air is added either by surface agitation or via diffusers using compressed air. The aeration has a dual

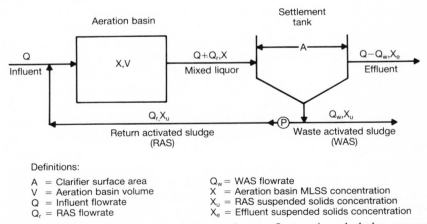

Definitions:

A = Clarifier surface area
V = Aeration basin volume
Q = Influent flowrate
Q$_r$ = RAS flowrate

Q$_w$ = WAS flowrate
X = Aeration basin MLSS concentration
X$_u$ = RAS suspended solids concentration
X$_e$ = Effluent suspended solids concentration

FIG. 4.31 Schematic diagram of the continuous flow activated sludge process (Jenkins et al. 1984).

function, to supply oxygen to the aerobic micro-organisms in the reactor for respiration and to maintain the microbial flocs in a continuous state of agitated suspension, which ensures maximum contact between the surface of the floc and the wastewater. This continuous mixing action is important, not only to ensure adequate food, but also a maximum oxygen concentration gradient to enhance mass transfer and to help disperse metabolic end-products from within the floc. Ardern and Lockett (1923) originally developed a batch process known as the fill and draw method, with aeration and settlement taking place within the same tank. Subsequently, a continuous system was developed with no settlement allowed within the aeration tank. As the settled wastewater enters the aeration tank it displaces the mixed liquor (the mixture of wastewater and microbial biomass) into a sedimentation tank. Here, the flocculated biomass settles rapidly out of suspension to form a sludge with the clarified effluent, which is virtually free from solids, discharged as the final effluent. In the conventional activated sludge process (Section 4.2.3.1) between 0.5–0.8 kg dry weight of sludge is produced for every kg BOD_5 removed. The sludge is rather like a weak slurry containing between 0.5–2.0% dry solids, and therefore is pumped easily. As the solids content increases the viscosity rapidly becomes greater, although, as discussed in Chapter 2, activated sludge is difficult to consolidate to > 4% dry solids by gravity alone. However, most of the activated sludge is returned to the aeration tank to act as an inoculum of micro-organisms, ensuring that there is an adequate microbial population to fully oxidize the wastewater during the period of retention within the aeration tank. The excess sludge requires further treatment prior to disposal.

The most important function in the activated sludge process is the flocculent nature of the microbial biomass. Not only do the flocs have to be efficient in the adsorption and subsequent absorption of the organic fraction of the wastewater, but they also have to rapidly and effectively separate from the treated effluent within the settlement chamber. Any change in the operation of the reactor will lead to changes in the nature of the flocs, which can adversely affect the overall process in a number of ways, most notably in poor settlement resulting in turbid effluents and a loss of microbial biomass (Section 4.2.4).

Although some variations of the activated sludge process are used to treat raw sewage (Section 4.2.3), most activated sludge processes use settled sewage. The sludge produced from the process should not be confused with primary sludge as it is entirely microbial biomass and adsorbed particulate matter and does not contain coarse organic or inorganic solids. For this reason, the activated sludge process is similar to normal fermentation processes except, of course, it is not aseptic. Compared to industrial fermenters, activated sludge is different in: (i) that there is no product being synthesized while material is being broken down; (ii) only a small proportion of the activated sludge is actually viable and capable of reproduction; and (iii) the nutrient concentration in the aeration tank is much lower than the nutrient medium used in a conventional fermenter. Ideally, the activated sludge process should be operated as close to a food-limited condition as possible in order to encourage endogenous respiration when the micro-organisms utilize their own cellular contents, thus reducing the quantity of biomass produced. During the endogenous respiration phase the respiratory rate will fall to a minimum that is sufficient for cell maintenance only. However, under normal operating conditions the growth of the microbial population and accumulation of non-biodegradable solids results in an increase in the amount of activated sludge produced.

The removal mechanism, assimilation or mineralization, can be selected by using specific operating conditions with certain advantages and disadvantages. For example, the most rapid removal of nutrients is achieved by removing organic matter by assimilation only, where it is precipitated in the form of biomass. Such processes produce considerable surplus sludge, which requires a higher proportion of the operating costs to be spent on sludge separation and disposal. Complete oxidation (mineralization) of wastewater is much slower and requires long aeration periods. Thus, although much less sludge is produced, the saving on sludge handling and aeration costs will be much higher (Winkler 1981).

The process is more intensive than in fixed-film reactors, treating up to 10 times more wastewater per unit reactor volume. Although this makes the activated sludge process cheaper in terms of capital cost, they are considerably more difficult to operate than percolating filters, have high operating costs, and produce comparatively large quantities of surplus biomass or sludge. The two process are compared in Table 4.1. The design of the activated sludge processes has been extensively studied and has also been extensively modelled (Niku *et al.* 1979; Ouano and Mariano 1979; Vavilin 1982).

4.2.1 Flocculation

The basic operational unit of activated sludge is the floc. Under the microscope, activated sludge comprises of discrete clumps of micro-organisms known as flocs, which vary both in shape and size. Good flocculant growth is important for the successful operation of the process, so that suspended, colloidal, and ionic matter in the wastewater can be removed by adsorption and agglomeration, and subsequently for the rapid and efficient separation of sludge from the treated effluent. There is a rapid agglomeration of suspended and colloidal matter on to the flocs as soon as the sludge and wastewater are mixed (enmeshment), which results in a sharp fall in the residual BOD of the wastewater. The volatile matter content of flocs is generally high, between 60–90%, although this depends on the nature of the waste and the amount of fine suspended and colloidal inert matter present. The adsorption capacity of the floc depends on the availability of suitable cell surfaces. Once all the adsorption sites are occupied, the floc has a very reduced capacity for adsorbing further material until it has metabolized that already adsorbed. Breakdown and assimilation of the agglomerated material proceeds more slowly (stabilization). Therefore, if the hydraulic retention time of the wastewater (HRT) in the aeration tank is too short, there will be a progressive reduction in BOD removal as there will have been insufficient time for adsorbed material to be stabilized. In the sedimentation tank, only well-formed flocs settle out of suspension with smaller flocs, dispersed micro-organisms and solids, carried out of the tank with the final effluent. The process depends on continuous reinoculation with recycled settled sludge, so the system will only select floc-forming organisms that rapidly settle in the sedimentation tank. Thus, the process is self-regulating with the best flocs recirculated.

Individual flocs are complex biochemical units. Each floc is a cluster of several million heterotrophic bacteria bound together with some inert organic and inorganic material. There is a wide range of particle sizes in the activated sludge process ranging from individual bacteria of between 0.5–5.0 μm, up to large flocs which may be greater than 1 mm ($= 1000\mu$m) in diameter. Parker et $al.$ (1971) found that there was a bimodal particle size distribution of flocs in activated sludge (Fig. 4.32), and suggested that the smaller particles were either individual micro-organisms or small aggregates that had not floccu-lated or had been sheared off larger flocs. The maximum size of flocs is dependent on their physical strength and the degree of shear exerted by the turbulence caused by the aeration system in the aeration tank. The process of floc formation is far from being understood. Originally, it was thought due to the slime-forming bacterium $Zoogloea$ $ramigera$, however, many other bacteria and protozoa are now known to be associated with floc formation (McKinney and Weichlein 1953; Kato et $al.$ 1971). Bacteria, protozoans, and detritus are either attached to the surface of the floc or embedded in some form of material forming a matrix. The exact nature of this flocculating material is still not known although it appears largely bacterial in origin. The material can be readily extracted from activated sludge (Brown and Lester 1979) and constitutes a significant portion of the dry weight of the sludge, up to 10%. All the studies have shown that the material is a polymer, which can be composed of a number of organic compounds, such as polysaccharides, amino polysaccharides, and protein (Sato and Ose 1980). Lipids may also be present, but the exact nature of these flocculating polymers will depend on the species of bacteria or protozoans producing it. Each polymer will have varying surface properties and charges that will influence not only the settling characteristics but also the water binding properties of the floc. The polymer does not only give the floc components cohesion, it also allows suspended particles in the waste to bind to the floc by adsorption. Therefore, the polymer has a critical role in the operation of the activated sludge process (Harris and Mitchell 1973; Forster 1976; Unz and Farrah 1976; Farrah and Unz 1976;

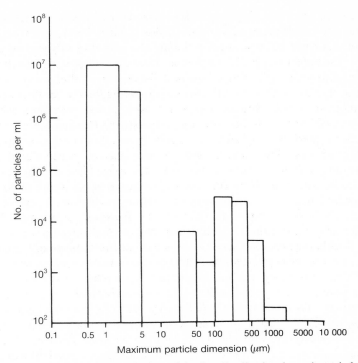

FIG. 4.32 Example of a typical particle size distribution in activated sludge (Parker *et al.* 1971).

Tago and Aiba 1977). These extracellular polymers (ECPs) are not food reserves, like poly-β-hydroxybutrate, and are not easily decomposed. The surface charges on the microbial cells and bridge formation by polyvalent cations also contribute to flocculation (Deinema and Zevenhuizen 1971; Steiner *et al.* 1976). Examination of flocs by electron microscopy has shown that granular and amorphous materials are present, and that fine cellulose fibrils are formed. Young flocs contain actively growing and dividing heterotrophic bacteria with a high rate of metabolism. Older flocs, in contrast, have a lower proportion of viable cells, being composed mainly of dead cells surrounded by a viable bacterial layer. Although the majority of these cells are no longer viable they retain active enzyme systems. Older flocs have a reduced rate of metabolism but as they are physically larger they settle far more readily than younger flocs, which are often associated with poor settleability. Weddle and Jenkins (1971) have suggested that the viability of micro-organisms making up flocs is quite low, estimating between 5–20% viability. As the floc ages, the slower growing autotrophs become established, especially the nitrifying bacteria, therefore, the concept of sludge age (Section 4.2.2) is important in terms of overall efficiency. Flocs undergo a secondary colonization by other micro-organisms such as protozoans, nematodes, and rotifers. The ciliate protozoans are considered particularly important as they feed on dispersed bacteria, thus reducing the turbidity of the final effluent, whereas the higher trophic levels present graze on the floc itself, which reduces the overall biomass (Section 4.3.5). A well-flocculated sludge is in a state of dynamic equilibrium between the flocs aggregating into larger flocs and being broken up into smaller flocs by the shear stress imposed by the aeration system (Tench 1979). Although, because flocs have strong binding forces, based on calcium, magnesium, or other multivalent cations, they can tolerate quite high shearing stresses. The surface area of flocs has been calculated by dye adsorption and found to range between 43–155 m^2g of floc. The radius of flocs can then be calculated from the specific surface area assuming them to be a homogenous mass of micro-organisms. Therefore, the lowest surface area of $43 \ m^2g^{-1}$ will have a floc radius of 0.58 μm. However, in practice, flocs are considerably larger than this, indicating that the high specific surface area is due to flocs being porous. This is confirmed by electron microscopy that shows flocs to have a spongy appearance. The porous nature explains why flocs are so good at adsorbing particulate matter, also why the diffusion rate of nutrients and oxygen into the centre of the floc is greater than if flocs were homogenous masses of bacteria (Coackley 1985). In general, good flocculation is associated with low-rate, and poor flocculation with high-rate processes. However, considerable variation is seen between flocs from different sludges and it is this variation that causes operational difficulties such as bulking, deflocculation, pinpoint floc, foaming, and rising sludge (Section 4.2.4). For example, toxic discharges, nutritional imbalance or changes in the microbial ecology of the process can all alter the surface chemistry of the flocs, which in turn will influence the settlement characteristics of the activated sludge.

4.2.2   Operating factors

4.2.2.1   Process control

Mixed liquor suspended solids
The concentration of suspended solids in the aeration tank, commonly referred to as the mixed liquor suspended solids concentration (MLSS), is a crude measure of the biomass within the aeration tank. It is measured in the same way as suspended solids in wastewater by filtering a known volume of the mixed liquor sample through Whatman GF/C filter paper and weighing it after drying in an oven at 105°C. The MLSS is a basic parameter used in the calculation of a number of other operating parameters and is expressed in $mg \ l^{-1}$ or gm^{-3}. Some of the MLSS may be inorganic, and under certain circumstances, this may represent a significant proportion of the solids

present. To remedy this, many operators estimate the organic fraction of the sludge by measuring the combustible matter present in the MLSS by burning the dried sludge in a muffle furnace at 500°C. This is also expressed in mg l^{-1} and is termed the mixed liquor volatile suspended solids (MLVSS). However, this does not distinguish between the biochemically active and the inert material and, therefore, a more complex technique must be employed to measure the biochemical sludge activity. The proportion of active micro-organisms in the MLVSS will vary depending on operating conditions of the activated sludge unit as well as the amount of volatile solids in the wastewater. Thus, it must be used with caution and not as an accurate estimation of microbial activity.

For day-to-day operational control, the MLSS is quite adequate with the MLVSS and the other measures of sludge activity used mainly in research and development work. Normal MLSS concentrations range from 1500–3500 mg l^{-1} for conventional activated sludge units rising to 8000 mg l^{-1} for high-rate systems. The MLSS concentration is controlled by altering the sludge wastage rate. In theory, the higher the MLSS concentration in the aeration tank the greater the efficiency of the process, as there is a greater biomass to utilize the available food. However, high values of MLSS are limited by the availability of oxygen in the aeration tank and by the capacity of the sedimentation unit to separate and recycle activated sludge (Hawkes 1983a).

Sludge residence time or sludge age
The sludge residence time (SRT) affects the character and condition of the activated sludge flocs within the aeration basin. It is calculated as the total amount of sludge solids in the system divided by the rate of loss of sludge which form the system. In practical terms, it is impossible to take into account all the sludge in the various stages of the activated sludge process, including the pipework and sedimentation tank as well as the aeration basin, so a simplified equation is used:

$$t_s = \frac{VX}{(Q_w X_w) + (Q_e X_e)}$$

where V is the volume of liquid in the aeration tank (m^3); X, the MLSS (mg l^{-1}); Q_w, the sludge wastage rate (m^{3}d^{-1}); X_w, the MLSS (mg l^{-1}) in the waste sludge stream; Q_e the effluent discharge rate (m^{3}d^{-1}); X_e the suspended solids concentration (mg l^{-1}), and t_s the SRT in days.

If the proportion of microbial cells in the MLSS is assumed constant, the SRT can be referred to as either the mean cell residence time (MCRT) or the sludge age. If the system is balanced with the amount of sludge in the aeration tank constant (VX), then the sludge wastage ($Q_w X_w$) will represent the net sludge production of the system (kg dry solids d^{-1}). SRT is an operational factor giving control over sludge activity because SRT is the reciprocal of the net specific growth rate of the sludge and thus can be considered as a measure of sludge activity. A low SRT (<0.5 d) indicates a sludge with a high growth-rate as used in high-rate units for pretreatment or partial treatment, a high SRT (>5 d) indicates a low growth rate sludge, such as extended aeration systems (Table 4.22). Conventional activated sludge has a SRT of between 3–4 d, and has good settling properties. However, at SRT >6 d or between 0.5–3 d, there is a reduction in settleability. SRT is controlled by altering the sludge wastage rate.

Plant loading
Three loading factors are used in the design and operation of activated sludge. The volumetric loading, which is the flow of wastewater in relation to the aeration tank capacity, organic loading, which is the BOD load in relation to the aeration tank capacity, and sludge loading, which is the BOD load in relation to the total biomass of activated sludge.

Volumetric loading. The retention time of the wastewater in the aeration tank, the hydraulic retention time (HRT), is expressed as:

$$\text{Volumetric loading} = \frac{V \times 24}{Q}$$

where V is the liquid capacity of the aeration tank (m^3) and Q the rate of flow of influent wastewater to the tank ($m^3 d^{-1}$), so that the hydraulic retention time is expressed in hours.

The HRT as measured here does not take into account the flow of recycled activated sludge returned to the aeration tank which may represent between 25–50% of the overall flow, making the actual retention time much less than calculated. For this reason, the HRT is often referred to as the 'nominal retention time'. The HRT must be sufficiently long to allow the required degree of adsorption, flocculation, and mineralization. In conventional activated sludge systems, the HRT will be at least 5 hours at DWF. However, because of fluctuations in flow and the recycling of sludge, the actual retention time is much less and during a storm, when maximum volumetric loadings are being received (3 DWF), the recycle of sludge is as much as 1.5 DWF, and the actual retention time may be as short as 1 h (Hawkes 1983*a*).

Organic loading. Wastewaters have different organic contents and it is useful to express loadings in terms of kg BOD per tank capacity per day:

$$\text{Organic loadings} = \frac{Q \times \text{BOD}_5}{V \times 1000}$$

where BOD_5 is the biochemical oxygen demand of the influent wastewater ($mg\ l^{-1}$). The organic loading is expressed as kg $\text{BOD}_5 m^{-3} d^{-1}$, therefore, an activated sludge unit with a retention time of 4.5 h and a settled sewage BOD

TABLE 4.22 Comparison of loading and operation parameters for different activated sludge treatment rates (Hawkes 1983a)

Treatment rate	Retention period (h)	BOD loading per capacity kg BOD m⁻³d⁻¹	Sludge loading (f/m) kg BOD kg⁻¹d⁻¹	Sludge age (d)	Sludge production kg dry sludge per kg BOD removed	Application
Median	5–14	0.4–1.2	0.2–0.5	3–4	0.5–0.8	Conventional treatment for medium and large works to produce 20:30 effluent with or without nitrification depending on loading within range
High	1–2	>2.5	>1.0	0.2–0.5	0.8–1.0	For pretreatment or partial treatment
Low	24–72	<0.3	<0.1	>5–6	0.4	Extended aeration for full treatment on small works. Effluent highly stabilized but may contain fine solids

of 240 mg l^{-1} will have an organic loading of:

$$\frac{240 \times 24}{4.5 \times 1000} = 1.28 \text{ kg BOD}_5\text{m}^{-3}\text{d}^{-1}$$

In essence, the higher the organic loading the greater the BOD$_5$ of the final effluent (Table 4.22).

The rate of BOD removal (g BOD$_5$d^{-1}) in the aeration tank can be calculated as:

$$\frac{\text{(Influent BOD} - \text{Effluent BOD)}}{\text{HRT}}$$

Although this works for complete mixed reactors that are continuously loaded, in plug-flow systems the influent wastewater is diluted by the recycled activated sludge. Thus, the BOD concentration at the inlet must be calculated separately as:

$$\frac{(Q \times \text{Influent BOD}) + (\text{Effluent BOD} \times Q_r)}{Q + Q_r}$$

where Q_r is the sludge recycle rate (m^{3}d^{-1}). The BOD removal rate can then be calculated in the normal way.

Sludge loading. With the biomass actively removing the organic fraction of the wastewater it follows that the BOD loading should be related to the amount of activated sludge in the aeration tank. The sludge loading is referred to as the food (f) to micro-organism (m) ratio and is calculated as:

$$\frac{f}{m} = \frac{\text{Organic loading rate}}{\text{Volume of sludge}} = \frac{Q \times \text{BOD}}{V \times X}$$

where X is the MLSS in the aeration tank. The f/m ratio is expressed as kilograms BOD$_5$ per day per kilogram MLSS (kg kg^{-1}d^{-1}). It can also be calculated using the equation:

$$\frac{\text{BOD}}{X} \times \frac{24}{\text{HRT}}$$

Thus, for a conventional activated sludge with a HRT of 4.5 h, influent BOD of 240 mg l^{-1} and a MLSS concentration of 2500 mg l^{-1}, the sludge loading is:

$$\frac{240 \times 24}{2500 \times 4.5} = 0.51 \text{ kg kg}^{-1}\text{d}^{-1}$$

When the f/m ratio is high, the micro-organisms are in the exponential growth phase. With excess food the rate of metabolism is at a maximum with large BOD removals achieved. However, under these conditions the micro-organisms do not form flocs but are generally dispersed making it difficult to settle and form a recyclable sludge. Because food is in excess not all the organic material will be utilized and the remainder will pass out in the final effluent giving a high BOD. In contrast, low f/m ratios put the micro-organisms into a food-limited environment even though the rate of metabolism may be high when the recycled micro-organisms are first mixed with the incoming wastewater. Once food becomes limiting, the rate of metabolism will rapidly decline until the micro-organisms are in the endogenous respiration phase with cell lysis and resynthesis taking place. Under low sludge loadings there is almost complete oxidation of organics resulting in a high quality effluent with the micro-organisms flocculating and settling rapidly (Fig. 4.33).

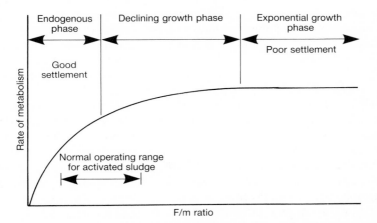

FIG. 4.33 The rate of metabolism of the micro-organism in the activated sludge aeration tank is controlled by the f/m ratio (adapted from Viesmann and Hammer 1985).

In theory, the sludge loading is related to SRT because sludge activity increases if the organic loading is increased, resulting in an increased rate of sludge growth. Therefore, in order to maintain a constant sludge concentration (MLSS) the sludge wastage rate must be increased, thus reducing the SRT. Therefore, it appears that the f/m ratio is approximately inversely proportional to SRT, although the work of Ekama and Marais (1979) does not support this.

Sludge settleability
Most problems associated with the activated sludge process involve poor settleability. Therefore, a rapid method of assessing settleability is vital if good separation of the sludge in the secondary tank is to be maintained, ensuring adequate sludge recycle and a final effluent with a low suspended solids concentration. Two indices can be used to assess the settling qualities of activated sludge, the sludge density index (SDI) and the sludge volume index

(SVI), the latter being the most widely used (Keeper 1963; Dick and Vesilind 1969). Both methods have been described in detail in Section 2.2.3.

The SDI and SVI are both calculated by settling the mixed liquor in a cylinder for 30 minutes and then measuring the volume of settled sludge. However, it has been demonstrated that within 30 minutes of quiescent settlement the process of sedimentation will have proceeded past the hindered settlement stage in which all the sludge flocs are evenly distributed (Type III settlement), and that transitional or even compression settlement phases may have started (Type IV) (Section 2.2.1). To minimize the effects of consolidation under compression, shorter periods of settlement for the test have been proposed (Finch and Ives 1950), or, taking a series of readings over a slightly longer period, and plotting the settlement curve. However, the problem has been adequately overcome by a modification of the SVI where settlement is measured while the mixed liquor is gently stirred at a constant temperature. This more accurate assessment of sludge settleability, known as the stirred specific volume index (SSVI), is described and compared with the SVI test in Section 2.2.3 (White 1976; Rachwal et al. 1982).

Sludge settleability is dependent on the microbiological, biochemical, and physico-chemical properties of the sludge. For example, a direct relationship between the bound water content of flocs and the SVI has been reported by Heukelekian and Weisberg (1956), whereas Forster (1968, 1971) has shown that the surface charge on flocs is directly proportional to the SVI. Under normal operating conditions, activated sludge settleability depends on the sludge loading, although good settling sludges are obtained from both conventional and high-rate loaded units. In the USA a functional relationship between sludge loading and SVI has been developed (Fig. 4.34), although the ranges and values quoted are *very* approximate.

Activated sludge will settle well if the $SVI < 100$ ($SDI > 1$), whereas a $SVI > 150$ ($SDI < 0.66$) indicates settling problems and possibly bulking (Fig. 4.35). In terms of SSVI, a good sludge has a value < 120 ml g^{-1}, whereas a bulking sludge or one with poor settleability has a value > 200 ml g^{-1}.

Sludge activity
Although theories and working models of activated sludge behaviour and performance have been developed using measures of microbial biomass such as MLSS or MLVSS, it has not been possible to relate bacterial numbers directly to performance. The problems of estimating bacterial numbers, of differentiating viable and non-viable cells, and of estimating levels of activity due to age of the cell, have all proved extremely difficult with the mixed populations of bacteria associated with activated sludge. Therefore, other methods of assessing biological activity have been developed using chemical analyses rather than direct counting techniques. The most widely used biochemical measure of biological activity is adenosine triphosphate (ATP) which is used to measure the number of viable cells present in activated sludge

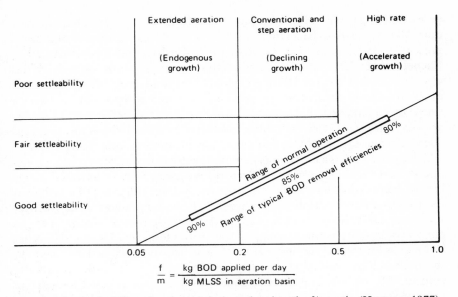

$$\frac{f}{m} = \frac{\text{kg BOD applied per day}}{\text{kg MLSS in aeration basin}}$$

FIG. 4.34 Settleability of activated sludge related to the f/m ratio (Hammer 1977).

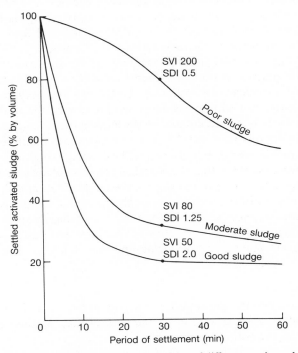

FIG. 4.35 Comparison of settling characteristics of different activated sludges using the sludge volume index and sludge density index (Hawkes 1983a).

(Roe and Bhagat 1982). Deoxyribonucleic acid (DNA), although less sensitive than ATP, appears to be a function of the type and the number of organisms present in activated sludge (Gentelli 1967; Bardtke and Thomanetz 1976; Raebel and Schlierf 1980). Enzymatic activity, especially protease activity, has also been used (Thiel and Hattingh 1966; Sridhar and Pillai 1973; Vankova *et al.* 1980). The electron transfer process associated with oxidation offers a range of co-enzymes that could be used as potential indicators of sludge activity. The rate of electron transfer is reflected by the activity of nicotinamide adenine dinucleotide (NAD) and flavin adenine dinucleotide (FAD), both of which can be used as a measure of metabolic activity in activated sludge (Jones and Prasad 1969; Dickson 1983). Dehydrogenase activity in activated sludge can be measured using the dye triphenyl tetrazolium chloride (TTC) which is reduced to a red dye triphenyl formazan (TF) in the presence of dehydrogenase enzymes. It is readily extracted from activated sludge and its concentration determined using a spectrophotometer. Unlike the other techniques for measuring sludge activity, which are complicated and based in the analytical laboratory so that they cannot be used for routine operational control, TF analysis is more straightforward and can be used for routine activated sludge operation (Coackley and O'Neill 1975). Also, the dehydrogenase response to changes is very rapid and Coackley and O'Neill (1975) observed that the increase in dehydrogenase activity, in response to the addition of organic matter to activated sludge, occurs much faster than can be accounted for simply by an increase in bacterial numbers.

An interesting new method of determining sludge activity is based on the assimilation of glucose. A standard glucose solution is added to the sample of activated sludge, which is continuously aerated in a small reactor. A glucose selective electrode is used to measure its assimilation by the activated sludge and sludge activity is then calculated on the basis of the reaction time required for 50% assimilation of glucose. There is a problem with interference with the dissolved oxygen concentration, therefore the dissolved oxygen concentration also has to be measured so that the electrical response of the glucose electrode can be corrected with that of the oxygen electrode (Olah and Princz 1986).

By being able to estimate the biological capacity of the activated sludge, system loadings including toxic substances, can be based on the functional relationship between the concentration of the waste and sludge activity, thus optimizing performance and reducing the possibility of overloading or inhibition.

Recirculation of sludge
Sludge is returned to the aeration tank in order to maintain sufficient microbial biomass for oxidation of the wastewater. Assuming no sludge wastage this relationship is expressed as:

$$\frac{Q_r}{Q + Q_r} = \frac{V}{1000}$$

where Q is the mean flow of influent wastewater to the aeration tank (m^3d^{-1}), Q_r the sludge recycle rate (m^3d^{-1}), and V the volume of settled solids after 30 minutes in a 1000 ml graduated cylinder. Using this relationship the volume of recirculated activated sludge (Q_r) can be calculated:

$$Q_r = \frac{V \times Q}{1000 - V}$$

as can the solids concentration in the recirculated sludge (X_r) and in the wasted sludge (X_w):

$$X_r = \frac{1\,000\,000}{SVI}$$

The ratio of sludge returned to the aeration tank (Q_r) is normally expressed as a percentage of the influent wastewater (Q):

$$\frac{Q_r}{Q} = \frac{V}{1000 - V} \times 100$$

Thus, if the return activated sludge rate is 20% and the influent wastewater into the aeration tank is $1\ m^3s^{-1}$, then the recycled sludge is equivalent to $0.2\ m^3s^{-1}$.

For example, for a conventional plant with a retention time of 4.5 h and an influent wastewater flow rate of $1\ m^3s^{-1}$, an influent BOD of $240\ mg\,l^{-1}$, a MLSS of $2500\ mg\,l^{-1}$, a sludge loading of $0.51\ g\,g^{-1}d^{-1}$, and a volume of settled solids in a 1000 ml cylinder after 30 minutes (V) of 240 ml, the SVI, volume of recirculated activated sludge (Q_r), solids concentration of the sludge (X_r) and the ratio of returned sludge to influent wastewater (Q_r/Q) are:

$$SVI = \frac{240 \times 1000}{2500} = 96\ ml\,g^{-1}$$

$$Q_r = \frac{240 \times 86\,400}{1000 - 240} = 27\,284\ m^3d^{-1}$$

$$X_r = \frac{1\,000\,000}{96} = 10\,416\ mg\,l^{-1}$$

$$\frac{Q_r}{Q} = \frac{240}{1000 - 240} \times 100 = 32\%$$

4.2.2.2 Factors affecting the process

The activated sludge process is affected by the nature of the wastewater being treated as well as environmental, climatic, and hydrological factors.

Biological activity of sludge flocs and their settling characteristics are affected by wastewater composition. In conventional activated sludge a BOD:N:P ratio of 100:6:1 is required to maintain the optimal nutrient balance for heterotrophic activity which is equivalent to an operational range of 0.03–0.06 kg nitrogen and 0.007–0.01 kg potassium per kg BOD. The presence of toxic or inhibitory substances affects the metabolic activity of aerobic heterotrophs, although activated sludge does become acclimatized to low concentrations with time, and can be used to treat potentially toxic wastes such as phenolic wastewaters. Unlike fixed-film reactors, activated sludge systems are sensitive to fluctuations in the organic strength of the wastewater, which seriously affects sludge characteristics and the SVI, with carbohydrates in particular encouraging filamentous growth in the sludge, causing a rapid increase in SVI. Thus, an important aspect of operational management of activated sludge systems is to ensure that strong organic loads due to the circulation of stormwater or the return of supernatant liquor from sludge consolidation or digestion are prevented. When hard detergents were being used widely, deep layers of foam would accumulate on the surface and around the aeration basin, especially those using diffusers. Foaming due to detergents made operation difficult and dangerous, with a loss of MLSS from the aeration tank as the flocs became entrained in the foam. Since the introduction of soft detergents, detergent foaming has been markedly reduced except where hard detergents still need to be used for specialized industrial purposes. Other chemicals, such as polyglycols and alkyl phenoxy compounds can also cause foaming and where it occurs foam suppression is required, by spraying with recycled effluent or the application of oil-based anti-foam compounds.

Increase in the temperature of the mixed liquor during the summer results in an increase in metabolic activity. In the activated sludge system, all the biochemical reaction rates, such as organic substrate stabilization, production of cellular material, maintenance energy requirements, oxygen utilization, auto-oxidation of cellular mass, and nitrification, follow the Arrhenius relationship over the 5–20°C range. However, except for the substrate utilization rate, they do not conform to this relationship above 25°C (Randall et al. 1982). Whereas increases in the temperature of the mixed liquor enhances the BOD removal of activated sludge systems, in practice it leads to problems of increased oxygen demand, which can lead to the system becoming oxygen-limited. Also, in warmer weather there is a significant increase in the incidence of bulking (Section 4.2.4). Lin and Heinke (1977a, b) considered the temperature to be of major importance in explaining the variation in performance of activated sludge systems and found that aeration tank values should be larger for lower temperature operation. They suggested that a much higher perforance could be achieved if hot compressed air was used for aeration, although whether this would be economically feasible at normal treatment plants is uncertain.

The oxygen requirement in conventional activated sludge depends on the SRT and the operating temperature. Benjes (1980) gives a typical value of 1.1 kg of oxygen per kg BOD at a SRT of 7 d at 10–20°C. If the SRT is long enough for nitrification to occur then the oxygen requirement will be greater due to nitrogenous demand. Therefore, at a SRT of 5 d, 1 kg BOD requires

1 kg O_2 for carbonaceous oxidation, and a further 4.3 kg of oxygen will be required to oxidize 1 kg ammonical nitrogen (Lister and Boon 1973). In the aeration tank, aerobic activity is independent of dissolved oxygen concentration above a minimum 'critical' concentration, < 1.0 mg l^{-1} for carbonaceous oxidation, below which metabolism is limited by a reduced oxygen supply. Minimum critical concentrations range from 0.2–2.0 mg $O_2 l^{-1}$ depending on the MLSS and other operating factors. As biological activity is just as great at low as at high dissolved oxygen concentrations and the oxygen transfer rate is proportional to the oxygen deficit in the mixed liquor, it is logical to operate the aeration tank as close to the critical minimum dissolved oxygen concentration as possible (Fig. 3.22). There is a reduction in concentration gradient as oxygen diffuses into sludge flocs. However, in practice, the dissolved oxygen concentration should not be allowed to rise to > 2.0 mg $O_2 l^{-1}$ but should never fall below 0.5 mg $O_2 l^{-1}$ for carbonaceous oxidation, whereas for nitrification the mixed liquor should be maintained as near to the 2.0 mg $O_2 l^{-1}$ concentration as possible. To maintain the critical oxygen concentration in aeration tanks, the dissolved oxygen concentration is automatically maintained by oxygen electrodes. In this way the aeration system is turned on or off when the dissolved oxygen concentration falls below or exceeds the critical oxygen limits set. The oxygen electrodes are normally positioned in the corner of the tank where lowest oxygen concentrations can be expected, or near the inlet where the oxygen demand is greatest (Lewin and Henley 1972). This type of control system is vital for optimum and economic operation (Barnard and Meiring 1988). Supplying extra oxygen for nitrification is very expensive and can be overcome by the use of an anoxic zone. Here, the recycled nitrified effluent is denitrified while in admixture, with the settled sewage in a separate anoxic area before entering the main aeration tank. The low oxygen conditions allow faculative anaerobes to use nitrates as a source of oxygen.

A sudden increase in the hydraulic loading to the aeration tank, due to storms or recirculation of wastewater within the plant, will increase the discharge of mixed liquor to the sedimentation tank. Where there is a constant rate of return of sludge, this will result in a reduction of the MLSS in the aeration tank, with more sludge stored within the sedimentation tank. This will result in the sludge being stored for longer periods within the sedimentation tank before being recycled, which may adversely affect the viability of the micro-organisms comprising the sludge. Increased flows also reduce the effectiveness of the sedimentation tanks by increasing the upward flow rate, which will extend the sludge blanket towards the surface with the possibility that some of the sludge will be discharged with the final effluent. This will be particularly significant if the SVI is high. Sudden storms will scour out the deposited stale accumulations in the sewerage system and carry it to the treatment plant. This will result in a shock load of organically rich wastewater to the aeration tank, which will adversely alter the characteristics of the sludge. In practice, the flow to an activated sludge aeration basin is normally greater than the design loading (i.e. 3 DWF for large and 6 DWF for small treatment plants), thus, any increase in the hydraulic or organic loading will have a serious effect on operation.

4.2.3 Modes of operation

By using different combinations of the main operating parameters, various different rates and degrees of treatment are possible. This flexibility in design, allowing operation over a wide range of loadings to suit specific treatment objectives, is the major advantage of the activated sludge process over other treatment processes. The relationship between the substrate and sludge concentration, the f/m ratio, is a fundamental one in the operation of activated sludge and is summarized in Fig. 4.36. The different regions of this growth curve are analogous to activated sludge systems operating at different loading levels with corresponding degrees of substrate removal. In general, high-rate processes have a high sludge loading, a short SRT, a high sludge activity, and a short retention time. Although large weights of BOD_5 can be removed per unit volume of reactor, there is a relatively high concentration of organic matter remaining in the final effluent. In contrast, low-rate processes have a low sludge loading, a long SRT, a low sludge activity, and a long retention time. The sludge is in the endogenous respiration phase so food is limited, resulting in a low residual concentration of organic matter in the final effluent, and as

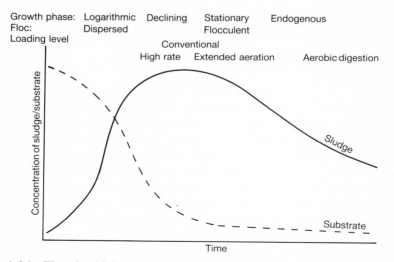

FIG. 4.36   The microbial growth curve showing the operational growth phase used in different modes of activated sludge (Winkler 1981).

the rate of microbial decay is high compared with the rate of microbial growth there is little excess sludge produced. Conventional operation falls between these two extremes (Winkler 1981).

Activated sludge operation can be categorized as either conventional, high-rate, or extended aeration, although the delineation between these categories is by no means precise and the terms are used only in their broadest sense.

4.2.3.1 Conventional activated sludge processes

Conventional activated sludge is the most widely used operational mode as it gives full treatment to wastewaters, particularly domestic sewage, producing a 20:30 effluent with or without nitrification, suitable for discharge to inland waters. The term conventional comes from the medium-rate of treatment with a BOD loading range between 0.5–1.5 kg BOD $m^{-3}d^{-1}$, a sludge age of between 3–4 d, a HRT of 5–14 h, a sludge loading rate (f/m) of 0.2–0.6 kg BOD $kg^{-1}d^{-1}$, and a sludge concentration in the aeration tank (MLSS) of between 2–3 kg m^{-3} ($=200$–300 mg l^{-1}). These figures are for domestic wastewater: the loadings for dairy or maltings wastes will be at the lower end of the range and those containing highly biodegradable substrates, such as simple alcohols, carbohydrates, and organic acids will be in the upper loading range (Calley et al. 1977). The sludge production is about 0.5 kg per kg BOD removed and a dry solids content of $<1\%$. This represents over 50 kg wet sludge produced per kg BOD removed, indicating the problem of handling waste sludge in the activated sludge process. The HRT is critical for nitrification with only about 6 h required for adequate carbonaceous oxidation and up to a further 4 h required for nitrification. However, the more recalcitrant the waste the longer the HRT required. For example, Cox (1977) found that phenolic wastewater required an HRT of 75 h to obtain a 96% reduction in the organic matter. In general, conventional activated sludge is expected to produce BOD reductions of between 90–95%.

Within this category is a wide variety of systems both in structure and operation. The most important difference is the mixing regime used (Bode and Seyfried 1984), with conventional activated sludge being nominally plug-flow or nominally completely mixed. Plug-flow conditions are difficult to maintain as the mixed liquor has to be aerated, which results in a significant degree of mixing, while the problem of completely mixing liquid, sludge flocs, and gas bubbles makes a truly completely mixed system difficult to achieve. Therefore, activated sludge systems have a tendency towards a particular mixing regime rather than exactly fulfilling the criteria for a particular mixing regime. However, although different modes of operation of conventional activated sludge are preferred by certain workers, comparative studies show that there is very little difference in performance of such systems when operated at particular sludge loadings (Knop 1966).

Plug-flow system

In the conventional plug-flow system, both the influent wastewater and the returned sludge are added at the end of an elongated rectangular aeration tank, typically 6–10 m wide, 30–100 m long, and 4–5 m deep (Fig. 4.37). The tank is equipped either with diffusers or, less commonly, surface mechanical aerators to provide oxygen to the mixed liquor along the length of the tank. As

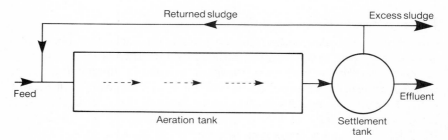

FIG. 4.37   Schematic flow diagram of a plug flow activated system.

the mixed liquor proceeds along the tank the organic matter is utilized, with the desired level of removal being controlled by the time it takes to reach the outlet at the far end. In theory, the sludge growth curve should be discernible as the mixed liquor moves along the tank, as in a batch system, with an initially rapid rate of removal becoming progressively slower as it makes its way along the tank. However, because of longitudinal mixing, the plug-flow effect is hidden so that the removal rate remains fairly uniform over the first third of the tank at least. There is also a discernible BOD concentration gradient along the tank. The rate of oxygen utilization also changes along the length of the aeration tank and the oxygen supply may be deficient at the inlet where demand is greatest, and be in excess (>5 mg $O_2 l^{-1}$) at the outlet, where the demand is lowest. This is overcome by a modification known as tapered aeration. The advantage of plug-flow systems is the low risk of short-circuiting of the tank. The risk can be reduced even further by incorporating dividing walls or baffles into the design of the tank, although this may increase localized mixing within the divided chambers. Toxic and shock organic loads are not diluted or buffered as in completely mixed systems, but pass through the tank as a discrete plug, resulting in serious deterioration of effluent quality. Plug-flow systems produce a sludge with good settling properties, although as the degree of mixing increases sludge settleability deteriorates.

Tapered aeration

Where constant aeration is provided along the length of a plug-flow system problems will arise due to the decreasing oxygen demand gradient that occurs along the length of the tank. The result is under-aeration at the inlet and over-aeration at the outlet. This can be overcome by tapering the aeration according to the respiratory requirement of the mixed liquor. Although the term tapered aeration suggests a gradual reduction in the aeration along the tank (Fig. 4.38), in practice, it occurs as a series of steps or movements. Lister and Boon (1973) estimate that the oxygen requirement in a plug flow tank is twice as high in the first half, thus, where three steps are used, about 45% of the aeration is supplied to the first third of the tank, 30% in the next third, and 25% in the final third. It is vital to ensure that the aeration at the outlet end of the tank is also sufficient to maintain the sludge in suspension as well as supply the required oxygen. It is easiest to introduce tapered aeration where surface aerators are used, where the different aeration rates are achieved by altering their depth of immersion or by their speed of rotation.

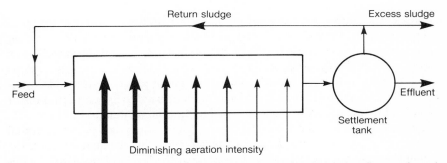

FIG. 4.38   Schematic flow diagram of tapered aeration in a plug flow activated sludge system.

Contact stabilization

In the contact stabilization, or biosorption process, adsorption and oxidation are carried out in separate tanks. The influent wastewater enters the contact tank where it is mixed with mixed liquor with a MLSS concentration of between 2000–3000 mg l^{-1} for between 0.5–1 h (Fig. 4.39). During this short contact period, the organic material present is adsorbed on to the activated sludge flocs. The mixed liquor is then settled in the sedimentation tank and the separated sludge pumped into the aeration tank where it is aerated at a high MLSS concentration of between 4000–10 000 mg l^{-1} for 5–6 h so that the adsorbed material can be fully oxidized. By having such a short HRT in the contact tank it is not necessary to provide aeration tank capacity for the entire wastewater flow, as is the case with other conventional activated sludge processes. In fact, approximately 50% less capacity is required and considerable savings in both capital and aeration costs are made. In some designs, the aeration tank is of the same capacity as a conventional plant, which allows a long SRT which gives a greater degree of oxidation. With the

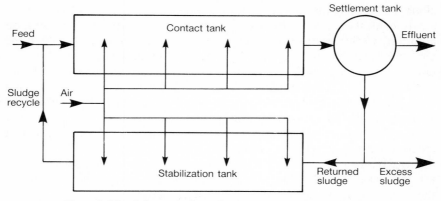

FIG. 4.39   Schematic flow diagram of contact stabilization.

sludge in the endogenous respiration phase less sludge is produced that is of better quality, although the larger aeration capacity will result in higher aeration costs. Moore and Todd (1968) obtained very low sludge production figures of 0.39 kg DS per kg BOD applied, which is significantly less than the 0.5–0.8 kg kg^{-1} BOD expected from conventional plants. Once oxidation is completed in the aeration tank, which is often referred to as reconditioning or reactivation, the mixed liquor is returned to the contact tank to be mixed with the incoming wastewater. The aeration and contact tanks can be incorporated into a single unit, with the influent wastewater entering the tanks towards the outlet zone but the recycled sludge entering at the far end (Fig. 4.40). In this way, two distinct zones are created within the aeration tank. The sludge loading rate (f/m) for contact stabilization is at the lower end of the range for conventional processes, between 0.1–0.2 kg BOD kg^{-1}d^{-1}. The process is particularly suited to strong industrial effluents having a large proportion of the biodegradable BOD present as colloidal or suspended matter. This process can also be used to uprate or expand existing activated sludge systems.

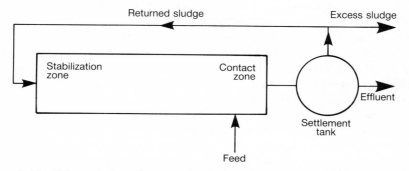

FIG. 4.40   Schematic flow diagram of a single basin contact stabilization activated sludge system.

Incremental feeding

The same 'evening out' of oxygen demand along the length of the aeration tank achieved by tapered aeration can also be done by introducing the influent wastewater incrementally at several points along the length of the tank (Fig. 4.41), with all the recycled sludge still introduced into the aeration tank at the inlet end. This results in a more even distribution of the BOD load and therefore a more even oxygen demand along the tank. Thus with a uniform

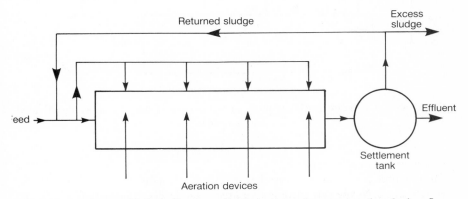

FIG. 4.41 Schematic flow diagram of step aeration in a conventional plug flow activated sludge system.

aeration system the BOD:dissolved oxygen ratio should remain more constant ensuring efficient use of the air supply (Hawkes 1983*a*). A useful advantage of this system is that the proportion of influent wastewater entering the tank at each stage can be varied according to changes in the organic or hydraulic loadings. This gives the process a considerable degree of operational flexibility. Even though the oxygen demand still rises at the inlet and falls at the outlet end of the tank, the conditions in the aeration tank are approaching those of a completely mixed system, and is generally considered as intermediate between a plug-flow and completely mixed process. Therefore, although the effects of shock loads are less than in strictly plug-flow reactors, the chance of short-circuiting is increased and the sludge settleability is slightly reduced. The process is known as incremental or stepped feeding but is also widely known by the confusing term 'step aeration', which would appear to be a more appropriate description of tapered aeration.

Incremental sludge feeding

A similar effect to tapered aeration or incremental feeding can be achieved by feeding the returned sludge incrementally at several points along the length of the tank, while the influent wastewater enters at a single point at the inlet and aeration is uniform along the length of the tank (Fig. 4.42). The theory is that the sludge activity is balanced with the dissolved oxygen concentration resulting in maximum oxygen utilization (Balmer *et al.* 1967). When compared with the other incremental modifications it appears to be only advantageous when the system is oxygen-limited at the inlet stage.

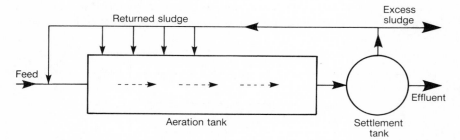

FIG. 4.42   Schematic flow diagram of an incremental sludge feed aeration activated sludge system.

Completely mixed system

The theoretical basis of completely mixed systems is that the influent wastewater and the returned sludge are immediately amalgamated with the mixed liquor as they are introduced into the aeration tank and are therefore diluted with purified effluent, which ensures uniform loading conditions throughout the tank (Fig. 4.43). The immediate dilution effect buffers the system to shock loadings of strong organic or toxic wastes. However, although short-circuiting is possible, completely mixed systems are considered to produce low density sludge which is difficult to settle compared with plug flow systems. In terms of operation, the most important factor in a completely mixed reactor is efficient mixing and aeration of the mixed liquor within the aeration tank. The MLSS of completely mixed reactors are generally higher ($3000-6000$ mg l^{-1}) than conventional plug flow reactors ($2000-3000$ mg l^{-1}) allowing for considerably higher BOD loadings. Critically low dissolved oxygen concentrations can result in some parts of the tank because of under-aeration or overloading, which prevents nitrification and also subsequently causes anoxic conditions in the settling tanks resulting in problems of rising sludge due to denitrification. Some of these disadvantages are overcome by attempting to utilize such reactors in a plug-flow manner by using a number of completely mixed tanks in series providing a 'stepped' plug-flow system. This is really a series of independent treatment stages that can be individually controlled in terms of loading and aeration. The most widely used adoption of such systems is as a modification to tapered aeration (Fig. 4.44) and step aeration (Fig. 4.45) systems.

Although other designs are used, completely mixed reactors are usually

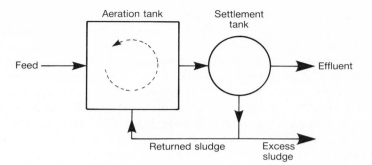

FIG. 4.43   Schematic flow diagram of a completely mixed activated sludge system.

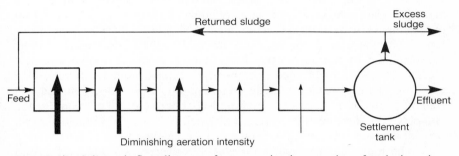

FIG. 4.44   Schematic flow diagram of step aeration in a number of tanks in series.

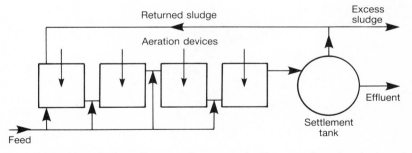

FIG. 4.45   Schematic flow diagram of step aeration in a number of tanks in series.

single tanks, square in cross-section, with a flat or slightly centrally inclined
bottom. There are many different types of mechanical surface aerators, but
those used for completely mixed conventional activated sludge systems are of
the vertical shaft type, i.e. rotating in a horizontal plane. Essentially, these are
inverted metal cones with blades welded or cast on to the surface and turned
by large electric motors using a suitable reducing gear. The whole apparatus is
mounted on a permanent support structure over the aeration tank with the

cone immersed in the mixed liquor. The cone is operated on a vertical shaft at between 35–60 rpm and is normally positioned at the centre of the tank to ensure maximum efficiency of action. A strong upward flow of mixed liquor from the centre of the tank is induced, which is thrown over the surface of the tank as a dense spray entrapping air and causing considerable turbulence that maintains the flocs in suspension. The oxygen transfer rate is approximately proportional to the power absorbed at the aerator shaft, which in turn is dictated by the size, speed, and immersion depth of the aerators. The rate of aeration is normally increased to compensate for an increase in organic loading by varying the level of the mixed liquor in the aeration tank by raising the outlet weir, or sluice gate, of the tank thus increasing the depth of immersion of the cone. Where the motor and cone are attached to a floating platform, which is widely used in lagoons and for the re-oxygenation of rivers and lakes, the rate of aeration is accelerated by increasing the rate of rotation of the cone. Two systems are widely used. A simple cone for medium depth tanks or for those using floating platforms, such as the 'Simcar' aeration cone manufactured by Simon–Hartley Limited (Fig. 4.46). This cone does not require a fixed draught tube and therefore can be used for a variety of functions. The diameter of the aerators can vary from 0.4–3.6 m and can transfer between 1–18 kg O_2h^{-1}. For deeper tanks of 10 m or more the 'Simplex' cone developed by Ames, Crosta, and Babcock, is mounted above a vertical draught tube which extends to just above the floor of the aeration tank. The updraught of mixed liquor caused by the cone ensures aeration and complete mixing (Fig. 4.47). These aerators will supply between 1.5 and 2.3 kg O_2 kWh, although the siting of the cone is critical so that maximum aeration and mixing is ensured.

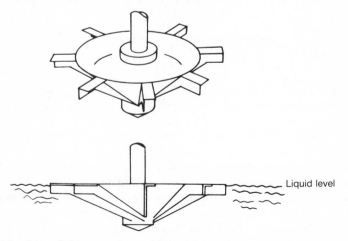

Liquid level

FIG. 4.46   One of the most frequently used aeration cones the 'Simcar' aerator manufactured by Simon–Hartley Ltd (Winkler 1981).

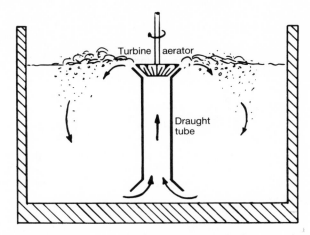

FIG. 4.47 The 'Simplex' cone aeration system employing a updraught tube to stabilize the flow pattern within the aeration tank (Winkler 1981).

4.2.3.2 Extended aeration

In extended aeration, the sludge loading level is very low, between 0.03–0.15 kg BOD $kg^{-1}d^{-1}$, which results in the process being food-limited, forcing the micro-organisms, which comprise the activated sludge, into the endogenous respiration phase of activity (Fig. 4.36). Therefore, the waste is fully oxidized (95% BOD reduction) and less sludge is produced (0.2–0.4 kg DS kg^{-1} BOD removed), which is also more stabilized and more easily dewatered than conventional activated sludge. Organic loadings range from 0.24–0.32 kg BOD $m^{-3}d^{-1}$. This type of low-rate activated sludge process has a long HRT of between 1–3 d and as the sludge activity is very low it requires less intensive aeration than conventional or high-rate processes. The sludge age of extended aeration processes as expected is quite long, >5–6 d (Table 4.22).

The first extended aeration system was developed by Pasveer in 1953. Known as the Pasveer or oxidation ditch, it is a simple, single-stage oxidation process with no primary sedimentation (Pasveer 1959). In its basic design, it is an oval ditch, usually with a central island, with the sewage agitated by horizontal paddles or cages that aerate the wastewater as well as giving it directional flow (Fig. 4.48). The ditch is normally between 1.0–3.5 m in depth and is either rectangular or trapizoidal in cross–section. This basic system represents 95% of the oxidation ditches operating in the UK. The influent wastewater enters the ditch upstream of the aerator ensuring maximum dilution; therefore, oxidation ditches are completely mixed systems, as well as supplying the oxygen where oxygen demand will be greatest. The mixed liquor is drawn off by displacement into the sedimentation tank and, after separation,

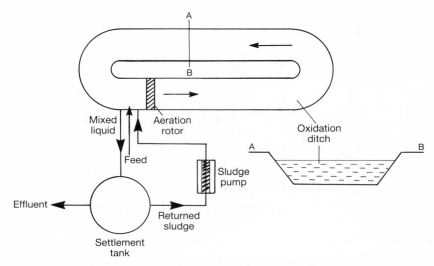

FIG. 4.48   Typical layout of a Pasveer oxidation ditch, with a cross-sectional profile
across the main channel between points A and B.

most of the sludge is returned. The aeration rate is controlled by the depth the
horizontal paddles or cage are immersed, and is altered by controlling the
depth of the mixed liquor in the ditch by using an adjustable weir. Two
developments of this system are also used.

The continuous split channel system has the channel on one side of the ditch
widened and split into two so that, by using control gates, the divided channels
can be used alternatively as a settlement tank and flow channel (Fig. 4.49). By
shutting the gate, the channel acts as a horizontal sedimentation tank with the
final effluent discharged over the end weir, and the settled sludge is resuspended
once the control gate is re-opened. Channels are operated alternatively every
4 hours. Excess sludge is removed by means of a trap, or a small sedimentation
tank, through which a small portion of the flow continuously passes.

In intermittent flow systems the ditch acts as the aeration and sedimentation
tank. It is used for small populations of less than 500 or from strong
wastewaters with an intermittent pattern of flow. The operation of the ditch is
quite straightforward and is fully automated. The influent enters a balancing
tank where it is stored, when full, the ditch aerator switches off and the sludge
is allowed to settle for 30 minutes. Influent wastewater is then pumped into the
ditch causing the level to rise and a syphon to operate and draw off the same
volume of settled final effluent at the same rate. Once the balancing tank is
emptied the pump switches off, the syphon breaks, and the aerator restarts so
that the whole cycle recommences. These systems are designed to operate on a
4–6 hours cycle.

The lower loading level gives a far greater reserve of dissolved oxygen to

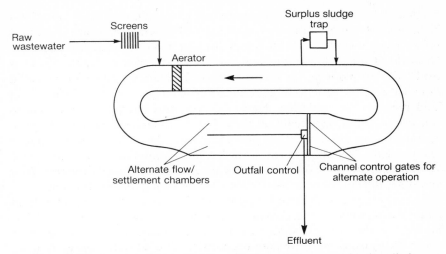

F IG. 4.49   Layout of a continuous split channel Pasveer oxidation ditch.

cope with surges in organic load, needs little maintenance compared to conventional or high-rate activated sludge and suffers less from sludge bulking or odour problems. Although the settleability of activated sludge tends to improve with increased SRT, after a critical period so much of the microbial mass has been broken down during endogenous respiration that the floc particles become too small to settle efficiently, thus causing problems. Basic design is based on simple formulations, for example, in the calculation of ditch capacity 4.73 m^3 of ditch is required per kg BOD load (PE = 0.055 kg BOD). The HRT (h) is calculated as:

$$\frac{\text{Ditch capacity (m}^3) \times 24}{\text{DWF (m}^3\text{d}^{-1})}$$

which is normally in excess of 24 h. In terms of aeration, about 2 kg of oxygen is required to remove 1 kg BOD d^{-1}. Using a horizontal cage aerator at 70 rpm immersed to a depth of 127 mm, a 1 m length of cage will remove 29.4 kg BOD d^{-1}. Thus, the length of aerator cage (m) required is:

$$\frac{\text{Daily BOD load (kg)}}{29.4}$$

The power requirement (kW) is very low, and at maximum immersion it can be approximated as:

$$\text{Length of cage aerator (m)} \times 1.6$$

The rate of flow in the channel is important and should be maintained at between 0.3–0.5 ms^{-1}, in order to prevent settlement of solids. Flows in excess of 0.6 ms^{-1} are undesirable because excess turbulence affects the stability of the flocs. The sludge loading should be <0.15 kg BOD $kg^{-1}d^{-1}$, and if too high, organic matter in the form of sludge will be removed from the channel before being completely oxidized resulting in a surplus sludge production. However, if sludge loading is kept low then most of the organic matter is oxidized, thus reducing the sludge volume. The normal range for MLSS in extended aeration systems is between 2000–6000 mg l^{-1}. Although the area of land required for oxidation ditches is large, this is offset by their cheap construction cost, especially as there is no primary sedimentation. Also, operating costs are low with regard to reduced sludge handling and power requirement. This makes them ideal for small communities and for Third World installations. Oxidation ditches are also widely used for treating dairy and other difficult organic wastewaters where a long SRT is desirable.

The problem of using extended aeration to treat domestic wastewaters from large populations of more than 10 000, when loaded at <0.3 kg BOD $m^{-3}d^{-1}$, is the cost of maintaining the sludge in suspension. This has been overcome by using two rotor aerators positioned equidistant from each other. A development of this system is the carrousel-type of extended aeration process which comprises of two rectangular tanks with an aerator at each end (Fig. 4.50) (Zeper and De Man 1970). The tanks are up to 4 m deep and 8 m wide with aeration and turbulence supplied by 'Simcar'-type vertical cone aerators. By siting one at each bend, the rotary motion is diverted to provide linear motion. There are a number of carrousel-type plants in operation throughout Europe, with the performance reported as being excellent. For example, the Ash Vale plant of the Thames Water Authority treats a PE (population equivalent) of 25 000. At a sludge loading of between 0.05–0.15 kg $kg^{-1}d^{-1}$, a 10:10 final effluent was produced (Pay and Gibson 1979). A carrousel-type system in the Cotswold division of the Thames Water

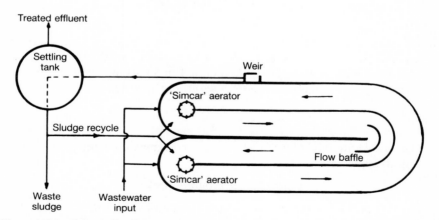

FIG. 4.50   Schematic diagram showing plan for the carrousel type of oxidation channel developed by Simon–Hartley Ltd (Winkler 1981).

Authority has been studied over a number of years and fully examined by Rachwal *et al.* (1983).

Three features of oxidation ditches in general, and of carrousel-type plants in particular are: (1) ease of operation; (2) low production of a highly stabilized sludge; and (3) the ability of such systems to nitrify and denitrify within a single tank. Denitrification occurs by allowing the part of the ditch, most remote from the aerators, to become anoxic (Johnstone and Carmichael 1982; Matsui and Kimata 1986).

Packaged plants
In Ireland, as in other parts of Europe, small sewage treatment plants serving isolated communities, which have traditionally used a septic tank and percolating filter, are now installing package activated sludge units. These self-contained treatment units, which are prefabricated either in concrete or steel, are designed to treat screened sewage which is directly fed into the aeration tank, therefore a primary sedimentation unit is not required. Sludge separation and recirculation facilities are built into the unit, and aeration is supplied by coarse bubble diffusers that also ensure adequate mixing (Fig. 4.51). With the HRT >24 h the activated sludge is in the endogenous respiration phase so that the minimum amount of surplus activated sludge is produced, thus, a sludge loading of between 0.05–0.15 $kg\,kg^{-1}d^{-1}$ is required. Approximately 0.23 m^3 of aeration tank volume is allowed per person, which is equivalent to a volumetric loading of about 0.24 kg BOD $m^{-3}d^{-1}$. Although each manufacturer has slightly different design features, sludge separation normally takes place in a small chamber within the aeration basin, with the settled sludge returned to the main tank through a slot at the base (Fig. 4.51), or by using an air-lift pump, although this makes the overall operation and management more difficult. Units used for villages, schools or hospitals have a very exaggerated diurnal variation in flow with often no flow

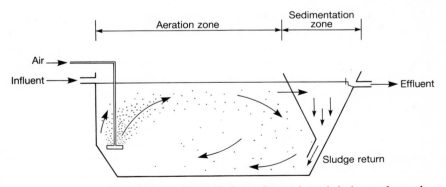

FIG. 4.51   Schematic diagram of extended aeration activated sludge package plant with a clarifier incorporated into the same tank.

at all between 01.00–06.00 h. Although flow balancing may be desirable, the long HRT of such units ensures that they can accept intermittent loads without upsetting the overall operation. These units are relatively easy and quick to install, and require only periodic desludging to remove surplus activated sludge. This is done by shutting off the aeration and allowing the solids to settle and then simply pumping the surplus sludge out of the tank using a gully sucker and tanker. The MLSS concentration ranges from 1000 mg l^{-1}, after desludging, to a maximum of 10 000 mg l^{-1}, when desludging is required. It is estimated that the MLSS will increase in concentration by about 50 mg l^{-1} per day, so that desludging is only required approximately every six months, if loaded correctly. As the MLSS builds up, there may be periodic losses of sludge in the effluent and many manufacturers recommend installing tertiary treatment to intercept any suspended solids before discharge to the receiving water. The nature of these units makes them ideal in many ways for isolated communities. However, experience in Ireland has shown that it is these very communities that often become cut off in winter because of snow and have frequent and often prolonged cuts in electricity. Although short periods without power to operate the aerators will not adversely affect the operation of package activated sludge units, long periods of more than 24 hours will begin to kill off the floc micro-organisms. Where this is a problem, the mechanically operated septic tank and percolating filter system is probably the best treatment option. Other activated sludge processes, such as contact stabilization are also available as package plants.

4.2.3.3   High-rate activated sludge processes

High-rate treatment is used, as is high-rate filtration, as a partial or roughing treatment for medium to strong wastes prior to further biological treatment or discharge to a sewer (Boon and Burgess 1972). They are widely used in the food processing and dairy industries (Bruce and Boon 1971), although they are also used for partial treatment of domestic wastewaters.

The loading levels are several times greater than those in conventional activated sludge processes (Table 4.22), which give a rapid but only partial removal of BOD. The BOD loading is 1.5–3.5 kg BOD m^{-3}d^{-1}, with a sludge loading of between 1.0–2.5 kg BOD kg^{-1}d^{-1} (Emde 1963). To maintain such a high sludge loading rate the MLSS concentration must be maintained at between 5–10 000 mg l^{-1}. This is a major design factor requiring a generous separation provision allowing for returned sludge facility of up to 100% of the influent wastewater. With a short SRT of < 0.5 d, the micro-organisms are in an environment where both food and oxygen are non-limiting, resulting in a rapid growth rate and a high sludge production rate of between 0.8–1.2 kg DS per kg BOD removed. These conditions tend to favour the dispersed bacteria that are able to grow faster than the floc-forming species and results in a less flocculent activated sludge, which may cause problems in the separation stage.

The dewatering characteristics of high-rate activated sludge are considered to be inferior to conventional activated sludge (Gale and Baskerville 1970), although, in practice, there is very little difference between the two. Where problems do exist Bruce and Boon (1971) have suggested that coagulants will help to dewater the sludge. In the aeration tank, the MLSS concentration is higher than in conventional plants, and with the increase in sludge activity there is a greatly enhanced BOD removal rate requiring a higher critical concentration of dissolved oxygen (>2 mg $O_2 l^{-1}$). The high oxygen demand of the mixed liquor combined with the higher MLSS concentration requires a highly effective aeration and mixing system. Although vertical cones can be used, fine or coarse bubble diffusers are favoured. With a HRT of only 1–2 h, BOD reduction is limited to 60–70% in high rate systems, although this represents a very high BOD removal per unit volume of aeration tank per unit time compared with other processes. The process is similar to contact stabilization, with an initial rapid adsorption of organic waste over the first 10–15 min of aeration, followed by a second period of adsorption after about 1 h, giving a considerable degree of BOD removal almost entirely by adsorption on to the flocs (Kehr and Emde 1960). Because of this, the HRT is largely unimportant so long as it is >1 h. However, even with an HRT of 20–30 min, BOD reductions are only reduced by 10–15% (Hawkes 1983a).

4.2.3.4 Advanced activated sludge systems

The main objective of advanced activated sludge systems has been to increase the amount of dissolved oxygen available for biological activity. This has been achieved by increasing the rate of oxygen transfer from the gas phase to the liquid phase by increasing the partial pressure of oxygen in the gas phase and so increasing the saturation concentration of dissolved oxygen. Two methods have received considerable attention: the deep shaft process in which the total pressure of the system is increased; and the use of pure oxygen instead of air to increase the proportion of oxygen in the gas phase. These advanced processes, such as the ICI deep shaft process, the Wimpey Unox, BOC Vitox, and the Megox Processes, have all intensified the activated sludge process making it even more compact both in terms of design and operation.

Deep shaft

The deep shaft system, developed by ICI as a spin-off from their single cell protein production, is perhaps one of the most interesting avanced treatment systems in operation today. By using a deep shaft or well (30–220 m), the hydrostatic pressure in the base of the unit reaches pressures of between 5–10 atmospheres, resulting in high oxygen transfer efficiencies without a high energy consumption. The increase in pressure results in an increase in oxygen solubility by as much as a factor of 10. Although there is a high degree of mixing in the process there is little dispersion or back mixing, making the process essentially plug-flow in nature (Dunlop 1978).

The shaft is between 0.8–6 m in diameter and is divided vertically to produce a central downflow pipe and an outer upflow section that are connected at the base (Fig. 4.52); although alternative arrangements have been used, such as a more simple U-tube configuration (Fig. 4.53) (Cox *et al.* 1980). Raw sewage is screened and degritted, and before it enters the shaft it is mixed with returned activated sludge (Fig. 4.53) (1). Once in the top chamber above the actual shaft it is mixed with the mixed liquor already in the unit (2), before it is drawn down the central shaft. Compressed air is added at about a third to a half of the depth of the shaft, and as the downflow liquor velocity exceeds the bubble rise rate the air is carried down the shaft and gradually dissolves as the pressure increases (3). The high pressure near the base of the shaft ensures that almost all the air present is in solution, providing an oxygen-rich environment for the activated sludge micro-organisms (4). At the bottom of the shaft mixing is induced by the configuration of the base but a minimum current velocity of $1.5 \, \text{m s}^{-1}$ must be maintained to prevent settlement (5). The MLSS concentration can be as high as $5000 \, \text{mg l}^{-1}$ so oxidation is rapid (6). The

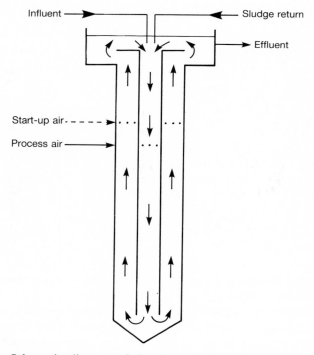

Fig. 4.52 Schematic diagram of the deep shaft system using a central tube configuration.

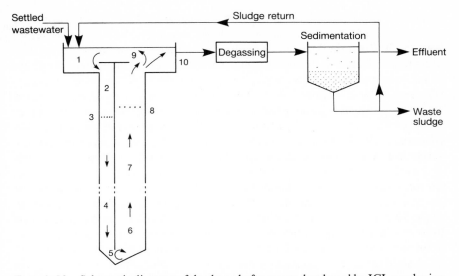

FIG. 4.53   Schematic diagram of the deep shaft process developed by ICI, employing a U-shaped reaction shaft.

turbulence due to the high liquid circulation rate results in a high liquid Reynolds Number ($>10^5$) and an increase in the rate of mass transfer. The contact time between the air bubbles and the liquid is also very much higher being up to 5 minutes compared to 15 seconds in a conventional diffused air system (Hemming *et al.* 1977). The pressure decreases as the mixed liquor moves up the upflow section releasing the air from solution, which is mainly carbon dioxide, nitrogen, and some residual oxygen forming bubbles. As the bubbles near the surface they increase in size thus increasing their upward velocity. The presence of the bubbles lowers the specific gravity in the upflow section forcing the mixed liquor to rise, which gives the process its overall circulation (7). The upflow air injection is used to start up the system and to control the flow rate; however, once circulation is established the release of air from solution as the pressure changes in the upflow section is enough to keep the circulation in operation (8). Once at the surface, the air escapes to the atmosphere (9), but due to the amount of air in solution it may be necessary to use vacuum degassing to ensure that the air in the mixed liquor does not interfere with the separation of sludge from the purified effluent in the sedimentation tank (10).

Pilot-scale trials have shown that the deep shaft process is able to achieve a good BOD removal rate (90%), but little nitrification. The normal organic loadings range between 3.7–6.6 kg BOD $m^{-3}d^{-1}$, with a sludge loading rate (f/m) of 0.8–0.9 kg $kg^{-1}d^{-1}$, a sludge age of 4–5 d, an HRT of 1.17–1.75 h, which is even shorter than in high-rate processes, and a MLSS of between

2000–6000 mg l^{-1} for the treatment of mixed industrial and domestic wastewater. The sludge has good settling characteristics, with SSVI values between 30–100 ml g^{-1}, which is readily dewatered. In terms of sludge activity and microbial ecology, sludges from deep shaft units are very similar to those from conventional plants, with sludge flocs about 0.1 mm in diameter (Dunlop 1978; Hemming 1979). The process is started by using activated sludge from a conventional plant and whereas the microbial population takes 2–3 weeks to stabilize, a good BOD removal can be expected within 1–2 days (Hemming 1979). However, the effluent does contain high concentrations of suspended solids therefore overall the sludge production is lower than conventional activated sludge plants at 0.50–0.85 kg kg^{-1} BOD removal (Bolton and Ousby 1977; Hemming et al. 1977; Cox et al. 1980). Clearly, the deep shaft process provides an excellent first stage for a two-stage process with the second stage providing nitrification. Cox et al. (1980) used a percolating filter as a second stage in their pilot trials which achieved a 95% reduction in ammonia when loaded at 1.7 m^{3}m^{-3}d^{-1}. Full-scale plants have been used to partially treat strong wastes. Collins and Elder (1980, 1982) describe a plant treating a strong domestic wastewater, with a mean BOD of 1060 mg l^{-1}, for discharge directly into the Thames Estuary. This plant was of the concentric tube design, 130 m deep and 1.86 m in diameter. Operated with a HRT of 1.5 h and at an MLSS concentration of 5000 mg l^{-1}, the average BOD reduction was >85%, with good sludge settleability (SSVI 40–70 ml g^{-1}). By filtering the final effluent to remove the suspended solids, the BOD reduction was increased to 95.5%.

In terms of loading, the deep shaft system falls between conventional and high-rate processes. It has a small space requirement, taking up 50% less land area than conventional activated sludge processes because of not having primary sedimentation and the aeration tank volume being largely underground. The power requirement is low, requiring approximately 0.85 kWh kg^{-1} BOD, which includes both aeration and recirculation of sludge. But in order to be competitive with other activated sludge processes the deep shaft process has to overcome a number of fundamental problems. For example, constructing deep shaft systems is difficult, and the stability of the soil and the bedrock is extremely important. Two options are available, to build a deep narrow shaft or a shallow wide shaft. The former gives higher hydrostatic pressures and increased oxygen transfer, but increased excavation and constructional costs, whereas the latter option provides a lower hydraulic resistance, which requires a lower specific power requirement than the narrower shafts and is also easier and cheaper to build. Other problems include foam production (Wheatland and Boon 1979), degassing, presence of fine organic solids in the final effluent that significantly increases the BOD, the possibility of a reversal of flow within the shaft, and the accumulation of dense solids in the bottom of the shaft due to inadequate screening or grit removal.

Pure oxygen systems

Atmospheric air contains only 21% oxygen and using pure oxygen instead of air will increase the saturation concentration of oxygen by a factor of five, thereby significantly increasing the oxygen transfer rate. In theory, this should increase treatment capacity of the aeration stage, or by switching from atmospheric air to pure oxygen at times of high organic loading it can be used to temporarily increase the capacity of the aeration tank at peak loadings.

The theory is quite straightforward. The rate oxygen is absorbed by water from air is controlled by the rate of diffusion across the air–water interface. The rate of oxygen transfer can be expressed as:

$$\frac{dc}{dt} = \frac{K_L A (C_s - C)}{V}$$

Where K_L is the liquid film mass transfer coefficient; A, the interfacial area; V, the volume of mixed liquor; C_s, the equilibrium saturation concentration of dissolved oxygen; and C, the actual concentration of dissolved oxygen in the mixed liquor (mg $O_2 l^{-1}$). When the mixed liquor is aerated the total interfacial area (A) cannot be realistically measured, therefore the overall oxygen transfer coefficient [$K_L a$] is used:

$$[K_L a] = \frac{K_L A}{V}$$

thus, a represents the interfacial area per unit volume of mixed liquor.

Thus,

$$\frac{dc}{dt} = K_L a (C_s - C)$$

The oxygen transfer coefficient [$K_L a$] in mixed liquor varies with the type of wastewater being treated, with detergents and the method of aeration adopted of particular importance. The effect on the oxygen transfer coefficient is complex. For example, in diffused air systems detergents tend to reduce [$K_L a$] as the interfacial area per unit volume a is increased by affecting the surface tension and bubble size, whereas the concentration of surface active material at the interface decreases the mobility of the surface film thus reducing K_L (Lister and Boon 1973; Boon 1976). With high-shear surface aerators the presence of detergents has the opposite effect on [$K_L a$] by increasing a to a greater degree, then K_L is reduced (Winkler 1981).

Therefore, in air systems, the maximum rate of oxygen transfer is achieved by maximizing [$K_L a$] by optimal use of the aeration system and by maintaining a maximum oxygen deficit ($C_s - C$) by operating the aeration tank at the critical dissolved oxygen concentration ($C < 2.0$ mg $O_2 l^{-1}$). Higher oxygen concentrations can only be obtained by increasing the rate of aeration, increasing the [$K_L a$] rate, but this will be expensive in terms of energy and in loss of settleability due to shear damage to the floc structure as turbulence is increased. The advantage of pure oxygen systems is that the oxygen deficit $(C_s–C)$ is so great that high operating oxygen concentrations can be maintained $(6–10 \text{ mg } O_2 l^{-1})$ without increasing $[K_L a]$. In fact, the $[K_L a]$ value is decreased in practice in order to achieve the same mass transfer, which results in a lower intensity of agitation (Lewandowski 1974).

As lower $[K_L a]$ values result in less disruption of the floc due to shear forces, which will improve sludge settleability, ideal designs have aimed to reduce the level of agitation in the aeration tank to the minimum so that the sludge is just maintained in suspension. The main advantages are: flexibility in operation to cope with surge flows; reduction in energy usage to drive aerators; higher organic loading per unit volume of reactor able to be treated; and better performance. Enhanced performance including nitrification, is due to the higher oxygen concentration, which ensures maximum penetration of oxygen into the floc as well as a smaller proportion of the activated sludge deprived of oxygen. Pure oxygen systems have been adopted because of three important advantages over air systems: (1) improved rates of BOD removal; (ii) reduced sludge yields; and (iii) improved settling characteristics of the sludge (Sidwick and Lewandowski 1975). However, comparative studies between pure oxygen and air systems have not always supported all these claims (Huang and Mandt 1978; Suominen 1980). Therefore, pilot-scale trials are necessary before the adoption of such an expensive system. The cost of producing oxygen is the limiting factor for this advanced process and, in recent years, new methods of producing oxygen have been continuously developed so that the cost of oxygen has steadily fallen (McWhirter 1978a,b).

There are two main types of pure oxygen activated sludge systems, those that operate in closed oxygen-rich atmospheres that generally employ surface mechanical aerators, and open systems that employ fine bubble diffusers.

Closed systems. By enclosing the aeration tank, a pure oxygen atmosphere can be maintained above the mixed liquor. There is, of course, an increased risk from having a pure oxygen atmosphere as many materials that may occur in wastewaters are highly combustible under these conditions, even though the gas phase is saturated with water, which reduces the hazard (Baker and Carlson 1978). Also, the materials used for the construction of such systems must be safe for use in a high oxygen environment. Safety equipment to monitor hydrocarbon accumulation in the space above the mixed liquor, and for rapidly venting the oxygen and replacing it with air, is standard. These systems operate at low pressures and at normal ambient temperatures. Like all enclosed treatment systems, odours are greatly reduced and problems associated with spray are contained.

Although marketed in Britain and Ireland by Wimpey Unox, the Unox Process was originally developed by the Union Carbide Corporation in the USA. Most plants are in the USA and Japan, and treat a variety of domestic, industrial, and chemical wastewater (McWhirter 1978c). The enclosed aeration tank is split up into a number of compartments by baffles, each served by a mechanical surface aerator (Fig. 4.54). The number of compartments depends on the level of treatment required and the nature of the wastewater, but the optimum number for domestic and municipal wastewaters appears to be between four and six. However, the greater the number of compartments

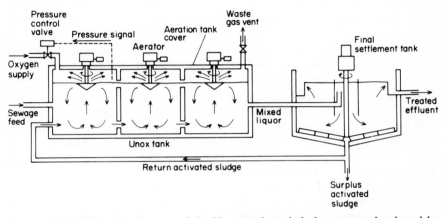

FIG. 4.54 Schematic diagram of the Unox activated sludge system developed by Wimpey-Unox Ltd which uses pure oxygen (Hawkes 1983a).

the greater the oxygen utilization (McWhirter 1978b). Using submerged inlets, the settled wastewater and returned sludge are fed into the final compartment, and the oxygen is pumped into the space above the mixed liquor. The oxygen and mixed liquor flow from one compartment to another so that the aeration tank becomes a series of completely mixed units. The oxygen concentration in the gas phase falls in each compartment as it is utilized by the micro-organisms, with 80–90% of the oxygen supplied utilized overall, which reduces the volume of gas to about 20% of the original volume. In the last compartment, the gas only contains 40–50% oxygen as it has been diluted by carbon dioxide, and is finally vented to the atmosphere. As the nutrient concentration in the mixed liquor also falls from stage to stage the design ensures that oxygen transfer is greatest where the oxygen demand from the mixed liquor is greatest. When treating typical municipal wastewater, the MLSS is maintained at 5000–6000 mg l^{-1}, which is about 50% higher than conventional activated sludge processes using air. The working concentration of dissolved oxygen in the mixed liquor is between 4–8 mg l^{-1} compared to only 1–2 mg l^{-1} in air operated plants. The high oxygen concentration limits the sludge loading even at high BOD loadings, which are 3–4 times higher than air systems at 2.5–4.0 kg BOD m^{-3}d^{-1}. The sludge loading rate (f/m) can be double the conventional loading rate (Table 4.22) at 0.4–1 $kg\,kg^{-1}d^{-1}$. Pure air systems have up to 40% shorter retention times and generally produce less sludge per unit mass of BOD removed with better settling properties. A most interesting Unox pilot plant study has been undertaken by the Scottish Development Department (1977) using a mixed domestic and industrial effluent.

Other closed systems include the 'Oases' system developed by Air Products and Chemicals Inc (USA) and the 'Forced Free-Fall' oxygenation system by Airco Inc (USA) (Wyatt *et al.* 1975; McWhirter 1978*a*).

Open tank systems. The use of pure oxygen in open systems requires a highly effective mechanism to ensure maximum dissolution of oxygen so that wastage is minimized. The Vitox aeration system, developed by the British Oxygen Company, was originally devised for re-aerating lakes and rivers that had become oxygen depleted. The potential of the system to uprate overloaded activated sludge plants was quickly realized and many new specifically designed systems are now available. The main advantage of these systems is that they can be used with existing aeration tanks with no need to construct additional tanks or even modifying existing ones. The use of pure oxygen can be used to replace the existing air system of aeration or can be used to supplement the existing aeration to provide extra oxygen at peak flows or when overloaded because of seasonal increases in loading. Other advantages of open systems include: no hazard because of an enclosed oxygen-rich atmosphere; no monitoring of potentially combustible compounds; and better access to the tank for maintenance.

The British Oxygen Company has developed two high-efficiency oxygenation devices known as Vitox 1 and Vitox 2. The Vitox 1 system involves increasing the pressure of the settled sewage flow to 2–3 bars and then pumping it through a vent, when it is then injected with oxygen. The oxygen is entrained in the flow as fine bubbles that are then broken up into micro-bubbles as the sewage is discharged into the mixed liquor within the aeration tank at a high velocity through an expansion nozzle (Fig. 4.55). The high velocity of the discharge causes considerable turbulence ensuring that the oxygen is rapidly mixed with the contents of the aeration tank (Boon 1978). The system has been adopted at a number of holiday resorts to cope with seasonal increases in the organic loading (Rees 1978). Under these circumstances, the cost of installing a Vitox 1 system will be only a few percent of the cost of extending a conventional activated sludge plant.

A modification of the Vitox 1 system is the use of a downflow bubble contactor, a process known as Vitox 2. The oxygen is pumped into the main wastewater stream, which then flows downwards through the contactor at a low velocity. The reduced velocity ensures that only very small bubbles with an extremely low upward flow velocity are carried out of the contactor and injected into the aeration tank. The larger bubbles remain in the contactor

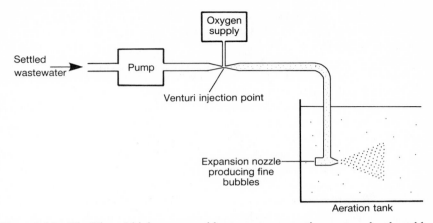

FIG. 4.55 The Vitox 1 high pressure side-stream oxygenation system developed by BOC Ltd.

until their size is reduced sufficiently by oxygen dissolution to be carried out of the wastewater stream (Fig. 4.56).

The Vitox 2 oxygenation system is used in the Megox treatment process, which has been particularly designed to treat difficult and nutrient-rich wastewaters, especially those from the food-processing industry. The Megox process incorporates both the biological and sludge separation phases into a single tank. The influent wastewater is mixed with the recycled activated sludge and injected with oxygen via the downflow bubble contactor. This oxygen rich mixed liquor is discharged into the central well of the tank, and as the oxygen concentration is already very high only a minimum amount of mixing is required to maintain the microbial flocs in suspension, which is provided by a rotating sludge consolidator with a very slow rotation speed. There is an outer concentric zone separated from the central reaction zone by a deep baffle, where the sludge is discharged at the base of the tank and recirculated (Fig. 4.57). The amount of oxygen injected into the influent stream is strictly controlled using a dissolved oxygen sensor linked to the injection system. This allows the process to cope with shock loads as well as conserving oxygen usage.

4.2.4 Sludge problems

The most common operational difficulties associated with activated sludge are those concerned with the separation of sludge from the clarified wastewater in the sedimentation tank. The ability of activated sludge to separate is normally measured by an index of settleability, such as the sludge volume index (SVI), the sludge density index (SDI) or the stirred specific volume index (SSVI)

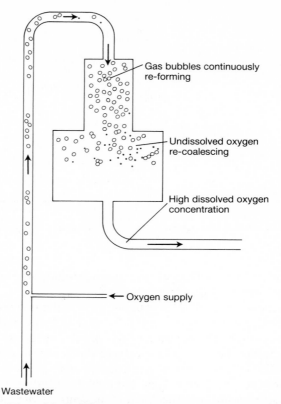

Gas bubbles continuously
re-forming

Undissolved oxygen
re-coalescing

High dissolved oxygen
concentration

← Oxygen supply

Wastewater

FIG. 4.56 The Vitox 2 bell diffuser oxygenation system developed by BOC Ltd
(Winkler 1981).

(Section 4.2.2.1). Problems in sludge settlement can be caused by bulking, deflocculation, pin-point flocs, foaming or denitrification. These terms describe the effects, although their definitions are rather imprecise and there is some overlap.

With the exception of denitrification, which has been dealt with in Section 4.4, all settleability problems can be traced back to the structure of the activated sludge floc. There is a wide range of particle sizes in the activated sludge mixed liquor ranging from individual bacteria of between 0.5–5 μm up to large flocs which may be greater than 1 mm (1000 μm) in diameter. Parker *et al.* (1971) found that there was biomodal particle size distribution of flocs in activated sludges (Fig. 4.32) and suggested that the smaller particles represented individual micro-organisms or small aggregates that have not flocculated or have been sheared off larger flocs. The maximum size of a floc is dependent on their cohesive strength and the degree of shear exerted by the turbulence within the aeration tank. Floc structure has been subdivided into

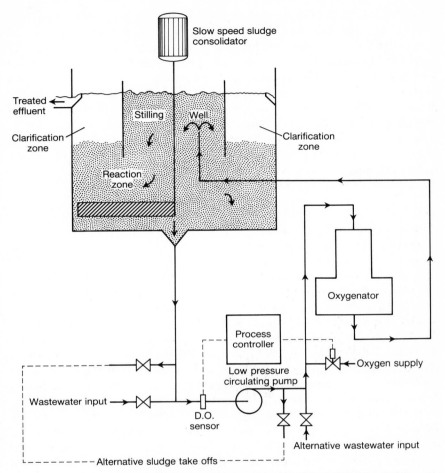

FIG. 4.57   The Megox process developed by BOC Ltd (Winkler 1981).

two distinct categories, micro- and macro-structure (Sezgin *et al.* 1978). Micro-structure is where the flocs are small (< 75 μm in diameter), spherical, compact, and relatively weak. They are composed of floc-forming bacteria and formed by aggregation and bioflocculation where individual micro-organisms adhere to one another to form large aggregates. The structure of such flocs is termed weak because in the turbulent conditions of the aeration tank they can easily be sheared into smaller particles. Although such flocs rapidly settle, the smaller aggregates that have been sheared from the larger flocs, which take longer to settle, may well be carried out of the sedimentation tank in the final effluent increasing the BOD and giving the clarified effluent a high turbidity. When filamentous micro-organisms are present, the flocs take on a macro-structure, with the micro-organisms aggregating around the filaments making larger flocs of irregular shape and able to withstand high shear forces within the aeration tank because of this extra support. This distinctive micro- and macro-structure of activated sludge flocs was demonstrated experimentally by Lau *et al.* (1984*a*) who grew a floc-forming bacterium *Citromonas* sp. in dual culture with the filamentous bacterium *Sphaerotilus natans*. When grown on their own, each formed compact spherical flocs with typical micro-structural properties, whereas when grown together in roughly equal proportions irregularly shaped flocs were formed with *S. natans* acting as the support structure on which the other bacterium aggregated (Fig. 4.58). A useful checklist of sludge settleability problems has been produced by the Water Research Centre (Chambers and Tomlinson 1981), which gives a step-by-step guide to identifying and solving sludge problems.

Deflocculation
Deflocculation or dispersed bacteria is a phenomenon caused by the micro-structure of flocs failing either by the flocs becoming less stable and breaking up, as would be the case in severe turbulence caused by over-aerating (Parker *et al.* 1972), or the bacteria failing to aggregate into flocs. The micro-organisms are present as individual cells or as very small clumps which remain in suspension in the liquid phase in the sedimentation tank and are continually washed out with the final effluent. Although some of the sludge continues to settle the final effluent becomes progressively turbid as the remaining flocs break up. It is mainly a problem of activated sludge with a microstructure character, but it is also associated with bulking. If deflocculation is not controlled, it will become gradually more difficult to maintain the MLSS concentration in the aeration tank, as the microbial biomass is washed out, with the sludge loading (f/m) gradually increasing as a consequence. Deflocculation is thought to be due to low dissolved oxygen concentrations, low pH or shock loads, although Pipes (1979) suggests that sludge loadings $> 0.4 \text{ kg kg}^{-1}\text{d}^{-1}$ will tend to lead to deflocculation. It would appear that if the sludge has a microstructure, then a high sludge loading will eventually cause deflocculation. Certain toxic wastewaters can also cause flocs to disintegrate into small aggregates resulting in very turbid effluents and a total breakdown of operations as the MLSS is rapidly reduced. High population densities of free-swimming protozoans, such as *Colpoda* sp. and *Paramecium* sp., are also known to cause the final effluent to appear turbid, although sludge settleability is unaffected.

Pin-point floc
Activated sludge flocs without any filamentous micro-organisms present have a pure micro-structure. Because of the lack of macrostructure the flocs are small, compact, and spherical, with a weak structure so that they are readily broken up into smaller flocs within the aeration tank. This total dependence on microstructure and the cohesive forces of individual bacteria to hold the

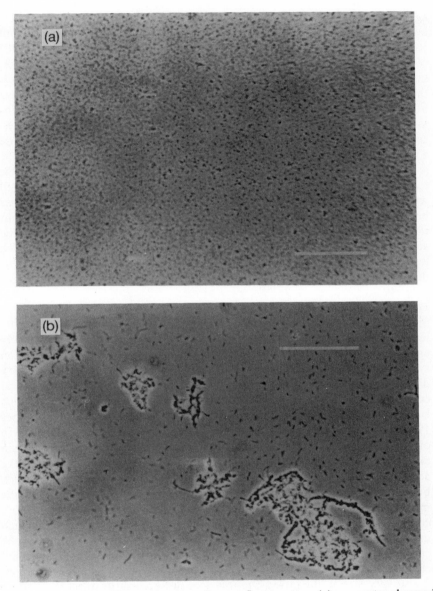

FIG. 4.58 Effect of filamentous organisms on floc structure: (a) aggregates observed for a pure culture of an activated sludge floc former; (b) aggregates observed for a dual culture of a single floc former and a single filamentous organism. Photographs taken at × 100 using phase contrast with the bar representing 100 μm (Lau *et al.* 1984).

agglutinated mass of cells together, means that the floc becomes progressively unstable as it grows. Thus, there is a high proportion of flocs with a small particle size that take much longer to settle than the larger flocs and can be carried out of the sedimentation tank in the final effluent. These small flocs, known as pin-point or pin floc, do not cause high turbidity in the effluent, as is the case with deflocculation, because the particles are much larger and visible to the naked eye as discrete flocs in the final effluent. The SVI usually remains low with much of the activated sludge continuing to settle, although the continued loss of microbial biomass due to pin-point floc may eventually pose a problem of maintaining sufficient MLSS in the aeration tank. Pin-point floc is mainly associated with a long sludge age > 5–6 d and with low organic loadings < 0.2 kg m^{-3}d^{-1} (Pipes 1979) and is therefore a problem associated with extended aeration systems (Forster 1977). It has been suggested that as pin-point floc is associated with long sludge ages and low organic loadings that the flocs are most likely to be the non-biodegradable fraction of the aerobically digested floc. This is supported by reports that the presence of such flocs in the final effluent did not significantly increase the BOD. However, this does not appear the case with all continuously loaded activated sludge plants.

Foaming
Since the ban on the use of non-degradable (hard) detergents, white frothy foam, forming deep banks often covering the aeration tanks, is now rarely encountered at activated sludge plants. However, the intensity of the aeration system produces small quantities of light, transient foam that quickly disperses. A denser foam, very similar to chocolate mousse both in texture and colour, occasionally occurs on aeration tanks and it is this phenomenon which is referred to as activated sludge foaming (Al-Diwamy and Cross 1978; Pipes 1978b; Dhaliwal 1979). This type of foaming is caused by the excessive growth of certain filamentous micro-organisms of the genus *Nocardia*, which entrain air bubbles from the aeration system and become buoyed to the surface of the aeration tank where a dense stable foam or thick scum is formed. Although bubble entrainment is possibly how denitrification causes rising sludge, the mechanism by which air bubbles become attached to *Nocardia* appears more complex. The mycelium of these species are generally hydrophobic, which makes it very amenable for the attachment of air bubbles. The foam can be carried over to the sedimentation tank and be discharged with the final effluent. Although the population density of this filamentous micro-organism is also very high in the mixed liquor it does not necessarily affect the settling qualities of the sludge which generally remain good. This is because *Nocardia* is highly branched (Fig. 4.59), giving the flocs a strong macrostructure that produces large firm flocs which readily settle. Unlike other filamentous micro-organisms the hyphae are largely retained within the floc and do not extend from its surface, which aids settlement. The depth of foam can be considerable, covering walkways, mechanical equipment, overflowing the outlet weirs, and

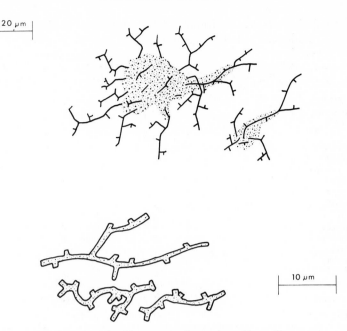

FIG. 4.59   The filamentous bacterium *Nocardia* sp.

generally making maintenance difficult and dangerous. Foaming in closed
aeration tanks reduces the available head space and has even reduced the
available hydraulic head for gravity flow through the tank. The MLSS
becomes trapped in the foam giving it a dark brown colour, and between
30–50% of the total activated sludge can be entrained in the foam making it
operationally difficult to maintain adequate MLSS in the mixed liquor in
order to continue treatment. The foam contains considerable air, as well as
Nocardia and activated sludge flocs, and has a bulk density of about
0.7 g ml^{-1} (Lechevalier and Lechevalier 1975). As expected, the density of
Nocardia is enriched in the foam with up to 10^{12} microcolonies ml^{-1}
compared to only 10^6 microcolonies ml^{-1} in the mixed liquor (Wheeler and
Rule 1980). Other problems include odour production in the summer, and if
the wasted sludge is digested the *Nocardia* can subsequently cause foaming in
the anaerobic digester (Jenkins *et al.* 1984).

Although the exact incidence of *Nocardia* foaming is not known, the
actinomycete is one of the most commonly observed filamentous micro-
organisms in activated sludge in the USA, South Africa, and Europe
(Eikelboom 1977; Gray 1982*b*; Richard *et al.* 1981; Wagner 1982; Strom and
Jenkins 1984; Blackbeard and Ekamma 1984; Ekamma *et al.* 1985; Byron
1987) (Table 4.23). The actinomycetes of the genus *Nocardia* occurring in
activated sludge are very diverse, with *N. amarae* the most commonly

TABLE 4.23   Ranking of dominant filamentous organisms recorded in major activated sludge surveys (Ekama *et al.* 1985)

Filamentous micro-organisms	S. Africa (Ekama *et al.* 1985)	USA (Jenkins *et al.* 1979)	The Netherlands (Eikelboom 1977)	W. Germany (Wagner 1982)
Type 0092	1	11	4	—
M. parvicella	2	7	1	2
Type 1851	3	9	12	—
Type 0675	4	6	—	—
Type 0914	5	—	—	—
Type 0041	6	5	6	3
Norcardia spp.	6	1	14	5
Type 0803	7	7	9	10
Type 1701	8	2	5	8
N. limicola	8	8	11	7
Type 021N	—	3	2	1
H. hydrossis	9	9	3	6
S. natans	—	6	7	4
Thiothrix spp.	9	4	17	—
Type 0581	9	12	8	—
Type 0961	—	10	10	9
Beggiatoa spp.	—	12	—	—
No. of plants	56	190	200	315
No. of samples	56	300	1100	3500

occurring species in the USA (Lechevalier and Lechevalier 1974), and like the fungus *Subbaromyces splendens* in percolating filters, it has not been isolated from any other environment except activated sludge. Other species isolated from activated sludge include *N. rhodochrus*, which is common in West Germany and Switzerland (Lemmer and Kroppenstedt 1984), *N. asteroides*, *N. caviae*, and strains of *Mycobacterium*. *Microthrix parvicella*, which is also thought to be an actinomycete, although not positively classified taxonomically so far (Slijkhuis and Deinma 1982), has also been associated with foaming (Jenkins *et al.* 1984). It should be remembered that many of these species are pathogenic and so there may be a risk of infection to humans through aerosol formation (Section 6.2.4.3). Causes of *Nocardia* growth are not known, with contradictory evidence relating to such factors as temperature ($> 18°C$), high loading levels, and long sludge ages (> 9 d) (Pipes 1978*a,b*). Experience in Europe does suggest that the presence of emulsifiable fatty material may be significant (Eikelboom 1975; Pipes 1978*b*; Matsche 1977). In Ireland, there certainly appears to be an association between foaming and discharges from abbatoirs and butchers, especially blood. The actual formation of foam depends on the mixed liquor containing adequate suspended solids (Wheeler and Rule 1980), with the amount of foam produced directly proportional to the amount of *Nocardia* present on a total filament length basis (Vega-Rodriguez 1983). Increases in temperature or in the air flow will both increase the height of the foam layer.

The most widely used method to control *Nocardia* growth in activated sludge is to reduce the sludge age. By increasing the sludge wastage rate it is possible to wash *Nocardia* out of the system, although the effective sludge age is a function of temperature, with the higher the temperature the lower the sludge age required. If nitrification is carried out in the aeration tank then this approach must be ruled out as the minimum sludge age for the retention of the nitrifying bacteria is much greater than that required to wash out *Nocardia*. Wilson *et al.* (1982) have successfully achieved *Nocardia* wash-out at a sludge age of 1 d. Anaerobic digester supernatant is toxic to *Nocardia* (Lechevalier and Lechevalier 1975; Lemmer and Kroppenstedt 1984) and some success has been achieved in controlling it by adding the supernatant to full-scale plants (Lechevalier *et al.* 1977), although this is in contrast to the findings of Wheeler and Rule (1980). Antifoaming devices employing water sprays that burst the air bubbles are ineffective against these stable foams and merely dilute it. However, by diluting it and reducing the thickness of the foam in the sedimentation tank it may be possible for the scum trap to cope with the foam and effectively contain and possibly even remove it. Another problem is that the activated sludge sedimentation tanks have small scum traps drained by small diameter pipes that do not allow the foam to freely drain away. It would be advantageous if such tanks could be fitted with full-width scum traps with larger diameter pipework. Some control has been obtained by removing the foam from the sedimentation tank via the scum traps and ensuring it is not fed into the system. The control options for *Nocardia* growth and foaming has been reviewed by Wheeler and Rule (1980) and Jenkins *et al.* (1984).

Filamentous bulking
Bulking is a phenomenon where filamentous organisms extend from the flocs into the bulk solution, interfering with settlement and subsequent compaction of the activated sludge with a SVI > 150 ml g^{-1} (Pipes 1967; Anon 1979*b*). Although bulking sludges settle more slowly than normal sludges they are normally quite efficient in purifying the wastewater and so produce good effluents. Even when bulking is quite severe a very clear supernatant is obtained as the larger number of extended filaments filter out the small particles that cause turbidity. Whereas poor settleability extends the sludge blanket so that larger flocs are carried out from the sedimentation tank with an increase in both suspended solids and BOD of the final effluent, the major problem associated with bulking is poor sludge compaction. This results in much thinner sludges being returned to the aeration tank with a low MLSS that leads to difficulty of maintaining the desired operational MLSS in the aeration tank with a subsequent fall in effluent quality. Attempts to control the height of the sludge blanket within the sedimentation tank by wasting more sludge than normal results in the MLSS concentration in the aeration tank rapidly declining.

In structural terms, bulking is due to flocs having a strong macro-structure, so much so that filamentous organisms are present in large numbers. In the ideal floc, where the SVI is between 80–120 ml g^{-1} and the final effluent is largely free from suspended solids and turbidity, the filamentous and floc-forming organisms are balanced. The filamentous organisms are retained largely within the floc giving it strength and a definite structure. Although a few filaments may protrude from the floc they are sufficiently scarce and of a sufficiently reduced length not to interfere with settlement. In contrast, the flocs comprising a bulking sludge have large numbers of filaments protruding from the floc, with two types of bulking flocs discernible. Fairly compact flocs with long filaments growing out of the floc and linking individual flocs together (bridging) forming a meshwork of filaments and flocs; or alternatively, flocs with a more open (diffuse) structure, which is formed by bacteria agglomerating along the length of the filament forming rather thin, spindly flocs of a large size. The type of floc formed, the type of compaction and settling interference caused, depends on the type of filamentous organisms present. Bulking is predominantly caused by bacterial species, with bridging caused by type 021N, *Sphaerotilus natans*, type 0961, type 0803, *Thiothrix* sp., type 0041 and *Haliscomenobacter hydrosis*. Open floc structure is associated with type 1701, type 0041, type 0675, *Nostocoida limicola*, and *Microthrix parvicella* (Anon 1979b).

About 25 different filamentous bacteria are known to be able to cause activated sludge bulking. A number of fungi and algae are also known to cause bulking, although fungi and algae are not normally found as dominant organisms in activated sludge (Farquhar and Boyle 1971b). The fungi *Geotrichium candidum* (Hawkes 1963) and *Zoophagus insidians* (Cooke and Ludzack 1958) have been identified (Tomlinson and Williams 1975), and the blue-green alga *Schizothrix calciola* was observed in an activated sludge unit treating a wastewater rich in acetate (Sykes *et al.* 1979). The same bacterial species has been observed world-wide, causing bulking, with approximately 10 bacterial species accounting for at least 90% of all bulking incidents. The frequency of occurrence and the frequency of dominance of filamentous micro-organisms in bulking sludge has been measured in the USA (Richard *et al.* 1981; Strom and Jenkins 1984), in South Africa (Blackbeard and Ekama 1984; Blackbeard *et al.* 1985; Ekama *et al.* 1985), and in Europe (Eikelboom 1977; Wagner 1982; Byron 1987). The most frequently encountered species observed in different countries are summarized in Table 4.23. Using the data on dominant filamentous micro-organisms in bulking sludge, the top 10 species can be compared (Table 4.24): from 525 samples taken from 270 treatment plants in the USA (Richard *et al.* 1984; Strom and Jenkins 1984), from 1100 samples taken from 200 plants in The Netherlands (Eikelboom 1977), 3500 samples from 315 treatment plants in West Germany (Wagner 1982), and 60 samples from 50 plants in South Africa (Blackbeard and Ekama 1984). Sixteen species were recorded in all, with only three species, types 021N, 0041, and *M. parvicella* occurring in the top 10 species from all four countries, and a further

TABLE 4.24 The 10 most frequently occurring filamentous micro-organisms recorded in activated sludge plants in South Africa, the USA, The Netherlands, and West Germany (based on the data in Table 4.23)

Ranks	S. Africa	USA	The Netherlands	W. Germany
1	Type 021N	Type 0092	*Nicardia*	M. parvicella
2	*M. parvicella*	Type 0041	Type 1701	Type 021N
3	Type 0041	Type 0675	Type 021N	*H. hydrossis*
4	*S. natans*	*Nocardia*	Type 0041	Type 0092
5	*Nocardia*	*M. parvicella*	*Thiothrix*	Type 1701
6	*H. hydrossis*	Type 1851	*S. natans*	Type 0041
7	*N. limicola*	Type 0914	*M. parvicella*	*S. natans*
8	Type 1701	Type 0803	Type 0092	Type 0581
9	Type 0961	*N. limicola*	*H. hydrossis*	Type 0803
10	Type 0803	Type 021N	Type 0675	Type 0961

six species were recorded in the top 10 from three countries examined. Using a mean ranking procedure where the top three values only are used, the most frequently occurring species can be identified. In order of occurrence the top eights species are : type 021N, *M. parvicella*, type 0041, *Nocardia*, types 0092, 1701, *S. natans*, and *H. hydrossis*.

According to Jenkins *et al.* (1983), the following filamentous micro-organisms are usually observed in domestic wastewater treatment plants at conventional organic loading rates: *S. natans*, *Thiothrix* I and II, *Beggiatoa* spp., *Nocardia* spp., *N. limicola* II, *H. hydrossis*, types 1701, 021N, and 1863. Plants treating industrial wastewaters or domestic plants operated a low organic loading included: types 0041, 0675, 021N, 0914, 1851, 0803, 0092, 0961, 0581, *Thiothrix* I and II, *Beggiatoa* spp., *M. parvicella*, *Nocardia* spp., *H. hydrossis*, *N. limicola* I, II, and III. Those species infrequently observed include fungi, Cyanophyceae, types 1702, 1852, 0211, 0411, *Bacillus* spp., and *Flexibacter* spp. Detailed descriptions of all these filamentous micro-organisms are given by Eikelboom (1975) and Eikelboom and van Buijsen (1981). There are a number of keys to these micro-organisms including Farquhar and Boyle (1971*a*), Eikelboom (1975), Eikelboom and van Buijsen (1981) and Jenkins *et al.* (1984). The latter two keys are beautifully illustrated. Each identification key is based on a number of morphological features that can be identified using simple staining techniques. For example, the key of Eikelboom and van Buijsen relies on just three staining techniques, the Gram, Neisser, and sulphur storage test (Fig. 4.60). Jenkins *et al.* (1984) have extended this key making use of three more simple stains (India ink reverse, polyhydroxbutyrate (βH), and crystal violet sheath stains) to aid identification (Fig. 4.61, Table 4.25).

There is a direct relationship between the density of filamentous organisms in the mixed liquor and the settling properties of activated sludge (Finstein and Heukelekian 1967; Pipes 1979). This can provide a more sensitive

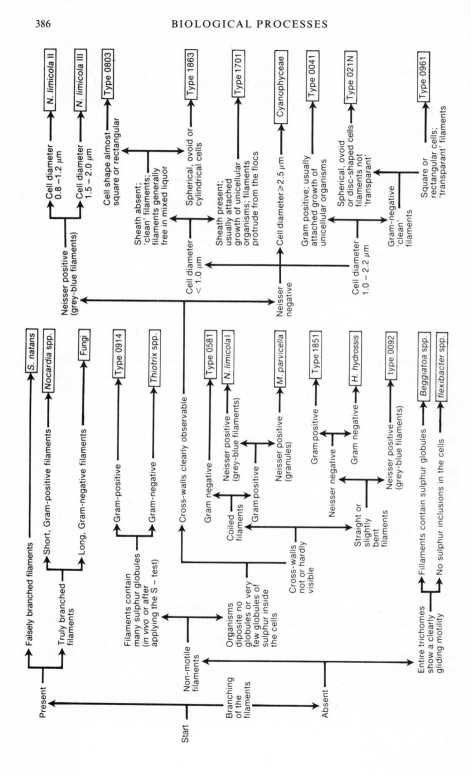

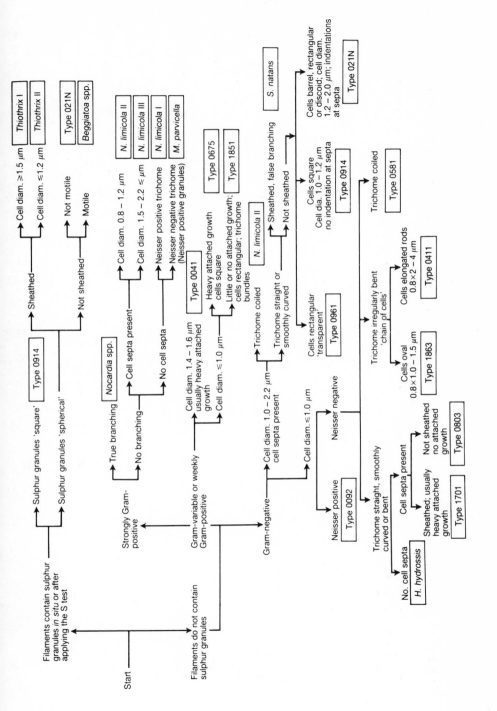

TABLE 4.25 Morphological and staining characteristics of filamentous organisms associated with activated sludge bulking (Jenkins *et al.* 1984)

Filament type	Bright field observation						Phase contrast observation 1000×									Notes
	Gram stain	Neisser stain trichome	Neisser stain granules	Sulphur granules In situ	Sulphur granules S test	Other cell inclusions	Trichome diameter μm	Trichome length μm	Trichome shape	Trichome location	Cell septa clearly observed	Indentations at cell septa	Sheath	Attached growth	Cell shape and size μm	
S. natans	−	−	−	−	−	βH	1.0–1.4	>500	St	E	+	+	+	−	Round-ended rods 1.4×2.0	False branching
Type 1701	−	−	−	−	−	βH	0.6–0.8	20–80	St,B	I,E	+	+	+	++	Round-ended rods 0.8×1.2	Cell septa hard to discern
Type 0041	+,V	−	−,+	−	−	−	1.4–1.6	100–500	St	I,E	+	−	+	++,−	Squares 1.4×1.5–2.0	Neisser positive reaction occurs
Type 0675	+,V	−	−,+	−	−	−	0.8–1.0	50–150	St	I	+	−	+	++,−	Squares 1.0×1.0	Neisser positive reaction occurs
Type 021N	−	−	−,+	−,+	+	βH	1.0–2.0	50–>500	St,SC	E	+	+	−	−	Barrels, rectangles, discoid 1-2×1.5–2.0	Rosettes, gonidia
Thiothrix I	−,+	−	−,+	+,−	+	βH	1.4–2.5	100–>500	St,SC	E	+	−	+	−	Rectangles 2.0×3–5	Rosettes, gonidia

Organism						βH	Diam (µm)	Length (µm)	Trichome shape	Location					Cell shape & size	Comments
Thiothrix II	–	–	–,+	+,–	+	βH	0.8–1.4	50–200	St,SC	E	+	–	+	–	Rectangles 1.0×1.55	Rosettes, gonidia
Type 0914	–,+	–	–,+	–,+	+	βH	1.0	50–200	St	E,F	+	–	–	–	Squares 1.0×1.0	Sulphur granules square
Beggiatoa spp.	–,+	–	–,+	+,–	+	βH	1.2–3.0	100–>500	St	F	–,+	–	–	–	Rectangles 2.0×6.0	Motile: flexing and gliding
Type 1851	+ weak	–	–	–	–	–	0.8	100–300	St,B	E	+,–	–	+	–,+	Rectangles 0.8×1.5	Trichome bundles
Type 0803	–	+	–	–	–	–	0.8	50–150	St	E,F	+	–	–	–	Rectangles 0.8×1.5	
Type 0092	–	–	–	–	+	+	0.8–1.0	20–60	St,B	I	+,–	–	–	–	Rectangles 0.8×1.5	
Type 0961	–	–	–	–	–	–	0.8–1.2	40–80	St	E	+	–	–	–	Rectangles 1.0×2.0	'Transparent'
M. parvicella	+	+	–	–	–	βH	0.8	100–400	C	I	–	–	–	–	–	Large 'patches'
Nocardia spp.	+	+	–	–	–	βH	1.0	10–20	I	I	+,–	–	–	–	Variable 1.0×1–2	True branching
N. limicola I	+	+	–	–	–	–	0.8	100	C	I,E	+	+	–	–	–	Incidental branching; Gram and Neisser variable
N. limicola II	–,+	+,–	–	–	–	βH	1.2–1.4	100–200	C	I,E	+	+	–	–	Discs, ovals 1.2×1.0	
N. limicola III	+	+	–	–	–	βH	2.0	200–300	C	I,E	+	+	–	–	Discs, ovals 2.0×1.5	
H. hydrossis	–	–	–	–	–	βH	0.5	20–100	St,B	E,F	–	–	+	–,+	–	'Rigidly straight'
Type 0581	–	–	–	–	–	–	0.5–0.8	100–200	C	I	–	–	–	–	–	
Type 1863	–	–,+	–	–	–	–	0.8	20–50	B,I	E,F	+	+	–	–	Oval rods 0.8×1–1.5	'Chain of cells'
Type 0411	–	–	–	–	–	–	0.8	50–150	B,I	E	+	+	–	–	Elongated rods 0.8×2–4	'Chain of cells'

Key: + = positive; – = negative; V = variable; Single symbol invariant; +,– or –,+ variable, the first being most observed. Trichome shape: St = straight; B = bent; SC = smoothly curved; C = coiled; I = irregularly-shaped; βH = poly-β-hydroxybutyrate.
Trichome location: E = extends from floc surface; I = found mostly within the floc; F = free in liquid between the flocs.

measure of the onset of bulking. A useful assessment of filament density is the method developed by Sezgin *et al.* (1978) for estimating total extended filament length. Good correlations between total extended filament length (TEFL) and settleability indices have been obtained (Sezgin *et al.* 1978; Palm *et al.* 1980; Lee *et al.* 1982; Baker and Veenstra 1986), with SVI increasing rapidly above 100 ml g^{-1} when the TEFL value was $> 10^7 \mu$m ml^{-1}, resulting in filament to filament or floc to filament aggregates being formed that have loose structures of low densities, which do not settle rapidly (Fig. 4.62). This

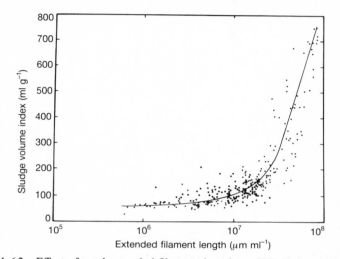

FIG. 4.62   Effect of total extended filament length on SVI (Palm *et al.* 1980).

method is, however, fairly time-consuming compared to more recent assessments of filament length (Walker 1982; Green 1982). For example, in the method developed by Walker (1982) a sample of mixed liquor is placed in a Lund Cell and using a Whipple eyepiece graticule, which superimposes a grid of 100 square on the image, the number of filaments is counted that cross each of the 10 squares of each diagonal. Only filaments that are in the square for at least two-thirds of a single square width are counted (Fig. 4.63). This is repeated five times giving a minimum count of 100 squares. The filament length is expressed in cm per mg of MLSS:

$$\text{Filament length:} \frac{N \times F \times 1000}{\text{MLSS}} \text{ cm mg}^{-1}$$

where N is the total count of filaments per 100 squares and F is the width of one square (cm)/area of 100 squares (cm^2) × cell depth (cm). F will, of course, remain constant for any specific microscope/cell combination at fixed magnification. However, for rapid routine asessment it is possible to use a

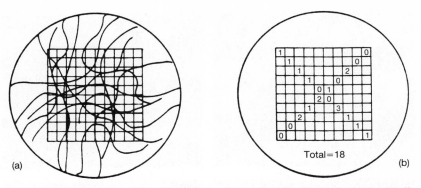

FIG. 4.63 Assessment of filament length using a Lund cell developed by Walker (1982). The diagram shows the microscopic field of view with (a) the eyepiece graticule superimposed on filamentous bacteria, and (b) the counts ascribed to this field by the diagonal counting technique.

subjective scoring index using an abundance scale (e.g. absent, rare, common, frequent or abundant), where the abundance of filaments can be compared to a set of drawings or photographs at × 100 magnification (Farquhar and Boyle 1971b; Rensink 1974; Forster and Dallas-Newton 1980; Eikelboom and van Buijzen 1981; Eikelboom 1982) (Fig. 4.64). This scoring system can also be correlated with a more precise assessment such as total extended filament length.

The major causes of bulking are thought to include low dissolved oxygen concentration due to insufficient aeration capacity in relation to sludge loading, low sludge loading rate in completely mixed systems, septic wastes, nutrient deficiency, and a low pH (< 6.5). By examining large numbers of bulking sludges and relating the dominant bulking organism to various operational parameters, including wastewater characteristics, correlations are gradually becoming evident, and as the data base grows more associations will emerge. Strom and Jenkins (1984) have summarized some of the major associations in Table 4.26, and by examining the filamentous organisms present in a bulking sludge some indications of the cause of bulking can be obtained. Some of these associations are extremely well correlated. For example, fungi indicates that the wastewater contains a strong acid discharge which has reduced the pH of the aeration basin. Type 1701 and *S. natans* both indicate the dissolved oxygen concentration is too low due to a high sludge loading, with type 1701 indicative of even more severe dissolved oxygen limitations than *S. natans*. Although the oxygen concentration may still appear within the critical limits, the presence of these filamentous organisms indicates that the sludge loading is too high for the existing oxygen conditions (Hao *et al.* 1983; Lau *et al.* 1984a,b). Other associations involving *M. parvicella* (Slijkhuis and Deinema 1982; Slijkhuis 1983) and type 021N (Richard *et al* 1984) have also been observed.

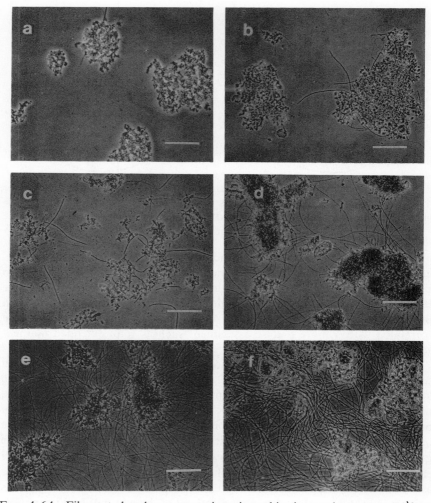

FIG. 4.64 Filament abundance categories using subjective scoring system: (a) few; (b) some; (c) common; (d) very common; (e) abundant; (f) excessive. (The photographs were taken by Jenkins *et al.* (1984) using phase contrast at × 100 magnification. The bar indicates 100 μm.)

Even if the filamentous organism responsible is known, there is no specific control method for bulking sludges. Indeed, many of the control methods suggested are contradictory and others are quite bizarre (Chambers and Tomlinson 1982a). Therefore, operators must try a series of corrective options, starting by correcting the possible cause as indicated by the filamentous organisms present, until the problem disappears (Waller and

TABLE 4.26   The use of dominant filament types as indicators of conditions causing activated sludge bulking (Strom and Jenkins 1984)

Suggested causative conditions	Indicative filament types
Low dissolved oxygen	Type 1701, *S. natans*, *H. hydrossis*
Low f/m	*M. parvicella*, *H. hydrossis*, *Nocardia* sp. types 021N, 0041, 0675, 0092, 0581, 0961, 0803
Septic wastewater/sulphide	*Thiothrix* sp., *Beggiatoa*, type 021N
Nutrient deficiency	*Thiothrix* sp. *S. natans*, type 021N, and possibly *H. hydrossis*, and types 0041 and 0675
Low pH	Fungi

Hurley 1982). Among the large number of possible control options that have been suggested the more commonly employed include:

1. *Controlling the sludge loading ratio:* Normal plant sludge loading is between 0.2–0.45 with bulking problems occurring if < 0.2 or $>$ 0.45 kg kg^{-1}d^{-1}, although this has been shown to depend on the mixing regime with the critical sludge loading in plug flow systems almost twice the value than for completely mixed reactors (Tomlinson and Chambers 1979). In order to avoid bulking, the sludge loading should be maintained within the operational range of 0.2–0.45 kg kg^{-1}d^{-1}.

2. *Nutrients:* A BOD:N:P ratio of 100:5:1 is required to prevent bulking (Pipes 1979). Filaments are able to store essential nutrients when the BOD:N or BOD:P ratios are high, making them available when the nutrient concentration falls. This gives filamentous organisms a strong competitive advantage over floc-forming bacteria.

3. *Oxygen concentration:* To prevent filamentous growth a minimum dissolved oxygen concentration of 2 mg O$_2$l^{-1} must be maintained (Pipes 1979). Although from the work of Palm *et al.* (1980) it is clear that the minimum dissolved oxygen concentration required to prevent bulking is a function of the sludge loading, increasing as the sludge loading increases (Fig. 4.65).

4. *Introduction of anoxic zone:* Although anoxic zones are principally used for denitrification they have been found to improve sludge settleability (Tomlinson and Chambers 1979; Price 1982). The reason for this is unclear but may be due in part to a reduction in the overall oxygen demand within the aeration tank.

5. *Mixing pattern:* Changing the mixing pattern in completely mixed aeration tanks to a more plug flow type, i.e. reducing the degree of dispersion, produces sludges with better settling characteristics (Chudoba *et al.*, 1973; Tomlinson and Chambers 1979; Chambers 1982; Humphries 1982). This also

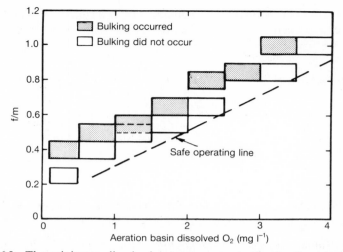

FIG. 4.65 The minimum dissolved oxygen concentration in the aeration tank required to prevent bulking as a function of sludge loading (Palm *et al.* 1980).

reduces the competitive advantage that filaments have over floc-forming bacteria due to their higher surface area to volume ratio which makes them more efficient in obtaining nutrients in conditions of low nutrient and low dissolved oxygen concentration.

6. *Chlorine, ozone, or hydrogen peroxide:* These chemicals are used to selectively destroy filamentous growth but not floc-flowing organisms. Much work has been done on this form of control using chlorine (Tapelshay 1945; Smith and Purdy 1936; Pipes 1974; Jenkins *et al* 1982, 1983, 1984). Many case studies have also been published using hydrogen peroxide (Cole *et al.* 1973; Keller and Cole 1973), although chlorine appears to be more widely used. Such control measures, however, can only be temporary while some other plant modification is carried out to permanently solve the problem. The costs of such treatments is reviewed by Chambers and Tomlinson (1982*b*).

7. *Polyelectrolytes:* Inorganic coagulants, such as lime or ferric chloride, or polyelectrolytes are used to improve flocculation and increase floc strength and settleability (Carter and McKinney 1973; Renskink *et al.* 1979; Thomanetz and Bardtke 1977).

Full details of control methods are given by Chambers and Tomlinson (1982*a*) and Jenkins *et al.* (1983, 1984).

Non-filamentous bulking
Bulking occasionally occurs without filaments being present. This is associated with deflocculation where the dispersed bacteria then produce a

zoogloeal matrix (Pipes 1979; Eikelboom and van Buijzen 1981). Known as zoogloeal or viscous bulking, it is due to a failure of the micro-structure of the floc with excess extra-cellular polymer (ECP) being produced, resulting in the mixed liquor having a slimey or even jelly-like consistency. It can be clearly seen by reverse staining using india ink. When normal flocs are stained in this way the ink penetrates deep into the flocs, whereas the extracellular material prevents the penetration of the ink so the flocs remain unstained.

Denitrification
In the sedimentation tank, the dissolved oxygen is rapidly utilized by the sludge flocs as it separates from the clarified effluent, so the oxygen conditions within the tank are usually anoxic. This can become a problem when the sludge residence time in the sedimentation tank is long and the effluent has been fully nitrified within the aeration tank, so that nitrates and nitrites can be reduced to nitrogen gas or nitrous oxide by denitrifying bacteria (Section 3.4.5). The gas released becomes entrained in the flocs making them buoyant so that they rise to the surface and are carried out of the tank with the final effluent. Known also as rising sludge, the rate of gas production can be extremely high in warmer weather causing significant turbulence within the tank inhibiting normal settlement. Apart from sludge being clearly visible on the surface of the sedimentation tank forming a thin scum, gas bubbles can also be seen rising to the surface. This problem is not associated with sludge structure but rather with operational practice. However, rising sludge will be more serious if flocs have a macrostructure because their irregular shape entrains gas bubbles more readily than the smaller, spherical flocs, and as flocs with macrostructure are considerably larger more sludge can be lost from the system.

Rising sludge can be overcome by ensuring that the settled sludge is not retained too long in the sedimentation tank before being recirculated or wasted. The problem can be the result of poor sedimentation tank design. Flat-bottomed tanks with central sludge take-off are particularly susceptible to denitrification problems as the sludge at the periphery of these tanks may remain there for long periods. If nitrification is not required, the easiest remedy is to prevent nitrification occurring in the aeration tank so that there is no oxidized nitrogen available in the secondary sedimentation tank. This can be done by either decreasing the aeration intensity or increasing the sludge loading to the aeration tank. Where nitrification is required, the oxidized nitrogen can be removed from the effluent by denitrification using an anoxic zone prior to aeration. Denitrification has also been controlled by increasing the sludge recycle rate from the sedimentation tank or ensuring the mixed liquor is well aerated before it enters the sedimentation tank (Chambers and Tomlinson 1981). If the sludge is retained for even longer periods in the sedimentation tank the liquid may become completely de-oxygenated resulting in the sludge becoming anaerobic and decomposing, producing hyrogen sulphide gas and an unpleasant putrescent sludge.

4.2.5   Ecology

Like all biological treatment processes, the activated sludge system relies on a mixed culture of bacteria to carry out the basic oxidation of the organic material present, with higher grazing micro-organisms also present, thus forming a complete ecosystem with various trophic levels. The activated sludge aeration tank is a truly aquatic environment, although the high BOD and high level of bacterial activity make it unlike any natural aquatic habitat. The constant aeration and recirculation of sludge makes it inhospitable for most aquatic species, especially those larger than the smaller mesofauna, such as rotifers and nematodes or those with long life-cycles. The main biological groups present are bacteria, fungi, protozoans, rotifers, and nematodes, with flocs a heterogeneous mixture of them all, together with organic and inorganic material. Other groups such as *Cyclops*, *Aelosoma*, and even larvae of some dipterans are occasionally observed, although they are not major components of the community structure. Algae are also present in the mixed liquor but rarely become established. In contrast to percolating filters (Section 4.1), where the active biomass is attached to a static support medium and the wastewater flows over the surface, activated sludge is a dispersed growth system with the microbial biomass present as discrete flocs that are suspended in the wastewater and mixed together by the aeration system. Due to the physical difference between a static and dispersed growth system fewer trophic levels are present in the activated sludge process (Fig. 4.66). The larger macro-

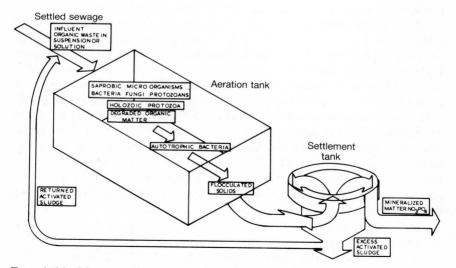

FIG. 4.66   Mass transfer processes in activated sludge showing the major trophic levels (Wheatley 1985).

invertebrate grazers associated with percolating filters are largely absent as they cannot be supported within the floc. Thus, apart from certain protozoans and nematodes, most of the larger grazers are found swimming in the liquid phase between the flocs. A simplified food web for conventional activated sludge is given in Fig. 4.67.

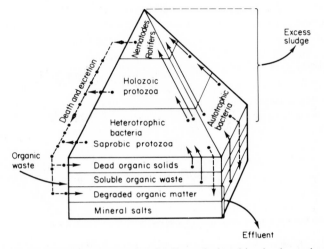

FIG. 4.67 Food pyramid representing feeding relationships in the activated sludge process: ——→, synthesis; −−−→, degradation (Hawkes 1983a).

As the floc ages it becomes colonized by bacteria-feeding organisms, such as ciliate protozoans, nematodes, and rotifers. Also, the proportion of the floc comprised of dead cells or accumulated inert solids, increases, with the living cells clustered in the outer surface of the spongy structure of the floc. Material continues to be adsorbed on to the floc and although complete oxidation is only possible by the living cells, the dead cells remain capable of enzyme secretion. However, as the floc ages the rate of oxidation gradually declines. A reduction in substrate removal with increasing sludge age has been reported by many workers (Keefer and Meisel 1953; Wuhrmann 1956). The slower growing nitrifying bacteria can only become established in the floc if long sludge ages are used (Section 4.2.6). Anaerobic bacteria are largely absent from activated sludge and those present have been introduced from the incoming wastewater or anaerobic activity either in the primary or secondary sedimentation tanks. As the floc increases in size the rate of diffusion of nutrients and dissolved oxygen into and the movement of metabolic waste products out of the floc becomes more difficult. Concentration gradients of both nutrients and oxygen occur throughout the floc, with the centre of larger flocs becoming anoxic or even anaerobic. In theory, it is possible that the activated sludge ecosystem can be completely balanced so that nearly all the available organic matter is utilized by the primary heterotrophic activity, which in turn is used by the grazers so that there is no excess microbial biomass, composed mainly of heterotrophic bacteria, which is surplus to that required to maintain the microbial population density in the aeration tank via the returned sludge. Therefore, the excess microbial biomass must be disposed of separately as unwanted sludge (Fig. 4.68). Under steady-state conditions the growth rate of the micro-organisms will be equivalent to the specific sludge wastage rate (Curds 1971*b*).

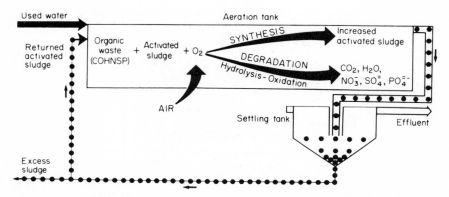

F IG. 4.68   Development and control of biomass in the activated sludge process (Hawkes 1983*a*).

Activated sludge is a complex ecological system made up of species forming several trophic levels, which compete for food resources, with predator–prey/ parasite–host relationships clearly discernible. The microbiology of activated sludge processes has been extensively studied and excellently reviewed (Pike and Curds, 1971; Curds and Hawkes 1975; Hughes and Stafford 1976; Pipes 1966, 1978*a*). Ecological considerations have also been dealt with by Hawkes (1963; 1983*b*), and Pipes (1966).

Bacteria
Bacteria in the activated sludge process are present as individual free-swimming cells dispersed in the liquid phase, floc-forming forms, and as dispersed non-floc-forming bacteria associated with the floc. As already discussed in Section 4.2.1, heterotrophic bacteria form the basis of flocs, which form the basic ecological unit of the activated sludge process, with individual cells agglutinating together or on to filamentous bacteria to form them. The biological condition of the flocs determines the rate of substrate removal, and their physical structure will determine the efficiency in being separated from the clarified effluent in the sedimentation tank. The principal species present will be those most able to reproduce in the activated sludge environment. In

terms of bacteria, the species making up the mixed liquor are significantly different to those present in the incoming wastewater (Dias and Bhat 1964). The process itself selects the most efficient flocculating bacteria by retaining those flocs which rapidly separate from the clarified liquor and settle in the sedimentation tank, which are then returned to the aeration tank to maintain the correct MLSS concentration. Those species associated with flocs or with other settleable material in the sedimentation tank will also be returned to the aeration tank and thrive. The rate at which bacteria multiply is also critical and, in order to survive, their reproduction rate must exceed the rate at which they are removed from the system by the wastage of excess sludge. The type of bacteria present also depends largely on the wastewater being treated, environmental factors in the aeration tank such as pH, temperature, dissolved oxygen, nutrient concentration, degree of turbulence, and the operating factors, in particular, sludge loading and sludge age. For example, high sludge wastage rates or dilution rates in the aeration tank will encourage a high growth rate. This will select faster-growing bacteria, suppress the higher trophic levels, and result in incomplete nutrient removal. Therefore, in high-rate activated sludge processes protozoans and rotifers are generally absent as will be the slower growing bacteria, such as the nitrifying species *Nitrosomonas* spp. and *Nitrobacter* spp. (Section 4.2.6). Dispersed, free-swimming bacteria are also present in large numbers but are constantly being removed from the system by either being discharged with the final effluent or by mesofaunal grazing mainly by protozoans. Earlier studies had assumed that floc formation was due to a particular bacterium *Zoogloea ramigera* (Pike 1975); however, bacteria from a large number of genera are able to form flocs, especially at the low nutrient concentrations associated with wastewater treatment (McKinney and Weichlein 1953; Kato *et al* 1971). The majority of the bacteria isolated from the process are Gram-negative species belonging to the genera *Pseudomonas, Flavobacterium, Achromobacter, Bacillus, Alcaligenes,* and *Micrococcus.* Many other genera are also found, but less frequently (Pike 1975). There is also a wide variety of filamentous bacteria isolated from activated sludge, which are associated with bulking problems (Section 4.2.4). Benedict and Carlson (1971) compared the aerobic heterotrophic bacteria from a laboratory and full-scale domestic plant. They recorded species from 11 different genera (Table 4.27) including a yeast *Debaromyces* sp., which occurred in significant numbers in the full-scale plant.

Bacterial growth kinetics have been used to model the activated sludge process and as bacteria are the most important group in terms of nutrient removal, some idea of bacterial density is often required. Three approaches have been used, direct counting, plate counts, and biochemical assessment of activity. Direct counting using a microscope is difficult because it is impossible to differentiate between living and dead cells, and a total rather than a viable count is obtained. Fluorescent dyes can be used to help identify active cells but other material can also absorb the dye leading to over-estimation. The

TABLE 4.27   Genera of aerobic heterotrophic bacteria and yeasts from activated sludge excluding zoogloeal strains (Benedict and Carlson 1971)

Genera	Laboratory activated sludge	Renton Metro sludge	Genera	Laboratory activated sludge	Renton Metro sludge
Acinetobacter	9[a]	4[a]	Debaromyces	0	4
Alcaligenes	4	2	Flavobacterium	7	1
Bacillus	0	5	Hyphomicrobium	0	2
Brevibacterium	17	7	Microbacterium	2	0
Caulobacter	2	0	Pseudomonas	8	16
Comomonas	7	5	Sphaerotilus	0	3[b]
Cytophaga	1	8	Unidentified	2	12[c]

[a] Number of isolates. [b] Only three colonies picked. [c] Did not grow in Difco nutrient broth or could not be keyed to genus.

counting procedure, using some form of counting cell such as a Helber chamber, is time-consuming and rather tedious, as a large number of replicates are required to reduce the confidence limits of the estimated population density to within a reasonable level. For the routine estimation of viable bacteria, plate counting has been widely adopted and appears far more reliable than direct counting methods. Before the mixed liquor can be plated out on to agar the individual cells have to be released from the floc matrix which is done by homogenization or ultrasonic irradiation (Pike et al. 1972). The use of homogenization increases the bacterial count by up to two orders of magnitude compared with unhomogenized samples. There has been much work done on finding the most suitable medium on which so many different bacterial species will grow (Prakasam and Dondero 1967; Gayford and Richards 1970; Lighthart and Oglesby 1969). Comparative studies have shown that nutrient agar or casitone–glycerol–yeast extract agar (CGY) is preferable to agar made from sewage or other nutrient-rich materials, with an optimum incubation period of 6 d at 22°C (Pike 1975). There is a large discrepancy between direct counts of total bacteria using a microscope and plate counts of viable bacteria in mixed liquors (Pike and Carrington 1972). This suggests that in conventional activated sludge processes, where the aeration tank is operated in the stationary or declining phases of the growth curve under low nutrient conditions, most of the bacteria in the mixed liquor will be either moribund or dead. As expected, the discrepancy between total and viable counts will decrease as the sludge loading rate or the specific sludge wastage rate increases. Biochemical methods of determining bacterial activity vary but as so many other living micro-organisms are present, specific enzyme analysis appears the most favoured, although problems of differentiating between living and dead cells is difficult as dead cells continue to secrete extracellular enzymes (HMSO 1971). The number of viable bacteria estimated to be in mixed liquor from activated sludge units varies between 0.02×10^{10} to 59×10^{10} cells g^{-1}, although a working figure of between $1-5 \times 10^{10}$ cells g^{-1} is normally taken (Banks *et al.* 1976; Takii 1977). The number of bacteria varies according to operational factors such as sludge age, environmental variables, and wastewater characteristics. Numbers of bacteria at different stages during the treatment process are summarized in Table 4.28. The number of viable bacteria in mixed liquor constitutes only between 1–2% of the total biomass.

Fungi
Fungi are rarely present as a dominant organism in activated sludge, although fungal hyphae is often seen associated with flocs. When the pH of the mixed liquor is lowered to below 6.0 then bacteria are inhibited and fungi will begin to dominate. Therefore, in general terms the presence of fungi as a dominant organism indicates acidic industrial effluents. Studies on the occurrence of fungi in activated sludge plants have found them to be rare with *Geotrichium candidum* and *Trichosporon* sp. being the most abundant (Cooke and Pipes 1968; Cooke and Pipes 1970) (Section 3.3.1).

A predacious fungus, which captures and consumes rotifers, *Zoophagous insidians* was identified by Cooke and Ludzack (1958) in a laboratory aeration tank, and by Pipes (1965) in a pilot-scale plant. These fungi can capture a range of mesofauna including protozoans and nematodes. Gray (1985*b*) found both endoparasitic and predatory nematophagous fungi from a number of activated sludge plants in Ireland, where they had a significant role both in floc formation and in the regulation of the nematode population density.

Protozoa
Protozoa are common components in activated sludge with population densities reaching up to 50 000 ml^{-1}, which can represent as much as 5–12% of the dry weight of the mixed liquor (Hopwood and Downing 1965; Pike and Curds 1971). This is similar to the total mass of bacteria, both viable and non-active, in the mixed liquor (HMSO 1971) and protozoans can represent a major proportion of the total biomass. Curds (1975) lists 228 species of Protozoa from activated sludge plants with 70% of them from the class Ciliatea. Flagellates, both Phyto- and Zoomastigophorea, are far more common in percolating filters, being found in small numbers in activated sludge units and being associated with overloaded plants. Although species diversity is dominated by ciliate Protozoa, testate, and naked forms of *Amoeba* sp. can occasionally be dominant numerically (Schofield 1971; Sydenham 1971).

In a survey of 56 activated sludge plants in the UK, Curds and Cockburn (1970*a*) found the commonest species were *Vorticella microstoma* (75% of plants), *Aspidisca costata* (69%), *Trachelophyllum pusillum* (64%), *Vorticella convallaria* (58%), *Opercularia coarctata* (54%), and *Vorticella alba* (38%). The most frequently observed Protozoa recorded by two other workers

TABLE 4.28 Mean numbers of total and viable bacteria recorded at various stages of domestic wastewater treatment (Pike and Carrington 1972)

Source (and number of samples)[b]	Bacterial counts[a]				Viability (%)	Total suspended solids (mg l^{-1})
	In samples (No. ml^{-1})		In suspended solids (No. g^{-1})			
	Total	Viable	Total	Viable		
Settled sewage (46)	5.6×10^8	6.3×10^6	3.0×10^{12}	3.4×10^{10}	1.1	190
Activated sludge mixed liquor, conventional rate (18)	5.9×10^9	4.9×10^7	1.3×10^{12}	1.1×10^{10}	0.83	4600
Activated sludge mixed liquor, high-rate (24)	1.4×10^{10}	2.4×10^8	3.0×10^{12}	5.0×10^{10}	1.7	4800
Filter slimes (18)	6.2×10^{10}	1.5×10^9	1.3×10^{12}	3.2×10^{10}	2.5	54 000
Secondary effluent (16)	5.4×10^7	1.1×10^6	1.9×10^{12}	4.1×10^{10}	2.1	28
Effluents, high rate activated sludge plants (24)	4.8×10^7	1.4×10^6	3.3×10^{12}	1.0×10^{11}	3.0	14
Tertiary effluents (11)	2.9×10^7	6.6×10^4	3.0×10^{12}	6.8×10^9	0.23	9.7

[a] Total counts obtained with Helber counting chamber, viable counts by plate dilution frequency method on CGY agar, incubation for 6 days at 22°C; viability expressed as percentage ratio of viable count to total count.

[b] Samples from sewage works and laboratory pilot plants; high-rate plants are pilot scale, working at loadings of 0.46–2.5 kg BOD removed/kg MLSS day, secondary effluents from nine filters and seven activated sludge plants, tertiary effluents from 10 lagoons, one grass plot.

(Brown 1965; Schofield 1971) are compared to the results of Curds and Cockburn (1970a) in Table 4.29. The protozoan fauna of activated sludge differs from that of percolating filters because of the significant difference between the two environments. For example, *Peritrichida*, mainly vorticellids, are far more common in activated sludge, with a greater diversity of suctorians also recorded (Fig. 4.69). The reason why opercularians are less common in activated sludge is that these stalked protozoans, unlike the vorticellids, do not have a contractile stalk and can be damaged when flocs come into close contact in the aeration tank. However, the static film environment in percolating filters is ideal for them.

Ciliate Protozoa in activated sludge can be categorized according to their habits. Three categories are discernible in the mixed liquor: sessile species, which are attached to the individual flocs (*Vorticella* spp., *Opercularia* spp., *Epistylis* spp.); crawling species, which move over the surface of the floc (*Aspidisca* spp., *Euplotes* spp); and free-swimming species, which live in the liquid phase of the mixed liquor and are not directly associated with flocs (*Paramecium* spp., *Colpidium* spp., *Litonotus* spp.). Whereas the Phytomastiophora are primary producers, the Zoomastigophora compete with the bacteria at the basic trophic level (Fig. 4.67). The holozoic forms of Protozoa, present mainly as ciliates, form the next trophic level. All three groups of ciliates feed predominantly on bacteria, although there are several attached (*Suctoria*) and free-living species (e.g. *Trachelophyllulm pusillum*), which are

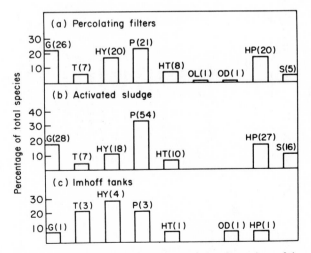

FIG. 4.69 Percentage distribution of species between the orders of the class Ciliatea found in used water treatment processes. G, Gymnostomatida; T, Trichostomatida; HY, Hymenostomatida; P, Peritrichida; HT, Heterotrichida; OL, Oligotrichida; OD, Odonstostomatida; HP, Hypotrichida; S, Suctorida. Numbers in parenthesis indicate actual numbers of species represented in each order (Curds 1975).

TABLE 4.29 The most frequently observed Protozoa in survey of activated sludge plants (adapted from Curds 1975)

Class	Brown (1965)	Curds and Cockburn (1970a)	Schofield (1971)
Phytomastigophorea		*Peranema trichophorum*	
Rhizopodea	*Euglypha alveolata*	Small amoebae *Arcella vulgaris* *Euglypha* sp.	*Amoeba* sp. *Cochliopodium* sp.
Ciliatea	*Aspidisca costata* *Aspidisca lynceus* *Aspidisca robusta* *Chilodonella cucullulus* *Epistylis plicatilis* *Euplotes aediculatus* *Litonotus fasciola* *Vorticella campanula* *Vorticella convallaria* *Vorticella microstoma* *Vorticella nebulifera* var. *similis* *Vorticella striata* var. *octava*	*Aspidisca costata* *Carchesium polypinum* *Euplotes moebiusi* *Opercularia coarctata* *Trachelophyllum pusillum* *Vorticella alba* *Vorticella convallaria* *Vorticella fromenteli* *Vorticella microstoma*	*Aspidisca costata* *Chilodonella uncinata* *Drepanomonas* sp. *Epistylis* sp. *Hemiophrys* sp. *Vorticella convallaria* *Vorticella microstoma*

predatory on protozoans. Although bacteria are primarily responsible for the removal of organic matter, protozoans also play an important role in the purification process as bacterial feeders. This has two major effects. First, it regulates bacterial density, preventing bacteria from reaching self-limiting numbers, both dispersed and floc-forming bacteria, causing enhanced bacterial activity in non-food-limited plants as well as reducing the overall bacterial biomass. Secondly, protozoans clarify the effluent by removing suspended material. When the nutrients are limiting in activated sludge, protozoal grazing will reduce the bacterial population, resulting in an increase in substrate concentration, which will cause either an enhanced nutrient uptake rate by the remaining bacteria or an increase in the substrate concentration of the effluent (Curds 1971b). Stalked protozoans attached to the flocs, and the free-swimming protozoans both feed on dispersed bacteria and organic particulate matter that is in suspension (Curds and Vandyke 1966). This was elegantly demonstrated by Curds et al. (1968), who compared the turbidity and chemical quality of the final effluent from six bench-scale activated sludge plants. When operated without Protozoa, the effluent was turbid with a high BOD and suspended solids concentration. However, when a mixed population of Protozoa was added to three of the plants, leaving the other three as protozoan-free control plants, a significant reduction in the turbidity, BOD, and suspended solids concentration was recorded (Fig. 4.70). Protozoa also feed on pathogenic bacteria which was demonstrated by Curds and Fey (1969) using *Escherichia coli*. They obtained a 95% removal of *E. coli* when Protozoa were present, compared to just 50% removal when absent, removal being due to flocculation and adsorption on to flocs (Fig. 4.71). Under certain food-limiting conditions, the dispersed bacteria may well be in

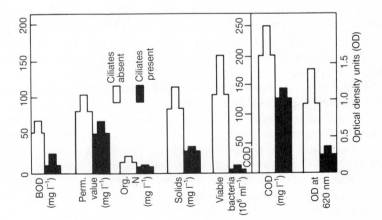

FIG. 4.70   Effluent qualities from experimental activated sludge units operating in the absence and presence of ciliated Protozoa. The 'shoulders' on blocks indicate ranges of means of final effluent quality (Curds *et al.* 1968).

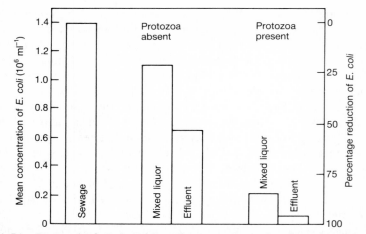

F IG. 4.71 Removal of *Escherichia coli* in experimental activated sludge plants operating in the absence and presence of ciliated Protozoa (Curds and Fey 1969).

competition with the floc-forming bacteria for the soluble substrate present. By removing the dispersed bacteria the protozoans will reduce the competition for the remaining substrate, allowing the floc-forming bacteria to produce larger flocs, which will improve settleability of the mixed liquor as well as reducing suspended material in the effluent, thereby enhancing overall quality (Downing and Wheatland, 1962; Curds *et al.* 1968).

Although no spatial variation in protozoan species is discernible in the completely mixed environment of the activated sludge aeration tank where the sludge is continuously recycled, which is in contrast to percolating filters, temporal succession of species has been well documented. The succession follows a similar pattern in all activated sludge plants as the mixed liquor matures with flagellates dominating initially followed by free-swimming ciliates, crawling ciliates, and attached ciliates (Agersborg and Hatfield 1929; Horosawa 1950; McKinney and Gram 1956; Curds 1966) (Fig. 4.72). Both the crawling and attached ciliates are closely associated with the flocs, and once established their population densities will be maintained by being recirculated from the sedimentation tank with the settled sludge. The wash-out rate for the flagellates and the free-swimming ciliates will be much greater than the other types of protozoans. This association has been shown by a computer model by Curds (1971*b*). Once the mixed liquor has matured, the Peritrichs and crawling species appear to be inversely related (Brown 1965; Curds 1966). As the mixed liquor matures and the protozoan population develops, the effluent quality gradually improves. A new mixed liquor or a plant that is in poor condition producing an inferior quality effluent contains mainly flagellates and rhizopods, with few ciliates present. In contrast, a satisfactory mixed liquor producing a reasonable quality effluent contains mainly ciliates with

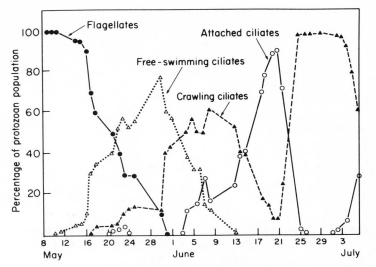

FIG. 4.72 Succession of Protozoa in the MLSS of an activated sludge plant (Curds 1966).

some flagellates and amoebae. The ciliates are dominated by species, such as *Chilodonella* spp., *Colpoda* spp., *Colpidium* spp., *Aspidisca* spp., and certain *Carchesium* spp. and *Vorticella* spp. Activated sludge plants producing nitrified effluents of excellent quality contain very few flagellates or amoebae and are dominated by ciliates, such as *Carchesium* spp., *Vorticella* spp., *Aspidisca* spp., *Loxophyllum* spp., and *Chaenea* spp. (Curds 1975). This association of certain species with the condition of the sludge and effluent quality has led to the use of protozoans to indicate the effluent quality of activated sludge plants. By utilizing the large data base collected in their survey of UK plants, Curds and Cockburn (1970b) have been able to correlate species diversity to effluent quality. Using four effluent quality categories based on BOD, very high (0–10 mg l^{-1}), high quality (11–20 mg l^{-1}), inferior (21–30 mg l^{-1}), and low quality (>30 mg l^{-1}), the frequency of occurrence of each species in the activated sludge plants sampled in the survey was recorded. The frequency of occurrence recorded in each of the four categories were then totalled and a value out of a total of 10 was awarded to each category depending on the proportion of the total occurring in each category (Table 4.30). This gave a weighting to the effluent quality categories that were most associated with a particular species. This was done for all the 288 species identified by Curds and Cockburn (1970a), a full list of which has been published by them (Curds and Cockburn 1970b). Effluent quality of activated sludge plants is then calculated by producing a comprehensive species list and adding up the various weightings for each species in each of the four effluent quality categories. The category with the highest total indicates the predicted

TABLE 4.30   Percentage frequency occurrence of some common Protozoa in activated sludge plants producing final effluents in four BOD ranges (Curds and Cockburn 1970b)

BOD range (mg l^{-1})	Frequency of occurrence (%) and points awarded (in brackets)			
	0–10	11–20	21–30	> 30
Vorticella convallaria	63 (3)	73 (4)	37 (2)	22 (1)
Vorticella fromenteli	38 (5)	33 (4)	12 (1)	0 (0)
Carchesium polypinum	19 (3)	47 (5)	12 (2)	0 (0)
Aspidisca costata	75 (3)	80 (3)	50 (2)	56 (2)
Euplotes patella	38 (4)	25 (3)	24 (3)	0 (0)
Flagellated protozoa	0 (0)	0 (0)	37 (4)	45 (6)

TABLE 4.31   Association ratings used to predict the final effluent quality from an activated sludge plant (Curds and Cockburn 1970b). In this example the highest points are awarded to BOD range 0–10 mg l^{-1}, which corresponds to the measured BOD of the final effluent which was 8 mg l^{-1}

Protozoa in sludge	Effluent BOD ranges			
	0–10	11–20	21–30	> 30
Trachelophyllum pusillum	3	3	3	1
Hemiophrys fusidens	3	4	3	0
Chilodonella cucullulus	4	4	1	1
Paramecium trichium	4	3	2	1
Vorticella communis	10	0	0	0
Vorticella convallaria	3	4	2	1
Vorticella fromenteli	5	4	1	0
Vorticella microstoma	2	4	2	2
Opercularia coarctata	2	2	3	2
Carchesium polypinum	3	5	2	0
Zoothamnium mucedo	10	0	0	0
Aspidisca costata	3	3	2	2
Euplotes moebiusi	3	3	3	1
Euplotes affinis	6	4	0	0
Euplotes patella	4	3	3	0
Total points	65	46	28	11

quality (Table 4.31). Curds and Cockburn (1970b) tested this method on their original data and on a further 34 plants not included in the original survey and found the predicted effluent quality was correct in 85% and 83% of the cases respectively. A lower rate of success has been achieved by the author who found that the prediction success rate was improved by weighting the scores of individual species obtained from the list compiled by Curds and Cockburn according to their relative abundance in the mixed liquor. In this way, in a very good quality mixed liquor the presence of a poor quality indicator species in low numbers is outweighed by an abundant good quality indicator. However, the method of evaluated plant performance using the mixed liquor fauna proposed by Curds and Cockburn is limited by two factors. The technique relies on identification of ciliate species, which requires some degree of specialist taxonomic expertise, also, the results will only suggest the BOD of the effluent. To some extent this has been overcome by Poole (1984), who compared nitrifying and non-nitrifying activated sludge systems, using a greater range of ciliate taxa including some non-ciliate protozoan and some metazoan groups. From this he was able to identify a number of reliable indicators of performance such as the holotrich *Prorodon* spp., shelled amoebae, and the Monogononta rotifers all of which are associated with low final effluent $BOD_{(ATU)}$ and ammonia–nitrogen concentrations with a high MLSS concentration. Conversely, the peritrichs, *Vorticella microstoma*, *Opercularia coarctata*, and *Opercularia microdiscum* are associated with high final effluent $BOD_{(ATU)}$ and ammonia–nitrogen concentrations and a low MLSS concentration.

The bacterial–protozoal relationship has been modelled by several workers (Curds 1971b; Curds 1973a,b; Canali et al. 1973), but although these models are of considerable interest ecologically, they do not appear to have been utilized in plant design or operation. This indicates perhaps that designers still regard the activated sludge process as an essentially bacterial process and model it in those terms, rather than as a more dynamic ecosystem functioning on several different trophic levels. In terms of modelling, it is certainly easier to consider it as a single trophic level system.

Other groups
The largest grazers in the activated sludge process are the nematodes and rotifers. Little is known of their role in the activated sludge process and generally they represent only a minor part of the microbial biomass. Nematodes are not abundant in mixed liquor as there is no suitable niche for them in the aeration tank. Also, the population doubling time is so long that the normal HRT and sludge ages of conventional plants do not allow nematodes to develop within the process, making them of little importance (Chaudhuri et al. 1965; Schiemer 1975). Rotifers are far more common in activated sludge than nematodes, with a wide diversity of species recorded (Godeanu 1966). They appear to have a number of important roles. For example, they break up flocs into smaller particles, which encourages new floc formation as well as contributing to floc formation directly by the production of faecal pellets consisting of undigested material held together by mucus, which form the basis of new flocs. Also, they clarify the effluent by removing dispersed bacteria which are in suspension. They appear more efficient than protozoans in filtering suspended material, not only because of their size but also because of their strong ciliary currents (Calaway 1968). Due to the longer development time, rotifers, unlike protozoans, are generally more abundant in

extended aeration plants or other processes with long sludge ages (Doohan 1975).

Other mesofauna are rarely recorded, although Copepods and the annelid *Aelosoma* sp. are both seen occasionally. Dipteran larvae are very occasionally recorded, but they develop most probably in the slime covering the walls of the aeration tank and are found in the mixed liquor when they become dislodged from their more stable environment.

4.2.6   Nitrification

Nitrifying bacteria have much lower specific growth rates than heterotrophic bacteria, therefore nitrification in the activated sludge process is dependent on the sludge age of the microbial biomass (Marais 1973; Jones and Sabra 1980). Like all species of micro-organisms, the nitrifying bacteria will only be able to grow and reproduce within the aeration tank if their specific growth rate in activated sludge is greater than the specific sludge wastage rate. Using *Nitrosomonas*, as it is the rate-limiting species in the two-stage nitrification process (Section 3.5.2), the reaction can be summarized under steady-state conditions as:

$$\frac{C_N - C_{NO}}{C_N} > \frac{\Delta S}{S}$$

where C_N is concentration of *Nitrosomonas* in the effluent mixed liquor, C_{NO} the concentration of *Nitrosomonas* in the inlet mixed liquor, ΔS the increase in MLSS, and S the MLSS at the inlet, so that $\Delta S/S$ is the specific sludge wastage rate. In operational terms, nitrification can only be expected below a critical sludge loading (f/m) rate of 0.15 kg kg^{-1}d^{-1} or a sludge age of >4 d. This may pose a problem for the operator because in order to provide sufficient time for nitrification to occur, the sludge age may no longer be optimal for settlement. Settlement may also be seriously affected if denitrification occurs in the sedimentation tank causing the sludge to rise. In this case, denitrification may be necessary by the provision of an anoxic unit before the aeration tank (Section 4.4). Downing *et al.* (1964) found that as the MLSS concentration increased there was a significant decrease in the period of aeration required to achieve nitrification (Fig. 4.73), so the greater the MLSS the better the degree of nitrification. The rate of ammonia oxidation in activated sludge is approximately 0.5–3.0 mg g^{-1}h^{-1} compared to 250 mg g^{-1}h^{-1} in pure culture studies (Painter 1977). Therefore, nitrifying bacteria comprise only a very small proportion of the microbial biomass, no more than a few per cent. Environmental factors are also important in nitrification once the bacteria have become established. For example, Painter (1978) suggests that the critical dissolved oxygen concentration should not be less than 2.0 mg O_2l^{-1} in the aeration tank in order to ensure that the oxygen

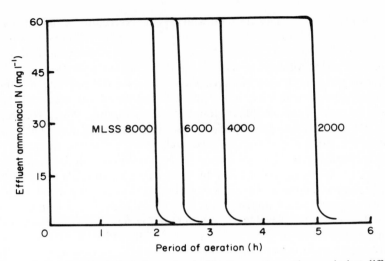

F IG. 4.73   Predicted relations between nitrification and aeration period at different
MLSS concentrations (Downing *et al*. 1964).

concentration for nitrifiers within the sludge floc should not fall below the
inhibitory 0.3 mg $O_2 l^{-1}$. The extra aeration costs to maintain this high critical
oxygen concentration as well as satisfying the high oxygen demand of
nitrification can be considerable. Linked to the larger tank capacity that is
required due to the longer aeration period compared to that needed for
carbonaceous oxidation only, the overall cost of nitrification is very high. The
rate of growth of the nitrifying bacteria increases considerably with
temperature over the range 8–30°C, with *Nitrosomonas* having a 9.5%
increase in growth rate per °C rise. This can be important in terms of the
aeration period, as the minimum period of aeration for nitrification can be
reduced by 50% if the temperature is increased by 7°C over the range 7–25°C
(DoE 1963*a*). It has been demonstrated that at high temperatures ($>20°C$)
nitrification is inevitable, even with short sludge ages. The minimum sludge
age required to achieve complete nitrification can be calculated by:

$$(SA)_T = 3.05 \times (1.127)^{20-T}$$

where *SA* is the minimum sludge age at temperature *T*. The effect of other
environmental variables, including pH (Wild *et al*. 1971; Haug and McCarty
1972) and inhibitory substances (Downing *et al*. 1964), has already been
discussed in Section 3.5.2.

In general, therefore, nitrification favours those biological processes that
are low-rate and have long MCRTs, which is associated mainly with attached
growth systems, such as percolating filters and fluidized beds. Although
extended aeration systems such as the Pasveer ditch and carrousel-type
systems have long MCRTs and so achieve a high degree of nitrification

(Stensel *et al.* 1978; Pay and Gibson 1979), conventional and high-rate activated sludge processes usually do not nitrify their effluents. In order to produce a fully nitrified effluent, a completely mixed conventional system requires a sludge loading of <0.25 kg kg^{-1}d^{-1} with a sludge age >7 d at 18°C. In open pure oxygen systems, good nitrification is possible, whereas in closed systems, nitrification is suppressed because of the accumulation of carbon dioxide in the gas phase above the mixed liquor, which reduces the liquid pH to 6, inhibiting the nitrifying bacteria. In terms of cost–effectiveness, nitrification can be carried out separately by a nitrifying filter after the activated sludge stage. This allows a smaller aeration tank to be used, reduces the aeration costs, reduces the sludge age, and also prevents denitrification occuring in the sedimentation tank ensuring good sludge separation. The use of a two-stage activated sludge process using two separate aeration tanks, one for carbonaceous removal and the second for nitrification has also been examined (Sutton *et al.* 1977; Young *et al.* 1979), although the system is less reliable and robust than nitrifying filters. Nitrification in the activated sludge process has been reviewed in considerable depth by Barnes and Bliss (1983).

Further reading

General: Hawkes 1983*a*; Bode and Seyfried 1984; Cech *et al.* 1985; Institute of Water Pollution Control 1987*b*; Chambers and Jones 1988.
Extended aeration: Barnes *et al.* 1983; Matsui and Kimata 1986.
Advanced systems: Hemming *et al.* 1977; McWhirter 1978*a*; Cox *et al.* 1980.
Sludge problems: Pipes 1979; Eikelboom and van Buijsen 1981; Chambers and Tomlinson 1982*b*; Jenkins *et al.* 1984.
Ecology: Curds and Hawkes 1975; Hawkes 1983*a,b*.
Bacteria: Pike 1975.
Fungi: Tomlinson and Williams 1975.
Protozoa: Curds 1975; Madoni and Ghetti 1981.
Nitrification: Poduska and Andrews 1975; Barnes and Bliss 1983.

4.3 STABILIZATION PONDS

Ponds have been widely used as a method of sewage disposal since ancient times. It is possible that castle moats, which received excrement directly from garderobes built into the walls, as well as night soil thrown from the battlements, although primarily defensive devices were also effective oxidation ponds (Porges and Mackenthun 1963). Because such ponds were highly productive they were often utilized for fish culture.

The development of waste stabilization ponds as a secondary treatment process has been largely accidental, with ponds initially constructed as simple sedimentation basins or as emergency holding tanks at treatment plants. It is only relatively recently that the design and operational criteria needed to successfully operate ponds has been established (Environmental Protection Agency 1983). They are now accepted as a major treatment process and are used throughout the world, serving populations ranging from < 1000 to > 100 000 (Gloyna 1971). Ponds are a popular alternative to other biological treatment systems in countries where there is plenty of sunshine, and where land is both cheap and readily available. They are particularly favoured in Australia (Parker *et al.* 1950, 1959); Central and Southern Africa (Stander and Meiring 1965; Marais 1970); India (Arceivala *et al.* 1970); Israel (Watson 1962) the USA (Oswald 1963*a*; Porges and Mackenthun 1963), and Canada (Townshend and Knoll 1987). They are becoming increasingly popular in Europe, especially in West Germany and France where 2000 and 1500 waste stabilization ponds are in operation respectively. In Europe they are used primarily for treating domestic wastewater for small communities (< 2000 PE), while larger sized plants (> 10 000 PE) are used for tourist areas where there is a large increase in population during the summer (Vuillot and Boutin 1987; Gomes de Sousa 1987). They are especially popular in the USA, where there is some 3500 treatment plants using waste stabilization ponds serving 7% of the population. However, as in other countries, they are used predominantly in rural areas, and in the USA, 90% of the communities served have a population equivalent of less than 10 000 (Hammer 1977). In Canada, there are over 1000 stabilization ponds in operation representing approximately half of the wastewater treatment systems in the country. Dependance on ponds increases to over 70% in Alberta, Manitoba, Saskatchewan, the Northwest Territories, and the Yukon Territory (Townshend and Knoll 1987).

The terminology and classification of ponds used in the literature is somewhat confused. Hawkes (1983*a*) classifies a waste stabilization pond as:

any natural, or more commonly, artificial lentic (i.e. standing) body of water in which organic wastes (either crude sewage, settled sewage, organic and oxidizable industrial effluents or oxidised sewage effluents) are treated by natural biological, biochemical and physical processes commonly referred to as 'self-purification' or 'stabilization'.

They are obviously similar to the activated sludge process but differ in that ponds have: (i) a much longer retention period; (ii) a lower loading rate; (iii) a less active microbial biomass; (iv) require less aeration due to a lower oxygen demand from the biomass; and (v) the lower aeration results in less mixing and agitation with the result that particulate solids settle and form a sludge layer in which anaerobic breakdown occurs. Stabilization ponds are most conveniently classified by the type of biological activity that occurs, for example, whether breakdown is primarily anaerobic or aerobic and the importance of algae in the process (Fig. 4.74).

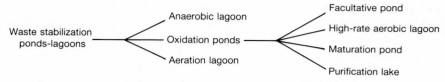

FIG. 4.74   The major types of waste stabilization ponds.

4.3.1   Anaerobic ponds and lagoons

Treatment in lagoons, which are designed to be predominantly anaerobic, relies on the development of a biologically active sludge layer. The sludge takes several months to build up before maximum biological activity is reached. The design criteria for anaerobic lagoons are different compared with other stabilization pond systems; the main difference being depth. Oxygen transfer through the air–water interface is not important in anaerobic ponds and, in fact, is undesirable. Therefore, deep basins are used up to 4.5 m in depth in order to reduce the surface area to volume ratio minimizing reaeration and heat loss. Three identifiable zones are observed in lagoons, the scum layer, the supernatant layer that contains about 0.1% volatile solids, and the sludge layer with 3–4% volatile solids (Fig. 4.75).

The scum or grease layer, which forms on the surface of the lagoon can become 40–60 mm thick and has a number of important functions. It insulates the pond, thus preventing heat loss, suppresses odours, and maintains anaerobic conditions in the supernatant by eliminating oxygen transfer between the air–water interface. Although the inlet pipe can be at the surface (Hawkes 1983c), anaerobic lagoons function most efficiently if the influent wastewater enters near the bottom of the basin. This prevents short-circuiting of the liquid and ensures that it mixes with the microbial solids in the active anaerobic sludge layer. The flow pattern of the liquid and the constant release

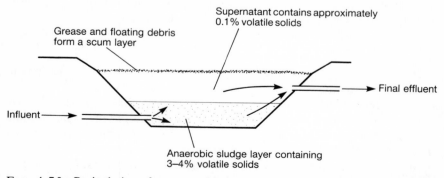

FIG. 4.75   Basic design of an anaerobic lagoon used for treating meat processing wastewater.

of methane and carbon dioxide resulting from degradation of organic matter ensures that some sludge particles remain in suspension with the wastewater. In this way, decomposition of organic matter continues in the liquid phase as well as in the sludge blanket, and a high mixing efficiency results in a high rate of BOD removal (Parker *et al.* 1950). The discharge pipe is located just below the scum layer at the opposite end of the basin to ensure an upward flow pattern. This encourages the bacterial flocs to settle out of suspension, ensuring a clarified effluent and also eliminating the need for sludge return by retaining the microbial biomass in the lagoon.

Anaerobic lagoons are generally used as a preliminary treatment for strong organic wastes. This means that the lagoon only partially stabilizes the wastewater and secondary aerobic treatment is required before discharge to natural waters, either in the form of facultative ponds in series, or activated sludge. However, anaerobic treatment can considerably reduce the organic loading to secondary treatment units, thus reducing the secondary treatment capacity required. If the subsequent aerobic treatment process is a facultative pond, then the sludge will be less likely to rise to the surface in warm weather as the sludge has already undergone anaerobic breakdown. Most industrial ponds operate as anaerobic lagoons and the process is widely used for the treatment of slaughterhouse waste, sugar, pulp, and food-processing wastewater (Porges and Brackney 1962; Barnes *et al.* 1984; Vuillot and Boutin 1987). Wastewaters most amenable to anaerobic breakdown in lagoons have a high organic strength, are rich in fats and protein, have a relatively high temperature, are free from toxic materials, especially heavy metals, and contain sufficient biological nutrients. Slaughterhouse and meat processing wastewaters are particularly suitable for treatment by anaerobic lagoons as they have a high BOD ($\simeq 1400$ mg l^{-1}), grease content ($\simeq 500$ mg l^{-1}), temperature ($\simeq 28°C$), and a neutral pH. Minimum pretreatment is required prior to discharge to an anaerobic lagoon, which should normally be limited to blood recovery, screening to remove coarse solids and dissolved air flotation or skimming to remove excessive grease. If pretreatment is too extensive then insufficient grease may remain in the wastewater to form a scum when discharged to the lagoon. Domestic sewage is not suitable for treatment by this method because of a relatively low temperature, BOD concentration, and grease content. Even then sewage has a high industrial content, anaerobic treatment often leads to serious odour problems due to lagoons becoming facultative instead of strictly anaerobic.

The design loadings for stabilization ponds are normally measured in BOD per unit area (BOD m^{-2}), but as light is not an important factor in anaerobic lagoons, loading is expressed as BOD per unit volume (BOD m^{-3}) or as a liquid retention time. Hammer (1977) gives mean loading values of 320 g BOD m^{-3}d^{-1} at a minimum temperature of 25°C to achieve 75% removal at a minimum retention time of 4 days. Similar figures are given by Gloyna (1971) who cites examples where the retention time ranges from 5 to 19 days for strong organic industrial wastewaters. The major advantages of an anaerobic system are that less sludge is produced compared with an aerobic system and no aeration equipment or power supply is required. The main operating limitations of the process are the temperature and strength of the wastewater. The temperature of the lagoon controls the BOD removal and below 15°C, purification will be due to physical settlement only with no anaerobic breakdown of organic matter occurring. In temperate climates, anaerobic lagoons act as settling tanks during the winter months and the accumulated sludge is subsequently degraded during the warmer summer months. This has been demonstrated by Parker *et al.* (1950) who recorded a 65–80% BOD removal in lagoons during the summer with a retention period of 1.2 days which fell to 45–65% removal in the winter even after 5 days retention. Therefore, special attention should be paid to those factors affecting the temperature of the lagoon such as ambient and wastewater temperature and insufficient scum development for insulation. Low BOD and grease content of wastewater will also reduce scum development and protection from wind is important in preventing mixing and the resultant break up of surface scum.

Under normal operating conditions, no odours are released due to complete anaerobiosis and adequate scum cover. However, problems of odour because of sulphide formation in low temperatures or high sulphate in the supply water can occur. Sulphate can also be a problem in some papermill wastes that use alum as a sizing agent. The odour nuisance can be alleviated by recirculating up to 40% of the flow from the secondary aerobic treatment system (Eck and Simpson 1966). However, in some installations a sulphide oxidation process has to be installed near the lagoon. No odour should arise from a lagoon if the liquid temperature is above 20°C. The scum layer can become a breeding ground for a variety of flies and in warmer climates this can become a serious nuisance. The operation of anaerobic lagoons in series is not recommended as it is difficult to maintain an adequate scum cover on the second lagoon. The reduced loading results in the surface layer of the lagoon becoming aerobic so that the system becomes facultative rather than strictly anaerobic. Under-loaded lagoons will also tend to become facultative ponds.

4.3.2 Oxidation ponds

The oxidation pond is designed to provide aerobic breakdown of organic matter in wastewater, with a significant proportion of the dissolved oxygen provided by photosynthetic algae. Three types of stabilization ponds fall into this category: facultative ponds; high-rate aerobic lagoons; and maturation ponds. All three are different from each other and are treated separately below. A fourth type of oxidation pond, the purification lake, is also considered.

Facultative ponds

Facultative stabilization ponds are characterized by having an upper aerobic and a lower anaerobic zone, with active purification occurring in both. They are able to treat completely both crude sewage and organic industrial wastewaters, and are the commonest type of stabilization pond in use. As organic matter enters the basin, the settleable and flocculated colloidal matter settles to the bottom to form a sludge layer where organic matter is decomposed anaerobically. The remainder of the organic matter, which is either soluble or suspended, passes into the body of the water where decomposition is mainly aerobic or facultative, although it is occasionally anaerobic. The principles of the pond are summarized in Fig. 4.76. Organic matter in solution or suspension is broken down principally by aerobic and facultative bacteria, which releases nitrogen and phosphorus compounds and carbon dioxide. Algae use these inorganic compounds for growth using sunlight for energy and releasing oxygen into solution. The dissolved oxygen is then utilized by the aerobic bacteria, thus completing the symbiotic cycle. Oxygen is also introduced by natural oxygen transfer, the rate of which is increased by turbulence caused by wind action.

In the sludge layer, the settled solids are anaerobically broken down, with methane, nitrogen, and carbon dioxide being released together with a variety of soluble degradation products. The gases escape to the atmosphere, and in a facultative pond up to 30% of the BOD load can be dissipated as gas production. The soluble degradation products, such as ammonia, organic acids, and inorganic nutrients are also released, and subsequently oxidized

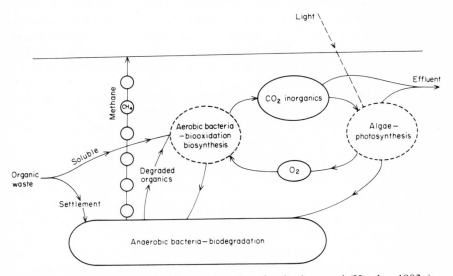

FIG. 4.76   Basic biological interactions in a facultative pond (Hawkes 1983*a*).

aerobically in the water layer. The hydrogen sulphide released from the deposited sludge is also oxidized in the water zone, thus preventing odours being released. Anaerobic degradation is temperature-dependent and no significant activity occurs in the sludge below a water temperature of 17°C. A four-fold increase in activity occurs in the sludge with every 5°C rise in the temperature, over the range of 4°–22°C (Gloyna 1971). Digestion rate of the sludge, expressed as the rate of gas production (K_s), at a particular temperature (T) can be estimated by:

$$K_s = 0.002 \ (1.35)^{-(20-T)}$$

Aerobic and facultative bacteria are the primary decomposers in the pond system. Although fungi are also present in ponds (Cooke 1963), they are not important and are thought to be restricted because of the high pH caused by the photosynthetic activity of the algae (Arceivala et al. 1970). The dominant bacteria found are similar to those isolated from other aerobic treatment systems, and belong principally to the genera Pseudomonas, Achromobacter, and Flavobacterium. Apart from obligate aerobes, other types of bacteria such as coliform bacteria, methane bacteria, sulphate reducers, and the purple sulphur bacteria, such as Thiocapsa floridana and Chromatium vinosum, are all common (Holm and Vennes 1970; Pike 1975). Laboratory-based studies using experimental ponds have demonstrated that the degree of illumination has no effect on bacterial activity and that BOD removal is related to bacterial density. According to Uhlmann (1969), the amount of solids synthesized, however, which was largely algal biomass, was proportional to the degree of illumination and an increase in algal biomass, resulted in an increase in nitrogen and phosphorus being incorporated into the sludge. In these experimental ponds, oxygen was never limiting, even in complete darkness, whereas oxygen is normally a limiting factor in full size ponds. Therefore, the full importance of algae in the euphotic zone was not demonstrated. The oxygen for aerobic degradation is mostly supplied by the photosynthetic activity of the phototrophs.

Although there is a succession of dominant algal species during the year, generally only one or two species will be dominant at any one time in a facultative pond. Porges and Mackenthun (1963) found a striking similarity between the algal species present in facultative ponds throughout the US and concluded that geographical location had little effect on speciation. The most commonly recorded genera are: Chlorella, Scenedesmus, Chlamydomonas, Micractinium, Euglena, Ankistrodesmus, Oscillatoria, and Microcystis. The dominant algal species is determined by the organic loading with those algae able to tolerate anaerobic conditions being recorded in ponds receiving heavy organic loads, e.g. Chlamydomonas sp. and Euglena sp. However, more recent research indicates that photosynthesis of many of the algae commonly found in ponds is inhibited by high ammonia conditions. The oxygenation capacity of different algae varies appreciably. For example, blue-green algae, e.g.

Anabaena variabilis, *A. cylindrica*, *Phormidium faveolarum*, and *Calothrix membranacea*, are less efficient than green algae in ponds, with *Chlorella* sp. and *Scenedesmus* sp. being desirable because of superior oxygenating capability. There is also some evidence that certain algae are capable of growth by photo-assimilation of organic compounds. Utilization of organic matter as a supplementary source of carbon normally only occurs when light or carbon dioxide are limited (Hawkes 1983c).

However, the main function of algae is as phototrophs, producing oxygen to maintain the aerobic condition of the pond. A supplementary role, but a very important one, is the removal of plant nutrients, such as nitrogen and phosphorus. The facultative pond is the most effective of all the biological treatment processes in removing these nutrients, and thus reducing eutrophication in receiving waters, without tertiary treatment. Apart from nutrient uptake to satisfy the normal nutritional requirements of algae, some species are able to take up nutrients, especially phosphorus, far in excess of their own requirements, which is known as 'luxury uptake'. Nutrients are also precipitated out of solution as a consequence of the pH change brought about by photosynthesis, which reduces the concentration of carbon dioxide in the water. Above pH 8, phosphates are precipitated out as calcium phosphate and at higher pH values nitrogen can be lost as ammonia. However, above pH 9 the conditions are no longer optimal for normal aerobic and facultative bacterial activity.

The purple photosynthetic sulphur bacteria of the family Thiorhodaceae can occur in significant numbers in facultative ponds. They oxidize sulphides using carbon dioxide as a hydrogen acceptor and therefore no oxygen is liberated. Unlike other bacteria, they do not contribute to the breakdown of organic matter, they synthesize it depositing sulphur granules in the cell. However, by reducing the sulphide concentration in the pond they do alleviate odours. As the group requires light, anaerobic conditions and the presence of sulphides, which are produced from protein degradation in the anaerobic sludge layer, they occur at the lowest depth in the pond where light can penetrate and as near to the anaerobic sludge blanket as possible. The photosynthetic bacteria generally utilize longer wavelengths than the green algae and so are found at depths below the algae zone. However, as the population density of the photosynthetic bacteria increase, they extend to the upper layers excluding the algae due to the increased turbidity. It is at this stage that the pond will change from the normal green to brown colour to pink, red, or brown (Pike 1975).

Unlike the other aerobic biological treatment processes, protozoans, except the phototrophic flagellates, do not appear to have a significant role in the process (Curds *et al.* 1968). Protozoa are unable to reduce significantly the number of algae in the final effluent and are out-competed for the dispersed bacteria by the filter-feeding Cladocera which also feed on the protozoans. Ponds have a range of zooplankton organisms including Rotifera, Copepoda, and Cladocera. However, Cladocera are by far the most abundant group found in facultative ponds although Rotifera and Copepoda dominate in maturation ponds.

Cladocera are found in ponds with a retention period in excess of 10 days, and under suitable conditions can develop large populations, which can significantly affect the operation of the pond, not always beneficially (Tschortner 1967). The cladoceran population affects the pond by three activities: (i) by feeding off the bacteria; (ii) by feeding off the algae; and (iii) by the formation of boluses. By feeding on bacteria and suspended particulate matter, Cladocera contribute directly to the stabilization process as well as clarifying the effluent by reducing the density of dispersed bacteria (Uhlmann 1969). Although they may help to maintain the bacterial population in an actively dividing phase, thus enhancing the rate of carbon metabolism in the pond, Cladocera do not appear to significantly reduce bacterial numbers overall (Loedolff, 1965). The rejection of food in the form of settleable boluses is probably more important in the clarification process than direct predation, especially when turbidity is high. Most cladocerans in oxidation ponds are filter feeders. Water is forced, by the rhythmic beating of numerous flat limbs, to pass between barbs on the setae fringing the limbs which act as a filtering apparatus. Particulate suspended matter, including bacteria and even colloidal solids, are filtered out of the water and collect in a food groove where it is concentrated before ingestion. If excess food is collected, which is normally the case when the pond is turbid, it is ejected from the food grooves as a bolus and rapidly settles because of its dense nature. Differences in the distance between the setae and barbs determine the size of particles collected by each species. For example, *Moina dubia*, which is a bacterial feeder, is more efficient at straining the bacteria *Escherichia coli* than *Daphnia magna*, an algal feeder, because of the difference in the coarseness of filtering setae (Loedolff 1965).

Although some Cladocera readily utilize small unicellular algae, such as *Chlorella*, it is doubtful if any species can feed on the larger algae such as diatoms, blue-green algae, or the filamentoids to any great extent. Unlike the filter feeding Cladocera, copepods seize and bite their food, being capable of utilizing even filamentoids. Effects of algal grazing can be beneficial by reducing the density of algal cells in the effluent and by reducing the biomass. Alternatively, it can be detrimental as reduced algal density will reduce the rate of reaeration of water, especially at higher temperatures when bacterial oxygen demand is greatest and solubility of oxygen is at its lowest. In India, blooms of Daphnids feeding on *Chlorella* have resulted in low dissolved oxygen concentrations due to reduced photosynthesis (Lakshminarayana 1965; Bodpardikar 1969). However, the Daphnid population was controlled by the addition of lime. The role of Cladocera and the other zooplankton in controlling algae is still unclear. Seasonal fluctuations in population and species dominance have been identified, and a general decrease in cladoceran numbers observed in the winter. It has been proposed that seasonal fluctuations in solar radiation and temperature, which affects algal activity, subsequently affects the cladocerans by altering the oxygen balance, pH, and food supply. A more detailed summary of the microbial ecology and the major biochemical interactions that occur in facultative ponds is given in Fig. 4.77.

Suggestions have been made to utilize the zooplankton either by harvesting the cladocerans for sale as fish food or by incorporating fish into the pond to feed directly on Cladocera, thus reducing the biomass and providing a more readily harvested end-product. Work in South Africa has shown that the fish *Tilapia mossambica* is particularly successful in facultative ponds as it feeds not only on the zooplankton and detritus but also controls mosquitoes and other flies.

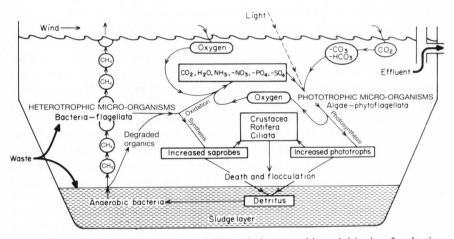

FIG. 4.77   Summary of the heterotrophic and phototrophic activities in a facultative pond that result in the complete stabilization of organic waste (Hawkes 1983*a*).

Several groups of invertebrates are present. In the sludge, benthic nematode worms, tubificid worms, and chironomid larvae are all common (Gloyna 1971). The worms contribute to the stabilization of the organic sludge, and both the worms and midge larvae increase the rate of exchange of materials between the sludge and water at the solids – water interface. The dominance and abundance of species is related to the dissolved oxygen conditions of the sludge (Kimerle and Anderson 1971) and the organic loading (Kimerle and Enns 1968).

It is important to prevent the development of surface mats of filamentous algae, or thin scum of encysted unicellular or blue-green algae (Hartley and Weiss 1970). Such growths seriously reduce the light penetration and interfere with the mass transfer of oxygen into the pond from the atmosphere. Infestations of surface-dwelling macrophytes such as *Lemna* have the same adverse effects (Hawkes 1983c). Surface material will eventually collect either on the banks or in the corners of ponds due to wind action and unpleasant odours are produced as it decomposes.

Apart from nutrient availability, including carbon dioxide, temperature and solar radiation are the major factors affecting algal photosynthetic activity. Most algae grow over a wide temperature range 4–40°C (Arceivala *et al.* 1970) with optimum growth for the green algae (Chlorophyceae) about 20°C but decreasing at 35°C. Above this temperature other algae become dominate, such as the blue-green algae (Cyanophyceae) and euglenoids. Although different algae have different light intensity optima, a high light intensity is not necessary for the successful operation of a facultative pond. For example, ponds operate successfully throughout the Indian subcontinent from latitudes 8°N to 34°N, although different organic loading rates are required for ponds from different latitudes to allow for the variability in temperature and solar radiation, both of which affect the rate of microbial action (Arceivala *et al.* 1970). Light penetration can be restricted by high densities of phytoplankton ($> 1 \times 10^6$m l) in the surface layer of the pond. In many ponds, optimum light intensity for the algae occurs at < 30 cm, although effective photosynthesis may not exceed 45 cm under such conditions. Therefore, most of the pond is in darkness, and although heterotrophic activity is independent of light, mixing is required to ensure that the rest of the pond depth is kept aerobic. Mixing, which can occur naturally in shallow ponds by wind or in deeper ponds by currents induced by temperature differences, also ensures the transfer of materials between the photosynthetic and bio-oxidation phases, which ultimately governs the rate of stabilization. In dense algal layers, the water quickly becomes supersaturated during the day with the excess dissolved oxygen being lost by transfer to the atmosphere unless mixed with the deeper unsaturated water. The circulation of algae in and out of the light ensures that a larger biomass of algae is supported and that more oxygen is made available to the bacteria. This is because algal cells fix a higher proportion of light energy under intermittent exposure to light compared to algae under constant exposure.

Aerobic bacterial breakdown is less sensitive to temperature than anaerobiosis, and can proceed effectively over a range of 10–35°C. However, if the water temperature is only a few degrees above freezing then the organic matter entering the pond will rapidly accumulate. Microbial activity will be reduced even further by ice and snow cover, which reduces light penetration and so effectively stops phototrophic production of oxygen, and also prevents oxygen transfer from the atmosphere, normally helped by wind action. The pond can become totally anaerobic and during the spring the odours will be produced until the algal population is re-established, which may take up to 4 weeks depending on temperatures, solar radiation, and the amount of accumulated organic matter (Porges and Mackenthun 1963). As with the anaerobic pond, the extra sludge accumulated in the pond during the winter is broken down by the enhanced anaerobic bacterial activity during the warmer months.

The major design function of facultative ponds is to maintain dissolved oxygen in the basin, provided by photosynthetic activity of the algae, to allow heterotrophic activity to proceed at an optimum rate. The site for such a pond should be at least 0.5 to 1 km from residential areas and, because of the production of odours during the spring, it should be sited preferably downwind. It should be an exposed site to ensure maximum wind sweep over the surface of the pond, which is a function of wind speed and distance, to encourage mixing. Thus, planting trees or landscaping the site to shield the pond from view will severely reduce mixing. Leaf fall will also reduce light penetration. The shape of the pond is not important as long as the influent mixes rapidly with the pond supernatant. Uniform flow throughout the pond is vital, therefore peninsulas and inlets should be avoided and corners of ponds should be rounded. If a rectangular pond is used then the length should not be greater the four times its breadth. Depth is a critical factor. Ponds range from between 0.7–1.5 m in depth with a freeboard of 0.9–1.5 m above the high water level to accommodate wave action. A minimum depth of 0.7 m has to be maintained in order to prevent the growth of rooted aquatic weeds that not only damage the lining of the pond and affect circulation, but also encourage mosquitoes and other flies. The recommended depth is between 1.0–1.5 m. Depths in excess of 1.5 m may create odour problems due to excessive anaerobiosis. The sides should slope gently, a maximum slope of 1 in 3 is recommended, to prevent earth movement, and should be covered with paving slabs to prevent erosion by wave action and the growth of marginal vegetation. If the soil is not impervious then it must be lined with bentonite clay or plastic to prevent groundwater pollution. The influent enters ponds horizontally at about 0.5 m above the bottom to induce a mixing current. The outlet is positioned to maximize mixing and avoid short-circuiting. Simple baffles can be used to enhance the plug flow nature of the pond and to minimize short-circuiting. Grit and screenings are generally removed from the wastewater before entry into the pond, which ensures that the sludge accumulates slowly over the years. Accumulation rates range from 10 to 90 mm per year, and depending on the rate, ponds will only require desludging after long intervals. If ponds are operated in series, then the majority of sludge will accumulate in the primary pond. As accumulation will reduce the capacity of the pond over the years, the rate of accumulation must be taken into account when designing the pond depth and retention period.

Photosynthetic activity is determined by the light entering the basin, which in turn is a function of the surface area of the pond. Organic loading is therefore expressed as kg BOD per hectare per day (kg BOD $ha^{-1}d^{-1}$) or g BOD per square metre per day (g BOD $m^{-2}d^{-1}$). The loading is limited by specific environmental factors such as insolation, temperature, and wind, which vary according to latitude. Therefore, different loading rates are recommended in temperate zones compared with tropical zones (Table 4.32). The maximum loading rate is limited by the need to maintain aerobic conditions in the pond during the most difficult season of the year when insolation and temperature are at their lowest and in temperate zones when ice formation occurs. Although ice formation reduces the radiation reaching the algae, snow accumulation will rapidly eliminate all light until photosynthesis is inhibited and the pond becomes anaerobic. Severe odours are produced as

TABLE 4.32   Loading and design criteria for facultative ponds constructed in different climatic zones (Gloyna 1971)

Surface loading (kg BOD ha^{-1}d^{-1})	Population per ha	Detention time (days)	Environmental conditions
<10	<200	>200	Frigid zones, with seasonal ice cover, uniformly low water temperatures and variable cloud cover
10–50	200–1000	200–100	Cold seasonal climate, with seasonal ice cover and temperate summer temperatures for short season
50–150	1000–3000	100–33	Temperate to semi-tropical, occasional ice cover, no prolonged cloud cover
150–350	3000–7000	33–17	Tropical, uniformly distributed sunshine and temperature, and no seasonal cloud cover

the ice breaks up due to anaerobic activity and although the odours become less severe they continue to be given off until the algae are re-established and the pond becomes aerobic once more. The duration of odour production is a function of BOD loading and duration of ice cover (Svore 1968). To minimize this period the organic loading can be reduced and the retention period increased which, of course, results in the pond being underloaded for the rest of the year (Marais 1970). This problem, as well as a reduction in BOD removal as the temperature falls due to reduced bacterial activity, can be partially overcome by slowly reducing the pond level in the autumn and early winter when there is adequate dilution flow in the receiving waters. This allows the discharge from the pond to be minimal or stopped completely during winter so that the influent is stored until spring when discharge from the pond can be recommenced, as the aerobic activity is able to reduce the BOD of the effluent to an acceptable level. Little anaerobic activity occurs in the sludge during the winter and no soluble degradation products are released to enhance the overall BOD of the stored wastewater. However, the sludge gradually accumulates as sewage is continued to be discharged to the pond, and it is not until the summer at the enhanced temperatures that there is a net reduction in sludge volume due to the increased rate of degradation. The increase in soluble degradation products released from the sludge layer results in a rise in the BOD of the supernatant, although no fluctuation in effluent BOD is recorded due to the enhanced heterotrophic activity at the warmer temperatures, which utilizes the increase in organic load. The BOD of the stored wastewater in the pond remains almost constant throughout the year, whereas the actual oxygen demand on the pond varies considerably as it is a product of both the BOD and the rate of degradation. Therefore, at higher temperatures the oxygen demand can increase several fold due to the higher degradation rate although the BOD remains constant. This provides a problem to the operator, because at temperatures in excess of 20°C the oxygen requirement of the pond cannot be determined by the standard BOD_5, nor can BOD_5 be used as an indicator of the onset of anaerobiosis (Hawkes 1983a). The use of stabilization ponds in cold climates has been the subject of a workshop organized by the Environment Protection Directorate of Environment Canada (Townshend and Knoll 1987).

In the northern states of America, where the climatic conditions are similar to those of Western Europe, the maximum loading permissible is 2.2 g BOD $m^{-2}d^{-1}$ if an odour nuisance is to be avoided in the spring. Where ice coverage of the pond does not occur this loading can be increased, and in the southern and south-western states of America loadings of up to 5.6 g BOD $m^{-2}d^{-1}$ are used. However, this is only equivalent to a volumetric loading of 0.0015–0.0037 g $m^{-3}d^{-1}$ for a 1.5 m deep ponds, which is a much lower organic loading compared to the activated sludge or percolating filtration process. Retention time depends on the load, depth of the pond, evaporation rate, and loss by seepage, but periods of 3–6 months are common for ponds in non-tropical locations. Ponds normally achieve a 90% reduction in BOD, which is less efficient compared with other biological treatment processes. Larger ponds are more effective because of the greater wind sweep over the surface, which enhances mixing. Generally, ponds are operated in series to prevent short-circuiting, to provide a greater reduction in pathogens (Section 6.3) and an increased BOD reduction. The first pond in series has a higher organic load and sludge accumulation compared with the other ponds, which may lead to excessive odour production. If ponds are operated in parallel, odours are prevented as the BOD load is distributed more evenly, although such an arrangement does allow short-circuiting. Facultative ponds rarely achieve a final effluent suspended solids concentration of 30 mg l^{-1}, even when in series. The suspended material is predominantly algae and zooplankton, which may be present at concentrations of between 50–70 mg l^{-1} in the final effluent. Tertiary treatment is required to remove the plankton biomass and any other colloidal or particulate matter if the receiving water is unable to assimilate the material. The three most popular options are chemical flocculation followed by a gravity filter, upflow filter, or by spray irrigation on to land (Section 2.4).

The depth of the aerobic zone fluctuates according to the photosynthetic activity and the degree of mixing, therefore, the depth of the aerobic zone will always be less during the hours of darkness. As the organic loading increases so the depth of the aerobic zone tends to decrease, with a greater portion of the organic matter being degraded anaerobically (Fig. 4.78). Apart from currents

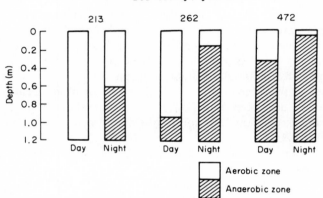

FIG. 4.78   The effect of organic loading on the depth of the aerobic zone during the day and night in facultative ponds at Nagpur, India (adapted by Hawkes 1983a from Arceivala et al. 1970).

induced by the influent wastewater entering the pond, circulation is also caused by wind action on the surface or currents caused by temperature differences within the pond. In shallow ponds, wind action will normally be sufficient to induce mixing as long as the sides of the pond are not too steep or trees have been planted, which may interfere with the wind sweep. In deeper ponds, and under calm hot conditions, ponds can become stratified. This is due to the surface water being warmed, which reduces its density so that it forms a distinct layer (the epilimnion) which is isolated from the lower mass of cooler denser water (the hyperlimnion) by a thin layer of water called the thermoclime through which the temperature gradient falls rapidly. Once stratification has occurred then mixing becomes restricted to the epilimnion only. As the density change per degree is greater at higher temperatures, the resistance to mixing will increase as the temperature rises. The hyperlimnion rapidly becomes anaerobic, even though the epilimnion may be super-saturated with oxygen. Although the non-motile algae tend to sink in the less dense epilimnion and to subsequently die because of lack of light in the hyperlimnion; motile forms, such as *Euglena* are able to maintain a position in the pond where the light conditions are most favourable and eventually become dominant, which is a characteristic of stratified ponds. Stratification can lead to short-circuiting of the wastewater through the epilimnion with a resultant reduction in BOD and pathogen removal. Under certain climatic conditions where hot days are followed by cold nights, daytime stratification will be followed by night-time mixing. However, in most cases, mechanical mixing is required using compressed air to break up the stratification, although the use of weirs to induce an under and over pattern of flow has been successful (Hartley and Weiss 1970). Mechanical mixing should not be continuous as the increase in suspended solids from the sludge layer will reduce the light penetration and reduce photosynthesis. Therefore mixing should be confined to short periods at night.

Facultative ponds are best suited to isolated communities where land is cheap and no industrial expansion is planned. When treating sewage, ponds are capable of producing odour free effluents throughout the year, except if ice formation is a problem in the spring. Industrial or strong organic wastes normally require pretreatment before discharge to a facultative pond to reduce the organic content, preventing overloading and permanent odours. The use of an anaerobic pond initially to maximize BOD removal followed by the facultative pond may be adequate. Incremental feeding, where some of the influent to the first pond in series is transferred to the second pond, or recirculation from the second pond to the primary pond may also improve efficiency and prevent odours (Marais 1970). Facultative ponds are most popular in areas where there is sufficient solar radiation throughout the year. Where the climatic conditions are suitable, oxidation ponds are the cheapest form of sewage treatment in constructional and operational terms. The advantage of the system is the ease of operation, no skilled labour is required, and no mechanical parts or power are required. However, among the disadvantages are a number of operational problems that include poor assimilative capacity for industrial wastes, production of odours, and meeting minimum standards for discharge.

New ponds are easily commissioned once construction is complete and the basin cleared of any loose debris, soil, and vegetation. Obviously, the pond cannot be operated at its full design load until a balanced community of bacteria and algae has been established. This is most effectively achieved by allowing raw wastewater into the centre of the pond to a depth of about 200 mm. Where the area of the basin is large, then a smaller area should be partitioned off by constructing an embankment at one corner or in the centre and filled with wastewater. Bacteria become established first, followed more slowly by the algae. Colonization occurs naturally, although the speed of colonization can be increased by the addition of algal-rich water from a pond or tank. Additional wastewater must be added at intervals to maintain the level in the pond, which reduces because of evaporation. Once an algal bloom has become established the pond can be gradually filled with wastewater, trying to maintain the rich algal bloom until the pond is full. The pond should be allowed to stand for a further 3–4 days before full continuous-flow operation is commenced. In temperate climates this process will take up to 6 weeks even in the summer months. Care must be taken to prevent weed growth developing during the start-up process and to this end it may be necessary to increase the initial depth to 1 m if it becomes a problem. Full details of commissioning all types of stabilization ponds are given by Arceivala *et al.* (1970) and Gloyna (1971).

The biologist plays a major role in the successful operation of facultative ponds by conducting regular microscopic examinations of the algae in order to give a clear quantitative assessment of the community structure within the pond (Arceivala *et al.* 1970). The overall photosynthetic activity of the pond can be determined either by measuring the biomass of the algae or the chlorophyll *a* concentration (American Public Health Association *et al.* 1985). Changes in the colour of the pond are also a useful guide to changes in condition. Healthy ponds are green or brownish due to the presence of green algae. However, a change to blue-green indicates a change in algal dominance from green to blue-green algae, and as the latter are less efficient than the green algae this indicates a deterioration in conditions. Changes in colour to pink, red or reddish-brown in the summer or autumn are caused by blooms of photosynthetic sulphur bacteria because of an increase in sulphate or sulphide concentration resulting from increased anaerobiosis. This is generally associated with overloading, thermal stratification or another operational problem. The commonest forms associated with sewage are *Chromatium* and *Thiospirillum*, whereas *Thiopeda rosea* and smaller spiral and rod forms are associated with industrial wastes (Gloyna 1971). A blue-green algae *Merismopedia tenuissina*, which is also associated with anaerobic conditions where sulphides are abundant, also imparts a pink colour to the pond. When the pond has broken down and gone completely anaerobic then it takes on a grey or black colour. This is normally accompanied by rising sludge due to excessive gas production in the sludge layer. Such a condition is normally due to a rapid change in temperature, severe overloading, change in wastewater characteristics or an input of a toxic waste.

High-rate aerobic ponds

These ponds are not designed for optimum purification efficiency but for maximum algal production. The algae is harvested for a variety of uses and is often sold to defray operational costs. The ponds are shallow lagoons 20–50 cm deep, with a retention period of 1–3 days. The whole pond is kept aerobic by maintaining a high algal concentration and by using some form of mechanical mixing. Mixing, which is normally carried out for short periods at night, prevents the formation of a sludge layer. Mixing may also be required for short periods during the day to prevent a rise in pH in the surface water due to photosynthesis. The pond is commissioned in the same way as a facultative pond except that continuous loading should not be permitted until an algal bloom has developed. Loading depends on insolation, and in California the average loading throughout the year is 134 kg BOD $ha^{-1}d^{-1}$ reaching an optimum summer loading of 366 kg BOD $ha^{-1}d^{-1}$ (Oswald *et al.* 1964). Strong organic sewages inhibit the photosynthetic action due to high ammonia concentrations, which results in the pond becoming anaerobic. Algal productivity, and uses of the algal crop are considered in Section 7.3.

Maturation ponds
Maturation ponds are widely used throughout the world as a tertiary treatment process for improving the effluent quality from secondary biological processes. Effluent quality is improved by removing suspended solids, reducing ammonia, nitrate and phosphate concentration, and by reducing the number of enteric micro-organisms. They are the same depth as facultative ponds, 1.0–1.5 m, although shallower ponds are used where insolation is high in order to achieve maximum light penetration. Unlike high-rate aeration ponds, maturation or tertiary lagoons are specially designed for maximum purification and not algal production (Hawkes 1983a). The retention period is normally 10–15 days, although shorter periods can be used for suspended solids (4 days) or phosphate removal (7–10 days) (Gloyna 1971; Toms *et al.* 1975). Unless specifically removed, removal efficiency remains low due to the initial low organic loading (Toms *et al.* 1975). By acting as a buffer between the secondary biological phase and the receiving water, maturation ponds can reduce the effects of toxic loads and fluctuations in the performance of the secondary biological phase by dilution (Potten 1972).

Different planktonic organisms vary in their efficiency in removing nitrogen and phosphorus. For example, the algae *Scenedesmus obliquus* is particularly efficient in utilizing organic and ammonical nitrogen, whereas *S. quandricauda* var. *alterans* is more efficient at using nitrate (Kalisz and Suchecka 1966). *Daphnia* reduces turbidity and enhances nitrogen removal by ingesting particulate matter and algae.

Work on the maturation ponds at Rye Meads sewage treatment works, which discharges to the River Lee in the UK, has shown that phosphate removal is directly correlated with the algal biomass, reaching a maximum removal of 73% in May and a minimum of 2% in January (Toms *et al.* 1975). However, only a small proportion (<20%) of the phosphate removal can be attributed to nutritional uptake by the phytoplankton as the majority is removed by precipitation at enhanced pH values (>pH 8.2), which is due to photosynthetic activity. For every unit increase in pH, the concentration of phosphate remaining decreased by a factor of 10. In the temperate climate of the UK, it is only possible to maintain sufficient algal biomass to remove a reasonable proportion of the phosphate from the pond during the spring and autumn by having a long retention time. In winter, this is not feasible, although the higher dilution factor, reduced temperature, and insolation all significantly reduce the risk of eutrophication in rivers. It has been reported that phosphorus can be regenerated during the night or during the winter at lower pH values (Hemens and Mason 1968). However, this was not observed at Rye Meads by Toms and his co-workers. Removal of nitrate is not linked with the algal biomass and although small amounts are utilized by the algae, the majority is lost by denitrification at the sludge–water interface (Toms *et al.* 1975). Optimum nitrate removal is achieved by ensuring the development of a sludge layer and by using shallow ponds to give the maximum sludge–water surface contact. In this way, up to $0.8 \text{ g N m}^{-2}\text{d}^{-1}$ can be removed. It has been reported that nitrate levels tend to be much higher in the effluent than in the feed and that nitrate concentrations can build up during the autumn and winter (Van der Post and Englebrecht 1973). In conclusion, although algae are beneficial in terms of phosphate removal and to a lesser extent nitrate removal, their presence in the final effluent from a maturation pond increases the suspended solids and turbidity. It has also been suggested that algae can reduce the mortality of faecal bacteria (Potten 1972).

River purification lakes

The development of large impounded lakes to remove and degrade the residual pollution from a river is becoming increasingly popular. Such lakes are very similar to maturation ponds. Five lakes have been constructed in the Ruhr area (Germany) with Lake Baldeney, south of Essen, the largest with a retention period of 60 hours at low flows (Imhoff 1982, 1984) (Fig. 4.79). Together they provide an area of 4488 km^2, with Lake Baldeney alone providing a purification capacity in the order of 100 000 BOD PE (population equivalent) per day. Lake Baldeney is also particularly efficient in removing dissolved phosphorus from the River Rhur (Fig. 4.80). The phosphorus is removed by algal growth and subsequent algal settlement. Up to 170 tonnes of phosphorus per annum, together with heavy metals, is removed by sludge settlement in the lake (Table 4.33). The lakes had not been desludged by 1988 and it is estimated by 1984, Lake Hengstey, constructed in 1929, had accumulated some 400 000 m^3 of sludge and in Lake Harkort, which was impounded in 1930, some 350 000 m^3 has been deposited. In Lake Baldeney, some 60 000 m^3 of sludge accumulates each year, but at flows 10 times greater than the average flow rate the sludge becomes mobilized. The sludge is subsequently flushed out of the lake when the river is in flood and flows are in excess of 15 times the average flow rate (Imhoff 1982). Because of the reduction in the depth of these lakes desludging is now underway. Some 750 000 m^3 of sludge is to be removed from Lake Baldeney alone by hydraulic dredging at a cost of 15×10^6DM at 1983 prices. However, even with these large desludging costs, it has been estimated that treatment by purification lakes is 30% cheaper than conventional treatment.

On the River Tame, UK, a polluted tributary of the River Trent in the West

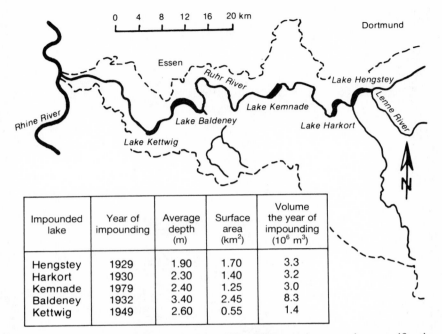

FIG. 4.79 Impounded lakes on the River Rhur (Germany) operated as a purification lakes system (Imhoff 1984).

Impounded lake	Year of impounding	Average depth (m)	Surface area (km²)	Volume the year of impounding (10⁶ m³)
Hengstey	1929	1.90	1.70	3.3
Harkort	1930	2.30	1.40	3.2
Kemnade	1979	2.40	1.25	3.0
Baldeney	1932	3.40	2.45	8.3
Kettwig	1949	2.60	0.55	1.4

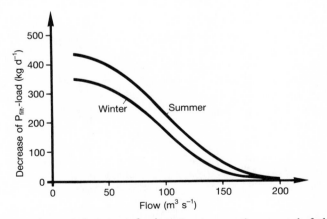

FIG. 4.80 The effect of flow rate ($m^3 s^{-1}$) and season on the removal of phosphorus in the impounded Lake Baldeney (Imhoff 1984).

TABLE 4.33   Comparison between sediments collected from the purification lakes at Lea Marston (UK) and the Rhur (Germany) catchment (Woods et al. 1984)

Lake site	Period	Dry solids (%)	Volatile solids (%)	Metal concentrations (mg kg^{-1} dry solids)					
				Ni	Cu	Cr	Zn	Cd	Pb
Lea Marston (sed. tank)	1969–70	10.5	40.5	554	1743	1275	8000	44	617
Lea Marston (exp. lake 1)	1972–74	11.4	35.1	1000	2350	1340	5160	70	950
Lea Marston (exp. lake 2)	,,	7.5	32.7	970	2230	1650	5400	80	950
Lea Marston (dredged sludge)	1981–82	4.5	24.9	410	1335	510	3570	39	780
Hengstey	,,	39.4	12.3	261	1240	440	4030	18	450
Harkort	,,	45.8	11.0	250	830	280	3120	29	480
Baldeney	,,	45.6	14.0	312	730	400	3540	36	525

Midlands, a purification lake has also been built. The River Trent receives vast quantities of sewage and industrial effluent reaching in excess of 80% of the total volume of the river at low flows. Lea Marston purification lake was built at a disused gravel works after the last major conurbation and sewage input. It was opened in 1980 and aims to remove settleable solids from the river that are especially a problem during storms. The solids are removed from the main body of water by settlement in the lake and which are subsequently removed by dredging. In contrast to the Rhur lakes, the Lea Marston lake is continuously desludged although it takes about 12 months to cover the whole flow area of the lake. The lake also reduces the need for tertiary treatment at Coleshill and Minworth sewage treatment plants and acts as a buffer in case of accidental spillages. The results so far have been very encouraging with the water quality in the Rivers Tame and Trent significantly improved. A coarse fishery has been established and fish are being caught in the River Tame for the first time in 100 years. The river enters the system and passes through large booms that remove floating debris (Fig. 4.81). Grit traps and grit channels remove > 2000 tonnes of dry matter per annum from the river with the lake itself removing a further 15 000 tonnes of sludge each year. The lake provides a retention time of 12 h at the daily average flow, falling to 1–2 hours at the design flood flow. The effect on water quality is quite dramatic, with the suspended solids reduced by 56% and the BOD by 34% overall (Woods *et al.*

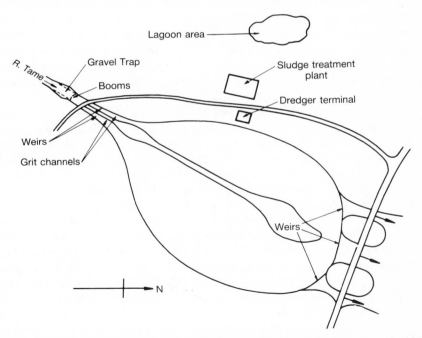

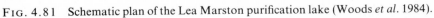

FIG. 4.81   Schematic plan of the Lea Marston purification lake (Woods *et al.* 1984).

1984). The quality of the sludge is compared with that removed from the Rhur lakes in Table 4.33.

4.3.3 Aeration lagoons

Aeration lagoons are used for the first-stage treatment of sewage or the pretreatment of industrial wastewater before secondary treatment, such as a facultative pond or activated sludge. The basins are deeper than other types of stabilization ponds, 3.0–3.7 m, with oxygen being provided mechanically and not by the photosynthetic activity of algae. Oxygen is normally provided by mechanical pier-mounted or floating surface aerators, which ensures that the microbial biomass is kept in suspension and that sufficient dissolved oxygen is provided for maximum aerobic activity. Bubble aeration, provided by compressed air pumped through plastic tubing laid across the bottom of the pond is also used.

A predominately bacterial microbial biomass develops and as there is no provision for settlement or sludge return, the process relies on sufficient mixed liquor developing in the basin. Retention times vary between 3–8 days depending on the degree of treatment required, the temperature, and strength of the influent wastewater. A retention time of 5 days at 20°C provides an 85% reduction in the BOD of domestic sewage, although a fall in temperature by 10°C results in a reduction in BOD removal to 65% (Hammer 1977). Bubble aeration is most successful in keeping ponds aerobic in locations where pond surfaces are frozen for several months in the winter. Under these conditions surface aerators cannot operate due to ice formation. In this type of climate unaerated facultative ponds are generally unsuccessful and tend to become anaerobic.

As a unit process they fall between the activated sludge process and facultative oxidation ponds. If the degree of agitation is sufficient then all the solids may be kept in suspension under aerobic conditions corresponding to the activated sludge process. In systems with less efficient agitation, the solids may tend to settle and form an anaerobic sludge layer similar to facultative oxidation ponds. Where the aeration is inadequate, then enhanced deposition of particulate solids and the reduced dissolved oxygen concentration in the basin will lead to anaerobic conditions, a loss of efficiency and the production of odours. Therefore, sufficient agitation and mixing coupled with a strict loading regime are required for the successful operation of aeration lagoons.

Aeration lagoons are susceptable to large inputs of biodegradable or toxic waste which can severely reduce efficiency. High infiltration and storm water result in reduced retention times and flush the microbial biomass out of the basin, thus reducing the MLSS and MCRT. Where this is a potential problem provision should be made to divert a portion of the enhanced flow directly to the second stage.

Further reading

General: Arceivala *et al.* 1970; Gloyna 1971; Hawkes 1983a; Middlebrooks *et al.* 1982; Environmental Protection Agency 1983; Townshend and Knoll 1987; Mara and Marecos do Monte 1987.
Design: Arceivala *et al.* 1970; Gloyna 1971; Heuvelen *et al.* 1960; Thirumurthi 1974; Middlebrooks *et al.* 1982; Environmental Protection Agency 1983.
Ecology: Hawkes 1983a.
Operation and maintenance: Arceivala *et al.* 1970; Gloyna 1971; Svore 1968; Middlebrooks *et al.* 1982; Environmental Protection Agency 1983.
Uprating: Middlebrooks *et al.* 1982; Environmental Protection Agency 1983.

4.4 ANAEROBIC UNIT PROCESSES

4.4.1 Introduction

Strong organic wastes (BOD > 500 mg l^{-1}) generated by the agricultural and food industries, often in large quantities, provide a particularly difficult wastewater treatment problem. These wastes contain large quantities of biodegradable organic matter and conventional aerobic treatment is beset with numerous operational difficulties. For example, the difficulty of maintaining aerobic conditions, especially if the wastewater has a high concentration of suspended solids, sludge bulking, inability to take high BOD or COD loadings, high operational and energy costs, and a high production of biomass as wasted sludge that requires subsequent disposal. Anaerobic treatment, although slow, offers a number of attractive advantages in the treatment of strong organic wastes: a high degree of purification; ability to treat high organic loads; production of a small quantity of excess sludge, which is normally very stable; and the production of an inert combustible gas (methane) as an end-product. The final effluent produced by anaerobic treatment contains solubilized organic matter that is amenable to quick aerobic treatment, which indicates the potential of combined anaerobic and aerobic units in series (Mathur *et al.* 1986). Unlike aerobic systems, complete stabilization of organic matter is not possible anaerobically and subsequent aerobic treatment of anaerobic effluents is normally necessary.

The basic difference between aerobic and anaerobic oxidation is that in the aerobic system, oxygen is the ultimate hydrogen acceptor with a large release of energy, but in anaerobic systems the ultimate hydrogen acceptor may be nitrate, sulphate or an organic compound with a much lower release of energy. Briefly, the process of anaerobic decomposition involves four discrete stages (Fig. 4.82). The first stage is the hydrolysis of high molecular weight carbohydrates, fats, and proteins that are often insoluble, by enzymatic action into soluble polymers. The second stage involves the acid-forming bacteria

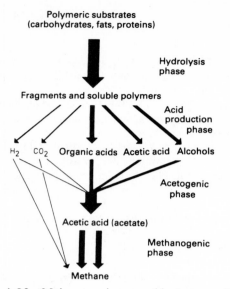

FIG. 4.82　Major steps in anaerobic decomposition.

which convert the soluble polymers into a range of organic acids (acetic, butyric, and propionic acids), alcohols, hydrogen, and carbon dioxide. Acetic acid, hydrogen, and carbon dioxide are the only end-products of the acid production that can be converted directly into methane by methanogenic bacteria. A third stage is present when the organic acids and alcohols are converted to acetic acid by acetogenic bacteria. It is in the fourth and final stage, which is perhaps the most sensitive to inhibition, when methanogenic bacteria convert the acetic acid to methane. Although methane is also produced from hydrogen and carbon dioxide, in practice about 70% of the methane produced is from acetic acid. Obviously, the methanogenic stage is totally dependent on the production of acetic acid and so it is the third stage, the acetogenic phase, that is the rate-limiting step in any anaerobic process. The biochemistry of anaerobic decomposition has been fully explored in Section 3.4.

A large number of anaerobic processes are available, including anaerobic lagoons, digesters, and filters. Present anaerobic technology can be divided into two broad categories. *Flow-through systems* for the digestion of concentrated wastes, such as animal manures or sewage sludges that have a solids concentration in the range of 2–10%. These include completely mixed reactors (Fig. 4.83*a*), which are used primarily for sewage sludges, and plug flow reactors used to a limited extent for the digestion of animal manures (Fig. 4.83*b*) (Hayes *et al.* 1979). Wastewaters with a lower solids concentration are treated anaerobically by *contact systems* in which the wastewater is brought

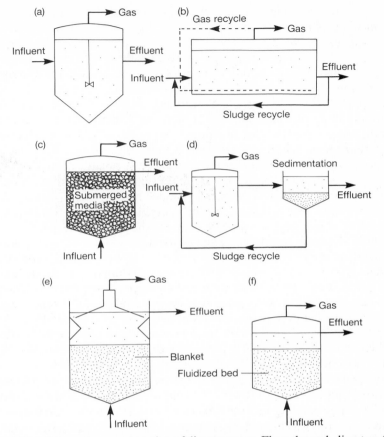

FIG. 4.83 Schematic representation of digester types. Flow-through digesters (a–b) and contact systems (c–f) (adapted Casey 1981).

into contact with an active microbial biomass that is retained within the system. In flow-through systems the residence time of the waste (hydraulic retention time, HRT) and of the microbial biomass (mean cell retention time, MCRT) in the unit are the same, whereas in a contact system the MCRT is far greater than the HRT of the wastewater. The biomass is retained within the reactor in a number of ways. For example, by allowing the anaerobic micro-organisms to develop as an attached film on a static medium, or to develop as flocs maintained in suspension either by mechanical mixing or by the upward flow of effluent through the reactor. By using a system similar to the activated sludge process, the anaerobic biomass is present as suspended flocs which are recovered in a separate settlement chamber and recycled back to the main reactor (Fig. 4.83d). This process can also take place within a single reactor where the depth of the sludge blanket is controlled by the upflow rate of the influent wastewater (Fig. 4.83e). Attached anaerobic films can be on a static filter medium of natural stone or plastic, similar to a percolating filter (Fig. 4.83c); whereas fluidized beds incorporate a fine grained medium such as sand (Fig. 4.83f). Both media reactors are completely flooded to ensure anaerobic conditions, and like the sludge blanket process are operated in the upflow mode (Casey 1981).

The basic kinetics of anaerobic treatment are fully discussed by Mosey (1983) and Hill (1983) (Section 3.1.2). Because so many different groups of bacteria are involved, the digestion process is not easily modelled, especially as the methanogenic bacteria have a much lower growth rate than the acid-producing bacteria. However, as the conversion of volatile acids to biogas is generally considered to be the rate-limiting step of the overall reaction, the methanogenic phase is usually used for modelling purposes. The majority of models are applicable to the continuously stirred tank reactor (CSTR) as they are the most widely used and operate closest to the steady-state, which is easier to model. However, the development of dynamic models has demonstrated the need for more accurate ways of examining operational problems of digesters. Steady-state models are usually derived from one or two differential equations that treat the digester contents as a single substrate and the bacteria as a single population. Among the best known models of this type are those using Monod kinetics (Lawrence and McCarty 1970; Grady *et al.* 1972; Pfeffer 1974). Of particular interest is the model based on contois kinetics developed by Chen and Hashimoto (1978, 1980). The dynamic models available all use Monod kinetics and normally consider the interactions between several substrates and bacterial populations. This results in a system of differential equations that can only be solved by computer. Andrews and Graef (1971) considered that relating substrate concentration to specific growth rate (the Monod function) was invalid for anaerobic reactors, as the volatile acids not only acted as substrate for methanogenic bacteria but were also inhibitory at higher concentrations. They replaced the Monod function with an inhibition function (Haldane 1930) so that:

$$\mu = \frac{\mu_m}{1 + K_s/s + s/K_i}$$

where K_i is the inhibition function. This is fully expanded and discussed by Andrews (1971, 1983). Whereas this early dynamic model only included a single substrate and organism, more recent models of this type consider more than one substrate (Hill 1982; Lavagno *et al.* 1983) or more than one bacterial population (Hill and Barth 1977; Hill and Nordstedt 1980; Lavagno *et al.* 1983). The use of models in anaerobic digestion is succinctly reviewed by Parsons (1984).

In plug-flow systems there is a substrate concentration gradient between the inflow and outflow ends of the reactor. As the rate of reaction is dependent on the substrate, the average rate of reaction in a plug-flow system is theoretically higher than for a CSTR. Casey (1981) illustrates the significance of this difference in terms of the reactor volume required for a first-order reaction process (Fig. 4.84). This shows that the advantage of the plug-flow reactor over the CSTR increases with the required degree of reaction completion. Although a number of contact systems can be considered as plug-flow processes, those treating domestic sewage sludge are largely of the completely mixed type.

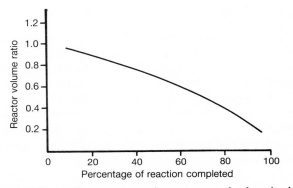

FIG. 4.84   Ratio of plug-flow reactor volume to completely mixed flow reactor volume for a first-order reactor process.

4.4.2   Flow-through systems (digestion)

Anaerobic digestion is a biological process in which the organic fraction of a wastewater sludge, proteins, carbohydrates, and lipids, are degraded in the absence of oxygen to methane and carbon dioxide by a variety of micro-organisms, principally bacteria. The process occurs quite widely in natural environments such as lake sediments, marshes, peat bogs, and even in the rumen of certain herbivorous animals.

The first direct application of anaerobic digestion to sewage sludge is attributed to Louis H. Mouras of Vesoul in France who developed a sealed cesspool (c. 1860) in which he claimed 'sewage solids were liquified'. By 1895, digester gas, also known as biogas or methane, was being collected and used as fuel. Donald Cameron constructed a large septic tank in Exeter (UK) and was able to use the gas for lighting the area around the treatment plant. By the turn of the century it had become common practice to incorporate a digestion chamber within sedimentation tanks, with the Travis hydraulic tank and the Imhoff tank the most widely adopted. The Imhoff tank, in particular, was installed at most small to moderate sized treatment plants and are still in wide use today in Ireland, although few have been constructed since 1950. The development of specific digesters for the anaerobic breakdown of sludges began in the early 1920s, with their use restricted to large cities. There has been a resurgence of interest in anaerobic digestion in the past 10–15 years, with much fundamental research and development work being done, which has led to the development of new digester and anaerobic reactor designs. The basic flow-through system can be separated into two categories, systems which combine settlement with digestion, and separate systems built for digestion only.

4.4.2.1   Combined systems

Combined systems of settlement and digestion are restricted to situations where there is only a small volume of wastewater to be treated. The combined system provides primary settlement and then sludge stabilization by digestion, all within a single chamber. By reducing the volume of sludge produced, and storing it in such a way as not to impede the settlement process, such systems are ideal for single houses or small communities. Anaerobic lagoons can also be categorized as combined systems, although their use in Europe is largely restricted to strong organic wastes from the food-processing and agricultural industries (Section 4.3).

Septic tanks

Septic tanks are often confused with cesspools, the latter being an underground chamber constructed solely for the reception and storage of wastewater with no treatment taking place. It is important to differentiate between the two as in many areas these two terms are used interchangeably. A cesspool is a storage tank which requires periodic emptying and is not intended as a septic tank in which decomposition of the settled material occurs.

   Cesspools can be constructed out of concrete, plastic or fibreglass and according to the British Code of Practice (British Standards Institution 1972, 1983) must be: (a) impervious; (b) not able to overflow; (c) have a minimum capacity of 18 m^3 or 45 days retention for two people assuming a per capita water usage of 180 l d^{-1}; (d) be constructed so that it can be completely emptied; (e) adequately ventilated; and, most importantly, (f) adequately covered to ensure safety. Clearly, cesspools are only used where no other form of treatment is possible, and the need to have them regularly emptied means that they are the most expensive form of treatment for domestic dwellings in terms of both capital and operational costs. The only improvement in this system can be achieved by adopting water saving improvements that reduce the volume of wastewater discharged. There are advantages, however, such as no power requirement, no quality control required, no mechanism that can go wrong, the process is not injured by intermittent use, and as there is no effluent discharge, there is no immediate environmental impact. The major limitation on the use of cesspools is the cost of emptying, although the construction of large underground storage tanks can be both difficult and expensive. For example, the land in a recent housing development in County Dublin was found after the construction of the houses to be unsuitable for septic tank usage, and cesspools were considered. The development had a person equivalent of 58, thus, assuming a daily per capita water usage of 180 l the daily flow was:

$$\frac{180 \times 58}{1000} = 10.44 \text{ m}^3\text{d}^{-1}$$

In this case, the council was faced with three tank sizes, 31 m^3 to give a minimum retention time of 3 days to cover a weekend, 73 m^3 for 7 days retention to cover the Christmas period, or 475 m^3 for the recommended minimum 45 days retention period. Although the land ultimately proved to be unsuitable for the construction of such a tank, the removal of 3810 m^3 of sewage each year from the site using a 9 m^3 tank would have required over 420 trips per annum and at an average cost of £24 per load, which would have amounted to over £10 000 each year. The advantages and disadvantages of cesspools have been examined by Mann (1979). One area where storage tanks are still widely used is on farms. Storage of farm effluents is expensive and, ideally, raw effluent should be spread immediately on to the land. However, this is impossible because of seasonal crop cycles, and at certain times of the year because of the risk of surface runoff causing pollution. The use of slurry tanks, both of underground and above ground construction, is widespread and the design of such systems has been reviewed by Weller and Willetts (1977).

Septic tanks are essentially a chamber in which the settleable solids settle out of suspension to form a sludge which undergoes anaerobic breakdown. The process provides partial treatment only and cannot produce an effluent of Royal Commission standard, and further treatment is required either by a percolating filter or by the provision of a percolation area in which the sewage percolates into the soil via a system of underground distribution pipes. If the effluent from a septic tank is discharged directly to surface waters then a minimum dilution factor of 300–400 is required. Although most commonly used for individual houses they can be used to serve small communities of up to 500, and are commonly used in combination with a percolating filter for small rural villages. Septic tanks require no power and where subsurface drainage is used quality control can be confined to suspended solids removal only. They are not adversely affected by intermittent use, have a very small head loss and can achieve a 40–50% BOD and 80% suspended solids removal. In its simplest form the septic tank consists of a single chamber (Fig. 4.85) with a single input and output. They can be of any shape although they are traditionally rectangular and made out of concrete. New prefabricated units in plastic or fibreglass come in a variety of shapes, designs, and even colours. The tank consists of three separate zones, a scum layer on top of the clarified liquid,

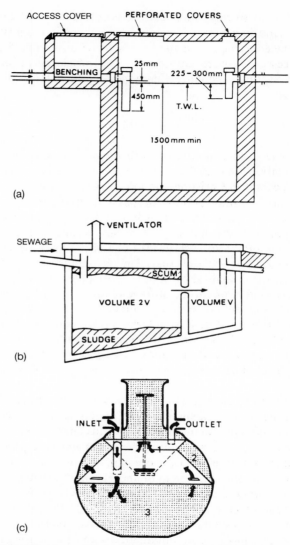

FIG. 4.85 Examples of manufactured septic tanks. (a) Single-chamber system; (b) two-chamber system; (c) GRP septic tank (Mann, 1979).

with a sludge layer in the base. Wastewater enters and leaves the tank via T-shaped pipes that prevents disturbing the scum layer or allowing solids to be carried out of the tank. As the wastewater moves through the chamber, settleable solids gradually settle forming a sludge, and any fats or buoyant material floats to the surface where they are retained by the baffle edge of the T-pipe and form the scum layer. The scum layer, although not vital to the successful operation of the tank, helps its operation in three ways: it prevents oxygen transfer through the air–water interface; insulates the anaerobic chamber by preventing heat loss; by attracting and retaining fats, floating material, and raised solids from the sludge layer. It is in the sludge layer where anaerobic decomposition takes place, with the organic fraction slowly being degraded. Often, decomposition is incomplete and the settled material is only hydrolysed and not broken down to methane. This results in intermediate products of anaerobic digestion, such as short-chain fatty acids, being produced and slowly being diffused back into the clarified liquid to be discharged from the tank. This incomplete anaerobic activity results in unpleasant odours. Methane production is inhibited by a number of factors in septic tanks, most notably low temperatures and insufficient sludge storage capacity. Methane production in septic tanks is low even in the summer, whereas in winter it may cease completely resulting in an overall increase in sludge accumulation. However, gas formation occurs even in the absence of methane formation, which can carry solids back into the liquid phase and cause an increase in the rate of scum accumulation. Most wastewaters contain a significant proportion of non-degradable solids, and even if anaerobic digestion is highly efficient there will be a gradual increase in solids in the tank. As these build up, the volume of the liquid zone in the chamber is reduced, thus reducing the retention time of the wastewater and the degree of settlement that is possible. If discharge of solids to the next treatment phase is to be minimized, then the septic tank must be regularly desludged. The floor of the chamber is sloped towards the inlet end above which a manhole is situated into which a suction pipe can be lowered (Fig. 4.85). Ideally, septic tanks should be desludged every 12 months, with only the sludge removed leaving about 5–10% of the sludge to re-seed the new sludge layer to ensure rapid anaerobic activity. In practice, the sludge is pumped out until it turns into water and then 5–10 litres are returned. Many operators feel that enough sludge will be left behind in the corners of the chamber and adhering to the walls to ensure seeding (Shetty 1971; Mann 1974). The scum should be disturbed as little as possible and the tank, which is designed to operate full, should be refilled with water as soon as possible after desludging. When commissioning septic tanks, they should be filled with water before use and seeded with a few litres of sludge. Although it is possible to purchase a commercial seed, the former is cheaper and probably better. There are a number of materials advertised for septic tank problems. A common claim is 'use chemical X and never again will you have to clean out your septic tank'. Schwab et al. (1975) examined a number of these chemicals and found that none had any effect on septic tank action and a few even had adverse effects on the percolation area. There is no short-cut to effective tank care and the addition of yeast, enzymes, and bacteria are not necessary for digestion within the tank. It is a common fallacy that odours produced by septic tanks can be reduced by cleaning the chamber using a proprietary cleaner. Under no circumstances should a tank be disinfected, as this will only result in even worse odours being produced on recovery and inhibiting the degradation processes already taking place in the chamber. If odours are a problem then the tank should be desludged.

The design and constructional details of septic tanks are given by the British Standards Institution (1972) and the Institute of Industrial Research and Standards (1975). The two most important design criteria for septic tanks are the suitability of land and the capacity of the chamber. Many by-laws insist on minimum distances for the siting of tanks and percolation areas from houses, wells, streams, and boundaries. The capacity and hence the retention time for settlement and sludge digestion is calculated according to the number of people discharging. The minimum capacity of a septic tank must not be less than 2720 l with the actual capacity calculated as:

$$c = (180P + 2000)\ l$$

where c is the capacity in litres and P the number of people discharging into the system. It is quite common for people to purchase a country cottage in which an elderly couple had lived for many years only to find that once modernized with mains water, a new bathroom, and a kitchen full of labour-saving devices, that the septic tank system can no longer cope. For example, a disposal unit for household kitchen waste (garbage grinder) results in a 30% increase in the BOD and a 60% increase in suspended solids. Where these are installed the capacity of the septic tank must be calculated as:

$$c = (250P + 2000)\ l$$

Therefore, in the design of a septic tank the future size of the family, number of bathrooms, visitors, and other potential developments must be included in the design. It is unacceptable to use half value for children in such calculations as their water demands are just as great as adults. Some wastewaters are not suitable for discharge to septic tanks, one lady who started a pottery at her country retreat discharged the wash water from her wheel directly to the septic tank. Within a year the excess clay had compacted so hard in the base of the chamber it had to be drained and manually dug out.

Septic tanks are easy to up-rate by adding extra chambers in series. This reduces the effects of surge flows and excessive sludge accumulation in the first chamber. The commonest faults associated with septic tanks include: *Leaking joints* when tanks have been constructed from concrete panels or concrete rings. It is important that septic tanks should be watertight to prevent contamination of the groundwater and should be constructed, if possible, without joints. *Non-desludging* is the commonest fault, resulting in a reduced retention time and a stronger effluent due to less settlement. The loss of sludge under these circumstances can block pipes, percolation areas or filters. *Blocked outlet pipes* are frequently a problem because of the scum layer becoming too thick or sludge physically blocking the outlet pipe, which causes the tank to overflow. In a review of Irish septic tanks, Keenan (1982) found that all the systems he visited had inadequate percolation areas and only a few had access to the tanks for the sludge tanker, with pipes often having to pass through peoples houses to reach the tank itself. The problems associated with soil percolation are examined in Section 4.5.3 and have been reviewed by Laak (1986).

Little information is available on the operation of septic tank systems, but on the whole they are very robust. However, care should be taken when using the following. *Disinfectants* should be used moderately as their bactericidal properties kill off the anaerobic bacteria, which can result in awful odours being produced during recovery. It is best to use disinfectants having free chlorine, as it reacts with the organic matter in the sewage rendering it harmless by the time it reaches the tank. *Caustic soda*, which is often used to remove grease from drains, can cause the sludge to flocculate and rise. It can result in sludge passing out of the tank and blocking the percolation area. Small amounts of acidic or alkaline cleaners do no lasting harm and should be used in preference. Some strong cleaners can upset the pH of the tank, which should be as near to neutral as possible. Kleeck (1956) suggests that the pH of the tank should be controlled by the addition of hydrated lime, but this should only be used as a last resort. Under good operational practice the tank will buffer itself. *High sodium* concentrations in the water does not affect the septic tank system directly but can impair the drainage properties of the soil. Those who have water softeners and use soil percolation as a secondary treatment process of septic tank effluents should take advice. *Detergents*, especially alkyl benzene sulphonate, are known to inhibit the digestion process, although, providing the tank capacity is sufficient, if normal concentrations of detergents are used the performance of the system will not be impaired. Enzyme-based washing powders have no effect on septic tank systems. *Large flushes of water* to the tank should be avoided if possible, to prevent scouring of the sludge, unless the tank is large enough to withstand them or a dual chamber system has been installed. Wherever legally permissible, bathwater should be diverted to a soak away, as should rainwater and melted snow from roofs and paved areas. *Solid material* such as disposable nappies, tampons, sanitary towels, coffee grinds, bones, cigarette ends, and cat litter will not degrade in the tank and should not be discharged to the septic tank as they will reduce the volume of chamber very quickly and can be difficult to remove. Fats and greases should also be disposed of separately whenever possible.

Maintenance of the septic tank, like all sewage treatment units, is important in maximizing treatment efficiency and preventing pollution. The scum and sludge accumulation should be inspected twice a year, depending on its volume. The depth of the sludge and the thickness of the scum should be measured in the vicinity of the outlet pipe. Records should be kept so that desludging frequency can be accurately predicted. The tank should be cleaned whenever the bottom of the scum layer is within 7.5 cm of the bottom of the outlet pipe or the sludge level is within 25–30 cm of the bottom of the outlet pipe (Public Health Service 1969). Scum thickness is measured using a hinged flap device, which is a weighted flap attached to long rod (Fig 4.86). Any device can be used which allows the bottom of the scum mat to be felt. The measuring device is pushed through the scum layer until the hinged flap falls into the horizontal position. It is then gently pulled upwards until the flap engages against the bottom of the scum layer. The handle is marked to correspond to a reference point on top of the tank. The same procedure is used to locate the lower end of the outlet pipe. The difference in height on the handle corresponds to the distance the scum is from the outlet. The depth of sludge is

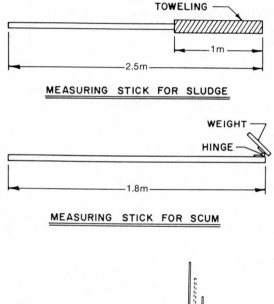

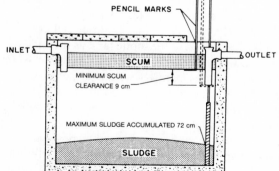

FIG. 4.86   Equipment and procedures for measuring the accumulation of sludge and scum in a septic tank (Schwarb *et al.* 1975).

measured by wrapping a long stick in rough white towelling which is tied securely. The stick is slowly lowered into the tank through the vertical piece of the outlet pipe to the bottom of the tank to avoid the scum. It is left for a few minutes and then slowly removed. The depth of the sludge can be distinguished on the towelling by black particles clinging to it. If the depth of the sludge is more than one-third of the total liquid and sludge depth at this point, desludging should be arranged.

Imhoff tanks

These combined settlement and digestion units are similar to septic tanks except they are specifically designed for sewage treatment plants rather than individual houses. They were widely adopted throughout the world and can still be seen in many small- and medium-sized sewage treatment plants in Ireland and the UK, although they have been superseded by separate sedimentation and digestion units in recent years. Imhoff tanks are rectangular in plan and comprise of a settlement tank with a digestion chamber below. Single or double units were made in a variety of sizes serving populations ranging from 300 to 10 000 (Fig. 4.87). Screened sewage enters the settlement chamber with a retention time of about 4 hours and as the solids settle they fall down the steep sides of the settlement unit and pass through the wide longitudinal slot that runs along the length of the settlement chamber and collects as a sludge in the digestion compartment. The overlapping lip of

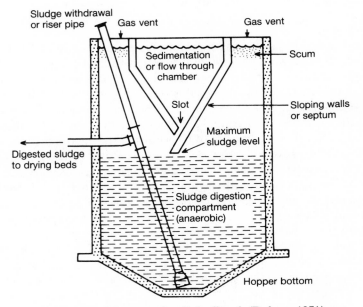

FIG. 4.87 Diagram of an Imhoff tank (Dabney 1971).

the base of the settlement chamber prevents solids buoyed up by gas or gas bubbles themselves entering the settlement chamber and interfering with the sedimentation process. The digestion compartment generally has a large storage capacity and sludge only needs to be removed at long intervals. The long sludge retention time allows digestion to take place with any gas produced vented to the atmosphere (Fig. 4.87). In some warmer climates, digestion proceeds at a more rapid rate allowing the gas to be collected. Imhoff tanks are operated in the same way as a primary sedimentation tank with 50–70% reduction in suspended solids and a 30–50% reduction in BOD being possible. They are compact treatment units that require little maintenance, making them ideal for small communities. However, as digestion is psychrophilic there is incomplete breakdown of the sludge during winter, which produces odours and poor gas production. In the summer, gas production can become excessive and interfere with settlement causing foaming and bubbling. Although simple in design they are not easy to construct in concrete, and soil stability is especially important as Imhoff tanks are fairly deep. The main problems associated with their operation in Ireland are: infrequent desludging, so that sludge is retained within the settlement chamber reducing the time for settlement; hydraulic overloading again reducing the retention time; the longitudinal slot at the base of the settlement chamber becoming blocked with large solids and in the summer gas accumulation under the harden crust of scum. Apart from proper operational management, the more interesting attempts to up-rate Imhoff tanks have included converting them into RBC units (Section 4.1.2.1).

Lagoons

The use of anaerobic lagoons has been fully reviewed in Section 4.3.1. They are used for strong organic wastes with digestion usually associated with the accumulated sludge in the bottom of the lagoon. Shallower lagoons are also used in some vegetable processing industries combining settlement of soil and waste vegetable matter with partial treatment. Such lagoons are commonly used in the sugar beet industry and although a 80–90% removal of suspended solids and a 60–70% reduction in BOD is possible, the final effluent produced will still be much stronger than a domestic wastewater (>500 mg l^{-1} BOD, >200 mg l^{-1} suspended solids), and requires further treatment. A number of problems are associated with lagoons. Vegetable-processing is generally seasonal and the operational period of the lagoons will be short, <100 days per annum, leaving little time for an active anaerobic biomass to build up. Retention time in the lagoons is too short to ensure complete anaerobic breakdown and in plug-flow systems, it is not possible for a sufficient MCRT to develop. In the treatment of sugar beet wastewater, incomplete anaerobic breakdown occurs with the carbohydrate waste rapidly hydrolysed to volatile acids. However, the volume of process water used in the industry is so great that the acids are present in only a low concentration making recovery or conversion to biogas via separate digestion very difficult. Complete degradation to methane is often inhibited in shallow lagoons because of partial reaeration of the water, and low temperatures. With vegetable wastes, inhibition is also often due to a high C:N ratio.

Anaerobic lagoons are particularly effective for stronger wastes and are widely used on farms for the storage and treatment of animal slurry. Distinct layers form in such lagoons with a bottom layer of settled solids forming a sludge, a liquid layer, and a floating crust. Cattle slurry rapidly forms a crust, excluding air and producing ideal anaerobic conditions, due to the high fibre content of cattle faeces. Pig slurry, in contrast, does not readily crust over thus allowing oxygen to diffuse into the lagoon so that the top 100–150 mm of the liquid is aerobic. Crust formation is quicker in summer and in a well-constructed anaerobic lagoon the BOD and suspended solids of the slurry will be reduced by 80–90%. During the summer, methane production can be high and large pockets of gas will be seen breaking through the crust. Desludging of farm lagoons should be done every 3–5 years depending on depth, total capacity, and degree of anaerobic degradation achieved, with the sludge spread directly on to the land without any of the associated pathogen problems. As most of these lagoons are between 2–3 m deep they are potentially dangerous and the crust rapidly becomes covered with grass, weeds, and even small bushes giving it a false appearance of stability. Although in certain circumstances they may be strong enough for an animal or even a child or a man to walk on, they cannot support the weight of a small vehicle. So for safety they should be securely fenced. The construction of a lagoon of this depth is not straightforward and advice will be required as they may need to be lined with welded butyl rubber or PVC sheeting in order to prevent seepage, which could cause groundwater contamination. Emptying a deep lagoon will require extremely stable banks from which heavy plant can operate, therefore earth banks may not always be suitable.

4.4.2.2 Digestion

Modern sewage treatment practice separates primary sedimentation from digestion. Digestion is carried out either in large open tanks or lagoons at ambient temperature (psychrophilic digestion), or more rapidly in covered tanks heated to between 30–35°C (mesophilic digestion). The former digesters are usually unmixed and as they are uncovered any gas produced is dispersed to the atmosphere. In winter, the rate of digestion will be extremely low or even zero, whereas in the summer digestion will be rapid. Therefore, long retention times are required of between 6–12 months to stabilize sludges under these conditions and to balance between sludge accumulation and sludge degradation. Psychrophilic digesters are restricted to smaller works where the output of sludge is low and land is readily available. Both septic tanks and Imhoff tanks fall within the psychrophilic range that operate from 5–25°C.

Heated digesters are much more cost-effective at larger treatment plants where there is sufficient sludge available to ensure a continuous operation, with gas collected and used to either directly or indirectly heat the digesters. In their simplest form, digesters are intermittently fed with no recycle. They can be a single-stage reactor (Fig. 4.88), but in the UK and Ireland, conventional digestion is usually a two-stage process with the primary digester heated to the desired temperature in order to allow optimum anaerobic activity, with acid

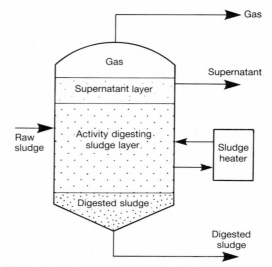

FIG. 4.88   Single-stage anaerobic sludge digester.

formation and gas production occurring simultaneously. The primary reactor is continuously stirred, unlike the earlier stratified digesters, which reduces the retention time from 30–60 d to < 15 d. The secondary digester is unheated and can be used for two functions. Either to continue digestion under psychrophilic conditions in which case it will be stirred to ensure complete mixing as in the primary digester, or it can be used for sludge separation (Fig. 4.89). However, as the evolution of gas interferes with settlement, the second reactor cannot be used for both functions. When used for sludge separation the secondary digester is not stirred and is generally uncovered. Under quiescent conditions, the solids separate leaving a strong supernatant liquor on top of the stabilized sludge. The liquor has a high soluble organic content with a high BOD which needs to be recirculated back to the inlet of the treatment plant. The settled solids are removed, dewatered, and disposed. When the secondary digester is used for digestion a third digester may be employed if thickening is required (Institute of Water Pollution Control 1979). This type of frequently fed reactor with no recycle of solids is used for the treatment of sludges with high solids content such as animal slurries and sewage sludge (Fig. 4.90) For example, the waste from a 300 pig unit can be treated in such a reactor with a HRT of 10 d

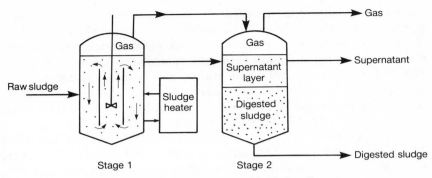

FIG. 4.89 Two-stage anaerobic sludge digester.

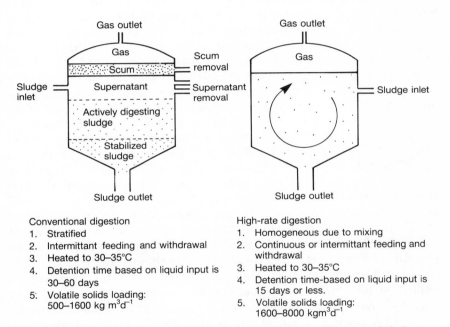

Conventional digestion
1. Stratified
2. Intermittant feeding and withdrawal
3. Heated to 30–35°C
4. Detention time based on liquid input is 30–60 days
5. Volatile solids loading: 500–1600 kg m³d⁻¹

High-rate digestion
1. Homogeneous due to mixing
2. Continuous or intermittant feeding and withdrawal
3. Heated to 30–35°C
4. Detention time-based on liquid input is 15 days or less.
5. Volatile solids loading: 1600–8000 kgm³d⁻¹

FIG. 4.90 Comparison of conventional and high-rate digestion of sludge.

and a gas yield of 0.3 m^3 kg total solids d^{-1} (Hobson 1984). Sludges and liquors with a lower solids concentration can only be digested in a conventional reactor where there is some form of solids recycle to prevent wash-out of bacteria. Generally, such systems have a short HRT of 0.5–5.0 d and an efficient secondary settlement system is vital to ensure an adequate SRT in the primary reactor (Section 4.4.3.1). The other type of flow through reactor is the plug-flow type (Fig. 4.83b), which is used mainly for animal manures, although still at the experimental stage they have also been used for primary sludge digestion in Dublin. However, adequate sludge recycle is required for seeding purposes, while mixing can be achieved by gas recycle. The major advantage of this sytem is that it allows simple tank configurations to be used, resulting in significant savings on capital cost (Casey 1981).

Digestion is used to stabilize both primary and secondary sludges having a solids content of between 20 000–60 000 mg l⁻¹ (2–6%). About 70% of the sludge is degradable and up to 80% of this will be digested reducing the solids content by about 50%. The remaining solids form a relatively stable sludge which has a characteristic tarry odour, although as discussed in Chapter 5, it can be difficult to dewater. The mixture of primary and secondary sludges fed into a digester is rich in carbohydrates, lipids, and proteins, which are ideal substrates for microbial degradation, and although the sludge is already rich in a variety of anaerobic bacteria, in order to obtain the correct balance of hydrolytic acid producers and methanogenic bacteria, the raw sludge needs to be seeded. This is particularly important as the main bacterial groups responsible for digestion are dependent on the end-product of the other. The particulate and high molecular weight organic matter is broken down to lower fatty acids, hydrogen, and carbon dioxide by facultative anaerobic bacteria (first stage) (Section 3.4.3.). The end-products are converted to acetic acid by acetogenic bacteria (second stage), which is subsequently converted in the final stage by methanogenic bacteria to methane and carbon dioxide. Obviously, anaerobic digestion will proceed most efficiently when the rates of reaction at each stage are equal. However, if the first stage is limited then the nutrient supply to the other stages will be reduced with the overall effect of suppressing the rate of digestion and biogas formation, but not inhibiting it. If either of the subsequent stages are restricted then the first-stage products will accumulate causing a gradual rise in the carbon dioxide fraction in the biogas (>30%) and a gradual fall in the pH as volatile acids accumulate until the pH falls below 7.0 and the whole process becomes stressed. All three stages proceed simultaneously within the same reactor with the bacterial groups in close proximity to each other. However, as you move along the anaerobic chain of reaction the bacterial groups become progressively more sensitive to their environmental conditions. Hydrolysis and acid formation are carried out by a diverse and large group of facultative bacteria that can tolerate a wide variation in temperature, pH, and a range of inhibitory substances. In contrast, the acetogenic and methanogenic bacteria are far more sensitive micro-organisms that are highly specialized and severely inhibited by even minor changes in operating conditions. Thus, it is more common to have operational problems involving the second and third stages of digestion.

Design

Primary digesters are normally covered with a fixed or floating top for gas collection (Fig. 4.91). Floating covers rise and fall according to the volume of gas and sludge, as they are the same design as conventional gas holders, with

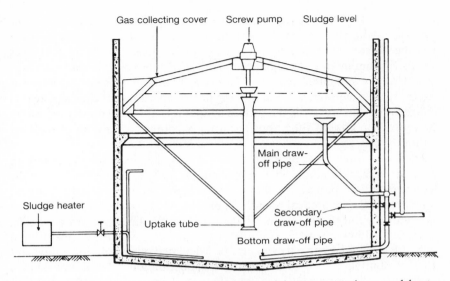

FIG. 4.91   Primary digestion tank with screw mixing pump and external heater
(Institute of Water Pollution Control 1979).

the weight of the floating cover providing the gas pressure. Fixed covers
require a separate gas holder that allows the gas to move freely in both
directions. This is particularly important if a vacuum is not to be produced
when sludge is removed from the digester, which would not only effect
operation but could affect the structural stability of a reactor with a fixed
cover. Methane is biochemically inert and it can be stored above the sludge
within the reactor without affecting or inhibiting the process. Digesters are
usually circular in plan with maximum diameters of approximately 25 m, and
to ensure the minimum depth the ratios of reactor height to diameter is
normally between 1:3 to 1:2, although smaller digesters tend towards a ratio
of 1:1. The floor of the tank is sloped between 12 and 30° with facilities to
withdraw heavier solids and grit from the base. Large tanks are built above
ground, and smaller units can be constructed below ground level or
surrounded by earth embankments to provide additional insulation. The
entire contents of the tank needs to be turned over once every 3–4 hours, with
the sludge warmed by heat exchangers using hot water from gas boilers or the
cooling systems of gas engines. Heating and mixing of digester contents can be
achieved by a number of techniques. The most common being a central screw
mixing pump with an external heat exchanger, a circulating pump, and
exchanger housed in a projecting chamber or internal gas-lift pumps. All these
methods are fully reviewed by the Institute of Water Pollution Control (1979).
The system shown in Fig. 4.91 has a screw pump mounted in the top of a
vertical uptake tube. Known as the 'Simplex System', it is designed and

manufactured by Ames Crosta Ltd. The pump circulates the contents of the tank by drawing sludge from the bottom of the reactor and then spraying it out over the surface of the main body of sludge, providing surface scum control as well as efficient mixing. At regular intervals, the pump is reversed for a short period to prevent blockages and improve mixing. The heat-exchanger is external to the digester, the sludge being drawn off from near the base, heated and then returned towards the surface so that it is not immediately drawn up the central updraught tube, thus ensuring maximum utilization of the heat within the reactor.

The basic design criterion is the provision of sufficient capacity to ensure an adequate sludge retention time. It is common practice to use 25–30 d as the design retention period to allow for variation in daily sludge loading. However, as the theoretical retention time can be as low as 7–10 d, many feel that with the improvements in mixing and heating digesters now available, that 15 d allows an adequate safety margin to cover most operational difficulties. Digester capacity can also be calculated on population equivalent (PE). Assuming a per capita sludge production of 1.8 l of raw sludge containing 4.5% dry solids, the capacity of a digester with a 25 d retention time is:

$$\frac{1.8 \times 25 \times PE}{1000} \, m^3$$

Although often quoted, organic loadings expressed as kg organic or volatile matter m^{-3} are not appropriate parameters for digester design. With the standard 25 d retention time, the typical sludge feed of 4.5% total solids with an 85% organic (volatile) matter content results in an organic loading equivalent to 1.5 kg $m^{-3}d^{-1}$. However, in the UK, organic loading ranges from 0.27–2.76 kg $m^{-3}d^{-1}$ indicating the varying water contents of the sludges being digested. Thus, in order to achieve the optimum organic loading of 1.5 kg $m^{-3}d^{-1}$ with a thin sludge, a much shorter retention time will result with incomplete digestion occurring (Swanwick *et al.* 1969). For this reason sludges should be characterized and loadings specified as a concentration of total (dry) solids (%).

Secondary digestion tanks are generally used for storage and separation, although they can be used for the further digestion and gas collection. Sludge is passed as frequently as possible from the primary digester into the uncovered tank where it cools and allows the liquor to separate from the solids so that each can be withdrawn separately. Temperature differences between the cool and warm sludge can cause convection currents within the digester that will hinder settlement, as will gas production, so that it becomes difficult to obtain a solids-free liquor. To overcome this, secondary digesters are relatively shallow with a maximum depth of 3.5 m. The capacity of older tanks were between 50–70% of the primary digester providing 15–20 d retention. Newer tanks are approximately the same size as the primary tanks and provide similar retention times for the sludge. Tank volume can be estimated on a PE basis using the equation:

$$0.035 \times PE \; m^3$$

In order to ensure adequate operating conditions for digestion the pH value, the concentration of carbon dioxide in the biogas, and the volatile acid concentrations should be continuously monitored. The normal operating conditions should be a pH of 7.0–7.2, alkalinity (as $CaCo_3$) of 4000–5000 mg l^{-1}, and a concentration of volatile acids (as acetic acid) of < 1800 mg l^{-1}. The carbon dioxide content of the biogas should not exceed 30%, and once any of these values are exceeded then immediate remedial action is required.

Digester operational Management

Successful operational management of anaerobic digesters depends on six major factors: composition of the raw sludge, method of addition of raw sludge to the digester, internal mixing and circulation, temperature, pH, and solids retention time.

(a) *Composition of raw sludge.* Like aerobic micro-organisms, those responsible for digestion require certain substrates, growth factors, trace elements, and nutrients for successful development. Mosey (1983) compares raw and digested sewage sludge and shows quite clearly that the major substrates such as lipids, cellulose, and some proteins are present as solids in suspension, and that a substantial fraction of the organic matter is either converted to microbial biomass or not metabolized at all as is the case with lignin (Table 4.34). Nitrogen and phosphorus are both vital for bacterial growth and are required at minimum concentrations of 2.5% and 0.5% of the dry organic matter content of the sludge respectively. This is equivalent to a C:N ratio of between 10–16:1 and an N:P ratio of 7:1, although higher C:N ratios of up to 30:1 have been cited (McCarty and McKinney 1961). What is clear is that anaerobic processes are far less demanding in terms of N and P than aerobic systems. Huss (1977) found the optimum BOD:N:P ratio to be 100:0.5:0.1, whereas the COD:N:P ratio was found to range between 42:0.7:0.1 and 150:0.7:0.1 by Henze and Harremoes (1983). In practice, sewage sludges have excess N and P ensuring adequate digestion, although certain trade wastes, including sugar beet and other food-processing wastes, may be deficient in these nutrients making them less amenable to the digestion process (McCarty 1964; Mosey 1974; DoE 1974). Sewage sludges are mixtures of complex organic materials and vary widely both in composition and strength from place to place. The average solids content in the UK for sewage sludge is often quoted as 4.5 g per 100 g wet sludge, although this will vary even from the same source depending on the operation of the settlement tank from which the sludge is removed.

TABLE 4.34 Typical composition of sewage sludges before and after digestion (g per 100 g total solids) (Mosey 1983)

Constituent or test	Raw sludge	Digested sludge
Suspended solids	95	97
COD	140	100
Organic carbon	40	31
Organic matter	60–80	45–60
Greases and fats	7–35	3.5–17
Cellulose	4	0.6
Hemicelluloses	3	1.6
Lignin	6	8
Protein	22–28	16–21
Anionic detergents	0.5–1.5	0.7–2.2
Zinc	0.09	0.14
Copper	0.035	0.055
Lead	0.016	0.026
Chromium	0.01	0.016
Nickel	0.0092	0.015
Cadmium	0.0022	0.0035

Secondary sludges from activated sludge or percolating filter units are less amenable to digestion than primary sludges as they contain a lower proportion of digestible matter per unit mass of solids as well as generally containing much more water than primary sludge. Only about 30% of the available organic and volatile matter is utilized in activated sludge compared with 50% in most primary sludges.

A wide variety of compounds normally present in sewage can inhibit the digestion process when present in excessive concentrations, most notably detergents, chlorinated hydrocarbons, heavy metals, and ammonia. The problem is that many toxic substances are separated at the primary settlement stage with organic compounds such as chloroform being heavier than water and thus collecting at the outlet hopper in the primary settlement tank, and many heavy metal salts being precipitated at this stage as hydroxides. All inhibitory susbstances will affect gas production as the methanogenic group of bacteria is made up of only a few sensitive species, unlike the diverse hydrolytic and acid-forming bacteria. Another important factor is that the hydrolytic and acid-forming bacteria are present in the raw sludge and are constantly replaced, in contrast, the methanogenic bacterial population is self-sustaining and once the population has been reduced it will take a long time for the population density to be restored and therefore re-seeding may be necessary.

Detergents. Non-ionic and cationic detergents have little effect on anaerobic digestion, even at high concentrations (Bruce *et al.* 1966), whereas anionic detergents inhibit the process. Household washing powders, which contain alkyl benzene sulphonates (ABS), both hard and soft, are non-degradable anaerobically (Little and Williams 1971; Klein and McGauhey 1965; Maurer *et al*. 1965; Punchiraman and Hassan 1986; Hernandez and Bloodgood 1960). As they are strongly adsorbed on to organic solids they are invariably present in sewage sludge. Anionic detergents at concentrations (expressed as Manoxol OT) in excess of 1.5% of the dry raw sludge solids inhibit anaerobic digestion even in heated digesters with relatively long retention times of up to 40 d, with gas production being particularly sensitive. Concentrations <1.5% will reduce the tolerance of the process to other operational variables, such as temporary overloading. The effects of ABS on the bacterial population are numerous and include direct toxicity (Swanwick and Shurben 1969; Meynell 1976). Although degraded aerobically (Huddleston and Allred 1963; Linden and Thijsse 1965; McKenna and Kallio 1965), it is clear that these detergents are not utilized in the anaerobic process because the enzymes responsible for their degradation, such as mono-oxygenase (Cain 1981), are not able to function due to the lack of oxygen (Little and Williams 1971; Alexander 1965). These concentrations are quite common in domestic sewage sludges and although bacterial populations can become acclimatized, serious inhibition will occur if the concentration is allowed to rise to 2.0% of the dry sludge solids. As digestion proceeds, the concentration of anionic detergents will increase so that at 1.5% the concentration of anionic detergents will rise to 2.5% dry solids in the digested sludge. If inhibition occurs then the anionic detergents can be rapidly neutralized by the addition of long-chain fatty amines, such as stearine amine, to the sludge prior to digestion (Swanwick and Shurben 1969).

Chlorinated hydrocarbons. Although high concentrations of these solvents are unusual in sewage sludges, they can cause a problem in certain trade wastes or where domestic sewage contains trade discharges. The most frequent source of this group of inhibitors is from dry-cleaning operations that discharge chlorinated hydrocarbons to the sewer. Although their effect on the process depends on a number of factors, it is the concentration in the sludge that is most critical (Swanwick and Foulkes 1971; Department of the Environment 1971). It appears that many chlorinated hydrocarbons act selectively on the methane bacteria, which is why they are so toxic to anaerobic digestion (Thiel 1969). The concentration at which 20% inhibition occurs is summarized in Table 4.35 for the most widely used chlorinated hydrocarbons, although their inhibitory effect will be increased if other inhibitory substances are also present.

Chloroform is the most toxic of this group and although most will be lost by volatilization in the sewer and during treatment, it can still cause detectable inhibition at concentrations as low as 10 mg kg^{-1} dry solids (Stickley 1970; Barrett 1972). Where chloroform is a problem it should be removed from the effluent by air stripping, prior to discharge to the sewer (Lumb *et al*. 1977).

Heavy metals. The most quoted cause of inhibition of sewage sludge digesters are heavy metals and, in particular, chromium, copper, nickel,

TABLE 4.35   Concentration of chlorinated hydrocarbons in digesting sludge at which 20% inhibition occurs (Institute of Water Pollution Control 1979)

Chemical	Concentration (mg kg^{-1} dry solids)
Chloroform	15
Trichlorethane	20
1,1,2-Trichlorotrifluoroethane	200
Carbon tetrachloride	200
Trichloroethylene	1800
Tetrachloroethylene	1800

cadmium, and zinc (Swanwick *et al.* 1969). However, particularly high concentrations of metals are required to have a significant effect (Institute of Water Pollution Control 1979) (Table 4.36). Synergistic effects have been noted with heavy metals and a number of other inhibitory substances, and under certain conditions even low concentrations of heavy metals may cause problems. Apart from the concentration of heavy metals in the raw sludge other factors such as solubility, pH, and the concentration of sulphide present will all effect their concentration in the digester. Anaerobic bacteria are able to withstand quite high concentrations of total heavy metals as a considerable percentage of the metal ions can be precipitated out of solution as either sulphides or carbonates. The concentrations of metals that can be present in sewage without adversely affecting anaerobic digestion of the resulting sludge have been summarized in Table 4.37 (Barth 1965). There is always more than one heavy metal present in a sewage sludge and the inhibitory effects are additive on an equivalent weight basis (Mosey 1976). Where the milligram equivalent weight (meq) per kg dry sludge solids (K) exceeds 400 meq kg^{-1} there is a 50% chance of digester failure which rises to 90% when K exceeds 800 meq kg^{-1}. To ensure a 90% probability that digestion will not be affected, the value of K should be < 170 meq kg^{-1}.

TABLE 4.36   Concentrations of heavy metals in digesting sludge that cause a 20% reduction in gas production in laboratory experiments (Institute of Water Pollution Control 1979)

Metal	Batch digesters: concentration (mg kg^{-1} dry solids)	Typical concentration in digested sludges (mg kg^{-1} dry solids)
Nickel	2000	30–140
Cadmium	2200	7–50
Copper	2700	200–800
Zinc	3400	500–3000

TABLE 4.37 Highest metal concentrations in sewage that will allow satisfactory digestion of sewage sludge (Casey 1981)

Metal	Concentration in influent sewage (mg l^{-1})	
	Primary sludge digestion	Combined sludge digestion
Chromium (hexavalent)	> 50	> 50[a]
Copper	10	10
Nickel	> 40	> 10[a]
Zinc	10	10

[a] Higher dose not studied.

K is measured using the concentrations of the most abundant heavy metals (mg l^{-1}) in sewage sludge, excluding chromium, using the equation:

$$K = \frac{((Zn)/32.7) + ((Ni)/29.4) + ((Pb)/103.6) + ((Cd)/56.2) + ((Cu)/47.4)}{\text{Sludge solids concentration (kg } l^{-1})} \text{ meq kg}^{-1}$$

Ammonia. Ammonia or ammonium ions are essential nitrogen sources for anaerobic digestion, but can be inhibitory when present at concentrations of > 150 mg N l^{-1} and 3000 mg N l^{-1} respectively. These concentrations are only occasionally found in very thick sewage sludges and undiluted farm slurries. However, as Mosey (1983) clearly explains, the system is largely self-regulating in that inhibition causes an accumulation of volatile solids, which in turn depress the pH value, converting dissolved ammonia (NH_3) to the less toxic ammonium ionic form (NH_4^+), thus alleviating inhibition.

Methane formation does not occur readily in the presence of electron acceptors, such as sulphate and nitrate. Sulphate can be a particular problem in the digestion process if present in sufficient quantities. The sulphate is reduced to sulphide by bacterial action (Section 3.4.4), with hydrogen sulphide eventually being formed. Sulphate concentrations > 500 mg l^{-1} can reduce methane production and generate up to 4% hydrogen sulphide in the biogas. The hydrogen sulphide will form insoluble compounds with heavy metals. Therefore, as long as hydrogen sulphide and heavy metals are present in equivalent proportions then hydrogen sulphide production will cause no problems and, in fact, will be beneficial. However, two problems can occur: hydrogen is consumed by sulphate reduction and is no-longer available for methane formation, which is inhibited, and hydrogen sulphide itself has a direct toxic effect on the methanogenic bacteria being toxic at concentrations > 200 mg l^{-1} (Mosey 1976). Remedial action involves precipitating the excess hydrogen sulphide out of solution as ferric sulphide by the addition of an iron salt, such as ferric chloride or ferric oxide, but not, of course, ferric sulphate. Nitrates can also cause problems because if denitrification occurs within the digester there will be a shift in the redox potential (Section 3.4.6), which will suppress methane production. Methanogens are strict anaerobes and can be completely inhibited by a dissolved oxygen concentrations as low as $0.01 \text{ mg } l^{-1}$ (Wolfe 1971). They require a reduced environment with a redox potential within the range -200 to -420 mv. This has been a serious problem at a number of sewage treatment plants in the UK and has been remedied by allowing denitrification to occur before the sludge enters the digester, with the nitrate being converted to nitrogen gas under anoxic conditions (Section 3.4.5).

(b) *Method of sludge addition to digester.* Digesters can be operated either as batch or continuous processes that may incorporate recycling of solids, gas, or both. For optimum performance sludge should be added to the digester as frequently as possible in order to avoid fluctuations in gas production or problems with scouring the active bacteria from the primary digester. It is not generally possible to feed sludge directly from a sedimentation tank on a continuous basis, as it requires consolidation, or thickening, before it is suitable for use. Although the majority of digesters are fed with raw sludge at least once a day, two or three times a day is preferable (Swanwick *et al.* 1969).

Commissioning digesters is a topic of some controversy. The most commonly used method is to gradually fill the digester with raw sludge together with the seed sludge. However, it is not possible to commence circulation or heating until the reactor is full, which can lead to problems in starting up due to compaction of solids at the base. Alternatively, the digester is filled with sewage so that circulation and heating can commence at once, with the raw and seed sludge added gradually. This method has the disadvantage that if the seed sludge is too diluted, so that the correct balance of micro-organisms is not maintained, incomplete digestion producing unpleasant odours will result. Although it is possible to use raw sludge which has been stored for several months as a seed, it is best to use sludge from another digester, preferably from the same plant. The ratio of seed sludge to raw sludge should be between 1:10–1:5. Once started, the percentage of seed can be increased but care must be taken to prevent excessive acid production from inhibiting the development of methanogenic bacteria. There will be little biogas production for the first 4–6 weeks and an alternative source of heating the reactor during commissioning will be necessary (Sambridge 1972).

(c) *Internal mixing and circulation.* Stratified and completely mixed digesters have already been discussed. The purpose of mixing and circulation within the digester is: to promote close contact between the raw and digesting sludges; to maintain a uniform temperature and solids mixture throughout the tank, and prevent localized accumulation of inhibitory substances; to discourage scum formation and settlement of grit and dense solids; and, most important of all, to encourage the release of gas from the sludge in the lower regions of the digester. Poor mixing will lead to stratification within the digester and will result in partially digested sludge being withdrawn (Institute of Water Pollution Control 1979). Efficient mixing turns the conventional plug-flow reactor into a high-rate digestion process by making use of the total reactor volume and ensuring a faster reaction rate due to the removal of mass transfer limitations of food and micro-organisms. Mixing can be done either mechanically or by the recirculation of biogas. Mixing is enhanced by the circulation of sludge through the heat exchanger that operates continuously. The evolution of gas, which reaches a maximum 2–3 hours after the addition of raw sludge, will also supplement the mixing effect. The actual mixing mechanism may not be operated continuously at all plants.

Scum, which is composed of grease, oil, and soaps, with floating material entrained, will tend to form on the surface of the digestion tank if mixing is not adequate. It causes a number of problems, such as reducing the effective capacity and thus the retention time of the reactor, interferes with mixing and the internal circulation of the sludge, interferes with the movement of the floating gas holder, and restricts the evolution of gas. Most mixing systems withdraw the sludge from the base of the tank and spray it over the surface preventing scum formation, and screw pumps can be reversed to carry any scum down into the body of the tank. Prevention is best, however, and the sludge should be adequately screened before entry to the digester and excessive grease and oil should be removed prior to discharge to the sewer. There are also a number of design features aimed at reducing scum formation including scum removal devices, which disperse sludge by discharging raw or digested sludge into the surface layer via jets or nozzles.

(d) *Temperature*. Anaerobic digestion can occur over a wide temperature range which has generally been subdivided into three separate ranges, psychrophilic (5–25°C), mesophilic (25–38°C), and thermophilic (50–70°C) digestion. The rate of anaerobic digestion and gas production is temperature-dependent (Fig. 4.92), with optimum gas production at the higher temperature ranges. Therefore, the warmer the reactor the shorter the MCRT needs to be for complete digestion (Table 4.38).

Septic tanks, Imhoff tanks, sludge lagoons, and unheated sludge digesters, which are located mainly at smaller treatment plants fall into the psychrophilic range. At ambient temperatures in temperate climates, the rate of digestion is so low that the residence time of the bacteria needs to be of the order of 3–12 months. Suitable bacteria for seeding such systems, acclimatized to these low temperatures, can be obtained from marshlands. The majority of heated sewage sludge digesters operate in the mesophilic range, usually between 30–32°C (Swanwick *et al.* 1969), with residence times of 20–40 days. Thermophilic digestion is technically feasible and a number of pilot plants have been built and operated successfully (Pickford 1984). However, the rate of gas production does not increase continuously with temperature but rapidly declines after reaching an optimum at 55°C (Fair and Moore 1934). A

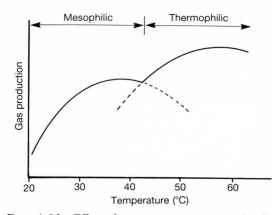

FIG. 4.92   Effect of temperature on gas production.

TABLE 4.38 Suggested mean
cell retention times (MCRT) for
the anaerobic digestion of sewage
sludge at various temperatures

Temperature (°C)	MCRT (days)
18	28
20	22
25	18
30	14
35	10
40	8–10

similar situation is found in the mesophilic range at 35°C (Fig. 4.92), indicating that different bacterial populations are responsible for thermophilic and mesophilic digestion, rather than thermo-tolerant mesophilic species. It appears that between these optima, erratic gas production will be encountered. Inhibition of methanogenic bacteria occurs at about 63°C (Pfeffer 1979). The effect of temperature is not great on the first stage of anaerobic digestion, as so many different species are involved that the operating temperature will always fall within the optimum range of some of the microorganisms present. However, the acetogenic and methanogenic bacteria are particularly sensitive to temperature with even a 2–3°C drop in a mesophilic digester adversely affecting biogas production.

Once operational, a heated digester will be adversely affected if the temperature is allowed to fluctuate by more than just a few degrees. It is general practice to ensure that the operating temperature is near the top of the preferred range before the onset of winter. However, a fall in temperature can be caused by a number of factors, including inadequate heating capacity, scaling of the heat exchanger surfaces, and the raw sludge having a low solids concentration. The temperature can be raised by reducing heat losses from the whole digestion unit by adequate insulation, increasing the heat input, reducing the water content of the raw sludge and thus reducing the total volume to be heated, and by descaling heat-exchanger surfaces.

(e) *pH*. Most anaerobic treatment systems have problems with pH control which arises from differences in the growth rate of the synergistic bacterial populations. The activity of the acid-producing bacteria tends to reduce the pH of digesting sludge from the optimal 7.0–7.5 range required by methanogenic bacteria. Under normal operating conditions, once a stable population of each of the groups has been established, an equilibrium is maintained by the buffering action of ammonium bicarbonate (the bicarbonate alkalinity), hence no external pH control is required. The bicarbonate ions are derived from carbon dioxide in the digester gas and the ammonium ions derive from the degradation of proteins in the raw sludge.

However, the digesting sludge does have a tendency to become acidic especially if the methanogenic bacteria are inhibited, or the digester is overloaded, which results in an excessive accumulation of volatile acids. Under these conditions, the buffering capacity may be exceeded with the pH rapidly decreasing to 6.0 causing the process to fail, resulting in a sudden decline in gas production (Fig. 4.93). Methanogenic bacteria exhibit a negative response when the pH shifts towards the acid region as they do when the temperature falls. Growth of methanogens is inhibited below pH 6.6, although the fermentative bacteria will continue to function until the pH has dropped to 4.5–5.0. It should be remembered that high concentrations of volatile acids that are likely to occur in digesters are not toxic to methanogenic bacteria in themselves, it is the pH that is inhibitory, with concentrations of acids up to 11 800 mg l^{-1} non-toxic to methanogens (Velsen and Lettinga 1979; Newell 1981). Therefore, any continuous downward trend in pH is an important warning sign requiring immediate attention. The measurement of pH must be done rapidly, as samples of digesting sludge once exposed to the atmosphere will rapidly lose carbon dioxide and cause erroneously high pH

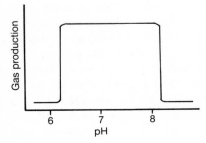

FIG. 4.93 The effect of pH on biogas production.

values. As long as the sludge has a fairly high alkalinity an increase in acid production will initially produce little effect on pH, and, in practice, the measurement of volatile acids is a better control factor of the buffering capacity within a digester. Any change in the loading of the digester must be gradual in order to ensure that the concentration of volatile acids does not exceed the normal buffering capacity of the system. Normal volatile acids concentrations in sewage sludge digesters are between 250–1000 mg l^{-1}, but values in excess of 1800–2000 mg l^{-1} indicate problems (Fig. 4.94). Determination of individual volatile acids is also very useful, as a shift to higher volatile

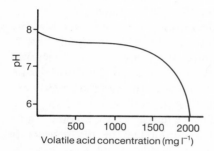

FIG. 4.94   Effect of volatile acid production on the pH in a digester.

acids, such as acetic to butyric is a sign of instability and remedial action is required. Provided adequate buffering capacity is available, higher volatile acids concentrations may be tolerated, although this will lead to incomplete conversion of biodegradable material to gas with a subsequent increase of the BOD$_5$ of the digester effluent. Depending on the chemical nature of the sludge it is possible to have rapid and effective digestion at all pH values between 6.2–7.8, although certain trade and sewage wastes restrict rapid digestion to much smaller ranges. Mosey (1983), in his excellent review on anaerobic processes, gives two interesting examples. Ferrous carbonate can become sufficiently soluble below pH 6.4 to release inhibitory concentrations of ferrous ions into solution, whereas above pH 7.5 an increasing percentage of ammonium ions will be converted to toxic dissolved ammonia gas. The latter example is particularly important in relation to strong sewage and farm wastes.

The pH value can be neutralized within the reactor by the addition of an alkali. The cheapest and most widely used alkali in full-scale digesters is probably calcium hydroxide (lime). Lime is extremely efficient at increasing the pH to about 6.4–6.5. However, further increases in pH can only be achieved by precipitating most of the dissolved carbonates as calcium carbonate. Therefore, at pH > 7.0 the lime reacts with carbon dioxide, which not only results in serious scale formation but reduces the pressure of the gas phase above the sludge by removing the carbon dioxide, which could seriously affect the structural stability of the reactor. Bicarbonates or carbonates of either sodium or potassium are generally used to raise the pH from 6.5 to 7.0, although excessive use of these salts can result in partial or severe inhibition at concentrations of 3500–5500 and >8000 mg Na l^{-1} and 2500–4000 and >12 000 mg K l^{-1} respectively. Other alkalis are less suitable, for example, sodium hydroxide also removes carbon dioxide from solution, and the use of ammonia or ammonium ions can inhibit the process. For laboratory-scale digestion units, the most useful reagents for adjusting the pH are hydrochloric acid and sodium bicarbonate.

If the sludge contains heavy metals in solution at concentrations likely to cause inhibition, the addition of alkali to raise the pH to 7.5–8.0 will precipitate most of the metal ions out of solution usually as carbonates, thus reducing the inhibitory effect. Sulphide can also be used to precipitate the heavy metal ions without causing a significant pH change.

Apart from chemical buffering, pH control can also be achieved by using other wastes or by increasing the sludge recycle rate where the reactor design allows, such as fluidized beds or downflow biofilters (Section 4.4.3) (Wheatley 1985).

(f) *Solids retention time.* The period that solids are retained in the digester is a crucial factor affecting performance. In flow-through systems the residence time of the waste (HRT) and of the microbial biomass (MCRT) will be the same. If complete digestion is to be achieved in heated mesophilic digesters then minimum retention times of more than 15 days are required to ensure that the methanogenic bacteria, which are slow growing, are able to accumulate to a sufficient population density. The design of digesters normally aims at a minimum retention period of 25–30 days, allowing for some loss in digester capacity due to accumulated non-biodegradable solids within the reactor. Clearly, any increase in the water content of the raw sludge will increase the HRT, and in flow-through systems this will cause an increased wash-out rate of the microbial biomass and an associated reduction in performance. As mentioned earlier, the minimum MCRT is dependent on the operating temperature of the reactor (Table 4.38).

Biogas and other products of digestion
There are three products of anaerobic digestion, the digested sludge, a waste liquor, and the sludge gas. Digested sludge is different to either primary or secondary sludges in a number of ways. It is pathogen free, stabilized, and far less inoffensive, having a tarry smell, and drying to an inert friable condition, thus making it ideal for disposal to agricultural land. The nature and utilization of digested sludge is fully discussed in Chapter 5. The waste liquor from anaerobic digesters has a high suspended solids (500–100 mg l^{-1}) and BOD concentration (400–800 mg l^{-1}), due to all the soluble organics present. The BOD of the liquor can be very high indeed reaching up to 10 000 mg l^{-1}, although up to 60% of the BOD is due to the suspended solids fraction.

Because of the degradation of organic nitrogen the liquor may have high concentrations of soluble nitrogen present. The characteristics and strength of the liquor makes it difficult to dispose of or treat separately, and it is returned to the works inlet where it is diluted by the incoming sewage and treated in admixture.

Sludge or digester gas is more commonly referred to as biogas. Gas production is the most direct and sensitive measure of the rate of anaerobic digestion, a decrease in production being the first indication that the process is unstable. Apart from trace amounts of water vapour, hydrogen sulphide, hydrogen, nitrogen, unsaturated hydrocarbons, and other gases, biogas is essentially a mixture of carbon dioxide and methane, the exact proportions of which determine its calorific value. A typical biogas contains between 65% and 70% methane by volume with the remaining 30–35% being carbon dioxide. Although hydrogen is an important precursor of methane formation in digestion, the concentration of hydrogen, which can be measured within the reactor, is usually very low, which suggests an immediate uptake by the bacteria (Zeikus 1979). When burnt, biogas produces water which complicates the determination of its colorific value. The net calorific value for combustion is where the water formed remains in the vapour phase, whereas the gross calorific value is where the water formed is condensed. The calorific values are calculated using the percentage of methane (M) present as the net calorific value $= (334 \times M)$ kJ m^{-3} (saturated with water), or the gross calorific value $= (370 \times M)$ kJ m^{-3} (saturated with water): both at 15.5°C and at 1 atmosphere. For normal biogas, the gross calorific value is between 24 000–26 000 kJ m^{-3}, with the net value being some 10% less.

The production of biogas can be related to the amount and type of organic matter utilized. For example, the removal of 1 kg of COD yields 0.35 m^3 of biogas at standard temperature and pressure (STP), whereas 1 kg of organic carbon would yield 1.87 m^3 at STP. The exact gas yield per kg of volatile solids removed depends on the composition of the waste but is approximately 1.0 m^3. Some food-processsng industries produce a pure substrate waste with 80–95% of the organic (volatile) matter removed, whereas only 40–50% of the organic matter in sewage sludge, and even slightly less in animal slurries, will be utilized with a proportionately lower gas yield in terms of m^3 of gas produced per kg of substrate supplied. Therefore, for a mixed primary and secondary sludge from a purely domestic sewage treatment works the gas production will be of the order of 0.5 m^{3}kg^{-1} organic (volatile) matter, or about 0.375 m^{3}kg^{-1} of total (dry) solids, added to the digester. The composition and quantity of gas produced by complete digestion can be theoretically determined using the equation:

$$C_c H_h O_o N_n S_s + 1/4(4c - h - 2o + 3n + 2s)H_2O = 1/8(4c - h + 2o + 3n + 2s)CO_2$$
$$+ 1/8(4c + h - 2o - 3n - 2s)CH_4 + nNH_3 + sH_2S$$

where $c, h, o, n,$ and s are the number of atoms of carbon, hydrogen, oxygen, nitrogen and sulphur respectively. This equation replaces the earlier equation developed by Buswell and Mueller (1952), which ignored the nitrogen and sulphur component of the waste that are generally utilized within the digester.

Thus, by neglecting the very small volume of other gases formed, and using the simplified formula (Section 3.4.3), carbohydrate will produce 73% CO_2 and 27% CH_4, lipid 52% CO_2 and 48% CH_4, and protein 73% CO_2 and 27% CH_4. The total volume of biogas produced from each of these basic substrates is 0.75, 1.44, and 0.98 m^3kg^{-1} dry matter respectively (Buswell and Mueller 1952). Actual values of gas yield and composition from the basic substrates have been given by Konstandt (1976) (Table 4.39). Lipids are only slowly degraded anaerobically and the measurement of their removal provides a very useful quality control parameter for the process.

TABLE 4.39   Gas yield and composition of biogas produced by the digestion of carbohydrates, proteins, and lipids (Casey 1981)

Substrate	Gas yield	Composition	
Carbohydrates	0.8 m^3kg^{-1}	50% CH_4	50% CO_2
Proteins	0.7 m^3kg^{-1}	70% CH_4	30% CO_2
Lipids	1.2 m^3kg^{-1}	67% CH_4	33% CO_2

A more accurate estimation of methane production per unit time can be obtained by using the equation:

$$G = 0.35 \, (L - 1.42 \, S_t)$$

where G is the volume of methane produced (m^3d^{-1}), L the mass of ultimate BOD removed (kg d^{-1}), and S_t the mass of volatile solids accumulated (kg d^{-1}). S_t can be estimated as:

$$S_t = \frac{aL}{1 + bt_s}$$

where a is the mass of volatile solids synthesized per kg of ultimate BOD removed, b is the endogenous respiration constant, and t_s is the solids retention time (Speece and McCarty 1964). Methane is the most valuable by-product of anaerobic digesters and it is useful to express the yield as $m^3CH_4kg^{-1}$ organic matter removed, or $m^3CH_4kg^{-1}$ COD removed. Typical methane yields from mesophilic digesters operated at 35°C are 0.86 m^3kg^{-1} organic matter removed for mixed sewage sludge, 0.55 m^3kg^{-1} for dairy manure, and 0.42 m^3kg^{-1} for pig manure. Under continuous reactor operation a mesophilic digester will produce between 12–16 l of methane per capita per day for primary sludge, rising to about 20 l d^{-1} for combined primary and secondary sludges (Casey 1981).

Biogas is usually burned on-site either to produce heat directly for the digesters using gas boilers, or to generate electricity using modified diesel (dual fuel) engines to drive alternators, with the cooling water from these engines used to heat the digesters. Some plants now use biogas for vehicle fuel, which is discussed in Chapter 7. Smaller digestion units are used widely in the Third World for cooking and lighting, and there are numerous small digesters operated specifically to produce energy. The simplest method of utilizing biogas, and the one requiring least capital investment, is using the gas to produce hot water in a boiler, giving the gas a value of 2p kWh^{-1}, compared to the cost of liquid petroleum gas (LPG). All the other methods involve the use of a generator to produce electricity (Parsons 1985).

In recent years, there has been considerable interest shown by the agricultural industry in using anaerobic digestion, not only to reduce the problem of disposing of animal slurry faced by many intensive farmers, but also as an economic source of energy (Hobson *et al.* 1981). A detailed feasibility study carried out by Parsons in 1985 on the economics of anaerobic treatment of dairy cow waste in the UK found that it was not economic when considered only as a source of energy. The cost of anaerobic treatment, after deducting the value of the energy produced ranged from £9–70 per cow y^{-1}, with the lowest costs (<£20 per cow y^{-1}) obtained only with herds of a minimum size of 200. Costs are reduced by reducing the water content of the sludge and by improving the gas yield at short SRTs. The sytem is only effective if there is a supply of slurry all the year round, and for animals only housed in the winter, slurry will have to be stored. At present, digestion is only economic for large intensive units where the disposal of slurry proves to be a particular and expensive problem. The economics of anaerobic digestion of farm animal wastes is reviewed by Parsons (1984). Other sources of biomass for the production of biogas include a wide range of fast-growing plants which are often grown specifically for biogas production (Section 4.5) (Table 4.40). However, there is a range of waste vegetable material that can also be used,

TABLE 4.40   Typical digestion times and gas yields at 30°C for common agricultural waste materials (Mudrack and Kunst 1987)

Starting material	For complete digestion Gas production relative to:				Gas yield after time stated as % of ultimate (days)		
	Total solids (ml g^{-1})	Organic solids (ml g^{-1})	Digestion time (days)	CH$_4$ content (%)	10	15	20
Cattle dung	237	315	117	80	24	36	48
Pig excreta	257	415	115	81	40	57	69
Straw 30 mm	357	383	123	80	29	38	45
Straw 2 mm	393	423	80	81	51	67	77
Potato haulm	526	606	53	75	85	90	92
Sugar beet leaves	456	501	14	85	99	100	100
Grass	490	557	24	84	87	96	99

such as sugar beet tops. One of the most effective substrates in terms of gas yield is waste silage.

4.4.3 Contact anaerobic systems

Because of the difficulties involved in degrading particulate organic matter and the slow growth of anaerobic bacteria, and of methanogens in particular, digester design has been traditionally based on flow-through stirred tanks with long retention times. Unlike sludge, the organic matter in wastewater is often in solution and is far more amendable to treatment. However, the primary problem with treating these wastes anaerobically has been how to retain sufficient biomass within the reactor and prevent wash-out of bacteria. Contact anaerobic systems are specifically designed to treat weaker sludges with low solids concentrations and strong effluents. Unlike the conventional continuously stirred digester, the HRT and MCRT are independent of each other, with the biomass either retained within the reactor or recycled after separation. Four major contact systems have been developed: anaerobic activated sludge; sludge blankets; static media filters; and fluidized media. Data on over 400 anaerobic plants, mainly treating vegetable-processing wastewaters, have been given by Demuynck *et al.* (1983), who give details on many contact systems having reactor volumes ranging rom 2000 to 20 000 m^3.

4.4.3.1 Anaerobic activated sludge process

This method is similar in design to the conventional two-reactor digester described in Section 4.4.2.2 (Fig. 4.89), with the primary tank being the main anaerobic reactor and the secondary reactor used for solids separation, with solids recycled back to the primary digester (Fig. 4.83d). Anaerobic activated sludge systems, also known as contact digesters, operate at relatively short HRTs of between 0.5–5.0 d. Therefore, it is essential that the settlement is efficient in order to ensure that digestion in the primary tank carries as high a solids concentration as possible and to maintain final effluent quality. This system is particularly suited to warm, dilute effluents, especially food-processing wastewaters. The design ensures a long MCRT (or SRT) and a short HRT, resulting in a high efficiency in spite of a comparatively small reactor volume. They have been used to treat a wide range of organic wastes and achieve high BOD$_5$ removal rates, for example meat packing wastewater 91% removal (Steffen and Bedher 1961) and sugar beet wastewater 93–95% (Frostell 1981). There are two large anaerobic activated sludge systems operating in the UK. One is Ashford, Kent, which is operated by Tenstar Products and treats starch waste; it was commissioned in 1976 (Morgan 1981). The other is a plant that was opened in 1982 by the British Sugar Corporation, Bury St. Edmunds, to treat sugar beet process water (Shore *et al.*

1984). Examples of anaerobic digestion systems used in the sugar-processing industry are given in Fig. 4.95.

The major operational problem appears to be poor settlement in the secondary reactor because gas bubbles become attached to the solids. This is usually overcome by vacuum degassing (Fullen 1953), although other modifications to commercially available digesters have also been successful (Stander 1967; Cillie *et al.* 1969). However, the process has many advantages over conventional completely mixed digesters: including ease of start-up; good process control; improved resistance to environmental shocks due to temperature; toxic or high loadings; and it can tolerate greater variation in influent quality. The main operational features of all the major anaerobic processes are compared in Table 4.41.

4.4.3.2   Sludge blanket process

In the up-flow anaerobic sludge blanket system, the biomass is retained in the reactor by flocculation using similar process technology to that used in sludge blanket clarification in potable water production. Originally developed in The Netherlands (Lettinga *et al.* 1980), it is in the experimental stage with only a few full-scale operational reactors. A heavily flocculated sludge develops within the special reactor (Fig. 4.83e), which acts as a separate fluidized bed, able to withstand high mixing forces. It should be stressed that no support medium is added to the reactor, the sludge that is produced having a granular nature. Mixing is achieved by pumping the feed sludge through the base of the reactor up through the sludge blanket. Above the blanket, finer particles flocculate and in the upper settlement zones they settle as sludge back to the blanket, thus preventing wash-out of biomass. The settlement zones occur between the gas collection bowls that slope at an angle of 50°, which promotes the return of the sludge from the settlement areas. The process is characterized by very high MCRT, which is even higher than anaerobic filters (Section 4.4.3.3). The process has the highest organic loading of any anaerobic reactor process because of two factors. First, because of the very high concentration of suspended sludge and secondly, because the process has only been used for treating easily degradable sugar refining wastes. A 200 m^3 digester of this type has successfully treated sugar beet waste at a rate of 16 kg COD m^{-3}d^{-1}, achieving a 90% removal at a HRT of only 4 h. However, little research has been done on more difficult wastes or wastes containing solids. Compared to the other contact processes, it is very difficult to operate and maintain the structure of the sludge blanket (Table 4.41). Sludge blanket reactors have been used as a separate methanogenic phase, with the first two stages of the digestion process being carried out in a separate reactor with increased biogas production (Morris and Burgess 1984).

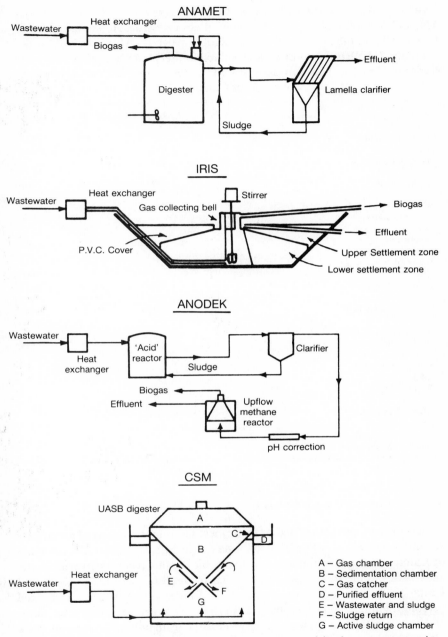

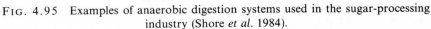

FIG. 4.95 Examples of anaerobic digestion systems used in the sugar-processing industry (Shore *et al.* 1984).

TABLE 4.41 Comparison of features of various anerobic reactors. ANFLOW, anaerobic flow-through digester without sludge recycle; ANCONT, contact digester with sludge recycle; ANBIOL, anaerobic biofilters; FANBIOF, fluidized anaerobic biofilter; UASB, upflow anaerobic sludge bed reactor. The more asterisks, the better the process (Oleskiewicz and Olthof 1982)

Feature	ANFLOW	ANCONT	ANBIOF	FANBIOF	UASB
Ease of start-up	*	*****	****	***	**
Ease of operation after proper acclimatization	**	**	*****	***	***
Good process control possible under transient influent conditions	*	*****	***	***	***
Resistance to shocks due to:					
temperature	*	***	*****	*****	*****
toxics	*	***	*****	*****	*****
high organic load	*	****	*****	*****	*****
Tolerance for influent quality variations	*	****	**	****	*****
Can tolerate high influent solids fluctuations	****	***	*	**	**
May incorporate sludges from pretreatment and aerobic polishing	*****	****	*	*	*

4.4.3.3 Static media filter process

Anaerobic reactors using fixed or static media to retain the active biomass are generally known as anaerobic biofilters (Fig. 4.83c). Although normally operated in the upflow mode, as all anaerobic contact reactors invariably are, anaerobic biofilters can also be operated in the downflow mode (Berg and Lentz 1979; Kennedy and Berg 1982). In design, they are submerged percolating filters, full of wastewater so that no oxygen can enter the reactor and are operated using a wide variety of media, from mineral to random pack plastics filter media. Apart from being operated as an anaerobic process, with some form of biogas collection device, submerged filters can be operated aerobically using aeration units or anoxically with a separate carbon input.

The design of anaerobic biofilters vary, although the shape and diameter of both the reactor and the media can have important effects on the stability of the attached film (Berg and Lentz 1980). Other factors, such as gas production, can also dislodge the film, although it may also reduce the chance of clogging within the filter medium. In cylindrical reactors the most satisfactory diameter to height ratio is 1:4. The entire reactor can be filled with the medium, and most anaerobic filters use plastic media (Anderson *et al.* 1984), or it may be restricted to the upper part of the reactor only, with the detached biomass settling to the lower chamber, thus providing a dual mixed and fixed system. Various media have been used. Of the mineral media, porous stones, gravel, and pottery fragments appear to be the favourites, although these have a large bulk density and relatively low surface areas. In contrast, plastic media have a much greater specific surface area (90–350 $m^2 m^{-3}$) and a lower density (50–100 kg m^{-3}), with a voidage in excess of 90%. Different media result in very different performance characteristics and unlike the traditional percolating filter, removal efficiency is not interrelated to specific surface areas of the media (Young and Dahab 1983; Wilkie *et al.* (1983) (Fig. 4.96). This is because in upflow filters, a significant portion of the active biomass will be present not as attached film, but as unattached dispersed growths in the interstices (voids) of the medium. The suspended solids flocculate as they travel up through the filter forming larger particles that finally settle back down the filter column. The movement of solids and gas bubbles being released introduces an element of mixing into this essentially plug-flow system. When operated in the downflow mode, all the dispersed

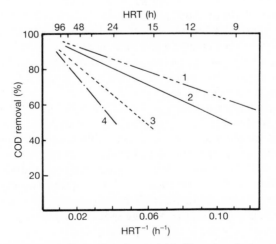

FIG. 4.96   COD removal efficiencies of anaerobic filters in relation to different types of media. 1, Large modular; 2, small modular; 3, pall rings; 4, spheres (Young and Dahab 1982).

solids are washed out of the reactor leaving only the attached biomass. In these reactors, the media must ensure stable film development as well as preventing excessive solids accumulation. Therefore, channelled or ordered media, as opposed to random media, are best (Wilkie *et al.* 1983). As all the biomass is attached in downflow anaerobic biofilters, the performance is directly related to the specific surface area of the medium. Submerged rotating biological discs have also been developed.

Anaerobic biofilters are ideal for relatively cold and dilute wastes as they have an extremely high MCRT:HRT ratio. This gives the process a high degree of stability, an excellent resistance to inhibitory compounds, and a satisfactory performance even at low temperatures (Table 4.41). They can treat similar organic loadings as anaerobic activated sludge systems, even though the MCRT is much higher (Table 4.42). Successful applications include: treating sewage; starch; whey; cellulose; distillery; pharmaceutical; and fish-processing wastes (Young and McCarty 1969; Taylor and Burm 1973; Witt *et al.* 1979; Genung *et al.* 1982; Rittman *et al.* 1982, Sachs *et al.* 1982; Bedogni *et al.* 1983; Mathur *et al.* 1986), with an average COD removal rate between 80–90% at a typical organic loading of 3.5 kg $m^{-3}d^{-1}$. A downflow biofilter containing a 35 l volume of fired pottery clay medium with a surface area of $157m^2m^{-3}$ was used to treat piggery waste at 35°C (Kennedy and Berg 1982). It was found to be able to treat higher loadings and have higher rates of methane production than partially or fully mixed reactors and also plug-flow reactors. The support medium was evenly coated with a thin microbial film 2–4 mm thick. Reactor performance, based on the amount of microbial film present, was between 1.1–1.4 g COD removed per g of film per day, which is similar to that reported by Berg and Lentz (1980) for other types of media. However, the biofilter system is limited to treating substrates with relatively low solids, as the interstices of the medium can easily become blocked. This is why they are often recommended to be operated in series, with an anaerobic activated sludge system, as the effluent from this process will be largely free from suspended solids, thus making it ideal for further anaerobic treatment using a biofilter.

At present, the major use of anaerobic biofilters has been for denitrification, i.e. removing nitrates from sewage effluents (Section 3.4.5). There is growing interest in denitrification throughout Europe, but especially in the UK. There has been a serious increase in the concentration of nitrates in both surface and groundwaters due to modern farm practice and to the use of artificial fertilizers, although the contribution from treated sewage effluents is significant in many areas (Greene 1980; Addiscott and Powlson 1989). In the past 20 years, the pressure on the farming industry to keep up yields has led to a continuing increase in the amount of nitrate leached from the soil. This has led to a steady increase in the concentration of nitrates in drinking water, and nitrates are thought to cause methaemoglobinaemia in babies, although the incidence of this is extremely rare and no direct evidence of nitrate in potable

TABLE 4.42 Features of various reactors compared with aerobic counterparts (Oleskiewicz and Olthof 1982). Key is given in Table 4.41

Feature	Anaerobic processes					Aerobic processes	
	ANFLOW	ANCONT	ANBIOF	FANBIOF	UASB	Activated sludge	Trickling filter
Loads practised (kg COD m^{-3}d^{-1})	0.5–3	2–8	2–10	0.5–12	1–15	0.5–2	1–3 roughing
Loads used in experimental scale (kg m^{-3}d^{-1})	0.5–10	0.5–100	0.5–25	1–40	1–60	1–10	2–15
HRT used (days)	8+	0.2–8	0.2–4	0.15–3	0.15–8	1–5	0.05–0.2
SRT resulting (days)	8+	15–80	20–300	20–100	30–300+	10–30	~30(+)
Temperatures used (C°)	35, 55	35, 55	15–35	35	5–35	15–25(+)	15–25(+)
COD removals attained at practised loads: similar waste assumed (%)	60	90+	90+	90+	90+	90+	60–80 in 1 stage. Multi-stages typical

supplies causing the condition has so far been found. Recent suggestions have been made that nitrates may influence the incidence of cancer in humans by the formation of nitrosamines in the digestive system (Royal Society 1983). The problem appears worse in East Anglia, where, in the Anglian Water Authority, some 500 000 people are presently receiving potable water containing concentrations of nitrate in excess of the EEC limit of 50 mg l^{-1} (= 11.3 mg l^{-1} as N) (Greene 1978). This concentration represents the maximum permissible concentration permitted in the directive on water quality for human consumption, which, although published in July 1980, did not have to be implemented by member countries until July 1985. The recommended guideline concentration is, in fact, half of the maximum value at 25 mg l^{-1} of nitrate (= 5.6 mg l^{-1} as N) (Economic Community 1980). In this intensively farmed region the most optimistic models predict that nitrate concentrations in aquifers could reach as high as 150–200 mg l^{-1} in certain areas by the end of the century. To some extent, the damage has already been done and even if farmers stopped using nitrogen fertilizers completely, the groundwater at least would continue to have very high nitrate concentrations for a minimum of 20–30 years. Therefore, the pressure on the Regional Water Authorities is to remove nitrates. In 1987, water cost about 30p per tonne (m^3) delivered to the consumer, which is about 100 times less expensive than the next cheapest material, salt. To denitrify potable water would add another 5–10p per tonne to the cost, and at present the capital investment for England and Wales is estimated (1986) at about £400m. There is growing interest in the possibility of farmers being forced to pay for this by using the 'polluter pays' principle to charge for the nitrate leached from their land. However, the dispersed nature of this leaching makes the implementation of such charges very difficult.

A wide range of facultative anaerobic bacteria can utilize nitrate as an alternative to oxygen for their terminal electron acceptor, releasing gaseous nitrogen. The stoichiometry of the process is outlined in Section 3.4.5. If a treated sewage effluent rich in nitrate was left to stand under anoxic conditions, denitrification would eventually occur as the population of denitrifying bacteria gradually increased. In order to make the process a part of the normal treatment operation, a more rapid and reliable method of denitrification is required. This is done by bringing the effluent into contact with a large and active population of denitrifying bacteria under ideal conditions. A source of readily degradable organic carbon is also required to act as a source of reducing agent. Some effluents will contain enough residual organic carbon to allow denitrification to proceed and, if not, untreated settled sewage can be used. However, the preferred source of organic carbon is methanol, which is also used for potable supplies, as the input of organic carbon into the reactor can be accurately controlled.

From the stoichiometric equations in Section 3.4.5 for denitrification, the amount of methanol, or organic carbon equivalent, can be estimated from:

$$C_m = 2.47 N_o + 1.52 N_i + 0.87 D_o$$

where C_m is the concentration of methanol required (mg l^{-1}), N_o the initial nitrate concentration (mg N l^{-1}), N_i the initial nitrite concentration (mg N l^{-1}), and D_o is the initial dissolved oxygen concentration, as the reaction will take place under anoxic as well as anaerobic conditions. The biomass produced can be estimated as:

$$C_b = 0.53N_o + 0.32N_i + 0.19D_o$$

where C_b is the biomass production (mg l^{-1}). Thus, when N_i and D_o are zero, the methanol:nitrate ratio is 2.47 and the biomass production is 0.53N (Barnes and Bliss 1983). In these calculations, only the amount of nitrogen present is used, thus for a sample containing 15 mg l^{-1} of nitrate ions at 0 mg l^{-1} nitrite ions and dissolved oxygen, the amount of methanol required will be $(2.47 \times 15 \times 14/62) = 8.4$ mg l^{-1}, with $(0.53 \times 15 \times 14/62) = 1.8$ mg l^{-1} of biomass produced. If, however, the sample contained 15 mg l^{-1} of nitrate–nitrogen, 37 mg l^{-1} of methanol would be required producing 8 mg l^{-1} of biomass.

It is important to provide the correct dose of methanol to the reactor as if it is under-estimated, denitrification will be incomplete and if over-estimated, methanol will remain in the final effluent and cause further pollution.

The major denitrification process is the denitifying filter which is a submerged filter containing an inert medium on which a film of denitrifying bacteria develop. This contact system provides the necessary MCRT for the development of these slow-growing bacteria and depending on the voidage of the medium the SRT can be very short, often less than one hour. In suspended growth systems, biological denitrification is a zero-order reaction, with respect to oxidized nitrogen and electron donor concentrations. There needs to be at least 2 mg l^{-1} of oxidized nitrogen and a C:N ratio of at least 3:1 for denitrification to occur. In fixed-film reactors, the rate of reaction is similar to suspended growth reactors except that the diffusion of both oxidized nitrogen and election donors within the biomass will modify the overall reaction. Where the film growth is thin, there will be no diffusion limitation on the rate of denitrification, but for thicker films a low dependence on these concentrations may be observed (up to half-order) (Riemer and Harremoes 1978). The overall effect of the denitrification reaction is to raise the pH by the formation of hydroxide ions. This replaces about 50% of the alkalinity consumed by the oxidation of ammonia during nitrification. For each mg of ammonia oxidized to nitrate, 7 mg of alkalinity are utilized, and 3 mg are produced during denitrification. The denitrification reaction is pH-sensitive, with an optimum range between 6.5–7.5, but falling to 70% efficiency at pH 6 or 8 (Environmental Protection Agency 1975; Moore and Schroeder 1970). Temperature is also important, with the reaction occurring between 0–50°C, but with the optimum reaction rate between 35–50°C. On the whole, the autotrophic nitrifying bacteria are more sensitive than the denitrifying heterotrophs and there will be little inhibition of denitrification of a particular wastewater if nitrification has already proceeded satisfactorily. Denitrifying filters can be operated in an upflow or downflow mode, although the former builds up considerably more biomass than the latter system. The most frequent operational problem is excessive film accumulation or entrapped gas bubbles, which alter the internal flow pattern, restricting the flow causing a significant head loss, or reducing the HRT by channelling the effluent through the filter. Nitrate removal efficiency increases with the accumulation of biological solids. However, removal efficiency sharply falls off as the flow is restricted and the pressure builds up until finally the liquid forces its way through the bed scouring away much of the film. Therefore, it is better to opt for a larger medium (15–25 mm diameter), with larger voids and a slightly reduced surface area so that film accumulation is less of a problem. In this way, the filter will operate more consistently (Tamblyn and Sword 1969). Better film control has been achieved using fluidized bed systems, although this has been designed primarily for potable water treatment (Jewell 1982; Croll et al. 1985).

4.4.3.4   Fluidized media

A modification of the anaerobic biofilter, still in the experimental stage, is the anaerobic fluidized bed reactor (Fig. 4.83f) in which fine grained inert (sand) or reactive (activated carbon) random media are used. The medium, which is very light becomes coated with bacteria and is fluidized by applying a high mixing velocity. However, it is difficult to maintain the optimum mixing velocity in order to ensure the medium remains in suspension without causing shear forces strong enough to strip off the accumulated biomass.

It may be that expanded bed reactors will be the more effective system, in which the bacterial-coated medium remains on the floor of the reactor and is expanded by the movement of the liquid forced up through the medium. This decrease in media particle size has resulted in a significant increase in the specific surface area available for biological growth. Although difficult to start up, fluidized beds are highly resistant to fluctuations in temperature, toxic compounds, and high organic loadings. In addition, they tolerate wide fluctuations in influent quality and tolerate solids in the influent (Table 4.41).

Further reading

General: Oleszkiewicz and Olthof 1982; Stafford et al. 1980; Hughes et al. 1981; Ferrero et al. 1984; Hawkes and Hawkes 1987.
Septic tanks: Environmental Sanitation Information Centre 1982; Laak 1986.
Imhoff tanks: Billings and Smallhurst 1971.
Digestion: Stafford et al. 1980; Ferrero et al. 1984.
Biogas: Hobson et al. 1981.
Sludge blanket: Lettinga et al. 1983; Pette and Versprille 1982.

Fixed-film ractors: Henze and Harremoes 1983; Salkinoja-Salonen *et al.* 1983.
Denitrification: Barnes and Bliss 1983.

4.5 PLANTS AND LAND TREATMENT

'Natural treatment systems' include anaerobic, high-rate, and facultative stabilization ponds; the controlled mass culture of higher plants and animals, the use of natural and artificial wetlands, land application, and composting. Stabilization ponds have already been considered in Section 4.3, and composting is discussed in Chapter 7. Such natural systems play an important role in the treatment of wastewaters, especially in warmer countries. For example, fish ponds have been used for many centuries throughout Asia, with the organic wastewater promoting algal growth which has resulted in increased yields of fish, feeding either directly on the algal biomass or on intermediate grazers. In recent years, the search for low-cost treatment systems yielding effluents low in nitrogen and phosphorus has renewed interest, especially in the southern USA, in the use of plants for treatment. Apart from adequately treating the wastewater, the use of plants results in the production of excess biomass, which can be used for a variety of purposes such as energy production, animal feed, and even protein production, thus offsetting the cost of treatment (Chapter 7). The use of wetlands, bogs (peatland), and other types of land for treatment is also being revived.

4.5.1 Plants

The controlled culture of aquatic plants is becoming widely used in wastewater treatment as a tertiary treatment system, for not only do plants accumulate nutrients and other soluble compounds from the water but they act as a food source for a variety of other organisms, such as zooplankton, small crustaceans (e.g. brine shrimp, *Artemia salina*), and a wide variety of fish. The excess plant biomass itself can be harvested and used for various purposes. The plants used in wastewater treatment systems can be loosely classified as emergent vegetation, submerged algae and plants, and floating macrophytes. The advantage of using plants is that the capital investment in the treatment works is low, and as little equipment, energy, and chemicals are required, the operational costs are also very low. The operational cost of harvesting either the plant or the secondary animal biomass is reduced by its value as a source of energy or protein (Dinges 1982). Much work has been done in evaluating plants for wastewater treatment in terms of treatment efficiency, biomass production, and value of the resultant biomass. The main criteria for such plants are: (1) ease of harvesting; (2) low water content; (3) high protein content; (4) low fibre and lignin content; (5) high mineral absorption capability; (6) extended growing and harvesting periods; (7) non-toxic either to humans or livestock; (8) capable of being processed into a useful by-product; and (9) have few natural pests. Selection of plants for wastewater treatment should, if possible, satisfy all of these criteria (Culley and Epps 1973).

Emergent vegetation
Although there are several hundred potential emergent species only a few have been selected as being potentially suitable for wastewater treatment, such as the bulrushes *Scripus lacustris*, *S. acutus*, and *S. validus*; the reeds *Phragmites communis* and *P. australis*; and the yellow flag *Iris pseudocorus*. Reeds have underground rhizomes and jointed stems that root at the nodes, and both reeds and rushes multiply vegetatively from adventitious buds located on the rhizomes. Rushes grow between 2–3 m in height and are restricted to shallow water < 0.5 m deep, whereas reeds are able to grow at depths of up to 2.5 m with some species growing to heights of 5 m (Dinges 1982). Emergent species are generally very hardy and resistant to both pests and diseases. They are able to withstand cold temperatures and even freezing, and are ideal for more temperate climates. These plants are perennials, and with proper management stands of rush or reeds can be maintained indefinitely. There is little information on the design of ponds, however, small experimental ponds have been constructed and operated successfully. Most reed–rush systems consist of unlined earthen ponds with the plants rooted in the soil, whereas others are sand and gravel filters planted with reeds or rushes. The latter system is expensive to construct, and Spangler *et al.* (1976) suggested that a single shallow gravel layer as a base for emergent plants is equally effective but much cheaper. This allows the construction of larger installations. This method of cultivation has now been widely adopted (Anon 1986). A major constraint of these systems is the availability of suitable land and access for harvesting.

In Flevoland (The Netherlands) an experimental rush pond of one hectare in area was constructed to cope with the seasonal influx of tourists at an isolated camping site (de Jong 1976). The primary function of the emergent vegetation is to provide suitable roots for the attachment of heterotrophic bacteria to purify the sewage, and the plants themselves also consume part of the available nitrogen and phosphorus. The experimental pond had a star-shape layout to ensure maximum utilization of the available area (Fig. 4.97) and was 0.4 m deep. In order to minimize maintenance this design was later modified to long narrow channels 375 m in length that could be maintained mechanically. Rushes were rooted in the mud of the unlined basin and the results showed that the BOD of the raw wastewater was reduced by 95.7%, COD by 80.7%, and faecal coliforms (MPN) by 99.9% (Table 4.43). Although the removal of nitrogen and phosphorus was rapid initially, the removal efficiency rapidly declined as the supply of nutrients exceeded the uptake rate, with mature rush stands removing between 300–500 kg N $ha^{-1}y^{-1}$ and 50–75 kg P $ha^{-1}y^{-1}$. De Jong (1976) observed that purification was achieved

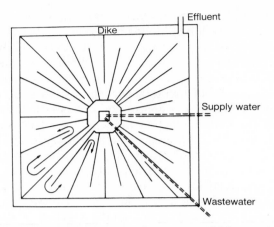

FIG. 4.97 Experimental rush pond covering an area of 1 hectare and 0.4 m deep. The star-shape was chosen to ensure maximum utilization of the available area although it made mechanical harvesting and maintenance difficult (Tourbier *et al.* 1976).

TABLE 4.43 Performance of experimental rush pond (de Jong 1976)

		Week number					Mean week 26–34
		26	28	30	32	34	
BOD	Influent	285	331	347	276	127	257
$(mg\,l^{-1})$	Effluent	12	8	18	17	7	11
COD	Influent	661	734	900	590	285	530
$(mg\,l^{-1})$	Effluent	48	54	94	83	66	70
Coliforms (MPN	Influent	43×10^4	40×10^4	38×10^4	52×10^4	41×10^4	36×10^4
$m\,l^{-1}$)	Effluent	1	64	2670	8	14	313

not only by the bacteria associated with plant roots but also by infiltration of the wastewater into the soil. In the channels the performance of reeds and rushes were compared, and whereas there was little difference in their ability to purify wastewater, the reeds required less maintenance. However, reeds are harvested annually during the summer and rushes can be gathered only every second or third year. Stem densities are about the same for both groups, but reeds have no commercial value, unlike rushes, which are highly sought after as wicker for basket and furniture manufacture and also can be used for feeding livestock (Pomoell 1976). The total biomass of rushes produced in a treatment pond can reach between 5–10 tonnes $ha^{-1}y^{-1}$.

Spangler *et al.* (1976) compared the efficiency of *Iris versicolor* (the blue flag), *Scripus validus* (soft-stem bulrush) and *S. acutus* (hard-stem bulrush) in bench-scale ponds over a winter period. They were planted in pea-sized gravel

70 mm deep in plastic lined basins (800 × 900 mm). Primary effluent was fed into the basins each day during Monday to Friday and retention times of 3, 5, and 15 days were studied. They found that there was no difference between the various ponds, including the control pond without the plants, in reducing BOD or COD. However, the presence of bulrushes made a significant difference in the removal of both total phosphorus and ortho-phosphate, and were more effective than the control of *Iris* ponds. Removal efficiency increased with the longer retention times with total phosphorus removal reaching 98% on occasions (average > 80%). The efficiency of four emergent species *Typha domingensis*, *T. orientalis*, *Phragmites australis*, and *Scripus validus* to treat the effluent from a poultry abattoir were compared by Finlayson and Chick (1983). The plants were grown in a gravel substrate in plastic-lined trenches. Using a retention time of between 2.7–3.6 d^{-1}, the effluent was allowed to percolate through the trenches. All three genera were able to significantly reduce the suspended solids (83–89%) and turbidity (58–67%) of the effluent, and *Phragmites* and *Scripus* were both able to oxygenate the anaerobic inflow. *Scripus* was more effective than the other genera in reducing the concentrations of total nitrogen and phosphorus, and also reduced the concentrations of sodium, potassium, and chloride. The genus *Typha* comprises of nearly 20 species, and three species in particular have been used for treatment studies: *Typha latifolia*, *T. domingensis*, and *T. orientalis* (Chick *et al.* 1983). The use of reeds is receiving much attention at present in the UK, where trials are underway by several Water Authorities, for use as a secondary and tertiary treatment process (Chalk and Wheale 1989).

Although emergent plants are an inexpensive treatment system for many types of wastewaters, some stronger effluents appear to be phytotoxic (Mitchell 1978). Piggery effluents, in particular, severely affect the vigour of emergent plants, especially *Typha domingensis* (Finlayson and Mitchell 1982; Bowmer 1985).

Submerged algae and macrophytes

The role of algae in wastewater treatment has already been discussed in Section 4.3. In facultative ponds, the algae use the inorganic compounds, released by aerobic and facultative bacteria, for growth using sunlight for energy. They release oxygen into solution, which in turn is utilized by the bacteria completing the symbiotic cycle (Fig. 4.76). The algae in facultative ponds are restricted to the euphotic zone, which is often only a few centimetres deep depending on the organic loading and whether it is day or night. High-rate aerobic stabilization ponds are not designed for optimum purification of wastewater but for algal production. Present research is aimed at producing sufficient biomass from such ponds to make them a useful and economic source of single-cell protein (Section 7.3.2). Mono-species cultures of green algae, such as *Chlorella* or *Scenedesmus*, have a protein content of 50%

(dry weight) compared to 60–70% for the blue-green alga *Spirulina*. Light availability is the most critical factor controlling algal growth and as it is uneconomical to use artificial light, ponds must be located in areas where there is maximum sunshine. Shallow lagoons are normally used for algal production because as the density of cells increases so the light penetration is reduced. Also, the cells nearest the surface can become damaged by constant exposure to sunlight. Ideally, deeper lagoons are used, which are gently mixed to maintain the algae in suspension, ensuring intermittent transport of the cells into the euphotic zone. Large-scale algal ponds are being used in several countries as an effective means of sewage treatment as well as producing valuable algal biomass which can be used for many different purposes such as biogas production (Table 4.44), fodder, and fertilizer (Oswald 1972; Taiganides *et al.* 1979; Shelef *et al.* 1978). The most recent use, and potentially the most valuable, is the removal and recovery of heavy metals from effluents. The ability of certain algae to absorb and accumulate metals suggests the possibility of removing the low concentrations of heavy metals found in normal domestic wastewaters (Filip *et al.* 1979; Harding and Whitton 1981; Nakajima *et al.* 1981). However, Becker (1983) indicates that removal of metals by algae may be possible where long retention periods are possible, for example, for the pretreatment of certain metal wastes. In ordinary algal ponds with relatively short retention times metal removal will only be marginal. Thus, complete removal and recovery does not appear feasible at present. The use of algae in stabilization ponds has been fully discussed in Section 4.3.

Submerged plants or macrophytes are used for tertiary treatment only as they require high clarity to ensure light penetration and an aerobic environment for plant respiration. Only a few of the many submerged aquatic plants found in natural waters can tolerate enriched waters, for example *Potamogeton berchtoldii*, *P. filiformis*, *P. foliosus*, *P. pectinatus*, *P. zosterformis*, *Ceratophyllum demersum*, *Elodea canadensis*, *Ludwigia repens*, and *Najas marina*. The macro-algae *Hydrodictyon* and the ubiquitous *Cladophora* are

TABLE 4.44   Comparison of digestibility and biogas production of various plant material (Braun 1982)

Material	Hydraulic residence time (d)	Volumetric loading (kg TS m^{-3}d^{-1})	Gas yield (m^{3}kg^{-1}TS)
Sea grass	20	2	0.1
Seaweed	20	2	0.17
Water hyacinth	10	4	0.1
Algae[a]	10	4	0.25
Algae[b]	3–30	1.4–11.2	0.26–0.5
Silage	15	6.7	0.45

[a] Freshwater algae; [b] Sewage lagoon algae.

also associated with wastewaters. They remove soluble nutrients and some other compounds from the water but also provide a solid substrate for the development of an active periphyton slime, including heterotrophic bacteria. Submerged plants grow well in the summer but die back in the winter. For example, in Michigan (USA), the plant biomass degenerated by 98–99% in the winter (McNabb 1976). Plants will compete with each other in ponds, and where mixed cultures are used *E. canadensis*, *C. demersum* or *P. foliosus* will dominate the others present.

In order that the euphotic zone extends to the bottom of the pond the depth should not exceed 2 m. Four experimental ponds 1.8 m deep and covering a total area of 16 hectares are being used to treat 1900 m^3d^{-1} of final effluent from an activated sludge plant serving Michigan State University (Bahr *et al.* 1977). The ponds contain *E. canadensis*, *C. demersum*, *P. foliosus*, and the epiphytic algae *Cladophora*. Harvesting of the filamentous alga *Cladophora* and the macrophytes is done four times a year during the growing season, resulting in a biomas production of 400 g (DW) m^{-2}. After two years of operation the nitrate concentration has reduced from 15 to 0.01 mg l^{-1} and the concentration of phosphorus from 4 to 0.03 mg l^{-1} by using all four ponds in series. The macrophytes can be used for fish food, feeding livestock or for composting. *Elodea canadensis* contains 16.3–16.9% ash, 2.89–3.86% nitrogen, and 0.092–1.180% phosphorus on a dry weight basis (Jewell 1971). Macrophytes can be harvested using a commercial weed harvester, although herbivorous fish can be used to crop the excess vegetation (Sutton 1974). Over-stocking with fish will lead to over-grazing and care must be taken with stocking levels. The fish can then be harvested annually and used for a much wider range of uses then the plant material (Dinges 1982).

Floating macrophytes

This group of aquatic plants use atmospheric oxygen and carbon dioxide but obtain the remaining nutrients they require from water. Three groups of floating macrophytes are used in wastewater treatment, the water hyacinth (*Eichhornia crassipes*), duckweeds (*Spirodela* spp., *Wolffia* spp., *Wolffiella* spp.), and water ferns (*Salvenia* spp. and *Azolla* spp.) (Jackson and Gould 1981; Mitchell 1978, 1979; Finlayson 1981).

Although originally discovered in the River Amazon basin, Brazil in 1824, the water hyacinth is now found growing wild throughout the warmer parts of the world including North America, Africa, Portugal, South and South-East Africa, Australia, and New Zealand (Holm *et al.* 1969; Pieterse 1978). The water hyacinth is the largest of the floating plants and the most studied macrophyte used in wastewater treatment. It is a rhizomatous plant with large glossy green leaves and a feathery unbranched root system. The pale lavender flowers with their beautiful splashes of yellow are very short-lived. The plant is able to reproduce asexually by the formation of stolons, a type of vegetative propagation, as well as sexually. New plants grow from the ends of each extended stolon which can grow up to 30 cm in length depending on the available space. The plants grow rapidly in both vertical and horizontal directions, although initially the hyacinths grow horizontally until the entire water surface is covered when they begin to grow vertically increasing the overall plant height. When hyacinths are grown in ponds containing wastewater they rapidly multiply and cover the available water surface. This crowding causes vertical growth with the plants growing in excess of 1 m in height. It is interesting that the more nutrient rich the water the smaller the root system that will develop, and in wastewater lagoons the root system may be restricted to only 10 cm in length. The biology of the plant is reviewed by Pieterse (1978).

Hyacinths are very much a tropical plant requiring a minimum light intensity of 1400 lm m^{-2} and prefer full sunlight for maximum growth, when their photosynthetic conversion efficiency approaches 4% (Penfold and Earle 1958; Wolverton and McDonald 1979). They have a growth range of between 10–35°C with optimum growth occurring between 25.0–27.5°C, but they cannot tolerate cold temperatures and the stem and leaves are killed by frost. The central structure of the plant, from which the stem, roots, and stolons arise, is the rhizome which is similar in shape to a carrot and grows up to 20 cm in length. The rhizome is critical to the plant and if just 4 cm of the rhizome tip is removed the plant dies. The tip floats just a few centimetres below the water surface and so it is very vulnerable to the cold, the tip is killed and the plant subsequently decays if the water temperature approaches freezing. If the stem and leaves are killed by the frost the plant rhizome will normally remain viable. However, as the dead vegetation dries the bulk weight of the plant is reduced so that the rhizome is displaced less and rises closer to the water surface making it more susceptible to possible damage from low temperatures. The northern limit of the plant in shallow waters has been shown to be where the mean air temperature for the coldest month never falls below 1°C. Therefore, in practice all the water hyacinth systems currently treating wastewater are located in tropical or warm climates (Middlebrooks et al. 1982). When hyacinth systems have been attempted in cooler climates they have had to be housed in heated greenhouses and the temperature maintained within the optimum range. Although it has been suggested that methane produced from digesting the waste biomass in anaerobic digesters could be used to heat the greenhouses, it is doubtful whether such a system could be cost-effective. Also, where the system only operates for part of the year there is the added expense of operating a culture unit to maintain sufficient hyacinths to introduce into the system each spring. Hyacinths are freshwater plants and are inhibited by salinities of > 1.6% (Haller et al. 1974), but they can tolerate a wide pH range of 4–10 (Haller and Sutton 1973). They are very disease resistant, with only three exotic insects being able to exert any significant effect on growth: the hyacinth weevils (*Neochetina eichhorniae, N. bruchi*), and the leaf minining mite (*Orthogalumna terebrantis*). These species have been introduced in some areas of the world to control the plant but other indigenous pests have had little effect on plant growth. However, insect-feeding on hyacinths is generally thought to be beneficial as it stimulates the formation of new growth.

The hyacinth is one of the most productive macrophytes in the world, with its biomass able to double every 7 days, which is equivalent to a daily gain in biomass of 108 kg (DW) ha^{-1}d^{-1}. Standing crops of hyacinth can reach 2500 g (DW) m^{-2}, although the exact biomass will depend on nutrient availability and the degree of crowding. When grown in domestic wastewater, the standing crop reaches over 3000 g (DW) m^{-2}, which is equivalent to 30 tonnes (DW) ha^{-1}d^{-1} (Wooten and Dodd 1976; Dinges 1982). With a moisture content of 95% this represents a vast quantity of biomass to be handled and eventually processed. The dense growth of hyacinths in a wastewater lagoon provides a unique and very stable environment. The extensive root system prevents horizontal movement, and particulate solids rapidly settle. Oxygen is released into the water via the roots resulting in an active heterotrophic and autotrophic (nitrification) community, and the lowest depths of the lagoon will be anoxic allowing denitrification. There is a complex micro- and macro-fauna associated with the root area of the lagoons but this has not been studied extensively (Dinges 1982).

Hyacinth lagoons are used for tertiary treatment and because they require warm temperatures they would seem particularly suitable for use as a tertiary treatment phase after facultative stabilization ponds that are widely used in warmer climates. In terms of treatment, most improvement in BOD and suspended solids is due to settlement, and nitrogen and phosphorus removal is by the plants, with the former unaffected by the season but the latter severely reduced in the winter. The plants exhibit luxury uptake of nutrients, far in excess of normal requirements, but also accumulate other components of wastewater, such as heavy metals and synthetic organic compounds. The uptake and accumulation of such compounds can be disadvantageous where the concentrations in the plant tissue prohibit the subsequent use of biomass for conversion into energy by digestion or use as a feed supplement.

The design of hyacinth lagoons is still in the experimental phase. Treatment objectives are important in order to ensure sufficient retention time but the limiting factors in design appear to be the periodic removal of settled solids and the harvesting of hyacinths. All the systems presently operating are <4 ha in surface area and the majority of systems, which are mainly experimental, are <1 ha. A few small units have been constructed to serve small communities of less than 1000 people. Long narrow channels with plug-flow appear most efficient as this prevents short-circuiting and allows easy harvesting of the macrophytes, although settlement is inefficient. Various modifications to this basic design have been made (Dinges 1978, 1982), for example, a zig-zag channel configuration has been used to increase settlement as well as ensuring adequate access for harvesting equipment (Fig. 4.98) (Wolverton *et al*. 1976).

Adequate root depth must be provided for the hyacinths so that the roots come into contact with the majority of the liquid passing through the lagoon. For example, a depth of 0.4 m ensures complete wastewater contact with the root system ensuring maximum nutrient removal. However, the depth can vary from 0.4–1.8 m, with the deeper lagoons allowing space for settlement and anoxic activity. Although deeper ponds reduce the total land area

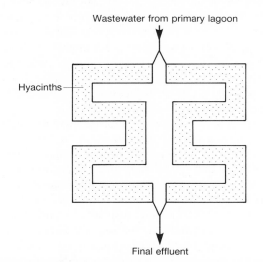

FIG. 4.98 NASA culture basin design for water hyacinth (Wolverton *et al*. 1976).

required, if they are completely covered with plants they tend to become completely anaerobic at night. Hydraulic loading is more critical than organic loading for wastewater that have already received some treatment. In practice, loading varies from 240–3570 $m^3ha^{-1}d^{-1}$ for domestic effluents. However, to produce a final effluent with a final BOD < 10 mg l^{-1}, suspended solids < 10 mg l^{-1}, total kejldhal nitrogen < 5 mg l^{-1}, and total phosphorus < 5 mg l^{-1} from a secondary effluent, then the loading rate should not exceed 1870 $m^3ha^{-1}d^{-1}$. Stronger effluents require lower loading rates. The retention time should not be less than 5 days. To achieve maximum nutrient removal, the hyacinth biomass necessary to absorb the nutrients is maintained by harvesting at a rate that balances net plant productivity. Frequent harvesting is necessary in order to ensure that the hyacinth population is kept in an active growing phase. Initial uptake of nutrients is rapid, with the amount absorbed related to the concentration of nutrients in the water. The removal rate of nutrients can be directly related to lagoon area. For example, a 2.1 ha hyacinth lagoon in Florida receiving 3800 m^3 of treated domestic wastewater each day removed 80% of the nitrogen and 44% of the available phosphorus (Cornwell *et al.* 1977). The productivity of the plant fluctuates seasonally according to temperature and other climatic factors, such as light intensity and day length, therefore, removal efficiency will fluctuate and harvesting has to be adjusted accordingly.

Numerous pilot studies on hyacinth lagoons have been conducted and there is a considerable amount of performance data available. The comparative pilot study by Hauser (1984) at Roseville, California is typical. Three separate ponds were operated using secondary effluent from the Roseville sewage treatment works. Pond 1 was unmanaged, with the hyacinths allowed to control their own density on the pond, whereas Pond 2 was aerated using a submerged perforated tubing connected to a compressor to ensure an aerobic aquatic environment, but the hyacinths were not harvested. Only the first two-thirds of the pond was aerated with the pipes spread three feet apart and the remaining third was left to allow settlement and filtration. In Pond 3, the plants were harvested but not aerated. Harvesting in Pond 3 was carried out every 2 weeks during the growing season with 20% of the surface area cleared each time. By alternating the harvesting location and not redistributing the plants after harvesting the period between disturbance of the hyacinths was maximized. Initially, all three ponds were operated in an unmanaged mode so that any difference in performance during maturation was due solely to unequal plant cover. All three ponds produced good final effluents with significant reduction in BOD and suspended solids; between 30–50% reduction in the winter and 70% for the warmer growing season (May–October). The final effluent concentration was independent of the influent concentration, suggesting that in these ponds the potential BOD reduction was only partially utilized. Efficiency in suspended solids removal varied between the ponds more than BOD removal. The unmanaged pond (Pond 1) produced a consistently good effluent with low suspended solids concentration throughout the year, whereas in the other two ponds occasionally high suspended solids concentrations were recorded. As the BOD was not high on these occasions it suggested that the high solids in Pond 2 were inert particles disturbed by the aeration system as they accumulated, and in Pond 3, as the high suspended solids concentration was associated with harvesting they were probably fragments of water hyacinths (Table 4.45). To control this 'harvest-induced' rise in suspended solids Hauser (1984) suggests that a portion of the pond near the effluent end should be left unharvested or, alternatively, a separate non-harvested polishing pond could be used. Dinges and Doersam (1987) have described a full-scale hyacinth treatment plant in Austin, Texas.

The two major mechanisms for ammonia reduction in hyacinth systems are bacterial nitrification and plant uptake. The primary function of water hyacinths in the bacterial nitrification process is to provide sufficient support for the development of the nitrifying bacteria. The nitrogen is converted from ammonia to nitrate with the possibility of denitrification in the lower anoxic

TABLE 4.45 Summary of the performance of three hyacinth ponds. Pond 1 was unmanaged, Pond 2 was aerated, and Pond 3 was harvested (Hauser 1984)

| Parameter | Influent concentration, mg l⁻¹ (except pH) | | Mean pond loading (all ponds) kg ha⁻¹ | Effluent concentration, mg l⁻¹ (except pH) | | | | | |
| | | | | Pond 1 (unmanaged) | | Pond 2 (aerated) | | Pond 3 (harvested) | |
	Mean	S.d.		Mean	S.d.	Mean	S.d.	Mean	S.d.
Nov 1981–Apr 1982									
BOD_5	11.6	6.5	18	4.9	3.6	3.6	2.5	3.5	1.9
SS	10.7	6.5	16	2.8	2.4	4.6	6.9	3.0	2.1
pH	7.8	0.2	—	6.9	0.3	6.9	0.3	7.1	0.3
NH_4 as N	14.1	2.7	21	6.1	4.6	1.8	2.4	9.5	3.3
Organic N	2.3	1.0	3.5	1.4	0.7	0.9	0.5	1.3	0.6
$NO_3 + NO_2$ as N	0.3	0.3	0.5	3.6	3.5	7.1	2.9	2.7	2.2
Total N	16.7	2.7	25	11.1	2.9	9.9	4.1	13.5	3.0
Alkalinity	132	14	200	97	35	64	17	109	22
May 1982–Oct 1982									
BOD_5	8.6	4.2	13	2.3	1.3	1.9	0.3	2.5	0.9
SS	7.1	4.2	11	2.0	2.0	2.6	1.0	5.4	7.0
pH	7.8	0.2	—	6.5	0.2	6.6	0.2	6.6	0.2
NH_4 as N	15.8	3.1	24	4.6	2.4	0.2	0.1	3.7	1.2
Organic N	2.3	1.9	3.5	1.1	0.5	1.2	1.3	1.2	0.6
$NO_3 + NO_2$ as N	0.2	0.1	0.3	2.7	2.7	3.8	2.4	3.2	2.4
Total N	18.2	2.5	28	8.4	3.0	5.2	3.0	8.0	3.1
Alkalinity	117	15	178	78	17	51	11	67	9

layers of the ponds. However, nitrogen is removed by plant uptake and is incorporated into the plant biomass. The aeration in Pond 2 enhanced nitrification resulting in the lowest final effluent ammonia concentration. Harvesting is thought to encourage plant uptake by eliminating the inhibitory effects of crowding on plant growth, although it may limit nitrification by reducing the availability of support structures as well as removing a portion of the nitrifying bacteria with the harvested plants (Hauser 1984).

There are numerous potential uses for harvested hyacinths, although these depend to some extent on the composition of the wastewater used to culture the plants (Table 4.46). Hyacinths grown on a domestic wastewater can be used for, e.g. livestock feed, a protein supplement, compost or soil amendment, paper manufacture, alcohol production, livestock bedding, pyrolytic conversion to oil, and incineration to produce steam for energy or for the generation of methane (Table 4.46). Chopped, pelleted hyacinths can be fed to cattle and sheep, or after the hyacinths have been ensiled. Ensiling requires the addition of about 4% citrus pulp or corn to act as a readily available carbohydrate source and can be fed directly to dairy cattle. The main value of hyacinth for livestock is as a source of roughage, minerals, and energy (Bagnall et al. 1974; Zerinque et al. 1979). A valuable compost can be produced from whole or chopped hyacinths by storing wet plants aerobically for 1–6 months followed by drying and grinding. Growth is poor in pure compost, but when mixed with sand at a ratio of sand to compost of 3:1 excellent growth is obtained, with the composted hyacinths able to retain water unusually well. Hyacinths grown on industrial wastewater accumulates high concentrations of heavy metals and are not suitable for consumption or disposal to land as a compost or amendment. As long as the heavy metal concentration does not inhibit digestion, methane can be produced with the residual sludge used for metal recovery of particularly valuable or toxic metals such as silver, gold, cadmium, mercury, lead, and all the base metals. Hyacinths can be readily digested in conventional mesophilic stirred digesters at low loadings and at high retention times (Klass and Ghosh 1980; Ghosh et al. 1980; Chin and Goh 1978) (Table 4.44). Better yields have been obtained using hyacinths mixed with primary sewage sludge and by using digesters in conjunction with upflow anaerobic filters (Chynoweth et al. 1981, 1982).

Water hyacinths are easily harvested as they float on the surface of the water. The leaves of the plant act as sails and where there is open water the wind blows the plants to the leeward side of the lagoon. Hyacinths are generally collected and chopped manually on experimental lagoons, although for larger installations there are a variety of mechanical harvesters commercially available. These normally operate from special platforms and combine a feed or collection system such as a rotary head, a chopper, and a conveyor to transport the material to trucks.

Two other groups of floating plants are also used for wastewater treatment, duckweed and the water fern. Duckweed (Lemnaceae) is widely distributed

TABLE 4.46 Conversion of vascular aquatic plants to useful products (Tourbier *et al.* 1976)

Pollution removal applications	Harvested plants: processing alternatives	Products
Removal of heavy metals from chemical and industrial wastewaters	Anaerobic fermentation $\longrightarrow$	Methane gas
	Residual sludge $\left.\begin{array}{c}\\\\\end{array}\right\}$ Metal extraction processes	Silver, gold, Cadmium, mercury, lead, etc., base metals
Removal of nitrates and phosphates from domestic sewage	Anaerobic fermentation $\longrightarrow$	Methane gas
	Residual sludge $\longrightarrow$ Dried: utilizing Methane gas or solar energy as source of thermal energy	Agricultural fertilizer (bagged or bulk)
	and/or	
	Chopped and dried plant material $\left.\begin{array}{c}\\\\\end{array}\right\}$ Animal-feed processing	Additive for cattle, swine and poultry feeds
	Potable food-processing	Protein supplement flour or meal cereal ingredient
	and/or Composted $\longrightarrow$	Yard and garden mulch (bagged or bulk)

and is found in all parts of the world except deserts and the polar regions. There are over 40 species in four genera *Spirodela, Lemna, Wolffia,* and *Wolffiella.* Duckweeds are comprised of a flattened leaf-like frond without a stem and many species even lack roots. Reproduction is primarily by vegetative multiplication, with biomass doubling times of between 1–5.3 days under optimum conditions. They are resistant to diseases or pests and will grow in slightly saline waters. They float on the surface of the water forming dense mats, but unlike hyacinths they do not provide much support for underwater micro-fauna. The fronds form a single layer on the water surface until crowding occurs, then the thickness of the surface layer may increase to form an extensive mat several centimetres thick. The water beneath a dense mat will have a lowered dissolved oxygen concentration and if the organic loading is high anaerobosis may occur. Duckweed will grow well even on raw wastewater, and will improve the water quality by direct plant uptake of organic and mineral material, suppression of algal growth, and by creating a suitable habitat for the growth of zooplankton, especially *Daphnia* (Dinges 1982). Harvey and Fox (1973) grew *Lemna minor* in laboratory-scale units in order to evaluate nutrient removal efficiency by using 80 aquaria filled with secondary wastewater to a depth of 450 mm, providing a surface area of 500 cm^2. The temperature was maintained at 24°C, with the light intensity held at 11 022 lm m^{-2} using 12 h photoperiods. Over a 10-day test period, frond doubling occurred every 4 days, with the harvested plants containing 4.59% nitrogen and 0.8% phosphorus (DW).

A high nutrient content is essential in order to maintain optimal uptake by the duckweed as well as maintaining biomass production. Removal rates for phosphorus remain fairly constant at between 90–95%. Experimental studies have indicated that duckweed may be more effective as a preliminary treatment phase, with hyacinths scavenging for nutrients in treated wastewaters.

Little is known of the design for duckweed ponds but as they readily grow on all types of natural water bodies, the design may not be critical as long as the water surface is sheltered and preferably still. The actual shape and depth of ponds will be primarily determined by the harvesting method employed. The plant is easily harvested using a skimming device similar to those used for removing oil from the surface of water. Microstrainers can also be used. Much interest has been shown in the use of these plants for poultry and livestock feeds as well as for the removal of nutrients from secondary treated wastewaters (Culley and Epps 1973; Harvey and Fox 1973; Hillman and Culley 1978). However, a high water content, the average water content for *Lemna* and *Spirodela* varies between 92–97%, and problems in removing the water have restricted their full development as an agricultural crop.

The water fern (*Salveniaceae*) consists of some 16 species in two genera, *Salvenia* and *Azolla* and little work has been done so far on this group. However, several workers have indicated that water ferns are less suitable than other plants because they produce large amounts of detritus within the pond and are not readily digested, although they compost well (Harvey and Fox 1973). This group of plants has been reviewed by Dinges (1982).

4.5.2 Wetlands

The position of natural wetlands has resulted in many of these habitats receiving wastewater for centuries. Since the early 1970s, however, considerable interest has been shown in the use of wetlands for complete, secondary or tertiary treatment of wastewaters. Natural wetlands include freshwater and saltwater marshes, swamps, fens, peat bogs, and cypress domes, all of which have been used for wastewater treatment (Dierberg and Brezonik 1983; Dolan et al. 1981; Good et al., 1978; Fetter et al. 1978; Tilton et al. 1976; Tourbier and Pierson 1976). In recent years, natural wetlands have come under considerable pressure from man for a variety of purposes, especially agriculture, when they have been drained or filled. With the increasing appreciation of the fragile nature and intrinsic value of wetlands, many wetland habitats are now considered too vulnerable to be modified by the addition of nutrients in the form of wastewater, therefore, artificial wetlands have been developed for treatment purposes (Gersberg et al. 1983, 1984, 1986; Seidel 1976; Sloey et al. 1978; Small 1976). Where natural wetlands have been adequately managed the damage has been minimal and in some particular wetland sites the addition of extra nutrients has increased the productivity of the habitat resulting in an improved wildlife habitat and an enhancement of the overall aesthetic appearance (Dinges 1982). It has been argued that the use of wetlands for this purpose may be a significant factor in the survival of these increasingly rare habitats that are under so much pressure from other sectors.

Natural wetlands
Wetlands are habitats where the water table lies above or close to the surface of the root substrate for a significant part of the year (Etherington 1983). Depending on their location they can be flooded with either fresh or saline waters, the former normally associated with, but not exclusively with, high rainfall regions, whereas the latter are situated on the coast or along the fringes of inland saline lakes. Freshwater wetlands can be subdivided into those where the decomposition rate is low and there is a continuous accumulation of organic matter (peat) to form bogs and fens, or those where negligible accumulation of organic matter occurs and where the rooting medium is mainly mineral, such as marshes and swamps (Jones 1986). Fens, marshes, and swamps are characterized by the presence of emergent aquatic species, and although the wetland communities of the tropics are floristically different

from those of the temperate zones, the actual diversity of emergent species is not great. The primary productivity of wetlands, excluding bogs, are among the highest for any natural community (Westlake 1963).

Wetlands are normally dominated by a single species that restricts the establishment and growth of other plants. Jones (1986) cites several examples from the papyrus swamps of East and Central Africa that are almost monospecific communities of *Cyperus papyrus*. There are also extensive stands of *Phragmites australis* or *Typha latifolia* that dominate many wetland areas in temperate regions, although these species extend into much warmer climates. Although some wetlands can have a high species diversity, species-rich communities are normally less productive than wetlands with few species (Grime 1979; Wheeler and Giller 1982). Three genera of emergent macrophytes dominate wetlands, *Typha*, *Phragmites*, and *Cyperus*, although other species, such as *Misacanthidium*, *Cladium*, and *Echmochloa* can be important locally. Examples of dominant emergent species in tropical, subtropical, and temperate zones are given by Jones (1986) (Table 4.47).

Marine wetlands include mangrove swamps that are restricted between the latitudes 30°N and 30°S, whereas elsewhere salt-tolerant grasses (*Spartina* spp.) are the primary vegetative cover (Valiela and Vince 1976; Whigham and Simpson 1976). Cypress domes are a common palustrine wetland found throughout the pine–palmetto woodlands of the Southern Atlantic and the Gulf coastal plain. They are topographic depressions that vary in size from 1–10 ha that are dominated by two species of tree, the tall pond cypress (*Taxodium distictium* var. *natans*) and the smaller black gum (*Nyssa sylvatica* var. *biflora*) (Monk and Brown 1965; Coultas and Calhoun 1975; Ewel 1983; Dierberg and Brezonik 1983).

The purification process in wetlands is complex and the most obvious biological component of the habitat, the emergent macrophytes, play an important, albeit minor role in the treatment of wastewater (Stephenson *et al.* 1980; Nichols 1983). As discussed in Section 4.5.1 the major removal mechanisms are bacterial and fungal transformations, and physico-chemical processes such as adsorption, precipitation, and sedimentation (Chan *et al.* 1982). The emergent plants have a number of important functions. The plant rhizomes provide a stable surface for heterotrophic growth as well as enhancing sedimentation of solids by ensuring flocculation and maintaining near quiescent conditions. Emergent macrophytes dominate wetlands due to the hostile root environment, which has a restricted oxygen supply so that facultative and anaerobic bacteria flourish (Armstrong 1982). The plants are able to translocate oxygen from their shoots to the roots making the rhizosphere (the root zone) an area where aerobic micro-organisms can survive. Here, heterotrophic and nitrifying bacteria flourish with nitrates diffusing to the oxygen limited zones (anoxic) of the wetland where it is removed from the system by denitrification (Sherr and Payne 1978; Iizumi *et al.* 1980; Gersberg *et al.* 1986). Nutrients are trapped in wetlands, which are

TABLE 4.47 Emergent species of some wetland areas (Jones 1986)

Location	Dominants
Tropical, subtropical	
Nyumba ya Mungu reservoir, Tanzania	*Typha domingensis, Cyperus alopecuroides*
Lake Naivasha, Kenya	*Cyperus papyrus,* Cyperus immensus, Cyperus digitatus* spp. *auricomus, Cyperus alopecuroides*
Kashambya swamp, Kigezi, Uganda	*Cyperus papyrus, Cladium mariscus* (pH > 6.3), *Mischanthidium violaceum* (pH 4.5), *Cyperus latifolius, Typha* sp.
S.E. Lango, lowland West Nile, Uganda. On alkaline clay-like soils in valleys draining into Lake Kyoga (seasonal swamps)	*Echinochloa pyramidalis,* Oryza barthii* (sometimes codominant with *E. pyramidalis), Leersia hexandra*
River Amazon, Manaus, Brazil (seasonal swamp)	*Echinochloa polystachya,* Oryza perennis, Paspalum repens, Hymenachne amplexicaulis*
Namiro swamp, northern end Lake Victoria, Uganda, landward zone (disturbed?)	*Miscanthidium violaceum,* Loudetia phragmatoides;* (towards water's edge), *Cyperus papyrus,* Miscanthidium violaceum*
Northern Sudd, Sudan	*Cyperus papyrus, Phragmites karka, Vossia cuspidata, Typha domingensis*
Temperate	
Huntingdon Marsh, Lake St. Francis, P.Q., Canada	*Carex aquatilis,* Carex lanuginosa,* Calamagrostis canadensis, Typha angustifolia*
Prairie glacial marsh, Eagle Lake, IA, USA	*Typha glauca, Scirpus fluviatilis, Scirpus validus, Sparganium eurycarpum, Sagitaria latifolia*
Freshwater tidal marsh, Augustine Creek, GA, USA	*Zizaniopsis miliacea,* Zizania aquatïca, Peltandra virginica, Pontederia cordata*
Norfolk Broadland, UK	*Cladium mariscus,* Phragmites australis,* Calamagrostis canescens, Carex acutifomis, Carex lasiocarpa, Carex paniculata, Glyceria maxima, Juncus subnodulosus, Phalaris arundinacea, Typha angustifolia*

* Most common species.

often referred to as nutrient sinks, not only in the sediment but in the actual plant biomass by nutrient uptake. The phosphorus in the wastewater does not move far in the soil as it is retained near the surface. The phosphate ions are chemically adsorbed on to the surface of hydrous oxides of iron and aluminium, and on to silicate clay minerals, where the phosphate binds to the aluminium atoms exposed at the edges of the clay particles. The chemical and physical adsorption of phosphate on to the surface of soil minerals is a rapid process, which in the laboratory occurs mostly within the first few minutes. In addition to this initial fast reaction, slower reactions continue to remove phosphate from solution for periods of several days to several months. Adsorption–precipitation by soils is not necessarily a permanent sink for the phosphorus in wastewater and is partially reversible by a reduction in the phosphate concentration in the water in contact with the soil, by plant uptake or by flushing or dilution with low phosphate concentration water, which will release some phosphorus back into solution. The soil therefore regulates the concentration of phosphate in solution. It is interesting that soils that adsorb phosphorus least readily are those very same soils that typically release phosphorus most readily. Peaty material, made up of plant remains, can adsorb minerals by ion exchange and retain them indefinitely. Organic soils are typically low in phosphorus and consequently have a high C:P ratio. The phosphorus content of peat is commonly less than 0.05–0.10%. Microbial immobilization has been suggested as a mechanism for retention of wastewater phosphorus where the C:P ratio is high. Initial immobilization of phosphorus will occur in response to wastewater application but as the C:P ratio of wastewater is low compared to peat, continued application would soon satisfy microbial requirement for phosphorus. Therefore, immobilization by micro-organisms is not likely to play a significant role in the long-term fixation of phosphorus by peat, but will be initially important.

The nitrogen cycle in wetlands is extremely complex. Nitrogen exists in a multitude of organic forms and as ammonia, nitrite, nitrate as well as gaseous forms of ammonia, nitrogen, and nitrogen oxides (Boyt et al. 1976). It is converted from one form to another in wetlands. Denitrification occurs in the anoxic sediments, although it occurs much more slowly under acid conditions compared with either neutral or alkaline conditions. Below pH 5, chemical rather than biochemical reactions convert nitrogen to gaseous forms. Algae and bacteria associated with wetlands can fix atmospheric nitrogen into available forms that can reduce the overall removal efficiency of wastewater nitrogen of wetlands. The application to a wetland of secondary wastewater effluent, which typically has a N:P ratio of less than 10 may stimulate nitrogen fixation, although this is dependent on the amount of wastewater applied and other nutrient sources.

The soil rather than the water is the major source of nutrients for emergent vegetation. Non-rooted plants, such as algae and duckweeds may also be present and obtain their nutrients directly from the water, and the subsequent incorporation of their detritus into the soil as a net transfer of nutrients from the water to the soil. Nutrient retention is greatest during periods of active vegetative growth and is low during the non-growing season. Unless the vegetation can be harvested, much of the stored nutrients will be released during the seasonal cycle of die-back and decomposition. Subsequently, there is often a release of nutrients from the decaying biomass in the winter followed by the rapid release into the water of 35–75% of the phosphorus from the plant tissue, although less of the nitrogen is released. A long retention time is an important factor in many of the physico-chemical processes and also ensures maximum removal of pathogens. Metals present in the wastewater become immobilized in the anaerobic mud, forming metallic sulphides. Release of carbon dioxide and methane from the anaerobic mud will also occur.

Nearly all the water, both precipitation and surface water, entering the wetland leaves it by surface discharge with little entering the groundwater strata. The quantity of water leaving the wetland will vary seasonally, with little or no discharge in the summer but a high discharge in the winter and spring. Most natural wetlands contain open channels to enhance the movement of water, and much of the water never comes into contact with either the soil or vegetation, whereas salt marshes may be regularly inundated by high tides. In a natural system there will be periodic flushes of accumulated solids and nutrients from the area. It is clear, therefore, that a natural wetland must be managed to ensure that the wastewater comes into maximum contact with the soil and the vegetation. During the winter, the loading must be reduced as the emergent vegetation dies back and the rate of biological activity of the associated micro-organisms in the rhizosphere is greatly reduced. Loading capacity varies, with a natural wetland able to treat the wastewater from about 100 people ha^{-1}, although the same area can remove the nitrogen produced by 125 people, but the phosphorus from only 25 people (Dinges 1982).

To ensure maximum removal, it is important that the wastewater comes into maximum contact with the soil and vegetation, which can be optimized by making a shallow depth < 30 cm. In essence, this is the basic design objective of artificial wetlands, which use long shallow channels or several ponds in series. Nichols (1983) has listed several natural wetlands that receive applications of secondary wastewater (Table 4.48). At low loading rates wetlands have the capacity to remove much of the phosphorus applied and continue to do so unless the loading rate is increased, when removal efficiency declines rapidly. It is not known how long a wetland can continue to remove phosphorus from wastewater, but if sufficient nutrient is added the adsorption capacity of the soil can become saturated. In addition, a wetland soil can release a proportion of the phosphorus previously adsorbed if the concentration of phosphorus in the water is reduced. The nitrogen pattern of wetlands is similar to that for phosphorus, with high removal efficiency, > 70%, at low loading rates (< 10 g N m^{-2}y^{-1}), rapidly declining as the loading rate increases. The hydraulic conditions in a wetland can also affect the removal of wastewater nutrients. Higher nitrogen and phosphorus loading rates are obtained by higher hydraulic loadings (Table 4.48), so that retention times in the wetland are reduced and less time for nutrient removal reactions is allowed. The morphology of wetlands is also important, for example, as the depth increases the chance for reactions to occur between wastewater nutrients and the underlying soil decreases. Deeper waters will, however,

TABLE 4.48 Removal of nitrogen and phosphorus from wastewater[a] applied to natural wetlands (Nichols 1983)

Types of wetland	Location	Size (ha)	Years wastewater applied	Hydraulic loading (cm y⁻¹)		Nutrient loading (g m⁻² · y⁻¹)		Nutrient removal	
				Wastewater	Other	Total P	Total N	Total P	Total N
(1) Shrub sedge fen	Michigan	1[b]	1[c]	70[c]	—	1.7[c]	1.9[c]	95[c]	96[cd]
(2) Forest shrub fen	Michigan	18.2	1[e]	36.8[e]	—	0.9[e]	1.5[ed]	91[e]	75[ed]
			2[f]	74.1[f]	205[f]	2.6[f]	6.5[fd]	88[f]	80[fd]
			3[g]	65.2[g]	183[g]	1.7[g]	9.3[gd]	72	80[gd]
			4[h]	55.7[h]	116[h]	1.8[h]	6.2[hd]	64[h]	77[hd]
			5[h]	57.3[h]	97[h]	1.7[h]	9.3[hd]	65[h]	75[hd]
(3) Blanket bog	Ireland	—	1	i	—	5.0	7.4[d]	96	82[d]
			2	i	—	13.1	15.4[d]	72	87[d]
			3	i	—	8.1	10.3[d]	43	68[d]
(4) Hardwood swamp	Florida	204	20	10.2	83	0.9	—	87	—
(5) Cattail marsh	Wisconsin	156	55	23.4	558	15.2	—	32	—
(6) Cattail marsh	Massachusetts	19.4	69	684	159	7.1	53.6	47	31
(7) Cattail	Massachusetts	2.4	69	5526	—	63.6	428	20	1
(8) Deepwater marsh	Ontario	162	55	231	5569	11.6	78.6	58[j]	41[j]
(9) Glycena marsh	Ontario	20	55	1870	—	77	404	24[j]	38[j]

[a] Secondary effluent; [b] area affected by study, entire wetland is 710 ha; [c] May–September; [d] inorganic N only, organic N not measured; [e] August–October; [f] March–November; [g] April–November; [h] June–November; [i] Chemical fertilizers, not wastewater applied; [j] Wastewater applied on a year-round basis but percent removal measured during the growing season only. Percent removed would likely have been more calculated on a year-round basis.

provide longer retention times than shallower wetlands at the same hydraulic loading.

Spangler *et al.* (1976) evaluated a natural wetland in Wisconsin. Using a 156 ha portion of a marsh with a total surface area of some 18 km^2, the marsh was loaded with a secondary effluent. The BOD was reduced from 26.9 to 5.3 mg l^{-1} (80%), with a 86.2% reduction in total coliforms and a 51.3% and 13.4% removal of nitrate and total phosphorus respectively. They observed that young shoots of emergent vegetation had the highest concentration of phosphorus which decreased as the tissue aged. They concluded that the quality of phosphorus removal could be greatly increased by frequent harvesting so that the tissue was not permitted to mature completely and subsequently die, decompose and so re-release the accumulated nutrients. Unlike artificial wetlands, it is not possible, or desirable in terms of conservation, to harvest natural wetlands.

Artificial wetlands
In contrast to natural wetlands, artificial systems are subject to close control and designed to maximize treatment efficiency and biomass production, with the shape of artificial wetlands made to facilitate harvesting. They can be used in all climates and provide a valuable habitat for wildlife.

In temperate climates, artificial wetlands function best in the warmer months and there is a need to reduce organic loading during the winter, and possibly even store effluent until the following spring. However, work in moderately cold climates, such as Ontario (Canada), has shown they can be operated successfully all the year round and produce good quality final effluents even when covered with ice (Wile *et al.* 1982; Reed *et al.* 1984). In terms of cost, they require: a large area of land; a high degree of labour to construct the bed and plant the vegetation; and periodic harvesting. However, work in the USA has shown that they are much more cost-effective on a long-term basis than natural wetlands.

Artificial systems can be lined to prevent exfiltration from the wetland, with the vegetation planted in a shallow layer of gravel. Where the substrate is suitable, emergent vegetation can be planted directly into the soil. Where wetlands are also constructed for wildlife use, it is important to include a diversity of habitats in the overall design, such as areas of deep, as well as shallow water, islands and adequate areas of open water for wildfowl. Such a system has been constructed at Martinez in California, where the wetland was created primarily for wildlife. The system is 6.1 ha in area and has a retention time of 10 days. Some 60% of the wetland is open water with stands of *Scirpus* and *Typha*. It has been necessary to use a further 2.1 ha around the edge of the existing wetland for growing plants, which the wildlife using the wetland can use as a food source.

Artificial wetlands function in the same way as natural ones except more nitrogen and phosphorus is lost from the system by harvesting. Such systems have been studied using plastic lined beds 18.5 m × 3.5 m and 0.76 m deep, with emergent macrophytes grown in gravel (Gersberg *et al.* 1986). Four beds were used to treat primary municipal wastewater from Santee, California. This is a particularly ideal climate for wetlands as the mean air temperature rarely falls below 12°C, and the plants continuously grow throughout the year. Three plants were compared for their ability to treat wastewater, *Scirpus validus* (bulrush), *Phragmites communis* (common reed), and *Typha latifolia* (cattail). During the experimental period, the mean ammonia concentration in the influent was reduced from 24.7 mg l^{-1} to 1.4 mg l^{-1} in the bulrush bed, 5.3 mg l^{-1} in the reed bed, 17.7 mg l^{-1} in the cattail bed, and 22.1 mg l^{-1} in the control bed, which did not contain any emergent plants. The ability of the bulrush and reed to remove ammonia from water is because they appear to translocate oxygen from the shoots to the rhizosphere, which stimulates nitrification–denitrification (Armstrong 1964). Similarly, BOD removal efficiencies were significantly higher in the bulrush and reed beds at 5.3 and 22.2 mg l^{-1} respectively, compared to the cattail bed at 30.4 mg l^{-1}, or the control bed at 36.4 mg l^{-1}, where the mean BOD of the influent was 118.3 mg l^{-1}. This showed that these two species were able to treat wastewater to an equal or even better quality than conventional secondary treatment and that the plants have a significant function in the removal of nitrogen separate to assimilation into plant tissue. The mean percentage removal of BOD was 96, 81, 74, and 69% for the bulrushes, reeds, cattails, and control bed respectively. The poor performance of cattails can be attributed to the fact that it is less able to transport oxygen to the root zone, although this ability is seen in many similar plants such as *Spartina* and *Zostera*, and has a shallow root zone with most of the root mass confined to the top 30 cm of the substrate compared to > 60 cm for bulrushes and > 75 cm for reeds. This provides less root surface area for the development of heterotrophic as well as nitrifying bacteria. It is clear from this study that at a hydraulic loading rate of 4.7 cm d^{-1} a 10:10 (BOD:SS) final effluent can be produced. At this loading rate 20 acres (8.1 ha) of wetlands would be required to treat 1×10^6 gallons (3785 m^3) of primary wastewater each day.

Earlier studies at the same site by Gersberg *et al.* (1983) had shown that the removal of nitrate from secondary effluents by denitrification could be enhanced by supplementing the carbon supply by the addition of methanol to stimulate bacterial denitrification activity. At loading rates of 16.8 cm d^{-1}, 95% of the total nitrogen was being removed. Mulched biomass was subsequently used, instead of methanol, to reduce operating costs, although lower nitrogen removals were achieved at reduced loading rates, 86% removal of total nitrogen at a loading 8.4 cm d^{-1} falling to 60% at 16.8 cm d^{-1}.

Bowmer (1987) has studied the effect of rhizosphere aeration and basin hydrology on nutrient removal in artificial wetlands in Australia. She found that the hydrology of such systems is driven primarily by gravity and transpiration. Also, that the macrophytes, *Typha orientalis* and *Schoenoplec-*

tus validus, were not introducing sufficient oxygen into the effluent for complete nitrification. Although this was due, in part, to inadequate outward radial diffusion of oxygen into the rhizosphere, the effect of channelling the effluent in preferential flow paths around the aerating root masses was also significant. To overcome these problems, Bowmer suggests using species that give a greater radial diffusion of oxygen from their roots into the substratum. To minimize preferential channelling of flow, suggested strategies include the use of a sprinkler system to distribute the effluent over a wide area, perforated baffles to improve mixing, and periodic draining and filling. The latter modification should provide both aerobic and anoxic conditions for complete nitrogen removal.

Reed beds are planned to help reclaim desert areas for agriculture in Egypt. The wastewater is settled before being pumped on to gravel beds lined with plastic. The beds are used to cultivate reed beds which surround agricultural plots protecting them from wind and sand. The central agricultural plots are used for growing cash crops with the sewage sludge used as a soil amendment and fertilizer (Anon 1986).

4.5.3 Land treatment

Land treatment of wastewater dates back to the sixteenth century and became widely used in Europe during the nineteenth century. At the time of the Rivers Pollution Prevention Act of 1876 new methods of sewage treatment were being investigated. At this time, land treatment remained the only alternative method of disposal of sewage apart from directly discharging it to a watercourse. However, land treatment itself had been improved. Instead of simply flooding land, the soil was ploughed deeply and the sewage allowed to flow into furrows. The waste matter decayed and made the soil very fertile, allowing vegetables of excellent quality to be grown on the tops of the furrows hence the term 'sewage farms'. Sewage farming became a lucrative business, with the rights to a town's sewage often costing farmers considerable amounts of money. However, because of the amount of land required, the occurrence of unpleasant smells, and the problem of flies in the warmer months, sewage farms were superseded by other biological treatment methods. Various improvements in land treatment were introduced in an attempt to reduce the area of land required. The most successful was the 'Latham soil-filter system', which comprised three layers of soil separated by layers of coke breeze overlying land drains. This method reduced the area of land required by up to a factor of 10 compared with the traditional sewage farm methods. However, even with these new developments the loading rate to land was restricted to a maximum of 0.05 $m^3m^{-2}d^{-1}$, which is roughly equivalent to 1000 persons per acre. In 1900, the City of Birmingham (UK) was using 2800 acres of land to treat its sewage, an area that stretched for 2.5 miles along the banks of the River Tame. The area available for treatment had to be increased at a rate of 1.5 acres per week to keep pace with the growing population (Bruce and Hawkes 1983). However, sewage farms were slow to disappear and several major European cities still partly relied on such systems until quite recently. The town of Greystones (Ireland) still employs land treatment using surface sprayers. The large areas of flooded land also proved an excellent feeding ground for many birds, especially waders, and recent studies have suggested that the demise of the sewage farm may be closely linked to the reduction in the population of certain birds (Fuller and Glue, 1978, 1980, 1981).

Although now not widely practised on a large scale in Europe for domestic wastewaters, spray irrigation of animal slurry and some food processing wastewaters is common. Spray irrigation of dairy wastewasters is widely practised in New Zealand, while in West Germany, at the Braunschweig Land Treatment Plant, the treated effluent from a population equivalent of 270 000 is re-used on 300 ha of farmland by sprinkler irrigation (Boll and Kayser 1986). Land treatment using percolation areas is still used widely in the UK and Ireland, in conjunction with septic tank systems. In areas of the world where water is in short supply, raw and treated wastewater is used for crop irrigation (Israel) or for watering public parks and green areas (southern USA) (Section 6.2.4.2). Agricultural land is a major recipient of sewage sludge in Europe and this is fully examined in Chapter 5.

Purification process
Wastewater is purified as it percolates through the soil by a range of physical, chemical, and biological processes. Suspended material, including micro-organisms, is physically filtered out of solution as the water percolates through the upper soil layer. Organic material trapped in the soil is rapidly utilized by the high density of heterotrophs present. For example, a single plate count using just one gram of soil can yield 10^7 bacteria, 10^6 actinomycetes, and 10^5 fungi, which in turn supports a large and diverse micro- and macrofauna (Miller 1974). The humus silt and clay particles that comprise the soil provide a very large surface area for ion-exchange with mineral ions, especially cations, becoming strongly bonded on to soil particles. In practice, this results in a high removal efficiency of metals, phosphorus, and ammonia from the water. Any nitrogen present is normally fully utilized by either plant uptake and subsequent harvest, the nitrification–denitrification process or volitization. Discharge of excess sodium (Na^+) ions to land disperses the soil particles and inhibits plant growth (Ellis 1974), and as it accumulates percolation is severely impaired and the tilth of the soil is reduced. In contrast, the divalent cations calcium (Ca^{2+}) and magnesium (Mg^+) will rectify the effects of excess sodium. The ratio between sodium and both calcium and magnesium is known as the sodium absorption ratio (SAR) which is used to characterize the suitability of

a soil for plant growth:

$$SAR = \frac{Na^+}{\sqrt{\left(\frac{Ca+Mg}{2}\right)}}$$

where all the cation concentrations are given as millequivalents l^{-1}.

Irrigation of soil with wastewater leads to complex chemical changes, such as precipitation and oxidation–reduction. Immobilization of cations is enhanced at a soil water pH > 7 but acidic conditions lead to some mineral solubilization and leaching. Overloading land treatment sites can result in the land becoming clogged with organic matter. This is unsightly, produces strong and unpleasant odours, often an associated fly nuisance, but as the soil is anaerobic the pH gradually falls until minerals are released and leaching occurs.

Land application
There are three types of land application of wastewater, irrigation (slow and rapid), overland flow, and infiltration–percolation, although there is some overlap between these categories (Environmental Protection Agency 1981). The important factors in the design of land treatment systems are: land availability; permeability and depth of soil; type of bedrock and nature of groundwater; climate including precipitation; evapo-transpiration and temperature; plant cover; topography; wastewater type; and required final effluent quality (Johnson 1973). All three types of systems are widely used in the USA, with irrigation systems most widely used, and rapid infiltration used primarily for groundwater recharge. In Europe, land treatment is less widely used and is generally restricted for secondary or tertiary treatment of wastewater from small communities.

Irrigation is the most widely used land treatment system and involves the disposal of screened and primary settled wastewater to agricultural or recreational land. Slow-rate irrigation is used to supplement rainfall and for maximizing crop production by utilizing the weak nutrient content of the wastewater as a growth promoter. This system is unsuitable for shallow soil or soil primarily comprised of clay. The ideal soil is either loam, or silt, has good permeability and a minimum depth of 1.75 m to the groundwater (Fig. 4.99). The rate of application of wastewater depends on soil permeability, evapo-transpiration rates, and precipitation. Normal hydraulic loading rates are in the range 0.6–6.0 m y^{-1}; for a population of 10 000, a surface area of between 23–227 ha is required (Pound *et al.* 1977). Particle size distribution of soil is important in the removal efficiency of wastewater constituents; smaller particles are most effective as they reduce the percolation rate (Tare and Bokil 1982). Soil profile and crop cover have been identified as major factors in nutrient removal. Complete treatment can be obtained with all the BOD and

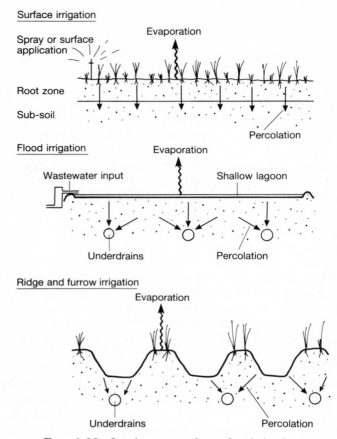

FIG. 4.99 Land treatment by surface irrigation.

suspended solids removed. The slow irrigation rate is important to ensure most of the phosphorus is immobilized and for the bulk of the nitrogen to be mineralized. Irrigation is not continuous and storage is required, unless sufficient area is available to allow the land to be rested between application periods. Also, irrigation may not be possible during periods of high rainfall or when the land is frozen. Wastewater is distributed on to the land either by surface flooding or by spraying. Surface flooding is only possible where the land is level and may involve flooding specific areas for several days and then resting it for a couple of weeks, or using conventional furrows as were used on the original sewage farms. Where deep furrows are preferred, each furrow is fed by a syphon connected to an open supply channel or simply filled with wastewater several times a day. In all surface flooding operations it is necessary to contain the wastewater on the site by surrounding it by a dike to prevent runoff. Spraying is done from either a fixed set of pipes or rotary sprayers, with different areas of the system rested intermittently, or by using a mobile rig which is moved manually from area to area. Spraying allows unlevel ground to be used, although there is a health risk from aerosols (Bausum *et al.* 1982). However, in the USA chlorination has been successfully used to reduce downwind levels of microbial and coliphage aerosols. Although soil damage is unlikely, unless severely overloaded organically, groundwaters are at risk. Studies have shown that the total dissolved solids concentration (alkalinity) and the nitrate concentration immediately below the site will be enhanced, although viruses and bacteria will be filtered out by the soil. However, prolonged application of wastewater to land may result in the translocation of some viruses through the soil (Shaub *et al.* 1982) (Section 6.2.4.2). Metals will be immobilized on to the soil particles, and the major metals, cadmium, copper, nickel, lead, and zinc in secondary effluents have been shown rarely to migrate deeper than 1.5 m (Brown *et al.* 1983). However, some persistent pesticides have been detected (Koerner and Haws 1979; Weaver *et al.* 1978). The largest slow-rate irrigation system in operation is Werribee Farm in Melbourne (Australia), which treats the wastewater from 800 000 people (546 000 m^3d^{-1}). This is sprayed on to pasture that is used for fattening livestock. Since it was opened in 1897, no adverse effects have been recorded to the soil, vegetation, or livestock (Dinges 1982).

Overland flow, which is also widely known as grass filtration, is a development of the irrigation system except that the underlying soil has a low permeability, being comprised of clay or clay loam soils. When wastewater is applied, the bulk of the water flows over the surface rather than percolates into the soil, resulting in the effluent having to be collected for re-use or discharge to a watercourse. Flat land is preferred so that it can be carefully regraded to give a slope of between 2–6%, although naturally sloping land can also be used. The wastewater is screened and the settleable solids removed by primary settlement. It is then applied to the top third of the gently sloping land either by overflow pipes or by spraying and allowing to slowly flow by gravity down the slope (Fig. 4.100). The ground is planted with a suitable species of grass so that purification occurs by filtration and bacterial decomposition as the wastewater moves through the vegetation. The grass is regularly cut to remove nutrients, taken up by the plants, from the system which could be released back if the vegetation is allowed to mature and die. The addition of lime has been found to enhance the removal of phosphorus from overland flow systems (Khalid *et al.* 1982). These systems are very dependent upon the weather and loading is normally kept between 1.5–3.5 m y^{-1}, and should not exceed 5 m y^{-1} (Pound *et al.* 1977). Performance of these systems is good for primary and secondary wastewaters, but unlike BOD or suspended solids, the removal of ammonia decreases as the wastewater application rate increases (Wightman *et al.* 1983). Problems have arisen when overland flow systems have been used as a secondary treatment phase for lagoon treated wastewaters, as the algae in the applied wastewater is not removed by the land system, which results in a

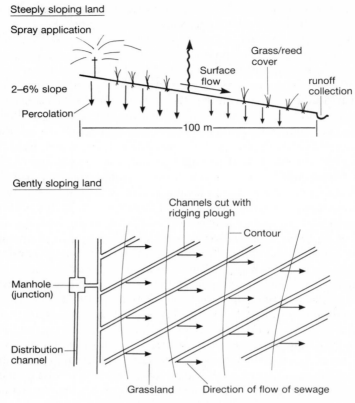

FIG. 4.100   Land treatment by overland flow.

high suspended solids concentration in the final effluent (Abernathy 1983; Witherow and Bledsoe 1983). The grass plots are normally 100 m in length and often several separate plots can be operated on a single slope. The use of grass plots in series has been beneficial in terms of enhanced effluent quality, but less efficient in overall removal per unit area of land. The design of overland flow systems has been reviewed by Hegg and Turner (1983) for use with animal wastes and by Smith and Schroeder (1983) for domestic wastewaters.

High-rate irrigation and infiltration–percolation systems are very similar except that the former is used on natural sandy soils that are irrigated at very high rates using sprayers, and the latter uses specially constructed basins in suitable soils so that the wastewater can soak away into the very permeable soil to the groundwater. High-rate irrigation uses loading rates between 6–170 m y^{-1}, with the dividing line between slow-rate irrigation and infiltration being 60 m y^{-1}, and an area 1.6–48.5 ha required to treat 10 000 people (Pound *et al.* 1977). Operation is continuous and due to the small area of land used, factors such as evapo-transpiration and precipitation are unimportant. Irrigation sites can have vegetation but as they are regularly ploughed to prevent accumulation of organic matter on the surface that could restrict percolation, they are normally bare. The small amount of clay, silt or humic material severely limits the capacity of the soil to remove metals, phosphorus, and other wastewater constituents, and the groundwater is much more vulnerable and receives a significant volume of mineral contamination. At these high loading rates there is only partial removal occurring, with all pathogens, especially viruses, reaching the groundwater, although both primary and secondary wastewaters have been successfully treated (Carlson *et al.* 1982). Although widely used in the USA as a secondary treatment process, high-rate irrigation is generally used for aquifer recharge using secondary treated effluents.

Irrigation–percolation is also restricted to sandy and sandy–loam soils that have a high permeability, so that the applied wastewater can pass quickly through the soil to the groundwater. Basins are specially constructed to enhance rapid percolation, however, depending on the organic loading, the top surface layer of sand will need to be renewed periodically and replaced with clean sand. Where partially treated wastewater is used, basins are flooded only for a couple of days before being rested for one or two weeks (Fig. 4.101). As with high-rate irrigation, the limiting factor for infiltration processes is the degree of groundwater contamination. For example, an infiltration system has been used in Milton, Wisconsin (USA) for the past 20 years. The wastewater receives primary treatment before being discharged into a set of four basins, which are continuously operated in series. The soil is comprised of glacial sands and gravels with inclusions of silt and clay. Test boreholes have shown that the quality of the groundwater below the site has become contaminated, with a BOD of 4.2 mg l^{-1}, ammonia–nitrogen at 13.7 mg l^{-1}, nitrate at

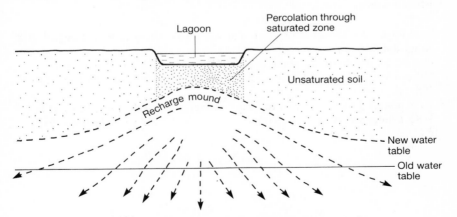

FIG. 4.101 Land treatment by infiltration–percolation.

1.5 mg l^{-1}, total phosphorus of 4.0 mg l^{-1}, and faecal coliforms reaching 258 per 100 ml (Anon 1979b). Groundwater contamination by trace organics have also been associated with rapid infiltration systems (Tomson et al. 1981). However, the most suitable natural sites are found in the flood valleys of rivers and there is little environmental damage. Where there is a risk of contaminating water supplies, this treatment process should be avoided.

Due to over-pumping, aquifers have become depleted in many countries which has resulted in attempts to recharge aquifers from other, usually less pure, water sources, such as rivers. There has been growing interest in using primary and secondary wastewater to replenish groundwater, relying on the soil and the aquifer itself to provide a high degree of purification (Idelovitch 1978; Idelovitch and Michail 1980; Roberts and McCarty, 1978). However, it has proved very difficult to predict or model the movement of the recharge effluent within the aquifer, and in practice the water quality has often become impaired. There may be a lag period of several years between commencement of recharge and contamination of observation boreholes, with a much longer period required for recovery of water quality once recharge has ceased. As recharge commences, a mixture of recharge effluent and natural groundwater will eventually be pumped from the supply boreholes, and regular monitoring of water quality is essential to ensure potable standards are maintained. Because of the potential long-term risks to important groundwater resources, where infiltration basins are used for groundwater recharge, it is advisable only to use wastewater treated to as near drinking water quality as possible, and to regard the basin system solely as a means of recharging the aquifer much in the same way that injection wells are used. Any benefits from the potential treatment capacity of the soil as the water percolates to the aquifer should be regarded as incidental (Idelovitch and Michail 1984).

Further reading

Plants: Pope 1981; Dinges 1982; Oron et al. 1985; Tourbier et al. 1976; Mitchell 1978; Athie and Cerri 1987.

Wetlands: Dinges 1982; Gersberg et al. 1984; Tilton et al. 1976; Sloey et al. 1978; Heliotis and De Witt 1983; Good et al. 1978; Butijn and Greiner 1985; Gershey et al. 1985; Brix 1987.

Land treatment: Environmental Protection Agency 1981; Loehr 1977; Tare and Bokil, 1982; Canter and Knox 1985; Hinesly et al. 1978; Oliveira and Almeida 1987.

5

SLUDGE TREATMENT AND DISPOSAL

5.1 SLUDGE CHARACTERISTICS AND TREATMENT

The treatment of sewage is essentially a separation process, a method of concentrating and converting suspended and soluble nutrients into a settleable form that can be separated from the bulk of the liquid. The removal of the settleable fraction of raw sewage at the primary settlement stage, and of the settleable solids produced by biological conversion of dissolved nutrients into bacterial cells at the secondary stage, continuously produces a large quantity of concentrated sludge. Although the liquid fraction of the wastewater can be fully treated and disposed of safely to surface waters, the accumulated sludge has to be transported from the wastewater treatment plant for disposal. Sludge separation, treatment, and disposal represents a major capital and operational cost in sewage treatment. Dewatering and disposal costs for a medium-sized activated sludge plant represents as much as 50% of the initial capital and 65% of the operating costs (Calcutt and Moss 1984). If one remembers that sludge is simply concentrated wastewater, then failure to provide adequate sludge treatment and disposal facilities can result in serious pollution.

Sludge is a complex material and is discussed in detail by Best (1980). At moisture contents greater than 90% sludges behave as liquids, whereas below 90% they behave as non-newtonian fluids exhibiting a plastic rather than viscous flow (Fig. 5.1). The water held within the sludge is either free or bound. In sludges with a moisture content of more than 95%, 70% of the water is in a free or readily drained form, and the remainder is bound to the sludge and more difficult to remove, with 20% present as floc or particle moisture, 8% chemically bound, and 2% as capillary water (Fig. 5.2). The more intimately the water is bound the more energy is required to remove it. Although most of the free water can be removed by gravity settlement, some of the bound water must also be removed if the sludge is to be made handleable. This is normally done by coagulation, using chemicals and dewatering equipment which removes moisture by altering the particle formation of flocs and the cohesive forces that bind the particles together, thus releasing floc and capillary water. The type of sludge produced depends on a number of factors, such as the type of sludge separation and the treatment processes employed, which are really a function of the size of the treatment plant and wastewater characteristics.

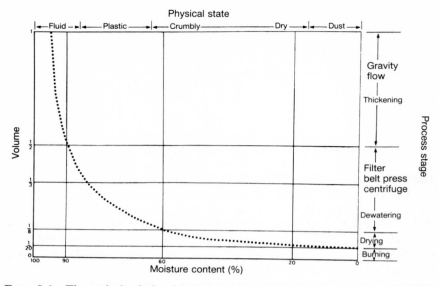

FIG. 5.1   Theoretical relationship between moisture content and the volume of sewage sludge as produced by various stages of sludge processing (Best 1980).

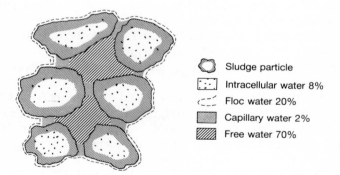

FIG. 5.2   Schematic diagram of a sludge floc showing the association of the sludge particles with the available water (after Best 1980).

Trade waters produce very different types of sludge that can be broadly categorized as either organic, such as those from slaughterhouse or fermentation effluents, or inorganic which may contain toxic materials, especially heavy metals. The presence of gross organic solids in sludge, such as offal from meat-processing wastes, pathogenic organisms or toxic compounds and heavy metals from pharmaceutical and metal industries may contaminate the sludge and restrict the disposal options. Primary or raw sludge is the settleable fraction of raw or crude sewage and is a rather unpleasant smelling thick liquid, which is highly putrescible, with a moisture content of between 94–98%. Secondary sludges contain the solids washed out of the biological treatment units, and either comprise of wasted activated sludge or sloughed microbial film. The secondary sludges are more stabilized than primary sludge with percolating filter sludge (humus) far more stabilized than wasted activated sludge having a dry solids content of 6–8% compared with 1.5–2.5% for activated sludge (Table 5.1). The quantity, volume, and solids content of sludge produced per capita each day from the most common sewage treatment units is shown in Table 5.2. It is common practice to mix primary and secondary sludges together for treatment and disposal. In general, the mean flow of wet sludge produced in a treatment plant and requiring treatment will be between 1–2% of the flow of raw sewage. White (1978) gives a useful formula from which the daily production of sludge can be estimated:

$$Q(K_1 S + (1 + K_2)YbK_3) \text{ kg solids d}^{-1}$$

where Q is the mean flow of sewage ($m^3 m^{-1}$), S and b the mean concentration of suspended solids and BOD respectively (mg l^{-1}), K_1 and K_2 the fraction of suspended solids ($\simeq 0.6$) and BOD ($\simeq 0.3$) removed at the primary sedimentation stage respectively, and K_3 the fraction of BOD removed at the biological oxidation stage ($\simeq 0.90$–0.95), and Y is the sludge yield coefficient

TABLE 5.1   Normal volumes of sludge and their solids content from primary sedimentation and secondary treatment process (Open University 1975)

Type of sludge component and source	Quantity (l cap^{-1}d^{-1})	Dry solids (kg cap^{-1}d^{-1})	Moisture content (%)
Primary sedimentation sludge	1.1	0.05 (4.5%)	95.5
Percolating filter sludge			
Low-rate filters	0.23	0.014 (6.1%)	93.9
High-rate filters	0.30	0.018 (6.0%)	94.0
Activated sludge, surplus sludge	2.4	0.036 (1.5%)	98.5

TABLE 5.2   Production of sludge by individual unit processes in sewage treatment (Casey and O'Connor 1980)

Process stage	Quantity (g cap^{-1}d^{-1})	Solids content (% by weight)	Volume (l cap^{-1}d^{-1})
Primary sedimentation	44–55	5–8	0.6–1.1
Biofiltration of settled sewage	13–20	5–7	0.2–0.4
Standard-rate activated sludge			
pre-settled sewage	20–35	0.75–1.5	1.3–4.7
Extended aeration raw sewage	22–50	0.75–1.5	1.7–6.7
Tertiary sand filtration	3–5	0.01–0.02	15.0–50.0
Phosphorus precipitation (Al or Fe)	8–12	1–2	0.4–1.2

for the conversion of BOD to secondary sludge in the biological oxidation unit ($Y=0.5$–1.0 for activated sludge and 0.3–0.5 for single pass percolating filter systems).

Some 30 million tonnes of sludge are disposed of annually in the UK, which is equivalent to 1.3 million tonnes of dry solids (National Water Council 1981). The cost of this is in excess of £180m per year (Collinge and Bruce 1981). As raw sludge is 94–98% water, the disposal problem can be significantly reduced by reducing the water content and enhance the volume of the sludge. The putrescible nature of sludge is also a problem, especially if it has to be stored before final disposal. Therefore, although raw liquid sludge can be disposed directly either to land or sea, it is normal practice for the sludge to receive further treatment in the form of thickening and dewatering. This may include chemical conditioning, which reduces the volume of the sludge, and digestion or lime stabilization to make the sludge more stable before final disposal (Fig. 5.3). Such treatment will alter the physical and chemical nature of the sludge, which is characterized by its water content and stability.

5.1.1   Treatment options

The treatment of sewage sludge follows a general sequence. The water content is reduced by some form of thickening that is either followed by a stabilization process and secondary thickening, and/or chemical conditioning and a dewatering process to reduce the water content even further (Fig. 5.3). The sludge is then ready for disposal.

Thickening
Untreated sludges from the primary and secondary sedimentation tanks have a high water content (Table 5.1), and in order to reduce the volume of sludge handled in the stabilization or dewatering processes the sludge needs to be concentrated or thickened. Thickening is achieved by physical means, such as flotation, centrifugation, and lagooning, but most usually by gravity settlement. Gravity thickeners can increase the sludge concentration in raw primary sludge from 2.5% to 8.0% resulting in a three-fold decrease in sludge volume, whereas a five-fold decrease in the volume of wasted activated sludge is not uncommon, with it being thickened from 0.8% to 4.0% solids (Table 5.3).

Gravity thickening takes place in a circular tank which is somewhat similar in design to a primary sedimentation tank (Fig. 5.4). The dilute sludge is fed into the tank where the particles are allowed to settle over a long period, which may be several days. The concentrated sludge is withdrawn from the base of the tank and pumped to the digesters or dewatering equipment. Sewage sludge is very difficult to dewater because of the small size of the particles that result in small interstices which retain the water. Some of the water in the sludge exists

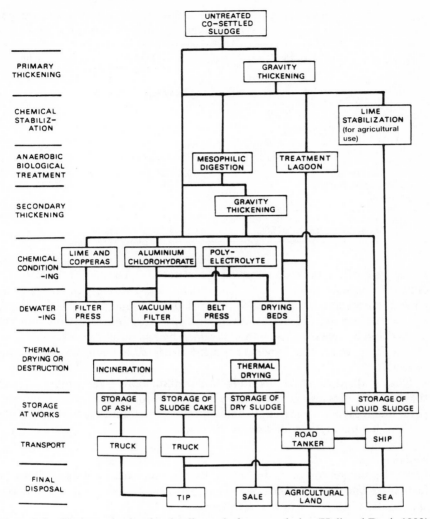

FIG. 5.3   Process selection for the disposal of sewage sludge (Hall and Davis 1983).

as a gel, and combined with strong capillary and electrostatic forces, water separation is difficult. Settlement is enhanced by slowly stirring the sludge with a rotating set of vertical blades or rods which are spaced approximately 100 mm apart. The rate of rotation is fairly critical, with optimum thickening occurring between peripheral blade velocities of between 0.5–3.0 m min^{-1}. The stirrer, which looks like a picket fence, gives its name to the unit which is universally known as a 'picket fence thickener'.

The sludge is thickened by the weight of the particles compressing the sludge

TABLE 5.3   Typical concentrations of unthickened and thickened sludges and solids loadings for gravity thickeners (Metcalf and Eddy 1984)

Type of sludge	Sludge concentration (%)		Solids loading for gravity thickeners ($kg\ m^2\ d^{-1}$)
	Unthickened	Thickened	
Separate			
Primary sludge	2.5–5.5	8–10	100–150
Percolating filter sludge	4–7	7–9	40–50
Activated sludge	0.5–1.2	2.5–3.3	20–40
Pure oxygen sludge	0.8–3.0	2.5–9.0	25–50
Combined			
Primary and percolating filter sludge	3–6	7–9	60–100
Primary and modified aeration sludge	3–4	8.3–11.6	60–100
Primary and air-activated sludge	2.6–4.8	4.6–9.0	40–80

at the base of the tank thus forcing the water out. The picket fence stirrer enhances particle settlement by encouraging particle size to increase by enhanced flocculation. It also encourages sludge consolidation in aiding water release by forming channels, preventing solids forming bridges within the sludge, and by releasing gas bubbles formed by anaerobic microbial activity. Picket fence thickeners can be operated either as a batch or continuous process.

The liquid released from the sludge will contain more organic material than the incoming raw sewage and it must be returned to the inlet of the treatment plant and recycled. The strength and volume of this liquid must be taken into account for the organic loading to the plant.

Stabilization
Once the sludge has been thickened, two options are available. It can be further dewatered to a solids content of between 30–40% or it can undergo stabilization before the dewatering stage.

Sludges are stabilized to prevent anaerobic breakdown of the sludge on storage (i.e. putrefaction), which produces offensive odours. There are other advantages depending on the stabilization process selected, including destruction of pathogens, partial destruction of sludge solids, increase in the concentration of soluble nitrogen, and improved flow characteristics. Specific sludge disinfection processes, specifically to kill pathogens, are becoming more widely practised (Bruce 1984). There are three categories of the stabilization process, biological, chemical, and thermal. Each process prevents the utilization of the volatile and organic fraction of the sludge during storage by different effects. Sewage sludge is only considered fully stabilized when it is humified, i.e. fully decomposed to humic substances that are non-

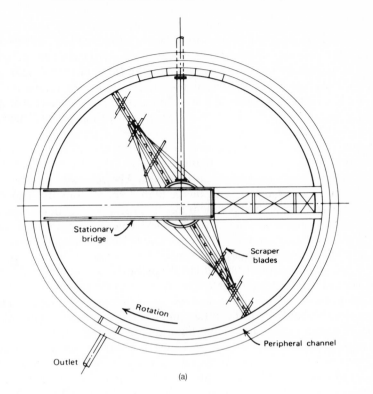

(a)

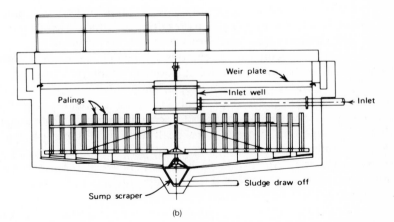

(b)

FIG. 5.4   A picket fence thickener in (a) plan and (b) in cross-section (Hammer 1977).

putrescible, odourless, and degrade further only very slowly (Hartenstein 1981). Confusion exists with this definition as the proposed EEC directive on sewage sludge disposal (European Communities 1982) defines stabilized sludge as simply as one that has undergone biological or chemical treatment, or long-term storage. Clearly, there is no universally accepted definition of stability and as yet no standardized test of sludge stability has been developed, although a number of possible methods have been proposed by Bruce and Fisher (1984) (Table 5.4).

Biological stabilization processes are the most widely practised, resulting in the utilization of the volatile and organic fraction of the sludge so that it will not undergo anaerobic breakdown on storage. The resultant sludge has a reduced volume, is odourless, and has a higher solids content. The most common stabilization method for medium- to large-sized treatment plants is heated anaerobic digesters, and at small works cold anaerobic digestion in tanks or lagoons is most common (Hall and Davis 1983). Anaerobic digestion utilizes up to 40% of the organic matter present in raw sludge that results in an increased nitrogen concentration of up to 5% of the dry solids, of which 70% is in the form of ammonical nitrogen. The sludge entering digesters does not have to be prethickened, but if it is, then the resultant digested sludge will have a higher dry solids and nitrogen content and there will be a saving in digester capacity. Thickening is required after digestion and can be achieved either by gravity settlement in a picket fence thickener or in lagoons over a period of several years with further digestions taking place. This results in a highly humified sludge with 8–10% dry solids (DS). The manurial value of lagooned sludge is reduced in terms of nitrogen and phosphorus content which is leached into the supernatant, with the proportion of total nitrogen present as ammonical nitrogen being reduced to <25%. Lagoons have a high potential storage capacity as the sludge volume is constantly being reduced by anaerobic digestion and water percolates into the underlying soil. Sludge is often stored in lagoons for many years resulting in a thick crust forming and becomes colonized by vegetation, with a dense sward of grass and even bushes developing. It is difficult to prevent this happening although it must be remembered that this gives a false impression of stability. The crust is not normally strong enough to prevent people, who inadvertently walk on to the lagoon, from falling in and most probably drowning. Therefore, a high level of maintenance is required to keep such sites securely fenced off. Both anaerobic and aerobic digestion processes are dealt with in Chapter 4. Another biological stabilization process is composting, with or without a bulking agent or recycled material. Much interest has been shown in the use of various composting processes to convert sewage sludge into a useful soil conditioner and fertilizer. This is fully explored in Chapter 7.

Lime stabilization is the most common form of chemical stabilization and is particularly popular in Norway. Much interest is developing in other European countries, although few treatment plants are employing the process

TABLE 5.4 Characteristics of sewage sludge stability (Bruce and Fisher 1984)

Basic parameter	Measurement	Possible applicability of each method of stabilization				Comments
		Anaerobic digestion	Aerobic digestion	Lime stabilization	Composting	
Odour emitted by the sludge	1. Odour intensity by dilution technique to give 'threshold dilution value'	✓	✓	✓	✓	Requires odour panel. Not related to odour 'quality'
	2. Gas chromatographic analysis (GCMS)	✓	✓	✓	✓	Expensive and difficult to interpret
Volatile solids	3. Volatile solids in the sludge as fraction of the total solids	✓	✓	×	?	Standard measurement
	4. Fraction of volatile solids destroyed (FVSD) by the stabilization process	✓	✓	×	?	Standard measurement
Residual, readily biodegradable matter	5. BOD$_5$ of filtrate	×	✓	×	×	Standard measurement
	6. Rate of increase of COD of filtrate with storage	✓	✓	×	×	Tentative basis for a 'stability index'
	7. Specific oxygen uptake rate (SOUR)	×	✓	×	×	Temperature-dependent
	8. Gas production during anaerobic incubation at 35°C	✓	✓	×	?	Standard control sludge required to detect inhibition
Chemical composition	9. Volatile fatty acids	✓	✓	×	×	Standard analysis
	10. pH and pH change during storage	✓	✓	✓	×	Standard test
	11. H$_2$S emission on storage	×	✓	×	×	Special test
	12. Nitrate concentration	×	✓	×	×	Standard analysis
Biological activity	13. ATP concentration	×	✓	×	×	Research application only
	14. Dehydrogenase concentration	×	✓	×	×	Research application only
Presence of putrescent matter	15. Attractiveness to house flies	✓	✓	✓	✓	Tentative. Under investigation

in the UK (Bruce and Fisher 1984). Unlike the biological digestion processes that utilize the organic fraction used by anaerobic micro-organisms, the addition of lime to untreated sludge until the pH is raised to > 11 creates an environment unsuitable for the survival of micro-organisms. Pathogens are readily killed by the process with lime stabilization at pH 12 for 3 h giving a higher reduction in pathogens than anaerobic digestion. Lime stabilization does not reduce the organic matter or provide permanent stabilization, but prevents putrefaction as long as the high pH is maintained. However, although the high pH suppresses the emission of volatile sulphides and fatty acids, the emission of amines and ammonia is increased making lime treated sludge less offensive than raw sludge but certainly not odourless. The elevated pH is normally maintained for several days or until incorporated into the soil. Obviously, lime addition has advantages to the farmer if sludge is disposed to agricultural land, and the addition of lime can help in the dewatering process. Two approaches to lime addition are used, the addition of hydrated lime to non-dewatered sludge and the addition of quicklime to dewatered sludge. The dosage is expressed as grams of lime per kg dry sludge solids (g kg DS^{-1}). The average dosage for a primary sludge is between 100–200 g Ca(OH)$_2$ kg DS^{-1}, and the mass of solids will increase after treatment. The pH of lime-treated sludges falls with time if the initial dosage is not sufficient, and it is important that sufficient lime is added to maintain the pH for the required period. The necessary lime dose required to maintain sludges at pH > 11 for at least 14 days is given in Table 5.5.

Thermal or heat treatment is a continuous process that both stabilizes and conditions sludges by heating them for short periods (30 minutes) under pressure. This releases bound water allowing the solids to coagulate, and proteinaceous material is hydrolysed resulting in cell destruction and the release of soluble organic compounds and ammonical nitrogen. The most recent modification, the Zimpro process, heats the sludge to 260°C in a reactor vessel at pressures up to 2.75 MN m^{-2}. The process is exothermic and results

TABLE 5.5   Lime dosage required to maintain the pH > 11 for at least 14 days at 20°C (Paulstrad and Eikum 1984)

Type of sludge	Lime dosage (g Ca(OH)$_2$ kg DS^{-1})*
Primary sludge	100–200
Septic tank sludge	100–300
Activated sludge	300–500
Mixed primary-chemical (Al, Fe) sludge	250–400
Mixed primary-chemical (Ca) sludge	None
Mixed activated-chemical (Al, Fe) sludge	300–500
Mixed activated-chemical (Ca) sludge	None

* DS = dry solids.

in the operating temperature rising. The solids and liquid separate rapidly on cooling with up to 65% of the organic matter being oxidized. The process sterilizers the sludge, practically deodorizes it and allows dewatering to be done mechanically without the use of chemicals. Because of the disadvantages of capital and operating costs, operational difficulties, and the large volume of very strong waste liquor produced, there are very few such plants operating in Europe (Institute of Water Pollution Control 1981). A pasteurization unit using submerged combustion of digester gas which heats the sludge to 70°C is used before the anaerobic digestion phase at the Colburn sewage treatment works in Yorkshire. This has been described in detail by Hudson *et al.* (1988).

Dewatering
Dewatering is a mechanical unit operation used to reduce the water content of sludge in order to obtain a solids concentration of at least 15%, and usually much more. It is normally preceded by thickening and conditioning by the addition of chemicals, which aids flocculation and water separation, and may be followed by further treatment. This reduces the total volume of sludge even further and thus reduces the ultimate transportation cost of disposal. The resultant sludge is a solid, not a liquid, and can be easily handled by conveyers or JCB tractors, although experience has shown that the dried sludge, known as cake, is more easily handled at solids concentrations of >20% (Institute of Water Pollution Control 1981). The resultant sludge is odourless and non-putrescible and can be stored without problems. Its solid nature makes it suitable for many more disposal options than liquid sludge.

The process is a physical one involving filtration, squeezing, capillary action, vacuum withdraw, centrifugal settling, and compaction. The most widely used European methods include drying beds, lagoons, filter and belt presses, and vacuum filtration. Centrifuges are uncommon at present in the UK and are more widely used in Ireland.

Drying beds are the cheapest and simplest form of dewatering. They are mainly restricted to smaller treatment plants because of the area of land required and the problems with odours and flies. Sludge drying beds are shallow tanks with a system of underdrains covered with layers of graded filter medium, normally a 100 mm of pea gravel covered with a 25 mm layer of sand (Fig. 5.5). The floor of the bed slopes towards the outlet at a minimum gradient of 1 in 60, and the sludge is pumped on to the bed to a depth of between 150–300 mm. Dewatering occurs mainly by drainage which is rapid for the first day or two and then progressively decreases until the solids have become so compacted on to the surface of the medium that drainage ceases. The surface liquid is decanted off and returned to the inlet of the treatment plant, so that evaporation can take over as the major dewatering process. Evaporation is affected by the weather, in particular the wind, humidity, and solar radiation, and at this stage rainfall will dilute the solids concentration. As the sludge dries, it cracks which increases the surface area of the sludge thus

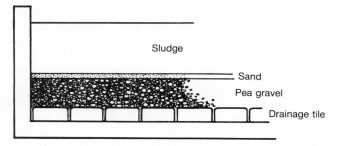

F<small>IG</small>. 5.5   The sludge drying bed (Institute of Water Pollution Control 1981).

increasing the evaporation. When the cracks reach the support medium, rainfall is no longer a problem as it can drain through the sludge via the cracks to the medium below. At this stage, the sludge can be lifted either manually or by machine. The performance of drying beds depends on the type of sludge, the initial solids content, the period of drying, the porosity of the medium, loading rate, and weather conditions. Solids contents of up to 40% are possible, although 35% is a more common attained value. They can only be used when drying conditions are good, which in the UK limits their use to 6–7 months of the year. In hot weather, as in the summer of 1976, 40% solids can be obtained after 10–15 days storage, but periods of 30–40 days are usual.

Sludge is not usually conditioned by the addition of coagulants prior to application to drying beds, although aluminium chlorohydrate and polyelectrolytes do improve drainage characteristics, so that more water is able to escape via the underdrains during the first few days after application. The total surface area of bed required is related to the population, although local factors, such as average rainfall, the dry solids concentration of the sludge, and its filtrability are also important. As a rough guide, 0.12–0.37 m² of bed are required per head of population for primary sludge, with 50–60% less area required for digested sludge (White 1978b). Secondary sludges (humus and activated sludge) are more difficult to dewater, and conditioning is required before application to drying beds. The use of transparent plastic roofing for sludge drying beds allows them to be used for much longer periods and makes them more efficient by increasing evaporation and protecting from rain. Full design and operational details are given in Institute of Water Pollution Control (1981) and Metcalf and Eddy (1984).

Filter presses are comprised of a series of metal plates, between 50–100 in number, suspended either from side bars (side-bar press) or an overhead beam (overhead beam press). Each plate is recessed on both sides and supported face-to-face so that a series of chambers are formed when the plates are closed. Filter cloths are fitted or hung over each plate that are then closed with sufficient force, using either hydraulic or powered screws, to form sealed chambers. The sludge is conditioned before being pumped into the chamber under pressure ($\simeq$700 KN m^{-2}) via feed holes in each plate. The pressure builds up and filtration occurs with the filtrate passing through the filter cloths and leaving the chamber via ports in the plates (Fig. 5.6). Pumping of sludge into the chambers continues until the chamber is filled with cake and the flow of filtrate has virtually stopped. Filter presses are normally situated on the first floor of a specially constructed building that allows access for trailers. The pressure is released to the press and the plates separated allowing the cake to drop out into the trailer parked directly below the press. In this batch process the operational cycle can vary from 3–14 h, with filling and pressure build up taking between 0.2–1.5 h, filtration under maximum pressure from 2–12 h, and cake discharge 0.1–0.5 h, although cycles normally take between 3–5 h to complete. Much work has been done to try and reduce the time taken for the second phase of the cycle, which is filtration under pressure. Important factors affecting the rate of filtration include the choice of filter cloth, pump pressure, filtrability of the dry solids content of the sludge, and condition of the filter cloth (Institute of Water Pollution Control 1981).

The process is primarily used for raw sludge, but is equally successful with digested, mixed, and thermally conditioned sludges. The final cake is quite thick, 25 or 38 mm, with a solids concentration of 30–40%. The advantage of the process is that it produces a drier cake than other dewatering devices, which results in a lower volume of sludge cake and reduced handling costs. The filtrate contains a relatively small concentration of suspended solids, although this is dependent on the type of sludge and filter cloth. Much has been written on the variation of filter cloths, including the material used, the

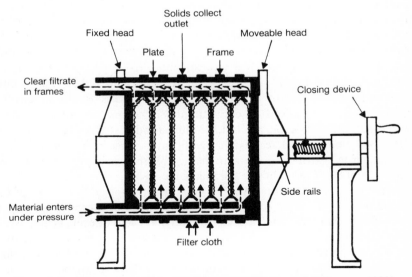

FIG. 5.6   Schematic outline of a filter press (Casey and O'Connor 1980).

weave, and the design. Gale (1975) found that cloths with an open weave had a higher concentration of solids in the filtrate than cloths with a tight weave, although the former were less susceptible to small particles packing into the spaces in the weave thus preventing the passage of filtrate. Filter cloths need to be washed occasionally between cycles using a high pressure water jet. The frequency depends on a number of factors, including the suitability of the cloth used and the chemical conditioner selected. For example, Pullen (1981) found sludge conditioned with lime and copperas gave 20 cycles between washing, whereas organic polyelectrolytes permitted 30 cycles before the filter cloths required washing. Ways of optimizing the performance of filter presses are reviewed by Hoyland *et al.* (1981) and Bruce and Lockyer (1982).

Conditioners aid the dewatering process by improving the filtration characteristics of sludges by increasing the degree of flocculation of the sludge particles so that the absorbed water can be more easily removed. It also prevents small particles clogging the filter cloths in filter presses. Apart from heat treatment, sludge conditioning normally involves the addition of one or more chemicals. Although expensive, the use of chemical conditioners are cost-effective because the increased solids content of the sludges produced reduces the sludge volume that has to be disposed. There are numerous chemical conditioners available including lime, ferrous sulphate, aluminium chlorohydrate, organic polymers or polyelectrolytes, and ferric chloride. All are commonly used in the UK with the exception of ferric chloride ($FeCl_3$), which is widely used in the USA.

Lime ($Ca(OH)_2$) is nearly always used in conjunction with copperas (ferrous sulphate $FeSO_4 \cdot 7H_2O$) unless iron salts are present in the sludge when lime can be used on its own (Swanwick 1973). The combination is used almost exclusively in association with filter presses, not being suitable for belt presses or centrifuges which require a more rapid flocculation that can only be obtained from using polyelectrolytes. Copperas is kept in a crystalline form and is dissolved as required. It is an acidic and highly corrosive liquid and special precautions have to be taken when used. The lime is delivered by bulk tanker and is normally kept in a silo ready for use. Lime is usually added after the copperas with doses of 10% copperas and 20% lime (dry solids) used for raw sludges, and 40% copperas and 30% lime for digested sludges. The large weights of chemicals used, results in a significant increase in the weight of solid cake that has to be disposed of, and in the case of lime and copperas this can be as much as 50% of the solids in the final cake.

Aluminium chlorohydrate ($Al_2(OH)_4Cl_2$) is just one of a number of the aluminium salts used for conditioning sludge prior to dewatering, and is delivered to the site as a concentrated solution that needs diluting before use. It is only suitable for filter presses and drying beds and has the advantage over lime and copperas of not significantly adding to the mass of sludge cake produced. Lime addition (5–10%) is sometimes used to control the emission of odours, especially if the sludge has been stored for longer than 4–5 days.

The mode of action of these conditioners, i.e. the charges on the metallic ions neutralizing the surface charges on the small sludge particles, is more controllable when using polyelectrolytes. These macromolecular organic polymers are water soluble and like the other conditioners are able to flocculate dispersed particles. They are available in a vast range of molecular weights and charge densities that allow the conditioner to be changed to cope with even subtle changes in sludge character. Polyelectrolytes are available either as a liquid, granules or powder, with the dry forms requiring to be dispersed in water before use so that they contain between 0.10–0.25% of active ingredients. Dispersion can take up to 2 h and, as their ability to flocculate particles deteriorates with storage beyond this period, they are usually made up as required. Dosage is expressed as either volumetric (mg m^{-3} of sludge) or more commonly as the weight of active ingredient added to a unit weight of dry sludge (kg tonne^{-1} DS). Normal doses for raw sludge are between 1–4 kg tonne^{-1} for belt presses, so there is no significant addition to the mass of sludge. Although they can be used in conjunction with filter presses, polyelectrolytes are most efficient when used with belt presses and centrifuges. They are particularly useful for use with dewatering systems that utilize shear force to aid water release. As a general rule, the higher the shear force required, low in drying beds and filter presses but high in belt presses and centrifuges, then the greater the molecular weight of the polymer used. Lime can be used with polyelectrolytes to suppress odours, but hydrogen peroxide is more efficient for this purpose and more easy to inject into the sludge. The optimum dosage of conditioner for a specific sludge is calculated by measuring its filtrability after conditioner is added. Numerous methods of assessing the dosage are used, including visual observation by the beaker test, gravity drainage test, capillary suction time (CST), standard shear test or the Buchner funnel test. These methods are fully explained elsewhere (Institute of Water Pollution Control 1981).

Unlike filter presses, belt presses are a continuous process. Instead of porous sheets of filter mesh being squashed between plates, the belt press comprises two continuous belts, one porous filter belt, and an impervious press belt, between which sludge is added, and are then driven by a series of rollers that also compress them, squeezing excess water from the sludge.

Conditioned sludge is fed evenly on to a continuously moving open meshed and endless filter belt that acts as a drainage medium. Dewatering then occurs in three distinct phases. First, before any pressure is added, water drains rapidly from the sludge due to the action of the polyelectrolyte. Sometimes a vacuum is used to increase the water removal at this stage. The second phase is the initial compression of the sludge, as a low pressure is applied by a series of rollers as the belt press converges with the lower filter belt. Water is squeezed out of the sludge as the pressure increases. In the final phase, the sludge is subjected to increased pressure and a shearing effect that rips open the sludge between the two belts producing new channels for more water to escape, while the increased pressure forces out the remaining free water. As the belts part, the sludge cake falls from the machine on to a conveyer belt or into a trailer ready for transportation to the disposal site. The filtrate from the process falls from the filter belt to drains under the machine and is pumped back to the inlet of the treatment plant (Fig. 5.7).

All belt presses are similar in principle, although the configuration of the rollers and the complexity of the machines can differ dramatically. The efficiency of these presses depend on the pressure applied to the sludge (i.e. clearance between rollers) and the retention time (i.e. length of compressed filter belt and belt speed). Other factors are also important, for example, with primary sludges fresh material dewaters more successfully than stored material or if insufficient polyelectrolyte is used then large conglomerate flocs are not formed so that complete free drainage does not occur (Institute of Water Pollution Control 1981). Organic polyelectrolytes are normally used with doses ranging between 1.3–7.7 kg tonne^{-1} (DS). Final solids concentrations of between 20–25% are normal, although surplus activated sludge and aerobically digested sludge are far less amenable to dewatering. New machine designs with vertical belts and spring loaded rollers (e.g. multistage and incremental pressure type filter-belt presses) can produce cake with up to 35% solids (Department of the Environment 1975; Institute of Water Pollution Control 1981). Whereas filter presses are the most widely used mechanical dewatering device in the UK, belt filter presses are more frequently used in Ireland, with small mobile units used at the smaller treatment plants.

Vacuum filtration and centrifugation are also used for dewatering sludges, although their use is largely restricted to industrial wastewaters in the UK, and are not considered in detail here. With vacuum filtration, water is sucked from the sludge under vacuum through a filter cloth which is carried on a slowly revolving drum partly immersed in the sludge (Nelson and Tavery 1978) (Fig. 5.8). Centrifuges consist of a rotating bowl into which sludge and polyelectrolytes are added. Centrifugal forces enhance the settling rates of the

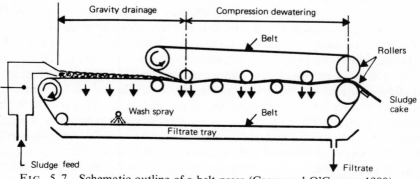

FIG. 5.7   Schematic outline of a belt press (Casey and O'Connor 1980).

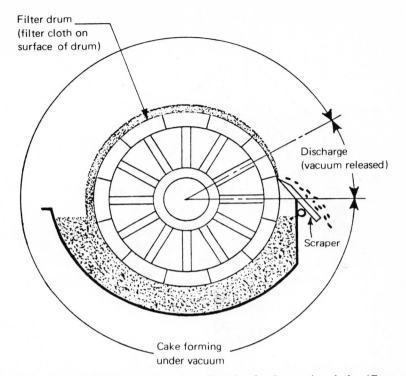

Filter drum
(filter cloth on
surface of drum)

Discharge
(vacuum released)

Scraper

Cake forming
under vacuum

FIG. 5.8   Schematic diagram of vacuum filtration for dewatering sludge (Casey and
O'Connor, 1980).

particles and cause the solids to separate out at the edge from where it is
removed (Fig. 5.9) (Egglink 1975).

5.1.2 Disposal options

In many respects, the final disposal options available dictate whether sludge
needs to be stabilized and dewatered, although the physical nature of the
sludge limits the choice of disposal methods. Factors such as sludge quality,
presence of toxic chemicals or pathogens, volume, location of treatment plant,
and cost are all involved in the selection of disposal methods. However, overall
cost and environmental factors are the foremost considerations. The
constraints on the Regional Water Authorities in the selection of the most
appropriate disposal route is summarized in Fig. 5.10. Disposal options can be
categorized into four groups: utilization on agricultural land; incineration;
landfill; and disposal to sea. The choice of the disposal methods varies from
country to country (Table 5.6), with agricultural use and landfill being most

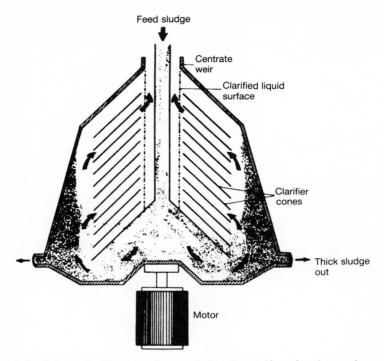

FIG. 5.9   Schematic diagram of a nozzle-bowl centrifuge for dewatering sewage
sludge (Institute of Water Pollution Control 1981).

frequently selected. However, dumping at sea is used for the disposal of sewage
sludge from most major cities located on the coast. There are significant
regional variations in sludge disposal methods (Department of the Environment 1978), with factors such as availability of suitable agricultural land,
proximity of plant to the sea, and quantity of sludge produced being limiting
factors on choice.

The most popular method of disposal is dumping on land. This
encompasses a number of options, such as using sludge for landfill, dumping
as cake on to municipal tips, allowing it to dry in shallow lagoons, or burying it
in trenches. The reuse of non-toxic sewage sludge as either fertilizer or a soil
conditioner is becoming increasingly popular in most European countries
(Section 5.2). Spreading sludge on agricultural land is an ancient practice and
it is used on grazing, arable, horticultural, and forestry lands, with digested
sludge also used on allotments. It is used in land reclamation and as a soil conditioner by composting with either refuse or straw (Section 7.3.3). Dumping
at sea has been adopted by most major cities located on the coast, e.g. London,
Liverpool, and Dublin, with the liquid sludge transported in purpose-built

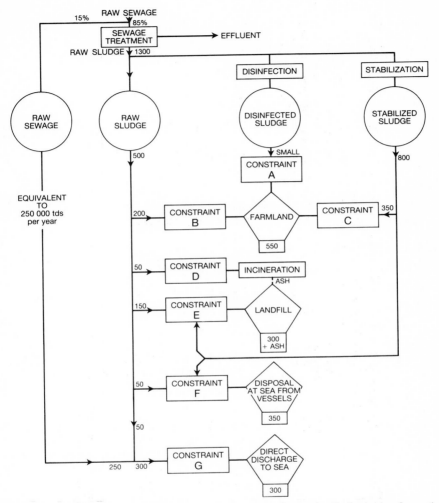

Constraint A, DOE (EEC) Guidelines – limits on metals; Constraint B, As A plus 6 month no-grazing period; Constraint C, As A plus 3 week no-grazing period; Constraint D, Planning permission required for incinerator construction; Constraint E, Control or pollution Act 1974 – licence required; Constraint F, Oslo Convention/Dumping at Sea Act 1974 – licence required; Constraint G, Paris Commission/Control of Pollution Act Part II – consent required. Units are '000 tonnes of dry solids per year.

FIG. 5.10 Current sewage sludge disposal routes in the UK and their legislative constraints. Constraints F and G will be modified according to the new EEC directive on urban wastewater treatment when adopted by the UK, when in practice these options wil no longer exist. New and impending EEC Directives now cover all these options although the national legislation enacting them is not yet in place (adapted Calcutt and Moss 1984).

TABLE 5.6　Total production of sewage sludge, treatment, and disposal methods employed by EEC member states, 1983 (Calcutt and Moss 1984)

Member state	Current sludge production		Amount stabilized (%)	Amount disinfected (%)	Use of disposal routes				
	DS per annum (tonnes × 10³)	% of estimated maximum			Agricultural land[a] (%)	Landfill[b] (%)	Incineration (%)	Dumping at sea (%)	Discharge via pipelines (%)
UK	1500	85	60	1	41	26	4	27	2
Belgium	70	45	80	1	45	53	2	0	0
Denmark	130	85	95	29	45	45	10	0	0
France	840	50	60	3	30	50	20	0	0
W. Germany	2200	80	75	17	39	49	8	0	0
Greece	3	<5	No information	No information	0	100	0	0	0
Republic of Ireland	20	20	40	No information	4	51	0	45	
Italy	1200	50	75	1	20[c]	55[c]	5	0	0
Luxembourg	11	70	Substantial	No information	90	10	0	0	0
The Netherlands	230	70	65	7	53	32	3	0	13

[a] Includes small quantities for horticulture, allotments, and gardens.
[b] Includes small quantities for land reclamation and forests.
[c] Unspecified.

vessels and dumped well out to sea. The capital investment is extremely large and this disposal method is restricted only to major centres of population. Among its many advantages is its independence of seasonal factors including the weather, as well as being economical and final (Section 5.3).

Sludge contains more volatile combustible matter and less fixed carbon than coal, and once dried it will burn and generate considerable heat. Because of high capital costs incineration has never been a widely adopted disposal route for sewage sludge except in France where 20% of the total sludge produced annually is incinerated (Table 5.6). Incineration has been selected primarily for industrial cities where heavy metal contamination of the sludge makes it unsuitable for other disposal routes, particularly disposal to agricultural land, where land for dumping is scarce or the sea disposal option is not available. Incineration destroys the organic and volatile components of the sludge, including toxic organic compounds, leaving a sterile ash in which all the toxic metals are concentrated. In comparison to the weight of the sludge cake incinerated only a small weight of ash is left which is disposed at tips. However, because of the high concentration of heavy metals present careful selection of disposal sites must be made to avoid leaching. The ash from the incinerator at Birmingham comprised 33.5% SiO_2, 25.0% Al_2O_3, 11.9% P_2O_5, 9.3% CaO, 8.8% Fe_2O_3, 1.2% K_2O, 1.0% Zn, 0.94% TiO_2, 0.85% MgO, 0.84% Na_2O, 0.69% Cu, 0.18% MnO, and 0.18% Pb (Institute of Water Pollution Control 1978).

The two most widely used types of incinerators used for disposing of sewage sludge are the multiple hearth incinerator and the fluidized bed incinerator. The amount of energy required to incinerate the sludge at the correct temperature in order to avoid odours and any possible harmful residues being formed, depends on the water content of the sludge cake. The amount of heat required per kg DS to burn sludge of various moisture contents is given in Fig. 5.11. If below 70% (>30% DS), little additional fuel is used as the energy released during combustion is greater than that required to evaporate the water present (i.e. autothermic). Raw sludge contains more energy than digested sludge as much of its potential energy is removed during the digestion process as methane. The calorific value of sludges varies considerably, with raw sludges having a higher calorific value (16 270–23 875 kJ kg^{-1} DS) compared with digested sludge (11 560–13 870 kJ kg^{-1} DS) (Grieve 1978). Where a town has a refuse incinerator then sewage sludge is often included, conversely refuse is occasionally mixed with sludge cake at treatment plant incinerators to increase the combustability of the sludge. New methods of sludge incineration, which have been developed largely in Japan, the Federal Republic of Germany, and the USA, are reviewed by Smith (1986).

With the growing awareness of air pollution, the incineration of refuse and sewage sludge have become less popular, although the flue gases from burning sewage sludge are mainly water vapour with small amounts of particulate solids that may contain metals. These contaminants, however, are easily

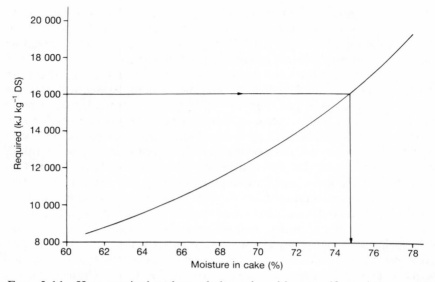

FIG. 5.11   Heat required to burn sludge cake with a specific moisture content
(Grieve 1978).

removed by conventional scrubbing processes. The price of supplementary
fuel for incinerators, such as oil and LPG has also made operating incinerators
much more expensive in recent years. Incineration of sewage sludge is re-
evaluated economically as a viable disposal method by Lowe (1988).

Further reading

General: Casey and O'Connor 1980.
Thickening: Institute of Water Pollution Control 1981.
Stabilization: Bruce 1984.
Dewatering: Department of the Environment 1975; Egglink 1976; Nelson and Tavery
   1978.
Disposal option: Institute of Water Pollution Control 1978; Calcutt and Moss 1984.
Incineration: Grieve 1978; Institute of Water Pollution Control 1978; Smith 1986;
   Lowe, 1988.

5.2   LAND DISPOSAL

5.2.1   Sludge disposal to land sites

Apart from the utilization of sewage sludge on agricultural land as a fertilizer
and soil conditioner, there are five other widely used methods of land disposal:

(1) irrigation; (2) shallow lagooning; (3) trenching; (4) disposal to refuse tips, (5) land reclamation.

The land used for these disposal methods can be categorized as either sacrificial or reclaimed. Irrigation, shallow lagooning, and continuous trenching are carried out permanently and intensively on a small area of land. Therefore, because of the level of contamination that inevitably results from such high loading rates of sludge the land is unsuitable for agricultural purposes, and probably never will be again. Such sites can only be used for further sludge disposal, hence the term sacrificial land. Sludge is also used for land reclamation, for bringing marginal land to its full agricultural potential, and reclaiming derelict land, including spoil heaps and refuse tips. Reclamation provides a major use of sludge cake, although trenching using liquid sludge is also employed. The disposal of sludge to tips can fall into either category. Where sludge is disposed to a special tip then it is likely that the metal concentrations in the soil and herbage will always be too high for that land to be used for agricultural purposes, although it may have a function for landscape or amenity. However, where sludge is incorporated into a refuse tip or used to cover an existing tip then it has a reclamation function. In a survey of 193 sewage treatment plants in the UK, only 37 did not dispose of sludge to agricultural land. The most frequent reason given was due to the unsuitable quality of sludge, normally in terms of heavy metal concentrations (Table 5.7) (National Water Council 1981).

Sacrificial land use
Irrigation is based on the established sewage farm principle except that liquid sludge is utilized and not raw sewage. Areas of ground are ploughed along the contours and then, using a ridging plough, deep furrows are made across the ploughed field down the slope at 9 m intervals. The sludge is allowed to flow down the ridge furrows and by using earth stops, it overflows and floods different sections of the ploughed field. Usually, areas close to the treatment

TABLE 5.7 Levels of metals in sewage sludges disposed to agricultural land in the UK (National Water Council 1981)

Metal	(mg kg^{-1}DS)				No. of works	No. of samples
	Min	Median	Mean	Max		
Zinc	199	1270	1820	19 000	193	2386
Copper	36	546	613	2889	193	2379
Nickel	5	94	188	3036	192	2343
Zinc equivalent	507	3440	4550	40 502	192	2343
Chromium	7	335	744	10 356	188	2310
Cadmium	0.4	17	29	183	180	2319
Lead	19	324	550	3538	164	2189

Note: Medians and means are weighted according to population served by works.

plant are kept specially for this purpose and are not used for agriculture. The sludge is allowed to dry periodically and is then ploughed into the soil. The land is generally reseeded and then left for several months, or even years, before being reused. The system is a particularly cheap method of disposal as there are no transport costs and the sludge need not be thickened, conditioned or dewatered. However, care must be taken to avoid runoff, contamination of groundwater, the production of odours, and in summer, the production of flies.

In their mode of action, shallow lagoons are very similar to irrigation, with exactly the same advantages and disadvantages. The lagoons are quite small in area and are formed by making parallel earth ridges, using a ridging plough, 9–15 m apart. These banks run along the contour of the land and similar ridges are then made at right-angles to them at similar intervals thus forming square lagoons. The banks and corners are finished off by hand to ensure that they are continuous and stable, and the lagoons flooded with sludge to a depth of about 300 mm. This is allowed to dry, although the process may be repeated before the banks are levelled and the ground deeply ploughed.

Deeper and larger lagoons of a permanent structure are also used for sludge storage and treatment, but not for disposal. Here, anaerobic stabilization takes place that reduces the volume of dry solids in the sludge by 40–60%. The sludge becomes highly stabilized, with digested sludges also undergoing further anaerobic digestion. Once the lagoon has become full the sludge is allowed to dry and the highly humified sludge is dug out for disposal.

Trenching also relies on soil percolation to reduce the water content of the sludge, which is then decomposed by the soil micro-organisms. Long trenches are dug 1.0 m wide and 0.6 m deep at intervals of between 0.6–2.0 m. Two approaches are then used. When trenching is used just once to bring marginal land back into full production, the trenches are filled with sludge at a rate of 860–1000 tonnes DS ha^{-1}, back filled and left for many months before being used for agriculture. The problem with this method is that the sludge can be displaced out of the trench during back filling and the land remains very unstable for some time afterwards, especially on heavy soils. In some extreme cases, vehicles are not able to get back on to the land for up to 12 months after treatment. The trenches are generally close together, 0.6–1.0 m, so that maximum application of sludge can be achieved. At permanent disposal sites, trenching is a continuous process. The trenches are spaced 2 m apart and after being filled with sludge are not immediately back filled. Instead, the water is allowed to separate from the sludge solids and permitted to percolate into the soil so that more sludge can be added to each trench. This is repeated until the trenches are full of semi-dry sludge, when the area is taken out of service. After the sludge has dried, new trenches are constructed between the old ones, using the excess soil to fill in the old trenches, and the process is then repeated. The next set of trenches are constructed at right-angles to the original trenches, and so on. Permanent trenching sites require at least three areas of land for successful and continuous operation. The land is used in strict rotation with one area receiving sludge, another out of use while the sludge is drying, and a third area that is being retrenched.

The majority of dewatered and dried sludge is disposed of to selected refuse tips. This is a safe method of disposal and is unlikely to add to the problems already arising from the presence of refuse. It is the single most popular disposal option in all the EEC countries except for the UK, Luxembourg, and The Netherlands. Forty-four per cent of all European sewage sludge is disposed of in this way, which is equivalent to 2.75×10^6 tonnes DS per annum (Calcutt and Moss 1983). When mixed with refuse, there is no problem with sludge disposal, although if the sludge is disposed on its own to special tips then care must be taken. The sludge cake can rapidly revert back to a semi-viscous state and will undergo decomposition, especially during the warmer months, producing very strong odours. Industrial sludge may contain very high concentrations of heavy metals that may be taken up by the vegetation once a tip is covered, therefore, these sites could be unsuitable for grazing livestock. If the sludge is stored at depth, then it will heat up due to bacterial activity, making it easy for vandals to set it alight, although spontaneous combustion can also occur. If this happens, the smoke and odours are very unpleasant, and such fires have proved extremely difficult to extinguish.

Use of sludge in land reclamation
The aim of land reclamation is to restore vegetation to areas without an established soil cover. Such areas include colliery waste, shale tips, mine waste, china clay waste tips, sand dunes, sides of embankments and cuttings, pits filled with overburden from open cast mining, and refuse covered with a layer of soil: in fact, any land that has suffered extensive disturbance so that the original soil structure has been destroyed. Most of these materials are deficient in soil nutrients and, more importantly, organic matter which results in a high soil bulk density. This means that the soil has a poor structure and lacks water holding capacity (Khaleel *et al.* 1981). The restoration procedure requires deep cultivation and mole drainage followed by applications of lime and fertilizer. Grass is then grown for several years to allow the turf to provide enough organic matter for a top soil to be eventually established (Bradshaw and Chadwick 1980). However, this takes a very long time, the resultant soil is unstable and is easily damaged, and the nutrients and lime are easily leached making their restorative effects short-term. Sludge is particularly valuable for land reclamation as the concentrations of plant nutrients and organic matter exactly complement the deficiencies of derelict soils. Also, the nature of the nitrogen and phosphorus, which are bound up with the organic matter, ensures that there is a slow release of nutrients over several years. In this way remedial treatment will have a permanent effect. By the time the nitrogen release has been severely reduced, the top soil will already be established and natural clovers will be supplying nitrogen. The use of sludge in reclaiming

derelict sites is almost totally restricted to dewatered sludge cake, due to the high proportion of dry solids, especially those which have been conditioned using lime or a mixture of lime and copperas. It has been used successfully to restore a variety of sites including colliery waste (Williams 1975; Brooker and Farnell 1979), mine tailings (Johnson *et al.* 1976), chromate smelter waste (Gemmell 1974), and motorway verges and banks (Matthews 1980). Coker *et al.* (1982) have shown that a single application of dewatered sludge can permanently restore fertility to derelict land at an application rate of 100 tonnes DS ha^{-1}. This is due mainly to the reduction in bulk density by the organic matter in the sludge which encourages root development. Lime-treated sludge helps to correct the low pH of acid soils, although very acid substrata, such as colliery shale, will still require extra lime. The level of heavy metals in a single application of sludge (100 tonnes DS ha^{-1}) will not cause any contamination of the soil. However, slightly elevated but quite safe concentrations will be recorded in the herbage in the first year but will decline over subsequent years.

In 1975, only 7% of the sewage sludge produced in the UK (equivalent to 60 000 tonnes DS per annum) was used for land reclamation. However, it is estimated that some 33 000 ha^{-1} of land in the UK would justify restoration, and at the same time provide a potential outlet of 3.3×10^6 tonnes of sludge at the application rate of 100 tonnes DS ha^{-1}. Of the 70×10^6 tonnes of refuse produced in the UK each year, about 80% is buried in the ground. The conventional restoration method used for refuse tips involves covering them with clay and top soil. Sewage sludge has been shown to be an excellent replacement for top soil, which is becoming increasingly scarce and expensive, and has been extensively used for restoring such tips back to agricultural use. In 1980, the cost of conventional restoration of a refuse tip was between £3000–£4000 ha^{-1}. Thus, the use of sludge as a partial or complete replacement for top soil will increase (Bennet 1981).

5.2.2 Sludge utilization to farmland

The manurial value of sewage has been recognized since the last century when raw sewage was disposed at sewage farms. However, it was during the Second World War that the high cost of artificial fertilizers and the need for increased crop production led to the widespread use of sewage sludge from drying beds on farmland. During the early-1950s, the demand for sludge exceeded the output from drying beds and road tankers were used to transport and spread liquid sludge. Since then, land disposal has become a major disposal method of sewage sludge, with many plants relying totally on the utilization of sludge by farmers as cheap manure as their sole method of sludge disposal. Most European countries dispose a significant proportion of their sewage sludge to land, with the majority of sludge in Luxembourg (90%) and The Netherlands

(53%) utilized in this way. In most other European countries, including the UK, land disposal is a major disposal method, accounting for between 30–45% of sludge output. Only in Ireland and Greece, where sludge output is low, is less than 5% of sludge used for agricultural purposes (Table 5.6). However, the utilization of sewage sludge on agricultural land is expected to increase in all European countries, and in Ireland it is estimated that by 1990, this will have risen from 4% to 18% (Dodd 1980).

In spite of the costs of processing and transport, disposal of sewage sludge on to agricultural land is an extremely economical method of elimination as well as providing the farmer with low-cost fertilizer. It contains plant nutrients, trace elements, and organic matter that have the potential for improving soil structure, crop yields, and the grazing value of the land. Unfortunately, sewage sludges also contain heavy metals and pathogenic organisms. Furthermore, the toxic metal content of some sludges makes them unsuitable for agricultural use and even domestic sewage contains heavy metals that become concentrated in the sludge (Table 5.8). Once a harmful concentration of metals has built up in the soil the damage to subsequent crops can be long lasting or even permanent. The potential toxicological and health hazards of using sewage sludge on agricultural land are reviewed below.

Manurial value of sewage sludge
Nearly 50% of the sewage sludge produced in the UK is utilized on agricultural land, which is equivalent to approximately 6 000 000 tonnes DS per annum. The general policy of the UK Regional Water Authorities is to expand this form of disposal as it appears to be the most economical and convenient method of sludge disposal for most treatment plants. Sewage sludge is rich in organic matter and nutrients, especially nitrogen (N), phosphorus (P), and potassium (K), and farmers are able to utilize it as a cheap soil conditioner and fertilizer (Williams 1979). The proportion of organic matter, the average nutrient content of sludges, and the availability of nutrients to crops depends on the type of sludge used and whether it has been dewatered and, more importantly, stabilized. Therefore, the agricultural value of sludge depends very much on the treatment it receives (Table 5.9). There is considerable variation in the fertilizer value of sludge produced at different sewage treatment plants, although individual plants produce a sludge of reasonably consistent quality as the inputs to a particular plant varies little (Table 5.8).

Many of the soluble nutrients in sewage, mainly potassium salts and ammonical nitrogen derived from urine, are lost in the final effluent. However, some of this source of nitrogen is recaptured when it is utilized in the biological treatment process and becomes eventually stored in the sludge. The organic matter and nitrogen in the primary sludge is largely contained in bacteria and excreta, whereas the secondary sludges contain a greater proportion of organic matter, a higher nutrient content, and usually less metals on a dry

TABLE 5.8　Major metal and nutrient content of 13 Irish sludges (Murphy *et al.* 1978)

| Treatment plant | Type | DS (%) | (mg kg^{-1}DS) | | | | | | | DS (%) | | |
			Mn	Pb	Cu	Zn	Cd	N	P	K
Kildare	D	16.4	550	350	3129	14 850	3.0	1.90	0.82	0.54
Michelstown	D	13.4	206	80	140	862	2.7	7.32	1.82	0.58
Macroom	EA	10.3	858	349	802	982	6.2	5.10	1.59	1.19
Newbridge	D	17.8	119	92	236	1100	3.3	2.78	0.52	0.22
Athy	D	17.2	131	486	455	8552	3.5	2.87	0.47	0.20
Mullingar	EA	8.1	232	345	774	475	4.2	7.15	1.61	0.50
Kilcoole	EA	6.8	441	250	558	605	2.8	5.28	1.52	0.72
Carrickmacross	EA	10.3	681	260	574	595	2.4	4.78	1.54	0.70
Letterkenny	EA	9.3	559	247	240	620	3.1	5.19	1.33	0.63
Leixlip	EA	2.9	422	178	370	1325	3.6	3.56	1.51	0.67
Killarney	EA	1.5	471	315	618	562	2.7	4.59	1.08	0.37
Ballincollig	D	15.1	612	175	432	690	4.4	6.78	1.43	1.20
Mountmellick	EA	1.7	540	228	330	440	3.8	3.78	0.45	0.73

EA, Extended aeration – activated sludge; D, anaerobic digestion – digested sludge.

TABLE 5.9 Typical nutrient concentrations of UK sewage sludge (Institute of Water Pollution Control 1978)

Element	Type of sludge (% DS)				
	Liquid primary	Liquid activated	Liquid, primary and secondary	Liquid digested	Air dried*
Nitrogen	2.1–7.6	3.8–7.6	1.0–6.5	0.9–6.8	1.5–2.5
Phosphorus	0.6–3.0	1.4–3.2	0.6–2.5	0.5–3.0	0.5–1.8
Potassium	0.1–0.7	Trace	0.1–0.7	0.1–0.5	0.1–0.3
Calcium	1.4–2.1	0.5–0.8	up to 2.0	1.5–7.6	1.6–2.5
Magnesium	0.6–0.8	0.5–0.8	up to 0.8	0.3–1.6	0.1–0.5

* Mainly digested sludges.

weight basis than primary sludges. The organic matter and nitrogen is present as bacterial flocs in surplus activated sludge and as the remains of grazers and sloughed biological film in humus sludges from percolating filters. Because the nitrogen is present in non-soluble forms bound up with the organic matter, dewatering has little effect on the nitrogen content of undigested sludges, with only 5–10% of the total nitrogen present in solution as ammonical nitrogen. Only soluble nitrogen, either ammonical nitrogen or nitrate, is available for uptake by plants. All sewage sludges are generally low in potassium as it is mainly soluble, and when sludge is used as a fertilizer a supplementary source of potash may be required. Although most arable crops and grass will require a potassium supplement, low quantities are desirable for dairy farmers, because a high concentration in herbage can cause hypomagnesaemia in cows. A single application of 100 m^3ha^{-1} of sludge with 5% dry solids content will provide 10–20 kg K ha^{-1}. The phosphorus content of sludges is not significantly affected by treatment because it is generally present in insoluble forms and is retained in the dry solids. The concentration of phosphorus remains constant at between 1.0–1.8% of dry solids, making the ratio of N:P low compared with artificial fertilizers. Sewage sludges are, therefore, phosphorus-rich manures, with 50–60% of the phosphorus readily available as in a super-phosphate fertilizer. Phosphorus as a plant nutrient appears less important than nitrogen as most soils have adequate reserves, and because the extent to which phosphorus is used is limited to the N:P ratio it is difficult to separate the effect of phosphorus on plant growth, which is generally controlled by nitrogen availability.

A major problem with modern arable farming is that little organic matter is returned to the soil. Although high yields are possible with low organic matter contents, these soils are difficult to cultivate, suffer waterlogging, and are particularly susceptible to drought. Organic matter helps to bind the soil particles thus forming aggregates, which create more air spaces so that water and air can penetrate, reduces bulk density, and increases the water holding capacity, Although 50–70% of the dry solids in sludge is organic, the amount disposed on land in liquid sludge is very small, between 1–5 tonnes ha^{-1}, so that improvement in soil quality can only be achieved after many applications. Dewatered sludge allow much higher rates of organic matter to be spread to land and can significantly improve the quality of sandy and free-draining loams (Table 5.10). The ability to increase the water holding capacity of such soils is of particular use in improving drought susceptible soils (Institute of Water Pollution Control 1978; Hall and Davis 1983). Sludge is also a rich source of trace elements. Zinc and copper are both essential elements for crops and animals, but in a survey of grassland used for grazing, 95% of samples were found to be deficient in zinc (<50 mg kg^{-1} DM) and 81% deficient in copper (<10 mg kg^{-1} DM). Clearly, a light dressing of sludge will not only supply useful nitrogen, phosphorus, and potassium, but also trace elements to deficient grassland. Secondary sludges are usually mixed with the primary sludge for disposal as surplus activated sludge is not suitable for land disposal because it has a very low dry solids concentration.

A significant proportion of sludge that is utilized on agricultural land is undigested primary and secondary liquid sludges. However, if sludge receives further treatment, such as chemical conditioning or stabilization then its manurial value to the farmer, especially its nitrogen content, is significantly altered. About 50% of the sludge spread on agricultural land receives some form of stabilization, usually either heated anaerobic digestion at medium to large treatment plants or cold digestion at smaller plants. Very little sludge receives either aerobic digestion or lime stabilization, although both these methods are becoming increasingly popular (Bruce and Fisher 1984).

Anaerobic digestion removes up to 40% of the organic matter in sewage sludge, converting it to gaseous methane. Therefore, on a dry weight basis the nitrogen content of digested sludges increases by about 5%, with up to 70% of the total nitrogen biologically transformed to ammonical nitrogen. The sludge is far more stable than undigested sludge and is far less offensive and has a dry

TABLE 5.10   The effect of digested sewage sludge application on a free-draining loam (Hall and Davis 1983)

Sludge applied tonnes (DS) ha^{-1}	Soil bulk density (g cm^{-3})	Water at field capacity (%)	Fresh yields of lettuce (tonnes ha^{-1})
0	1.13	30.1	4.5
19	1.10	32.2	8.5
38	1.11	32.4	9.6
76	1.09	35.2	12.8
152	0.99	39.4	15.1
250	0.87	43.1	n.d.
500	0.72	47.3	n.d.

Note: n.d. = not determined.

solids content of 2–4%. Whereas the nitrogen in undigested sludge is slowly released for plant growth over a period of several years as the organic constituents are degraded by soil micro-organisms, digested sludge provides a more instant supply of nitrogen for plants.

Crop response to liquid undigested sludge is related to the total nitrogen in the sludge of which only 10% is in the ammoniacal form. An important factor in the response of plants to fertilizers is the texture of the soil and their ability to retain applied nitrogen. For example, soluble nitrogen is more readily leached from sandy soils than those containing silt or clay. In undigested sludge, the nitrogen is in a form that is not easily leached as it requires further mineralization. Therefore it is less likely to be leached from soils thus making it ideal for use on sandy soils. Unlike artificial fertilizers, this type of sludge can be applied at any time over the winter period, with little loss of fertilizer value, as long as surface runoff is controlled. Therefore, the value of sewage sludge as a fertilizer depends more on soil type than on time of application. This has a particular practical advantage for the farmer. Because of the six-month non-grazing rule following the application of undigested sludge to grassland, early winter treatment allows the farmer to graze treated grassland much sooner than if applied in the spring, when a hay or silage crop would have to be taken first. Crops respond differently to undigested sludge application. When it is applied to arable land and ploughed in, the nitrogen is released throughout the year and as mineralization proceeds faster at warmer soil temperatures nitrogen availability reaches a maximum in summer and autumn. However, unlike grass, cereals absorb most of their nitrogen requirement by the end of June and so uses only a small proportion of the available nitrogen from the sludge. Although undigested sludge is about 30–40% as effective as artificial fertilizers on grassland, it is only 15–20% as effective with cereals because of the short growth periods. Thus, cereal crops are unable to take full advantage of the slow continuous mineralization of the organic nitrogen in the sludge.

Two problems are associated with liquid sludge utilization on farmland. If applied at a high rate on grass under dry conditions a papery mat can be formed, which smothers the grass and inhibits growth. Ideally, rain is required following sludge application in order to help wash solids through the turf on to the soil surface. Nitrogen immobilization can also occur if the C:N ratio in the soil exceeds 13.5:1, where C is the concentration of readily metabolizable carbon, which results in a poor crop response. However, this is only significant if the soil is very deficient in nitrogen with the effect usually disappearing after 1–2 months (Hall and Davis 1983).

Stabilization of undigested sludge with hydrated lime (pH > 11) inhibits bacterial action and its subsequent decomposition in the soil. Therefore, mineralization in the soil will be delayed until the soil pH has fallen to 8.5, which may take several weeks. Although liming reduces the nitrogen value to about half that of unlimed sludge, it has great potential for use on acid soils. The treated sludge contains about 20% (dry solids) of lime and an application of 100 m³ha⁻¹ of lime-stabilized sludge will provide in excess of 800 kg CaO ha⁻¹, and trials by the Water Research Centre have shown that long-term application will increase soil pH (Table 5.11).

Many rural areas in Britain and especially those in Ireland are not on main sewerage and still rely on septic tanks, and appreciable quantities of septic tank sludge needs to be disposed of. In the South-West Water Authority area in England, 10% of their sludge comes from this source. The dry solids of these sludges are variable, with mean nitrogen, phosphorus, and potassium concentrations in the order of 4.2%, 0.8%, and 0.7% dry solids respectively. Septic tank sludges have a higher ammoniacal nitrogen content at 18% of total nitrogen, and a higher potassium content than undigested primary sludges (3.5% N, 1.3% P, and 0.2% K). It can be used on grassland, although such sludges can produce very unpleasant odours (Carlton-Smith and Coker 1982).

TABLE 5.11   Effect of spreading limed undigested sludge on the pH of soil (Hall and Davis 1983)

Rate (m³ ha⁻¹)	Total limed applied (tonnes CaO ha⁻¹)	Soil pH after 2 years	Increase in pH
0	0	5.8	
70	1.16	5.9	+0.1
140	2.32	6.0	+0.2
280	4.64	6.3	+0.5

Surplus activated sludge contains between 1.5–2.5% dry solids, with 4–7% of nitrogen, mainly present as bacterial floc. The ammoniacal nitrogen concentration is very low, much lower than in primary undigested sludges, with the C:N ratio closer than for other sludge types. When applied to land it decomposes very rapidly, resulting in odours, with up to 50% of the total nitrogen becoming available within the first 12 months. Because of its lower dry solids content and the difficulty in dewatering such sludges it is generally mixed with primary sludge before treatment and/or disposal.

Liquid digested sludge contains more nitrogen on a dry solids basis than any other type of sewage sludge, with the major portion, up to 85% of total nitrogen, readily available for plant growth (Williams 1979). Little metaboliz-able carbon is left in sludge after digestion and the remaining organically bound nitrogen is mineralized much more slowly in the soil than undigested sludge, with 15% of the nitrogen being released in the first year, although the remainder may take many years to become available for plant growth. Nitrogen in sludge is not prone to leaching as it needs to be nitrified to nitrate before loss occurs. However, volatilization of ammonia can be high, resulting in a significant loss of nitrogen if sludge is applied to bare, dry soil on hot windy days. This can be avoided if sludge is applied to land with a moist soil, a full crop cover or if cultivated immediately after treatment. If the dry solids content of sludge is high then it sticks to the blades of grass, which increases volatilization. Therefore, it is best to apply thick sludges on wet days, whereas thin sludges can penetrate the turf mat areas on dry days. The nitrogen availability of liquid digested sludges can be estimated by:

$$(\text{Ammoniacal nitrogen}) + 15\% \text{ (organic nitrogen)}$$

The fertilizer value of liquid digested sludge can be estimated by multiplying the nitrogen availability by a factor C which is the proportion of ammoniacal nitrogen lost:

$$C(\text{NH}_4\text{-N}) + 15\% \text{ (organic nitrogen)}$$

when the formulae is applied to grassland in the British Isles $C = 1.0$, but if applied under adverse conditions then C can be 0.5 or less (Coker 1982). Fertilizer efficiency of liquid digested sludge can be 35–100% of that artificial fertilizer, as measured on a ammoniacal nitrogen basis, and is related to the proportion of ammoniacal nitrogen to total nitrogen which rises as the percentage of ammoniacal nitrogen increases. Applications of digested liquid sludge are most effective for both grass and cereals if applied during January and February, being 70–100% as effective as spring applied fertilizer on an ammoniacal nitrogen basis. Details of the amounts of available nutrients supplied at various rates of sludge application are given elsewhere (Institute of Water Pollution Control 1978).

Lagoon storage of digested sludge results in a more stabilized sludge with a lower proportion of ammoniacal nitrogen to total nitrogen, at 25%, although the total nitrogen increases with the dry solids concentration. The annual mineralization rate of organic nitrogen for this type of sludge is about half that for liquid sludge, at 9%.

The proportion of sludge cake being spread on land is declining with liquid sludge being far more economical to use as it can be pumped, does not require dewatering, has a reduced volume of dry solids, and is easier to apply. However, 37% of all mechanically dewatered sludge and 21% of cake from drying beds are used on agricultural land annually in Britain. The dry solids content of sludge cake is between 20–35%, although most of the ammoniacal nitrogen has been lost, therefore the main value of cake to the farmer is as a bulk organic manure or as a slow release nitrogenous fertilizer. Sludge cake contains significantly less nitrogen than non-dewatered sludges and the release of nitrogen from sludge cake due to mineralization of the organic nitrogen component in the soil decreases by about 50% each year. It is difficult to spread evenly, although spreading qualities can be improved by storage for several months before use to allow further stabilization and mineralization. This produces a more friable material with an improved nitrogen value. Sludge that has been chemically conditioned with lime and copperas can be of benefit to the farmer as it can increase the soil pH, but the volume of dry solids is increased by 20% due to the addition of chemicals. Aluminium chlorohydrate is still used, although polyelectrolytes are rapidly taking over.

Sludge cake from drying beds is more expensive to produce than mechanically dewatered sludge because of the large areas of drying beds required. The final cake is much drier and harder than the other types, with a dry solids content of between 50–70%. The sludge is much easier to handle due to its low moisture content and is used mainly as a soil conditioner for very poor soils.

Contamination by heavy metals
Metals originate from both domestic and industrial sources, with primary and secondary sludges containing between 44–96% of the total metal load of crude sewage (Davis 1980). All sewage sludges contain metals to a greater or lesser extent and certainly in higher concentrations than found in soil. Therefore, when sludge is disposed to land, heavy metals will accumulate in the soil. The effects of heavy metal accumulation can be long lasting or even permanent in nature with phytotoxic effects being the major problem, although there is a danger that metals may be introduced via domestic livestock or directly into the human food chain. For man, the main source of these elements is in meat and plants, with recommended maximum daily intake of the common sewage-associated metals of 30 $\mu g\, d^{-1}$ for Cd, 170 $\mu g\, d^{-1}$ for Pb, 5000 $\mu g\, d^{-1}$ for Cu, and 17 000 $\mu g\, d^{-1}$ for Zn (WHO 1971, 1984). The quantity of metals found in sludge depends on the source of wastewater, although Zn, Cu, Ni, and Pb are present in relatively large concentrations even in domestic sewage. The most toxic metal in sewage is probably cadmium. In a survey of UK sludges, a mean value of 7.15 mg Cd kg^{-1}DS was obtained for domestic sewage with no industrial inputs, whereas levels of > 10 mg Cd kg^{-1}DS indicated industrial inputs. Even very small concentrations in crude sewage can result in high concentrations in the sludge. For example, cadmium is normally present in crude sewage at concentrations between 0.008–0.01 mg Cd l^{-1}. Assuming an 85% removal efficiency of the metal during treatment and that 350 mg of primary and secondary sludge solids are generated per litre of sewage treated, then the sludge will contain 20 mg Cd kg^{-1}DS (Hall and Davis 1983). The normal range of metals found in agricultural soils is compared with the levels found in sewage sludge in Table 5.12, although a more comprehensive table is given by Davis (1980).

There is a general lack of agreement as to what constitutes acceptable levels of metals in soils for plant growth, and at what levels accumulated metals in plants present a danger either to man or animals. This is complicated by the fact that availability of metals to plants varies with soil and plant type, the element concerned, and the environmental factors that affect plant growth. Leeper (1978) has reviewed the effect of soil properties on plant uptake of heavy metals, with pH and cation-exchange capacity the most important

TABLE 5.12 The range (R) and common values (CV) of the major elemental contaminants in sewage sludge and the soil (mg kg^{-1} DW) (Davis 1980)

Contaminant	Sludge		Soil	
	R	CV	R	CV
Ag	5–200	25	1–3	1
As	3–30	20	1–50	6
B	15–1000	30	2–100	10
Be	1–30		0.1–40[a]	3[a]
Cd	2–1500	20	0.01–2.4	1
Co	2–260	15	1–40	10
Cr	40–14 000	400	5–1000	100
Cu	200–8000	650	2–100	20
F	60–40 000	250	30–300	150
Hg	0.2–18	5	0.01–0.3	0.03
Mo	1–40	6	0.2–5.0	2
Ni	20–5300	100	10–1000	50
Pb	50–3600	400	2–200	20
Sb	3–50	12	2–10[a]	—
Se	1–10	3	0.01–2	0.2
Sn	40–700	100	2–200	10
Tl	—	1[a]	—	0.1[a]
V	—	15[a]	—	100
W	1–100	20	—	1[a]
Zn	600–20 000	1500	10–300	50
Dieldrin	<0.03–300[a]	0.4[a]	—	—
PCB (Aroclor 1254)[b]	<0.01–20[a]	3[a]	—	—

[a] Tentative data.
[b] A WRC survey of 11 UK sludges found PCB (Aroclor 1260) concentrations of 0.06–2.4 mg kg^{-1} dry solids.

factors. Some elements, such as cadmium and zinc are rapidly absorbed and translocated to aerial plant pests, whereas others, e.g. lead, are largely unavailable. If the organic content and pH of the soil are high, then the availability of most metals is reduced. As most of the contaminants from sludge are absorbed to soil or organic matter, or form relatively insoluble precipitates at pH values normally associated with agricultural soils, heavy metals appear to be largely immobile and are not readily leached from the soil into the groundwater (Edworthy *et al.* 1978). Therefore, it would appear that lime-treated sludges help to reduce the availability of metals; however, care must be taken at high pH levels as molybdenum becomes more available and is accumulated by crops. Although high molybdenum levels are not toxic to plants it can result in copper deficiency symptoms in livestock. Lime-treated sludge can also cause trace element deficiency in plants under certain soil conditions. A single application of sludge with the common values (CV) shown in Table 5.12 at a rate of 5 tonnes ha^{-1} and cultivated into the top 20 cm of soil (density 1.0 g ml^{-1}) will produce very small percentage increases in soil metal concentrations, except for zinc and copper (Table 5.13). The

TABLE 5.13 Effect on soil concentration of potentially toxic metals of a single application of sewage sludge (Hall and Davis 1983)

Element	Concentration (mg kg^{-1})		Increase in soil concentration following dressing of sludge	
	Sludge	Soil	mg kg^{-1}	%
Cd	20	1	0.05	5
Pb	400	50	1	2
Cu	600	20	1.5	5.5
Ni	100	50	0.25	0.5
Zn	1500	50	3.75	7.5

levels of metals at which toxic effects in soils and plants are recorded are given in Table 5.14. Davis and Beckett (1978) suggest that as the metal concentration as measured in soil is so variable, because it depends on so many environmental factors, the metal accumulated in plant tissue is a more convenient and accurate indicator of metal levels in soils and that phytotoxic levels have been reached. They observed that critical concentrations of metals in the tissue of young barley plants occurred at between 18.2–20.3 mg kg^{-1} for Cu (median 19.1 mg kg^{-1}), 10.8–13.0 mg kg^{-1} for Ni (median 11.8 mg kg^{-1}), 124–220 mg kg^{-1} for Zn (median 199 mg kg^{-1}), and 6.0–10.0 mg kg^{-1} for Cd (median 8.0 mg kg^{-1}). Metals may be described as either phytotoxic or zootoxic depending on whether the concentration toxic to plants is lower or higher than the concentration toxic to animals that may eat them. Although uptake efficiency varies between plant species, and even varieties, the effect of contaminants, once in the tissue of the plant or the animal eating it, is relatively constant for most crop plants (Davis 1980). The background and upper critical concentrations of metals in plant tissue to give phytotoxic and zootoxic effects are summarized in Table 5.15, although no account has been taken of possible interactions between elements. Metals also reach soil from

TABLE 5.14 Normal and toxic concentrations of common metals in soil and plants (Dodd 1980)

Element	Concentration in soil (mg l^{-1})		Concentration in plants (mg kg^{-1})	
	Normal	Toxic	Normal	Toxic
Zn	1–50	>200	20–100	>200
Cu	0.5–5.0	>100	5–15	>25
Ni	1–5	>25	1–10	>50
Pb	0.5–5.0	>200	0.1–2.0	>10
Cd	0.2–2.0	20–50	0.2–0.5	>10
Hg	0.05–0.5	>50	0.02	>2

TABLE 5.15 Background and upper critical concentrations of elemental contaminants in plant tissue (mg kg^{-1} DW) (Davis 1980)

Contaminant	Background concentrations	Upper critical concentrations	
		Phytotoxic threshold	Zootoxic threshold[a]
Ag	0.06	4	—
As	<1	20	—
B	30	80	—
Be	<0.1	0.6	—
Cd	<0.5	10	—
Co	0.5	6	50
Cr	<1	10	50
Cu	8	20	30
F	8	>2000[b]	30
Hg	0.05	3	1
Mo	1	135	10
Ni	2	11	50
Pb	3	35	15
Sb	<0.1	—	—
Se	0.2	30	5
Sn	<0.3	60	—
Tl	<1[a]	20	—
V	1	2	—
W	0.07	—	—
Zn	40	200	500

[a] Tentative data: it is particularly difficult to assign zootoxic thresholds to Cd, Pb, and Hg.
[b] Applies only to F taken up from soil

other sources and although only 2% of the UK agricultural land receives sludge in any one year, all receive inputs from aerial deposition and phosphate fertilizers. The cadmium level in phosphate fertilizers ranges from 1–160 mg Cd kg^{-1}, depending on the origin of the rock phosphate used (Davis and Coker 1980). Hutton (1982) estimates that away from areas of localized contamination about 8 g Cd ha^{-1} are deposited annually, 5 g originating from fertilizers and 3 g from atmospheric deposition. Interestingly, the contribution from sewage sludge application is too small to be included either on a national or regional basis.

The UK Ministry of Agriculture, Fisheries and Food introduced the concept of *zinc equivalent* to characterize the phytotoxicity of the metals in sludge and to calculate tolerable rates of application. Copper is assumed to be twice as phytotoxic as zinc, and nickel eight times more phytotoxic. As the toxic effect of the three elements is also assumed to be additive, the amount of toxic metal in the sludge can be expressed as a single figure (the zinc equivalent) by adding the metal concentrations in the sludge in the form of:

$$Zinc\ equivalent = Zn + (2 \times Cu) + (8 \times Ni).$$

Where there is no previous contamination of the soil it is assumed that an addition of sewage sludge to the soil of 250 ppm of zinc equivalent per kg DS of sludge over a 30-year period is quite safe, provided that the soil pH value is close to 6.5. The concept is extremely useful as Zn, Cu, and Ni are common in sewage sludge and are useful indicators of the overall toxicity of the sludge (Chumbley 1971). However, concern that the toxicity of these three metals is additive has been expressed, so that the use of zinc equivalents may be greatly under-estimating the amount of metals in sludge that could be safely applied to land at pH ≤ 6.5 (Beckett and Davis 1982). Although zinc equivalents still remain in the current UK guidelines (National Water Council 1981), the proposed EEC directive on the use of sewage sludge on land seeks to treat Zn, Cu, and Ni separately (European Communities 1984). This is supported by recent work carried out by the Water Research Centre (Davis and Carlton-Smith 1984; Davis *et al.* 1985).

UK guidelines allow the application rate to be calculated for individual metals so that the maximum application rate of sludge can be controlled by the concentration and toxicity of the individual metals present. The application rate is calculated as:

$$\frac{A-B}{C} \times \frac{1000}{D} \text{ tonnes DS ha}^{-1}\text{yr}^{-1}$$

where A is the recommended limit for the addition of specific metals (kg ha^{-1}) (Table 5.12), B is the concentration of available metal already in the soil (kg ha^{-1} or $2.2 \times$ mg kg^{-1}DS), C is the total concentration of the metal in the sludge (mg kg^{-1}DS), and D is the application period, which is generally over 30 years. By using the formula, the application rate is calculated for all the major metals present and the lowest application rate indicates which metal is limiting and gives the recommended annual loading (Tables 5.16 and 5.17). Incremental loadings can be in excess of the recommended annual loading as long as the maximum loading in any one year does not exceed 20% of the total permissible loading, and that the total permissible loading is not exceeded during the 30-year period (National Water Council 1981). The proposed EEC directive is based on the maximum permissible level of metals in the sludge and in agricultural soils. Maximum annual application rates are also given based on a 10-year application period, and, as with all EEC directives, both mandatory and recommended values are given (Table 5.18). Fears have been expressed that the implementation of a single set of standards for all EEC countries, as with the bathing water directive, will be extremely difficult because there is such a wide range of sewage treatment, and farming practices, climate, and soil conditions within the community. Until the standards are adopted, member countries will continue to use their own guidelines for sludge disposal. These have been reviewed by Hucker (1979) and summarized in Table 5.19.

Contamination by pathogens
Pathogens are common in sewage and are concentrated in the sludge.

TABLE 5.16 Elemental application rates for various metal concentrations in sludge and soil (Forster 1985)

	Concentration in sludge (mg kg^{-1})	Concentration in soil (mg kg^{-1})	Elemental application rate (tonnes DS ha^{-1} year^{-1})
Zinc	1650	40	$\dfrac{560-(2.2 \times 40)}{1650} \times \dfrac{1000}{30} = 9.5$
Cadmium	34	1	$\dfrac{5-(2.2 \times 1)}{34} \times \dfrac{1000}{30} = 2.7$
Lead	150	30	$\dfrac{1000-(2.2 \times 30)}{150} \times \dfrac{1000}{30} = 207.6$
Copper	400	15	$\dfrac{280-(2.2 \times 15)}{400} \times \dfrac{100}{30} = 20.6$
Nickel	30	10	$\dfrac{70-(2.2 \times 10)}{30} \times \dfrac{1000}{30} = 53.3$
Zinc equivalent	2690	150	$\dfrac{560-(2.2 \times 150)}{2690} \times \dfrac{1000}{30} = 2.9$

Although there is a range of animal and plant pathogens that can be transmitted via sewage sludge utilization on farmland, two are of particular significance, *Salmonella* bacteria and the beef tapeworm *Taenia saginata*. However, the normal level of contamination by these pathogens in sewage sludge poses little risk of infection to livestock (Argent *et al.* 1977). The health hazards associated with land disposal of sewage sludge are fully reviewed in Section 6.2.4.2.

Application of sludge to land
The production of sewage sludge at a treatment plant is constant and continuous, whereas the requirement for sludge by farmers will fluctuate and be seasonal. It is important, therefore, that provision is made for the storage of sludge either at the treatment plant or at farms. However, other factors, such as bad weather, mechanical breakdown, industrial action, restricted movement into farms due to disease, will all slow up delivery and make storage at the treatment plant inevitable. Storage of dewatered sludge in the form of cake is relatively straightforward as it can be stacked until required. Liquid sludge must be kept in special tanks, lagoons or old drying beds. Storage can be advantageous as the separated liquor can be decanted before final transportation from the site, thereby increasing the dry solids content and reducing the volume of sludge to be transported. But care must be taken to ensure that valuable nutrients are not lost in the liquor, thus reducing its value as a fertilizer. This is less likely with undigested sludge as most of the nutrients are present in an insoluble form. However, after digestion much of the remaining nutrients are soluble and will be lost if the liquor is removed.

TABLE 5.17 Recommended limit of toxic metal added to the soil from the spreading of sewage sludge (Hudson and Fennel 1980)

Element	Recommended limit in the soil after 30 yr or more (mg kg⁻¹)	Normal range in soils (mg kg⁻¹DS)	Reason for control**	Possible source of sludge contamination
Zinc	250*	10–300	P	Food, cosmetics, galvanizing, rayon
Copper (×2 to give *ZnE*)	125*	2–100	P	Pig wastes, cable, and tube manufacture
Nickel (×8 to give *ZnE*)	30*	5–500	P (Toxic to sheep)	Plating, chemical and steel industries
Chromium	450	5–500	Low toxicity	Plating, tinning, steel industry
Cadmium	2.3	1	P & T	Plating, plastics, electronics
Lead	450	2–200	T	Leaded petrol, metal processing and finishing
Mercury	0.9	0.01–0.3	T	Pharmaceuticals, dye products
Molybdenum	2.3	2	T	Metallurgy, electronics
Arsenic	4.5	0.1–40	T	Pesticides, electronics
Selenium	2.3	0.2–0.5	T	Electronics
Boron		2–100	P	Detergents, glass
Annual limits				
Pasture				
1st year	3			
subsequent years	2.3			
Arable				
1st year	2			
subsequent years	1.6			

* Limits may be increased by a factor up to 2 for permanent grassland. If more than one of these metals is present, use the *zinc equivalent*.

**P, potentially phytotoxic; T, potentially toxic to animals/humans.

TABLE 5.18 Maximum permissible concentration of metals in soil and sewage sludge applied to soil allowed under the EEC directive (European Communities 1984)

Element	Limit in sludge (mg kg DS^{-1})		Loading limit (kg ha^{-1} yr^{-1})		Soil limit (mg kg DS^{-1})	
	G	I	G	I	G	I
Cd	20	40	0.10	0.15	1	3
Cu	1000	1500	10	12	1	100
Ni	300	400	2	3	30	50
Pb	750	1000	10	15	50	100
Zn	2500	3000	25	30	150	300
As	—	—	0.35	—	20	—
Cr	750	—	10	—	50	—
Hg	16	—	0.4	—	2	—

G, recommended; I, mandatory.

The procedures adopted by the Regional Water Authorities of England and Wales for the disposal of sludge to agriculture land are generally similar. For example, the Severn and Trent Water Authority produces $>4 \times 10^6$ tonnes of sludge per annum, which is equivalent to 260 000 tonnes DS, and disposes 51% of this to farmland. The Authority hopes to increase this to 62% by 1990. It also aims to phase out the use of unstabilized sludge on land altogether, because of the odour and potential pathogen problems. However, only 13 000 of the 260 000 tonnes DS is unstabilized (Severn and Trent Water Authority 1982). The sludge if usually supplied and delivered free in most areas to the farmer if within a reasonable distance, although farmers often collect the sludge themselves using their own equipment. Transportation charges depend on the demand for sludge because in some areas suitable land is at a premium. Although most sludge is still spread by the Authorities there is a growing trend to employ contractors to dispose of the sludge, and in some areas small firms have sprung up which market the sludge as vigorously as the major fertilizer companies. The disposal of sludge follows the Department of the Environment and National Water Council Working Party guidelines (National Water Council 1981), although many of the Water Authorities have their own guidelines, and contractors are monitored to ensure that they are followed. The guidelines are fully explained in the sections on contamination by heavy metals (Section 5.2) and pathogens (Section 6.2.4.2). Land and sludge are analysed to calculate annual application rates over a 10–30 year period and details are generally supplied to the farmer. Like the Severn and Trent Water Authority, the Regional Water Authorities keep detailed records of where the sludge is spread, monitor the soil and sludge, and reassess the application rates on a regular basis. In order to prevent the transmission of pathogenic organisms present in the sludge to livestock, a minimum of 90 days should

TABLE 5.19   Summary of the sewage sludge disposal guidelines to agricultural land (Hukker 1979; Dodd 1980; Webber *et al.* 1983; Bell 1988). (The guidelines for those countries within the EEC are those prior to the implementation of the new directive)

Sludge	EPA (USA)	Wisconsin, USA
Type acceptable for agricultural use	Sludges should be stabilized (chemical, physical, biological, or thermal treatment). Sludges with hospital wastes should be pasteurized, irradiated, composted, stored long-term, or pH raised to 12	Raw should not be used. Root crops or crops to be consumed raw not to be grown within 1 year of application. Milk cows should not be grazed until 2 months after application, other animals 2 weeks
Application recommendations	Avoid highly porous soils, direct contamination of crops, ingestion by animals and humans, and public nuisance. Limits vary according to cation exchange capacity of receiving soil; and apply to soil with a pH ≥ 6.5 and increase with increasing cation exchange capacity in the ranges <5, 5–15, >15 meq. $100\ g^{-1}$ soil.	At least 0.6 m soil above impermeable layer or water table. At least 305 m from public supply well. At least 152 m from private supply well. At least 152 m from nearest residence.
Maximum application	Crop demand and N in sludge Cd addition 2 kg ha^{-1}yr^{-1} (up to 1981) Cd addition 1.25 kg ha^{-1} yr^{-1} (1982–85) Cd addition 0.5 kg ha^{-1}yr^{-1} (from 1986)	Crop N demand. Metal addition limited to 10% of soil EEC Cd limit 2.4 kg ha^{-1} yr^{-1}

Metal additions	meq. <5	Max. total loading to soil (kg ha⁻¹)		Max. total loading to soil (kg ha⁻¹)
		5–15	>15	
Cd	5	10	20	24
Co				
Cr				
Cu	125	250	500	364
Hg				
Mn				
Mo				
Ni	50	100	200	
Pb	500	1000	2000	182
Se				
Zn	250	500	1000	
B				
To other		Max. cumulative loading on soils for consumed food, inc. tobacco, is Cd <0.5 kg ha⁻¹ and Pb 800 kg ha⁻¹. Rate of application 25, 100, and 10 mg kg⁻¹ (DW) for Cd, Pb and PCB's respectively		No differentiation for pasture
Application period				
Soil pH	Not specified	>6.5		Not specified >6.5
Zinc equivalent (ZnE)	Not used			1:2:4. ZnE addition limited to 10% of EEC Cd limit
Zn:Cu:Ni				

$To\ crops$ brace groups: Cu, Hg, Mn, Mo, Ni, Pb, Se, Zn, B

TABLE 5.19—*Continued*

Sludge	Canada	England and Wales (DoE)
Type acceptable for agricultural use	Only processed sludges acceptable (aerobic or anaerobic digestion or other stabilization process). Land not to be used for fruit, vegetables or grazing within 6 months of application	Sludges may be used for animal feed and crops to be cooked, but should preferably be treated (well digested, dewatered and stock-piled, lagooned or lime treated). On pasture, delay required between application of raw sludge and grazing: 6 months for untreated and 3 weeks for treated sludges. Salad crops not sown within 12 months of application. Only mesophilic anaerobic digested or heat-processed sludge used for seed potatoes, nursery stock (especially for export), and gardens or allotments
Application recommendations	No application to soils with organic content >17% or available $P > 60$ ppm	Avoid water contamination, odour, raingun drift and physical soil damage
Maximum application	Crop N demand. Maximum 133 kg ha^{-1} amm. N over 5 yr or 4 yr on turf. 121 m^3 ha^{-1} liquid per application varies according to province	Crop N demand + 50%. Up to 1/5 of total 30 yr metal addition in any one year and no more until running average falls to 30 yr average. Reduced where soil metal levels are high. Varies between water authorities

Metal additions	Max. concentration sludge (mg kg⁻¹)	Max. annual metal addition (kg ha⁻¹ yr⁻¹)	Max. total loading to soil (kg ha⁻¹)
As	75	0.33	15
Cd	20	0.09	4
Co	150	0.67	30
Hg	5	0.022	1
Mo	20	0.09	4
Ni	180	0.8	36
Pb	500	2.2	100
Se	14	0.06	2.8
Zn	2000	8.2	370

(To crops)

To pasture: No differentiation
Application period: Suggested 50 applications at 4 yr intervals
Soil pH: >6.0
Zinc equivalent Zn:Cu:Ni: Not used

	Max. total loading (kg ha⁻¹)		Max. annual metal addition (kg ha⁻¹ yr⁻¹)
	Non-calcareous	Calcareous	
As	10	10	0.333
B	3.25	3.25	3.5
Cd	3.5	3.5	0.167
Cr	600	600	33
Cu	140	280	9.3
F	500	500	—
Hg	1	1	0.067
Mo	4	4	0.167
Ni	35	70	33
Se	3	3	0.167
Pb	550	550	18.6
ZnE	280	560	—

Application period: 30 years
Arable 6.5
Pasture 6.0

Zinc equivalent Zn:Cu:Ni: 1:2:8

TABLE 5.19—Continued

Sludge	The Netherlands			W. Germany		
Application recommendations	Not used for crops for human consumption			Precautions to be taken against disease infections in agricultural use		
Maximum application	2 t DS ha^{-1}yr^{-1} (arable), 1 t DS ha^{-1}yr^{-2} (pasture) to maximise the N and P value of sludge or metal addition limit			5 t DS ha^{-1} over 3 yr or metal addition limit		
Metal additions — Metal	Total loading to soil (kg ha^{-1})	Max. annual metal addition (kg ha^{-1}yr^{-1})	Max. concentration in sludge (mg kg^{-1})	Max. concentration in sludge (mg kg^{-1})	Max. annual metal additon (kg ha^{-1}yr^{-1})	Max total loading to soil (kg ha^{-1})
As	2	0.02	10	20	0.03	0.4
Cd	2	0.02	10	1200	2	210
Cr	100	1	500	1200	2	210
Cu	120	1	500			
Hg	2	1.2	10	25	0.042	5.7
Mo						
Ni (To arable crops)	20	0.2	50	200	0.3	60
Pb	100	1	500	1200	2	210
Se						
Zn	400	4	2000	3000	5	750
To horticultural crops	as above			as above		
To pasture	half above			as above		
Application period	50 years			60 years		
Soil pH	Not specified			Not specified		
Zinc equivalent	Not used			Not used		
Zn:Cu:Ni						

Sludge	France	Scotland
Type acceptable for agricultural use	At least 70% domestic origin. At least 25% DS	Dried sludge not acceptable on grazing land. Raw sludge—no grazing for 6 m. Digested sludge—no grazing for 21 d. or until 100 mm growth of herbage
Application recommendations	Sludge is considered unsuitable for land application if the concentration of any heavy metal exceeds twice the reference values below, or if the sum of the concentrations of Cr, Cu, Ni and Zn exceed 4000 mg kg⁻¹ DW	Avoid pollution of watercourse, application on underdrained land, and physical soil damage. Sludge not spread on land used for growing potatoes, nor vegetables and fruit if eaten raw
Maximum application	Application rate allowed is 30 t ha^{-1} DW during a 10 yr period, sludge may also be applied to land with a pH <6	Max. loading 2 t DS ha^{-1} yr^{-1} or 10 t DS ha^{-1} 5 yr^{-1}, max. application rate $= 55$ m^3 ha^{-1}, not used on soils if Ph <5.5

Metal additions:

	France			Scotland	
	Max. concentration in sludge (mg kg^{-1})	Max. annual metal addition (kg ha^{-1} yr^{-1})	Max. total loading to soil (kg ha^{-1})	Max. concentration in sludge (mg kg^{-1} DS)	Max. total loading to soil (mg kg^{-1} DS)
As				150	12
Cd	20	0.06	5.4	20	1.6
Cr	1000	3.0	360	800	80
Cu	1000	3.0	210	1000	60
Hg	10	0.03	2.7	7.5	0.4
Mo				25	2
Ni	200	0.06	60	250	40
Pb	800	2.4	210	800	80
Se	100	0.3	—	40	2.4
Zn	3000	9.0	750	2500	150

To arable crops

	France	Scotland
Application period	Not specified	50 years
Soil pH	See above	6.5 (arable)
Zinc equivalent	Not used	
Zn:Cu:Ni		1:2:8

TABLE 5.19—Continued

Sludge	Denmark		Finland		
	Max. concentration in sludge (mg kg^{-1} DS)	Max. annual metal addition (kg ha^{-1} yr^{-1})	Max. total loading to soil (kg ha^{-1})	Max. annual metal addition (kg ha^{-1} yr^{-1})	Max. concentration in sludge (mg kg^{-1})
Type acceptable for agricultural use	Raw sludge not acceptable. Stabilized sludges only for parks, arable and industrial crops, and pasture. Sludges containing abattoir waste to be disinfected for pasture and crops for raw consumption		Raw sludge not acceptable. Bio or lime stabilized or composted acceptable for parks, arable, and industrial crops, but not recommended for pasture or crops for raw consumption		
Application recommendations	Avoid surface and groundwater pollution, public nuisance, and health hazards		Avoid surface and groundwater pollution, public nuisance, and health hazards		
Maximum application	Crop demand and release inorg. N or metal addition limit <0.01 kg ha^{-1} yr^{-1} Cd. Max. rate 1.5 t ha^{-1} yr^{-1}.		20 t DS ha^{-1} 5 yr^{-1}		
Metal additions					
As				20	30
Cd	8	0.010	0.2	100	
Co					
Cr				4000	1000
Cu				12 000	3000
Hg	6	0.010	—	100	25
Mn					
Mo				12 000	3000
Ni (Crops for animal or human consumption)	30	0.045	—	2000	500
Pb (Crops for animal or human consumption)	400	0.60	—	4800	1200
Se					
Zn				20 000	5000
B					
Application period	Not 'long-term' for consumable crop		Not 'long term' for consumable crops		
Soil pH	Not specified		Not specified		
Zinc equivalent Zn:Cu:Ni	Not used		Not used		

Sludge	Sweden	Norway
Type acceptable for agricultural use	Raw sludge acceptable for arable and industrial crops, but not for parks, pasture or crops for raw consumption. Stabilized sludge not acceptable for pasture or crops for raw consumption. Disinfected sludges acceptable for all uses	Raw sludge not recommended. Bio, or lime stabilized acceptable for parks, arable, and industrial crops, but 2 years required before grazing pasture or growing crops for raw consumption. Composted or heat-treated sludges generally acceptable
Application recommendations	Avoid surface and groundwater pollution, public nuisance, and health hazards. Spray irrigation not recommended. Sludge to be worked into soil within 1 day	Avoid surface and groundwater pollution, public nuisance, and health hazards
Maximum application	5 t DS ha^{-1} 5 yr^{-1} 1 t DS ha^{-1}yr^{-1}	50 t DS ha^{-1} once only (land reclamation only) 20 t DS ha^{-1} 10 yr^{-1} repeated 10 t DS ha^{-1} 5 yr^{-1} repeated 2 t DS ha^{-1} 1 yr^{-1} repeated

Metal additions

	Sweden			Norway		
	Max. concentration in sludge (mg kg^{-1})	Max. annual metal addition (kg ha^{-1}yr^{-1})	Max. total loading to soil (kg ha^{-1})	Max. concentration in sludge (mg kg^{-1})	Max. annual metal addition (kg ha^{-1} yr^{-1})	Max. total loading to soil (kg ha^{-1})
Cd	15	0.015	0.075	10	0.02	0.2
Co	50	0.05	0.25	20	0.04	0.4
Cr	1000	1	5	200	0.4	4
Cu	3000	3	15	1500	3	30
Hg	8	0.008	0.04	7	0.014	30
Mn				500	1	10
Ni	500	0.5	2.5	100	0.2	2
Pb	300	0.3	1.5	300	0.6	
Zn	10,000	10	50	3000	6	60

	Sweden	Norway
Recreation, forestry, flowers	According to case	According to case
Application period	5 years, but repeated application discouraged	Not 'long term' for consumable crops
Soil pH	Not specified	Not specified
Zinc equivalent Zn:Cu:Ni	Not used	Not used

elapse between the application of unstabilized sludge to grassland and the return of livestock. However, this can be reduced to 21 days in the case of stabilized sludge. As similar controls cannot be exerted on small areas of land, sewage sludge is not generally available to private gardeners.

Application of sludge to land inevitably involves transportation costs that can be reduced considerably if the sludge is thickened or dewatered to form cake. However, three options are available for transporting liquid sludge, the selection of which depends on the location of the treatment plant in relation to the farmland requiring the sludge. The least common is the use of temporary or permanent pipelines that transport sludge directly from the plant to the farmland. This system is only suitable for small- to medium-sized treatment plants that are located in rural areas and are surrounded by suitable farmland. A network of pipes are laid and the sludge pumped to fields when and where required. The disadvantage of this system is that a network of pipes is very restrictive, which can result in over-application of sludge. Tractor-hauled trailer tanks are generally employed at small- to medium-sized treatment plants where only small volumes of sludge are produced or sludge production is intermittent. These units are limited to short distances and the tanker capacity restricted to between 3.5–5.5 m^3 (normally 3.6 m^3). Road tankers (4.0–22.5 m^3) are used for transporting larger volumes over longer distances. Four-wheel drive tankers (4.0–5.5 m^3) are the most economical as they can also be used for spreading the sludge, whereas over distances of 15 km, larger articulated tankers (>20 m^3) are used, with the sludge transferred to storage tanks on the farm or smaller units for spreading.

The farmer requires the sludge spread evenly over the soil, without damaging the soil structure, causing pollution or having unpleasant odours produced. The control of these factors is dependent on the method of application selected, although essentially they have four principle objectives: (1) to prevent unacceptable contamination of agricultural land with potentially toxic elements (Table 5.20); (2) to prevent dissemination of human, animal, and plant diseases (Table 5.21); (3) to avoid public nuisance; and (4) to avoid water pollution.

There are three widely used methods of applying sewage sludge to land; surface spreading; irrigation; and injection. Sludge cake is transported by lorry and is then distributed on to the land directly by conventional tractor-towed manure spreaders. Liquid sludge or slurry is spread on to land either from slurry tanks with heavy-duty flotation-type tyres enabling them to operate even in wet weather, without damaging the soil, or from small road tankers that can carry up to 2 m^3 more sludge than the tractor-hauled slurry tanks. The small road tankers also have wide tyres to minimize vehicle pressure, reducing damage to soil and improving tractability. However, because of the weight of these vehicles they cannot be used during wet conditions as the soil structure can be damaged or the tanker could become

TABLE 5.20 Summary of the recommendations by the UK regarding potentially toxic elements and the use of sewage sludge on land (National Water Council 1981)

Element	Maximum permissible addition (kg ha^{-1})	Provisional maximum soil concentration (mg l^{-1})
Cd	5	3.5
Cu[a]	280	140[b]
Ni[a]	70	35[b]
Zn[a]	560	280[b]
Pb	1000	550
Hg	2	1
Cr	1000	600
Mo	4	4
As	10	10
Se	5	3
B	3.5	3.25[a]
F	600	500

[a] These elements are assumed to have additive toxic effects in the ratio $Zn + 2(Cu) + 8(Ni) = zinc equivalent$. The limits for the *zinc equivalent* are as for zinc.
[b] Extractable.

stuck. The tractor-drawn tanker is relatively expensive due to its smaller size and also completes less runs per day than a road tanker. However, it is often the only available option where the ground is not stable enough for the weight of a road tanker. The liquid sludge is discharged by gravity on to the land via fish-tail nozzles that ensure even distribution over a 3.0–4.5 m wide strip, with the rate of application being controlled by the speed of the vehicle. Other methods of spreading sludge from tankers involves a rotating disc distributor or a pair of contra-rotating discs. Sludge is first fed into a hopper from which it falls on to the discs which distribute the sludge over a much wider area. Liquid sludge can be also pumped, under pressure, through arms or sprayers so that lighter applications of sludge can be made over a wider area, up to 18 m or even 45 m wide strips, which is particularly useful for treating existing crops (Institute of Water Pollution Control 1978). Because of the weight of the tanker and sludge, access to the land must be restricted to a period April to November, although this depends on the weather and condition of the land. Failure to restrict access will result in severe compaction and rutting of the soil, which may take several years to recover. After application, grassland should be chain harrowed, especially if raw sludge is used. When arable land is treated, this should be ploughed and cultivated to incorporate the sludge into the soil, and if raw sludge is used this should be done at once to avoid odours.

TABLE 5.21 Recommendations on the application of sewage sludge to agricultural land to avoid potential pathogen problems (National Water Council 1981)

Sludge	General arable including forestry, land reclamation, conservation, crops	Seed potatoes and export nursery stock	Grazed crops	Salad and crops consumed raw	Park flower beds	Orchards, turf
1. Liquid raw	√	×	C	×	×	×
2. Liquid raw stored 2 weeks	√	×	C	×	×	D
3. Cold anaerobic digested	√	×	AB	E	√	√
4. Lagooned	√	×	AB	E	√	√
5. Mesophilic anaerobic digested	√	√	AB	E	√	√
6. Heat processed	√	√	AB	E	√	√
7. Non-limed cake	√	×	C	×	×	D
8. Non-limed cake stored	√	×	AB	E	√	√
9. Limed, cake or stabilized	√	×	C	E	√	√
10. Full treatment biological (inc. aerobic digested and extended aeration of whole sewage)	√	×	C	E	√	D
11. Partial treatment biological	√	×	C	×	×	D

√ Acceptable. × Unacceptable.

A 3 weeks grazing interval, but 5 weeks for cattle which produce milk which is consumed unpasteurized.

B Where 40% organic matter destroyed in digestion or storage period less than that recommended, the grazing interval should be increased accordingly from 3 weeks to 6 months for pigs or cattle.

C Grazing interval for pigs, cattle, 6 months, other animals as A.

D No fruit or turf harvested for 3 months after application.

E None to be sown until 12 months after application.

F Only mesophilic digested and heat treated sludges to be used in gardens and allotments – this practice to be phased out.

Liquid sludge can be sprayed on to agricultural land under pressure using rain or manure guns. The sludge is pumped via a pipeline to one or more rain guns with 15–50 mm openings. The guns are normally directional and able to cover areas of between 0.1–0.4 ha each. Coverage is not as even as the other methods, with blockages or wind effecting evenness of distribution. Care must be taken to ensure that slurry is not sprayed into local streams or on to roads or footpaths. The wind can take the slurry considerable distances and the operator must ensure that it is not allowed to drift into neighbouring fields or on to buildings. Aerosols which can be formed by these irrigation techniques have received much attention because of the risk of disease transmission, and precautions have to be taken (Section 6.2.4.3). This type of sludge application is labour-intensive, with the guns requiring frequent relocation that involves the operator walking over ground which has been already sprayed with sludge and moving the pipework, usually sectional aluminium pipes in 6 or 9 m lengths, and the guns that are both fouled by sludge. A method that is less labour-intensive and much less unpleasant for the operator to use is a sledge-mounted rain gun attached to flexible pipework wound on a tractor-drawn bobbin, known as a travelling irrigator. The sledge supporting the gun is anchored temporarily at the far side of the field and the pipework fully unwound from the bobbin. The end of the flexible pipework is attached to a single temporary sludge pipe running along the near-side of the field and secured. The sludge is released, and the tractor very slowly rewinds the flexible pipework, slowly hauling the gun over the field as sludge is sprayed over the ground (Fig. 5.12). The operator is not required to handle the dirty pipe or gun and does not have to walk over treated land, with the sequence repeated on the next section of fresh field. The use of travelling irrigators is reasonably cheap and because it involves little movement of vehicles on the soil and no

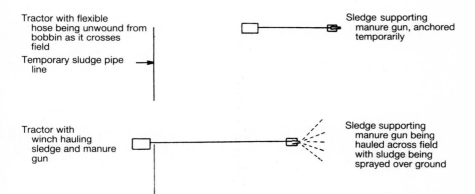

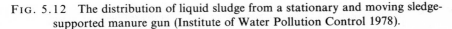

FIG. 5.12 The distribution of liquid sludge from a stationary and moving sledge-supported manure gun (Institute of Water Pollution Control 1978).

movement of heavy tankers, there is no risk of soil damage, and this allows operations to continue even when the soil is wet.

The final and least objectionable method of land application of sludge is direct sub-surface injection which uses a chisel plough to inject sludge 25–50 mm below the surface of the soil. The equipment can be fitted to most road tankers and tractor-hauled units, although custom-built vehicles are far more efficient and flexible in the depth of injection and rates of application that can be achieved. This method effectively eliminates odours and the risk of disease transmission, so that the long period between treatment and allowing animals back to the land to graze can be reduced considerably. High application rates are possible and up to 180 tonnes ha^{-1} can be achieved at a single pass, but at these levels of application the soil is too wet and unstable to allow animals or vehicles back on to the land for 4–6 weeks.

The major application methods are compared in Table 5.22 and are

TABLE 5.22   Comparison of the various methods employed to spread liquid sewage sludge on to agricultural soil in terms of operational performance and cost (Hall and Davis 1983)

Application method	Prevention of odours and disease transmission	Effective on wet soil and for many land uses	Evenness of spread*	Economy	General comments
Tanker direct	+	+	+ +	+ + +	Widely used. Economic. Allows operational flexibility
Field tanker	+	+ +	+ +	+	High capital cost. Should be kept continuously supplied with sludge
Tractor-drawn tanker	+	+ +	+ +	+	Widely used
Movable rain gun	+	+ + +	+	+ +	Uneven application, problems of moving dirty pipes
Travelling irrigator	+	+ + +	+ + +	+ +	No risk of damage to soil
Sub-soil injection using field tanker or tractor-drawn	+ + +	+ +	+ + +	+	Avoids odour and disease-transmission problems

* Evenness of spread does depend considerably on the type of ancillary equipment used and driver skill.

+, fair; + +, good; + + +, excellent.

reviewed more fully by Critchley *et al.* (1982). Caution must be exercised when applying sewage sludge to agricultural land. Groundwater contamination can occur if liquid sludge is applied directly to dry clay soils, which have cracked, as this allows the sludge to drain rapidly deep into the soil. Surface waters are also at risk if sludge is over-applied to wet land and surface runoff occurs. Soil damage is a major concern when tankers of sludge are being driven over land, even though extra-wide tyres may be used. As a general rule, no vehicles should be driven on land immediately after rain, other than on sandy soils that drain readily. The soil should be allowed to drain until it regains sufficient bearing strength to support vehicles of high weights.

Further reading

General: Best 1980; Vincent and Critchley 1983; Davis 1987.
Sludge disposal to land sites, general: Metcalf and Eddy 1984.
Sacrificial land use: Institute of Water Pollution Control 1978.
Use of sludge in land reclamation: Coker *et al.* 1982; Davis 1982.
Sludge utilization to farmland, general: Hudson and Fennel 1980; Hall and Davis 1983; Institute of Water Pollution Control 1986.
Manurial value of sewage sludge: Davis 1980.
Contamination by heavy metals: Davis 1980.
Contamination by pathogens: Watson *et al.* 1983.
Application of sludge to land: Institute of Water Pollution Control 1978; Hall and Davis 1983.
Legislative control: Matthews 1983.

5.3   SEA DISPOSAL

Only in the UK and Ireland is sea disposal an important method of disposing of sewage sludge, with most other EEC member countries making more extensive use of the landfill option (Table 5.6). In the USA, 40% of the sludge goes to landfill sites and some 15% is disposed to sea, including the sewage sludge from New York. Of the remainder, 25% is incinerated and 20% is applied to farmland (Rothman and Barlett 1977). Sea disposal is often the preferred disposal method in environmental, economic, and logistical terms, and where geographical position permits, it is the cheapest method of sludge disposal (Collinge and Bruce 1981). The UK is particularly well suited to this disposal method as nowhere in the country is further than 120 km from the sea, and the coastline is the longest of all EEC countries being, 9840 km in length of which 3790 km borders the Irish Sea. The tidal currents are fast (100–200 cm s^{-1}) and the tidal ranges moderate to high, reaching maxima of approximately 6 m off the east coast, 5 m in the English Channel, 7 m in the Irish Sea, and 10 m in the Bristol Channel. Therefore, the fast currents and high tidal ranges combine to provide excellent dispersal and dilution characteristics (Oslo and Paris Commissions 1984). Sludge disposal to the sea has a long history in the UK, being used to dispose of London's sludge to the outer Thames estuary as early as 1889. Currently, dumping at sea accounts for 29% of all the sludge produced in the UK, a total of 10.5×10^6 wet tonnes from 30 locations, which is equivalent to 300 000 tonnes DS per annum. The dependence of authorities on this disposal methods varies, with 9 of the 10 Regional Water Authorities in England and Wales disposing of some sludge by dumping at sea. Only the Severn and Trent Water Authority do not use this method. Strathclyde Regional Council, North West Water Authority, and Thames Water Authority dispose of 80%, 49%, and 30% of their sludge by this route respectively (Table 5.23). All of London's sludge is disposed to sea at the Barrow Deep by Thames Water Authority, which represents 49% of all the sludge disposed to the sea in the UK (Oslo and Paris Commissions 1984). The quantities of sludge disposed of at sea increased by 10% between 1976–81 (Table 5.24), with dumping from Edinburgh commencing in 1978 (350 000 wet tonnes per annum) and from Newcastle-upon-Tyne in 1980 (500 000 wet tonnes per annum) (Fig. 5.13). Two authorities, Northumbrian Water Authority and Lothian Regional Council, have made major capital investments in sea disposal equipment, as has Dublin Corporation, which treats the

TABLE 5.23 Amount of sewage sludge dumped at sea at licensed sites in 1981 by UK water authorities, see also Fig. 5.13 (Oslo and Paris Commissions 1984)

Location	Amount dumped (net tonnes)	% of UK total (nearest 1%)
England and Wales		
Northumbrian Water Authority	271 000	3
Yorkshire Water Authority	92 000	1
Severn Trent Water Authority	1 000	<1
Anglian Water Authority	225 000	2
Thames Water Authority	4 920 000	49
Southern Water Authority	267 000	3
South West Water Authority	148 000	1
Wessex Water Authority	266 000	3
Welsh National Water Authority	62 000	<1
North West Water Authority	1 693 000	17
Total	7 945 000	79
Scotland		
Strathclyde Regional Council	1 512 000	15
Lothian Regional Council	287 000	3
Total	1 799 000	18
Northern Ireland		
Total	279 000	3
United Kingdom		
Total	10 023 000	100

TABLE 5.24   Amount of sewage sludge dumped at sea by the UK
between 1976–81 (to nearest '000 tonnes) (Oslo and Paris
Commissions 1984)

Year	England and Wales	Scotland	N. Ireland
1976	7 011 000	n.a.	213 000
1977	7 572 000	n.a.	218 000
1978	7 844 000	n.a.	242 000
1979	7 625 000	1 965 000	262 000
1980	8 242 000	2 016 000	296 000
1981	7 945 000	1 799 000	279 000

sewage from a third of the population of Ireland. In Dublin, a major new primary sedimentation and sludge consolidation plant has been built and a larger sludge vessel, the Sir Basil Jet, has been purchased from Thames Water Authority. Clearly, sea disposal is recognized as a major disposal route by many water authorities and the level of investment indicates that it will continue to be exploited.

Sea disposal is mainly carried out by dumping liquid sludge into estuaries and coastal waters from specially designed ships, with < 5% being discharged from pipes (Fish 1983). This should not be confused with sewage outfalls from which 15% of all UK sewage is discharged. It is estimated that if the dumping of sewage sludge to sea was stopped it would cost the British Water industry £85 × 10^6 in capital plus a further £32 × 10^6 per year in operating costs (Water Authorities Association 1984) (see Preface).

5.3.1   Legislative control

Licences for dumping sludge at sea in Britain are issued annually for each source by the Ministry of Agriculture, Fisheries and Food in England, the Welsh Office, the Department of Agriculture and Fisheries for Scotland, and Department of the Environment for Northern Ireland in their respective areas. The licences are issued under the Dumping at Sea Act (1974) which makes it an offence to dump any material from a vehicle, ship, aircraft, hovercraft or other marine structure without a licence from the relevant licensing authority whose responsibility is 'to have regard to the need to protect the marine environment and the living resources it supports from any adverse consequences of dumping . . .' (Norton 1978). This act enables the UK to ratify the Oslo and London International Conventions that regulate dumping to the marine environment. Under these conventions, sludges can only be dumped if substances listed in Annex I of the Oslo Convention, which are considered extremely toxic (e.g. mercury, cadmium, and organochlorine compounds) (Table 5.25), are not present other than as *trace contaminants*. This is

FIG. 5.13   Licensed UK sludge dumping sites and the amount dumped each year in tonnes (Oslo and Paris Commissions 1984).

currently interpretated as not substantially in excess of levels in domestic or light industrial sludges. Similarly, for Annex II substances, which include potentially bio-accumulative and toxic substances (e.g. arsenic, lead, copper, and zinc) and toxic substances (e.g. cyanides, fluorides, and those pesticides excluded from Annex I) (Table 5.26), 'special care' is to be taken where sludges contain 'significant' quantities of these substances, which is currently defined as 0.1% of the weight of the wet waste (Institute of Water Pollution Control 1978; Norton 1980; Collinge and Bruce 1981; Oslo and Paris Commissions

TABLE 5.25 Summary of the substances listed in Annex I of the Oslo and London Conventions that may not be dumped at sea

Prohibited under Oslo and London Conventions	Prohibited under Oslo Convention only	Prohibited under London Convention only
Organohalogen compound[a,c]	Organosilicon compounds[a]	Oils taken on board for dumping[c]
Mercury and its compounds	Carcinogenic substances[b,c]	High level radioactive wastes
Cadmium and its compounds[c]		Materials of biological and chemical warfare
Persistent plastics and other persistent synthetic material[c]		

[a] Excluding those which are non-toxic or rapidly converted in the sea into substances which are biologically harmless.
[b] As agreed by the contracting parties.
[c] Does not apply to those wastes containing these substances in trace contaminants.

TABLE 5.26 Summary of the substances listed in Annex II of the Oslo and London Conventions that can only be dumped at sea with a special permit[a]

Included in both Oslo and London Conventions	Included in Oslo Convention only	Included in London Convention only
Arsenic ⎫ Lead ⎬ and their Copper ⎰ compounds Zinc ⎭ Cyanides Fluorides Pesticides not listed in Annex I Containers, scrap metal and other bulky wastes[b]	Non-toxic substances which may be harmful because of large quantities in which they had dumped	Organosilicon compounds Beryllium ⎫ Chromium ⎬ and their Nickel ⎰ compounds Vanadium ⎭ Radioactive matter not included in Annex I

[a] For the purposes of the London Convention all wastes other than those listed in Annex I and Annex II require a prior general permit. For the purposes of the Oslo Convention, the provisions of Annex II apply only to wastes containing significant quantities of the substances shown.
[b] For the purposes of the Oslo Convention, such substances may be dumped only in waters where the depth is greater than 2000 m and the distance from land is not less than 150 nautical miles.

1984). Licences issued under individual countries legislation to conform with the Oslo and London Conventions must have regard to the factors listed in Annex III (Table 5.27), which include the composition and properties of the waste, the characteristics of the proposed dumping area, and the method of disposal. In all cases, the decision to license dumping operations is solely a matter for the national licencing authority. This has led to disagreements about licensing policy between countries who share common seas, for example, the dumping of radioactive waste in the Irish Sea by the UK.

TABLE 5.27 Annex III of the Oslo Convention giving the provisions governing the issue of permits and approvals for the dumping of wastes at sea

---

1. *Characteristics of the waste*
   (a) Amount and composition.
   (b) Amount of substances and materials to be deposited per day (per week, per month).
   (c) Form in which it is presented for dumping, i.e. whether as a solid, sludge or liquid.
   (d) Physical (especially solubility and specific gravity), chemical, biochemical (oxygen demand, nutrient production) and biological properties (presence of viruses, bacteria, yeasts, parasites, etc.).
   (e) Toxicity.
   (f) Persistence.
   (g) Accumulation in biological materials or sediments.
   (h) Chemical and physical changes of the waste after release, including possible formation of new compounds.
   (i) Probability of production of taints reducing marketability of resources (fish, shellfish, etc.).

2. *Characteristics of dumping site and method of deposit*
   (a) Geographical position, depth, and distance from coast.
   (b) Location in relation to living resources in adult or juvenile phases.
   (c) Location in relation to amenity areas.
   (d) Methods of packing, if any.
   (e) Initial dilution achieved by proposed method of release.
   (f) Dispersal, horizontal transport and vertical mixing characteristics.
   (g) Existence and effects of current and previous discharges and dumping in the area (including accumulative effects).

3. *General consideration and conditions*
   (a) Interference with shipping, fishing, recreation, mineral extraction, desalination, fish and shellfish culture, areas of special scientific importance, and other legitimate use of the sea.
   (b) In applying these principles the practical availability of alternative means of disposal or elimination will be taken into consideration.

In 1985, a new Directive on the control of dumping waste at sea was proposed by the EEC (European Communities 1985) that supersedes the 1976 proposal (European Communities 1976b). This is critically evaluated by Matthews (1986) who feels that it is an unnecessary addition to the existing legislation, especially as it is more restrictive than the present Oslo Convention.

5.3.2 Dumping sites

The careful selection of dumping sites can minimize any detrimental effects sewage sludge may have on the environment and minimize the interference with other legitimate uses of the sea. Sites should either ensure a high degree of dilution and dispersion, thus preventing the build up of sludge or harmful concentrations of substances in sludge, or ensure sludge dispersion is contained so that any effects are restricted to as small an area as possible. Thus, sites are characterized by their different hydrographic qualities as either dispersive or accumulative. At dispersive sites dumping occurs in generally shallow well-mixed waters in areas with an open coastal aspect. The site is swept by high tidal currents to give rapid dispersion. Dispersal is enhanced by the release of sludge, at mid-tide, into the wake of the moving vessel that increases the area over which particles will settle. Most sites are of this type, although a few, such as Garroch Head on the Clyde are containment or accumulating sites. There is little dispersion at such sites which are characterized by low tidal velocities that are usually more physically restricted and are in deeper waters. Settlement is encouraged by rapid release of sludge from the bottom of the vessel while it is stationary above the site. Dumping is most effective in accumulative sites on or just before slack water, ensuring minimum dispersion.

Similar criteria have to be met for both types of site and these are listed in Annex III of the Oslo Convention (Table 5.27). This states that the potential effects of sludge dumping on all the other legitimate uses of the sea should be fully evaluated before a licence is issued. Norton (1978) has listed the most important uses that may be affected by dumping activities. These are:

Amenity: avoidance of contamination of bathing waters and beaches with sewage-associated solids and debris, and, in particular, with pathogenic micro-organisms. To achieve this, dumping sites should be located several miles from the nearest shore.

Shipping: sites should be away from busy shipping lanes, especially if the sludge vessel has to steam within the defined area for an extended period to dispose of its load.

Commercial fishing: dumping sites must avoid areas which are extensively fished whether by trawling, seining, lining or potting.

Spawning and nursery grounds: areas where eggs, larval, and juvenile stages of fish and shellfish abound should be avoided. The developmental stages are generally more sensitive to environmental changes than adults.

Mineral extraction: offshore mineral extraction is rapidly developing and dumping should avoid areas which may be suitable for future exploitation, especially for sand and gravel extraction. Unless there are very strong dispersive currents, settlement of sludge would make any future exploitation of the minerals very difficult.

Existing discharges: although sludge dumping is insignificant in terms of the amount of nutrients and certain metals that are discharged, combined with land based and atmospheric inputs, local environmental problems could result if the local dispersive capacity is exceeded.

Clearly, site selection will require extensive field evaluation in order to define an area suitable for sludge disposal, with specialist studies on the hydrography to measure current strengths and direction, direction of residual movement, vertical stratification of the water column, and water quality. Specialist studies are also required to locate the fisheries, to characterize the physical and chemical nature of the existing sediment, and the biological nature of the sediment in order to establish the susceptibility of the existing fauna to sludge. Once selection has taken place and dumping commenced then a permanent monitoring programme is required to determine the chemical, physical, and biological effects on the site. What is of most interest is the short- and long-term fate of the sludge, nutrients, and metals in the area, and the overall effect on fish and shellfish quality.

5.3.3   Environmental impact

Dumping sites in the UK are chosen to minimize environmental impact, although the major criterion for controlling sea discharges remains the protection of seafood for human consumption. Both the Ministry of Agriculture, Fisheries and Food, and the Department of Fisheries for Scotland monitor all major licensed sites. They employ a system of intermittent intensive studies at approximately five-yearly intervals, supported by less intensive routine monitoring. Disposal sites are also chosen to preclude contamination of commercial fish and shellfish, and some monitoring of fish for pathogens and metal content is also undertaken. Traditionally, the major impact of sludge dumping is assessed in terms of abundance and diversity of species plus the concentration of metals in the sediment. From their data, it would appear that sludge dumping has very little impact on the marine environment. At sites where the sludge is highly dispersed, such as the Thames Estuary and Liverpool Bay dumping sites, environmental damage is not detectable above background variations. However, at sites where sludge accumulates, such as Garroch Head, some localized effects have been detected (Davis *et al*. 1985).

High turbidity and low dissolved oxygen concentrations are rarely serious problems at sludge dumping sites. The dissolved oxygen concentration is not significantly affected at dispersed sites, as in open coastal situations any oxygen depletion in overlying waters is rapidly removed by dilution and dispersion, and only localized and transient reductions occur. For example, in the Thames Estuary and Liverpool Bay, the dissolved oxygen concentration rarely falls below 99% saturation with localized concentrations only falling as low as 97% in the surface waters at dumping sites (Norton 1978). The sediment oxygen demand at accumulative sites can often be considerable resulting in a reduced oxygen concentration in the overlying water, with subsequent reduction or elimination of sensitive species. Therefore, at sites where there is less mixing, reductions in dissolved oxygen will be larger and longer in duration. Density stratification of the water inhibits vertical mixing that can result in critically low oxygen concentrations.

Changes in environmental conditions will disturb the ecological balance of dumping sites resulting in the reduction of species diversity. Under extreme conditions, only those species tolerant of the extremes of the particular stress factors survive, although they will generally proliferate resulting in large population densities because of reduced competition. Where sludge does not accumulate, the benthic fauna is unaffected in terms of species diversity (Domenowske and Matsuda 1969). However, where deposition takes place, species diversity can be reduced over a wide area (Pearce 1969; Pearson and Rosenberg 1978). On the boundaries of dumping sites where the level of organic enrichment is low, a number of species will preferentially feed off the organic material with a subsequent increase in their numbers. At high levels of organic enrichment a reduction in species diversity is recorded as sensitive species are eliminated and only tolerant species are left. High biomasses of tolerant species are recorded at the centre of dumping sites and are dominated by the detritiophagous polychaetes such as *Capitella capitata* and also several species of nereids. These sites are characterized by low-diversity polychaete communities and are well documented (Bellan 1967; Kitamori 1971; Reish 1973). At Garroch Head, where the organic carbon concentration reaches 8% in the central area of the dumping site, the species diversity is severely restricted, although the animal biomass is very high (Topping and McIntyre 1972). In the surrounding areas, where the organic carbon is $< 3\%$, a normal mixed fauna is present. However, in the Thames and Liverpool Bay dumping areas, it is the mobility of the bottom sediments that appears to be the main factor controlling the benthos, with few adverse effects attributable to sludge dumping (Gould 1976). The stability of the sediment is an important factor affecting the burrowing forms of molluscs and crustaceans, and sludge dumping tends to eliminate many of these species. For example, at the Garroch Head dumping site, the sludge has made the sediment unsuitable for the Norway lobster (*Nephrops*) over a 10 km^2 area, so that the species is severely reduced in numbers (McIntyre and Johnston 1975). In localized areas in dumping sites where excessive accumulation has occurred and there is little mixing, low dissolved oxygen conditions can persist, eliminating even tolerant species. The response of organisms depends very much on the physical and chemical nature of the site as well as the level of dumping that occurs, and individual sites tend to be unique in terms of the response by the flora and fauna, making generalizations difficult.

The major environmental effects of sludge dumping are caused by nutrient enrichment, metal and pesticide accumulation, and the persistence of pathogens (Chen and Orlob 1972; Eppley et al. 1972; Jenkinson 1972; Halcrow et al. 1973; Watling et al. 1974; Caspers 1976; Eagle et al. 1979). The nutrients released during the degradation of sewage can lead to an enormous increase in algal productivity when discharged to the marine environment. The subsequent death of the biomass formed adds, and can even exceed, the load of organic material derived directly from the outfall or discharge, releasing further nutrients as the algal mass degrades. In extreme cases, this secondary degradation of organic matter can lead to anaerobic decomposition, with the release of hydrogen sulphide gas and severe stress on the existing biota. This is a particularly serious problem in the larger enclosed estuaries that receive large effluent loads, such as Dublin Bay which receives the settled sewage from almost 1 million people. The water quality at dumping sites changes because of the presence of organic matter, nutrients, trace metals, and suspended particulates in the sludge. Although changes in water quality are generally minimal because of dilution and dispersion, the input of nutrients is so significant that elevated nutrient concentrations occur on a local basis. However, at dumping sites located near major estuaries, such as the Mersey or the Clyde estuaries, the nutrient enrichment in the water at the dumping site is less than the adjacent inshore areas that receive nitrogen and phosphorus from the rivers and coastal sewage outfalls. Therefore, nutrient enrichment arising from sludge dumping may not in itself be important, but it can make a significant contribution to the total nutrient levels in some coastal waters.

Primary productivity of phytoplankton in marine waters appears limited by nutrient concentration rather than temperature or light, and any increase in nitrogen and phosphorus could result in increases in algal biomass as measured by chlorophyll *a*. However, no elevated primary productivity has yet been recorded at a sludge dumping site, even accumulative ones, and the problem of algal blooms and coastal eutrophication is most likely due to sewage outfalls and nutrients originating from riverine sources. Blooms generally appear in restricted waters where poor dilution and dispersion leads to locally high nutrient concentrations. Algal growth is enhanced in shallow waters where the water temperature is high and light penetration is at a maximum. These conditions cause an explosion in the growth of algae, which results in a very dense, low diversity or even mono-specific algal crop. If this bloom is composed of toxin-producing species, further damage may result from the concentration of toxins in the marine food web. Some blooms are promoted by organic enrichment as well as nutrients, such as the algae *Phaeocystes* and dinoflagellates, and sludge dumping may well encourage the formation of these blooms in existing nutrient-rich waters. These particular blooms are generally toxic to both fisheries and man via contaminated shellfish. Although the effects of nutrient enrichment in open coastal waters may be rather subtle and difficult to detect, phytoplankton will become concentrated at the shoreline as they are washed ashore. Luxuriant growths of *Ulva* and *Enteromorpha* in shallow enriched waters are particularly common and like the phytoplankton blooms, will cause secondary pollution upon decay, resulting in odour and discolouration of the water (Gould 1976). It has been suggested that sludge dumping acts as a marine fertilizer (Segar *et al.* 1985), as it does on land. There is, however, little evidence to support this and many authors feel that sea disposal represents a waste of a useful resource (Rothman and Barlett 1977; Norton 1978).

The disposal of sewage sludge from ships accounts for only a small proportion of the total quantity of pollutants reaching the marine environment. Metals, for example, come from three other sources: via rivers; direct discharge of sewage and industrial wastes; and from atmospheric pollution. Of the total quantity of cadmium in the North Sea, for example, 63% comes from aerial deposition, 34% via rivers, 2% from direct discharges, and only 1% from sludge dumping. Similar figures are available for mercury with 14%, 68%, and 4% respectively (Oslo and Paris Commissions 1984). Sixty per cent of sludge dumped from vessels is dumped in the North Sea. However, sewage sludge dumping in this area represents only 5% of the total BOD load and 1% of both the nitrogen and phosphorus loads. However, the southern North Sea, which is closest to the UK but is only 7.5% of the total area of the North Sea, receives over half the nutrient load and between a third and a half of all the metals. Therefore, if the southern North Sea is considered separately, then the proportion of nutrients supplied from sludge dumping rises to 12%, 2%, and 3% for BOD, nitrogen, and phosphorus respectively (Norton 1982; Hill et al. 1984). In some areas, however, such as the Clyde, Liverpool Bay, and the outer Thames estuary, sludge disposal accounts for 30–50% of the input of certain metals, whereas in other areas, such as the Severn and the Firth of Forth, the contribution from sludge rarely exceeds 5% (Table 5.28) (Collinge and Bruce 1981). Considerable efforts have been made by the Water Authorities to reduce the concentrations of Annex I substances and other less toxic substances listed in Annex II reaching treatment plants, and eventually being concentrated in the sludge and dumped at sea. The Thames Water Authority has been able to reduce the average concentration of mercury and cadmium in its sludge disposed at sea by 78.3% and 41.0% respectively, by setting new limits for metals discharged to sewers by the 800 firms discharging metals to the Beckton, Crossness, Riverside, and Deephams treatment plants that serve London (Fish 1983). This is reflected in the total quantities of metals disposed by the UK (Table 5.29). This shows that although there has been a continuous decline in the quantities of Annex I metals dumped in sludge since 1976, there has been little change in the Annex II metals. However, the weight of zinc has decreased from 939.6 tonnes per annum in 1979 to 439.3 tonnes in 1983 (DoE 1984).

Because nearly all sludge is dispersed, it is important to consider total inputs into marine waters, not only from dumping, but from the other sources as well. However, increasing pressure to restrict disposal of wastes to sea from EEC countries has now extended to sludge dumping, with the introduction of the Urban Wastewater Directive which prohibits dumping of sludge at sea from vessels or outfalls from 1998. While these are minor routes for entry of contaminants into the marine environment, they are the most easy to control.

At dispersive sites there is very little accumulation of organic matter and contaminants, whereas at accumulative sites the reverse is true. The concentration of metals and other contaminants is linked to the quantity of organic matter present. At two accumulative sites, the New York Bight and

TABLE 5.28 Annual discharge of heavy metals to five major UK estuaries and the percentage input from sludge dumping (Collinge and Bruce 1981)

Estuary	Cd	Cr	Zn	Ni	Cu	Pb
Clyde						
Total discharge (tonnes yr^{-1})	6	205	716	49	158	185
Percentage as sludge	16	59	18	13	30	23
Firth of Forth						
Total discharge (tonnes yr^{-1})	20	160	823	110	338	453
Percentage as sludge	2	4	2	2	3	2
Liverpool Bay						
Total discharge (tonnes yr^{-1})	53	324	1907	188	238	—
Percentage as sludge	—	28	14	8	46	—
Severn						
Total discharge (tonnes yr^{-1})	40	492	1652	107	294	458
Percentage as sludge	1	4	2	4	9	4
Thames						
Total discharge (tonnes yr^{-1})	42	142	1126	243	242	189
Percentage as sludge	12	36	40	21	42	54

TABLE 5.29 Quantities of heavy metals dumped annually (tonnes) at sea as sewage sludge, 1978–84 by the UK (DoE 1987)

Heavy metal	1978	1979	1980	1981	1982	1983	1984
Mercury	2.8	2.8	3.6	2.4	1.4	1.1	0.9
Cadmium	8.6	8.6	9.3	6.5	5.1	4.0	4.0
Copper	235.7	236.6	220.7	199.8	203.9	158.6	156.8
Lead	207.6	197.6	182.3	159.8	164.8	160.0	158.0
Zinc	867.3	939.6	684.8	640.8	442.0	439.2	500.3
Total amount of sludge ($\times 10^6$)	8.3	8.5	8.9	8.5	8.1	7.3	7.5

Garroch Head dumping grounds, the organic carbon levels reach 5% and 8% respectively, with correspondingly elevated levels of contaminants. After dumping, metals and organic compounds are solublized either from the sediment or from the suspended particulate matter, resulting in enhanced metal levels in the water. At the Liverpool Bay dumping site, high concentrations of mercury $(200 \, \mu g \, l^{-1})$ and Polychlorinated biphenyls (PCBs) $(1.5 \, \mu g \, l^{-1})$ have been recorded (Gardner and Riley 1973; Dawson and Riley 1977).

It is difficult to isolate the effect of sludge dumping from the effects of other dumping operations and other sources of metals. However, there is evidence of elevated heavy metal concentrations in sludge dumping grounds and also increased levels in some commercially fished species of fish and shellfish in some coastal areas. In the Clyde estuary, where there are no significant dispersion effects, the changes in the sediment and benthos is clearly attributable to sludge dumping, and when compared with the Liverpool Bay dumping site, where dispersion is good, significantly higher concentrations of metals have been recorded (Gould 1976).

Heavy metals, PCBs, pesticides, and pathogenic micro-organisms can all be accumulated by fish and shellfish, with some materials being passed through the food chain. Many of the species most at risk such as inshore fish and shellfish, are commercially fished for human consumption. Accumulation can be either direct by feeding on contaminated organic matter or indirect due to species feeding on organisms which already contain high concentrations of contaminants. Sub-lethal effects of heavy metals on the marine biota have been reviewed by Jones (1978). Primary productivity can be suppressed by quite low levels of heavy metals (Bernhard and Zattera 1975), although larval stages of marine animals are generally more vulnerable than the adults (Reish 1973). Work done by McIntyre (1975) has demonstrated that copper can be passed through the marine food chain with adverse effects noted at metal concentrations of just 2 or 3 times above the normal background levels, which is two orders of magnitude less than the LD_{50} values for copper on the species involved. The concentration of toxic substances, including metals and pesticides accumulated in fish, are monitored annually. Fish and shellfish around the UK shoreline have shown elevated levels of metals, especially mercury, from areas used for sludge disposal (Ministry of Agriculture, Fisheries and Food 1971). However, the level of mercury in cod caught in the Thames estuary has declined significantly, in line with the reduction of the quantity of the metal dumped at this site by Thames Water Authority.

The persistent organic contaminants in sludge, organochlorine pesticides, and PCBs, constitute a much greater potential threat to fish and human health than the other contaminants found in sludge, because their high persistence and toxicity. Although they are not abundant in sludge they are present in measurable concentrations in sewage sludge (Table 5.30), providing another route for these highly dangerous compounds to enter the environment. Levels in fish remain low enough not to be a hazard, although the concentration of both total DDT, which includes DDT, DDE, and DDD, and PCBs in cod caught in the Thames, Liverpool Bay, and South Bight dumping areas, are all significantly higher than concentrations found in deep sea fish.

The fate of pathogenic micro-organisms associated with sewage and sewage

TABLE 5.30 Breakdown of the major constituents (mg kg^{-1}DS) of sewage sludge from Manchester and Salford (Gould 1976)

Total P	16 000
Total N	30 000
Anionic surfactant	9500
Mineral oil	38 000
Metals	
Zn	2900
Cu	1900
Pb	530
Cd	56
Hg	37
Ni	500
Cr	930
Fe	22 000
Mn	650
Chlorinated solvents	
1,2-Dichlorobenzene	78
1,2-Trichlorobenzene	600
Chloroform	2.6
1,1,1-Trichloroethane	0.5
Carbon tetrachloride	0.3
Trichloroethylene	2.7
Tetrachloroethylene	2.2
PCBs as Arochlor 1254	5.1
Organochlorine pesticides	
BHC	0.1
DDE	<0.3
TDE	<0.3
DDT	<0.3

sludge in the marine environment is examined in Section 6.2.4. The contamination of bathing beaches is largely due to sewage outfalls with no evidence of such contamination arising from sludge dumping from vessels. However, the contamination of commercial fish and, in particular, shellfish has been closely associated with sludge dumping operations (Section 6.2.4.1). As long as good cleansing and heat treatment is practised then all the potentially hazardous bacteria can be removed from shellfish prior to consumption. However, the fate of viruses in the sea from dumping and other discharges is still relatively unknown and occasional food poisoning outbreaks still occur with the causative agents being viruses associated with contaminated seafoods (Anon 1985b). A major study carried out at two eastern American dumping sites in the mid-Atlantic region, the New York Bight, and the Philadelphia dump site, showed that viruses can survive for up to 17 months in the sediments. Also, that viruses could be isolated in the absence of faecal indicator bacteria, reinforcing the inadequacy of such bacteria in predicting the virological quality of water. Human enteric viruses were not only detected in the water and sediments at these sites but also in crabs (Goyal *et al*. 1984).

Monitoring the dispersal of organic material and contaminants is difficult at sludge dumping sites because of sea area and the rapid dilution and dispersion that normally occurs. Monitoring the dispersion of sludge can only be effectively done using radioactive tracers, although other techniques have been proposed, including characterization of the origin of organic material using the carbohydrate to organic ratio (Hatcher and Keister 1976), the use of faecal bacteria (Ayres 1977), and the use of tomato pips as an indicator of sludge-contaminated sediments (Shelton 1973). Many biological and biochemical methods are used to investigate the effects of pollutants on the fauna at dumping sites. Davis *et al*. (1985) used mussels in cages suspended from buoys at dumping sites and examined the digestive glands after exposure, for cellular damage caused by elevated metal levels. This method, and a range of other cellular and physiological indices, such as the lysosomal test (Moore 1980, 1985) and the 'scope for growth' test (Widdows 1985), are compared in an assessment of the biological effects of sludge dumping at the Plymouth sites by Lack and Johnson (1985).

Further reading

General: Calcutt and Moss 1984; McGlashan 1983.
Legislative control: Norton 1980; Oslo and Paris Commissions 1984.
Dumping sites: Norton 1978; Whitelaw and Andrews 1988.
Environmental impact: Head 1980; Watling *et al*. 1974; Wilson 1988; Whitelaw and Andrews 1988.

6

PUBLIC HEALTH

6.1 DISEASE AND WATER

The term pathogenic is applied to those organisms that either produce or are involved in the production of a disease. This direct and indirect action of pathogens allows water-related diseases to be classified into one of the following four categories (Bradley 1974).

Water-borne diseases are enteric diseases caused by organisms that are excreted in large numbers by infected persons and the route of infection is normally by oral ingestion. The diseases are usually transmitted when pathogens in water are drunk by an animal or a human which may then become infected. The classical water-borne diseases are mainly low-infective dose infections, such as cholera, typhoid, and leptospirosis. All the remaining diseases are high-infective dose infections and include infectious hepatitis and bacillary dysentery. All water-borne diseases can also be transmitted by other routes that permit faecal material to be ingested. For example, by faecal-oral contact via contaminated food, the most notorious case being that of the Irish cook who worked in North America in the late-nineteenth century and was known affectionately as 'Typhoid Mary' (Table 6.1).

Water-washed diseases are caused by a lack of personal hygiene because of water scarcity. The incidence of all these diseases will fall if adequate supplies

TABLE 6.1 The water-borne pathogens associated with wastewater important in European temperate areas (adapted Feachem 1977)

Group	Genus or species	Disease or effects on human health	Notes on distribution if restrictive
Viruses	Poliovirus	Fever, poliomyelitus enteritus	
	Coxsackievirus A	Headache, muscular pain	
	B	Nausea, meningitus	
	Echovirus	Diarrhoea, hepatitis	
	Adenovirus	Fever, respiratory infection, enteritis, conjunctivitis, involvement of the central nervous system	

TABLE 6.1—*Continued*

Group	Genus or species	Disease or effects on human health	Notes on distribution if restrictive
Viruses	Reovirus	Common cold, respiratory tract infections, diarrhoea, hepatitis	
	Hepatitis A	Infectious, hepatitus (fever, nausea, jaundice)	
Bacteria	*Salmonella* spp.	Typhoid fever, parahtyphoid fever, bacterial enteritis, Salmonellosis, food poisoning	
	Campylobacter spp.	Campylobacter enteritis, gastro-enteritis, acute diarrhoea, food poisoning	
	Shigella spp.	Bacillary dysentry	
	Escherichia spp.	Enteritis (pathogenic strains)	
	Vibrio cholera	Cholera, enteritis, food poisoning	Not established in Australia, New Zealand, Pacific Islands or Americas
	Leptospira spp.	Leptospiirosis	
	Mycobacterium spp.	Tuberculosis, skin granuloma	
	Clostridium spp.	Gas gangrene, tetanus, botulism, food poisoning	
	Brucella tularensis	Tularaemia	Mainly in N. America, Europe, USSR, and Japan
Protozoa	*Entamoeba histolytica*	Amoebic dysentry	
	Giardia intestinalis	Giardiasis	Epidemic from water supply reported in Colorado
	Balantidium coli	Balantidiasis	Epidemics reported in Brazil, Georgia, and USSR
Nematodes	*Ascaris lumbricoides*	Ascariasis (roundworm infestation)	
	Anchylostomum duodenale	Hookworm infestation	
Cestode	*Taenia* spp.	Tapeworm infestation	

of washing water, regardless of microbial quality, are provided. These are diseases of mainly tropical areas and include infections of the intestinal tract, the skin, and the eyes. The intestinal infections are all faecal in origin and include all the water-borne diseases that are contracted because of poor personal hygiene. There is growing evidence that Shigellosis is linked more closely to personal hygiene than water quality (Feachem 1977). Most of the intestinal infections are diarrhoeal diseases responsible for the high mortality rates among infants in hot climates. The infections of the skin and mucous membranes are non-faecal in origin and include bacterial skin sepsis, scabies, and cutaneous fungal infections (such as ringworm). Diseases spread by fleas, ticks, and lice are also included in this category, such as epidemic typhus, rickettsial typhus, and louse-borne fever.

Water-based diseases are caused by pathogens that have a complex life-cycle and which require an intermediate aquatic host. All these diseases are caused by parasitic worms with the severity of the infection depending on the number of worms infesting the host. The two commonest water-based diseases are Schistosomiasis carried by the trematode *Schistosoma* spp. and Guinea worm which is the nematode *Dracunculus medimensis*. *Schistosoma* worms use aquatic snails as intermediate hosts and are estimated as infecting as many as 200 million people, and the Guinea worm uses the small crustacean *Cyclops* spp. as its intermediate host.

Water-related diseases are caused by pathogens carried by insects that act as mechanical vectors and which live near water. All these diseases are very severe and control of the insect vectors is extremely difficult. The most important water-related diseases include two viral diseases, yellow fever, transmitted by the mosquito *Aedes* spp. and dengue, which is carried by the mosquito *Aedes aegypti* which breeds in water. Gambian sleeping sickness, trypanosomiasis, is caused by a protozoan transmitted by the riverine testse fly (*Glossina* spp.) which bites near water, and malaria is caused by another protozoan, (*Plasmodium* sp.) and is transmitted by the mosquito *Anopheles* spp. which breeds in water.

This chapter deals exclusively with water-borne diseases, caused by pathogens present in sewage from temperate regions of Western Europe, which, in terms of wastewater treatment, is the most important category of water-associated diseases. However, water-washed diseases are also important to those involved with the treatment, disposal, and re-use of wastewater, including farm wastes, treated effluents, and sewage sludge. A number of diseases of the skin or respiratory tract are directly related to contact with, or close-proximity to, contaminated wastes (Sections 6.2.2 and 6.2.4).

As contamination of water by faeces from people and animals suffering from enteric diseases is the cause of water-borne disease, the introduction of wastewater treatment to remove these pathogens and the disinfection of water supplies has led to the elimination of all the classical diseases. The development of the public health authorities with specialist inspectors and laboratories, linked with better sanitary conditions and a greater awareness of personal hygiene, has resulted in all the water-borne and water-associated diseases being brought under control throughout the developed world.

6.2 WATER-BORNE DISEASES

6.2.1 Introduction

Industrial wastewaters rarely contain pathogens, whereas pathogens are common in food-processing wastes. The pathogens found are directly related to the original plant or animal materials being processed. For example, in the case of potato processing, the effluent contains plant pathogens mainly specific to potatoes, such as the potato cyst eelworm (*Heterodera rostochiensis* or *H. pallida*) and a variety of other viruses, fungi, and bacteria all pathogenic to potatoes. In sewage, it is the diseases excreted by man via faeces and urine which are of primary importance to public health. The numbers and diversity of potential pathogens in sewage reflects the standards and the socio-economic levels of the community. For example, fewer pathogens are present in sewage from industrialized countries having high standards of living compared with those countries with little industrialization and a correspondingly lower standard of living.

Pathogens in sewage are able to infect man and animals by oral ingestion, via the skin or by respiratory routes. The commonest infection route for pathogens is oral ingestion, which generally causes gastro-enteric disorders. The pathogen multiplies and is excreted with the faeces and subsequently ingested by the next host via sewage-contaminated water or food (Table 6.1).

6.2.2 The water-borne pathogens

The most important water-borne pathogens associated with wastewater treatment are summarized in Table 6.1. Of these, the most important in terms of frequency of isolation in sewage-contaminated water, in sewage or sludge are strains of *Salmonella*, *Shigella*, *Leptospira*, enteropathogenic *Escherichia coli*, *Francisella*, *Vibrio*, *Mycobacterium*, human enteric viruses, cysts of *Entamoeba histolytica*, or other pathogenic protozoans, and larvae of various pathogenic worms. All these organisms are well characterized and have been fully reviewed elsewhere (Dart and Stretton 1977; Carrington 1980a). Therefore, only a brief summary of the major organisms is given below.

(Details of potential public health risks from pathogenic organisms in wastewater are discussed in Section 6.2.4.)

6.2.2.1   Bacteria

Salmonellosis
The genus *Salmonella* is now probably the most important group of bacteria affecting the public health of humans and farm animals in Western Europe. For humans, this is undoubtedly due to the elimination of the other classical bacterial diseases through better sanitation, higher living standards, and the widespread availability of antibiotic treatment. The greater awareness of animal hygiene linked with new drug thereapy has also largely eradicated many of the formally important diseases, such as bovine tuberculosis and brucellosis. *Salmonella* species are extremely widespread in nature being recorded as pathogens not only of man but of nearly all known animals. By reacting a *Salmonella* strain in the agglutination reaction with the antibodies produced in the serum of animals inoculated with standard strains of *Salmonella*, some 1800 subtypes or serotypes can be identified (Carrington 1980*b*). Some serotypes are largely specific to a single host, and these include the typhoid organism *S. typhi*, specific to man; *S. dublin*, specific for cattle; *S. abortus-ovis*, specific for sheep; and *S. cholerae-suis*, specific for pigs. Other serotypes are not host-specific, for example, *S. typhimurium* which can infect a wide range of animals, including man (Table 6.2).

Typical symptoms of salmonellosis are acute gastro-enteritis with diarrhoea, and is often associated with abdominal cramps, fever, nausea, vomiting, headache, and in severe cases, even collapse and possible death. In pregnant animals, abortion may occur. In comparison with farm animals, the incidence of salmonellosis in humans is low and, interestingly, shows distinct seasonal variation. A large number of serotypes are pathogenic to man and their frequency of occurrence varies annually from country to country (Table 6.3). Low-level contamination rarely results in the disease developing, because

TABLE 6.2   Number of incidents of *Salmonella* infection in animals caused by various serotypes in England and Wales during 1968–74 (Carrington 1980*b*)

Serotype	Cattle	Poultry and other birds	Sheep	Pigs	Other species	Total
S. dublin	15 446	48	302	78	55	15 929
S. typhimurium	3785	732	59	97	169	4842
S. abortus-ovis	0	0	243	0	0	243
S. cholerae-suis	1	0	3	309	1	314
S. gallinarum	0	44	0	0	0	44
S. pullorum	0	65	0	0	0	65
Other serotypes	1094	855	68	74	81	2172
Total incidents	20 326	1744	675	558	306	23 609

TABLE 6.3 The most common *Salmonella* serotypes pathogenic to humans isolated from the UK, 1965, compared to those isolated in Denmark during 1960–68 (Dart and Stretton 1977)

UK	Denmark
S. typhimurium	S. typhimurium
S. heidelberg	S. paratyphi B
S. newport	S. enteritidis
S. infantis	S. newport
S. enteritidis	S. typhi
S. saint-paul	S. infantis
S. typhi	S. indiana
S. derby	S. montevideo
S. oranienberg	S. blockley
S. thompson	S. muenchen

between 10^5–10^7 organisms have to be ingested before development. Once infection has taken place then large numbers of the organisms are excreted in the faeces ($> 10^8 g^{-1}$). Infection can also result in a symptomless carrier-state, in which the organism rapidly develops at localized sites of chronic infection, such as the gall bladder or uterus, and is excreted in the faeces or other secretions. Carriers comprise of between 1–4% of the human population depending on the country of residence, although the number of persons excreting *Salmonella* at any one time is never exactly known. Estimates for the USA and UK put the number of carriers at less than 1%, whereas in Sri Lanka it is nearly 4% (Dart and Stretton 1977). In the UK, the number of reported human cases varies annually with 5564 being reported in 1971 (Lee 1974).

The main source of infection for man is by eating infected food, particularly meat and milk, which has been contaminated during production and subsequently carelessly prepared and stored. The infection pathways of *Salmonella* are more fully explored in Section 6.2.4. Since the last major outbreak of typhoid in Britain, which occurred in Croydon, Surrey, during the autumn of 1937 when 341 cases were reported resulting in over 40 deaths, there have been five minor outbreaks of typhoid and three of paratyphoid fever. The number of reported cases of typhoid fever in the UK has fallen to less than 200 per annum, with 85% of cases contracted abroad. Of the remainder, few are the result of contaminated drinking water (National Water Council 1975). Although *Salmonella typhi* has been recorded from surface waters from around the British Isles (Public Health Laboratory Service 1978). There are less than 100 reported cases of paratyphoid fever reported annually (Galbraith *et al.* 1987). Typhoid has also largely been eliminated from the USA, although in 1973 there was an outbreak in Dade County in which 225 people contracted the disease from contaminated well water.

The incidence of *Salmonella* infection is much higher in farm animals than in humans. Table 6.2 summarizes the frequency of infection by the main serotypes involved in animal salmonellosis. Cattle (86%) and poultry (7.4%) are the most 'at-risk' groups of animals, with *S. dublin* responsible for 67%, and *S. typhimurium* for 21% of all reported cases. *Salmonella* can be isolated from perfectly healthy farm animals with 13.4% of pigs, 13.0% of cattle, and 15.0% of sheep being symptomless carriers (Grunnet and Nielsen 1969; Prost and Riemann 1967). Other commonly isolated serotypes of cattle include, *S. derby, S. oranienburg, S. java, S. anatum, S. infantis, S. abony, S. neurington, S. stanley, S. meleangridis*, and *S. chester* (Richardson 1975; Dart and Stretton 1977). Manufactured animal feed that has not been subject to pasteurization has been widely implicated in the transmission of animal salmonellosis (Skovgaard and Nielsen 1972; Williams 1975). Associated human salmonellosis has also been reported (Richardson 1975), the most recent case being the contamination of eggs by *S. enteritidis* in 1988. Due to the importation of contaminated protein used for animal feed, there has been a steady increase in the incidence of 'exotic' serotypes of *Salmonella*.

Campylobacter
Campylobacter are spiral curved bacteria 2000–5000 nm in length and comprise 2–6 coils. Although discovered in the late-nineteenth century, they were not isolated from diarrhoeic stool specimens until 1972. It is only since the development of a highly selective solid growth medium, allowing culture of the bacterium, in 1977, that its nature has been revealed (Kist 1985). Campylobacter species have been isolated from both fresh and estuarine waters (Pearson *et al.* 1977; Pearson *et al.* 1985), with counts ranging from 10 to 230 campylobacters per 100 ml in rivers in North-West England (Bolton *et al.* 1982, 1985). Although the epidemiology of human campylobacter infections remains to be fully elucidated, certain sources of infection are well established (Skirrow 1982). There were 27 000 reported outbreaks of campylobacter enteritis in the UK during 1987, causing severe acute diarrhoea, and it is now thought that campylobacter is the major cause of gastro-enteritis in Europe, being more common than *Salmonella*. (Andersson and Stenstrom 1986). The most important reservoirs of the bacterium are meat, in particular poultry (Brouwer *et al.* 1979; Skirrow 1982), and unpasteurized milk (Robinson and Jones 1981; Bates *et al.* 1984; Boer *et al.* 1984; Wright *et al.* 1983). Household pets, farm animals, and birds are also know to be carriers of the disease (Hill and Grimes 1984; Fox *et al.* 1983, 1981; Sticht-Groh 1982). Unchlorinated water supplies have been identified as a major source of infection. For example, 3000 of a total population of 10 000 developed campylobacter enteritis from an inadequately chlorinated mains supply in Bennington, Vermont (Vogt *et al* 1982); and 2000 people who drank unchlorinated mains water contaminated with faecally polluted river water in Sweden also contracted the disease (Mentzing 1981). Other incidents have

been reported by Rogol *et al.* (1983) and Taylor *et al.* (1983). Water is either contaminated directly by sewage, which is rich in camplylobacter (Marcola *et al.* 1981), or indirectly from animal faeces. Bolton *et al.* (1985) obtained a definite seasonal variation in numbers of campylobacters in river water, with greatest numbers occurring in the autumn and winter. Serotyping of isolates confirmed that *C. jejuni* serotypes common in human infections were especially common downstream of sewage effluent sites. This confirmed that sewage effluents are important sources of *C. jejuni* in the aquatic environment. Gulls are known carriers and can contaminate water supply reservoirs while they roost. Dog faeces, in particular, are rich in the bacterium (Blaser *et al.* 1978; Svedhem and Norkrans 1980). Wright (1982) isolated *C. jejuni* from 4.6% of 260 specimens of dog faeces sampled, whereas *Salmonella* spp. were isolated from only 1.2%. The incidence of *C. jejuni* is low compared to other studies on dog faeces, with infection rates ranging from 7–49% (Bruce *et al.* 1980; Holt 1980). Dog faeces can cause contamination of surface waters during storms, as surface runoff removes contaminated material from paved areas and roads. An outbreak affecting 50% of a rural community based in northern Norway was identified to contaminated faecal deposits from sheep grazing the banks of a small lake that were washed into the water during a heavy storm and which melted the snow on the banks. The water supply for the village came directly from the lake without being chlorinated (Gondrosen *et al.* 1985).

Natural aquatic systems in temperate areas are usually cool, and research has shown that campylobacters can remain viable for extended periods in streams and groundwaters. Rollins *et al.* (1985) found that survival decreased with increasing temperature but, that at 4°C, survival in excess of 12 months was possible. The incidence of campylobacter in water can be estimated by an MPN technique developed by Bolton *et al.* (1982). The topic of campylobacter infections has been reviewed by Skirrow (1982).

Shigellosis
Shigella causes bacterial dysentery and is the most frequently diagnosed cause of diarrhoea in the USA, where it accounted for 19% of all the cases of water-borne diseases reported in 1973 (Hughes *et al.* 1975). The genus is similar in its epidemiology to *Salmonella*, except that it rarely infects animals and does not survive quite so well in the environment. When the disease is present as an epidemic it appears to be spread mainly by person-to-person contact, especially between children, as shigellosis is a typical institutional disease occurring in overcrowded conditions. However, there has been a significant increase in the number of outbreaks arising from poor quality drinking water contaminated by sewage. Of the large number of serotypes (> 40), *Shig. sonnei* and *Shig. flexneri* account for >90% of isolates. The number of people excreting *Shigella* are estimated as 0.46% of the population in the USA, 0.33% in Britain, and 2.4% in Sri Lanka (Dart and Stretton 1977). In England and

Wales notifications of the disease rose to between 30 000 to 50 000 per year during the 1950's, falling to < 3000 per annum in the 1970s. However, in the 1980s, notifications have doubled to nearly 7000 per annum (Galbraith *et al.* 1987).

Leptospirosis
All the species of the genus *Leptospira* are pathogenic to man except for *L. biflexa*, which is a heterotroph found in rivers and lakes. Although there are over 100 serotypes known, it is *L. icterohaemorrhagia* that causes Leptospira jaundice or Weils disease. This is probably the most widely known form of the disease as it is transmitted to man via infected rats. The bacteria are characterized by being motile and comprised of very fine spirals wound so tightly that they are barely distinguishable under the microscope (Dart and Stretton 1977). Entry to the body is via minor cuts or the mucous membranes where they enter the blood stream and affect the kidneys, liver, and central nervous sytem.

Infection is widespread in domestic and wild animals, especially rats, and the bacteria can be transferred to man either directly or indirectly via water into which infected animals have urinated. In Israel, 2.3% of cattle and 27% of dogs are carriers, although the level of infection in the human population is only 0.37% (Torten *et al.* 1970). The disease occurs throughout Europe, North and South America, the Middle and Far East. However, in general, the incidence of leptospirosis reported is low (< 1%) except for those in high-risk occupations. Those at high-risk include persons handling animals, involved with meat-processing, or who are in contact with sewage or polluted waters, including fish farms, where the contamination occurs from the urine of infected rats. In these groups, the annual rate of infection can be as high as 3%, and up to 20% have a positive antibody reponse to the organism. *Leptospira* survives best in cold waters with low levels of organic contamination and it is from these situations that the low-risk category people are infected. It would appear, from a number of studies, that the majority of low-risk category people become infected when swimming in water heavily contaminated by the excreta of infected animals. For example, in The Netherlands, people who had swum in a canal contaminated by animal effluent became infected. During 1933–48, 5% of the 983 reported cases of leptospirosis in the UK were associated with bathing, accidential immersion in water, or water sports. During the six years from 1978 to 1983, the proportion of cases associated with the recreational use of water or accidental immersion increased to 25%, of a total of 177 reported cases. These were mainly bathers in freshwater ponds and streams and those engaged in water sports, such as canoeing (Waitkins 1985) (Table 6.4). Since 1986 there has been a four-fold increase in reported cases of Leptospirosis in the UK, reaching 130 cases in 1988. Nineteen of these died, thirteen having contracted the disease during water sports. The increase in the incidence has been linked to (a) the explosion in the population density of rats

TABLE 6.4   Most frequently reported diseases, organism responsible, and incidence of infection, from the recreational use of water in the UK from 1937–86 (Galbraith *et al.* 1987)

Disease	Organism	Incidence
Skin infections	*Pseudomonas aeruginosa*	Outbreaks associated with swimming pools and whirlpools, probably common
	Mycobacterium marinum	Probably about 100 cases since the 1960s
Conjunctivitis	Various	Not known, probably common
Gastro-intestinal infections	*S. typhi*	
	(i) contact with sewage polluted river water	12 + cases
	(ii) drinking polluted river water	61 + cases
	(iii) sea bathing	2 cases
	S. paratyphi	
	(i) contact with sewage polluted river water	12 cases
	(ii) sea bathing	7 cases
	Campylobacter	One outbreak of 4 cases
	Non-specific	One outbreak of 21 cases
Respiratory infections	Adenovirus	Not known
	Mycobacteria	5 reported infections
	Legionella	One outbreak, 26 cases of legionnaires' disease and 7 of Pontiac fever
Primary amoebic meningoencephalitis	*Naeglaria fowleri*	6 cases
Leptospirosis	*Leptospira icterohaemorrhagia*	45 cases, 1978–83 About 200, 1937–86

and (b) to the increasing popularity of all water sports and of windsurfing, on both inland and coastal waters, in particular.

Enteropathogenic Escherichia coli
There are 14 distinct serotypes of *Escherichia coli* that cause gastro-enteritis in man and animals, being especially serious in the new-born and in children under five years of age. It is common throughout Europe and is also thought to be the cause of 'traveller's tummy', the bout of diarrhoea that affects so many tourists who visit the warmer areas of Europe. The symptoms are profuse watery diarrhoea, with little mucus, nausea, and dehydration. The disease does not cause any fever and is rarely serious in adults. Up to 2.4% of children in England and Wales are thought to be carriers, although much higher percentages are found in people engaged in high-risk occupations, such as food handling. Enteropathogenic *E. coli* is commonly isolated from sewage but probably represents less than 1% of the total coliforms present in polluted

waters (Table 6.5). However, only 100 organisms are required to cause illness. Survival of the organism is the same as for other serotypes of *E. coli*, and in warm nurient-rich conditions they are able to multiply in water. The outbreak of gastro-enteritis in Worcester in the winter of 1965–66 affected 30 000 people, was thought to be due to contamination of the water supply because of flooding. An incident in north-east Leeds in July 1980, caused by a leaking sewer contaminating a borehole in a limestone aquifer, affected 3000 people.

TABLE 6.5 The level of *E. coli* found in various animals and birds, compared with sewage and sewage effluents (Jones and White 1984)

	Faecal production $\mathrm{g\ d^{-1}}$	Average number of *E. Coli* per g faeces	Daily load *E. coli*
Man	150	13×10^6	1.9×10^9
Cow	23 600	0.23×10^6	5.4×10^9
Pig	2700	3.3×10^6	8.9×10^9
Sheep	1130	16×10^6	18.1×10^9
Duck	336	33×10^6	11.1×10^9
Turkey	448	0.3×10^6	0.13×10^9
Chicken	182	1.3×10^6	0.24×10^9
Gull	15.3	131.2×10^6	2.0×10^9

	E. coli concentration 100 ml^{-1}
Sewage	$3.4 \times 10^5 - 2.8 \times 10^7$
Sewage effluent	$1 \times 10^3 - 10^7$

E. coli survival in freshwater = mean T_{90} 62.3 h
E. coli survival in seawater = mean T_{90} 2.3 h

The failure of the chlorination system to deal with this pollution was traced to complaints from consumers close to the borehole who found the water too chlorinous to the taste, so the chlorine dose had been kept too low to deal with the resultant pollution (Short 1988). More recently, the water tanks on a cruise liner became contaminated by sewage resulting in an outbreak of the disease that affected 251 passengers and 51 crew (O'Mahony *et al.* 1986).

Tularaemia
This disease is only endemic in North-west America, although cases have also been reported in the USSR and Czechoslovakia (Dart and Stretton 1977). The bacterium *Franciscella tularensis* enters the body via abrasions and mucous membranes, leaving an ulcer at the site of contact. Infection results in progressively worsening symptoms: chills and fever, swollen lymph nodes, and general malaise. Without treatment, patients suffer delirium coma and possibly death. Most cases occur in people who have handled infected wild animals, especially rodents and rabbits, and in the USA ground squirrels are the natural vector, although contact with wood ticks can also result in infection (Bow and Brown 1946). The disease cannot be transmitted from

person-to-person and multiple infection, such as the case in the Soviet Army reported by Schmidt (1948), is generally caused by contact with, or drinking water contaminated by, urine, faeces, or the corpses of infected animals.

Cholera

Cholera is thought to have originated in the Far East, where it has been endemic in India for many centuries. In the nineteenth century, the disease spread throughout Europe where it was eventually eliminated by the development of uncontaminated water supplies, water treatment, and better sanitation. It is still endemic in many areas of the world especially those which do not have adequate sanitation and, in particular, situations where the water supplies are continuously contaminated by sewage. However, over the past 10 to 15 years the incidence and spread of the disease has been causing concern that has been linked to the increasing mobility of travellers and the speed of travel. Healthy, symptomless carriers of *Vibrio cholerae* are estimated to range from 1.9% to 9.0% of the population (Pollitzer 1959). However, this estimate is now thought to be rather low, with a haemolytic strain of the disease reported as being present in up to 25% of the population (Yen 1965). The holiday exodus of Europeans to the Far East, which has been steadily increasing since the mid-1960s, will have led to an increase in the number of carriers in their countries and an increased risk of contamination and spread of the disease. There have been nearly 50 reported cases of cholera in the UK between 1970–86, although no known cases have been water-borne (Galbraith *et al.* 1987).

Up to 10^8–10^9 organisms are required to cause the illness, and cholera is not normally spread by person-to-person contact. It is readily transmitted by drinking contaminated water or by eating food handled by a carrier or that has been washed with contaminated water. It is regularly isolated from surface waters in the UK (Lee *et al.* 1982). It is an intestinal disease with characteristic symptoms: sudden diarrhoea with copious watery faeces, vomiting, suppression of urine, rapid dehydration, lowered temperature and blood pressure, and complete collapse. Without therapy, the disease has a 60% mortality rate, and the patient dies within a few hours of first showing the symptoms. However, with suitable treatment, the mortality rate can be reduced to < 1% (Dart and Stretton 1977).

The bacteria are rapidly inactivated under unfavourable conditions, such as high acidity or high organic matter content of the water, although in cool, unpolluted waters *Vibrio cholerae* will survive for up to two weeks (Mara 1974).

Tuberculosis

Mycobacterium tuberculosis, *M. balnei* (*marinum*), and *M. bovis* all cause pulmonary tuberculosis. Infection is by inhalation or ingestion of bacilli released in the sputum, milk or other discharges, including the faeces, of infected animals. The source of infection is difficult to identify as the disease has a very long incubation period before clinical tuberculosis is diagnosed, which in some cases may be many years. However, *M. tuberculosis* is frequently isolated in wastewater from hospitals and meat-processing plants. Like *Leptospira*, the bacilli are able to survive for several weeks at low temperatures in water contaminated with organic matter. Clearly, drinking contaminated water must be a source of infection and there is considerable circumstantial evidence to support this.

Brucellosis

Brucellosis, or undulant fever, is rare in man but is common in cattle where it is known as contagious abortion. It is an extremely serious disease in cattle and can result in economic ruin for farmers whose herds become infected. The bacteria are excreted from cattle but not by humans and, therefore, *Brucella* is rarely found in sewage but is common in wastewaters from milk parlours and yards where infected cattle are kept.

6.2.2.2   Viruses

Infectious hepatitis, enteroviruses, reovirus, and adenovirus are all thought to be transmitted via water. Of most concern in Britain is viral hepatitis. There are three subgroups, hepatitis A which is transmitted by water, hepatitis B which is spread by personal contact or inoculation and is endemic in certain countries, such as Greece (Papaevangelou *et al.* 1976), and hepatitis C which is a non-A or B type hepatitis virus. Hepatitis A is spread by faecal contamination of food, drinking water, and areas that are used for bathing and swimming. Epidemics have been linked to all these sources, and it appears that swimming pools and coastal areas used for bathing that receive large quantities of sewage are particular sources of infection. The virus cannot be cultivated *in vitro* and studies are confined to actual outbreaks of the disease.

All the enteroviruses, poliomyelitis virus, coxsackievirus, and echovirus, cause respiratory infections and are present in the faeces of infected people. Poliomyelitis, in particular, is common in British sewage, but this is due to the vaccination programme and does not indicate actual infection (Carrington 1980a). Reovirus is thought to be associated with gastro-enteritis, and adenovirus type 3 is associated with swimming pools and causes pharyngo-conjunctival fever.

Warm-blooded animals appear able to carry viruses pathogenic to man. For example, 10% of beagles have been shown to carry human enteric viruses (Lundgren *et al.* 1968). Therefore, there appears a danger of infection from waters not only contaminated by sewage but by other sources of pollution, especially stormwater. Most viruses are able to remain viable for several weeks in water at low temperatures as long as there is some organic matter present.

6.2.2.3 Protozoa

Protozoan pathogens of man are almost exclusively confined to tropical and sub-tropical areas. However, with the increase in travel, carriers of the diseases are found world-wide and cysts of all the major protozoan pathogens occur in European sewages from time to time. All the diseases are transmitted by cysts entering water supplies via faecal contamination or by the host swimming in contaminated water.

The most important protozoan parasite conveyed by water is *Entamoeba histolytica* which causes amoebiasis or amoebic dysentery. The number of carriers of this disease are in the order of 10% of the population in Europe, 12% in the USA, 16% in Asia, and 17% in Africa. The highest rate of incidence occurs in those groups with unprotected water supplies, inadequate waste disposal facilities, and poor personal hygiene. As with so many other diseases, the incidence among high-risk occupations is much higher reaching 14% in sewage workers. The number of cysts in sewage is generally low, $< 5 \, 1^{-1}$ in raw sewage, although because of their poor settleability (Section 6.3.2) they often pass through treatment plants. The infection is centred in the large intestine and the symptoms include mid-abdominal pain, diarrhoea alternating with constipation, or chronic dysentery with a discharge of mucus and blood.

Giardia lambia, an intestinal protozoan causing giardiasis, an acute diarrhoeal illness, has a world-wide distribution. Transmission of the flagellate protozoan is by ingestion of cysts in contaminated water or food, and is often associated with salad crops irrigated with sewage. It is particularly common in the USA, where it is now considered to be endemic with a carrier rate of 15–20% of the population, depending on their socio-economic status, age, and location (Lin 1985; Craun 1979). The number of reported cases in England and Wales has risen from 1000 per annum in the late 1960s to over 5000 per year by the late 1980s, although these are almost exclusively associated with travel. There was, however, a significant outbreak of giardiasis in south Bristol (England) in the summer of 1985, when 108 cases were reported (Jephcote *et al.* 1986; Browning and Ives 1987).

Another protozoan found in sewage effluents is *Naegteria fowleri*. This causes the fatal disease amoebic meningo-encephalitis. The cysts enter the body through the nasal cavities and rapidly migrate to the brain, cerebrospinal fluid, and blood stream. Cases are rare but are exclusively associated with swimming in warm contaminated water, especially hot springs. Since it was first recognized in 1965 there have only been six cases reported in the UK. The last reported case was in 1978, an 11-year-old girl, who had been swimming in a municipal swimming pool filled with water from a natural hot spring in the City of Bath (Galbraith 1980).

6.2.2.4   Parasitic worms

The incidence of worm infection in European countries is generally low and is limited to *Taenia* and *Ascaris*. Infection is normally contracted from animals reared for food, although many incidents of infection from contaminated water have been reported. Ova of both genera are commonly isolated from sewage.

Taenia saginata, the beef tapeworm, is generally disseminated by polluted waters. Infected persons can excrete up to 10^6 ova per day, and in raw sewage 20 ova 1^{-1} is common. *Taenia* ova are abundant in meat-processing effluents and extreme caution should be taken if handling the effluent or sampling a river receiving this type of wastewater. *Ascaris* or roundworms also produce large numbers of ova and infected people can excrete $> 200\,000$ ova d^{-1}. Ascariasis is especially associated with young children, although the ova of *A. lumbricoides* may also enter water from soil and vegetables, as well as from faeces of infected people (Dart and Stretton 1977).

There are a number of other parasitic worms that can also be transmitted by faecal contamination of water supplies. These include several hookworms, schistosomes, and a number of tapeworms, such as the fish tapeworm *Diphyllobothrium latum*. Infection by the dog tapeworm (*Echinococcus* and *Canidae* spp.) is normally by direct contact with infected dogs. However, contamination of water supplies by dog faeces and by stormwater from paved areas has contributed to the significant increase in the incidence of this parasite in man. The frequency of occurrence of helminth ova in wastewater, sludge, and soil has been reviewed by Theis and Storm (1978) (Table 6.6).

6.2.3   Indicator organisms

Most countries require by law that drinking water is regularly examined to ensure that it is free from organisms that will cause disease (pathogens) and is therefore safe for consumption. However, acceptable standards vary from country to country (WHO 1971; Hendricks 1978).

The wide diversity of pathogenic micro-organisms found in wastewaters are normally present in low numbers compared with other non-pathogenic organisms that are also present. The isolation and identification of individual pathogenic organisms is often complex, being different for each species and extremely time consuming. It is, therefore, impossible to examine all water samples on a routine basis for the presence and absence of all pathogens. Also, the selection of individual pathogens may be misleading as each species can tolerate different environmental conditions and the presence of one may not indicate the presence of another. In order to routinely examine water supplies a rapid and preferably a single test is required. It is far more important to

TABLE 6.6 Summary of helminth life-cycles (Theis and Storm 1978)

Species	Free-living stages	Infective stage	Larval worm migrates through body of host	Adult worms (location)	Intermediate host(s) required	Normal definitive vertebrate host	Human infestation possible	Able to complete life-cycle in US
Ascaris lumbricoides	No	Ova with larvae *in situ*	Yes	Small intestine	No	Humans	—	Yes
Toxocara	No	Ova with larvae *in situ*	Yes	Small intestine	No	Dogs or cats	Yes	Yes
Toxascaris	No	Ova with larvae *in situ*	No	Small intestine	No	Dogs or cats	No	Yes
Oxyuris equi	No	Ova with larvae *in situ*	No	Colon	No	Horses	No	Yes
Trichuris trichiura	No	Ova with larvae *in situ*	No	Colon Ceacum Rectum	No	Humans	—	Yes
Necator	Yes	Larvae that burrows through skin	Yes	Small intestine	No	Humans	—	Yes warm, moist areas
Strongyloides larvae	Yes, may live as a nonparasite	Larvae that burrows through skin	Yes	Small intestine	No	Human, dogs, sheep	—	Yes warm, moist areas
Strongylus ova	No	Larvae that climb up grass blades and are eaten	Yes	Small intestine	No	Horses	No	Yes
Taenia	No	Cyst in meat of cow or pig	Yes in cow or pig	Small intestine	Yes cow or pig	Humans	—	Yes
Hymenolepis nana	No	Ova with embryo	No	Small intestine	No	Humans	—	Yes
H. diminuta	No	Cyst in tissue of insect	No	Intestine of insect	Yes	Rodents	Yes rare	Yes
Clonorchis	No	Cyst in flesh of fish	Yes in part	Liver	Yes snail & fish	Humans	—	No
Fasciola	No	Cyst on grass stems	Yes in part	Liver	Yes snail	Sheep	Yes uncommon	Yes cool, moist areas

examine a water supply frequently by a simple test, as most cases of contamination of water supplies occur infrequently, than occasionally by a series of more complicated tests (HMSO 1977, 1983b). This has led to the development of the use of indicator organisms to indicate the likelihood of contamination by faeces. The criteria for selection of indicator organisms are: (1) they are present whenever pathogens are present and in much greater numbers; (2) they should be consistently and exclusively associated with the source of pathogens; (3) they must be as resistant or even more resistant to disinfectants, treatment systems and the aqueous environment than pathogenic organisms; and (4) be suitable for routine isolation by growing readily on selective media and be easily identified and enumerated with reasonable accuracy and precision (Cabelli 1979; Lynch and Poole 1979). The non-pathogenic organisms that are always present in the intestine of man and animals are excreted along with the pathogens, but in far greater numbers. Several of these are easily isolated and are ideal for use as indicators of faecal contamination. The most widely used are the non-pathogenic bacteria, in particular coliforms, faecal streptococci, and *Clostridium perfringens*. There is now much interest in other indicator organisms and the ability to rapidly detect minute concentrations of faecal-associated compounds, which are unaffected by environmental factors. This has put the whole use of indicator organisms to test water for faecal contamination under review. There is considerable interest in the development of chemical indicators of faecal pollution and sterols are particularly promising (Murtaugh and Bunch 1967). For example, the faecal sterol coprostanol which is produced in the intestine by the microbial reduction of cholesteral has been shown to be an accurate indicator of faecal pollution (Dutka *et al.* 1974; Hatcher and McGillivary 1979; McCalley 1980; McCalley *et al.* 1981). Coprostanol has been successfully used to trace faecal pollution in the Clyde estuary, where it closely paralleled the faecal coliform counts (Goodfellow *et al.* 1977).

Three groups are normally used to indicate faecal contamination, *Escherichia coli* or faecal coliforms, faecal streptococci, and *Clostridium perfringens*. The three groups are able to survive for different periods of time in the aquatic environment. Faecal streptococci die fairly quickly outside the host and their presence is an indication of recent pollution. *E. coli* can survive for several weeks under ideal conditions and are far more easily detected than the other indicator bacteria. Because of this, it is the most widely used test although the others are often used to confirm faecal pollution if *E. coli* are not detected. *Clostridium perfringens* is a sporulating anaerobe that can exist indefinitely in water. When *E. coli* and faecal streptococci are absent, its presence indicates remote or intermittent pollution. It is especially useful for testing lake water, although the spores settle out of suspension. The spores are more resistant to industrial pollutants than the other indicators and it is especially useful in waters receiving both domestic sewage and industrial wastewaters. It is assumed that these organisms do not grow outside the host and, in general, this is true. However, in tropical regions *E. coli*, in particular, is known to multiply in warm waters and there is increasing evidence that *E. coli* is able to reproduce in enriched waters generally, thus falsely indicating an elevated health risk (Eliasson 1967; Henricks 1972; Dutka 1973, 1979). Therefore, great care must be taken in the interpretation of results from tropical areas, and the use of bacteriological standards designed for temperate climates are inappropriate for those areas (Mara 1974).

To successfully protect water supplies, especially those in rural areas, it is necessary to be able to trace the source of faecal pollution. Standard bacteriological techniques are traditionally used but fail to distinguish between human and animal faecal pollution. There are a number of modifications to traditional methods available such as estimating the ratios of faecal coliforms to faecal streptococci or the ratios of different types of streptococci (section 6.2.3.2). However, the use of specific indicator bacteria are perhaps the most promising.

Duncan Mara and John Oragui of the University of Leeds have suggested that *Rhodococcus coprophilus* and *Streptococcus bovis* should be used for detecting animal faecal pollution, while sorbitol-fermenting bifidobacteria should be used for human faeces. Faecal streptococci and *E. coli* are excreted by both animals and humans, and so cannot be used as a specific indicator. Also, the ratio of faecal coliforms to faecal streptococci is not always reliable. Therefore specific indicators are best. *Streptococcus bovis* is often present in significant numbers in faeces from people living in tropical areas including India and parts of Africa, while *R. coprophilus* is only found in animal faeces. Bifidobacteria have also been isolated in animal dung in low numbers, however the sorbitol-fermenting strains isolated using YN-17 and Human Bifid Sorbitol Agar media appears restricted to humans.

6.2.3.1 Escherichia coli *and coliforms*

Coliform bacteria are part of the Enterobacteriaceae and are defined by Bonde (1977) as conforming to the type genera *Escherichia*, *Citrobacter*, and *Klebsiella* (*enterobacter*). A more specific definition describes coliforms as Gram-negative, oxidase-negative, non-sporing rods capable of growing facultatively on agar medium containing bile salts and of fermenting lactose within 48 h at 37°C, with the production of both acid and gas (HMSO 1977). Microscopically, all the species of Enterobacteriaceae appear as Gram-negative rods of between $2–3 \times 0.5$ μm, and are impossible to distinguish by visual examination. Both the quantitative tests measure all the coliforms present in the water sample, which is known as the total coliform count. However, only *E. coli* is exclusively faecal in origin, being present in fresh faeces in numbers $>10^8 g^{-1}$, whereas the other coliforms are normal inhabitants of soil and water even though they can occur in faeces. Therefore, the presence of coliforms does not imply faecal contamination is present, although in practice, it is assumed that the coliforms are of faecal origin unless proved otherwise. It is important to confirm whether *E. coli* are present and this is done by a series of simple biochemical tests. The tests characterize the major coliforms present and are known by the acronym IMViC which stands for:

I, Indole. The amino acid tryptophan is converted by certain coliforms to indole which accumulates in the medium. The presence of indole is detected colorimetrically with Kovac's reagent.

M, Methyl red. This test detects the fall in pH of the medium due to the production of acids by the mixed acid fermentation pathway.

Vi, Voges–Proskauer. The detection of acetoin (acetylmethylcarbinol) indicates butylene glycol fermentation. Detected colorimetrically with creatine and potassium hydroxide.

C, Citrate. The utilization of citrate as the sole carbon source.

In temperate waters, the ability to produce gas at 44°C by indole-producing organisms is restricted to *E. coli* (Table 6.7), although in warmer climates non-faecal coliforms also show similar heat tolerance (Bonde 1977). Specific identification of coliforms is more complex and requires the use of one of the multi-test differential identification systems (HMSO 1983*b*). Full details of the IMViC and other bacteriological tests are given in the text by Mara (1974) who has used excellent colour photographs to highlight the important colour changes in the tests. Other useful methodologies and sampling techniques are given elsewhere (Bonde 1977; HMSO 1983*b*; American Public Health Association *et al.* 1983). It should be noted that different methods and materials are used in the USA for nearly all the standard bacteriological tests compared with Britain, so care should be taken when selecting methods to check for the preferred methods in the bacteriological standards being followed.

Three methodologies for counting total coliforms are widely used. The multiple tube method, the pour plate method, and the membrane filtration

TABLE 6.7   Identification of coliforms to species using the response to the IMViC tests and the ability to ferment lactose at 44°C

Species	IMViC pattern				
	I	M	V	C	44°C
Escherichia coli	+	+	−	−	+
Citrobacter freundii	−	+	−	+	−
Klebsiella pneumoniae	+	−	+	+	−
Enterobacter cloacae	+	−	+	+	−

technique. Reliable estimates of coliform densities in water and effluent samples can be obtained by any of these techniques. The methods outlined below are those most widely used in the UK and Ireland. However, more recent adaptations of these methods have been proposed by the UK Public Health Laboratory Service (HMSO 1983*b*) and reference should be made to this for specific instructions on conducting the tests.

The multiple tube method
This is the most widely used method of estimating total coliform and *E. coli* densities. It is especially useful for routine analysis of drinking waters where very low coliform counts are expected.

The principle of the test is that various volumes of the sample water are inoculated into a series of tubes containing a medium which is selective for coliform bacteria. From the pattern of positive and negative growth responses an estimate of the number of coliforms and subsequently *E. coli* can be made. The technique is carried out in two discrete stages. The first estimates the number of coliforms present on the assumption that all the tubes which show acid and gas production contain coliform organisms. Because this assumption is made, this is known widely as the presumptive coliform count. The second stage tests for the presence of *E. coli*. The first stage is completed over 48 hours, and the second takes a further 24 hours, the whole test taking three days to complete.

A range of small volumes of the water sample are added to tubes containing the selective medium Fig. 6.1. A range of volumes are used to ensure that the optimum numbers of coliforms are inoculated into the medium. For example, with good quality waters one 50 ml and five 10 ml volumes should be inoculated into the medium. If water of doubtful quality is tested then one 50 ml, five 10 ml, and five 1 ml volumes should be used making 11 tubes for the one sample. If effluents are tested then the water sample will have to be diluted using 0.25% Ringer's solution. With heavily polluted waters, such as raw sewage, dilution of 10^{-2}, 10^{-3}, or even higher may be required to give some negative reactions and thus obtain a finite figure for the most probable number (HMSO 1983*b*). Minerals Modified Glutamate Medium is now used in Britain and has superseded MacConkeys Broth which was made from peptone and bile salts and were complex compounds that varied from batch to batch in their nutrient and inhibitory properties respectively (Mara 1974). The media used for the multiple tube test contains lactose as the principle carbon source and a pH indicator to show if acids are produced from the fermentation of the lactose by coliforms. Minerals Modified Glutamate Medium is sufficiently selective for the coliform group and, unlike MacConkeys Broth, does not include a compound specifically to inhibit non-coliform bacteria. An inverted Durham tube (a small vial) is used to detect the formation of gas, and the inoculated tubes are incubated for 48 h at 37°C. At this temperature, coliforms can ferment lactose producing acid and gas, and any tubes showing a pH

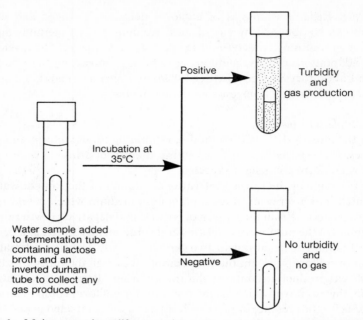

FIG. 6.1 Major steps in coliform testing using the multiple tube method where fermentation tubes containing either lactose or lauryl tryptose broth are used.

reaction and with a bubble of gas trapped in the inverted Durham tube, are considered as positive. The results are expressed as the most probable number (MPN) of presumptive coliforms per 100 ml of water. The pattern of positive and negative tubes in decreasing volume, or increasing dilution, are used to obtain an estimate of the presumptive coliforms using special probability tables with 95% confidence limits (Table 6.8). This allows for the lower and upper limits between which the real density of coliforms in a particular water sample can be expected to occur. A worked example is given in Table 6.9.

The second stage of the multiple tube method is to test the positive tubes in the presumptive coliform test for the presence of *E. coli* (faecal coliforms). A small subsample (<0.01 ml) is taken from each tube that showed a positive reaction in the first stage presumptive coliform test, using a sterile wire loop to inoculate each of two tubes, one containing Tryptone water the other containing either Lactose Ricinoleate Broth or Brilliant Green Bile Lactose Broth. Both tubes are incubated at 44°C, but anaerobic lactose fermenting bacteria can also grow in it at this temperature and cause false results. Therefore, the sodium ricinoleate or brilliant green dye inhibits the growth of these spore-forming anaerobic bacteria, which might have been responsible for false positive reactions in the presumptive test. Only *E. coli* can reduce indole and ferment lactose at this temperature and so *both* tubes must show a positive reaction to confirm its presence. From the pattern of positive and negative tubes the number of *E. coli* or faecal coliforms can be estimated from the MPN tables.

Pour plate method

This technique is often used when counting coliforms in very polluted waters. Small subsamples of water are inoculated directly on to Lactose Teepol Agar.

TABLE 6.8   Tables for the calculation of the most probable number (MPN) of coliform bacteria in 100 ml of original water sample for various combinations of positive and negative results, with 95% confidence limits. *Table A* is used when five 10 ml, five 1 ml, and five 0.1 ml subsamples are used. *Table B* when one 50 ml and five 10 ml subsamples are used, and *Table C* when one 50 ml, five 10 ml, and five 1 ml subsamples are used (HMSO 1983*b*)

Table 6.8A

No. of tubes giving positive reaction out of			MPN per 100 ml	95% confidence limits	
				Lower limit	Upper limit
5 of 10 ml each	5 of 1 ml each	5 of 0.1 ml each			
0	0	1	2	<0.5	7
0	1	0	2	<0.5	7
0	2	0	4	<0.5	11
1	0	0	2	<0.5	7
1	0	1	4	<0.5	11
1	1	0	4	<0.5	11
1	1	1	6	<0.5	15
1	2	0	6	<0.5	15
2	0	0	5	<0.5	13
2	0	1	7	1	17
2	1	0	7	1	17
2	1	1	9	2	21
2	2	0	9	2	21
2	3	0	12	3	28
3	0	0	8	1	19
3	0	1	11	2	25
3	1	0	11	2	25
3	1	1	14	4	34
3	2	0	14	4	34
3	2	1	17	5	46
3	3	0	17	5	46
4	0	0	13	3	31
4	0	1	17	5	46
4	1	0	17	5	46
4	1	1	21	7	63
4	1	2	26	9	78
4	2	0	22	7	67
4	2	1	26	9	78
4	3	0	27	9	80
4	3	1	33	11	93
4	4	0	34	12	96

TABLE 6.8—*Continued*

Table 6.8A

No. of tubes giving positive reaction out of			MPN per 100 ml	95% confidence limits	
5 of 10 ml each	5 of 1 ml each	5 of 0.1 ml each		Lower limit	Upper limit
5	0	0	23	7	70
5	0	1	31	11	89
5	0	2	43	15	114
5	1	0	33	11	93
5	1	1	46	16	120
5	1	2	63	21	154
5	2	0	49	17	126
5	2	1	70	23	168
5	2	2	94	28	219
5	3	0	79	25	187
5	3	1	109	31	253
5	3	2	141	37	343
5	3	3	175	44	503
5	4	0	130	35	302
5	4	1	172	43	436
5	4	2	221	57	698
5	4	3	278	90	849
5	4	4	345	117	999
5	5	0	240	68	754
5	5	1	348	118	1005
5	5	2	542	180	1405
5	5	3	918	303	3222
5	5	4	1609	635	5805

Table 6.8B

No. of tubes giving positive reaction out of		MPN per 100 ml	95% confidence limits	
1 of 50 ml	5 of 10 ml each		Lower limit	Upper limit
0	1	1	<0.5	4
0	2	2	<0.5	6
0	3	4	<0.5	11
0	4	5	1	13
1	0	2	<0.5	6
1	1	3	<0.5	9
1	2	6	1	15
1	3	9	2	21
1	4	16	4	40

Table 6.8C

No. of tubes giving positive reaction out of			MPN per 100 ml	95% confidence limits	
				Lower limit	Upper limit
1 of 50 ml each	5 of 10 ml each	5 of 1 ml each			
0	0	1	1	<0.5	4
0	0	2	2	<0.5	6
0	1	0	1	<0.5	4
0	1	1	2	<0.5	6
0	1	2	3	<0.5	8
0	2	0	2	<0.5	6
0	2	1	3	<0.5	8
0	2	2	4	<0.5	11
0	3	0	3	<0.5	8
0	3	1	5	<0.5	13
0	4	0	5	<0.5	13
1	0	0	1	<0.5	4
1	0	1	3	<0.5	8
1	0	2	4	<0.5	11
1	0	3	6	<0.5	15
1	1	0	3	<0.5	8
1	1	1	5	<0.5	13
1	1	2	7	1	17
1	1	3	9	2	21
1	2	0	5	<0.5	13
1	2	1	7	1	17
1	2	2	10	3	23
1	2	3	12	3	28
1	3	0	8	2	19
1	3	1	11	3	26
1	3	2	14	4	34
1	3	3	18	5	53
1	3	4	21	6	66
1	4	0	13	4	31
1	4	1	17	5	47
1	4	2	22	7	69
1	4	3	28	9	85
1	4	4	35	12	101
1	4	5	43	15	117
1	5	0	24	8	75
1	5	1	35	12	101
1	5	2	54	18	138
1	5	3	92	27	217
1	5	4	161	39	>450

The plates are incubated at 30°C for 4 h followed by 20 h at either 37°C or 44°C to yield coliform or *E. coli* counts respectively. When coliforms are present the green agar is changed to a translucent yellow colour due to acid production. The colonies of coliforms are counted directly from the plate. It is always advisable to do a series of dilutions to ensure that at least one plate contains a countable number of distinct colonies.

TABLE 6.9 Calculation of the number of coliform
bacteria (MPN) from samples collected in the field

(i) The Coliform count for a sample of river water taken below an
outfall from a sewage treatment plant at Osberstown, Co. Kildare on the
River Liffey was calculated using the most probable number technique
(MPN). Five replicate tubes were prepared using 10 ml, 1 ml, and 0.1 ml
of sample.

Sub sample used (ml)	Replicates	Positive tubes
10	5	3
1	5	1
0.1	5	0

Using Table 6.8(A) the combination of 3–1–0 gives an MPN value of
11 coliforms per 100 ml of sample.

(ii) Stronger samples such as wastewaters and effluents will need to be
diluted before setting up the tubes. When calculating the coliform count
this dilution factor must be taken into account. For example the final
effluent from Carlow sewage treatment plant was examined and the
results were:

Sub sample used (ml)	Replicates	Positive tubes
0.1	5	4
0.01	5	3
0.001	5	1

Using Table 6.8A the combination of 4–3–1 gives an MPN value of 33,
this is multiplied up by the dilution factor of 100 to give a coliform count
of 3300 coliforms per 100 ml of sample.

Membrane filtration technique
This technique, originally developed in the 1940s (Waite 1985), is rapidly
growing in popularity. It is replacing the multiple tube method because the
preparation of large quantities of tubes with media and inverted Durham
tubes is very time consuming. This method is far more rapid, essentially
disposable, has a smaller percentage error than other methods, is simpler to
use, and does not require specialist training, but it can be comparatively
expensive if pre-prepared plates are purchased (Table 6.10).

Known volumes of water are passed through a membrane filter (pore size
0.45 μm that retains the bacteria present. The filter is placed on Membrane–Enriched Teepol medium, which contains the detergent Teepol to
inhibit non-intestinal bacteria, and is then incubated (Geldreich et al. 1965)
(Fig. 6.2). The technique assumes that each bacterium retained by the filter
will grow and form a small visible colony. During the incubation period, the
nutrients diffuse from the medium through the membrane and the coliforms
are able to multiply and form recognizable colonies. Membranes are

TABLE 6.10 Comparison of the advantages and disadvantages of the multiple tube (MPN) and membrane filtration methods of coliform analysis (Hutton 1983)

Multiple tube MPN	Membrane filtration
Costs Large quantities of culture media and glassware and large autoclave Capital costs fairly high	*Costs* Smaller quantities of media Membrane expensive. Disposables (petri dishes, pipettes) expensive Capital costs of proprietary equipment high
Accuracy Statistically based estimate Possibly large errors especially at low levels	*Accuracy* More accurate especially to low levels (less than 100 colonies per 100 ml)
Field use Needs static base	*Field use* May be operated in the field and in transport if portable incubator used. Transport media can be used prior to incubation
Suspended matter May be used for turbid samples	*Suspended matter* Not suitable for turbid waters due to membrane clogging
Convenience Large amount of material to be prepared prior to analysis and disposed of after incubation	*Convenience* Less manipulation and hence lower chance of contamination Disposable, pre-sterilized equipment can be purchased
Incubation times Up to 48 h or 72 h	*Incubation times* 24 h (or even 7 h in some special cases)

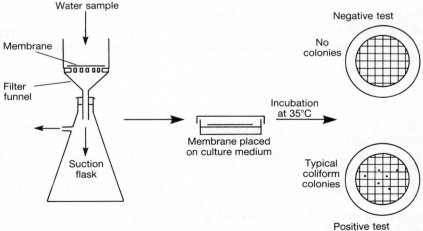

FIG. 6.2 Major steps in coliform testing by the membrane filtration technique.

incubated at 30°C for 4 h, followed by 14 h at 37°C for total coliforms, or 14 h at 44°C for *E. coli*, the colonies of which are a distinctive yellow colour. The volume of water filtered should yield between 10 and 100 colonies, and a series of water volumes should be filtered. Problems have been reported using this media with chlorinated waters, and Mara (1974) describes an alternative isolation technique that should be used to prevent false results. Different media are used in the USA; for total coliform Membrane–Endo Medium (commonly abbreviated to ME Medium) and for *E. coli* Membrane–Faecal Coliform Medium (or MF Medium). No preliminary incubation is recommended, just 24 h at either 37°C (total coliforms) or 44°C (faecal coliforms).

The actual number of coliforms is determined by counting the number of colonies and is expressed as the number per 100 ml of water. The filters have a grid printed on the surface to aid counting. Like all the bacteriological tests good sterile practice is essential. With the membrane filtration technique, both the forceps used to pick up the filters and the filter unit have to be sterilized between successive filtrations. The most rapid system is a UV sterilizer and a second unit is required if large number of samples are routinely processed. The method is not suitable for turbid effluents which are difficult to filter. Smaller volumes can be used but it is wise to increase the number of replicates; prefiltration through a series of coarser filters overcomes the problem of filterability but the estimation of coliforms will always be greatly reduced. Full details of the method are given in many standard reference texts including Mara (1974), HMSO (1983*b*), American Public Health Association *et al.* (1983), and Viessman and Hammer (1985). A useful working guide for those not familiar with the multiple tube and membrane filtration techniques has been produced by Hutton (1983) who explains how they can be used under field conditions. Recent research indicates that the brand of membrane and the pore size of membrane filters may have significant effect on the counts obtained (Tobin and Dutka 1977; Tobin *et al.* 1980). The effect of temperature on this test has also been under scrutiny and those who are undertaking comparative studies on the group are advised to read the excellent review in Dutka (1973).

6.2.3.2   *Faecal streptococci*

The streptococci normally occurring in human and animal faeces are the enterococcus group which includes *Str. faecalis* (including var. *liquefaciens* and var. *zymogenes*), *Str. faecium*, *Str. durans*, and the viridans group which comprises of *Str. mitis*, *Str. salivarius*, *Str. bovis*, and *Str. equinus*, all of which are commonly found in contaminated waters (Stanfield *et al.* 1978). *Streptococcus bovis* and *Str. equinus* are not found in man and can be used as a specific indicator of animal wastes or meat processing wastewaters. However, because all the species are indicative of faecal pollution, it is not necessary to identify individual species, as a single test for the whole group is sufficient.

Faecal streptococci are Gram-positive cocci approximately 1 μm in diameter that occur in chains of varying lengths, and are non-motile and non-sporing. Numerically, they can be equally as abundant as coliforms, especially in stormwater and in the effluents from intensive pig-rearing units, but under normal circumstances they are slightly less abundant.

Methods for isolation and enumeration are the same for coliforms, except that different media and incubation conditions are used, with the multiple tube and membrane filtration methods the most widely used (HMSO 1977). Glucose Azide Broth is used in the multiple tube methods, and as faecal streptococci can ferment lactose and other sugars but do not produce gas, no inverted Durham tubes are necessary. Growth occurs at 45°C in the presence of sodium azide or potassium tellurite, which inhibit Gram-negative bacteria, including coliforms. Faecal streptococci can be distinguished from non-faecal species by their tolerance to 40% bile salts, their ability to grow in 6.5% sodium chloride, and the ability to withstand heating to 60°C for 30 minutes (Mara 1974; Buchanan and Gibbons 1975).

The tubes are set up in the same way as for coliforms and incubated at 37°C for 72 h, and if the pH indicator in the medium has changed from purple to yellow, indicating acid production, then an inoculum from each positive tube is transferred to fresh medium and reincubated at 45°C for 48 h. Tubes showing acid production are then used to calculate the MPN (Table 6.8). Membrane filtration is the same as for the coliforms except that filters are plated out on Membrane Enterococcus Agar and incubated for 4 h at 37°C followed by 44 h at 45°C. Colonies appear red or maroon in colour and are counted directly. An excellent comparative study on the specificity and selectivity of five membrane filtration methods for isolating and enumerating faecal streptococci has been conducted by Stanfield et al. (1978) and this should be consulted before conducting such tests.

Faecal coliform–faecal streptococci ratio
Different animals excrete different quantities of faecal coliforms and faecal streptococci, and the ratio of one to the other remains remarkably constant for specific sources. Therefore, the ratio of faecal coliforms (FC) to faecal streptococci (FS) can be used to identify the source of faecal contamination in a water body. The FC:FS ratio in fresh human faeces is > 4, whereas for other animals it is < 0.6 (Table 6.11) (Geldreich 1972). Intermediate values are difficult to interpretate, but FC:FS ratios between 2.0 and 4.0 indicate a predominance of human waste in mixed pollution, whereas 1.0–2.0 indicates equal levels of human and animal pollution, and 0.7–1.0 a predominance of animal waste in mixed pollution. The ratio is especially useful in tracking down the source of pollution and ratios of > 4.0 or < 0.6, warns of the particular danger of human or animal disease transmission respectively. However, some caution if required when using the FC:FS ratio (Section 6.2.4.1). For example, the variability between samples can be significant

TABLE 6.11   Densities of faecal coliforms (FC) and faecal streptococci (FS) in animal faeces and the resultant FC:FS ratios (Lynch and Poole 1979)

Faecal source	Densities g^{-1} of faeces (median values)		Ratio FC:FS
	Faecal coliform	Faecal streptococci	
Man (USA)	13 000 000	3 000 000	4.33
Cow	230 000	1 300 000	0.177
Pig	3 300 000	84 000 000	0.039
Sheep	16 000 000	38 000 000	0.421
Horse	12 600	6 300 000	0.002
Duck	33 000 000	54 000 000	0.611
Chicken	1 300 000	3 400 000	0.382
Turkey	290 000	2 800 000	0.104
Cat	7 900 000	27 000 000	0.293
Dog	23 000 000	980 000 000	0.024
Field mouse	330 000	7 700 000	0.043
Rabbit	20	47 000	0.0004
Rat	180 000	78 900 000	0.0023
Chipmunk	148 000	6 000 000	0.002
Elk	5100	760 000	0.007
Robin	25 000	11 700 000	0.002
English sparrow	25 000	1 000 000	0.025
Starling	10 000	11 800 000	0.0009
Red-winged blackbird	9000	11 250 000	0.0008
Pigeon	10 000	11 500 000	0.0009

making replicates vital. Both coliforms and streptococci are killed by extreme pH and industrial pollutants and care must be taken to ensure that samples are free from contaminants and that the pH is between 4–9. The death-rate of the two groups of micro-organism vary, for example *Str. bovis* and *Str. equinus* die off rapidly, whereas *Str. faecalis* can persist far longer than faecal coliforms, so that the FC:FS ratio will change significantly if micro-organisms have been free in the environment for > 24 h. This can only be avoided by taking samples as close to the suspected pollution source as possible and processing samples quickly. Finally, it must be stressed that only the faecal coliform count obtained at 44°C should be used and *not* the total or presumptive coliform count (Mara 1974; HMSO 1977).

From the research by Tobin and Dutka (1977) it would appear that the ratio of FC:FS can alter according to the membrane filter used or the incubation temperature used in the estimation of the faecal coliform count, raising the 4:1 ratio for human waste to as high as 8–15:1 (Dutka 1979).

6.2.3.3   Clostridium perfringens

Clostridium perfringens is a Gram-positive anaerobic spore-forming rod that is found in both human and animal faeces, and its presence in water is a positive indicator of faecal contamination. The bacterial rods are 5×1 μm and although it sporulates readily in nature it does not on laboratory medium. *Clostridium perfringens* is less abundant than the other bacterial indicators, being found in maximum numbers of 10^5–10^7 per 100 ml of water. Unlike coliforms or streptococci, its spores can survive indefinitely and it is particularly useful in identifying old or remote contamination. Spores are very resistant and are able to withstand heating at 75°C for up to 15 minutes. It can ferment lactose, producing large quantities of gas, and is able to reduce sulphite to sulphide.

A two-stage MPN technique is used to routinely isolate and enumerate *C. perfringens*. First, the water sample is heated to 75°C for 10 minutes and after cooling is inoculated directly into Differential Reinforced Clostridial Broth and incubated at 37°C for 48 h. The inocula are placed in screw-topped bottles partially filled with medium, which are then filled to the top with excess medium and then tightly closed to ensure anaerobic conditions. The medium contains sulphite and if clostridia are present then it will be reduced and precipitated to ferrous sulphide which turns all the medium in the bottle black. However, as any clostridium can cause blackening of the medium the presence of *C. perfringens* has to be confirmed. This second stage requires an inoculum from each positive bottle to be transferred to freshly prepared tubes of Litmus Milk which are then incubated at 37°C for 48 h. The redox potential of the medium is reduced, to encourage anaerobic growth, by adding a small length of iron wire or a steel nail which is sterilized immediately before use by heating until red-hot. If *C. perfringens* is present a 'stormy clot' is formed in the tube. The acid production causes the milk to clot and the coagulated protein is forced up the tube towards the top by the gas produced. The purple medium now appears a yellow colour which is the whey of the milk with the reddish clot which is normally broken up on the top of the whey taking on the appearance of a thunder cloud, hence the term 'stormy clot'. The MPN is then obtained by reference to Table 6.8. This confirmatory test is not totally accurate as other clostridia can also produce this effect in Litmus Milk. Therefore, although this method is adequate for routine analysis a more specific test may be required, details of which are given by Mara (1974).

The membrane filtration method can also be used. After the preliminary heat treatment, to destroy the vegetative bacteria, a volume of the sample is filtered through a membrane and incubated anaerobically on a sulphite-containing agar medium. A count of the sulphite-reducing clostridia are made by counting the black colonies which have formed. The presence of *C. perfringens* is confirmed by subculturing each colony to be tested into a tube or bottle of Litmus Milk as described previously (HMSO 1983*b*).

6.2.3.4 Other indicators

Problems with the accuracy and reproducability of coliform and faecal coliform counts have led to a continuing review of the techniques to identify and enumerate the group. These problems have also led to the search for new indicators of faecal contamination (Dutka 1979). With the increase in the re-use of water for human consumption a more rapid routine test is needed, which allows water to be checked for faecal contamination before reaching the consumer. Ideally, such a test should be automated.

Candida albicans

The most exciting new potential indicator to be evaluated in recent years is the yeast *Candida albicans* (Grabow *et al.* 1980). It is extremely widespread in developed countries with 80% of the adult population having low levels of infection and detectable levels of the yeast in their faeces. Although it lives at low levels of activity in the rectum of most people, it can result in mouth, vagina, groin, and skin infections. The latter are especially common in swimmers (Buckley 1971), where there has been a steady increase in the incidence of serious infections (Briscou 1975). The yeast cannot exist for prolonged periods without a natural host and does not exist or grow independently in water. Therefore, the presence of *C. albicans* is a direct result of faecal contamination and makes it ideal for monitoring water, sewage, and estuarine waters, and if it could be successfully enumerated would provide very accurate details of the level of contamination. The yeast is effectively removed by sewage treatment (Buck 1977), although concentrations of up to 100 *C. albicans* per 100 ml have been reported in treated effluents (Dutka 1979). There are no universally recognized standard enumeration procedures, although Dutka (1979) has evaluated a membrane filtration system, and a medium, developed by Buck which he found to be both rapid and extremely reproducible. For example, typical numbers for beaches with high faecal coliform counts can be up to 20–25 *C. albicans* l^{-1} but falling to 0–2 l^{-1} for relatively unpolluted beaches.

Pseudomonas aeruginosa

Although coliform and faecal coliform counts can indicate the presence of faecal contamination in water they do not necessarily reflect the risk to those using such waters for recreational purposes. The most frequent diseases associated with this type of activity are skin infections, and diseases of the nasopharyngeal and upper respiratory tract.

One micro-organism which has been widely implicated in such diseases is *Pseudomonas aeruginosa* (Hoadley 1977). It is the only species of the pseudomonads that is pathogenic to man and causes a range of diseases including otitis externa and infections of the urinary tract. It is particularly resistant to many anti-bacterial drugs which makes it a difficult pathogen to control. The bacterium is aerobic, forming Gram-negative rods $0.5 \times 2 \ \mu m$, which are motile with montrichous or lophotrichous flagella but non-sporing and non-capsulate. It has interesting growth properties, forming both oxidase and catalase, growing at $42°C$ but not at $4°C$. It can reduce nitrates and nitrites, produces ammonia from the breakdown of acetamide, and is able to hydrolyse casein but not starch. An important characteristic of this pseudomonad is that it can produce the blue-green pigment pyocyanin or the fluorescent pigment fluorescein, or both. Fluorescent pseudomonads are common in water with *Ps. fluorescens* and *Ps. putida* forming slimes in water supply pipes. Although *Ps. aeruginosa* is relatively short lived it is an extremely useful indicator of faecal contamination, especially applicable in small areas of waters, such as swimming pools and lidos as well as for monitoring in the natural environment. Pseudomonads are capable of multiplying in water containing suitable nutrients. It is, therefore, not recommended as a routine monitor of faecal pollution (HMSO 1983*b*). The micro-organisms enter the water directly from contaminated people, either from the skin of bathers, skin surfaces contaminated by faecal material, urine or from contamination by faeces.

Isolation and enumeration is by the MPN method using Drakes' Asparagine Medium (American Public Health Association *et al.* 1983), which recovers greater numbers of the organisms than the comparable membrane filtration technique (Mara 1974; Dutka and Kwan 1977) for samples taken from rivers and sewage treatment plants (Hoadley 1977). The colonies are green, blue or reddish-brown and fluoresce under UV light. However, a confirmatory test is required for *Ps. aeruginosa* that checks for growth at 42°C, casein hydrolysis, and production of either pyocyanin or fluorescein (Brown and Foster 1970; HMSO 1983*b*).

Staphylococcus aureus

There is a much higher incidence of staphylococci infections in swimmers than non-swimmers. Much concern has been expressed that the normal enteric indicators do not provide an index for these infections as contamination is via the mouth, nose, throat, and skin of bathers, rather than urine and faeces. *Staphylococcus aureus* is the major pathogen of concern. It is commensal in the nostril, armpits, and groin in about 40% of healthy adults, but can cause a variety of diseases of the upper respiratory tract, eyes, and ears. Among the commonest diseases are osteomyelitus, impetigo, minor boils, and carbuncles. *Staphylococcus aureus* forms characteristic clusters of Gram-positive cocci, 1 μm in diameter, which are non-motile, non-sporing, and non-capsulated. Colonies appear as yellow colonies when isolated by the membrane filtration technique and grown on a mannitol salt medium, such as Membrane Staphylococcus Medium.

There appears to be an urgent need to measure the incidence of staphylococci in swimming pools and natural waters that are used for swimming or other recreational pursuits. As the organisms are more resistant to chlorination and other environmental factors than *E. coli*, staphylococci would appear ideal indicators, although enumeration and isolation techniques for *Staph. aureus* are not fully effective. The techniques used for the group are described elsewhere (Mara 1974; Alico and Palenchar 1975; American Public Health Association *et al.* 1983).

Bacteriophages
These are viruses that are pathogenic to specific bacteria, for example, coliphages are specific to strains of *E. coli*. They are widely used as models for viral studies as they are technically much easier to handle than viruses themselves. The ratio of coliforms to coliphages varies, being 87:1 in raw sewage, 4.2:1 in lagoon effluent, and 0.15:1 in river water (Bell 1976). Rapid tests are being developed using specific bacteriophages to indicate not only the presence of specific faecal bacteria but their relative abundance (Kenard and Valentine 1974; Bell 1976). However, bacteriophages are much smaller than bacteria and behave differently in water, being more readily adsorbed on to suspended solids. Bacteriophages may not directly imitate enteric bacteria in wastewater systems, hence the variable ratio during the treatment process (Section 6.3.2). These ratios do not remain constant because the concentration of coliforms fluctuates with environmental conditions, although the concentration of bacteriophages and enteric viruses remains contant. Therefore, because bacteriophages are so similar to enteric viruses, being of the same size and behaving in the same way in water, it has been proposed that bacteriophages could be used as indicators of enteric viruses (Kott *et al.* 1974). Although early attempts proved disappointing, more recent research has shown that coliphages giving rise to plaques of 3 mm in diameter are positively related to enteric viruses in secondary effluents (Funderburg and Sorber 1985). Jofre *et al.* (1986) evaluated bacteriophages infecting *Bacteroides fragilis* as a possible surrogate indicator of human viruses in water, while Grabow *et al.* (1984) have evaluated coliphages. Bacteriophages are plentiful in wastewater and can be isolated without concentration, and assay procedures are now cheap and can be completed within a single day (Wentsel *et al.* 1982). These tests could greatly increase the sensitivity and speed of faecal monitoring, although it would be naïve to think that such tests will not require a high level of expertise and high-quality laboratory facilities.

6.2.4   Hazards associated with wastewater and sludge

6.2.4.1   *Water pollution*

A wide variety of enteric pathogens can be detected in sewage (Table 6.2), the frequency and population density of which is a reflection of the degree of infection within the community. Therefore, special care should be taken at treatment works when there are outbreaks of diseases within the community, and precautions should be taken when treating wastewaters from hospitals. Occasionally, wastewaters from isolation hospitals are sterilized or disinfected before being discharged into the main sewer in order to minimize the chance of infection spreading. Other wastes are also potentially hazardous. There is a wide range of pathogens, especially *Salmonella* and intestinal parasites, that are associated with meat-processing. Thus, care should be taken when coming into contact with wastewaters, even after large dilution, from abattoirs, meat and poultry processing, and intensive animal and poultry units.

Diseases that are commonly associated with faecal contamination of water in temperate regions include *Salmonella* (typhoid, paratyphoid, food poisoning), *Shigella* (bacterial dysentry), *Mycobacterium* (tuberculosis), and *Leptospira icterohaemorrhagia* (Leptospira jaundice). Viruses are excreted in large numbers in faeces, for example, poliovirus, coxsackievirus groups A and B, echovirus, infectious hepatitis virus, reovirus, and adenovirus (Pike 1975). The most serious of these, in terms of human infection in temperate climates, is infectious hepatitis and there is a direct link between faecal contamination of water supplies and bathing water, and infection. After mass vaccination of children at schools with live attenuated vaccine, large increases in the numbers of poliovirus can be recorded in the sewage from the community. Children are often symptomless excreters of poliovirus as well as other enteric viruses, although the number varies according to socio-economic as well as environmental factors (Kollins 1966). The level of infection will depend on a number of factors, such as the density of the pathogen in the water, the invasiveness and toxigenicity of the pathogen, the degree of contact between the pathogen and the host, and the innate and acquired immunity of individuals. Although sewage workers appear to be among the healthiest in the community, which suggests that they are immunized by regular exposure to low levels of pathogens, mainly via aerosols, they still run a high risk of infection by certain water-borne diseases (Benarde 1973; Clark *et al.* 1984*a,b*). The incidence of Leptospira jaundice is particularly high amongst sewage workers as is the incidence of intestinal parasites, such as *Entamoeba histolytica*.

Most pathogens of mammalian origin are highly specialized parasites which grow best at mammalian body temperatures (37°C) and in a suitable nutrient environment. Once excreted, pathogens are in a hostile environment and their numbers rapidly decline (Fennell 1975). However, the chance of infection decreases after treatment, after discharge and dilution in the receiving water, and with time, so the chance of infection from faecal contamination by treated effluents is extremely low.

However, certain areas are at particular risk of contamination: *bathing waters*, notably coastal waters that receive large volumes of untreated or, at best, settled sewage, where the dilution effect may be reduced by currents and wave action; *edible shellfish* growing in polluted coastal and estuarine waters also accumulate pathogens; *groundwater* contamination from septic tanks, leaking sewers or agricultural wastes; *reservoirs* from roosting gulls; and *water which is re-used* several times for supply purposes.

Bathing waters

The microbial quality of water is comparatively unimportant except where a risk of infection exists. Infection normally only occurs if there is intimate human contact with contaminated water or it is swallowed. Therefore, in water-based recreational activities that require participants to enter the water, such as swimming, wind surfing or water ski-ing, a risk of infection does exist and so these activities require water that is free from faecal contamination. There is only minimum contact with water with other water-based activities, such as boating and fishing, and because there is a much lower risk of infection, much higher levels of contamination can be tolerated. This is supported by studies that have shown a higher incidence of water-borne diseases among swimmers than non-swimmers, with skin irritation, gastro-intestinal infections, and infections of the eyes, ears, nose, and throat all associated with contaminated coastal waters (WHO 1975). A wide range of infections are known to be transmitted from person to person at swimming pools, especially skin infections caused by *Pseudomonas aeruginosea* and *Mycobacterium marinum*, conjunctivitis caused by *Chlamydia trachomatis*, and respiratory infections (Table 6.4). However gastro-intestinal infections have been reported in people bathing or becoming immersed in water either accidentally or during water sports. For example, a study of swimmers who took part in a snorkel swimming contest in Bristol during 1983 showed that 21% experienced gastro-intestinal symptoms within 48 h of entering the untreated water (Phillip *et al.* 1985). Other examples of more serious diseases contracted during the recreational use of both fresh and sea water including typhoid, paratyphoid, polio, and leptospirosis are cited by Galbraith *et al.* (1987). In fact, 25% of the reported cases of leptospirosis in the UK during 1978–83 were associated with the recreational use of water or accidental immersion (Waitkins 1985) (Table 6.4). A study by the Environmental Protection Agency in the USA shows a strong link between bacterial counts in coastal waters at the time of bathing and the number of cases of severe gastro-intestinal disease among the swimmers. With an enterococcus bacterial concentration of 10 per 100 ml there were 8 cases per 1000 swimmers, at 100 per 100 ml there were 30 cases, which rose to 50 cases as the concentration of bacteria reached 1000 per 100 ml (Cabelli 1980; Pearce 1981). However, the risk of contracting such diseases from bathing waters is still extremely small, unless gross pollution is present, and under those conditions it is unlikely that it would be aesthetically acceptable to the bather for swimming (Moore 1977; Barrow 1981).

In England and Wales, there were 333 sewage outfalls discharging to coastal waters in 1972, and at that time only 49 of these (15%) outfalls extended further than 90 m below the low water mark at ordinary spring tides (Agg and Stanfield 1979). Between 1974 and 1981, only long outfalls were built. In 1981, 190 beaches in England and Wales were at risk from sewage pollution, including some of the major resorts. For example, Scarborough, with a summer population of 100 000, was discharging raw sewage at the low water mark (Pearce 1981). However, length of outfall is not the only important criterion in the design of outfalls, with depth and currents also of importance (Grace 1978). The majority of the population in Ireland live in coastal towns and settlements, all of which discharge their sewage directly to the sea via short outfalls. Therefore, with so many coastal towns in the British Isles discharging either raw or primary treated sewage directly to the sea, often with very short outfalls, it is not surprising that so many bathing beaches are contaminated by faecal pathogens. Although beaches can show visible signs of such pollution in the form of plastic strips from disposable toiletries or actual faecal matter, normally, contamination can only be detected by microbial examination of the water. Sea outfalls are not the only source of faecal contamination on beaches and in coastal waters. There is a considerable pathogen load from rivers that have received effluents from treatment plants serving inland towns, with higher levels of contamination at estuaries. Gull droppings are also rich in coliforms and *Salmonella*, and enhanced levels of these faecal bacteria are often recorded at sites near to roosting colonies (Section 6.2.4.1) (Gameson 1975*b*). However, there are so many factors affecting the level of contamination of beaches, such as dilution, currents and tide movements, environmental factors affecting die-off of bacteria, length of outfall, quality of effluent, and concentration of pathogens in sewage, that it is impossible to generalize.

The problem of sewage contaminated bathing waters has been realized for a long time, although many traditional holiday resorts, famous for their beaches and '*safe swimming-areas*', are often those very towns that discharge their untreated sewage to coastal waters via short outfalls. In the USA, legal standards have been in force for a long time with a maximum microbial standard of 200 faecal coliforms per 100 ml generally used for all direct-contact recreational waters (i.e. bathing waters), with public health authorities posting '*sewage-polluted beach*' warning signs advising against bathing in areas exceeding this limit (Waldichuk 1985). In 1976, the risk to those bathing in faecally contaminated water in Europe was recognized by the introduction of the EEC directive on the quality of bathing waters (European Communities 1976). The directive cites mandatory standards for total coliforms, faecal coliforms, *Salmonella*, and enteroviruses in bathing waters. It states that samples of water must be taken from bathing areas at fortnightly intervals during the bathing season and that they must contain < 10 000 total coliforms and < 2000 faecal coliforms per 100 ml in 95% of cases. More stringent guideline values are also given but these are not mandatory (Table 6.12). There are many difficulties with the interpretation of this directive, for example, which bathing areas are covered by the legislation, the choice of sampling sites on a particular beach, sampling frequency, and the time of sampling in relation to tide and currents (Moore 1977; Water Research Centre 1977).

In 1985, there were only 27 designated beaches in Britain, with 19 in the

TABLE 6.12 Microbiological quality requirements specified for bathing waters within EEC countries. Percentage values in parentheses are the proportion of samples in which the numerical values stated must not be exceeded (European Communities 1976*a*)

Parameters	G	I	Minimum sampling frequency	Method of analysis and inspection
Microbiological:				
Total coliforms 100 ml^{-1}	500 (80%)	10 000 (95%)	Fortnightly (1)	Fermentation in multiple tubes. Subculturing of the positive tubes on a confirmation medium. Count according to MPN (most probable number) or membrane filtration and culture on an appropriate medium such as Tergitol lactose agar, endo agar, 0.4% Teepol broth, subculturing and identification of the suspect colonies
Faecal coliforms 100 ml^{-1}	100 (80%)	2000 (95%)	Fortnightly (1)	
				The incubation temperature is variable according to whether total or faecal coliforms are being investigated
Faecal streptococci 100 ml^{-1}	100 (90%)	—	(2)	Litsky method. Count according to MPN (most probable number) or filtration on membrane. Culture on an appropriate medium
Salmonella 1 l^{-1}	—	0 (95%)	(2)	Concentration by membrane filtration. Inoculation on a standard Medium. Enrichment— subculturing on isolating agar—identification
Enteroviruses PFU* 10 l^{-1}	—	0	(2)	Concentrating by filtration, flocculation or centrifuging, and confirmation

G, recommended; I, mandatory.

* PFU = Plaque Forming Unit – a method of counting viruses; one PFU may be regarded as one infective virus particle.

(1) When a sampling taken in previous years produced results which are appreciably better than those in the Directive when no new factor likely to lower the quality of the water has appeared, the competent authorities may reduce the sampling frequency by a factor of 2.

(2) Concentration to be checked by the competent authorities when an inspection in the bathing area shows that the substance may be present or that the quality of the water has deteriorated.

south-west between Weston-super-Mare, Avon to the Isle of Wight. In contrast, the French have designated thousands of beaches both on the coast and on rivers (Anon 1985a). Beaches are designated as 'Eurobeaches' in Britain if the threshold of 1500 bathers per mile ($= 1000$ per km) is exceeded. If this threshold is not reached then there is no need to comply with the standard. However, the degree of freedom allowed in calculating the bathing density has resulted in some of the major resorts in the UK being originally excluded from the list, including such resorts as Blackpool, Brighton, and Eastbourne. For example, Pearce (1981) asks the pertinent question if there are 2000 bathers on a 2 mile beach all congregated in a 0.5 mile stretch around the pier is the bathing density 1000 or 4000 per mile? The fact that Blackpool has 8 miles of beaches may well provide the answer why it was not designated. Some smaller resorts have very severe beach pollution. For example, Bridlington discharges sewage from a population of 75 000 at the low watermark in a single outfall, which resulted in total coliform counts of 100 000 per 100 ml (10 times the EEC limit) and 42 000 faecal coliforms per 100 ml (21 times the EEC limit) in 1979. Margate, a designated beach, discharges the screened sewage from 275 000 people 100 m below the low water mark and regularly fails to meet the standard (Pearce 1981). For some resorts, the only way to achieve the standard is to disinfect their effluent, although solids may still become stranded on the beach. Huge discharges of crude sewage enter coastal waters via the major estuaries, such as the Humber and Severn, with the effluent ending up at resorts on the estuaries. Disinfection of these effluents does reduce the level of bacterial contamination (e.g. Weston-super-Mare) but chlorination of raw sewage has been shown to interfere with the migration of Salmonids (Waldichuk 1985). The quality of beaches in Britain and Ireland has not improved since the introduction of the EEC Directive. In 1984, a survey of 120 beaches in Wales showed that only 38 met the EEC standards. The bathing water quality of British beaches in 1985 is summarized in Table 6.13. Member states were required to conform to the limits set by the EEC directive for all designated beaches by December 1985, except where a derogation had been granted. In the case of the two designated beaches at Scarborough derogations have been granted until 1990. At Ryde, a new outfall is near completion and trial plants are in operation at Sandown and Shanklin. New treatment plants and outfalls have been proposed for Southend and Weston-super-Mare (DoE 1987). The number of designated beaches in Britain was increased from 27 to 379 in 1986. A complete list of British beaches, including all those designated under the Directive, including details of discharges has been produced by the Marine Conservation Society (1987).

The criteria for selecting designated beaches are so broad that in Ireland none of the existing bathing beaches come within the directive. However, seven beaches were originally designated, which was increased in 1988 to 51 designated areas two of which lie within Dublin Bay that receive almost a third of the sewage from the country's total population of three million. Both sites

TABLE 6.13 Bathing water quality survey of designated beaches in the UK during 1985 (DoE 1987)

'Identified' bathing waters	No. of samples	Faecal coliforms 100 ml⁻¹				Total coliforms 100 ml⁻¹			
		Median	Maximum	No. of samples exceeding		Median	Maximum	No. of samples exceeding	
				2000 100 ml⁻¹	100 100 ml⁻¹			10 000 100 ml⁻¹	500 100 ml⁻¹
Bournemouth Pier	15	50	320	—	5	100	740	—	3
Bridlington, North beach	12	25	130	—	1	65	680	—	1
Bridlington, South beach	12	75	590	—	5	190	700	—	3
Christchurch, Avon	16	70	2000	—	6	230	8000	—	4
Margate	14	115	2000	—	8	240	17 000	1	6
Newquay, Fistral	8	41	130	—	1	105	450	—	—
Newquay, Towan	8	70	250	—	3	142	770	—	1
Paignton, Broadsands	6	79	1600	—	3	101	2300	—	1
Paignton, Goodrington	17	98	530	—	6	129	950	—	3
Paignton, Paignton beach	6	151.5	470	—	4	545	600	—	4
Penzance, Sennen Cove	12	114	420	—	7	283	1580	—	2
Poole, Shore Road	15	10	3400	—	3	10	980	—	1
Ryde	24	2550	80 000	13	24	5450	121 000	10	23
St. Ives, Porthmeor	12	199	420	—	9	442	940	—	4
St. Ives, Porthminster	12	180	1410	—	9	420	2590	—	4
Sandown	24	79	6100	2	9	130	7600	—	2
Scarborough, North bay	18	205	4500	1	10	225	9500	—	8
Scarborough, South bay	18	305	18 800	2	13	510	18 800	1	9
Shanklin	12	180	13 200	1	6	470	14 300	1	5
Southend, Thorpe	22	335	16 100	3	21	800	40 000	2	12
Southend, Westcliff	22	528	17 600	3	19	820	25 500	1	14
Swanage, Central	15	20	380	—	2	50	590	—	1
Torquay, Meadfoot	6	22.5	48	—	—	32.5	82	—	—
Torquay, Oddicombe	6	8.5	56	—	1	15.5	81	—	—
Torquay, Torre Abbey	6	28	290	—	1	68.5	1030	—	2
Weston-super-Mare	61	360	4300	4	48	800	13 600	2	39
Weymouth, Central	14	45	370	—	4	60	430	—	—

have failed to meet the 95 percentile limit of the EEC standards since monitoring began, although the directive only came into force in Ireland in 1985. The EEC standards for total and faecal coliforms are generally felt to be too high; it is alleged that this is in order to make it politically acceptable to all the member states. However, some member states have set more severe limits and in the case of Ireland, is half the EEC values (Fegan 1983). The same standards apply to all European waters from the Atlantic to the Mediterranean, with an enormous variety of physical and meterological conditions. Bathing habits also vary in terms of duration and frequency. They are much higher in the Mediterranean than elsewhere, resulting in much higher water–skin contact times and therefore a higher risk of infection. Problems in the selection of techniques for faecal coliform analysis result in variability between counts, and even the risk that the results could be manipulated by the careful selection of specific techniques and sampling sites.

The directive is based on coliforms that are only directly related to the incidence of bacterial gastro-intestinal infections. Although the ratio of this indicator to other pathogens is reasonably good (Bonde 1977), it is less effective in indicating risks of skin infections or those of the eyes, ears, nose, or throat, which are most frequently associated with swimmers. It may be that more specific and longer lived indicators, such as *Staphylococcus aureus* would be more effective.

In the long term, the answer to this problem, and that of coastal eutrophication, resulting in algal blooms, is to limit the discharge of pathogens and nutrients by adequate treatment at coastal sites (Baalsrud 1975). However, the authorities have responded by building longer (> 1000 m) and deeper outfalls with diffusers, thus ensuring maximum dilution and dispersion of the sewage entering the sea (Gray 1986). For example, the new outfall at Bray, County Wicklow, is 1690 m in length. A useful case study of the 1000 m long Kirkcaldy sea outfall in the Forth estuary explains in detail the various environmental, biological, and hydrographical investigations that need to be carried out, as well as highlighting some of the design and constructional problems (Moore *et al.* 1987). This particular project was of special interest as the sea-bed did not shelve as steeply as in other areas where such outfalls have been constructed. Therefore, the available water depth was marginal, resulting in very weak currents to disperse the discharged waste. The design, construction, and maintenance of long sea outfalls is examined in depth elsewhere (Institution of Public Health Engineers 1986).

Shellfish

Enteropathogenic diseases are known to be transmitted by the consumption of contaminated shellfish, in particular, mussels, oysters, clams, cockles, and to a lesser extent, scallops (James and Wilbert 1962; Ohara *et al.* 1983). This particular hazard has been known since the latter part of the nineteenth century when typhoid was shown to be transmitted via contaminated shellfish and since then the list of associated diseases has grown to include cholera, paratyphoid, infectious hepatitis, and other gastro-enteric disorders (Pike and Gameson 1970). All the shellfish listed are filter feeders that ingest food indiscriminately and extract faecal bacteria along with the other debris. However, the faecal micro-organisms are not digested by the shellfish or even inactivated, instead they accumulate at gill surfaces and in the alimentary canal (Wood 1979). These shellfish are common in estuarine and coastal waters throughout Europe that receive organic pollution. Also, many species are commercially farmed in suitable coastal and estuarine locations.

The major hazard to health arises from eating shellfish raw. The risk can be minimized by either cooking them at a sufficiently high temperature for a suitable length of time or by immersion in tanks of clean sea water to allow self-cleansing to occur (Carrington 1980b). Water quality standards for growing shellfish that will be uncontaminated have been adopted in both Canada and the USA. There is a maximum permissible faecal coliform count < 14 per 100 ml with < 10% of samples exceeding 43 per 100 ml, and a maximum permissible faecal coliform count of 230 per 100 g of shellfish meat (Environmental Protection Agency 1976; Waldichuk 1985). However, the suitability of shellfish for consumption in the UK is based on the bacterial content of the flesh, not the water, with 0.5 E. coli per ml of flesh considered safe (Evison 1979). An EEC directive on the quality required for shellfish waters has recently come into force (European Communities 1979).

Groundwater
Chemical pollution can travel up to twice the distance of bacterial pollution in soil, while both viruses and bacteria are able to move through the soil and contaminate wells and groundwater (Allen and Geldreich 1985; Filip *et al.* 1988; Polprasert *et al.* 1982). The enteric organisms originate from sewage irrigation, grass-plot treatment of final effluents, septic tank percolation areas, and land disposal of sewage sludge. The movement of these pathogens, which travel in both a vertical and horizontal direction, depends on factors such as the size and shape of the micro-organism, the soil, type, the surface characteristics of the soil, and the viscosity of the interstitial water. In fact, so many factors influence the survival and passage of viruses and bacteria in the soil that it is difficult to generalize for all soil types. However, bacteria can travel faster in highly permeable soil (> 3.0 m d^{-1}) than in soils with slow permeability (< 1.2 m d^{-1}). Faecal coliforms have been reported as travelling long distances horizontally through soils (> 100 m) (Dappert 1932) and at flow rates of up to 15 m h^{-1} (Bouma *et al.* 1974), with the distance travelled related to the soil properties and hydrological variables (Reneau 1978) (Table 6.14). There is a reduction in the number of pathogenic organisms with depth and distance form the source of pollution, although this varies with the adsorptive properties of the soil (Carrington 1980a). For example a column of garden soil 0.9 m high reduced the E. coli concentration by 75% and

TABLE 6.14 Examples of the distance travelled by faecal micro-organisms in soil (Polprasert and Rajput 1982)

Type of organism	Distance transported (m)	
	Vertical	Horizontal
E. coli	3.04–9.14	70.71
E. coli		24.38
E. coli		121.92
'Lactose fermenters'	0.76	0.61
Coliform bacteria	0.61–0.91	3.048–121.92
Coliform bacteria		54.86
Coliform bacteria	45.72	
Clostridium welchii	2.13–2.43	
Bacteria		609.6

Coxsackievirus by 50%. Reneau et al. (1975) have suggested that the removal of pathogens is a function of the distance from the source of pollution, for that particular soil type. Vertical contamination is increased much more than horizontal contamination by rain. Viruses have been reported at depths of 2.6 m below the surface after heavy rain (Wellings et al. 1975). Viruses, because of their smaller size, are able to penetrate deeper and farther than bacteria, although they are less able to survive for long periods. The retention of viruses decreases proportionally to the increase in the flow rate. Adsorption is greater at lower pH values, with some metal complexes (e.g. magnetic iron oxide) rapidly adsorbing viruses. Cations can reduce the repulsive electrostatic potential between soil particles and viruses thus enhancing adsorption, and the high ion exchange capacity and large surface area of clay soils gives a high retention capacity. The presence of soluble organics has been shown to reduce adsorption capacity (Gerba et al. 1975; Goyal and Gerba 1979). This topic is reviewed by numerous authors (Hutchinson 1974; Allen and Geldreich 1975; Carrington 1980a; Seppanen and Wihuri, 1988).

Re-use of wastewater for human consumption
Not all areas of a country will have access to upland reservoirs or groundwater sources of good quality water for human consumption. In order to meet the demand for water in highly populated areas alternative sources have to be utilized. The commonest source of water in such situations are rivers, and some major cities in Europe rely almost exclusively on rivers that have already received sewage from other towns upstream to supply all their water needs. Therefore, a consumer downstream of other towns could be drinking water that has already been used, treated, and returned to the river on several previous occasions.

The volume of sewage effluent discharged into a river may remain constant, however the ratio of sewage effluent to river water (the dilution factor) will vary. It is possible at time of drought for the volume of effluent discharged to exceed the volume of clean water in the water course. The River Thames in the UK comprises up to 14% sewage effluent under normal conditions and the River Lea is even higher. The only available surface water in Rotterdam is the River Rhine that receives over 70% of the total sewage of West Germany (Dart and Stretton 1977). During periods of drought in Agra (India), the water supply is comprised entirely of sewage effluent from New Delhi, 190 km away (WHO 1973).

Although it is possible to purify almost any river water, no matter how polluted, to drinking water standard (Table 6.15), the costs can be prohibitive. The most effective method of re-use is to combine maximum sewage treatment of wastewaters before discharge to a river with adequate treatment of abstracted water. Although this is comparatively easy to ensure in a small catchment, such as the River Thames where there is a single controlling authority, it becomes increasingly difficult when more than one authority is involved and near impossible in the case of international rivers like the Rhine, which travel through several countries. The problem that normally occurs is that one authority or country may wish to discharge wastes to the river and interfere with another who needs to re-use the water.

Provision of a clean supply of drinking water is the major cause of the decline in the incidence of water-borne diseases. Therefore, extreme care must be taken to ensure maximum removal of pathogens during sewage treatment so that only a small population of pathogens exist in the river. When water is re-used, full treatment is necessary no matter how clean it is bacteriologically. Water supplies taken from lowland rivers are normally stored for several days prior to treatment to enhance the die-off of pathogens. The water is then

TABLE 6.15 Microbiological quality requirements specified for surface waters intended for the abstraction of drinking water within EEC countries (European Communities 1975)

Parameter		A1 G	A1 I	A2 G	A2 I	A3 G	A3 I
Total Coliforms. 37°C	100 ml^{-1}	50		5000		50 000	
Faecal coliforms	100 ml^{-1}	20		2000		20 000	
Faecal streptococci	100 ml^{-1}	20		1000		10 000	
Salmonella		Not present in 500 ml		Not present in 1000 ml		—	

I, mandatory; G, recommended.

Category A1: waters treated by simple treatment and disinfection, e.g. rapid filtration and disinfection.

Category A2: normal physical treatment, chemical treatment, and disinfection, e.g. prechlorination, coagulation, flocculation, decantation, filtration, disinfection (final chlorination).

Category A3: intensive physical and chemical treatment, extended treatment, and disinfection, e.g. chlorination to breakpoint, coagulation, flocculation, decantation, filtration, adsorption (activated carbon), disinfection (ozone, final chlorination).

subjected to full treatment, i.e. microstraining, sand filtration, and, finally, disinfection. This ensures that the water is free from all pathogens. There is, of course, an increased risk of infection when water is used more than once, and routine and regular bacteriological monitoring is required (HMSO 1983*b*). Not all the chemical compounds that are added to water during consumption can be subsequently removed either at the sewage treatment or water treatment stages, resulting in them being supplied to the next consumer downstream. Some of these chemical compounds may be toxic or carcinogenic and particular concern is being expressed about a number of organic pollutants from domestic sewage.

Numerous surveys have indicated that there may be long-term effects in re-using water after being consumed, with higher incidences of cancer among people drinking river water already used for supply purposes compared with those drinking groundwater (Diehl and Tromp 1953; Stocks 1973; Page *et al.* 1976; Kuzma *et al.* 1977). However, a more recent study carried out in London over the period 1968–74 showed that there was no significant difference between the mortality rate between those drinking re-used River Thames water and groundwater (Beresford 1980). This work did, however, provide epidemiological evidence, consistent with the earlier studies, that a small risk to health existed from the re-use of drinking water. In this study, the percentage of domestic sewage effluent in the water was positively associated with the incidence of stomach cancer and bladder cancer in women (Beresford 1983; Beresford *et al.* 1984).

High levels of viruses are recorded in rivers waters used for supply purposes, but as long as sufficient water treatment is used then no public health problems will occur. However, such supplies must be subject to very strict and regular microbiological, toxicological, and chemical testing.

Pollution by gulls
Although migratory waterfowl, which roost at reservoirs, have no serious deleterious effect on water quality (Brierley *et al.* 1975), gulls, which feed on contaminated and faecal material at refuse-tips and sewage works, have been shown to excrete pathogens. All five British species of *Larus* gulls have been rapidly increasing in recent years with the Herring gull (*Larus argentatus*) doubling its population size every 5 to 6 years. This has resulted in a very large increase in the inland population of *Larus* gulls throughout the British Isles, both permanent and those over-wintering (Gould 1977). These birds are opportunist feeders and have taken advantage of the increase in both the human population and its standard of living, feeding on contaminated waste during the day and then roosting on inland waters, including reservoirs, at night (Hickling 1977; Gray 1979). Faecal bacteria especially *Salmonellae* spp., have been traced from feeding sites, such as domestic waste-tips to the reservoirs, showing that gulls are directly responsible for the dissemination of bacteria and other human pathogens (Williams *et al.* 1976; Riley *et al.* 1981;

Benton *et al.* 1982). Many reservoirs have shown a serious deterioration in bacterial quality because of contamination by roosting gulls (Fennell *et al.* 1974; Jones *et al.* 1978; Benton *et al.* 1983). Those situated in upland areas, where the water is of a high quality and treatment is minimal before being supplied to the consumer, are particularly at risk from contamination. Recent research has shown that numerous *Salmonella* serotypes can be isolated from gull droppings (Fenlon 1981; Girwood *et al.* 1985) (Table 6.16), as well as faecal coliforms, faecal streptococci, and spores of *Clostridium welchii* (Gould

TABLE 6.16 The most frequently isolated *Salmonella* serotypes obtained from gulls during 1982–83 compared to the frequency and ranking from humans, cattle, and sheep. The number of isolates are given in parentheses (Girdwood *et al.* 1985)

Salmonella serotype	Seagull	Human	Cattle	Sheep
S. virchow	1 (271)	3 (571)	3 (78)	—
S. typhimurium	2 (175)	1 (2487)	1 (1002)	2 (83)
S. bredeney	3 (49)	9 (64)	—	—
S. infantis	4 (17)	—	—	—
S. hadar	5 (16)	—	—	—
S. meunchen	6 (15)	—	—	
S. stanley	6 (15)	—	6 (9)	—
S. mbandaka	8 (14)	—	—	—
S. montevideo	9 (13)	—	6 (9)	—
S. agona	10 (11)	8 (65)	4 (28)	—

1977). Densities of these indicator bacteria are given in Table 6.17. Ova of parasitic worms have also been isolated from gull droppings (Crewe 1967) and birds are thought to be a major cause of the contamination of agricultural land, with the eggs of the human beef tapeworm (*Taenia saginata*) (Crewe and Owen 1978).

There is a clear relationship between the number of roosting gulls on reservoirs and the concentration of all the indicator bacteria, including *E. coli*, faecal streptococci, *C. welchii*, and *Salmonellae* spp. Fennell *et al.* (1974) reported maximum densities of gulls on their reservoir of 16 000 individuals that resulted in high levels of faecal contamination (*E. coli* 4000, faecal streptococci 600, and *C. perfringens* 16 per 100 ml of reservoir water). The results in Table 6.17 are particularly interesting as they show that the faecal coliform (FC) to faecal streptococci (FS) ratio for gulls is higher than the 4:1 ratio proposed by Geldreich (1972) to indicate contamination of human origin (Section 6.2.3.2). In a study by Gould (1977), 65% of all gull droppings had a FC:FS ratio in excess of 4:1, with the percentage of values exceeding this limit for the major *Larus* species being *L. argentatus* (94%), *L. fuscus* (100%), *L canus* (33%), and *L. ridibundus* (35%). Gould went on to suggest that the FC:FS ratio could be replaced by the ratio of faecal coliforms to *Clostridium*

TABLE 6.17 Number of droppings, total weight, and bacterial loads produced in a 24 h period by individual gulls of four different species. Total coliforms (TC), faecal coliforms (FC), faecal streptococci (FS), and spores of *Clostridium welchii* (CW) are given. The estimated 24 h nutrient loads for the four species of gull are also given (Gould 1977)

Larus species	Total no.	Total wt (g)	TC (10^8)	FC (10^8)	FS (10^6)	Ratio FC:FS	CW (10^3)	Org. C.	NH$_3$	Kj·N	Org. N (mg^{-1})	Sol. P	Total P
L. argentatus	39	17.5	5.1	5.2	<5.9	87	<0.98	1392	402	1819	1416	92	>115
	34	24.9	17.6	17.7	<20		<3.4						
L. fuscus	41	13.4	47	50	14.7	340	11.8	705	211	919	708	47	58
L. canus	62	11.8	7.2	6.2	1.1	585	<0.63	698	134	829	689	42	50
L. ridibundus	52	11.2	2.9	3.0	2.2	135	<0.20	502	113	608	495	30	38

welchii as an effective indicator of human pollution. Gull droppings can also contribute significant amounts of nitrogen and phosphorus to reservoirs which can lead to eutrophication (Table 6.17).

Although contamination of small water storage reservoirs has been eliminated by covering them, this is impracticable for large upland reservoirs serving major cities and towns. A considerable degree of success has been achieved by using bird-scarers and, more recently, by using techniques developed for the control of birds at airports (Bremond *et al.* 1968). Roosting can be discouraged by broadcasting species-specific distress calls of *Larus* gulls, which has led to dramatic reduction in bacterial contamination. Such a control option is far more cost-effective than installing and operating more powerful disinfection treatment systems (Benton *et al.* 1983).

6.2.4.2   Land Pollution

Sewage sludge disposal

Sewage sludge arises from the separation and concentration of the settleable fraction of wastewater as it passes through the treatment plant. Sludge is essentially of two types, raw and treated, with raw sludge from primary settlement tanks usually mixed with the less noxious secondary sludges, and treated sludge which has undergone some form of stabilization process, such as digestion, lime treatment, chlorination, composting, or drying (Section 5.1). Sludge disposal is a major operational cost in the treatment of wastewaters and the water industry tends to dispose of sludge *at minimum cost consistent with acceptable environmental impact and with flexibility and security* (Davis 1980). Sewage sludge is rich in organic matter, nitrogen, phosphorus, and trace elements. Therefore, combined with the philosophy of utilization of waste rather than disposal, spreading on agricultural land is steadily growing throughout Europe. However, the utilization of sewage sludge on agricultural land has been shown to affect the growth of plants and the health of animals because of a range of contaminants. Inorganic and organic contaminants are dealt with in Section 5.3, however, the third category of contaminants, the pathogens, are considered below.

The density and diversity of pathogens in sewage sludge depends on the nature of the wastewater, which is determined by the general health of the community and the presence of wastewater from hospitals, abattoirs, and meat-processing factories. Other factors, such as the diurnal and seasonal variation inflow and sewage constituents, and dilution by infiltration and stormwater are also important. The majority of the pathogen content of crude wastewater is concentrated in the primary sludge (Engelbrecht 1978). The effectiveness of the secondary treatment processes in removing pathogenic micro-organisms determines their concentration in secondary sludges (Table 6.18). However, as primary and secondary sludges are usually combined for disposal it is the choice of stabilization process that will ultimately control the

TABLE 6.18 The removal of pathogenic micro-organisms by sewage treatment (Carrington 1978)

Micro-organism or pathogen	Range of percentage removals of organism and number of reports reviewed (in parentheses)		
	Percolating filter	Activated sludge	Anaerobic digestion
Total bacteria	70–95 (4)	70–99 (3)	
Salmonella	84–99 (2)	70–99 (2)	84–92.4 (1)
Escherichia	82–97 (4)	91–98 (1)	
Mycobacterium tuberculosis	66–99 (2)	88 (1)	69–90 (2)
Poliovirus		Most (1)	
Coxsackievirus A	60 (1)		
Tapeworm ova	62–70, 18–26 (2)	Little effect (1)	97 (1)
Entamoeba histolytica	88–99.9 (1)	No reduction (2)	Removed (1)
Ascaris	70–99.8 (2)	93–98 (1)	45 (1)
Hookworm	100 (1)	81.5–96 (1)	

pathogen load being spread on to the land. The efficiency of the various stabilization methods of sewage sludge are compared and fully discussed in Section 5.1. The traditional methods of treatment, such as digestion and drying, are unable to eliminate pathogens, and although numbers are reduced, it appears that digestion, in particular, has little effect on the removal of pathogens such as *Salmonella* (Argent *et al.* 1977). Methods, such as composting and heat treatment, however, that involve heating the sludge to about 60°C, are able to totally eliminate or reduce pathogens to very low levels (Table 6.19).

In the UK, and Europe generally, the major hazard is associated with *Salmonella* and *Taenia* infections (WHO 1981). The bacterium *Brucella abortus* has also been indicated as a potential hazard for brucellosis-free attested herds feeding on sludge-treated grassland. Infected humans do not excrete Brucellosis and it is rarely found in sewage. However, it is encountered in farm and dairy wastes, although infection is normally transmitted by the aborted discharges of acute cases and by the milk of chronically infected animals. Sewage sludge can occasionally contain *Brucella* if it contains contaminated wastes from abattoirs or intensive farming units (Bell *et al.* 1978), although the bacterium cannot survive for long in the sludge. If an institutional source of infection exists in the community then *Mycobacterium tuberculosis* may enter and remain in the sewage treatment plant and even survive sludge digestion and drying. However, as there is such a low frequency of infection of the disease within the community it is unlikely to be normally present in sewage and sludge (Argent *et al.* 1977). Viruses are extremely host-specific and it is unlikely that any human enteroviruses in sewage sludge would present a hazard to animals or vice versa. In farming terms, the only viruses of

TABLE 6.19   Relative efficacies of various methods of sludge treatment in reducing of different pathogens or their time of survival (Carrington 1978)

Treatment	Relative reduction		
	Poor	Moderate	Good
Raw sludge storage	*Ascaris* ova	Viruses	
	Taenia ova	Bacteria	
Digestion	*Ascaris* ova	Hookworm ova	Viruses
	Taenia ova		Bacteria
			Entamoeba cysts
Composting			Viruses
			Bacteria
			Fungi
			Helminth ova
Lime treatment	*Ascaris* ova		Bacteria
Heat treatment			Viruses
			Bacteria
			Helminth ova
Irradiation	*Ascaris* ova		Bacteria
			Viruses

importance are foot and mouth and also swine vesiccular disease, both of which are controlled by veterinary inspection so that they are unlikely to find their way into sewage sludge (Hudson and Fennell 1980).

The utilization of sewage sludge on agricultural land is subject to strict controls in the form of guidelines (National Water Council 1981) and an EEC directive (European Communities 1982) for controlling the spread of diseases, with greater restrictions imposed on the use of raw than for treated sludges (Anon 1982a). In essence, the UK guidelines state that untreated sludge should preferably be ploughed-in or, if applied to pasture, grazing should be prohibited for a minimum of six months. This sludge should only be used on arable crops that are to be consumed by animals or only after processing if eaten by humans. However, Watson (1980) found that *Salmonella* was no longer detectable on cabbages seven weeks after the application of contaminated sewage sludge, indicating that a ban on the consumption of all vegetables may be over-cautious. There does appear to be some evidence that contamination of vegetable salad crops occurs when the soil has been treated with sludge (Wright *et al.* 1976). However, as Watson points out, this must be weighed against the dosage required to produce clinical symptoms. It would appear that if accurate assessments of the possible health risks of sludge disposal to land are to be made, then the concentration of pathogens being applied must be known. The only restriction for treated (digested) sludge is that if used on grassland or pasture a minimum of three weeks must be allowed before grazing recommences in order to ensure that pathogen concentrations are below the minimum infective dose. The minimum infective dose is the number of pathogens that have to be ingested before the disease is contracted. This varies with the pathogen and the health of the animal (Pike and Davis 1984). For example, in healthy cattle the minimum infective dose of *S. typhimurium* or *S. dublin* is between 10^7–10^8.

The overall concentration of *Salmonella* in sewage sludge applied to grassland is low in relation to the minimum infective dose for cattle, with between 3–24 000 *Salmonella* bacteria per 100 ml of wet raw sludge ($\bar{x} = 4500$), compared with 3–210 per 100 ml ($\bar{x} = 60$) for digested sludge (Gray 1979). The cycle of infection between man and grazing animals involving sludge is complex (Fig. 6.3) and has been reviewed by Pike (1981). No clear association between sludge utilization and salmonellosis in grazing animals has been established in the UK (Hall and Jones 1978; Williams 1978), although there have been a number of reported cases (Jack and Hepper 1969; Bicknell 1972; Burnett *et al.* 1980). However, there is some evidence from other European countries, such as The Netherlands, Germany, and Switzerland, that infection can occur via this route (Pike 1981). There are also several cases where contamination has been linked to contaminated water, leaking sewers, and septic tank discharges (Carrington 1980*a*). Over 50 serotypes of *Salmonella*

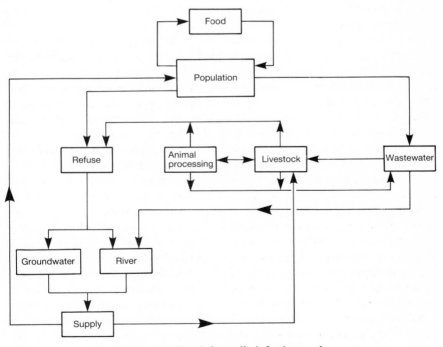

FIG. 6.3   The *Salmonella* infection cycle.

that can infect either man or animals have been isolated from sewage sludge, and if animals are unusually susceptible or if there is heavy contamination of grazing land by sludge then infection could readily occur (Hall and Jones 1978; Haugaard, 1984; Linklater *et al.* 1985). *Salmonella montevideo* was recorded as the most common serotype isolated from sewage sludge and abattoir effluents in a survey conducted in Scotland by Linklater *et al.* (1985). This serotype has caused frequent outbreaks in animals throughout eastern Scotland and causes abortion in sheep that has resulted in heavy losses in flocks from that part of the UK. This is of particular interest as the transfer of this serotype between flocks and herds by the gull *Larus argentatus* has also been demonstrated (Coulson *et al.* 1983). Methods of isolation and identification of *Salmonella* spp. from sewage sludge are summarized by Carrington (1980*b*).

Apart from salmonellosis the dissemination of the human beef tapeworm, *Taenia saginata*, is the other major microbial hazard associated with the agricultural use of sewage sludge. The cestode worm infects both cattle (*T. saginata*) and pigs (*T. solium*), but it is the former that is of particular concern. The primary host is man where it grows in the intestine into an adult worm and continuously sheds ova which are passed out with the faeces. In the sewage works, the ova, that are denser than water (density = 1.1), readily settle out in the settlement stages and concentrate in the sludge. Once the sludge is spread on to grassland the secondary host, cattle, ingest the ova that hatch, forming cysticeri (*Cysticerus bovis*), which develop in muscle tissue. The cycle is completed if humans consume infected and inadequately cooked beef. The ova are very resistant and can remain viable for long periods, making the strict adherence to the six-month ban on grazing absolutely necessary. The frequency of infection in humans has decreased since the Second World War because of two main factors. First, meat is rigidly inspected at the abattoir and infected carcasses are either frozen to destroy the cysts or if infection is heavy the carcass is destroyed, and secondly, the greater awareness of adequately cooking meat, which destroys the cysticeri. There are less than 100 confirmed cases of infection in humans per year in England and Wales, with 50–65% of these being in foreign visitors from countries where infection is far more prevalent. The frequency of infection in cattle has remained constant for a long time at <0.05–0.10% (Blamire *et al.* 1980), although infection in cattle can only occur by ingestion of ova shed by infected humans. However, there is growing evidence that birds feeding at sewage treatment works or on infected material at refuse tips may ingest tapeworm proglottides and disseminate eggs in their droppings (Crewe and Owen 1978). Other modes of transmission, such as defecation by motorists near lay-bys and discharge of stormwater to watercourses used by cattle for drinking, may also be possible (Pike and Davis 1984). A study on the outbreaks of *Cysticerus bovis* in Scotland has shown that only 7% of the 218 outbreaks reported occurred on farms using sewage sludge. The geographical pattern of outbreaks, however, approximated to the major road routes suggesting that the sanitary habits of motorists may well be involved (Reilly et al. 1981). Other parasites of veterinary importance that occasionally occur in sewage sludge and effluents are *Ascaris lumbricoides*, *Trichuris trichiura*, *Necator americarius*, *Ancylostoma duidenate*, *Ascaris suum*, *Fusciola hepatica*, and *Monieza* spp. The life-cycles and the factors that govern the survival of these parasites in sewage sludge are discussed by Watson et al. (1983) and also by Kiff and Lewis-Jones (1984).

The survival of pathogens on agricultural land depends on a number of environmental factors and will vary according to whether the herbage or soil is contaminated. Once sludge is spread there is a continuous die-off of pathogenic micro-organisms, the rate of which depends on the environmental conditions, although they can persist for a long time. The meterological factors are most important as pathogens are highly sensitive to sunlight and desiccation. However, rainfall will help in the dispersion of pathogens. Those micro-organisms on the herbage are most exposed and are short-lived in comparison with those that find their way into the soil (Table 6.20). Both the intensity and duration of daylight are important. For example, coliform bacteria on green plants are inactivated within 10 h in summer daylight, whereas in wet cold weather, this is extended to 28 h. Within the soil, bacteria will survive for shorter periods in more permeable drier soils, at higher temperatures, in acid conditions, at the soil surface exposed to sunlight, when the organic matter content is low and re-growth is prevented, and, finally, where there is an active soil microfauna that can predate on the pathogens (Thunegard 1975). The availability of organic matter in the immediate vicinity of the bacterium is important for its survival as it can only be utilized as food for maintenance as well as for protection. The pH of the soil is also important as the nearer the pH is to neutrality the better the bacterial survival rate (Watson 1980). *Salmonella* survival is very variable, but reports of bacteria surviving on grass and soil for up to 27 weeks is not unusual. Field trials on arable land has indicated that although the survival of *Salmonella* is linked with the environmental conditions and the concentration of bacteria in the

TABLE 6.20 Median bacterial counts (per gram) in four replicate samplings of a single grassland site at varying times after application of digested sludge (Carrington 1980a)

Days after application	Soil			Grass		
	E. coli	Total coliform bacteria	Faecal streptococci	E. coli	Total coliform bacteria	Faecal streptococci
15	250	770	17	689	4720	0.8
22	4602	1774	20	464	>1415	1.3
29	8550	950	23	12	144	1
36	1200	1225	250	32	175	<0.2

sludge, under summer conditions, *Salmonella* will only survive for about 6 weeks in the soil (Watson 1980). When sludge is buried in trenches and covered with 0.3 m of soil, *Salmonella* can still be isolated 9 months later. The ova of *Ascaris* remain viable in the soil for up to 5 months, but if contaminated grass has to be used for silage then all the ova should be destroyed after 80–90 days storage. The persistence of pathogens in the soil is reviewed by Carrington (1980*a,b*).

Unlike the chemical contaminants of sludge, the hazards presented by the presence of pathogens are short-term, depending less on their numbers but rather on the method of application to the soil. Four methods of applying sewage sludge to land are widely used: surface spreading, trenching, irrigation using sprayers, and direct injection (Section 5.3). All contaminate the soil, although trenching and direct injection prevent the herbage becoming contaminated, thus reducing the risk of disease transmission to grazing animals. The problem of using irrigation via sprayer rain guns, is that the herbage is more effectively covered with sludge, thus increasing the distribution of pathogens. There is also a possible risk of disease transmission from the formation of aerosols.

Apart from human and animal pathogens, sewage also contains vegetable waste and soil that harbour a wide variety of plant pathogens, including viruses, fungi, bacteria, and nematodes. The use of sewage for irrigation, and especially the use of sewage sludge on agricultural land, helps the dissemination of plant pathogens back to agricultural soils. The most important plant pathogens are the cyst-forming nematodes belonging to the genus *Heterodera*. Seven species are regularly isolated from sewage sludge (Table 6.21), the most important being the potato cyst eelworms, *H. rostochiensis* and *H. pallida* that are endemic in Europe. In the soil, the nematode is present as cysts which contain the eggs of the parasite. Even in the absence of the host plant these cysts can remain viable for many years and infection is particularly difficult to eradicate. The parasite is easily dispersed via contaminated soil, and in wet weather, when vegetables retain a greater proportion of soil on their surface, there is a particular risk of spreading the disease. Cysts accumulate in the soil washed from potatoes, sugar beet, and carrots during processing, the sludge from vegetable-processing factories must be carefully disposed of, if *Hetero-*

TABLE 6.21   Species of *Heterodera* associated with sewage sludge

Species	Name	Host crops
H. avenae	Cereal cyst-nematode	Oats, barley, wheat
H. carotae	Carrot cyst-nematode	Carrot
H. cruciferae	Cabbage cyst-nematode	Brassicas and other plants
H. goeltingiana	Pea cyst-nematode	Peas and beans
H. pallida	Pale potato cyst-nematode	Potato, tomato
H. rostochiensis	Potato root eelworm	Potato, tomato
H. schachtii	Beet cyst-nematode	Sugar beet and fodder beet

dera contamination of cyst-free soil is to be avoided. Potato growing is excluded from heavily infected soils, but the danger that sludge could introduce the pathogen to cyst-free soils is a real one. The EEC directive that concerns potato growing, stipulates that potatoes grown for export can only be produced on certified cyst-free soil.

Several species of cyst nematodes, including *H. rostochiensis* and *H. pallida*, can also parasitize the roots of tomatoes. If tomato plants are allowed to grow on sludge beds or on sludge dumps, or even as weeds, after the application of sludge to land, this will greatly increase the number of cysts in the sludge or soil.

Sludge can be examined for cysts, using a simple flotation method, and as long as the level of infection in the soil remains below 25 cysts per 100 g of dry soil then no significant losses in production will occur (Linfield 1977).

Sewage irrigation
Since the development of the sewage farm, when sewage was allowed to flow on to deeply ploughed land, sewage has been disposed of by land irrigation. This type of treatment has been almost completely replaced in the British Isles, although some European cities have retained part of their sewage farm system. For example, the sewage farm developed between 1890–1920 in Paris is still in use, although since then the city has grown dramatically, with an enormous increase in the total flow of sewage, the majority of the sewage is now treated by large biological units before being returned to the River Seine (Dean 1978). The 100-year-old sewage farm at Wroclaw (Poland) is also still in use. The sewage, after a 6 h settling period, is used for the irrigation of grazing land for dairy cattle, fodder, and trees (Cebula and Kutera 1978) (Sections 4.5.3 and 7.2).

Sewage irrigation of agricultural land is practised throughout the entire arid tropical and semi-tropical areas of the world, being particularly common in the Southern Mediterranean, South-west USA, South America, Australia, and India. Contamination of plants and soil by faecal micro-organisms and pathogens readily occurs, allowing pathogens to complete life-cycles by re-infecting animals. Sewage irrigation also introduces plant pathogens into the soil and spreads viruses and bacteria that will either infect grazing animals or find their way into surface or groundwater. Bacteria are fairly persistent in the soil with *Salmonella* surviving for up to 6 weeks after irrigation (Joshi *et al.* 1973). Plant pathogens are spread by the re-use of infected water or sewage, and plant-pathogenic bacteria, fungi, and nematodes can remain viable for many years in the soil (Steadman *et al.* 1975). The greatest proportion of irrigation water is applied to pasture or to grain crops that need to be processed and cooked before being eaten, and cotton also responds favourably to irrigation with sewage (Dean and Lund 1981). Treatment requirements for sewage used for crop irrigation varies significantly and depends on the quality of the raw sewage and the type of crop to be irrigated.

In California, for example, primary effluent may be used for fodder, fibre, and seed crops, and secondary treatment is the minimum treatment required before sewage can be used on food crops or on pasture used for dairy herds (Pettygrove and Asano 1984). The use of sewage for irrigation of vegetables and fruit, however, is discouraged (Sheikh *et al.* 1984). But in desert areas, where water is the limiting factor for plant growth, sewage may have to be used for these types of crops, with the nutrients in sewage being an extra economic incentive. Fresh vegetables and fruit in arid areas are a rich source of income to farmers and it is inevitable that sewage will be used. However, the dangers are very real, for example, a recent cholera outbreak in Jerusalem was traced to lettuce grown by a farmer who had illegally diverted raw sewage to irrigate his crop (Shuval and Fattal 1980; Fattal *et al.* 1986*b*). Poliovirus was isolated from lettuce sprayed with sewage up to 4 weeks after application, although survival appeared to be weather-dependent (Larkin *et al.* 1976). The virus was shown to persist for between 93–123 days in the winter when sprayed on the soil used to grow vegetables, but this was reduced to 8–11 days in the spring (Tierney *et al.* 1977). The degree of contamination depends on the quality of the wastewater used. Guidelines have been proposed that state that sewage effluents must be treated and disinfected before use and contain <100 coliforms or <20 *E. coli* per 100 ml (Krishnaswami 1971). Chlorinated secondary wastewater effluents are widely used in the southern USA for the irrigation of golf courses, school grounds, parks, and cemeteries. The bacterial standards for such effluents vary from state to state, but are becoming more stringent (DeBoer and Linstedt 1985).

6.2.4.3   Atmospheric pollution

Aerosols are particles with a diameter of between 0.01–50 μm that are suspended in the air. Only particles <5 μm are medically important and these are readily released during the treatment of sewage, especially from the aeration unit of the activated sludge process and from the distributor nozzles on percolating filters (Adams and Spendlove 1970). The aerosols from sewage treatment plants can contain any of the pathogenic viruses and bacteria associated with sewage (Dart and Stretton 1977), and can infect animals or humans by inhalation. The degree of inhalation is dependent upon particle size, with particles >2–5 μm in diameter being retained in the upper respiratory tract, but smaller ones able to enter the alveoli of the lungs (Sorber and Guter 1975). Such small droplets of water rapidly evaporate so that any micro-organisms in the aerosol would probably be dehydrated within seconds. Therefore only those people who inhale aerosols within close proximity of the treatment plant (<20 m) are at risk. At distances >20 m from units, the risk of contamination rapidly falls, so that there is negligible risk to people who live or work adjacent to treatment plants. However, aerosols that contain micro-organisms resistant to dehydration or contain organic matter which can delay the dehydration of micro-organisms, can increase the effective distance at which aerosols can be hazardous. This may be a particular problem when sewage sludge is sprayed on to agricultural land.

Aerosols are primarily a problem associated with activated sludge aeration basins (Raygor and Mackay 1975), although aerosols are also formed as sewage is poured on to percolating filters, being especially serious when spray nozzles are used in conjunction with plastic media. The concentration of bacteria in aerosols and the distance they can travel from source depends on a number of factors, the most important being wind velocity, temperature, solar radiation, and humidity. The spread of micro-organisms from treatment units follows the Gaussian-plume equation with counts of up to 2.4×10^6 *E. coli* per 50 m^3 recorded by Grunnet (1975) in the air next to an activated sludge aeration tank, and with *S. paratyphi* B also occasionally isolated. Goff *et al.* (1973) examined the air around two percolating filters and recorded coliforms extending 100 m from the filters in concentrations up to 90 coliforms per m^3.

When sewage is used for irrigation, contaminated aerosols can be produced if the effluent is either poured from tankers or sprayed on to the land. This is particularly a problem in public areas where sewage is used for the irrigation of golf courses and parks. Crop spraying employing high-pressure jets produce visible aerosols that carry viable pathogenic bacteria up to 650 m downwind. Therefore, in order to minimize any risk to public helath, a 1000 m zone around the sprayer is cleared of people and animals (Shtarkas and Krasil'Schchikov 1970). Attempts have been made to reduce the production of aerosols by using low-pressure jets, but, surprisingly, research has indicated that such sprayers may increase aerosol production (King *et al.* 1973), although this has yet to be fully evaluated. There is some evidence that an irritant contact dermatitis from airbourne contamination from spreading sewage sludge is also a problem for those engaged in this activity (Nethercott 1981).

6.2.4.4   *Antibiotic resistance in enteric bacteria*

The widespread use of antibiotics in medical and veterinary treatment, and in the control of infection in intensively reared livestock by the medication of food and drinking water has led to the emergence of antibiotic-resistant strains of bacteria. Although resistant strains arise by genetic mutation within the cell, which then produce resistant bacteria by normal cell division, resistance to antibiotics is also transferable either *in vivo* or *in vitro* to other enterobacteria. Not all resistant bacteria have the ability to transfer antibiotic resistance, only those processing R-factors. The R-factors, and the genetic factors conferring resistance, are carried on extra-chromosomal elements of nucleic acids (plasmids) that can replicate autonomously during cell division. Genetic material conferring resistance can be transferred during conjugation between pairs of bacteria without R-factors. For example, an R$^+$ strain of *E. coli* could transfer antibiotic resistance to a R⁻ strain of *E. coli* or *Salmonella* sp. (Pike 1975; Carrington 1979).

Antibiotic-resistant enteric bacteria are particularly hazardous as they are extremely difficult to treat medically, especially where strains have acquired multiple resistance to antibiotics, which is a feature of transferable resistance. This has been highlighted by the rapid spread of multi-resistant strains of *Staphylococcus aureus* in hospitals world-wide, resulting in numerous deaths. Once established in a host, such bacteria have a distinct competitive advantage over similar sensitive strains when particular antibiotics are administered. Transmissable R-factors have been found in 61% of resistant strains (Linton *et al.* 1972), and a high proportion of the resistant isolates from rivers and coastal waters examined by Williams-Smith (1970; 1971) were able to transmit resistance to *Salmonella typhi* and *S. typhimurium*. It is estimated that about 50% of healthy adults in Britain excrete some antibiotic-resistant coliform bacteria (Linton *et al.* 1972), with higher percentages recorded in adults, having close contact with farm animals.

Like other bacteria, antibiotic-resistant strains are reduced by sewage treatment with 90% removal achieved by full biological treatment. In the absence of the relevant antibiotic, resistant bacteria have no advantage over sensitive organisms in the aquatic environment, which is shown by the ratio of antibiotic-resistant to antibiotic-sensitive coliforms remaining constant before and after treatment. The release of antibiotic-resistant bacteria into the environment is best controlled by restricting the prescribing and use of antibiotics (HMSO 1969). The distribution of antibiotic-resistant bacteria in sewage and the effect of sewage treatment is reviewed by Carrington (1979).

Further reading

Pathogens: Dart and Stretton 1977; Carrington 1980; Galbraith *et al.* 1987; Andersson and Stenstrom 1986.
Indicator organisms: Burman *et al.* 1969; Bonde 1977; Mara 1974; Hoadley and Dutka 1977; Dutka 1979; HMSO 1983*b*; WHO 1984.
Hazards: Piechuch, 1983; Lundholm and Rylands 1983; Clark *et al.* 1984*a,b*; Sridhar and Oyemade 1987; Galbraith *et al.* 1987; Clark 1987.
Bathing waters: Moore 1977; Agg and Stanfield 1979; Waldichuck 1985; Jones and White 1984; Fattal *et al.* 1986*a*; Brown *et al.* 1987.
Shellfish: Wood 1979; Wilson 1988.
Groundwater: Allen and Geldriech 1975; Seppanen and Wihuri 1988.
Reuse of wastewater: Dean and Lund 1981; Beresford *et al.* 1984; Fattal *et al.* 1986*c,d*; Shuval *et al.* 1984, 1986*a,b*.
Pollution by gulls: Gould 1977.
Sewage sludge: Pike 1981; Strauch *et al.* 1985; Schwartzbrod *et al.* 1987.
Atmospheric pollution: Sorber and Guter 1975.
Antibiotic resistance: Dart and Stretton 1977; Carrington 1979.

6.3 REMOVAL OF PATHOGENIC ORGANISMS

6.3.1 Environmental factors and survival

All pathogens are able to survive for at least a short period of time in natural waters, both fresh and saline. Usually, this period is extended when there are cooler temperatures and if organic pollution is present. The major environmental conditions affecting the survival and viability of pathogens are: sedimentation, solar radiation, predation, bacteriophages, nutrient deficiencies, algal and bacterial toxins, and physico-chemical factors. Adsorption and sedimentation can be significant factors in the removal of enteric bacteria from surface waters, depending on the nature of sewage, the degree of treatment, and the effect of vertical dispersion. In raw sewage, 50–75% of the coliforms are associated with particles with settling velocities > 0.05 cm s^{-1}. Therefore, in conventional primary sedimentation at a treatment plant significant removals of enteric bacteria are expected (Section 6.2.2). If discharged to coastal waters other conditions, such as enhanced flocculation, adsorption to marine and estuarine sediments combined with sedimentation, may result in significant removals of bacteria. However, Irving (1977) demonstrated that adsorption of bacteria on to particulate matter was not significant in sea water. By comparing the mortality rate (T_{90}) at different concentrations of suspended solids (0–10 000 mg l^{-1}) under dark and light conditions he was able to show that although the value of T_{90} varied with turbidity in the light, under dark conditions it remained stable showing that solar radiation was the major factor controlling inactivation, whereas adsorption to particulate matter was insignificant (Table 6.22). These processes are hindered by vertical mixing because of wind and bottom scouring, which can maintain in

TABLE 6.22 Effect of particulate matter on the mortality rate of coliform bacteria expressed as the time in hours required for 90% mortality to occur (T_{90}) in light (50 cal cm^{-2}h^{-1}) and dark conditions. The correlation coefficients for the regression lines calculated from the mortality data in each case are also given (Irving 1977)

Suspended-solids content (mg l^{-1})	Light		Dark	
	T_{90}	Correlation coefficient	T_{90}	Correlation coefficient
0	1.0	0.98	116	0.94
300	1.3	0.99	155	0.89
600	6.5	0.77	382	0.61
1000	12	0.86	123	0.91
5000	20	0.92	107	0.88
10 000	37	0.64	108	0.96

suspension or re-suspend settleable bacterial (Mitchell and Chamberlin 1975). Normally, sewage receives some degree of treatment before being discharged and even those effluents that are discharged directly to coastal waters generally receive primary settlement. Therefore, the effect of sedimentation in the natural environment is not generally significant.

The rate of *Escherichia coli* mortality in sea water has been shown to be proportional to the size of the marine microfauna (Mitchell *et al.* 1967). Mitchell (1971) implicated three groups of micro-organisms, cell-wall lytic marine bacteria, marine amoebae, and marine bacterial parasites. Although bacteria in the natural environment competes with the unadapted enteric bacteria, especially for nutrients (Jannasch 1968), the concentration of predators such as, protozoans and bacteriophages, is normally too low to have a significant effect on pathogen survival (Pike and Carrington 1979). In sea water, the diversity of predators is lower and bacteriophages are inactivated. Toxins and antibiotics produced by algae and other bacteria can cause a significant increase in the death-rate of enteric bacteria in enclosed water bodies such as, ponds and lakes, but the dilution effect in rivers, estuaries, and sea water renders them harmless (Carlucci and Pramer 1962; Foxworthy and Kneeling 1969). The concentration of organic matter in the form of carbon, nitrogen, and phosphorus can aid survival and, in some instances, encourage growth of enteric bacteria, although in sea water the concentration of the important nutrients is too low (Won and Ross 1973).

Some idea of the effect of time on the survival of pathogenic micro-organisms can be obtained by examining the removal efficiency of storage lagoons that are used at some waterworks to improve water quality prior to treatment and supply. Up to 99.9% reduction of enteric bacteria can be obtained by storage, although this is dependent on temperature, retention time, and the level of pollution of the water (Kool 1979). For example, a 99.9% reduction in pathogenic viruses by storage requires a retention time of 20 d at 20–27°C, but up to 75 d at 4.8°C. The half-life or die-off rates of the commonest pathogenic bacteria have been assessed by McFeters *et al.* (1974) who suspended samples in well water enclosed in membrane chambers (Table 6.23). These results indicate that the half-life of pathogenic bacteria and coliforms are of the same order of magnitude, except for *Shigella* sp. Using a similar technique, the percentage survival of bacteria in flowing river water was calculated after 4 d: for *E. coli* (78%); faecal streptococci (46%); and *Salmonella* sp. (74%) (Smith *et al.* 1973). However, pathogenic bacteria have been reported as surviving for much longer periods, especially *Salmonella* Sp. (Geldreich *et al.* 1972; Leclerc *et al.* 1976). Viruses can survive longer than bacteria in natural waters, with some viruses surviving for up to twice as long as certain indicator bacteria (Table 6.24) (Kool 1979).

Temperature is an important factor in the survival of viruses and at low temperatures (4–8°C) survival time will increase (Table 6.24). For example, echovirus type 6 is inactivated more readily in sea water at 22°C (8 d) than at

TABLE 6.23 Half-life of enteric bacteria in well
water (McFeters *et al.* 1974)

Species	Half-life (h)
Salmonella paratyphi B	2.4
Streptococcus bovis	4.3
Vibrio cholerae	7.2
Streptococcus equinus	10.0
Salmonella paratyphi A	16.0
Salmonella typhimurium	16.0
Coliform bacteria	17.0
Salmonella paratyphi D	19.2
Enterococci	22.0
Shigella dysenteriae	22.4
Shigella sonnei	24.5
Shigella flexneri	26.8
Aeromonas	72.0

TABLE 6.24 Effect of temperature on the inactivation of enteric micro-
organisms during storage. Time taken for 99.9% reduction of micro-
organisms (Kool 1979)

Species	Temperature (°C)	Time (d)	Temperature (°C)	Time (d)
Poliovirus	4	27	20	20
Echovirus	4	26	20	16
E. coli	4	10	20	7
Str. faecalis	4	17	20	8
Poliovirus type 1, 2, 3	4–8	27–75	20–27	4–20
Coxsackie virus type A2, A9	4–8	12–16	20–27	4–8

3–5°C (91 d) (Won and Ross 1973). Coliphages survive for only 7 d in pond
water at 20°C in the absence of hosts, but will survive for up to 14 d when
bacteria are present (Scarpino 1975). Many studies have demonstrated that
the death-rate of enteric and indicator bacteria is accelerated at higher
temperatures in natural waters (Table 6.25) (Jamieson *et al.* 1976). This is also
illustrated by Evison and James (1973) who compared the survival of indicator
bacteria over a 5 mile distance in two rivers with similar flow rates in East
Kenya ($\bar{x} = 18.5°C$) and Britain ($\bar{x} = 2°C$). There was a 96.5% reduction of
indicator bacteria in the warmer river compared with only a 56% reduction at
the lower temperature.

Both UV radiation and short-wave visible light are lethal to bacteria, with
the rate of death related to light intensity, clarity of the water, and depth. In the
dark, the death-rate of coliforms follows first-order kinetics over the initial
period, however, predation by protozoans causes a departure from the log-
linear relationship (Morgan 1985). The death-rate is measured as the time for
90% mortality to occur (T_{90}). In the dark, at 20°C, the T_{90} for coliforms is

TABLE 6.25   Effect of temperature and salinity on the survival of pathogenic micro-organisms in sea water, expressed as the time for a 99% reduction of the original inoculation of $1.5 \times 10^7 ml^{-1}$ (Jamieson *et al.* 1976)

Temp. (°C)	Salinity (g kg⁻¹)	Period for a 99% reduction in concentration (d)					
		E. coli	*Myco.* tuberculosis	*L. interrogans*	*S. typhi*	*Shig.* dysenteriae	*V. cholerae*
4	5	5	7	1	6	5	1
	20	5	7	1	6	4	1
	35	5	7	1	6	1	1
25	5	4	6	1	5	1	1
	20	3	6	1	5	1	1
	35	3	6	1	5	1	1
37	5	1	6	1	3	1	1
	20	1	6	1	1	1	1
	35	1	6	1	1	1	1

49 h, whereas under midday sunlight, the T_{90} is reduced to 0.3 h. *Salmonellae* sp. have mortality rates comparable with the faecal coliforms (Gameson 1984). The death-rate is also temperature-dependent, decreasing by a factor of 1.97 for each 10°C rise in temperature, and is proportional to the total radiation received, regardless if this is continuous or intermittent. The amount of radiation required to kill 90% (S_{90}) of coliforms is estimated as 23 cal cm^{-2} (11.3 cal cm^{-2} for *E. Coli*). The S_{90} ratio of faecal streptococci to coliforms varied from 10–39 indicating that the die-off rate for faecal streptococci is appreciably slower than for total coliforms (Gameson and Gould 1974). The effect of temperature is far less marked under light conditions and although variations in salinity have no effect on the death-rate it is substantially slower in fresh and brackish waters (Gameson 1985a). The daily solar radiation in southern Britain is between 50–660 cal cm^{-2}, and there appears to be ample radiation to inactivate all faecal coliforms. Wavelength is also important, with about 50% of the lethal effect of solar radiation attributable to wavelengths below 370 nm, 25% to near visible UV (370–400 nm), and 25% to the blue-green region of the visible spectrum (400–500 nm). The effect of wavelengths > 500 nm is negligible (Gameson and Gould 1974). The amount of radiation reaching the sea is dependent on a number of factors, such as solar elevation and weather conditions. The lethal radiation is rapidly absorbed by suspended and colloidal solids, which rapidly reduce its effect, with depth being effective only up to 5 m in sea water (Pike 1975). In sea water, the mortality rate of bacteria rapidly decreases as the turbidity of the water increases (Fig. 6.4). The die-off rate is related to depth, with an effective attenuation coefficient of approximately 0.22 m^{-1}. As this die-off rate for coliforms is proportional to the light intensity and is therefore essentially a first-order relationship, it can

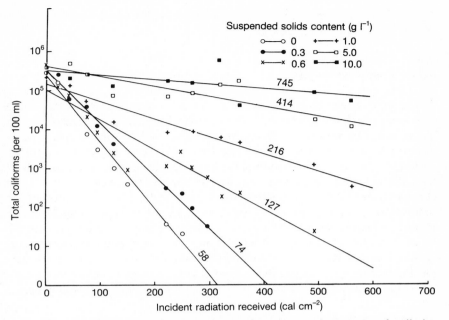

FIG. 6.4 The effect of suspended solids concentration on the amount of radiation required to inactivate coliform bacteria in 1 litre beakers. The amount of radiation required to inactivate 90% of the coliforms originally inoculated (S_{90}) in cal cm^{-2} is shown against the plotted regression (Irving 1977).

be expressed as:

$$\frac{dC}{dt} = -kl_0 \cdot e^{-\alpha z}C$$

where C is the coliform concentration at time t at a depth Z, k is a proportionality coefficient, l_0 is the light intensity just below the water surface and α the effective attenuation-coefficient (Mitchell and Chamberlin 1975).

Since the introduction of the bathing waters directive by the EEC (European Communities 1976) there has been considerable interest and research into the survival of pathogens in estuarine and coastal waters. For example, the series of technical reports produced by the Water Research Centre in Britain entitled 'Investigations of sewage discharges to some British coastal waters', largely written by Dr A. L. H. Gameson. Sea water does not favour the survival of pathogens and *E. coli* in particular, with osmotic effects, pH, and specific ion toxicity, such as heavy metals (Jones 1964), all important factors effecting the survival of enteric bacteria (Mitchell and Chamberlin 1975). However, only a slight increase in the death-rate of pathogens is recorded at increased salinities (Table 6.25), although the mean T_{90} for coliforms in river water is double the value for sea water (Carrington 1980a).

Viruses also show a slight increase in death-rate as the salinity increases (Table 6.26).

6.3.2 Treatment processes

The primary objective of wastewater treatment is to reduce the BOD_5 of the wastewater so that it can be safely discharged without causing deoxygenation in the receiving waters. A secondary objective, and in some respects more important than the first, is the removal of enteric bacteria and other pathogens. It is the introduction of the biological treatment of sewage, linked with the chlorination of domestic water supplies, that has eliminated the classical water-borne diseases from Western Europe. The effectiveness of the removal of pathogenic micro-organisms from wastewaters depends on a number of factors, such as the diversity and numbers of pathogens present, the type of treatment processes used, loading rates and plant efficiency, and seasonal factors, e.g. temperature and the chemical nature of the wastewater. During treatment of sewage the microbial flora changes from predominately faecal in character to that found more commonly in enriched fresh waters.

Methods of removal
The removal of pathogenic micro-organisms is effected by a combination of physical, chemical, and biological processes. Physically pathogens are removed by adsorption and settlement, and the overall concentration is reduced by dilution. The chemical nature of the wastewater will determine

TABLE 6.26 Effect of temperature and salinity on the survival of enteroviruses inoculated into sea water at a concentration of 10^5–$10^6 ml^{-1}$, expressed as the time for a 99% reduction in virus concentration and the time in weeks until no virus is recovered (Lo *et al.* 1976)

Incubation temp (°C)	Salinity (g kg^{-1})	Period for 99% reduction (wks)			Period until virus no longer recovered (wks)		
		Polio	Echo	Coxsackie	Polio	Echo	Coxsackie
4	10	18	18	32	46	46	>53
	20	12	20	46	46	53	>53
	34	10	18	53	53	53	>53
15	10	40	12	46	53	24	>53
	20	10	12	20	22	32	53
	34	6	14	32	22	32	46
25	10	4	6	6	8	12	12
	20	2	4	4	6	8	10
	34	4	4	6	8	6	10
37	10	<2	<2	<2	<2	<2	<2
	20	<2	<2	<2	<2	<2	<2
	34	<2	<2	<2	<2	<2	<2

whether the environmental conditions are suitable for the survival or even the growth of pathogens. However, factors, such as hardness, pH, ammonia concentration, temperature, and the presence of toxic substances, can all increase the mortality rate of the micro-organisms. Biologically, death of pathogens can occur because of a number of reasons including starvation, although predation by other micro-organisms and grazing by macro-invertebrates are important removal mechanisms. Bacteriophages and the predatory bacterium *Bdellovibrio bacteriovirus* are common in sewage. However, if they were active during treatment than a rapid increase in their numbers would be expected as treatment progressed. Infected bacterial host cells will release hundreds of phage progeny some 20–30 minutes after infection, and host cells of *Bdellovibrio* result in a 3–5-fold increase of the organism 4–5 hours after infection. However, during sewage treatment this does not occur, indicating that they are inactivated by contact with sewage (Pike 1975). This has been confirmed by studies on rivers receiving sewage (Fry and Staples 1976) and activated sludge plants to which coliphages have been added (Balluz *et al.* 1978). Enteric viruses tend to survive longer than bacteria in wastewater as bacteria are more susceptible to environmental factors than viruses because of their cellular nature. Also, viruses are much smaller than bacteria and so behave as colloids in water, carrying a charge on their surface and thus are more readily adsorbed to solids than bacteria. There is also evidence to suggest that adsorbed viruses are protected from inactivation and will remain viable longer than unadsorbed viruses and bacteria (Funderburg and Sorber 1985).

It is convenient to look at the wastewater treatment plant as an enclosed system with inputs and outputs. It is a continuous system and, therefore, the outputs, in the form of sludge and a final effluent, will also be continuous. Although a comparison of the number of pathogens in the influent with that of the final effluent will provide an estimate of overall efficiency, it will not give any clues as to the mechanism of removal. Essentially, pathogens are either killed within the treatment unit, discharged in the final effluent, or concentrated in the sludge. The latter will result in secondary contamination problems if disposed either to agricultural land or into coastal water (Section 6.2.4). An estimation of the specific death-rate $(-\mu)$ of a pathogen can be calculated by accurately measuring all the inputs and outputs of the viable organism (Pike and Carrington 1979). All the individuals of the population of the pathogen (x), which are assumed to be randomly dispersed within the reactor, have the same chance of dying within a specific time interval $(t–t_0)$. Under steady-state conditions, the causes of death can be assumed to remain constant in terms of concentration (in the case of a toxic substance) or number (of a predator), and the rate of death will be proportional to the number of survivors (x_t) of the original population (x_0). This can be represented by a first-order, exponential 'death' equation:

$$x_t = x_0 \cdot e^{-\mu(t-t_0)}$$

The problem with using this equation in practice is obtaining accurate estimates of the viable micro-organism present in the outputs, especially if the cells have flocculated or are attached to film debris (Pike *et al.* 1972; Pike 1975). The type of reactor is also important in the estimation of the death-rate. In an ideal plug-flow system (e.g. a percolating filter) in which first-order kinetics apply, the fraction of pathogens surviving (x_t/x_o) can be related to the dilution rate or the reciprocal of the retention time (D) as:

$$\frac{x_t}{x_o} = \frac{-\mu}{D}$$

In a continuous-stirred tank reactor (CSTR), the specific death-rate is calculated by assuming the rate of change in pathogen concentration within the reactor (dx/dt) equals the input concentration (x_o) minus the output concentration (x_t) minus those dying within the reactor under steady-state conditions, then:

$$\frac{dx}{dt} = Dx_o - Dx_t - \mu x$$

Primary Settlement

Counts of indicator bacteria and viruses are not greatly reduced by primary settlement (Pike 1975), although some removal of pathogens must be occurring as primary settled sludge contains the whole range of pathogens found in raw sewage. Some reports suggest that the concentration of bacteria that is particularly associated with particulate matter is reduced by settlement, with removals of 10% for *E. coli* and 60% for *C. perfringens* being recorded by Bonde (1977). Numbers of some indicator bacteria have been reported to increase during primary settlement (Harkness 1966), although viruses are generally not effectively removed at this stage. Berg (1966) found that 33–67% removal of poliovirus type 1 occurred after 24 h, whereas normal retention periods were ineffective.

Ova (eggs) and cysts of parasites are only significantly removed at this stage in wastewater treatment plants. Settlement efficiency is dependent on the size and density of the ova and cysts, and as their free-falling settling velocities are not much greater than the theoretical upflow velocity, near quiescent conditions are required for optimum removal (Section 2.2). The rate of removal of these pathogens is easily affected by currents, eddies, reduction in retention time, and any other factor that reduces settling efficiency. The larger, denser ova of *Ascaris lumbricoides* and *Taenia saginata* are more efficiently removed than the smaller cysts of *Entamoeba* spp. The settling velocity of *T. saginata* is about 0.6–0.9 m h^{-1}, although much less if detergents are present, resulting in 68% removal after 2 h and 89% after 3 h settlement (Liebmann 1965). No significant settlement of *E. histolytica* occurred in a settlement tank

0.67 m deep after 3 h, although the ova of *A. lumbricoides* settled out rapidly (Cram 1943). It has been estimated that up to 97% of the major genera of worms (*Ascaris*, *Trichuris*, *Enterobius*, *Diphyllobothrium*, and *Taenia*) are normally removed by primary settlement (Silverman and Griffiths 1955). However, ova of *Taenia* and *Ascaris*, and cysts of *Entamoeba* are regularly detected in settled sewage and occasionally in the final effluent.

Activated sludge

The activated sludge process is highly efficient in the removal of pathogenic bacteria and viruses, achieving a 90% removal efficiency or more. Among the many references to removal efficiency, Geldreich (1972) found a reduction of 88–99% for *Salmonella*, *Shigella*, and *M. turberculosis*. The major removal mechanism of bacteria in the activated sludge process is predation by a variety of amoebae, ciliate protozoans, and rotifers. In reactors without sludge return predator–prey oscillations are discernible. However, the return of wasted sludge, which is normally practised, tends to dampen out these oscillations so that under steady-state conditions the overall removal of bacteria by predation remains constant (Curds 1971a,b). The ciliate protozoans and rotifers feed only on the freely suspended bacteria and not on flocculated forms. The ability of ciliates to clarify turbid effluents caused by high concentrations of suspended bacteria was demonstrated by Curds *et al.* (1968) (Figs 4.71 and 6.5). Amoebae occur in similar numbers to ciliates and have similar yield coefficient biomass and generation times. They also play a significant role in the removal of bacteria by predation and are able to feed on flocculated forms as well as the freely suspended bacteria. Clearly, the amoebae may have a particularly important role in grazing on the pathogens that have become adsorbed on to sludge flocs.

Protozoal feeding is a major process in the removal of *E. coli* cells (Curds and Fey 1969). Using the data of Curds and Fey, Pike and Carrington (1979)

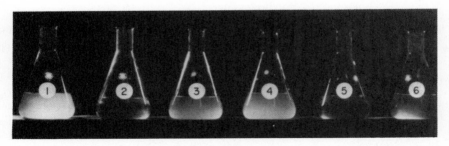

FIG. 6.5 Photographs of effluents from replicate pilot-scale activated sludge plants. Effluents from plants 2, 5, and 6 contained ciliate protozoans, whereas plants 1, 3, and 4 did not, resulting in more turbid effluents (Curds 1975).

estimated the specific growth rate of *E. coli* to be $-7.9 \, d^{-1}$ in the presence of protozoans compared with $+0.12 \, d^{-1}$ in the absence of protozoans. This suggested that *E. coli* is capable of very slow growth in sewage when ciliates are absent. In general terms, the percentage removal of coliform bacteria is directly related to the specific sludge wastage rate, with 90% removal at a sludge wastage rate of $0.65 \, d^{-1}$ that rapidly decreases as the wastage rate increases (Pike and Carrington 1979).

Experiments on predation have suggested that physical adsorption is not important in the removal of pathogenic bacteria. For example, Curds and Fey (1969) found that the half-life of *E. coli* introduced into a plant lacking protozoans was 16 h compared with 1.8 h in a parallel plant with protozoans, indicating that removal mechanisms other than protozoan feeding are insignificant in comparison. However, this may not be the case, as rapid adsorption of *E. coli* on to sludge flocs have been shown to occur within the first hour of sewage entering the aeration tank (Drift *et al.* 1977). After an hour, the removal of coliforms being primarily by predation, was no longer as rapid or constant. The adsorbed bacteria is retained in the solids fraction where mortality by predation, starvation, or other means continues as in the liquid phase. However, the wasted sludge is rich in pathogenic bacteria that have been concentrated mainly by adsorption.

The prime removal method of viruses is adsorption onto sludge flocs, predation having a negligible effect on concentration in the liquid phase. Both viruses and bacteria are adsorbed according to the empirical Freundlich adsorption isotherm, where the count of particles adsorbed per unit mass of sludge (Y/m) is proportional to a power (n) of the count of particles (x) in the liquor at equilibrium:

$$\frac{Y}{m} = kx^n$$

This adsorption model is for unreactive sites, whereas Michaelis–Menton or Monod kinetics are for reactive sites. However, as Pike and Carrington (1979) point out, the successful fitting of an adsorption model to data may not demonstrate that adsorption is the only factor operating because, for example, protozoans attached to the activated sludge flocs and feeding off freely suspended bacteria will quantitatively behave as a continual adsorption site.

Upwards of 90% of enteroviruses are removed by the activated sludge process (Pike 1975), with high removal efficiencies of poliovirus type 1 (90%) and coxsackievirus type A9 (98%) being recorded (Geldreich 1972; Varma *et al.* 1974), although removal efficiencies are erratic (Berg 1973). Apart from the indicator viruses, other viruses, such as coxsackievirus type B, echovirus, and adenovirus, have also been shown to be adsorbed on to sludge flocs (Lund and Rønne 1973), but are also present in the final effluent (Irving and Smith 1981). Adsorption has been shown to be very rapid, for example poliovirus is adsorbed from solution within three minutes of being added to liquor from a waste stabilization pond (Sobsey and Cooper 1973). This rapid adsorption of viruses was confirmed by Malina *et al.* (1975) who used radioactivity counts of tritium-labelled poliovirus to monitor removal mechanisms. They found that the initial rapid adsorption was followed by a period of equilibrium which, they suggested, was due to the maturation of available adsorption sites on sludge flocs. There is a significantly higher concentration of viruses in secondary sludge than primary sludge (Lund and Rønne 1973). This is due to the mixing action of the aeration tank that allows contact between available adsorption sites and viruses. This is confirmed by the increase in the adsorption of viruses by increasing the mixing rate of sludge flocs after the initial uptake of viruses, by breaking up flocs and providing more adsorption sites. A lack of mixing is probably why percolating filters have so little effect on virus removal. Many workers have noted that the concentration of viruses in the sludge is directly related to the concentration of viruses in the influent. The addition of alum to aeration tanks to precipitate phosphates also enhances the removal of viruses and faecal bacteria (Davis and Unz 1973, 1975). Other physico-chemical processes will also greatly reduce the virus and bacterial load by increasing flocculation (Berg 1973). For example, the addition of lime to sewage thus raising the pH > 11.5 results in 99.99% removal of poliovirus type 1 (Satter *et al.* 1976) and *E. coli* (Morrison *et al.* 1973).

Viruses in the activated sludge process are also inactivated, provided that the aeration of the mixed liquor continues (Glass and O'Brien 1980). Inactivation rates of poliovirus of -1.47 h^{-1} have been recorded. Inactivation also occurs in settlement tanks (Balluz *et al.* 1977, 1978). They also recorded that poliovirus is mainly associated with the solids phase of mixed liquor (85%), whereas coliphages are associated mainly with the supernatant (83%).

Ova and cysts of parasites are able to survive the activated sludge process and are not effectively removed.

Fixed-film reactors

Percolating filters are extremely effective in removing coliform bacteria with normal removal efficiencies of >95% (Pike 1975). Removal is achieved by similar mechanisms as in the activated sludge process, except that filters are plug-flow systems with a fixed and not a mixed microbial biomass, so that opportunities for contact between pathogens and adsorption sites in the biomass are probably reduced.

Removal of bacteria is directly related to the bacterial count of the sewage and at low-rate loadings to the surface area of the filter medium (Tomlinson *et al.* 1962). In a comparison between high- and low-rate filtration, Bruce *et al.* (1970) found that the removal of coliforms in high-rate filters with hydraulic loadings of 6 or 12 m^{3}m^{-3}d^{-1} was only 6–74% compared with 90–99.7% in a control low-rate filter operated in parallel with a hydraulic loading of only $0.4 \, m^3 m^{-3} d^{-1}$. They also observed that removal of pathogenic bacteria was directly related to BOD removal, thus linking removal of pathogenic bacteria with adsorption. A similar situation was observed using faecal streptococci and *E. coli* as indicator organisms (Bruce and Merkens 1973). An earlier study had shown that removal efficiency fell off during winter (Allen *et al.* 1944), which suggests that maximum removal of pathogens occurs when the film is most actively growing and under maximum grazing pressure: this is when maximum availability of adsorption sites will occur.

This is further confirmed by the low efficiency of percolating filters in removing viruses and bacteriophages (Berg 1966; Malherbe and Strickland-Cholmley 1967; Sherman *et al.* 1975; Carrington 1980*a*), which are known to be removed by adsorption, which in the activated sludge process is more effective due to the mixing action in the aeration tank. In the percolating filter, there is far less chance of viruses being attached to available sites, especially under conditions of ponding and short-circuiting, when only a small portion of the available surface area of the medium is being used, or under high-rate loading when the retention time (or film contact time) is very short.

Ciliate protozoans can ingest pathogenic bacteria as can rotifers, nematodes, and annelid worms. However, in percolating filters there is a much larger range of macro-invertebrate grazers feeding directly on the film and thus indirectly feeding on the pathogens. Once pathogens have been adsorbed on to the film they are essentially 'removed' and their subsequent ingestion by a grazing organism may not be significant. The major limitation of percolating filtration in the removal of pathogens is their physical adsorption from suspension.

Percolating filtration is not very effective in the removal of parasite ova and cysts, although the nature of the film does allow some retention of ova and removals of up to 30% have been observed (Silverman and Griffiths 1955). Geldreich (1972), in a review on water-borne pathogens, quotes higher removal rates, 18–70% for tapeworm ova and 88–99% for cysts of *E. histolytica*.

Rotating biological contactors are also extremely efficient in the removal of pathogenic bacteria, with median removal of *E. coli* normally >99.5%. Although the numbers of bacteria in raw and settled sewage varies diurnally, the removal efficiency remains constant and unaffected by factors such as temperature, bacteria concentration in the influent, or the organic loading of disc surfaces over the range $1.59–9.47 \, g \, BOD \, m^{-2} d^{-1}$ (Pike and Carrington 1979).

Waste stabilization ponds

In general, all stabilization ponds and lagoons are extremely effective in removing bacteria, viruses, and other parasites. A variety of removal mechanisms has been reported, including settlement, predation, inactivation because of solar radiation which is also linked with temperature, increase in pH due to daytime assimilation of carbon dioxide which can reach in excess of pH 9, and, finally, anti-bacterial toxins produced by algae (Pike 1975).

It is most appropriate to consider ponds as CSTRs, so that the survival of pathogens (x_t/x_o) can be calculated as:

$$\frac{x_t}{x_o} = \frac{1}{\left(1 - \frac{\mu}{D}\right)}$$

Ponds are normally in series, and if they have similar dilution rates the survival of pathogens can be calculated as:

$$\frac{x_n}{x_o} = \frac{1}{\left(1 - \frac{\mu}{D}\right)^n}$$

where n is the number of ponds. A problem arises if the number of ponds in series is large (i.e. > 5) because then the system behaves more like a plug-flow reactor. In this case the relationship becomes:

$$\frac{x_n}{x_o} = \exp-\left(\frac{\mu}{D_n}\right)$$

where D_n is the dilution rate of the complete system.

In temperate climates, only maturation ponds are commonly used in the treatment of domestic sewage. They are very effective in the removal of bacteria and viruses (Windle Taylor 1966; Adams *et al*. 1972; Toms *et al*. 1975). The removal efficiencies of *Streptococci* sp. and *Salmonella* sp. are equal to or better than the removal rate for *E. coli*, although *C. perfringens*, both as active cells and spores, are less effectively removed (Toms *et al*. 1975). Removal efficiency in maturation ponds is related to retention time, with removal rates falling rapidly as the retention time is reduced. For example, high removals of *E. coli* have been reported in Britain at retention times of 8.9 d (99.9%) and 17.0 d (99.5%) (Windle Taylor 1966; Adams *et al*. 1972), whereas at 3.5, 3.0, and 2.3 d retention, the removal rates were 94.0, 92.4, and 81.0% respectively (Fish 1966; Windle Taylor 1966). Retention time is critical and any short-circuiting within ponds will reduce removal efficiency.

Solar radiation is a major removal mechanism in maturation ponds. In South Africa the mean removal efficiency of *E. coli* during the summer was 99.6% compared with 96.9% in winter (Drews 1966). This is more clearly illustrated in Britain where the difference between solar radiation in the summer and winter is more pronounced. At a retention time of 3.5 d, the removal rate of *E. coli* was > 90% in the summer but fell to 40% in the winter (60% for faecal streptococci) because of the seasonal difference in light intensity (Toms *et al.* 1975). In the British Isles, the algal density is never sufficient to significantly shift the pH so that the pathogens are killed. In fact, the mortality of bacteria and viruses can be reduced by algae because of the reduction of light intensity by shading. The depth is also an important factor with removal rates being reduced with depth.

In tropical and sub-tropical countries, the removal of pathogens is as important as BOD removal is in temperate zones, and in many countries more so. Facultative ponds are very efficient in the removal of pathogens (Arceivala *et al.* 1970; Gloyna 1971), with removal rates of coliform and streptococci bacteria >99%, and viruses inactivated by light so that >90% removals are achieved. The greater efficiency of facultative ponds compared with other systems is because of the much longer retention times (Coetzee and Fourie 1965), and ponds in series achieve greater removals than single ponds (Marais 1970). Major removal mechanisms are the high pH values created by photosynthesis and the higher zooplankton predation rate. Anaerobic treatment followed by facultative treatment gives a higher die-off rate of pathogenic bacteria than a facultative pond followed by a maturation pond in series (Davis and Gloyna 1972). Pond systems appear extremely effective in the removal of parasites because the long retention time allows maximum settlement. Cysts of the protozoans *Entamoeba histolytica* and *Giardia lamblia* are almost completely removed, and the helminth parasites, such as *Schistosoma*, *Ascaris*, *Enterobius*, *Ancylostoma*, and *Trichuris* are also effectively removed (Gloyna 1971). Maximum removal occurs in the first pond, and as the nematode eggs in particular are highly resistant, extreme care must be taken with the disposal of the raw sludge from the pond if contamination is to be prevented. Removal efficiencies for three ponds in India have been given by Veerannan (1977). For helminth ova these were *Ascaris lumbricoides* (89–93%), hookworm (>92%), *Trichuris* (>68%), *Enterobius* (>95%), *Hymenolepis* (>83%), and for protozoan cysts removals were *Entamoeba coli* (39–77%), *Entamoeba histolytica* (>62%), and *Giardia lamblia* (>98%).

Digestion

There are very conflicting results relating to the efficiency of anaerobic digestion and composting of removing pathogens from sewage sludge. Anaerobic digestion certainly reduces the numbers of pathogens considerably but not always completely (Bates 1972; Carrington 1978). Although many bacteria, fungi, and viruses are rapidly killed by air-drying they can survive anaerobic digestion at 20°C or 30°C for long periods. For example, *S. typhi* can survive digestion for 12 d at 20°C or 10 d at 30°C, *Ascaris* ova 90 d at 30°C, and hookworm ova 64 d at 20°C or 41 d at 30°C. Both the latter ova can also withstand air-drying. Some researchers have indicated that complete destruction of pathogens is only possible by heating the sludge to 55°C for 2 h or treating with lime (Smith *et al.* 1975).

6.3.3 Sterilization and disinfection methods

Sterilization and disinfection of final effluents to remove any disease-causing organisms remaining in effluents is not widely practised in Europe, but is common in the USA with up to 50% of the total treatment works employing chlorination (Thoman and Jenkins 1958). The increasing need to re-use water for supply after wastewater treatment will mean that the introduction of such methods to prevent the spread of diseases via the water supply is inevitable (Dean and Lund 1981). The two processes are distinct from each other. Sterilization is the destruction of all the organisms in the final effluent regardless of whether they are pathogenic or not, whereas disinfection is the selective destruction of disease-causing organisms. There are three target groups of organisms, viruses, bacteria, and amoebic cysts, and each is more susceptible to a particular disinfection process than the other. The main methods of sterilization and disinfection are either chemical or physical in action.

Chemical
This group includes the most widely used methods of removing pathogens from wastewater effluents. The oxidizing chemicals are the commonest and safest chemicals to use, with chlorine, ozone, and hydrogen peroxide all widely used. There are a variety of other chemical methods used to disinfect wastewaters and these are fully reviewed by Venosa (1983) and Metcalf and Eddy (1984).

The factors affecting the efficiency of chemical disinfectants are contact time, concentration and type of chemical agent, temperature, types and numbers of organisms to be removed, and the chemical nature of the wastewater. The most widely used chemical disinfectant is chlorine. However, unlike drinking water, the disinfection of treated effluents or crude and settled sewage, if discharged to coastal waters, requires a high chlorine dose. Chlorine is normally used in its elemental form or as hypochlorite, and depending on the pH of the water and the presence of ammonia, the chlorine may take the form of $HOCl$, OCl^-, Cl_2 or chloramines when in solution. At a neutral pH, free chlorine is converted almost instantaneously in the presence of ammonia to monochloramine so that free chlorine is not found in sewage effluents, although some free chlorine residuals may exist in highly nitrified effluents (White 1978*a*). The reaction of chlorine with a compound containing a nitrogen atom will continue forming a range of N-chloro compounds:

$$NH_3 \leftrightarrow NH_2Cl + H_2O$$
(monochloramine)

$$NH_2Cl + HOCl \leftrightarrow NHCl_2 + H_2O$$
(dichloramine)

$$NHCl_2 + HOCl \leftrightarrow NCl_3 + H_2O$$
(nitrogen trichloride)

At higher pH values nitrogen trichloride can be slowly oxidized to nitrogen gas:

$$NH_2Cl + NHCl_2 \leftrightarrow N_2 + 3H^+ + 3Cl^-$$

With time, inorganic chloramines formed in sewage will be converted to organic chloramines that are less reactive and thus less efficient in the inactivation of bacteria or viruses. The chemistry of chlorination has been explained in most standard texts and will not be discussed here. A particularly good summary of the process is given by Irving and Solbe (1980). Normal dosage rates for treated effluents range from 2–15 mg $Cl_2 l^{-1}$ depending on the degree of treatment with a minimum contact time of 20–30 minutes. Bradley (1973) reported chlorine doses, using a contact time of 15 minutes, of 6–24 mg l^{-1} for raw sewage, 3–18 mg l^{-1} for settled sewage, and 3–9 mg l^{-1} for an activated sludge effluent.

There is a noticeable difference in the resistance of bacteria, viruses, and protozoan cysts to chlorine (Kool and Kranen 1977). Chlorine is a strong oxidizing agent and attacks the chemical constituents of bacterial cells and viruses (Kool 1979). The exact method of inactivation in bacteria is still not fully understood, although chlorine is suspected of interfering with the sulphydryl groups of the enzyme triose-phosphate-dehydrogenase. Although chlorination may not eliminate all potentially pathogenic bacteria, for example, lactose non-fermenters can survive normal chlorination, certain bacteria are extremely sensitive to the process, *Salmonella* sp. for example are more sensitive than *E. coli*. Bacterial cells require longer contact times with chlorine as they age, and free residual chlorine is required to eliminate many important viruses. Poliomyelitus virus type 1, hepatitis A, and coxsackievirus type A2 both require much higher concentrations of chlorine, or longer contact times for their destruction than *E. coli* (Botha 1984) (Table 6.27). Viruses also have various sensitivities to disinfectants (Table 6.28).

In order to obtain sufficient free residual chlorine (breakpoint chlorination) higher concentrations of chlorine are required, resulting in free residual chlorine remaining in the final effluent. This residual chlorine is extremely toxic to organisms, especially fish, and can react with organic compounds in wastewater to form toxic compounds, such as chlorinated biphenyls, some of which are carcinogenic (Jolley 1975; Gaffney 1977; Irving and Solbe 1980). Fawell *et al.* (1987) have studied the reaction of Cl^- with various organic

TABLE 6.27 Comparison of the concentration ($mg\,l^{-1}$) of the most frequently used disinfectants required to inactivate the major microbial groups within 10 minutes (White 1978a)

Disinfectant	Enteric bacteria	Viruses	Bacterial spores
HOCl	0.02	0.40	10.0
OCl$^-$	2.0	> 20.0	> 1000
NH$_2$Cl	5.0	100.0	400
Free Cl$_2$ (pH 7.5)	0.04	0.8	20.2
O$_3$	0.001	0.10	0.2

TABLE 6.28 Time required for 99.99% inactivation of enteric viruses by a $0.5\ mg\,l^{-1}$ free chlorine residual at pH 7.8 at 20°C in natural river water (White 1978a).

Species	Time for 99.99% inactivation (min)	Species	Time for 99.99% inactivation (min)
Polio type II	36.5	Coxsackie virus type B1	8.5
Coxsackie virus type B5	34.5	Adenovirus type 12	8.1
E. coli type 29	18.2	Coxsackie virus type A9	7.0
E. coli type 12	16.7	E. coli type 7	6.8
Polio type III	16.6	Adenovirus type 3	4.3
Coxsackie virus type B3	15.7	Reovirus type 2	4.2
Adenovirus type 7a	12.5	Reovirus type 3	4.0
Polio type I	12.0	Reovirus type 1	2.7

chemicals produced during water treatment chlorination (Table 6.29). Many of these compounds have been shown to be potentially carcinogenic using a biassay technique with *Salmonella typhimurium* strain TA100. Some regulatory agencies in the USA have specified a maximum residual chlorine concentration of $0.1\text{--}0.5\ mg\,l^{-1}$ in undiluted effluents to prevent potential toxicity in receiving waters. Once discharged, the residual chlorine is too reactive to persist for long in most environments (both fresh and saline waters). This has allowed the permissible level of chlorine in natural waters to be set nearer to the lethal limits for most organisms than would otherwise be advisable. The European Inland Fisheries Advisory Committee (1973) have proposed a limit of $0.004\ mg\ HOCl\ l^{-1}$ for European rivers, but even at this level, some freshwater and marine organisms will be susceptible. If chlorine exceeds these limits dechlorination is required to detoxify the discharge. This is normally achieved by adding sulphur dioxide or using activated carbon. The level of residual chlorine needed to destroy viruses in final effluents will mean that such effluents will always have to be dechlorinated before discharge to the

TABLE 6.29 Organic compounds produced during water treatment chlorination (Fawell *et al.* 1987)

Benzaldehyde	Chloropicrin
Benzylcyanide	Dibromoacetonitrile
Bromoethane*	Dibromoiodomethane
Bromobutane*	Dibromomethane*
Bromochloroacetonitrile*	Dichloroacetic acid
Bromochloroiodomethane	Dichloroacetonitrile*
Bromochloromethane*	Dichlorodibromomethane
Bromochloropropane isomers (4)	1,2-dichloroethane*
Bromodichloromethane*	Dichloroiodomethane
Bromoform*	Dichlorophenol
Bromopentachloroethane	Dichloropropene*
Bromopropane*	Hexachloroethane*
Bromotrichloroethylene	Hexachloropentadiene
Carbon tetrachloride	*p*-Hydroxybenzyleyanide
Chloral*	Iodoethane*
Chlorobutane	Methylbromodichloroacetate
Chlorodibromomethane*	1,1,1-Trichloroacetonitrile
Chlorodiiodomethane	Trichloroacetonitrile
Chloroform	Trichlorophenol

* Mutagenic to strain TA100 (without S9 activation).

natural environment. The toxicity of free chlorine and chloramines to freshwater and marine organisms is reviewed by Evins (1975) and Irving and Solbe (1980) respectively. Chlorination of crude and settled sewage discharged to coastal waters is not a practical substitute for efficient treatment nor the provision of a suitably long sea outfall. However, in problem areas the use of chlorination can protect bathing areas or shell-fisheries from bacteriological contamination. The major problem is to break up small suspended particles in the sewage that will contain bacteria, before chlorination. However, laboratory studies have indicated that although chlorination is effective in destroying bacteria, small numbers survive and this can result in regrowth of pathogens, especially if the water is stored before discharge (Irving 1980). However, regrowth is governed by the degree of dilution, so when the water is discharged to the sea it is assumed that regrowth will be minimal.

Ozone has similar bactericidal properties to chlorine and is generally considered more effective in destroying viruses (Venosa 1972, Katznelson and Biederman 1976). As with chlorine, enteric viruses are much more resistant to ozone treatment than coliform bacteria (Table 6.27) (Sproul *et al.* 1982). Ozone has a more powerful action than chlorine, by a factor of 10–100 depending on the form of chlorine used. For example, ozone has a higher oxidation potential (-2.07 v) when compared with HOCl (-1.49 v), Cl_2 (-1.36 v) or NH_2Cl (-0.75 v) (Kinman 1972). The effectiveness of ozone is not significantly affected between normal wastewater pH values (pH 6.0–8.5), although it has a reduced mass transfer efficiency at higher water temperatures (White 1978*a*). Typical dosage rates are > 50 mg $O_3 l^{-1}$ for raw sewage and

15 mg O l^{-1} for treated effluents, with 5 minutes contact for complete virus inactivation. However, although ozone is expensive, it is completely and rapidly converted to oxygen on addition to water:

$$2O_3 \rightarrow 3O_2$$

Therefore, there are no persistent toxic chemical residuals remaining in the final effluent. Organic matter impedes the inactivation ability of ozone and it is only suitable for use with good quality effluents. It is possible that all disinfection processes would be more efficient and more cost-effective if used after a physical tertiary treatment process, such as sand filtration, micro-straining or treatment on grass plots, all of which help to reduce the organic as well as the bacterial load. Ozone has another advantage over chlorine in that it aerates the final effluent as well as removing phenols and chlorophenols from water. Ozonation of wastewater is widely practised in France, and because it is such a strong oxidizing agent it must be generated on site at the point of use. Hydrogen peroxide and the other oxidizing chemicals can also be used in the collection (for the control of odours, hydrogen sulphide production and subsequent corrosion of sewers, and in the control of sewer slime) and the treatment of wastewater (in the control of ponding, filter flies, sludge bulking, and in the removal of grease and for BOD$_5$ reduction) as well as for the disinfection of final effluents. There is considerable literature on the treatment processes for the removal of micro-organisms from public supplies (White 1978a; Kool 1979) and Carrington (1980a) has reviewed the effectiveness of the various disinfection procedures in removing pathogens from crude and treated wastewaters.

Physical
Sunlight is known to be a good disinfectant, and in wastewater stablization ponds it is sunlight that is largely responsible for reducing the concentration of pathogenic bacteria. A modification of this principle is the growing use of UV radiation to sterilize effluents. Ultraviolet radiation acts on the cellular nucleic acids destroying bacteria, viruses, and any other organisms present. Although expensive in terms of energy, such systems can be highly effective. The greatest effect occurs at a wavelength of 265 nm, with low pressure mercury arc lamps (254 nm) most widely used. The major operational problem is to obtain maximum penetration of the rays to ensure that even turbid effluents are fully sterilized. Numerous systems have been evaluated to obtain maximum exposure of wastewater to the radiation, but the most effective system to date is the use of thin film irradiation (<5 nm) (Venosa 1983). Ultraviolet irradiation is particularly useful in preventing contamination by pathogens of lakes and coastal waters that are popular for bathing. The introduction of the bathing waters directive in Europe (European Communities 1976a) has made it very difficult for local authorities to meet the high bacterial standards required, therefore UV sterilization appears to be an attractive solution. This

is particularly so as chlorination of sewage is not fully effective when used with primary treated effluents in the destruction of viruses and protozoal cysts. Furthermore, chlorinated discharges appear to have deleterious effects on the marine and estuarine environments near the outfalls (Irving and Solbe 1980). Gamma rays have also been used to sterilize wastewaters (Melmed 1976; Metcalf and Eddy 1984).

Further reading

Environmental factors: Gameson 1975a; Mitchell and Chamberlin 1975; Carrington 1980a; Gameson 1985b.

Treatment processes: Pike 1975; Pike and Carrington 1979; Carrington 1980a; Yaziz and Lloyd, 1979; Grabow 1986.

Sterilization and disinfection methods: White 1978a; Kool 1979; Venosa 1983; Fawell *et al.* 1987.

7

BIOTECHNOLOGY AND WASTEWATER TREATMENT

7.1 THE ROLE OF BIOTECHNOLOGY

A particularly clear definition of biotechnology is given by Rothman *et al.* (1981), as the '*exploitation of living organisms, generally micro-organisms, or biological processes in an industrial or commercial situation to provide desired goods or services*'. Such a definition encompasses a large number of areas, including wastewater treatment. Biotechnology is not a distinct discipline but rather a result of four traditionally separate disciplines, biochemistry, microbiology, engineering, and chemistry, interacting with each other (Sikyta 1983). Although the term biotechnology has only become widely studied as a subject in its own right since the mid-1970s, many biotechnological processes have been used by man for centuries, such as the exploitation of yeasts in baking and brewing, or bacteria to ripen cheese. The recent progress and upsurge in interest in biotechnology has been due to a number of incentives. First, the increased cost of energy led to a re-examination of the processes producing fuel from renewable sources, and the increasing scarcity of raw materials, especially minerals, has made them more expensive as the poorer resources are exploited. Micro-organisms can be used not only in the extraction of metals from low grade ores but also to reduce the pollution load on the environment. The recent advances in biotechnology are closely associated with developments in general and molecular genetics. New strains of micro-organisms can be developed that produce a desired substance in far greater quantities than the original parent strain. Although this can be achieved by careful culturing and selection, the most important advance is the development of genetic manipulation, known ubiquitously as genetic engineering. This involves the transfer of genetic material from one cell to another in order to modify that cell's behaviour to benefit some industrial process. Turning the bacterial cell into a factory capable of producing enzymes, vaccines, amino acids, steroids or other cellular products, such as insulin and interferon, has revolutionized the biomedical and food industries (Old and Primrose 1980). The application of genetic manipulation to wastewater treatment is varied, ranging from the inoculation of specially cultured bacteria to enhance the performance of an existing conventional effluent treatment plant to the biodegradation of recalcitrant compounds.

Most of the investment in biotechnology to date has been in those areas with most capital return, which have been the food and biomedical sectors. However, it is estimated that *environmental biotechnology* will be the market which will develop most in the next decade (Anon 1982*b*) (Table 7.1). In financial terms, the most recent forecasts of world markets in biotechnology in the year 2000 indicate that energy and food products will be the major sectors, with pollution control at the bottom of the estimates (Table 7.2) (Williams 1982). The lack of financial return in comparison with other sectors, linked with the conservative outlook of the water technology industry itself, which is very slow to move away from the familiar treatment methods, is why biotechnology has so far had little real impact on wastewater treatment (Anon 1982*c*; Gray 1985*b*).

TABLE 7.1   Markets for microbes and enzymes, where AGR is the average annual growth rate estimated for 1981–91 (Anon 1982*c*)

Application	1975	1977	1979	1981	1986	1991	(AGR)
Environmental clean-up	5	7	9	10	22	48	17
Biomass conversion	0.5	3.3	5.5	7.0	8.5	10.0	3.8
Sweeteners	7.8	28.0	29.5	30.5	42.5	60.0	7
Processed food beverages	35.7	39.6	42.4	45.0	52.9	64.0	3.6
Biomedical	15	18	29	31	45	67	8
Laundry products	7.0	5.0	6.0	6.0	6.3	6.6	1
Tanning and textiles	18.0	18.0	9.0	8.6	8.4	7.8	−1
Miscellaneous (insecticides, mining, feed, biosensors, etc.)	0.4	0.4	1.1	2.1	3.3	5.6	10
Total ($m)	89.4	119.3	131.5	140.2	188.9	269.0	

TABLE 7.2   Forecast of world markets for biotechnology in the year 2000 (estimated biotechnology products 1981 = $25m) (Williams 1982)

Market	$m
Energy Ethanol, methanol, enhanced oil recovery, hydrogen generation	16 350
Food High fructose, sugar, single cell protein	12 655
Health care Vaccines, antibiotics, hormones, blood products	9080
Chemicals Ethylene oxide, glycerol, methanol, isopranol, amino acids	10 550
Agriculture Modified crops, fertilizer	8546
Metal recovery Especially copper and nickel	4570
Pollution control	100
Miscellaneous	3000
Total	64 851

Effluent treatment is undoubtedly the largest controlled application of micro-organisms in the manufacturing and service industries with $9 \times 10^6 m^3$ of domestic and $7 \times 10^6 m^3$ of industrial effluent treated daily in the UK alone. With 80% of the population served by 7795 treatment plants, 17 of which serve populations $> 300\ 000$, the cost of pollution control is currently running in excess of £110m per annum (Water Data Unit 1979). In the USA, it has been estimated that $400 000m was spent on pollution control in the decade 1976–86 (Congressional Office of Technology Assessment 1981). The size of this market is therefore very large and with the installations constructed during the early expansion of wastewater treatment in the late-nineteenth and early-twentieth centuries now nearing the end of their useful lives, the opportunities for the biotechnologist to apply new technologies to pollution control, such as genetic manipulation combined with new reactor designs, are enormous. The fact that the total value of crude fats, proteins, and metals in wastewater disposed to sewers in the UK is estimated as being worth about £150m annually is an added incentive (Clapham 1980). In the future, cheaper, more efficient and more compact processes will be developed, with the traditional aims of removing organic matter and pathogens to prevent water pollution and protect public health, replaced with a philosophy of environmental protection linked with conservation of resources and by-product recovery. Unlike the other sectors of biotechnology, development in environmental biotechnology is unlikely to be stimulated or supported by normal market forces. However, the introduction of the 'polluter pays' principle for industrial wastewater (Deering and Gray 1986) and water charges for domestic wastewater, linked with the possible privatization of the UK water industry, may go some way towards this goal. Any economic deficit can be rectified by legislation on pollution and government subsidy for the conservation of vital resources. For example, in Italy an effluent treatment plant which produces biogas from waste receives a 70% grant, as the country is short of indigenous energy (Wheatley et al. 1983). In the future, the operating costs of wastewater treatment plants will be offset by the various by-products or resources that are recovered. Although it is unlikely that this will normally result in a clear operating profit, taken with its service role of environmental protection then there is no reason why wastewater treatment should not become increasingly cost-effective. It is unrealistic to think that the by-products of environmental biotechnology, which are mainly of low to intermediate value, would ever recoup capital investment. It is likely, therefore, that the initial cost of wastewater treatment plants will always have to be funded centrally by governments. At present, this concept is more widely seen in relation to solid waste disposal, especially by incineration. Plants in the UK and Ireland are to be operated by private companies with the major income coming from the local authorities who not only pay to have their refuse disposed of, but also for the energy generated which is generally used for group heating schemes.

The major areas of biotechnology that have the greatest potential in wastewater treatment are resource recovery (Section 7.2), biological conversion (Section 7.3), and environmental protection (Section 7.4).

Further reading

General: Bull *et al.* 1982; Cooke 1983; Sikyta 1983; Wheatley *et al.* 1983; Wheatley 1985; Sidwick and Holdom 1987.

7.2   RESOURCE REUSE

7.2.1   Fertilizer value

Sewage and agricultural sludges are rich in organic matter and the major plant nutrients, nitrogen, phosphorus and potassium, as well as all the important trace elements. Farmers are able to utilize these sludges as an effective but cheap soil conditioner and fertilizer. The exact manurial value of sewage sludge depends on the nature of the sludge, whether it is domestic, industrial or agricultural and whether it is primary or secondary sludge, and also whether it has been dewatered or stabilized (Table 5.9). Liquid digested sludge in the UK is mainly used on agricultural land. There are problems with contamination of the soil and vegetation by heavy metals and pathogens. This has led to the introduction of legislation to control the level of contamination as well as to prevent the transfer of human, animal, and plant diseases. The whole topic of sludge utilization as a fertilizer is dealt with in Section 5.2.2.

7.2.2   Reuse of effluents

With mains water in the UK currently costing £1 per tonne (or cubic metre) and effluent disposal to sewers between £1–3 per tonne (t^{-1}), it is rarely economical to treat wastewater for reuse within a particular production process. Recyling of water is only economic when the quality of the water required is unimportant, as with industries such as power generation, steel making, and coal washing. Thus, the water can be continuously recycled with only minimum treatment required to remove gross particles or to cool the water. However, two situations can make the reuse of effluent economic. Where local water capacity is insufficient to meet the needs of industry and the effluent is the only other source of water available, and where effluent treatment costs levied by water authorities make it more economic for industries to treat their own waste before discharging effluent. In these circumstances it may be cheaper to reuse their own treated effluent rather than pay for mains water and even minimum disposal charges. Usually, water is reused several times within the factory before eventually being discharged, starting with processes that require clean water and finishing up being used for processes which only require low grades of water, such as vegetable washing (Shore *et al.* 1984). With the introduction of water supply and disposal charges, the conservation and the multiple use of water, industry has greatly reduced water usage and alleviated water pollution. An interesting example of water reuse can be seen in the sugar beet processing industry. The total water requirement for processing a tonne of sugar beet is between 9–19 m^3 and until recently most beet factories operated on a once through basis with little or no recirculation of water which resulted in vast quantities of wastewater being generated. Wastewater from sugar beet is particularly polluting as it contains low molecular weight carbohydrates, usually sucrose or volatile fatty acids, in high concentrations (2000–5000 mg l^{-1} BOD). These can cause severe sewage fungus outbreaks and deoxygenation in receiving waters (Gray 1987, 1988). However, with careful water management, the surplus water can be reduced to as little as 0.5–1.0 m^3t^{-1} (McNeil 1984). The Bury St. Edmonds factory operated by British Sugar processes more than 11 000 t of beets daily during the processing season. They have been able to reduce the quality of effluent produced from >100 t $m^{-3}d^{-1}$ to only 4.8 t m^3d^{-1} by introducing an extensive water re-use programme (Fig. 7.1). The high water content of sugar beets (78%) makes it impossible, however, for such systems to operate as a closed loop, with no excess water produced and so no discharge at all (Shore *et al.* 1984).

In areas where water is scarce, treated sewage effluent may be reused after sufficient disinfection for uses not associated with human or animal consumption. At one time, it was proposed to have a dual water supply system in UK towns, one with fully treated water and the other with a lower grade of water, usually untreated river water. The idea was to conserve water of the highest quality for consumption and household uses while also saving money by not having to treat all the water being supplied. Although this system is in use in some other countries it was never feasible in the UK because of the high cost of providing two separate mains, having two separate plumbing systems in buildings, plus the danger of people mistakenly using the lower grade of water for consumption. However, in arid areas it is common to use sterilized treated effluents for non-consumable activities such as car washes, flushing toilets, and even washing clothes (Fewkes and Ferris 1982).

In the UK at present about 30% of the raw water used for public supply is obtained from recycled effluent (Water Data Unit 1979). This is a mean value for the whole of England and Wales, and in areas where supplies of upland water are very restricted, such as the South-East of England this figure may be as high as 70%. Treated effluent is discharged from one consumer area into a lowland river and abstracted for reuse at the next urban area downstream. It is incredible, but true, that the Rivers Thames and Lee consist of 95% effluent

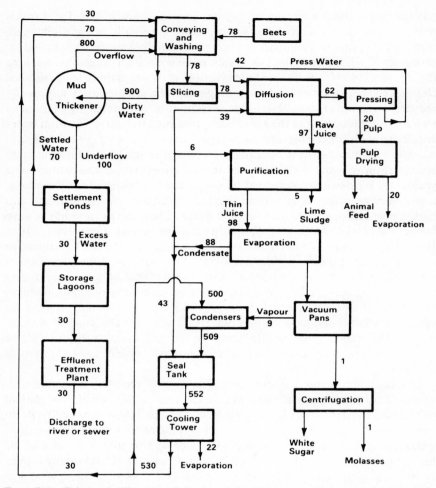

FIG. 7.1   Schematic diagram of the water flow in a sugar beet processing factory operated with maximum recirculation. Figures are the nominal water % w/w on beets (Shore *et al.* 1984).

during dry summers (Pacham 1983). With the major areas of population centred in the Midlands and the South-East, it is unlikely that new upland supplies will be made available in the future and any subsequent increase in demand will have to be met by using groundwater or reclaimed water. The Thames River Basin in England is an example of open cycle reuse, when the sewage from one community is converted to drinking water in another. Overall, the population of London is in excess of 10 million, with water being supplied from bore holes, the River Thames, and its tributaries, including

$1.2 \times 10^6 m^3 d^{-1}$ of sewage. The recycle rate is, on average, 13% but during the 1975–6 drought it exceeded 100% (Blackburn 1978). The sewage receives full biological treatment followed by nitrification and denitrification when required (Cooper *et al.* 1977), whereas the water supply is stored for seven days prior to treatment, which normally involves slow sand filtration followed by chlorination. The re-use of the River Thames water is shown schematically in Fig. 7.2. It is interesting that one community, Walton Bridge, actually discharges its effluent upstream of its own intake (Eden *et al.* 1977).

All municipal wastewaters discharging to rivers used for public supply are fully treated biologically and the abstracted water is then subjected to full water treatment (which may also include treatment with activated carbon, membrane filtration or de-ionization if necessary). However, when water is recycled many times dissolved salts will accumulate in it, particularly the anions from biological oxidation. Nitrate, sulphate, phosphate, and chloride all accumulate in the water supply which causes unpleasant tastes and a fall in

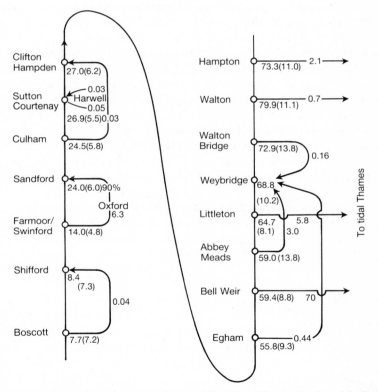

FIG. 7.2   An example of the reuse of water. The River Thames basin, with river flow (as averages) given in $m^3 s^{-1}$, and in parentheses the percentage of flow under average flow conditions that comprises sewage effluent (Dean and Lund 1981).

quality, corrosion and scaling in pipes, and even toxicity. If there is no alternative source of supply then these inorganic salts may have to be removed by advanced water treatment techniques such as ultrafiltration, reverse osmosis or ion exchange, which makes the water expensive (Botto and Pawlowski 1983; Hedges and Pepper, 1983). Some pollutants are not easily removed and are very persistent. Although they are normally only present in trace amounts they are still toxic or in some cases carcinogenic. These compounds, known widely as recalcitrants, can be degraded or scavenged by specialized micro-organisms (Section 7.4.1). The short- and long-term health effects of consuming re-used water is examined in Section 6.2.4.1.

Sewage contains a wide range of inorganic plant nutrients, although they are in much lower concentrations than either liquid sewage sludge or synthetic plant feeds (Table 7.3). Even though sewage contains small quantities of nutrients its main potential is only realized if the effluent is also needed as a water supply. Therefore, irrigation with sewage is not widely practised in Western Europe as the nutrients are so dilute, that large quantities of sewage would be required to meet plant needs, making it an expensive source of nutrients. However, in drier and more arid areas, sewage irrigation is common, especially in Israel, India, Australia, and southern USA. However, as sewage contains pathogens and heavy metals, care must be taken to prevent contamination or accumulation of toxic materials in the crops. This is done by careful selection of crop species and restricting the use of certain effluents to particular categories of crops. This is fully discussed in Section 6.2.4.2. Sewage irrigation is widely used in south western USA where primary effluents are used as fodder, fibre, and seed crops, and secondary treated effluents on food crops and on pasture used for dairy herds (Pettygrove and Asano 1984). Much interest is being shown in using sewage effluents in soil-less nutrient-film hydroponics (Winfield and Bruce 1981). The use of sewage on vegetables and

TABLE 7.3 Comparison of inorganic nutrient concentrations ($mg\ l^{-1}$) in domestic sewage with synthetic nutrient solutions used to culture micro-organisms (Wheatley 1985)

Nutrient	Domestic sewage effluent	Synthetic nutrient solution
Nitrate (as N)	20–25	40–45
Ammoniacal nitrogen (as N)	0.1–5.0	0.1
Phosphorus (as P)	5–12	90–100
Potassium	10–18	90–100
Calcium	80–90	125
Magnesium	5–9	85
Iron	0.01–0.1	8–12
Manganese	0.1	1
Sodium	80–90	60
Chloride	60–100	50
Boron	1–2	—

fruit is discouraged, especially if eaten raw, because of the long survival of bacteria and viruses. Chlorinated secondary effluents are widely used throughout southern USA for the irrigation of all large areas of grass, such as golf courses and parks.

Irrigation with sewage removes much of the nitrogen and phosphorus from the effluent, which prevents eutrophication, and land treatment is widely used as a tertiary treatment method.

7.2.3 Metal recovery

The consumption and therefore the value of metals world-wide has risen dramatically over the past two decades. This is not only because of demand but also to scarcity, and increasing costs of production, as less metal-rich ores are worked. Also restrictions, both natural and international, on the discharge of metals into the environment, because of toxicity and the dangers associated with bioaccumulation and metals entering the food-chain, has led to a growing awareness of the importance of their recovery and possible reuse. Nearly all wastewaters contain some metals but industrial wastewaters and, in particular, mining wastewaters can be rich in metals. It is not generally wastewaters that contain high concentrations of metals that pose the greatest environmental risk, as such wastes can be economically treated by conventional physical and chemical recovery systems. The wastewaters that pose the greatest problems are those where the concentrations may be low enough to make conventional chemical recovery uneconomical but at the same time high enough to cause environmental damage. Certain micro-organisms are well known to be able to remove metals from solution by precipitation (*Sphaerotilus*, *Leptothrix*, and *Gallionella*), by precipitating soluble metals in the form of insoluble sulphides (many anaerobic bacteria) and by oxidation (*Thiobacillus*). Recovery of metals using micro-organisms would only be a fraction of the cost of physical or chemical recovery processes, therefore, the active utilization of suitable micro-organisms to remove metals from wastewaters appears very attractive both economically and environmentally.

Water exposed to sulphides or sulphur-rich coal will have associated with it several types of bacteria which derive their energy from the oxidation of inorganic substances, such as ferrous iron, sulphur, and soluble and insoluble sulphides. These bacteria can be used to extract these elements from process wastewaters (Fig. 7.3) (Kelly *et al.* 1979; Kleins and Lee 1979). Sulphur is a major constituent of many effluents, including gypsum wastewaters, excess sulphuric acid, coal sulphurization by-products, acid mine waters, and general metallurigcal effluents. Cork and Cusanovich (1978) studied the quantitative conversion of sulphate to elemental sulphur using *Desulfovibrio desulfuricans* to convert sulphate to sulphide and photosynthetic bacteria, such as *Chromatium thiosulfatophilium* and *C. vinosum* to convert sulphide to

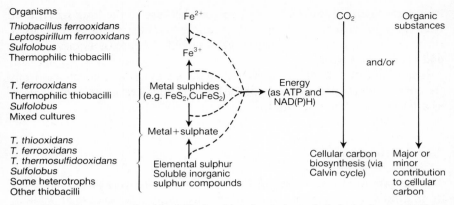

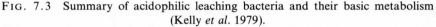

FIG. 7.3   Summary of acidophilic leaching bacteria and their basic metabolism
(Kelly *et al.* 1979).

elemental sulphur, and concluded that a viable industrial process could be developed from such a system. In the laboratory, mixed cultures of *Thiobacillus ferrooxidans* and *T. thiooxidans* are effective in the removal of pyritic sulphur from 20% slurries of a commercial grade of pulverized coal, even though neither species is effective on its own (Dugan and Apel 1978). Although further work remains to be done on the use of micro-organisms to solubilize the organic sulphur fraction of coal, as these were not removed in the laboratory-scale trials, Dugan and Apel are confident that commercial-scale operation is feasible.

The principle contaminants of coal mining wastes are ferrous (Fe^{2+}) iron and sulphuric acid. Wichlacz and Unz (1981) were able to remove >90% of the ferrous iron using a rotating biological contactor, regardless of the ferrous iron concentrations or the hydraulic residence times tested. The biological film that developed on the discs was found to contain the chemolithotrophic iron oxidizing bacterium *Thiobacillus ferrooxidans*, as expected, and also at the higher mass loading concentrations of ferrous iron, acidophilic heterotrophic bacteria. They concluded that the practice of treating soil and strip mine overburden with organic inhibitors to prevent oxidative weathering may be pointless, as a heterotrophic bacterial population may be induced to grow on the organic molecules.

The runoff from mine tailing dumps are acidic and rich in metals. This observation has led to the development of the microbial leaching process, which is the major method of recovering metals using micro-organisms. The process has been known since Roman times and the practice of percolating acidified water through heaps of low-grade ore to remove the metal sulphide formed by the bacterial activity within the heap, was carried out in Anglesey as early as the sixteenth century. However, it is only in the latter half of this century that bacterially assisted leaching has been practised on a large scale

(Rothman *et al.* 1981). Ore deposits contain a wide range of metallic sulphides which are naturally oxidized and solubilized by a complex microbial community. Initially, the autotrophic oxidizing bacteria, such as *Thiobacillus* spp. are able to oxidize both the sulphide ores and the lower oxidation states of the metals (Kelly *et al.* 1979). The leaching process is quite simple. Low-grade ore and other waste rock are heaped on to an impermeable base. Little preparation is required except the grading and shape of the ore particles should allow sufficient voidage to permit maximum recirculation within the heap ensuring long residence times, unimpaired drainage, and aeration. The leaching solution is sprayed on to the surface of the heap and percolates slowly through, dissolving metals released by microbial oxidation. The metal salts are extracted and concentrated by electroprecipitation, solvent extraction or conventional precipitation. The leaching solution is recycled after the metal salts have been extracted, although it needs to be replenished with acid and bacteria. Although copper and uranium are the principal metals extracted in this way commercially, the leachate contains a variety of other valuable metals. Iron, arsenic, antimony, cadmium, lead, silver, gold, and chromium have all been extracted by this method (Ferraiolo and Del Borghi 1987).

Between 10–15% of the annual primary copper production in the USA is by microbial leaching. In Utah, a massive heap of 250 000 tonnes of mineral ore produces 150 tonnes of copper per day, with the liquid heavily charged with bacteria. The bacteria form copper sulphate from which the copper can be readily recovered with the acid solution returned to the leaching process. Microbial leaching operations run continuously and require only a small workforce; the plant described above in Utah only requires six people to keep it running. Similar units of equal size are recovering copper in Chile and Romania, and uranium is being extracted by microbial leaching in Canada (Rothman *et al.* 1981).

Using *Thiobacillus* spp. Ebner (1978) leached metals from sulphidic dust, acidic fly ash from a copper process, slag from a lead smelting process, and javosite from zinc electrolysis. He found that leaching efficiency depended on material, treatment, and bacterial species, with maximum outputs for zinc and copper of 95% and 70% respectively (Table 7.4). Ebner recorded high concentrations of other metals in the leachate, especially cadmium, arsenic, and lead. Torma *et al.* (1970) were able to leach zinc rapidly using *Thiobacillus ferrooxidans*, producing leachate concentrations of up to 72 g l^{-1} of zinc, concluding that at this level of recovery the process was commercially competitive with other forms of mining. Electron micrographs of bacteria from a mixed culture of *T. ferrooxidans* and *T. thiooxidans* revealed that silver was being concentrated by the bacteria without inhibition (Pooley 1982). Small silver sulphide granules growing on the surface of bacteria collected at different stages of a batch leaching process, which finally accumulated to form large lumps (Fig. 7.4). The cells containing the silver could be readily collected making viable the bacterial leaching of precious sulphide minerals, such as

TABLE 7.4 Maximum leaching efficiencies (%) from different wastes using either *Thiobacillus ferrooxidans* (F) or *Thiobacillus thiooxidans* (T) (Ebner 1978)

	Sulphidic dust concentrate a			Sulphidic dust concentrate b			Fly ash			Javosite		Lead slag	
	Cu	Zn	As	Cu	Zn	As	Cu	Zn	As	Cu	Zn	Cu	Zn
Bacterial leaching	12	38	3	35	94	14	42	85	81	69	52	33	50
Sterile control	0.1	16	—	0.4	10	0.2	10	16	—	5	10	0	0
Thiobacillus strain	F-E2 F-E21	F-E2	F-E2	F-E2	F-E2	F-E2	T-E2	T-E2	T-E2	T-E2	T-E2	T-E2	T-E2

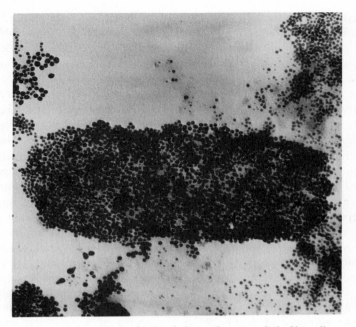

FIG. 7.4 Electron micrograph of a simple bacterium emptied of its cell contents and showing silver sulphide particles adsorbed on to its surface (Pooley 1982).

silver and possibly gold, from previously uneconomic ores. More recent leaching methods employ stirred reactors with finely crushed ore kept in constant suspension (Lundgren and Malouf 1983). Much attention is being focused on developing a more efficient strain of *Thiobacillus ferrooxidans*, the bacteria mainly used in this process, by using genetic engineering.

A number of heterotrophic bacteria are capable of accumulating heavy metals in rivers and streams where the concentration of metals is extremely low (Friedman and Dugan 1968b; Patrick and Loutit 1976; Norris and Kelly 1980; Gray and Clarke 1984; Gray 1985a). Biological wastewater treatment units are also effective in removing metals from domestic and industrial wastewaters, with the metals being concentrated in the biomass and sludge (Brown and Lester 1979). The concentration of metals accumulated in the sludge can be so high that it restricts the disposal options for the sludge (Section 5.1). For example, the biomass produced from a plant treating acetate rayon wastewater which contained between 50–100 mg l^{-1} of zinc, accumulated 12% by weight of zinc without interfering with the BOD removal (Kiff and Brown 1981). The major removal mechanism of metals is adsorption which is controlled by physico-chemical factors, such as temperature and competing ions. The micro-organisms and their extracellular polymeric secretions are highly charged due to the presence of carboxyl, amine, and hydroxyl groups in their protein, carbohydrate, and phospholipids. Adsorption, which is extracellular binding of the metal ion on to the microbial cell, occurs because of the electrostatic attraction between positively charged metals and the negatively charged surfaces of micro-organisms or their secreted polymers. Wheatley (1981) has demonstrated with some excellent electron photo-micrographs that the active bios in biological treatment units of wastewater treatment plants is largely composed of extracellular bacterial polymer rather than bacterial cells. This is not altogether surprising as the bacteria rely on the polymer to adsorb and transfer nutrients from solution to the parent cell. The area and structural nature of the polymer varies according to the substrate with carbohydrates, and especially oligosaccharides, generating most polymer under controlled conditions. To the process engineer, the very large active surface area supplied by microbial surfaces and microbial polymers is cheaper and more flexible for adsorption than chemical surfaces, such as activated carbon. Heavy metals appear to be taken up by specific uptake mechanisms by micro-organisms along with useful metals necessary for metabolism. This property and the mechanisms by which micro-organisms may resist metal toxicity, such as polymer traps, enzymatic oxidation, precipitation, and efflux pumps, may well provide the genetic engineer with the means to modify existing species to enhance metal accumulating properties (Wood and Wang 1983). The ability of *Sphaerotilus natans*, in particular, to accumulate metals in the mucilaginous layer outside its sheath makes it ideal for concentrating and recovering metals that are present in low concentrations in wastewaters. Similarly, metals are bound in the extracellular slime of the zoogloeal growths that are a major component of the bios of all aerobic biological treatment units. Hatch and Menawat (1979) found that *S. natans* grew well in the presence of sulphates of iron, magnesium, copper, cobalt, and cadmium, accumulating these metals in the outer layer of its sheath. However, growth was inhibited in the presence of nickel and chromium sulphate, and in chlorides of iron, magnesium, copper, and cobalt. So far, the ability of the *Sphaerotilus–Leptothrix* group of bacteria to recover metals has not been fully exploited, although the potential is enormous. The bacteria could be particularly useful in removing low-level metal contamination from wastewater or even reducing the metal load of rivers *in situ*. For example, Gray and Clarke (1984) found that *S. natans* was able to accumulate cadmium when grown at concentrations ranging from $0.001–0.1$ mg l^{-1}, reaching maximum concentrations in the bacteria of between $12.5–19.6$ μg g^{-1} dry weight. Another potential area is in the treatment of radioactive contaminated water used in the processing of nuclear fuel. A number of heavy metals are present in this wastewater, in particular, uranium. Although conventional methods are generally ineffective or very expensive, experiments using the yeast *Saccharomyces cerevisiae* and the bacterium *Pseudomonas aeruginosa* were both able to recover the metals from solution. The bacterium was more effective than the yeast with the uranium concentration in solution approaching the equilibrium value after a contact time of only 10 minutes (Schumate *et al.* 1979). Initially, uranium binds to an active site but then additional accumulation occurs by crystallization on to the bound molecules (Beveridge and Murray 1976). The rate and degree of uranium recovered from solution by the bacterium suggests that such a system could be used to decontaminate process wastewaters from the nuclear fuel cycle and allow the uranium to be recycled. Like all metals bound to a microbial surface, uranium can be elutriated by acid hydrolysis. Both yeasts and fungi can also accumulate metals (Norris and Kelly 1980; Gray and Clarke 1984), with metal accumulation in the yeast *Saccharomyces* being particularly effective, reaching 10–15% of its cell weight. Two particular species have demonstrated tolerance to particularly toxic metals. The fungus *Chrysoposporium* is able to absorb mercury (Williams and Pugh 1975), and the bacterium *Citobacter* is able to accumulate cadmium (Macaskie and Dean 1984). Such species could be particularly useful for detoxifying wastewaters.

Further reading

Fertilizer value: Davis 1980.
Re-use of effluents: Shuval 1977; Dean and Lund 1981.
Metal recovery: Murr *et al.* 1978; Ferraiolo and Del Borghi 1987.

7.3 BIOLOGICAL CONVERSION

7.3.1 Bio-energy

Energy produced from the bioconversion of wastewater sludges is mainly confined to the production of biogas from anaerobic digestion, although the production of alcohol from various other types of wastes is also important. Energy can also be produced from waste material by a variety of other processes including combustion, pyrolysis, liquefaction, and gasification.

The simplest method of obtaining energy from organic waste and biomass is to burn it, with combustion in suitable devices being one of the most efficient methods of utilizing the energy potential of these substances. For example, in direct fired furnaces and steam boilers it is possible to achieve thermal efficiencies of up to 85%, although in practice they are much less efficient. The combustion of biomass ($C_xH_yO_z$) can be expressed as:

$$C_xH_yO_z + (x + y/4 - z/2)\,O_2 \rightarrow xCO_2 + (y/2)H_2O.$$

with the heat produced varying between 16–24 GJ t^{-1} of oven dried biomass.

The presence of water in the biomass does not reduce the thermodynamic heat yield of the combustion reaction, although in practical terms the efficiency of the reaction is severely reduced because of the need to heat the water and evaporate it off. A moisture content >30% will prevent direct burning so that the material must be either dried, or a supplementary fuel added. Recent advances in incinerator design have increased efficiency, with fluidized bed incinerators able to accept material containing up to 55% water. Wood and straw are the most widely used biomass materials for combustion, although solid animal wastes, sewage sludges and composted sludge containing up to 75% water have all been utilized. However, preliminary drying is essential (Boyle 1984), and this is normally done by using the flue gases from the actual combustion process (Pedersen 1982).

Pyrolysis, liquefaction, and gasification are upgrading processes, converting biomass into a stable and transportable fuel that can be substituted for conventional fuels. Solids, liquids, and gases can be produced from biomass including wastes, by a variety of processes, which have similar properties to coal, oil and natural gas. When heat is applied to biomass, water is released, and above 100°C the biomass begins to decompose. Between 250–600°C the main products are charcoal and an oily acidic mixture of tar with variable quantities of methanol, acetic acid, acetone, and traces of other organic molecules. Before the wide availability of fossil oils the destructive distillation of wood provided the major source for these chemicals. The reactions are very complex and are summarized in Fig. 7.5. There are various adaptations of this process which are reviewed by Boyle (1984), but it is clear that domestic and

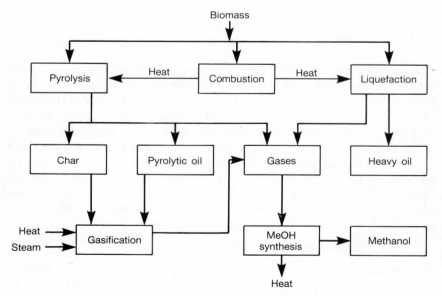

FIG. 7.5   Thermal upgrading of biomass.

agricultural wastes have not been fully exploited by these techniques, with the potential for the production of fuel oils being particularly encouraging.

Biogas

Biogas recovery is the most widespread bioconversion process used for wastewaters and for domestic sewage and animal slurries, in particular. Since its introduction in 1901, anaerobic digestion has developed from being a process for stabilizing sludge with the evolved gas often vented to the atmosphere, to a highly sophisticated energy producing process. The digestion process is fully explained in Section 4.4, but suffice it to say that anaerobic digestion is based on a complex microbial community made up of three distinct groups. The hydrolytic fermentative bacteria hydrolyse the complex organic material to organic acids, alcohols, esters, and sugars, generating carbon dioxide and hydrogen. The second group is dependent upon the first, and is comprised of hydrogen-producing and acetogenic bacteria. Finally, the methanogenic bacteria convert the acetate and hydrogen into biogas which is a mixture of methane and carbon dioxide.

Biogas produced from domestic wastewater comprises 65–70% methane, 25–30% carbon dioxide, and small amounts of nitrogen, hydrogen, and other gases. Biogas has a specific gravity of about 0.86 compared with air. With the conventional digesters built at treatment plants up to the early 1970s, a 30-day retention time was typical, with only a 50% conversion of the organic matter achieved. New reactor designs today ensure efficient and rapid digestion with maximum production of methane, with digesters continuously fed and residence times as short as 10 days. Modern digesters tend to treat the wastewater rather than the sludge because in the wastewater the organic component is more likely to be in solution and so more amenable to digestion (Section 4.4). Apart from sewage treatment, the most common application of anaerobic digestion is for the treatment of vegetable and food-processing wastewaters, with reactors of between 2000–20 000 m^3 in use. There are at present fourteen permanent major installations in the UK that are being used for the dual role of high-rate pretreatment, and the production of energy (Table 7.5). The best wastes for anaerobic digestion are the strong, warm wastewaters from the food- and drink-processing industries, where the majority of the organic material is in a soluble form. Using such ideal wastes, maximum loads of up to 25–30 kg COD $m^{-3}d^{-1}$ can be achieved (McInerney *et al.* 1980). The rate of conversion of organic matter to carbon dioxide and methane is discussed in Section 3.4, although in practice, each kg of COD should yield about 350 l of methane. The total gas production is usually estimated from the volatile solids loading to the digester, with the volume of gas produced fluctuating over a wide range, depending on the volatile solids concentration in the sludge feed and the level of microbial activity within the digester. Although various formulae are available for estimating the volume of biogas produced (Metcalf and Eddy 1984), it is possible to roughly estimate

TABLE 7.5 Major UK industrial anaerobic treatment plants (Dr A. D. Wheatley, Cranfield Institute of Technology)

Owner	Waste	Location	Type of reactor	Contractor	Date commissioned
J. Sturge	Molasses	Selby	CSTR	Ames Crosta Esmil	1970 Rebuilt 1988
Tenstar Products	Starch	Ashford	CSTR	Biomechanics	1976
British Sugar Corporation	Sugar beet	Bury St. Edmunds	CSTR	Sorigana	1982
		Peter-borough	CSTR		1986
		Lincoln	CSTR		1988
McCains	Potato	Peter-borough	UASB/ Lagoon	ADI	1981
Swizzels Matlow	Confec-tionery	Stockport	Filter	Prototype ETA	1982
Caernarvon Creameries	Dairy	Chwilog, Caernarvon	UASB	Hamworthy	1982
Distillers	Yeast production	Stirling	Filter	Biomass	1986
Davidson	Paper mill	Aberdeen	UASB	Pacqs	1986
Wrexham Lager	Brewery	Wrexham	CSTR	Biomechanics	1987
Callard & Bowser	Confec-tionery	Bridgend	CSTR	CLEAR	1985
Hall & Woodhouse	Brewery	Blandford Forum	CSTR	CLEAR	1987
Cricket & Malherbie	Dairy	Mether-stowie, Somerset	CSTR	CLEAR	1987
General Foods	Starch/ coffee	Banbury	UASB	Biwater	1988
Tunnel Refineries	Starch	London	CSTR	Biomechanics	1988

the volume on a per capita basis. For primary sludge, typical yields range between 15–22 m^3 per 1000 persons, whereas secondary sludges can produce up to 28 m^3 per 1000 each day. However, the methane content of the biogas is correlated to the chemical composition of the substrate, so the methane content of biogas produced from carbohydrate may be only 50% rising to 75% for alcohol substrates. Although methane production by micro-organisms occurs over a wide temperature range, 0–97°C, two distinct optima exist. Most digesters are mesophilic (35°C), although there is growing interest in thermophilic (60°C) digestion. The cider makers', Bulmers, have developed a thermophilic version of a fully mixed reactor to digest 'still bottoms' at an operating temperature of between 60–70°C (Pickford 1984; Anon 1984b). Furthermore, the potential of digesters has been demonstrated by Caernarvon Creameries, who have perfected a high-rate digester that they intend to use to treat $20–22 \times 10^6$ l of whey produced at the creamery each year, thus producing 775 000 m^3 of biogas (Anon 1983a).

In 1980, the gross energy content of sewage, wastewater from intensive livestock production, and industry-generated wastes in the then nine member states of the EEC was equivalent to 33×10^6 tonnes of oil (Anon 1982b). Most potential for biogas production appears to be from effluents from intensive animal rearing (Gibbs and Greenhalgh 1983). Although not all the 1.47×10^8 tonnes of animal waste produced in the EEC is recoverable for biological conversion, as not all the animals are housed, this still represents an appreciable resource. Farm animals in the USA produce about 2 billion tonnes of waste annually, with about half of this produced by intensive animal production systems and therefore easily recoverable (Bryant et $al.$ 1977). Although agricultural wastes of this kind are of variable concentration, they are generally considered to be highly polluting and a serious threat to water resources. In terms of biogas production, the stronger the effluent the greater its energy potential (Delaine 1981). Therefore, food industry wastes of all types present a potentially very rich source of methane production. However, effluents with high fat and protein contents are even more productive than those containing high percentages of carbohydrates (Table 7.6).

Specific examples of anaerobic treatment plants converting food-processing and agricultural wastewaters to biogas are given by Dodson (1981), who also gives figures for treatment efficiency and gas production rates. Typical values are a 99% reduction in BOD and a gas production of 0.85 m^3kg^{-1} BOD for milk-processing wastes at a retention time of 6 days. Although pea canning liquor has a shorter retention time of 3.5 days, it has a similar gas production at 0.87 m^3kg^{-1} BOD but only a 75% BOD reduction. The effectiveness of anaerobic digestion and biogas production in treating strong effluents has been extensively reviewed. Isaac and McFiggans (1981) give details of the treatment of effluents from the malting, brewing, and distillery industries, with

TABLE 7.6 Gas yields and methane content from the anaerobic digestion of various wastes (Delaine 1981)

Material	Gas yield (m^3kg^{-1} dry solids)	Methane (% volume)
Sewage sludge (municipal)	0.43	78
Dairy waste	0.98	75
Abbatoir:		
paunch manure	0.47	74
blood	0.16	51
Brewery waste sludge	0.43	76
Potato tops	0.53	75
Beet leaves	0.46	85
Cattle manure	0.24	80
Pig manure	0.26	81

patented processes all claiming effective treatment and plentiful gas production (Newell 1982; Anon 1983*b*, 1984*b*). For example, the 'Bio-energy' system has been used in the treatment of effluent produced in the processing of wheat into starch, gluten and glucose, ham-processing effluent, cheese-processing effluent, and a range of other food-processing effluents (Anon 1983*b*). It is claimed that for every tonne of BOD removed, 0.45 tonnes of heavy fuel oil is recovered. The disposal of whey is the most serious problem facing cheese manufacturers, with about 10 tonnes of whey resulting from the production of each tonne of cheese, and the strength of whey varying from 32 000 to 60 000 mg BOD l^{-1}. Hickey and Owens (1981) suggest that, on average, 35% of the operating costs of a cheese manufacturing plant could be recovered by the fermentation of the whey to methane with its subsequent use on site. The patented system, 'Bio-process', is claimed to be able to recover up to 1 tonne of oil equivalent in methane for every 130 tonnes of whey treated (Anon 1984*c*).

It is the anaerobic treatment of domestic and municipal sewage that has received most attention and from which most of the expertise in digestion has been developed. Simplified versions of fully mixed reactors have found widespread applications in many Third World countries, particularly China and India. Alaa El-Din (1980) estimates that about 5 million individual anaerobic digesters had been set up in China by 1978. About 5 m^3 of gas per day is generated by one 10 m^3 biogas plant. This is sufficient to supply a Chinese family with enough gas for cooking and lighting, with the sludge and effluent produced used as fertilizer and irrigant respectively. It is interesting that the average family does not produce enough waste to keep a digester of this size working at full capacity, so it is customary for each family to have at least one pig to help keep the digester fed with waste. Some larger collective digesters have also been built serving a number of families and even small villages. In Fu-Shang City near Shanghai, 45 m^3 capacity septic tanks each generate about 230 m^3 of gas per day which is converted to electricity. In India, extensive use of firewood for energy has led to widespread deforestation and apart from burning dung, which results in lower crop yields as there is no other source of fertilizer for the soil, there are no other locally available energy sources. Therefore, attention has been turned to locally produced biogas from animal dung and agricultural by-products. Parikh and Parikh (1977) describe in detail the potential of biogas production in India. They suggest that village plants of 170 m^3 capacity should be constructed to serve about 100 families, thus supplying all the energy for cooking and lighting. With 576 000 suitable villages in India, the widespread introduction of such a scheme could supply up to 45% of the total energy requirement of that continent.

Although it is generally agreed that small methanogenic systems are potentially viable, such as those designed for small communities, individual industries or in areas where there is no fuel resource, doubts exist on the ultimate economy of using wastes in this way (Hungate 1977; Loll 1977; Kirsop 1981). At present, large-scale digestion of domestic and municipal wastes appears to be non-economic on capital costs alone. However, improvements are expected in many areas of anaerobic digestion and biogas production, including the types of organisms used, the rate of methanogenesis, temperature of operation, the rate of feeding the waste substrate into the digester, physical separation of the groups of organisms involved in the different stages of the process, adoption of more multi-stage processes, and the improvement of mixing in the reactor (Gallo *et al.* 1979; Stuckey and McCarty 1979; McCornville and Maier 1979; Hashimoto 1981; Ghosh 1981; Crueger and Crueger 1984).

There are three main options for the use of biogas: burning it to produce heat, or to generate electricity, or to fuel vehicles. The energy can be supplied to the National Grid, either in the form of gas or electricitiy, when sufficient biogas is produced. Each of these options requires further intermediate processing after production, ranging from simple storage until required, to cleaning and compression in the production of fuel gas (LPG). Methane has a high calorific value of 35.8 MJ m^{-3} at standard temperature and pressure compared with 37.3 MJ m^{-3} for natural gas, which is a mixture of methane, propane, and butane. However, digester gas is only 65% methane, so without cleaning to remove the carbon dioxide, biogas has a typical calorific value of 23 MJ m^{-3}. At the sewage treatment plant the gas is generally used in gas heated boilers to heat sludge digesters, and the excess is used to generate electricity for use on the plant. The electricity production at some intensive agricultural plants may be so high that they are able to sell the surplus energy via the National Grid.

The most exciting and cost-effective use of biogas to date has been its use to fuel vehicles. Ortiz-Canavate *et al.* (1981) have described the use of biogas in both spark ignition and compression engines. In a pilot study at the Modesto wastewater treatment plant (California), compressed biogas has been used to fuel both cars and lorries. A similar study in the UK has been conducted at the Avonmouth Sewage Treatment Plant of the Wessex Water Authority where biogas has been used to generate electricity at the works for many years. Here, a full-scale trial was conducted to operate their eight vehicles on biogas rather than petrol. Before the raw gas can be used in high-efficiency internal combustion engines the carbon dioxide content of the gas, which can be up to 45%, must be removed. The purified gas must then be compressed. A biogas processing plant was constructed (Fig. 7.6), which is able to convert up to 30 m^3 of raw gas per hour. It is a two-stage compression process, with the methane/carbon dioxide mixture scrubbed with the clarified effluent from the plant in order to remove carbon dioxide, and then further compressed to 198 bar for storage in cylinders. The cylinders are able to store 240 litres of compressed gas, which is 99% methane and produced at between 15–25 m^{3}h^{-1}. The cylinders are connected to vehicle filling bays with standard LPG valves and snap-on hose connectors. Approximately 0.7 m^3 of gas is equivalent to 1 litre of petrol and the Avonmouth plant is producing up

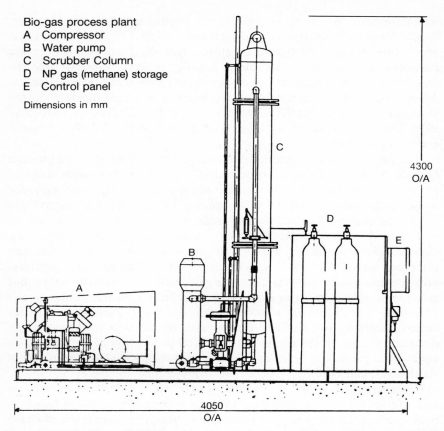

Bio-gas process plant
A   Compressor
B   Water pump
C   Scrubber Column
D   NP gas (methane) storage
E   Control panel

Dimensions in mm

FIG. 7.6 The biogas (LPG) process plant at Avonmouth Sewage Treatment Plant which is able to convert up to 30 m^3h^{-1} of 60% methane and 40% CO_2 biogas to high performance LPG with 99% methane and an output of between 15–25 m^3h^{-1} (Anon 1984*e*).

to 23 litres of petrol equivalent each hour, which is enough to satisfy 95% of their fuel requirements. In the winter, they have not been able to produce enough gas from their existing digesters, although during the summer there is a massive surplus, which at present is wasted. Modifications to the existing digesters will ensure an adequate supply of gas throughout the year in the future. Although the Water Authority has to pay tax on the gas at £0.135 m^{-3}, they estimate that the plant is saving them about £10 000 per annum in petrol costs. This figure does not include the surplus fuel that could be sold. Each converted van has two cylinders of gas located transversely behind the driver's seat. The only other modifications required are the gas carburettor and the gas–petrol change-over switch. The cylinders in each van provide enough fuel for 160 km and if this is exhausted then the engine can be switched instantly over to petrol–drive without even stopping. The performance of gas is claimed to be as good as petrol, with claims of a cleaner engine which requires less maintenance, prolonged life of the engine oil, oil filter and the spark plugs, and a cleaner exhaust. The cost of the project at Avonmouth was £30 000; approximately £25 000 for the plant and £5000 for the vehicle conversions (Anon 1984e). A similar scheme is being operated by the Anglian Water Authority at a sewage treatment plant in Essex (Anglian Water Authority 1982). The plant also cost £30 000, and at a conservative estimate of a plant life of 10 years and only 5 years for the vehicle conversions, they estimated that the cost of the gas equivalent of a gallon of petrol is £1. The Anglian Water Authority has also been working on compression engines, which are simpler to convert than spark-ignition engines. The gas can be used in mixture with the normal diesel fuel at a possible saving of 30% (Anon 1984e).

Fuel–alcohol
There has been increasing interest in the idea of obtaining chemical and lipid fuels from cellulose (Gallo *et al.* 1979). The production of fuel–alcohol, in particular, has received considerable attention because of the pioneering work by the Brazilian Government who are attemping to replace all imported petrol by the year 1990 (Gochnarg 1979). However, it is clear that with present fermenter technology, fuel–alcohol can only be competitive with petrol when the world price of oil is more than $40 a barrel.

Cellulose is converted to ethanol (ethyl alcohol) by a two-stage reaction (Kirsop 1981; Crueger and Crueger 1984). The crude cellulose, in the form of oligosaccharides and polysaccharides, generally requires to be hydrolysed to monosaccharides in a separate reaction before fermentation. Hydrolysis can be either chemical, using acids, or enzymatic, using cellulases obtained from bacteria and fungi, such as *Cellulomonas* spp., *Trichoderma vivide*, *T. lignovum*, *T. koningii*, *Chrysposporium lignonvum*, *C. pruninosum*, *Penicillium irieusis*, and *Fusarium solani* (Mandels 1975). Starch hydrolysis is relatively easy by using both acid and enzyme methods, whereas cellulose requires pretreatment to free the associated lignin before enzymatic hydrolysis. If the percentage of lignin in the crude cellulose is high, as is the case with a number of wastewaters, enzymatic hydrolysis is much less effective. Therefore, chemical or mechanical methods must be used instead of enzymes, which results in high energy costs. However, a number of biological alternatives are being investigated. Using genetic engineering techniques, improved strains of the fungus *Trichoderma veesei*, may provide a suitable source of an enzyme able to cope with lignin wastes (Gallo *et al.* 1979). Kent-Kirk (1975) has reported on lignin-degrading enzymes produced by the white rot fungus that destroys wood (*Polyporus vesinosus*). He feels that whole organism rather than isolated enzymes may be more effective because the highly complex nature of lignin. Cooney *et al.* (1979) carried out the two processes of hydrolysis and fermentation in a single operation using the thermophilic bacterium *Clostridium thermocellum*.

However, the reaction is slow and the yields poor, although they aim to eventually obtain an ethanol return of 25% of the total solids fed into the reactor. The major commercial problem with enzymes is that the specific rates of cellulose hydrolysis achieved is low in comparison with acids (Boyle 1984). The hydrolysis reaction of cellulose or starch can be summarized as:

$$-(C_6H_{10}O_5)n- +nH_2O \rightarrow nC_6H_{12}O_6$$

The monomeric sugars are then fermented, anaerobically, by yeast such as *Saccharomyces* spp. to alcohol.

$$C_6H_{12}O_6 \rightarrow 2C_2H_5OH + 2CO_2$$

Crude cellulose can be obtained either as plant material or as waste. Although the latter is cheaper, the cellulose content of domestic and most agricultural wastes varies enormously. The main carbohydrates obtained from various wastes and plant material are summarized in Table 7.7. The yield of carbohydrate varies considerably, for example, wood is composed of 60% cellulose (dry weight), where sugar beet and cane both contain between 15–20% of sucrose.

Only 2% of Brazil's land will be required to supply enough cellulose to produce sufficient fuel–alcohol for their own requirements. The cellulose comes almost exclusively from sugar cane, which because of the high photosynthetic rate in the tropical climate of Brazil, grows so rapidly that three harvests a year can be taken. Alcohol is a non-polluting, anti-knock fuel that can be used instead of, or in combination with petrol (Humphrey 1975). On its own, it can also be used as boiler fuel with a thermal efficiency approaching 80%. However, compared with fuel oil its calorific value is only 57% by volume or 66% on a weight basis. Alcohol can be used directly as a substitute for petrol except that the performance is lower. Gashol is a blend of 99.9% alcohol and petrol, with some blends containing up to 20% alcohol. Engines using gashol require only minor modifications compared to engines adapted

TABLE 7.7   Sources of carbohydrates found in wastewaters (Boyle 1984)

Source	Carbohydrate
Monosaccharides and oligosaccharides	
Sugar cane and beet	Sucrose
Molasses	Sucrose, glucose, fructose
Dairy wastes	Lactose, galactose
Sweet sorghum	Sucrose, glucose
Polysaccharides	
Wood and crop residues	Cellulose, hemi-cellulose
Municipal and paper wastes	Cellulose
Maize and other cereals	Starch
Cassava and potatoes	Starch

to use alcohol alone. A major problem associated with using alcohol is that it is more corrosive than petrol so that storage tanks, pumps, and vehicle storage tanks all require to be lined with a protective material. Engine parts also wear out more quickly and engines using pure ethanol will require modification. The development of a suitable engine is continuing with current designs requiring additional lubrication and a higher compression ratio. Diesel engines do not function well on alcohol or alcohol–diesel mixtures because up to 20% by volume of additives are required to achieve the necessary octane ratio (Rothman *et al.* 1981; Boyle 1984).

At present, the alcohol, which is a waste product of yeast metabolism, is removed by constantly removing a portion of the fermenter liquid and distilling off the alcohol, and returning the unused substrate to the reactor. The alcohol inhibits the yeast at concentrations approaching 8–10% and it must be continuously removed if the biological metabolism is not to be severely inhibited or even halted. In the production of alcohol, more energy is required to produce a unit of fuel–alcohol than the energy it contains. Many stages in the process of converting cellulose to alcohol are energy-intensive and none more so than the distillation stage. Kirsop (1981) has established that the energy costs of distillation to remove ethanol from the treated waste increases as the concentration of ethanol diminishes. Distillation is likely to be uneconomical, unless the ethanol concentration is greater than 5%, and to ensure this the carbohydrate concentration of the substrate should be in the order of 10%. An interesting proposal is to replace distillation with reverse osmosis to separate the alcohol which will save considerable energy. Clearly, the production of ethanol from wastewaters is only economic when the wastewater has a high concentration of cellulose or other carbohydrate, is free from toxic materials, and when the cost of alcohol production is offset against the cost of pollution control. The waste from the fermenter, which can be more than 10 times the volume of alcohol, is very polluting. The present policy in Brazil is to pump this effluent back to the cane fields without any treatment. It may be possible, however, to devise a more integrated system where the effluent could be digested anaerobically to produce methane gas which could be used to fuel the distillation process as well as producing a useful and more handleable fertilizer. Plants are already in operation in Brazil and Australia, and orders have been placed for plants in Pakistan, France, and Germany that will produce 30 000 l of ethanol per day from cane molasses, and 90 000 l and 45 000 l of ethanol each day from beet molasses respectively (Anon 1984*d*).

7.3.2 Single-cell protein and biomass

The link between biomass for food production and wastewater treatment arose from two particular developments. First, large crops of protein-rich algae grow on oxidation ponds (Section 4.5) and secondly, the conversion of wastes from various food-processing industries to yeast resulted in the purification of the wastewater as well as the production of a useful by-product. This has led to the development of numerous schemes for the utilization of wastewater as substrates for the production of biomass or single-cell protein, and at the same time purifying the effluent (Samuelov 1983). The conversion of soluble and suspended nutrients to microbial biomass during the biological unit processes in conventional wastewater treatment has already been discussed and these gross solids in the form of activated sludge flocs or sloughed film from percolating filters have been used directly as an animal feed supplement (Grau 1980; Beszechits and Lugowski 1983). The macro-invertebrates washed from fixed-film reactors have also been utilized as a source of food for non-intensive fish farming.

Single-cell protein (SCP) is microbial biomass produced by some form of fermentation process and can be used as a food or a food additive. In its simplest form, a suitable organism, such as a yeast, is cultured in a suitable substrate, normally a carbohydrate, such as a molasses solution, and under suitable conditions. The yeast or biomass is recovered from the fermentation by filtration, washed and dried to produce a free-flowing powder, rich in protein. The SCP is cheap to produce, versatile, and depending on its quality, can be used for either animal or human consumption. Its most promising applications are in the animal feed trade, although it is widely used in fortifying poor diets and in adding flavour and protein to processed foods (Rothman *et al.* 1981).

Various groups of micro-organisms have been used for the production of SCP, including bacteria, yeasts, fungi, and algae (Table 7.8). The easiest micro-organisms to be cultured and used for SCP production are the bacteria, but they are less acceptable in terms of palatability than yeasts and fungi, more difficult to separate, and contain considerably more nucleic acids. A wide range of bacterial species have been utilized, especially the photosynthetic bacteria, and these are reviewed by Kobayashi (1977) and Ensign (1977). The most widely investigated groups are the yeasts and the *fungi imperfecti*. A number of higher organisms are also utilized for biomass and protein production, although they are not classed as SCP. These include worms from composting wastes and sludge (Section 7.3.3), fish as the top of the food chain in oxidation ponds, and the fast growing water hyacinth and other plants that are effective methods of wastewater treatment as well as being a useful by-product.

Although a wide variety of substrates can be utilized, few have been successfully exploited on a commercial scale (James and Addyman 1974; Rose 1979a: Atkinson and Mavitune 1983) (Tables 7.7 and 7.8). The problem is that in order to produce a well-defined end-product, a pure culture fermentation of a substrate with constant quality is required. Fermentations of substrates of consistent quality, such as molasses, whey or starch can successfully use pure cultures to ensure a constant product, wastewaters, however, are rarely of

TABLE 7.8 Commonly used micro-organisms for single-cell protein (SCP) production (Atkinson and Mavitune 1983)

Micro-organism	Substrate
Algae	
Chlorella sovokiniana	Carbon dioxide
C. regularis S-50	Carbon dioxide
Spirulina maxima (synthetic medium)	Carbon dioxide
S. maxima (sewage)	Carbon dioxide
Bacteria and actinomycetes	
Acinetobacter (Micrococcus) certificans	n-Hexadecane
Cellulomonas sp.	Bagasse
Methalomonas clara	Methanol
Methylophilus (Pseudomonas) methylotrophus	Methanol
Thermomonospora fusca	Pulping fines
Yeasts	
Candida lipolytica (Toprina)	n-Alkanes
C. lipolytica	Gas oil
C. utilis	Ethanol
Hansenula polymorpha	Methanol
Kluyveromyces (Saccharomyces) fragilis	Cheese whey
Saccharomyces cerevisiae	Molasses
Trichosporon cutaneum	Oxanone wastes
Moulds and higher fungi	
Agaricus campestris (white var.)	Glucose
A. campestris (brown var.)	Glucose
Aspergillus niger	Molasses
Fusarium graminearium	Starch
Morchella crassipes	Glucose
M. crassipes	Sulphite waste liquor
M. esculenta	Glucose
M. hortensis	Glucose
Paecilomyces vaviota (Pekilo)	Sulphite waste liquor
Trichoderina viride	Starch

constant quality. Domestic sewage, for example, is composed of multi-carbon substrates and a mixed culture fermentation is required to ensure that all the carbon is utilized, thus providing adequate pollution control. Although mixed culture systems are more stable, easily maintained, and less susceptible to contamination, the product is less easily defined with an inconsistent nutrient value and possibly containing toxic metabolites and residues. Few demonstration plants have successfully produced SCP from sewage, although it is theoretically and technically feasible. The problem of contamination of the substrate by metals and toxic organic compounds that can inhibit microbial growth and become accumulated within the microbial biomass is a major drawback when municipal wastes are considered for SCP recovery. Also, inconsistency of the product and the relatively low value of the recovered protein suggests that although SCP production from sewage could be a useful by-product to partially offset wastewater treatment and sludge disposal costs, it will probably never be a major or economic source of protein. Processes that produce large volumes of dilute carbohydrate wastes free of toxic materials are the most promising substrates for SCP production. These include milk and cheese-processing, confectionery manufacturing, and food canning plants (Forage and Righelato 1979). These effluents have a high biochemial and chemical oxygen demand and are costly to treat, so the development of waste recovery using SCP production would offset the cost of treatment. The development of SCP processes using these effluents is widespread. For example, Bassetts of Sheffield are using a protein recovery process based on the dilute wastes from Liquorice Allsorts manufacturing which was developed by Tate and Lyle Process Technology Ltd. The Swedish 'Symba' process is based on the production of yeast protein from starch waste, especially from potato processing. Other investigations have used a range of wastewaters including paper mill wastes (Holderby and Moggio 1960) and cellulose materials (Callihan and Clemmer 1979), milk wastes (Wassermann et al. 1961; Meyrath and Bayer 1979; Wheatley et al. 1982), coconut and pineapple wastewater (Smith and Bull 1976; Prior, 1984) citric acid wastes (Braun et al. 1979), distillery-type wastes (Quinn and Marchant 1980; Tauk 1982), and many more (Tomlinson 1976a,b; Rose 1979a). However, economic SCP production is limited to a small number of specific substrates (Norris 1981). A number of SCP production processes have been patented and are being operated commercially, whereas others remain at the demonstration stage (Table 7.9). The fall in the value of SCP in relation to other sources of protein, such as fish meal and soya suggests that there will be little further development in this field for the present. Research is continuing, however, with particular emphasis on dual SCP and biogas production. Cheaper methods of SCP production using existing systems are also being examined. Significant reductions in both capital and running costs can be made by avoiding sterilization and controlling contamination by using high inoculum concentrations, cell recycling or a low pH. Methods of recovering the protein are generally expensive, especially those involving centrifuging, filtration, and spray or drum drying, and alternative methods of separation, such as flocculation, flotation, and screening should be used (Wheatley 1985). There has been a general movement away from dried SCP to a liquid end-product. This means that the whole fermentation broth can be used after pasteurization, which is sufficient if the protein is to be used as a supplement to ordinary animal feeds, although is probably less acceptable for use in human foodstuffs.

Although SCP is of excellent nutritional value, it is very variable in quality and has a number of drawbacks that limit its widespread use. Yeasts, such as *Saccharomyces cerevisiae* and *Candida utilis*, and most fungi, are quite acceptable to animals and man, whereas algal and bacterial biomass are less so, being less palatable and containing undesirable levels of certain cellular materials. Possible health hazards include a high nucleic acid content and toxic or carcinogenic substances absorbed from the growth substrate by the

TABLE 7.9 Plants producing single-cell protein (SCP) of 'feed' grade (Atkinson and Mavitune 1983)

Plant class	Company	Plant location	Substrate	Type of Organism	Plant size (tonne y^{-1})
Demonstration	British Petroleum	UK	n-Alkane	Yeast	4000
Demonstration	Dianippon	Japan	n-Alkane	Yeast	1000
Demonstration	Chinese Petroleum	Taiwan	n-Alkane	Yeast	unknown
Demonstration	ICI	UK	Methanol	Bacteria	1000
Demonstration	Kanegafuchi	Japan	n-Alkane	Yeast	5000
Demonstration	Kohjin	Japan	unknown	Yeast	2400
Demonstration	Kyowa Hakko	Japan	n-Alkane	Yeast	1500
Demonstration	Milbrew	US	Whey	Yeast	5000
Demonstration	Shell	Netherlands	Methane	Bacteria	1000
Demonstration	Svenska-Socker	Sweden	Potato starch	Yeast	2000
Semi-commercial	British Petroleum	France	Gas oil	Yeast	20 000
Semi-commercial	ICI	UK	Methanol	Bacteria	50 000
Semi-commercial	United Paper Mills	Finland	Sulphite waste	Yeast	10 000
Semi-commercial	USSR State	USSR	unknown	Yeast	20 000
Commercial	British Petroleum	Italy	n-Alkane	Yeast	100 000
Commercial	Liquichemica	Italy	n-Alkane	Yeast	100 000
Other systems	LSU-Bechtel		Cellulose	Bacteria	
Other systems	Tate and Lyle	Belize	Citric acid	Fungi	
Other systems	ICAITI	Guatemala	Coffee waste	Fungi	
Other systems	IFP		CO_2/Sunlight	Algae	
Other systems	General Electric		Feedlot waste	Bacteria	
Other systems	Mitsubishi		Methanol	Yeast	
Other systems	United Paper Mills	Finland	Pulp waste	Fungi	

microbial biomass. Also, as SCP is generally digested slowly in the digestive tract, allergic reactions or indigestion are possible (Pokvovsky 1975; Sinskey and Tannenbaum 1975; Garattini *et al.* 1979). There have been many reports of adverse reactions in humans following the consumption of microbial biomass (Scrimshaw 1975). For example, amounts of SCP comprised of *Aerobacter aerogenes* or *Hydrogenomonas eutropha* of between 12–25 g d^{-1} caused nausea, vomiting, and diarrhoea when fed to young female volunteers, although these bacteria caused no adverse effects when fed to animals (Waslien *et al.* 1968, 1969). *Candida utilis* is also known to have caused gastro-intestinal disorders when fed in quantities as little as 15 g d^{-1} (Scrimshaw 1975). If microbial biomass is to become widely used as a form of protein then more research on the physiological effects is urgently required.

Yeasts and fungi
The yeasts and fungi have been exploited more than any other microbial group for SCP production (Table 7.8). The filamentous fungi are easier to separate and dry than the yeasts, and have the advantage of having a better food texture than other micro-organisms. Full-scale trials have been conducted using *Trichoderma* and *Geotrichum* for the treatment of vegetable-processing wastes (Church *et al.* 1973), and the latter fungus has also been used with distillery wastes (Quinn and Marchant 1980) and for the treatment of milk wastes, along with *Fusarium* (Wheatley *et al.* 1982). *Fusarium* is also used with many carbohydrate wastes (Munden and King 1973). However, it is the yeasts that have made the most significant impact in the wastewater treatment industry in terms of SCP production, especially strains of *Candida* and *Saccharomyces*. Although the idea of producing SCP from fungi is relatively new, producing yeast for food dates back to 1910, with both *Candida utilis* and *Saccharomyces cerevisia* used for this purpose. These yeasts are capable of using a range of carbohydrates, *Candida* using both pentoses and hexoses, whereas *Saccharomyces* can only utilize hexoses for growth. Sulphide liquor, for example, contains both pentoses and hexoses and in North America some 50 000 tonnes y^{-1} of yeast are produced from this source alone. World-wide yeast production is more than 5000 tonnes per week and is used for both animal and human consumption. One particular use for this type of yeast is the production of the human foodstuff 'Incaparina'. The yeast is used to enrich maize flour with protein (3% yeast) and is widely available in Central and South America (Norris 1981).

Carbohydrate is the most useful energy and carbon source available for SCP production. However, much of the available carbohydrate in waste-waters is in the form of large polymers, such as cellulose, and it cannot be utilized directly by many bacteria and fungi, especially on a large scale. Therefore, much attention has been focused on those wastewaters that are rich in fermentable substrates, such as sucrose-rich effluents from the sugar industry, sulphite waste liquor from paper manufacture, and potato starch wastes. Although fermentable carbohydrates produced by agricultural industries are potentially very attractive, many are seasonal and large-scale development is unlikely to be economic.

Each year, thousands of tonnes of fruit and vegetables are wasted at packaging and canning factories. This led to the development of a novel process by Tate and Lyle to increase the protein content of such wastes, thereby converting it into a valuable product. The process involves solid substrate fermentation, using the mould *Aspergillus niger*. The company claims that the protein content of such waste with an adequate level of fermented carbohydrates can be increased to 20–30% using what is essentially a low-technology process (Davy 1981). One of the largest SCP plants treating waste carbohydrate is that owned by the confectionery firm George Bassett and Co in Sheffield. This is a batch fermentation following heat sterilization of the effluent (Fig. 7.7). A pure culture of *Candida utilis* is grown in the wastewater which is recovered following centrifugation and drying. The SCP is bagged and sold as a high-protein additive for animal feed. The present output for this plant is about 140 tonnes y^{-1} and it is able to reduce the COD of the wastewater by 65% and it is therefore weak enough to be discharged directly to the sewer. However, with the fall in the value of protein in recent years the development of similar installations appears unlikely at present, although with the adoption of the 'polluter-pays' principle system of effluent charges, it may be a cost-effective way of reducing the wastewater treatment bill for manufacturing and processing industries (Deering and Gray 1986). There are many other examples: Romantschuk (1975) describes the 'Pekilo' process, which is a continuous process that uses large 360 m^3 fermenters to produce SCP from the polysaccharide-rich effluents from wood pulp production. The SCP yields obtained in this Finnish plant range from

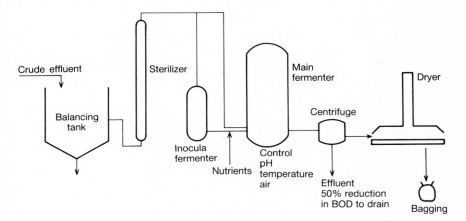

FIG. 7.7   Single-cell protein (SCP) from confectionery effluent using *Candida utilis* (Wheatley 1985).

2.7–2.8 kg m^{-3}h^{-1} using a mould of the genus *Paecilomyces* at a dilution rate of 0.2 h^{-1}. With a solution containing 32 g l^{-1} of reducing sugar, up to 55% of the sugar can be converted to biomass. A major process that has been developed in Ontario, Canada, is the 'Waterloo' SCP process (Moo-Young *et al.* 1979; Moo-Young 1980) (Fig. 7.8). This is based on the mass cultivation of the cellulolytic, heat-tolerant fungus *Chaetomium cellulolyticum* or the yeast *Candida utilis* in an anaerobic fermenter using waste carbohydrate as the main carbon nutrient source. The yeast is grown in a liquid-substrate fermentation on sugars produced by acid hydrolysis of the original waste material, whereas the fungus is grown in a solid-substrate fermentation on the original waste which is pre-softened and partially delignified by thermal or thermochemical treatment. Supplementary nutrients are required as is sterile air and a low pH. Anaerobically predigested manure is used for energy conservation, with the methane gas produced used to supply the energy required for the process. By recycling the processing water, the pollution potential of the final effluent is greatly reduced (Moo-Young 1980). The process appears to be particularly well suited for the treatment of wastewaters from intensive animal rearing.

Two SCP processes are of particular interest. The first is the treatment of distillery wastewaters from the manufacture of whisky. In 1978, the annual production of liquid wastes generated by whisky distilleries in Britain and Ireland was estimated to be 9.6×10^6m^{3}y^{-1} with a BOD equivalent to the wastewater generated by 6.1 million people (Quinn and Marchant 1980).

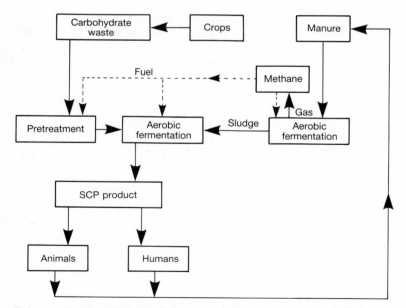

FIG. 7.8   Application of the 'Waterloo' SCP process to the utilization of cattle slurry (Moo-Young 1980).

Traditionally, distillery wastes were disposed to rivers, but now they are usually discharged directly to the sea, applied to derelict land, or evaporated and mixed with spent grain to produce distillers' dark grains that are sold as low grade animal feeds (Quinn and Marchant 1980, Quinn *et al.* 1981; Barker *et al.* 1982). The problems of safe disposal of large volumes of highly polluting wastes have increased in recent years because of high energy costs and increasing stringent anti-pollution legislation (Quinn and Marchant 1979). The spent wash produced by whisky distilleries is extremely strong, with a BOD of up to 43 000 mg l^{-1}, and are rich in carbohydrates and protein (Table 7.10). However, because of the high organic loading and the low pH,

TABLE 7.10   The limit values of the main constituents of spent wash produced by whisky distilleries in Ireland and the UK during 1976–77 (after Quinn and Marchant 1980)

Total carbohydrate	6.7–21.2 g l^{-1}
Protein	15.1–31.0 g l^{-1}
Glycerol	4.4–7.5 g l^{-1}
Total titratable acidity (as lactate)	4.9–18.4 g l^{-1}
Free amino acids	2.1–4.3 g l^{-1}
BOD	10 500–43 000 mg l^{-1}
COD	15 000–58 000 mg l^{-1}
TOC	4920–35 000 mg l^{-1}
Total dissolved phosphorus	740–1960 ppm
Copper	0.2–4.8 ppm
pH	2.9–3.9

conventional sewage treatment methods are not suitable for the treatment of distillery wastes. Quinn and Marchant (1980) investigated the possibility of devising a low-technology microbial treatment process which would not only remove the bulk of the organic loading from the spent wash and produce an effluent that could be more easily disposed of, but which would transform the organic material into high protein biomass to serve as a dietary supplement for animals. They selected a suitable micro-organism by enrichment of inoculum from a variety of natural habitats from which they found that the predominant organism in all instances was the fungus *Geotrichum candidum*. Batch cultures of the fungus grown in 25 ml samples of spent wash, diluted 5-fold, in 25 ml Erlenmeyer flasks gave a BOD reduction of 92.2%, a COD reduction of 80.6%, and a TOC reduction of 63.7%, with a productivity of 0.106 g $l^{-1}h^{-1}$ over the 34 h period prior to the onset of a stationary growth phase. The yield of organisms in terms of organic carbon utilized was 126.4%, with a TOC removal rate of 0.084 $l^{-1}h^{-1}$.

Batch cultures of undiluted samples obtained maximal TOC removal rates of 0.54 g $l^{-1}h^{-1}$ and a biomass production of 0.68 g $l^{-1}h^{-1}$ after 40 h incubation. The BOD removal varied from 63.5–91.4%, COD reduction from 31.7–77.8%, and TOC reduction from 22.5–75.6%. Continuous culture trials were also carried out. Single-stage continuous culture at a predetermined optimum temperature of 22°C and an optimum dilution rate for biomass productivity resulted in 50%, 30%, and 40.7% reductions for BOD, COD, and TOC respectively. With the addition of a second fermenter in line with the first giving an overall retention time of 19.75 h, the BOD removal achieved was 87%, with 69.8% TOC removal and a biomass yield of 34.0 g l^{-1}, and an overall productivity of 1.72 g l^{-1}h^{-1}. Biomass from the second fermenter had a protein content of 45.5%. Quinn and Marchant concluded that the process of organic matter removal by SCP production by an organism that had already been previously cultured for animal and human consumption and was readily harvested, was certainly economically viable. The process also produced an effluent of superior quality to that of the evaporation methods widely employed at that time.

Barker *et al.* (1982) found that two strains of yeast were normally isolated from spent wash samples along with *Geotrichum candidum*. Both yeasts were isolated and identified as *Hansenula anomala* and *Candida krusei* both of which had previously been used for the production of microbial protein from methanol and whey respectively. They tested the growth and substrate assimilation of all three organisms on whisky distillery spent wash in both pure and mixed, batch and continuous culture. Their results showed that although there was no difference between batch and continuous culture in terms of yield or productivity, mixed culture of all three organisms gave greater protein assimilation. The reduction in COD from batch cultures of individual organisms ranged from 44–49%, being 55% for the combined batch cultures. Nutrient examination of the protein produced indicated that it would be suitable as a dietary protein source for non-ruminant animals. Figures of BOD removal were not included in the results so that the protential of the treatment to reduce the pollution load of the waste cannot be assessed. This work has led to a new approach to the treatment of this traditional waste. In Northern Ireland, a pilot plant for conversion of distillery wastes into microbial biomass was set up at the Old Bushmills Whisky Distillery in County Antrim. The plant operated successfully and as a result the company are in the process of changing over from the traditional evaporation treatment method to a new biomass recovery system. The produce from the plant will be used as a food supplement by a local pet food company. The new plant came on-stream early in 1986.

A number of processes producing SCP from starch-containing substrates have been commercialized, including those that treat effluent from the processing of potatoes, corn, and other starch containing foods (Forage and Righelato 1979). The potato-processing industry is another example of an industry that has to contend with the disposal of substantial volumes of highly polluting wastes. During processing, about half of the potato is lost in various forms of waste, such as peel, trim, filterable particulates, processing water and blancher water (Lemmel *et al.* 1979). The larger solids can be used as cattle feed, and the more liquid wastes are subjected to primary and secondary waste treatment. Potato-processing wastes contain a lot of starch and the purification of such wastes is difficult and expensive by conventional wastewater treatment methods. As with the distillery industry described previously, more stringent water quality standards and increasing costs of treatment have prompted research into alternative waste treatment processes, with attention focused on the bioconversion of waste into microbial biomass for feed or food.

The 'Symba' process was developed by the Swedish Sugar Company and the Chepmap Company to treat effluent and solid wastes from potato processing, and has been in full-scale operation since 1973 (Skogman 1976). The process is based on the symbiotic culture of the yeasts *Endomycopsis* (= *Saccharomycopsis*) *fibuliger* and *Candida utilis*. The basic substrate for SCP production in the wastewater is starch, which is not readily assimilated by most microorganisms, including *C. utilis*, so it has to be hydrolysed to simple carbohydrates. Although starch can be hydrolysed using mineral acids, enzymatic hydrolysis is preferred as the former requires highly corrosive-resistant equipment. Also, acid hydrolysis can result in the formation of compounds that can not be used by the yeast in the second phase of the process and gives lower yields of biomass when calculated on the basis of total starch supplied. The starch is hydrolysed by the enzyme amylase, which is produced in large quantities by *E. fibuliger*, to low molecular weight sugars, principally glucose and maltose. These are then used by *C. utilis* which has a high nutritional value and is suitable for a wide variety of uses as a food supplement (Lawford *et al.* 1979). The amylase-producing yeast *E. fibuliger*, which does not have a high nutritional value, is much slower growing than *C. utilis* and is much smaller in size, and the final fermentation product is predominantly *C. utilis* (98% by weight).

A flow diagram of the 'Symba' process is given in Fig. 7.9. The wastewater is strained to remove any large particles and then sterilized by heating in order to destroy any microbial contaminants that may interfere with the fermentation process. The sterile substrate passes into a small preliminary fermenter in which *E. fibuliger* is grown to supplement its population in the main fermenter, as it is much slower growing than *C. utilis* (Table 7.11). This is particularly important as the growth rate of *C. utilis* is governed by the rate at which *E. fibuliger* can hydrolyse the starch to produce the low molecular weight sugars,

TABLE 7.11 Composition of a mixed culture (cells ml^{-1}) as used in the 'Symba' process (Norris 1981)

Cultivation time (h)	Candida utilis	Endomycopsis fibuliger
0	4×10^7	5.5×10^7
20	1.2×10^9	1.4×10^8

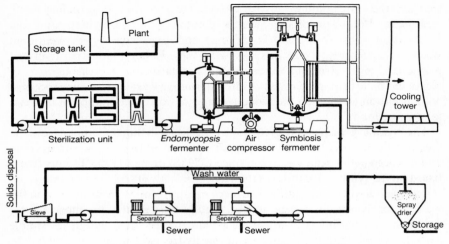

FIG. 7.9   The 'Symba' process (Norris 1981).

and if a single fermenter was used then the growth rate of *C. utilis* would be severely limited as would the rate of biomass production. Trace elements have to be added to the small fermenter to ensure that a steady supply of *E. fibuliger* cells and reduced sugars pass into the main fermenter, which has a normal capacity of about 300 m³, where the symbiotic culture develops. Like all fermenters of this type, sterile air is required which is supplied by pumping air through a series of sterile filters. The culture within the fermenter is vigorously stirred to maintain adequate aeration. The resultant microbial activity produces large quantities of heat that is removed by cooling towers or heat exchangers. Excess biomass passes through several mechanical purification stages to bring the biomass up to the required quality, and it is then concentrated in continuous flow centrifuges in two steps, being washed in between. The concentrated biomass is then spray or drum dried before being packaged ready for use.

Skogman (1976) describes the original 'Symba' plant that reduces the BOD of the wastewater from 10 000–20 000 mg l⁻¹ to 1000–2000 mg l⁻¹ (90% removal), with 50% reduction in the concentration of both nitrogen and phosphorus. The process is able to treat 20 m³h⁻¹ of wastewater, yielding between 250–300 kg dry yeast h⁻¹ at 3% dry solids, which is equivalent to a biomass production of 45% based on the dry weight of the substrate supplied. The yeast contains about 45% protein and is low in nucleic acids but rich in vitamins, especially vitamin B. With the addition of a small amount of methionine, the protein has a nutritional value comparable to that of casein. Feeding trials have been carried out using the yeast as a skim milk replacement for young stock. It is estimated that up to 40% of milk protein used to feed calves could be replaced by 'Symba' SCP. All the trials carried out to date have yielded positive results without any adverse effects. The use of the 'Symba' process in food for human consumption is also being investigated and it seems likely that it will be used in the production of potato powders, bread, and milk products. It is a continuous process and highly automated and can be operated by one person. Although it operates at a level below the economically viable carbohydrate SCP fermenters, it is certainly cost-effective for those industries faced with the problem of disposing of a difficult and expensive waste.

Algae

The proposition that the growth of photosynthetic algae could be used for SCP synthesis, with energy derived from sunlight and carbon from carbon dioxide in the air, has been considered for a long time. In fact, algae form part of the staple diet of the native tribes which live near Lake Chad in Africa, who collect matted algae from pools around the shore of the lake and dry it in the sun. As Norris (1981) points out, the equation of 'free sunlight + free carbon dioxide + cheap inorganic nitrogen = valuable protein' is an attractive one, and it is bound to continue to attract attention.

Algae have a high nutritive value, with green algae, such as *Chlorella* and *Scenedesmus*, containing 50% protein (dry weight), and the large blue-green alga *Spirulina*, 60–70% protein. These figures are only approximate and depend on the nature of the substrate. For example, Milner (1953) found that by altering the growth conditions he could control the cellular constituents of *Chlorella* over a range of 7–88% protein (dry weight), 6–36% carbohydrate, and 1–75% true fat. The protein content in algal cells decreases and the lipid content generally increases under conditions of nitrogen limitation, and other nutritional limitations will also have strong effects on the composition of micro-algae (Waslien 1975; Benemann *et al.* 1979). Table 7.12 summarizes the chemical components of the most important algae in SCP production: *Chlorella*, *Scenedesmus*, and *Spirulina*. However, SCP production using algae is frought with difficulties. Cultivation has to take place in dilute aqueous solution and as cell yields are generally low, recovery is difficult. This makes the resultant protein rather expensive in comparison with other protein sources. Light availability is the critical factor controlling algal growth and although artificial light could be used it is not economically viable. Sunshine is the only practical light source and this restricts the location of such plants to tropical areas. Furthermore, the algae have to be grown in thin layers to ensure maximum penetration and utilization of sunlight, unless constantly mixed. The most economic systems employ very shallow lagoons but in hot climates this results in a high evaporation rate of the water, which is normally a rare and expensive medium in those parts of the world best suited to algal lagoons. Finally, as the algal cultures are not grown under strict aseptic conditions they often become contaminated by other micro-organisms. A pilot plant has been constructed in Mexico to produce algal SCP using

TABLE 7.12 Chemical composition of selected micro-algae (Rose 1979a)

Component (% DW)	Chorella sp. 7–11–05	Scenedesmus sp.	Chlorella–Scenedesmus (10:1, sewage-grown)	Spirulina maxima
Protein*	55.5	53.0	41.8	55.5
Fat	7.5	13.0	7.2	12.7
Carbohydrate	17.8	13.5	27.4	17.4
Ash	8.25	6.5	19.1	7.4
Moisture	7.0	6.0	4.5	7.0
Crude fibre	3.1	8.0	—	—
Ascorbic acid	1.46	—	3.96	1.03
β-Carotene	5.02	—	6.02	2.25
Pantothenic acid	1.12	1.2	0.46	—
Pyridoxin	0.3	—	0.11	0.043
Thiamin	0.77	0.17	0.115	0.138
Riboflavin	—	0.42	0.269	0.285

* Protein contents were calculated by multiplying the value for the total nitrogen content by 6.25. True protein contents are about two-third of the values quoted.

sunlight and carbon dioxide for energy and substrate respectively. The plant uses the alga *Spirulina platensis*, which has helically coiled cells that readily intermesh to form large aggregates which can be easily and cheaply separated, and is able to produce about a tonne of algal biomass daily. However, the limitations of high capital costs, moderate yields, poor product quality control, and a lack of markets have restricted the practical applications of algal cultures for food or feed production (Benemann *et al.* 1977*a*; Prentis 1984).

When algal culture is combined with wastewater treatment, the production of SCP is no longer the primary objective but merely a part of the treatment process that produces a useful by-product which can make a contribution to the economics of the process. Growing algae in sewage has received much attention in recent years, and several plants have been built to take advantage of this 'cash crop'. In hot and arid areas, the culture of algae may be an economically competitive use of land and certainly lagoons of this type are the most cost-effective method of treating wastewaters under such climatic conditions. These systems offer a degree of flexibility so that either algal production or treatment efficiency can be optimized. What is particularly important in some arid coastal regions is that the system operates equally well with wastewater or a mixture of wastewater and sea water. Algal lagoons are very efficient in terms of BOD and nutrient removal and in arid regions the final effluent, after algal separation, may be more valuable that the algae because of the scarcity of water. In Haifa, Israel, a pilot algal lagoon system was found to be as effective as any other treatment process, more so if demand existed for irrigation water and SCP simultaneously with wastewater treatment (Table 7.13) (Shelef *et al.* 1977). Dubinsky *et al.* (1979) summarized the problems of exploiting algal mass culture that must be overcome before it becomes an economic reality (Fig. 7.10). Although physical configuration of algal lagoons is not critical, with circular, square, and rectangular tanks all successfully employed; the important design criteria include water depth, mixing intensity, retention time, and organic loading. However, there are a limited number of interrelated operational variables that can be adjusted during pond operation (Table 7.14). The design and operation of such plants is examined further in Sections 4.3 and 4.5.

The major restriction of algal lagoons for sewage treatment is the relatively large area of land required, about 1 hectare per 1000 people, and the dependence on continuous sunlight throughout the year. Their use to date, therefore, has been restricted to smaller, isolated communities, mainly in Australia and southern USA. Between 40–100 tonnes (dry weight) of algae can be produced per hectare per year. Shelef *et al.* (1977) harvested over 110 tonnes per hectare per annum from their high-rate wastewater lagoon. Over 70% of the dry matter was algae with a protein content of between 42–48%. In Bezonbagh, Nagpur, India, a yield of 123.5 tonnes per hectare per annum was obtained from an experimental pond which was a very high yield on an area

TABLE 7.13   Performance of a high-rate oxidation pond at Technion, Haifa (Shelef *et al.* 1977)

	Raw sewage (mg l^{-1})	Pond effluent (mg l^{-1})	Final effluent (mg l^{-1})*
Suspended solids	240	268	515
BOD$_5$			
total	330	106	10
dissolved	—	12	5
COD			
total	750	670	148
dissolved	—	64	46
Nitrogen			
total	86	71	20
dissolved	—	18	12
Phosphorus			
total	16	10	1
Coliforms per 100 ml	6×10^7	3.5×10^5	8×10^3

* Following flocculation and flotation.

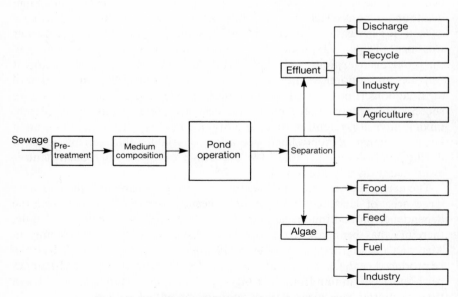

FIG. 7.10   The exploitation of algal mass culture.

TABLE 7.14 Major operational variables in lagoons for the production of algal biomass (Rose 1979a)

Parameter	Control method(s)	Normal limits
Algal concentration	Harvesting, dilution, recycle, inoculation	$150-700 \text{ mg l}^{-1}$
Depth	Dilution, harvesting	$20-50 \text{ cm}$
Hydraulic retention time	Dilution	$1.5-6.0 \text{ days}$
Phytoplankton retention time	Biomass recycle, dilution	$1.5-6.0 \text{ days}$
Zooplankton retention time	Harvesting–recycling with DSM screen	$0.5-6.0 \text{ days}$
Hydraulic loading	Dilution (applicable to oxidation ponds)	$2-20 \text{ cm d}^{-1}$
pH value	Carbon dioxide, dilution	$6.0-10.5$
Nutrient additions	Add with dilution water or independently	Should not be limiting
Oxygen tension	Mixing, carbonation	$0-25 \text{ mg l}^{-1}$
Light absorption	First four parameters, mixing	Absorbtion of 99–99.9% of incident light

basis, compared with other agricultural crops grown in the region (Arceivala *et al.* 1970). Few species of algae are suitable for SCP production because of their indigestibility, toxicity, presence of silicone or carbonate shells, and the lack of economic cultivation technology (Benemann *et al.* 1979). The major problems of successful algal culture are harvesting the algae and maintaining the desired balance of species in the pond, which is normally achieved by separation and recirculation of algae. Harvesting the algae has proved the most difficult problem to resolve. Centrifugation is the most reliable and effective method but is not economically viable. Filtration and screening are usually not effective, especially for the single-celled green algae, although they may be satisfactory for the filamentous species. Algal settling and flotation using chemicals, such as lime, alum or polyelectrolytes, is effective but is expensive both in capital and operating costs. Also, the algal chemical mixtures cannot be used for SCP and have limited value for agricultural use. Benemann and Weissman (1977) have obtained 90% removal efficiencies of algae, resulting in a 10-fold increase in concentration, by using micro-strainers. The device is a rotating drum covered with a fine mesh screen ($25-100$ μm). The pond effluent enters the bottom of the drum and as it slowly rotates, the excess water drains through the mesh leaving the algae held to the inside of the drum. A pressurized jet at the top of the drum dislodges the algae into a collecting trough. This system costs 10% of chemical treatment and leaves the algae free from contamination so that it can be used for SCP. Ideally, strainable algae, especially the filamentous blue-green species, should be cultured on ponds, although such control is difficult.

Algae can be used to supplement the diet of intensively reared animals, including fish (Shelef *et al.* 1977). For example, Carp (*Cyprinus carpio*) and St. Peter's fish (*Tilapia galila*) in which 30% of the algae in the fish's diet replaced 85% of the fish-meal portion; which is usually 15% of the fish commercial diet (Hepher *et al.* 1975). In controlled experiments, weight gain of these fish was consistently higher when fed on the algae-supplemented diet. Even when algae have been chemically separated and contain 4% aluminium, they can be successfully used to replace 25–40% of the soya bean and fish-meal protein fed to chickens (Mukadi and Berk 1975) and fish (Hepher *et al.* 1975).

The production of algae on wastewater can be linked to biogas production (Weissman *et al.* 1979). Nitrogen and phosphorus can be removed from the effluent emanating from another secondary treatment unit by using a tertiary lagoon containing algae. Blue-green algae are especially suitable as they are filamentous and comparatively easy to harvest, and also are a suitable substrate for methane production by anaerobic digestion, with the final digested sludge being rich in nitrogen and phosphorus (Benemann *et al.* 1977*a,b*) (Fig. 7.11). In algal fermentations for biogas, 60–65% of the potential energy content of algae can be converted into methane, with the blue-green algae *Osciallatoria* and *Spirulina* being slightly more digestible than single-cell algae, such as *Euglena* and *Scenedesmus* (Benemann *et al.* 1977*b*). Anderson *et*

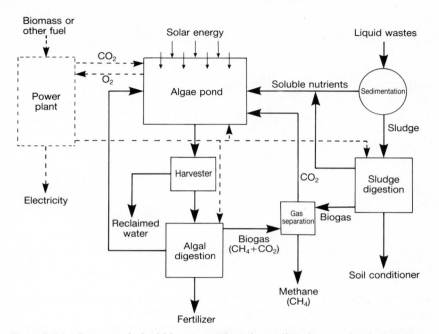

F IG. 7.11 Integrated algal biomass and methane digestion production system.

al. (1980) have proposed a scheme that is almost totally integrated from the recovery of biogas from fermenter effluents to almost complete recycling of nitrogen and other nutrients to the utilization of solar energy for thermophilic digestion to produce ethanol (Fig. 7.12). Although the scheme is purely hypothetical, progress has been achieved on different elements, and this may well be the type of integrated fermentation system used for the treatment of wastewater in the future.

Although algae have only normally been exploited in facultative and high-rate lagoons, much attention has also been paid to the exploitation of algae growing in maturation ponds in the UK. Although algae in maturation ponds are beneficial in terms of phosphorus and nitrogen removal, their presence in the final effluent increases its suspended solids and turbidity, reduces the mortality rate of faecal bacteria, and because of the increase in pH caused by their photosynthetic activity, there is an increased risk of ammonia toxicity to fish (Hawkes 1983a). Suggestions have been made that cladocerans could be used as grazers to remove the algae from the effluent and that the cladocerans could then be harvested as a commercial product as food for aquarium fish. The harvesting would prevent overcrowding, encourage maximum growth rates, and minimize fluctuations in the cladoceran population (Green and Watts 1973). In practice, this has proved very difficult to operate and a rather unreliable means of removing algae, although the presence of a large

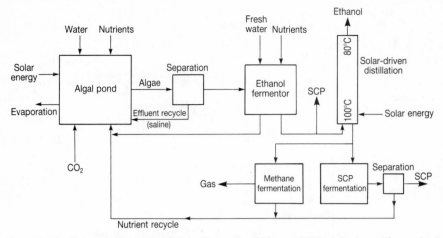

FIG. 7.12   Integrated process for the recovery of biogas, SCP and ethanol from algal fermentation.

cladoceran population does play a significant role in reducing the density of algal cells in ponds and final effluent (Hawkes 1983a).

Bellamy (1975) reviewed the increasing use of biological treatment systems incorporating many different trophic levels. Nutrients released during the degradation of organic matter by heterotrophs are utilized by algae which in turn are consumed by zooplankton, and so on up the food chain to the top trophic level that is usually occupied by a species of fish. The fish can then be harvested for human consumption, with coarse fish being especially popular in Central Europe. This form of treatment has been popular in China for centuries and is increasing in popularity in both developing and developed countries around the world. However, the system requires careful management in order to ensure that the rate of organic loading does not disturb the ecological balance. European wastewater lagoons can provide ideal conditions for fish culture, especially roach, carp, and, to a lesser extent, chub and perch. Experiments on carp culture have shown that a high standing crop can be maintained in British maturation ponds, with good growth and production rates. However, problems with disease were encountered at high stocking densities, and the fish proved very difficult to recover because of the layout of the ponds and nature of the banks. The inclusion of fish in maturation ponds has a significant effect on the ecological balance. In their original model for the control of algae by Cladocera, Green and Watts (1973) did not include predation of the grazers. Unfortunately, the fish, especially the young fingerlings which give such high productivity, feed on the zooplankton which, in turn, feed on the phytoplankton. Therefore, as the fish population increases so does the density of algae in the pond, with a corresponding increase of algae in the final effluent (Williams et al. 1973) (Fig. 7.13). This may not be a

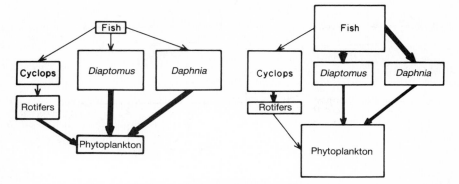

FIG. 7.13 The effect of overstocking with fish on the food web in a fish pond system (Hawkes, 1983c).

problem in countries where the quality of the final effluent is less important in relation to the production of fish, or if the effluent is to be used for irrigation.

The effluent from stabilization ponds in India which is rich in both phyto- and zooplankton, the natural food of so many fish species, has been profitably exploited for rearing fish (pisciculture). As fish are extremely sensitive to oxygen depletion, even for short periods, and the dissolved oxygen concentration in stabilization ponds fluctuates diurnally often reaching zero at night (Section 4.3), fish are rarely stocked in the main stabilization pond itself but normally in secondary ponds (Fig. 7.14). The major species of fish cultured in lagoons and ponds are all carp species, with *Catla catla*, *L. rohita*, *L. calbasu*, and *C. mrigala* all grown successfully in both the stabilization and the subsidiary lagoon. In Bhilai, 2000 fingerlings of *L. rohita* and *C. catla*, with mean weights of 18.7 and 5.5 g respectively, were stocked in a stabilization pond and after 12 months their weight had increased to 140 and 990 g respectively (Chatterjee *et al.* 1967). Fish fry grown in a pond using sewage effluent in Bhopal has been shown to have a higher rate of growth than regularly grown nursery pond fry. It is clear that the main species of carp grow best in the subsidiary rather than the main stabilization ponds, and the conditions can be more carefully controlled for pisciculture in the former and dissolved oxygen depletion is prevented. However, *Ophiocephalus* is a popular fish in some regions of India because of its high iron content and reports have indicated that it may be particularly suited to intensive rearing in stabilization ponds as it is able to utilize both dissolved and atmospheric oxygen; hence its local name 'the surface-breathing fish'. Normal fish yields from freshwater ponds have also been reported as doubling when effluents from stabilization ponds are added in the ratio of 1–2 volumes of effluent to each volume of pond water (Arceivala *et al.* 1970).

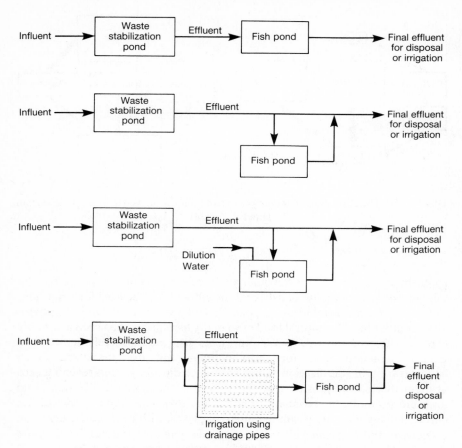

FIG. 7.14   Fish ponds in waste stabilization pond systems.

7.3.3   Composting

Composting is usually applied to solid or semi-solid materials and is the biological decomposition and stabilization of organic substrates, with putrescrible organic material converted to a stabilized compost by microbial action. Compared to other stabilization processes, composting is a fast, simple, and safe approach to the bulk treatment of organic wastes. A wide variety of materials are suitable for composting including sewage sludge, animal and agricultural wastes, and household refuse (Gasser 1985). Almost all the widely used composting systems are aerobic, with the main products being water, carbon dioxide, and heat. Significantly less heat is produced during anaerobic composting, with methane, carbon dioxide, and low molecular weight organic acids being produced. Due to the nature of the intermittent products of anaerobic breakdown there is also a high possibility that odours will be produced. Fermentation and anaerobiosis also lead to products that will undergo further decomposition when exposed to air. In contrast, aerobic systems have a greatly reduced odour potential because the organics that remain are relatively stable so that further decomposition will continue only slowly. Energy, or heat production, is essential to the success of the composting process. Two distinct temperature phases occur during composting. First, a mesophilic phase in which microbial action raises the temperature up to 40–55°C, forcing the process into a thermophilic phase in which the mesophilic micro-organisms are inactivated and are replaced by thermophiles. At these higher temperatures, both animal and plant pathogenic micro-organisms are inactivated, as well as insect pests, their eggs, and weeds. Composting also reduces the water content of the composted waste thereby reducing the volume of material that has to be eventually disposed of.

Composting effectively recycles the organic matter and nutrients in wastewater and converts wastes into a useful material. Composted waste can be used as a soil conditioner to reduce the bulk density of the soil; to increase the water holding capacity and encourage proper soil structure by the addition of organic matter, thus encouraging a healthy soil microfauna/flora and healthy plant root development. Compost replaces the organic matter lost each year due to current farming practices. Indeed, the problem of organic matter restoration in soil is becoming increasingly urgent as field size increases and desertification becomes more common (De Bertoldi et al. 1985). Composted waste is also a fertilizer that can improve crops by the addition of nitrogen, phosphorus, and trace elements. The nitrogen is organically bound and is therefore slowly released throughout the growing season, with minimum losses due to leaching in contrast to when soluble inorganic fertilizers are used. Composting is used world-wide, with over 100 plants operating in the USA alone (Willson and Dalmat 1983).

Methods of composting
Wet substrates are difficult to compost because their high moisture content makes aeration difficult. As a general rule, the higher the moisture content of the substrate the greater is the need to maintain a high voidage to ensure adequate ventilation. Dewatered sludge cake contains between 70–80% water and lacks both bulk and porosity, therefore its dense nature makes diffusion of oxygen difficult and prevents aerobic composting. When stored, unstabilized sludge cake will undergo anaerobic breakdown and as the sludge is plastic in nature, it readily compacts under its own weight, reducing the available voidage even further. Because of these factors, dewatered sludge cake is rarely successfully composted on a commercial scale, although on the pilot-scale level, composting has been achieved with constant mixing so that new surfaces are continuously exposed for oxygen transfer (Haug 1980). However, a recent full-scale demonstration project in Norway has shown that heap composting

of dewatered sewage sludge, without the addition of a material to reduce bulk density, is possible although it is highly dependent upon environmental factors (Paulsrud and Eikum 1984). This is fully discussed below.

To successfully compost dewatered sludge cake the bulk density and porosity need to be reduced so that air can penetrate through the cake more efficiently. Three approaches can be used to overcome the problem of oxygen transfer when composting wet substrates, such as dewatered sludge cake (Fig. 7.15): (1) recycling of compost and mixing with the substrate before composting; (2) addition of an organic amendment to the sludge; and (3) the addition of a bulking agent which is recovered after composting and re-used.

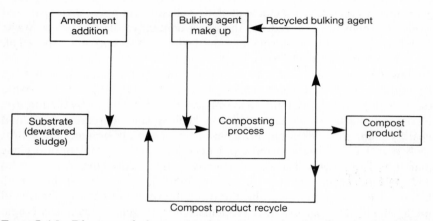

FIG. 7.15 Diagram of the composting process showing inputs of substrate, amendment, and bulking agent.

These methods are generally used separately, although the use of amendments and recycling are occasionally used together. The most popular of the three methods is the addition of an amendment to the sludge, as it allows greatest control over the quality of the final product as well as disposing of a secondary waste material. The amendment is an organic material that is added to the substrate to reduce its bulk density and to increase the voidage, so air can penetrate providing adequate aeration. Sawdust, straw, peat, rice husks, manure, refuse and garbage, lawn, and tree trimmings have all been successfully used. The choice of amendment is generally limited to what is available locally, but ideally it should be dry, have a low bulk density and be degradable. The advantage of carefully selecting the amendment is that it can significantly increase the quality of the final compost. Bulking agents can be either organic or inorganic particles of sufficient size and shape to provide structural support to the sludge cake as well as maintaining adequate aeration. The bulking agent provides a matrix of interstices between particles in which sludge is trapped and undergoes decomposition. Enough space is left between particles to ensure sufficient ventilation. After composting, the bulking agent is recovered, normally by screening, and reused. Inert bulking agents have the longest life, although degradable materials can be used to improve the organic quality of the compost and are less of a problem as the former is not totally removed by the screens. Wood chips are the most widely used bulking agent, although other suitable materials include pelleted refuse, shredded tyres, peanut shells, tree trimmings, and graded mineral chips.

Because composting is an exclusively biological process, those factors that affect microbial metabolism either directly or indirectly are also the factors that affect the process as a whole. The most important operational factors are: aeration, temperature, moisture, C:N ratio, and pH level (Suler and Finstein 1977; MacGregor et al. 1981; De Bertoldi et al. 1983; Finstein et al. 1983). In the compost pile, the critical oxygen concentration is about 15%, below which, anaerobic micro-organisms begin to exceed aerobic ones. Oxygen is not only required for aerobic metabolism and respiration, but also for oxidizing the various organic molecules that may be present. During composting, the oxygen consumption is directly proportional to the microbial activity, and a direct relationship between oxygen consumption is expected, with the high oxygen consumption occurring at temperatures between 28–55°C (Haug 1980). High temperatures in composting result from heat produced by microbial respiration. Composting material is generally a good insulator and as the heat is only slowly dispersed the temperature in the pile increases. Heat loss to the outside of the pile is a function of the temperature difference and the rate of microbial activity. As the rate of activity is limited by the rate that oxygen can enter the pile, heat production is affected by oxygen availability. In an aerated pile, 60°C can be reached within 1–2 days, whereas in unaerated windrow systems where oxygen is more limited, similar temperatures are reached after 5 days (Audsley and Knowles 1984). The compost will remain at this maximum operating temperature until all the available volatile solids have been consumed (Fig. 7.16). Excessively high temperatures inhibit the growth of most micro-organisms present in the compost severely reducing the rate of decomposition, although a number of thermophilic bacteria can withstand temperatures > 70°C (e.g. *Bacillus stearothermophilus*, *B. subtilis*, *Clostridium* sp., and *Thermus* sp.) (De Bertoldi et al. 1983). Therefore, for rapid and efficient composting, long periods of high temperatures must be avoided, except as an initial phase for the destruction of heat-sensitive pathogens. Moisture content is linked to aeration of the pile, and is a difficult factor to quantify. Too little moisture results in dehydration of the pile, which will reduce or even stop biological activity, resulting in a physically stable but biologically unstable compost. Too much moisture fills the interstices of the compost, excluding air, so that aerobic growth is replaced by anaerobic activity, which will result in odour production. The bulking agent can act as a reservoir for water, with straw, for example, capable of absorbing up to three times its own weight of water (Gifford 1972). The

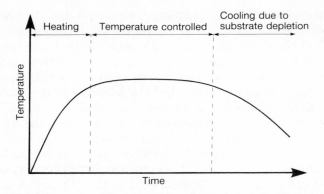

FIG. 7.16   The three stages of the composting process in a static pile system (adapted from Stentiford *et al.* 1985).

optimum moisture content for composting is discussed by Audsley and Knowles (1984) who concluded that a moisture content of at least 50% was required. The ratio of available carbon to nitrogen (C:N) should be about 12 for good microbial growth (Vogtmann and Besson 1978). All the carbon present may not be readily available, especially when bulking agents, such as wood chips are used, which are only slowly degraded and only a small proportion of the carbon is then available. In these circumstances, an optimal C:N ratio for composting may be as high as 25 (De Bertoldi *et al.* 1982, 1985). A higher C:N will slow up decomposition until excess carbon is oxidized, whereas a low C:N ratio also slows decomposition and increases the rate of nitrogen loss through ammonia volitization, especially at the higher pH and temperatures so typical of the composting process. The final compost product must have a C:N ratio < 10–12 otherwise the soil microbial population will immediately utilize the available nitrogen, which will limit crop growth and defeat the purpose of using compost to improve soil quality. Although the optimum pH for composting is between 5.5–8.0, organic matter with a pH range of between 3–11 can be successfully composted. The pH will begin to drop as soon as composting commences because of the acid-forming bacteria breaking down complex carbohydrates to organic acids, which tends to favour fungal development. As stated before, a high pH can result in nitrogen loss if the temperature is also high.

The most critical factor in composting is the supply of oxygen and the various systems used for providing adequate aeration ranges from the relatively simple to the very complex. Various techniques have been developed to supply oxygen and these are used to classify the various composting systems. There are two main methods of composting. Open systems do not require a reactor but employ piles and windrows, whereas closed systems take place within a specially constructed reactor and involve a high degree of mechanization.

Open systems are of two types: turned piles or windrows (long rows of compost triangular or trapezoidal in cross-section) and static piles. In the former, oxygen is provided by natural diffusion and periodic turning. In the latter, oxygen is supplied by forced aeration or possibly by natural ventilation.

Turned or agitated systems require that the composting mixture, which is placed in piles or windrows, is broken up periodically during the composting cycle to allow oxygen into the pile. The compost is not fully mixed each time but rather turned or tumbled over and rebuilt back to its original shape. This can be done manually, although larger scale operations require this to be done by special machines. The height, width, and shape of the piles or rows depend on the nature of the compost mixture and the type of equipment used for turning. Generally, rows are narrower at the top than at the base to ensure stability. Oxygen is mainly supplied by gas exchange during turning, although oxygen also enters the pile by diffusion and natural ventilation caused by the warm gas escaping which induces convection currents through the pile. Although they are simple to operate and the cheapest form of composting, turned systems have severe limitations. The problem is that the pile is only fully oxygenated when turned, which results in a cyclic variation in the oxygen concentration within the pile. In practice, this means that biological oxidation can never be maintained at maximum efficiency as the oxygen concentration is normally limiting (De Bertoldi *et al.* 1982) (Fig. 7.17). Also, pile turning requires more space than static pile systems especially if they are moved laterally. During the final stages of composting when the material is nearly dry, turning releases quantities of dust containing spores of *Aspergillus fumigatus* that are hazardous to the operator (Millner *et al.* 1977) The spores or conidia are present in soil at very low densities and rarely constitute a hazard. However, within the compost, vast densities of spores accumulate as they are heat tolerant, and are released when the pile is turned. There does not

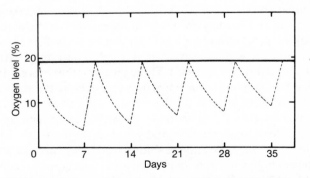

FIG. 7.17 Cyclic variation of oxygen during composting within a pile that is periodically turned. The straight line represents the oxygen demand of a composting mass, and the dotted line the oxygen availability within the pile during each period between being turned (Stentiford *et al.* 1985).

appear to be a problem with aerated static piles or closed systems of composting, with the release of spores being worst when the compost is dry. Health-associated problems are restricted to plant operators who, if they are hypersensitive or suffer from pulmonary disease, may display a severe allergic response. It is possible that the spores will germinate after inhalation and that the fungus may invade lung-parenchyma to produce typical aspergillosis (Clark *et al.* 1984*a,b*; Vincken and Roels 1984).

Windrowing is more popular than piles as it is more efficient in terms of space utilization. They can be constructed either on a soil base or on concrete floors with underdrainage. Soil-based systems are generally used by farmers composting animal wastes by using straw as a bulking agent. They are generally constructed at the periphery or headland of fields and are left to rot slowly without turning, which allows pathogens to survive in the outer part of the pile. The final quality of the compost produced in this way is much poorer compared to piles that have been turned periodically. At sewage treatment works, concrete bases are normally employed with long windrows 1 m high and 4 m wide constructed at regular intervals so that they can be regularly turned mechanically with ease. This ensures that all parts of the compost mixture, especially the sludge cake, spends an adequate period in the centre of the pile at 55°C, so that total pathogen destruction occurs. There are a variety of turned systems in use, employing a variety of amendments and bulking agents. The Bangalore process was developed in 1925 in India and is used extensively there today. A 0.5–1.0 m deep trench is dug in the ground and filled using alternative layers of refuse, night soil, earth, and straw. The material is then turned by hand as frequently as possible, with composting completed after 120–180 days.

Static piles are not turned for aeration, instead, oxygen is supplied by natural ventilation or more usually by forced aeration. By using a forced air method the amount of oxygen entering the pile can be controlled, as can the moisture content and temperature of the compost. Static piles take up less area than turned piles and savings are made in manpower as the compost is not touched until the process is completed. Air can be sucked or blown through the pile, or used alternately to give overall oxygen and temperature control (Haug 1980; De Bertoldi *et al.* 1985). There are a number of well-documented systems using forced aeration in operation throughout the world. A widely used system in the USA is the Beltsville process in which air is sucked through the pile at a rate of about $0.2 \text{ m}^3 \text{ min t}^{-1}$, providing an oxygen concentration of 15% in a windrow of sludge cake and wood chips. A base of wood chips is laid down on a concrete base and the sludge/wood chip mixture is laid on top to form a 300 cm high windrow which is covered with screened compost from earlier batches. Air is pulled through the pile by a perforated pipe laid in the wood chip base. The air which has been sucked out of the pile is dried and vented off through a pile of screened compost which eliminates any odours. Residence time in the pile is between 14–21 d, with adequate pathogens killed to allow the final product to be spread on land without causing health risks. As with all composting methods that employ a bulking agent, the wood chips are screened from the composted waste and reused (Fig. 7.15). The system devised by Rutgers University (New Jersey) essentially prevents the pile overheating. Most micro-organisms do not survive at temperatures $> 60°C$, especially fungi that degrade cellulose and lignin, thus by controlling the temperature in this way composting can continue at maximum microbial efficiency. Once the temperature rises above 60°C, microbial degradation begins to fall off rapidly and eventually stops, and composting will not continue until the temperature falls and the micro-organisms can re-invade the affected area. The system uses thermocouples placed within the pile that can activate a blower automatically if the temperature reaches critical levels. Air is continued to be blown through the pile until the temperature returns to the required optimum. Unlike suction methods, blowing air through the compost enhanced evaporation and produces a highly stabilized end-product with a low moisture content (Finstein *et al.* 1980; Willson *et al.* 1980).

It should be remembered that high temperatures destroy pathogens, and some methods combine an initial phase of suction to ensure oxygenation without any heat loss. This allows the temperature to rise for a few days so that the pathogens are killed, before blowing is introduced to control the temperature. The effect of blowing (positive pressure) and sucking (negative pressure) aeration on the temperature gradients within piles is clearly shown by Stentiford *et al.* (1985) who composted sludge using refuse as an amendment (Fig. 7.18). The shape of the pile gradually slumps as composting progresses (Fig. 7.19), with the problem of pathogen removal in the outer layer of the pile overcome by using a thick insulating cover of mature compost (Fig. 7.20). This layer of mature compost also reduces the amount of odour produced by the piles.

Closed system composting takes place in specially constructed reactors or vessels. There are essentially two types, vertical flow reactors which are towers of up to 9 m in height, and horizontal or inclined flow reactors. Neither of these types of reactors are widely used for composting sewage sludge and are extremely difficult to manage on a continuous basis. Because of their high capital and operational costs they are not a cost-effective method of stabilizing sewage sludge under normal conditions. Closed systems of composting have been extensively reviewed by Haug (1980) and De Bertoldi *et al.* (1985). An interesting closed composting system has been in operation at the Bekkelaget wastewater treatment plant in Oslo since 1980. It is a BAV-type plant that is capable of composting up to 30 m³ of dewatered sludge each day. A mixed activated sludge is used that is dewatered by centrifuge to 20% and then mixed with raw compost and sawdust before being fed into the 500 m³ reactor, which is cylindrically shaped having an overall diameter of 9 m (Fig. 7.21). The nominal retention time in the reactor is 10–12 d. However, problems with

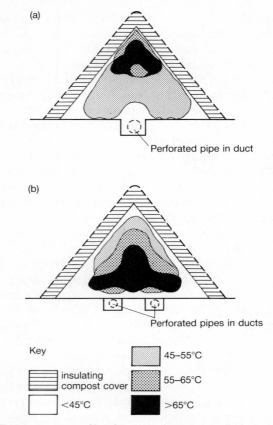

FIG. 7.18 Temperature profile of composting sewage sludge where (a) is positive pressure (blowing) aeration and (b) negative pressure (sucking) aeration (Stentiford *et al.* 1985).

short-circuiting have been reported. The excess compost which is not recirculated is stockpiled for 6–9 weeks before being used by the Oslo Parks' Department (Paulsrud and Eikum 1984).

Research has centred in Europe on open systems of composting. For example, in the UK much research has been conducted into composting sewage sludge with straw. There is approximately 4.3×10^6 tonnes of wheat straw and 2.1×10^6 tonnes of barley straw surplus to requirements in the UK each year (Larkin 1982), which at present is burnt in the fields. However, with increasing pressure on farmers to curtail straw burning on environmental and conservation grounds, they are left with a serious disposal problem. There is a reluctance to plough straw into the soil because this has been shown to reduce yields by between 3–20%, depending on the weather during autumn, because of immobilization of soil nitrogen and retardation of germination and seedling

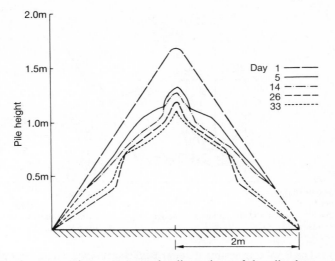

FIG. 7.19 As composting progresses the dimensions of the pile changes. Here the sludge is being composted with domestic refuse (Stentiford *et al.* 1985).

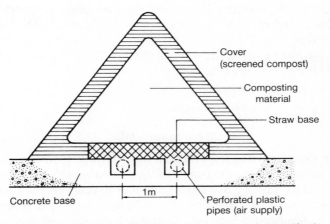

FIG. 7.20 Diagram of the most effective pile structure used (Stentiford *et al.* 1985).

growth as it decomposes. However, in many areas of the UK there is also a surplus of sewage sludge (Table 7.15) that could be composted with the surplus straw to provide a valuable soil conditioner and fertilizer of superior quality to either straw or sludge on their own. Research at the National Institute of Agricultural Engineering (NIAE) has shown that straw and sewage sludge can be successfully composted (Audsley and Knowles 1984). The straw can be used whole or chopped and the sludge can be either raw or digested, with the dry solids content between 4–20%, and dewatering is not

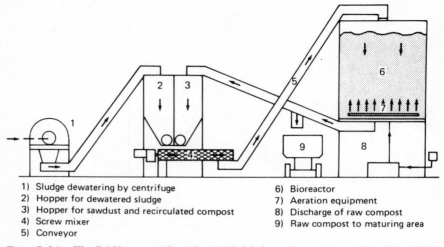

1) Sludge dewatering by centrifuge
2) Hopper for dewatered sludge
3) Hopper for sawdust and recirculated compost
4) Screw mixer
5) Conveyor

6) Bioreactor
7) Aeration equipment
8) Discharge of raw compost
9) Raw compost to maturing area

FIG. 7.21   The BAV composting plant at Bekkelaget sewage treatment plant, Oslo (Paulsrud and Eikum 1984).

essential although preferable. Dewatering removes about 30% of the nitrogen in raw sludge and up to 50% in digested sludge (Table 7.16). Therefore, in terms of nitrogen, assuming only half of the remaining nitrogen in sludge is available for plant growth, the value of the four types of sludge as a fertilizer can be expressed as £ per tonne dry matter:

Raw sludge	14.9
Digested sludge	17.6
Dewatered raw sludge	8.4
Dewatered digested sludge	11.6

Clearly, dewatered sludges, with their lower nitrogen content, have a lower fertilizer value. All composting systems can be used with sludge and straw, but three systems have been particularly successful and cost-effective. The *on-field* system composts the material where it is to be used the following year. Straw is taken to the headland of each field while the sludge is stored in a lagoon until harvest. An open windrow is constructed on the headland which is not turned. The costs are between £15–21 per tonne of compost produced, depending on farm size and transport costs (not including the cost of either digestion or straw). As the pile is not turned, pathogens will not be killed in the outer surface of the pile, and digested sludge only should be used. If dewatered sludge is used, the volatile solids content may be too low to maintain the critical temperature (Table 7.16). With *on-farm* systems, the straw is baled and stored centrally, with windrows constructed on a concrete base as the sludge is delivered. The hard base allows the farmer to use his tractor to turn the windrows regularly, and maximum pathogen removal is achieved allowing

TABLE 7.15 How surplus straw and sewage sludge is distributed in the UK (by county). The sludge is expressed as 4% dry matter (Audsley and Knowles 1984)

County	Wheat straw (tonnes)	Barley straw (tonnes)	Sludge solids (tonnes)	Surplus as tonnes sludge solids		Closest county	Final surplus sludge (tonnes)
				straw	sludge		
London	7950	3404	8740	0	7141	Herts	0
Bedfordshire	107 896	46 328	12 462	9248	0		
Cambridgeshire	396 978	115 451	10 188	64 411	0		
Essex	419 581	110 287	30 606	47 114	0		
Hertfordshire	106 436	51 847	5233	16 774	0		
Lincolnshire	533 379	235 894	13 337	94 649	0		
Northamptonshire	166 321	67 396	9366	23 715	0		
Norfolk	338 657	233 655	4730	71 886	0		
Suffolk	370 601	151 892	10 415	63 462	0		
Berkshire	42 527	26 042	21 845	0	12 540	Wilts	0
Buckinghamshire	51 247	27 188	10 727	82	0		
East Sussex	38 880	14 229	2053	5533	0		
Hampshire	136 013	79 966	13 224	16 214	0		
Kent	126 535	40 897	25 944	0	1772	E Sussex	0
Oxfordshire	133 962	90 660	12 656	17 481	0		
Surrey	15 408	13 405	22 607	0	18 854	Sussex, Hants	0
West Sussex	66 467	29 436	9036	4424	0		
Cheshire	1615	1429	13 232	0	12 836		12 836
Derby	20 575	35 026	20 041	0	13 386	Lincs	0
Greater Manchester	292	849	19 710	0	19 581		19 581
Hereford & Worcester	55 042	25 055	8808	2404	0		
Lancashire	2276	2671	39 312	0	38 691		38 691
Leicestershire	132 141	61 697	21 726	5339	0		

TABLE 7.15

County	Wheat straw (tonnes)	Barley straw (tonnes)	Sludge solids (tonnes)	Surplus as tonnes sludge solids		Closest county	Final surplus sludge (tonnes)
				straw	sludge		
Merseyside	1011	1257	8573	0	8290		8290
Nottingham	113 114	67 349	29 917	0	5353		5353
Shropshire	78 073	27 245	7897	7210	0		
Staffordshire	3083	4515	41 296	0	40 396	Leics	35 029
Warwickshire	51 300	18 216	6154	3802	0		
West Midlands	2441	2493	16 515	0	15 885	Warks	0
Avon	4593	2451	3989	0	3018	H & W	614
Cornwall	164	1007	4412	0	4289		4289
Devon	5064	8615	6749	0	5111	Dorset	2042
Dorset	44 191	14 460	5389	3069	0		
Gloucester	105 926	68 358	3181	20 329	0		
Somerset	14455	5976	8372	0	5485	Wilts	2124
Wiltshire	136 376	72 777	12 905	15 901	0		
Cleveland	6661	320	620	476	0		
Cumbria	0	0	2834	0	2834	Cleve	2834
Durham	13 312	798	3571	0	1364		887
Humberside	228 128	92 593	5383	40 006	0		
Northumberland	29 390	34 683	7826	205	0		
North Yorkshire	170 249	145 539	20 447	20 764	0		
South Yorkshire	38 695	21 996	5466	2836	0		
West Yorkshire	16 836	14 581	49 090	0	44 996	Yorks	21 395
Tyne & Wear	768	2272	5262	0	4921		4921
Total	4 334 609	2 072 205	601 846	557 347	266 724		158 889

TABLE 7.16 Typical characteristics of raw, digested, and dewatered sewage sludge (Audsley and Knowles 1984)

	Raw sludge (%)	Digested sludge (%)	Dewatered, raw sludge (%)	Dewatered, digested sludge (%)
Solids content	4	4	20	20
Carbon content (DS)	35.5	31.7	35.5	31.7
Nitrogen content (DS)	5.2	6.7	3.6	3.4
Volatile solids (DS)	75	50	52.5	35

raw sludge to be used. The compost costs about £8–20 per tonne, depending on farm size and transport costs (the cost of straw is not included). The final system is *central-site* composting where the straw is bailed and stored at the farm and brought to the treatment plant as required. As with the previous system, windrows are constructed on a concrete base and turned regularly, with the compost returned to the farm for spreading. Raw sludge can be used, and the cost of the compost at the central site can be as little as £5–8 per tonne. However, because of transportation back to the farm its eventual cost is between £11–15 per tonne, depending on farm size and distance from the treatment plant, but excluding the cost of the straw (Audsley and Knowles 1984).

Maintaining an adequate C:N ratio is difficult as straw has a typical C:N ratio of 48 and sewage sludge 5–9. In order to obtain the optimum C:N ratio of 12 for good microbial growth it is necessary to mix sludge dry matter to straw dry matter in a ratio of 1.3:1, 0.7:1, 4.7:1, and 4.0:1 for raw (4% DS), digested (4% DS), dewatered raw (20% DS), and dewatered digested sludges (20% DS) respectively. More dewatered sludge is required as it contains significantly less nitrogen than non-dewatered sludges. To ensure that a high temperature is maintained for a sufficient time, large piles with a minimum cross-sectional area of 5 m² are required to ensure that the heat generated exceeds the heat lost. As heat is produced from microbial degradation of volatile solids, a high volatile solids content in the compost mix is desirable (Table 7.16).

In conclusion, the NIAE study found that outdoor composting required care and was liable to fail if the winter was very cold. Forced aeration improved the chance of successful composting, whereas enclosed forced aeration would, they felt, always succeed. However, methods involving forced aeration are comparatively expensive and would probably make this type of composting economically not feasible.

Experiments in Norway have been conducted on the composting of primary sludge after dewatering using a centrifuge to 20–25% dry solids. The dewatered sludge is dumped in piles on the composting site which should have good drainage during autumn to spring. During the early summer, the water content of the sludge decreases further due to evaporation, with a dry crusty layer forming with a dry solids content $> 30\%$. In midsummer, the pile is turned for the first time so that the dry crusty surface layer is mixed with the wetter sludge in the centre, and so acts as a bulking agent. At this stage, the porosity is sufficient for composting to start, and with further turning, the composting process continues for the rest of the summer and autumn (Fig. 7.22) (Paulsrud and Eikum 1984).

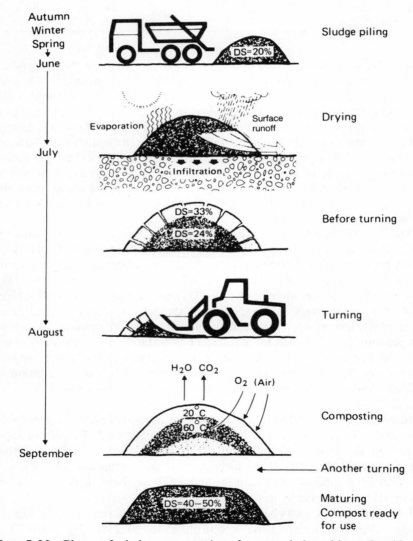

FIG. 7.22   Phases of windrow composting of sewage sludge without the aid of a
separate bulking agent (Paulsrud and Eikum 1984).

Product quality

Fully composted sludge closely resembles peat, being dark brown in colour with a crumbly texture, and having a pleasant earthy smell. The fertilizer value of such compost is generally low compared with inorganic fertilizers, although it does reflect the original fertilizer value of the substrate and amendments used (Table 7.17). It is possible to fortify the compost and increase its nutrient value by adding nitrogen, phosphorus, and potassium, although the cost of

TABLE 7.17 Fertilizer value of raw and digested sewage sludges before and after composting (Parr *et al.* 1978)

	Raw sludge	Composted sludge	Digested sludge	Composted sludge
Water (%)	78	35	76	35
pH	9.5	6.8	6.5	6.8
Organic carbon	31	23	76	35
Total nitrogen	3.8	1.6	2.3	0.9
Ammonia	1540	235	1210	190
Phosphorus	1.5	1.0	2.2	1.0
Potassium	0.2	0.2	0.2	0.1
Calcium	1.4	1.4	2.0	2.0

this probably exceeds its value (Hileman 1982). Compost has been successfully marketed as a soil additive for householders (Heaman 1975), as an ingredient for potting compost, where it is comparable to a sand–peat mixture, and for general horticulture and landscaping (Gonin 1982), where compost mixed with fertilizer produces better plant growth than the addition of fertilizer alone (Diez and Weijelt 1980). Most compost producers are spurred to improve their final product in the knowledge that 250 000 tonnes of peat are sold annually in the UK alone, at a basic price of £80 per tonne. With proper marketing, it is possible that some of that market could be replaced by compost. Straw produces a coarser compost than other types of amendments and bulking agents, and is only suitable for use as an agricultural soil conditioner (Audsley and Knowles 1984).

Heavy metals may be a problem with composted sludge as with sewage sludge, when disposed to land. Therefore, compost must be treated as sewage sludge under the existing legislation and guidelines (Section 5.2.2). Composts are generally pathogen free. The major cause of inactivation of pathogens in composting is exposure to high temperatures ($> 55°C$) for a number of days. The Environmental Protection Agency in the US have produced standards to ensure pathogen destruction during composting. In aerated piles, the coolest part of the pile must be at least $55°C$ for 3 days, and in the windrow system, the centre must be at least $55°C$ for 15 out of the total of a 21–30 day composting period. During this period, the pile must be turned five times (Haug 1980). In forced aeration systems, the volatile solids can be utilized very quickly, and although a high temperature is achieved it may not last as long as the slower and slightly cooler turned system, and so give poor pathogen inactivation (Audsley and Knowles 1984). Pathogen inactivation is greater if the pile is turned, as cool spots develop within the pile, especially in the corners where pathogens can survive and even increase in numbers. Regular turning ensures all parts of the pile are subjected to the high temperatures. It is now common practice in forced aeration piles to build the new compost mixture on top of a layer of mature compost and for the pile to be sealed by a thick layer of the material, which acts as insulation. Under these conditions, high pathogen kills are possible without turning and there is no outer layer in which the pathogens can survive. Composting crop residues is an extremely efficient way of destroying all plant pathogens including fungi, bacteria, nematodes, and viruses. However, inactivation is caused not only by heat generated during the thermophilic phase but also by microbial antagonism and the toxicity of conversion products formed during decomposition (Bollen 1985; Lopez-Real and Foster 1985).

Other uses for compost have been investigated. Composting has been described as biological drying, and certainly composting is an excellent means of removing water from sewage sludge, thus reducing the volume to be transported to disposal sites. This ability could also be exploited to convert moist organic wastes into fuel (Finstein and Miller 1985). Some research has been done on direct heat recovery during composting, although so far success has been limited (Baines *et al.* 1985; Thostrup 1985).

Vermiculture

A promising approach to sludge management is the use of earthworms to convert waste sludges into a useful compost, with the worms themselves being utilized as a valuable source of protein. The process is known as vermiculture or vermicomposting depending on the emphasis placed on the final product.

It is well known that earthworms break down organic matter while feeding on the micro-organisms present (Edwards and Lofty 1977), and composting systems using windrows, piles, and boxes have been developed by worm breeders for fish bait (Tomlin 1983). Work by Hartenstein *et al.* (1979), Neuhauser *et al.* (1980), and Edwards (1982, 1983) has demonstrated that earthworms can break down a variety of wastes and sludges including sewage sludge, pig and cattle solids and slurries, waste from chickens, broilers, turkeys and ducks, potato wastes, spent mushroom compost, and paper pulp. However, not all the wastes are equally acceptable to earthworms, with the production of worm biomass varying with substrate (Fig. 7.23). Similarly, not all sewage sludges are suitable for vermiculture and so far only dewatered activated sludge has been widely utilized. Hartenstein *et al.* (1979) report that *Eisenia foetida* grew faster in activated sludge than in manure, their natural habitat, although the rate of reproduction may not be as high. For this reason

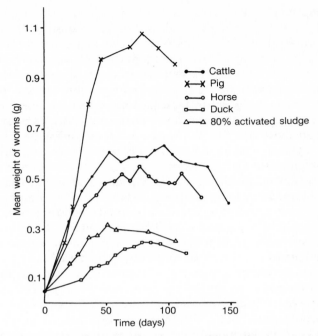

FIG. 7.23 The growth of *Eisenia foetida* on various animal wastes and surplus activated sludge (Edwards *et al.* 1985).

they suggested that a separate culture of worms should be maintained using manure in parallel with the composting mixture and used to constantly reinoculate the composting sludge. Although a number of species of earthworm have been tested, the *tiger* or *brandling* worm (*Eisenia foetida*) is the most efficient natural species for use with vermiculture as it has rapid growth and reproductive potential. This species takes 5–7 weeks to reach sexual maturity but after that it can produce up to five cocoons per week from which up to seven hatchlings are produced per cocoon. Thus, each adult worm has the potential to produce 15–20 young per week over an average reproductive period of 22 weeks. However, the search for an even more efficient exotic species continues. *Eudrilus eugeniae*, the African night crawler, grows about twice as fast as *E. foetida* but requires a greater space per individual. It can tolerate higher temperatures than temperate worm species, having an optimum growth rate at 28°C. Below 12°C, however, the worm population is rapidly dessimated (Hartenstein *et al.* 1979). In the Philipinnes, where commercial production is already underway, the worm *Pheretima asiatica* is used (Anon 1984*a*).

The sludge undergoes normal composting with the high density of worms grazing on the micro-organisms. *Eisenia foetida* requires a mixed population of bacteria, fungi, and protozoans to sustain maximum growth and reproduction, with the fungi and protozoans constituting the major proportion of its diet. Therefore, sewage sludge and agricultural wastes are ideal media for growing worms. The production of tunnels by earthworms increases aeration, allowing the material to compost more efficiently. The earthworms also accelerate the process by fragmenting organic matter and producing casts both of which increase the surface area available for microbial growth. At 25°C, an earthworm can convert up to twice its body weight of waste into compost each day, the exact amount depending on the density of the solids ingested. The casts not only decompose more rapidly than the original material, but also dry twice as rapidly. Minerals as well as organic matter are assimilated by the worms which results in an overall decrease in the ash content of the waste being composted. At the end of the composting period the worm-worked compost is superior to normal compost. It is an odourless material resembling peat, rich in available nutrients, with a good moisture holding capacity and excellent porosity. The action of the worms is such that most of the nitrogen is converted to nitrate by enhanced microbial activity. The amount of soluble potassium, phosphorus, and magnesium is also increased (Edwards *et al.* 1985).

The earthworms themselves contain between 60–70% protein, 7–10% fat, 8–20% carbohydrate, and 2–3% minerals, and have a gross energy of 4000 kcals kg^{-1}. In nutritional terms, worm tissue is excellent being equal to meat or fish. It is particularly rich in amino acids, especially lysine and the vitamins niacin, riboflavin, and vitamin B_{12}, which makes worm tissue valuable as an animal feed. Worms have already been successfully fed to a wide range of animals including fish (Guerreo 1983; Tacon *et al.* 1983), chickens and pigs (Sabine 1978; Jin-you *et al.* 1982). The tissue also contains plenty of long-chain saturated fatty acids, such as linoleic acid, dihomo-linoleic acid, and arachidonic acid that are essential to non-ruminant animals who cannot synthesize them. Their conversion efficiency of agricultural wastes into worm tissue on a dry weight basis can be as high as 10%, so that for every tonne of suitable waste composted 100 kg of worms can be produced.

Processing of worm compost falls into five stages: preparation of waste, composting/culture chamber, harvesting, worm processing, and processing of compost (Fig. 7.24). The preparation of the waste depends very much on the water content of the sludge and whether additional material, such as straw or wood chips are to be added. Worms grow most rapidly at moisture contents of between 80–90%, with the optimum pH being 5, although they can tolerate a pH range 4–9. Temperature and moisture content are the two most important environmental factors, with little growth below 10°C and temperatures above 35°C being lethal. Optimum temperature for growth, the number of cocoons produced and the number of young worms hatching per cocoon is 25°C, although there is a rapid decrease at temperatures >25°C. The operating temperatures for maximum productivity is usually between 15–20°C. The

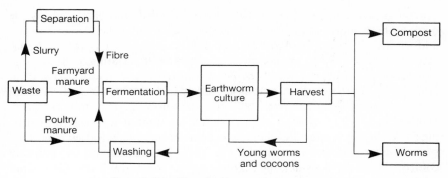

FIG. 7.24   Vermiculture system processes.

compost material is laid approximately 0.5 m deep, depending on the density of the material, on a solid floor of concrete, bricks, or railway sleepers. The larger the composting chamber the better the economics, although the waste material has to be built up slowly in successive layers. Once the compost is active, slurries and other liquid wastes can be sprayed on top. Vermiculture cannot be carried out during October to March in most temperate climates unless there is sufficient insulation and heating. In order to maintain optimum moisture conditions the compost may require periodic watering and some reports indicate that if worms migrate from the chamber then lighting may be required. Best yields of worms and most efficient waste conversion rates have been obtained under strictly controlled environments using polythene or fibreglass insulated tunnels (Anon 1984a). Composting is normally completed after 2 to 3 months if straw or wood chip based wastes are used, less if the separated waste only is used, with the length of time depending on the type of waste, the inoculation level of worms at the commencement of the process, and the environmental factors. Harvesting presents two problems. First, the separation of the worms from the compost and secondly, the separation of the young worms from the adults. The juvenile worms need to be recovered so they can be used to inoculate the next batch of waste to be composted. Harvesting methods are still being developed and rotary sieves have been widely used, although attempts to use the response of worms to certain stimuli such as heat, light, moisture, and chemicals (including ammonia and formalin) to separate them from the compost have also been made with limited success. Once separated, the worms are cleaned by washing and soaking in water to remove debris from the surface and the intestines respectively. They are generally then sterilized by heat to destroy any pathogens and to help prevent deterioration of the tissue. They are subsequently preserved by freeze drying, air or heat drying, freezing or pickling and processed into either a powder or paste of uniform quality before being sold. If the compost is to be used directly on land as a soil conditioner no further treatment is necessary. However, much worm-worked waste is marketed through garden centres for horticultural and general garden use and so needs to be dried, blended, and packaged. As the compost has a moisture of 75% it will not store without further decomposition and drying is necessary. Also, as a standardized product is required that cannot be guaranteed by the composting system itself, blending of the product is necessary. The economics of the process have been reviewed by Fieldson (1985) who estimates that from case studies in the UK, 63% of the total benefits come from the compost, 29% from the worms, and 8% from the waste disposal saving. As vermiculture can be used as a treatment process for cattle, pigs, and poultry wastewater then the benefits in waste disposal savings alone may in some circumstances make this an attractive method of pollution control. There are considerable markets for vermiculture. Rank, Hovis and McDougall produces some 40 000 tonnes a year of spent mushroom compost, and Bowater is currently disposing of 30 000 tonnes of paper pulp per annum to landfil sites. Both of these wastes are particularly suited to vermiculture. Another advantage of worm composting is that not only is it odour-free itself, but the compost can be used to remove odours produced from other treatment processes or from intensive rearing units. For example, the air stream extracted from a pig unit, which was in danger of closure due to odours, was vented through the worm compost which absorbed the odours completely. Estimated market value for worms is hard to gauge but is estimated to be a minimum of £350–400 per tonne, although for certain specialized uses, such as feeding young eels, their value may be 5–10 times greater, and the compost is worth the same as peat, at £80 tonne. If just 10% of the available waste material in the UK was composted, assuming £400 per tonne for the worms and £80 per tonne for the compost, then its market value could be in the order of £1800m annually (Anon 1984a). At present, there are no plans to produce worm protein for human consumption in the UK, but in the Philippines they are producing a worm protein in the form of a powder which is to be used as an extender in beefburgers for the Japanese market. Also, according to recent media coverage the Japanese, who consider worm protein to be an aphrodisiac, are also interested in the idea of marketing whole worms.

At present, there is only one major UK-based firm commercially providing vermiculture equipment, British Earthworm Technology Limited, based in St. Ives in Cambridgeshire. They have exploited the research carried out by the NIAE and Rothamsted Experimental Station who have developed two types of beds for worm compostings. The first is a batch process using a simple static bed, and, more recently, a continuous bed has been developed. Fresh waste is loaded on the top surface of the worm-worked waste and the worms move slowly upwards against the flow of the waste, with the worked compost being discharged through a special floor. This system retains all the worms within the bed so that the need for worm separation is eliminated. Other advantages of the continuous system, compared with the batch process, includes a faster working rate, better quality product, lower labour requirement, and easier management (Phillips 1986).

Further reading

Bio-energy: Boyles 1984; Lewis 1983, 1988.
Single-cell protein and biomass: Rose 1979*a,b*; Benneman *et al.* 1979; Norris 1981; Samvelov 1983; Goldberg 1985; Wheatley 1987.
Composting: Haug 1980; De Bertoldi *et al.* 1983, 1988; Gasser 1985; Anderson and Smith 1987.

7.4 ENVIRONMENTAL PROTECTION

Wastewater biotechnology has an important role to play in environmental protection. Apart from removing organic matter and nutrients from wastewater before it is discharged to a watercourse, thus preventing pollution, biotechnological advances have been made in four important areas. These are: the breakdown of recalcitrant compounds, the removal of metals from effluents, bioscrubbing, and in the use of packaged micro-organisms for seeding or uprating biological reactors. These areas are considered separately below, except for the removal of metals which has been reviewed in a previous section (Section 7.2.3).

7.4.1 Breakdown of recalcitrants

It has often been reported that certain compounds are not degraded following release to the environment, even when conditions appear adequate for microbial growth (Slater and Somerville 1979). Such compounds are termed recalcitrant and are mainly synthetic organic chemicals. Slater and Somerville advise caution in placing compounds into this category as there may be any number of relatively trivial reasons why biodegradation does not occur under a given set of conditions. However, many compounds, the so-called foreign or xenobiotic compounds in particular, are known to be recalcitrant (Leisinger *et al.* 1981). A list of recalcitrants currently under investigation is given by Leisinger (1983). Recalcitrant molecules persist for extended periods in all natural environments regardless of whether the compound is or is not inherently biodegradable (Lynch and Poole 1979). They include detergents, pesticides, and a number of common polymers, although the latter group is of little interest in aquatic systems. Vast quantities of recalcitrant organic chemicals are produced, about 150×10^6 tonnes annually, which can find their way into surface and ground waters as constituents of industrial effluents (Mitchell 1974; Hutzinger and Veerkamp 1981). However, two groups of recalcitrants are particularly significant, the aromatic compounds and pesticides.

Many micro-organisms can metabolize particular aromatic compounds (Fewson 1981) (Table 7.18). Some green, brown, and red algae contain dioxygenases allowing them to partially oxidize hydrocarbons such as naphthelene, whereas many mesophilic organisms such as protozoans, rotifers, and nematodes use the hepatic cytochrome P-450 to detoxify and eliminate aromatic xenobiotics. In the natural environment, the degradation of xenobiotics is brought about by communities rather than individual species, and this observation has been of primary importance in improving the microbial degradation of xenobiotics and recalcitrants (Fewson 1981). Two basic methods are used to obtain suitable microbial cultures to degrade xenobiotics: enrichment culture and gene manipulation. Enrichment culture is the standard method used and involves taking an inoculum of micro-organisms from a habitat that preferably has already been exposed to the compound of interest. The inoculum is then cultured and exposed to progressively larger concentrations of the xenobiotic compound until it eventually becomes a source of essential nutrient. Batch culture can be used to screen for suitable organisms, whereas with continuous culture, selection pressure is exerted resulting in the culture becoming dominated by the micro-organism best able to adapt to the available nutrients. It is possible to increase the rate at which adaptation takes place by increasing the mutation rate using radioactive or chemical mutagens (Harder 1981; Powledge 1983). Although this technique is both slow and random, with the selected organisms

TABLE 7.18   Major micro-organisms involved with the degradation of aromatic compounds (Fewson 1981)

Organism	Substrate
Procaryotes	
Gram-negative bacteria	
E. coli	3- or 4-Hydroxyphenylacetate
Alcaligenes spp.	Aromatics
Azotobacter spp.	Aromatics
Flavobacterium spp.	Aromatics
Gram-positive bacteria	
Bacillus megaterium	Salicylate
B. brevis	Salicylate
B. stearothermophilus	Phenols, cresols, 4-hydroxyphenylacetate
Actinomycetes	
Nocardia spp.	Aromatic carboxylic acids, phenol, and substituted phenols
Rhodococcus spp.	Naphthalene, biphenyl
Methanotrophs	
Methosinus trichosporium	Benzene, toluene, ethylbenzene, cresols, 1-phenyl-theptane, phenylacetate, and others
Methanogens	Aromatics
Eukaryotes	
Trichosporon cutaneum	Phenol
Cuiruinghainella elegans	Biphenyl (partial metabolism)

discovered by accident, genetic manipulation offers a faster and more controllable development of useful organisms. Gene manipulation using plasmid transfer, gene cloning, and transposon mutagenesis have all been used with some success. For example, the breakdown of the alkanes, toluene and xylene is now understood quite well (Chakrabarty 1980; Williams 1981), with all the information required by the micro-organism contained in the plasmids which can be transferred by genetic manipulation. Gene cloning techniques using *Pseudomonas aeuroginosa* enables genetic manipulation procedures to be carried out on a wide range of Gram-negative bacteria that will improve degradation of aromatic compounds, such as toluene, xylene, 3-chloroben-zoate, and 2,4-dichlorophenoxyacetate (Christopher *et al.* 1981). The General Electric Company have patented a *Pseudomonas* bacterium that can degrade octane, xylene, metaxylene, camphor, and salicylate (Davis 1979). Strains of bacteria able to decompose aromatics can be added to a normal bioreactor, usually activated sludge type systems, where they will either persist or donate their special abilities to other strains (Dagley 1981; Knackmuss 1981). The main agents for degrading crude soil, which is a mixture of many different aromatic and aliphatic hydrocarbons, are bacteria such as *Pseudomonas*, *Micrococcus*, *Coryuebacterium*, and *Mycobacterium*, with yeasts also able to degrade petroleum hydrocarbons to a lesser extent. In natural waters, this degradation is, almost without exception, an aerobic process. Davis (1979) describes an extended activated sludge process, with added nutrients, used for the biological treatment of waste machinery coolants. The reactor was originally seeded with activated sludge from a nearby municipal wastewater treatment plant and suitable micro-organisms gradually developed.

Pesticides are extremely persistent in soils and are usually extremely resistant to degradation (Table 7.19). There are no specific pesticide-decomposing micro-organisms, with degradation only occurring if the micro-organism has the ability to synthesize the appropriate enzyme. Degradation is often dependent on the sequential action of two or more micro-organisms (Mitchell 1974). Golovlena and Skvyabin (1981) examined the pathway of DDT degradation by *Pseudomonas aeuroginosa* (strain 64OX). They found that degradation depended on the metabolism of co-substrates and that the succession of these different substrates and the aeration conditions required were probably too complex to occur under natural conditions. However, Gasner (1979) felt that recalcitrant wastes, such as pesticides and halogenated aromatics, are often of sufficient concentration in wastewaters to support microbial growth and to be possibly biodegradable within the normal treatment process.

7.4.2 Bioscrubbing

Waste gases are a major pollution control problem for many industrial processes, with unpleasant odours particularly difficult to remove because of

TABLE 7.19   Persistence of pesticides in the soil (Alexander 1965)

Common name	Chemical structure	Persistence
Aldrin	1,2,3,4,10,10-Hexachloro-1,4,4a,5,8,8a-hexahydro-endo-1,4-exo-5,8-dimethanonaphthalene	> 15 yr
Chlordane	1,2,4,5,6,7,8,8-Octachloro-2,3,3a,4,7,7a-hexa-hydro-4,7-methanoindene	> 15 yr
DDT	1,1,1-Trichloro-2,2-*bis*(*p*-chlorophenyl)-ethane	> 15 yr
Dicamba	3,6-Dichloro-*o*-anisic acid	4 yr
Diuron	3-(3,4-Dichlorophenyl)1,1-dimethylurea	> 15 mth
2-(2,4-DP)	2-(2,4-Dichlorophenoxy)propionic acid	> 103 days
Endrin	1,2,3,4,10,10-Hexachloro-6,7-epoxy-1,4,4a,5,6,7,8,8a-octahydro-endo-1,4-endo-5,8-dimethanonaphthalene	> 14 yr
Fenac	2,3,6-Trichlorophenylacetic acid	> 18 mth
Fluometuron	N'-(3-Trifluoromethylphenyl)-N,N-dimethylurea	195 d
Heptachlor	1,4,5,6,7,8,8-Heptachloro-3a,4,7,7a-tetrahydro-4,7-endomethanoindene	> 14 yr
Lindane	1,2,3,4,5,6-Hexachlorocyclohexane	> 15 yr
Monuron	3-(*p*-chlorophenyl)-1,1-dimethylurea	3 yr
Parathion	O,O-Diethyl O-*p*-nitrophenyl phosphorothioate	> 16 yr
PCP	Pentachlorophenol	> 5 yr
Picloram	4-Amino-3,5,6-trichloropicolinic acid	> 5 yr
Propazine	2-Chloro-4,6-*bis*-(isopropylamino)-*s*-triazine	2–3 yr
Simazine	2-Chloro-4,6-*bis*-(ethylamino)-*s*-triazine	2 yr
2,4,5-T	2,4,5-Trichlorophenoxyacetic acid	> 190 d
2,3,6-TBA	2,3,6-Trichlorobenzoic acid	2 yr
Toxaphene	Chlorinated camphene	> 14 yr
Trifluralin	α,α,α-Trifluoro-2,6-dinitro-N,N-dipropyl-*p*-toluidine	> 40 wk

their low concentrations. For example, some mercaptans have an odour threshold of below 1 ppb. The control of odours and waste gases requires complex and very expensive technology, with incineration, dispersion, catalytic oxidation, scrubbing, and adsorption the major control options available. However, many of these physical and chemical processes are not very flexible if the volume, concentration or composition of the gas alters, resulting in a general reduction in performance. This is largely overcome by the biological control of gases, first proposed in 1930, which provides combined liquid and gas treatment with no chemical costs and, compared with other methods, negligible energy costs. Packed beds, similar to percolating filters, are used as the medium supports a large area of biofilm that provides a large gas–liquid surface area for adsorption (Le Roux and Mehta 1980; Le Roux 1982). Removal occurs by adsorption on to the biofilm followed by solubilization of the compound, which is subsequently degraded by microbial action. Bioscrubbing units need to be located at sites where there is a suitable substrate for the development and maintenance of an active biofilm. For this reason, Wheatley (1985) suggests that prototype units will most likely be part of existing wastewater treatment plants. At present, most systems are treating odours from other treatment units, such as digesters or composting systems. However, it is possible to seed reactors with specially cultured micro-organisms that are capable of removing specific gases. Wheatley concludes that although the area of bioscrubbing is in its infancy he envisages biofilm systems for oxidizing (aerobic) mercaptans, hydrogen sulphide, and cyanide, and reducing (anaerobic) sulphur dioxide and chlorinated hydrocarbons. Walshe (1988) describes a biofiltration unit, to remove the offensive odours produced at an animal rendering plant in South Tipperary, using a mixture of 25% peat and 75% heather. The odorous air is sprayed with water to lower its temperature and increase its humidity before it is pumped under pressure up through the filter. Efficiencies of such plants in West Germany are in the order of 95% reduction in odour. At sewage treatment plants most odour arises from high-rate percolating filters and during sludge treatment, especially sludge thickening and dewatering (North 1979; Frechem 1988). Although there are over seventy biofilters in operation to treat odorous gaseous emissions from treatment plants, the injection of such gases into activated sludge aeration tanks is also widely practised.

7.4.3 Packaged micro-organisms

Many wastes, although not necessarily recalcitrant, may be difficult to treat. Gasner (1979) describes how live micro-organisms, with specific activity for one specific component or one specific enzyme activity are commercially available. These packaged micro-organisms are of three types:

(1) Natural; such as rumen bacteria with high cellulase activity
(2) selected mixtures that have been propagated together in a single fermenter, or
(3) mixtures that have been propagated individually and combined in the final formulation.

The micro-organisms themselves can be naturally occurring species collected from special sites where natural selection has already favoured microbes adapted to unusual conditions. For example, up to half the micro-organisms in coastal waters can degrade hydrocarbons, while meat-processing plants and even septic tanks can yield microbes capable of degrading lipids and proteins. These organisms are grown in the laboratory and are cultured on a medium which is rich in the pollutant they are supposed to degrade. This enrichment technique can be developed by using mutagenesis to find more efficient degraders or by producing new custom-built microbes by gene manipulation (Powledge 1983) (Section 7.4.1). The microbial culture eventually produced is grown in large quantities and dried. Most companies market their microbes as a powder that also contains important additives, such as wetting agents, emulsifers to aid dispersion, and nutrients. To activate the microbes all that is required is to add warm water and stir.

Packaged micro-organisms can be divided into three broad categories according to their use. The largest of these is uprating or seeding existing wastewater treatment plants, although microbes are also sold for cleaning up oil spillages and for degrading toxic and recalcitrant wastes.

Little is known of the techniques used by five major UK companies that specialize in the production of packaged micro-organisms or indeed how widely their products are being used. They are Biolyte (London), Bactozyme (Worksop), Agrico (Stafford), Ubichem (Middlesex), and Interbio (London). Together they market a wide range of products for improving the BOD removal of all types of wastewater treatment plants, eliminating filamentous bulking in the activated sludge process, and improving methane generation in anaerobic digesters. Although most of their products are supplements for wastewater treatment they also produce a range of other pollution control products including dried microbial seeds for the BOD test. The seed normally comes in a capsule that contains 100 mg of specialized microbial culture made up of 15 to 20 different strains including *Pseudomonas* sp., *Nocardia* sp., *Streptomyces* sp., *Bacillus* sp., and *Micromonospora* sp. Each capsule provides enough seed for up to 500 BOD tests, with the culture being activated by adding the contents of a capsule to 1 litre of nutrient water and leaving overnight, but kept continuously stirred. Then, 5 ml of the activated culture is then added to each litre of dilution water used. Fitzmaurice (1986) compared the performance of one BOD seed with raw sewage. He found that the commercial seed produced BOD results with a reaction rate similar to that obtained using raw sewage as seed, but with a higher repeatability estimate. He concluded that, apart from the obvious convenience factor, its widespread use in water pollution laboratories could significantly reduce inter-laboratory variation (Fitzmaurice and Gray 1987a).

Much more is known about the three major companies in the USA which had a total annual turnover of between $5–7m in 1983. Polybac (Allentown, Pennsylvania) is a subsidiary of the Cytox Corporation and began operations in 1975. It achieved sales in excess of $1.5m in 1983 and specializes in engineered microbes for commercial wastewater treatment for food processing, chemical manufacturing, petroleum refining, petrochemical plants, the pulp and paper industry, and municipal sewage works. Like Polybac, the Sybron Corporation based in Birmingham, New Jersey, produces microbial cultures for industries ranging from food processing to steel coking. It was the first company to successfully patent a bacterium and among its more recent products is a microbe capable of degrading Arochlor 1260 which is one of the most highly chlorinated of the polychlorinated biphenyls (PCBs). The third company, Flow Laboratories (Inglewood, California) only uses organisms selected from the environment. It does not use mutation or gene manipulation which means that there are no problems about releasing new genetically engineered species into the environment. They specialize in municipal water and wastewater treatment and have a wide range of products. For example, New York City in 1982, paid $25 000 for bacteria to degrade the grease that clogged the sewers beneath the city's restaurants. The biological result was the same as the previous method, which was using augers to physically drill through the grease, but at a third of the price (Powledge 1983).

There has been considerable success in using specialized cultures to clean up spillages and also decontaminating soil. Packaged micro-organisms for cleaning up spillages of crude oil or hydrocarbons are produced by most companies. However, there is a wide range of microbial cultures that are able to deal with most of the major chemicals that are transported in bulk. An interesting example occurred in northern California where a spillage of 21 000 gallons of 50% formaldehyde solution was successfully treated by containing the liquid in a sealed drainage ditch and pumping it through a temporary bioreactor where it was aerated and mixed with 'Hydrobac', the trade name for a hydrocarbon degrader made up of several mutant bacteria and produced by the Polybac Corporation. The formaldehyde concentration was reduced from 1400 mg l^{-1} to <1 mg l^{-1} after just 14 days, with the whole operation costing about 10% of physically removing the spillage. As well as dealing with spillages to surface water, opportunities also exist for cleaning up contaminated wells and aquifers.

Specialized cultures have also been produced to degrade difficult and recalcitrant wastes, such as pesticides (Cooke et al. 1983). Professor A. M. Chakrabarty of the University of Illinois Medical Centre, Chicago has devised a Pseudomonas strain that can degrade 2-4-5-T a highly dangerous herbicide, better known as Agent Orange. Remarkably, the genetically manipulated bacterium is $>90\%$ efficient. Like all packaged micro-organisms, these organisms and their enzymes have to survive and function in the sub-optimal conditions of a complex ecosystem. The strains that are marketed are generally selected on their ability to metabolize certain pollutants, rather than on their ability to survive environmental stress. Although they are ill-adapted to cope with predation by protozoans, myxobacteria, and slime moulds, like all natural populations they will adjust to the predation level. However, other factors, such as abiotic stresses, including sunlight and temperature, may have a significant effect on numbers and hence efficiency. Clearly, in these situations enhanced cultures may be more successful than either mutated or gene manipulated cultures.

There can be no doubt that packaged micro-organisms can achieve success in treating many difficult and recalcitrant wastes as well as uprating and seeding wastewater treatment plants, providing the conditions are suitable and the correct culture of microbes is utilized. However, no independent studies have so far appeared in the literature and there still seems to be a certain unwillingness by the purchaser of packaged micro-organisms to be enthusiastic by their performance in the field. Gasner (1979) suggests that this lack of confidence is probably due to overly optimistic claims on the part of some suppliers, whereas the truth may be that the companies have failed to clearly present the product and its limitations to those who they hope will purchase it. The US companies realize why the industry has not developed as quickly as other biotechnological areas: the Environmental Protection Agency (EPA) regulates the release of mutagenic and genetically engineered organisms into the environment, and there is also considerable public concern about these organisms. This means that until suitable tests have been developed by the EPA to evaluate new organisms, there are going to be long delays before permission is given to allow these organisms to be used, and future research is going to be at a standstill. Without the prospect of newly developed products, the investment in pollution control will decline, even though biotechnological processes are much cheaper and safer than traditional disposal alternatives (Powledge 1983).

Further reading

General: Cooke 1983
Breakdown of recalcitrants: Leisinger *et al.* 1981, Leisinger 1983.
Bioscrubbing: Le Roux and Mehta 1980.
Packaged micro-organisms: Powledge 1983; Hakulinen 1988; Kobayashi 1983.

REFERENCES

Abernathy, A. R. (1983). *Overland flow treatment of municipal sewage at Easley, SC.* EPA-600/2-83-015, NTIS PB83-170985. US Environmental Protection Agency, Ada, Oklahoma.

Adams, A. P. and Spendlove, J. C. (1970). Coliform aerosols emitted by sewage treatment plants. *Science*, **169**, 1218–20.

Adams, I. M., Ayres, P. A., and Wood, P. C. (1972). Bacterial reduction during tertiary treatment of sewage effluents. *Institute of Public Health Engineers Journal*, 71, 108–25.

Addiscott, T. and Powlson, D. (1989). Laying the ground rules for nitrate. *New Scientist*, **112** (1662), 28–9.

Ademoroti, C. M. A. (1984). Short-term BOD tests. *Effluent and Water Treatment Journal*, **24**, 373–7.

Ademoroti, C. M. A. (1986). Models to predict BOD from COD values. *Effluent and Water Treatment Journal*, **26**, 80–4.

Adin, A., Bauman, E. R., and Warner, F. D. (1985). Evaluation of temperature effects of trickling filter plant performance. *Water Science and Technology*, **17** (2/3), 53–67.

AFF (An Foras Forbartha) (1984). *Development of a water quality model for Irish rivers.* Report: WR/R16. An Foras Forbartha, Dublin.

Agersborg, H. P. K. and Hatfield, W. D. (1929). The biology of a sewage treatment plant – a preliminary survey. *Sewage Works Journal*, **1**, 411–24.

Agg, A. R. and Stanfield, G. (1979). Effect of sewage outfalls on marine water quality in the U.K. In *Biological indicators of water quality*, (ed. A. James and L. Evison) 15.1–15.19. John Wiley, Chichester.

Alaa El-Din, M. N. (1980). Recycling of organic wastes in agriculture. In *Bioresources for development: the renewable way of life*, (ed. A. King, H. Cleveland, and G. Streatfield), pp. 184–203. Pergamon, New York.

Al-Diwamy, L. J. and Cross, T. (1978). Ecological studies of *Nocardia* foams and other actinomycetes in aquatic habitats. In *Nocardia and Streptomyces*, (ed. M. Modarski, W. Kurytowicz, and J. Jeljaszewiz), pp. 153–74. Fischer V, Stuttgart.

Alexander, M. (1965). Biodegradation: problems of molecular recalcitrance and microbial fallibility. *Advances in Applied Microbiology*, **1**, 38–40.

Ali, H. I. and Bewtra, J. K. (1972). Effect of turbulence on BOD testing. *Journal of the Water Pollution Control Federation*, **44**, 1798–1807.

Alico, R. K. and Palenchar, C. A. (1975). *Staphyloccoccus aureus* recoveries on various brands of membrane filters. *Health Laboratory Science*, **12**, 341–6.

Allen, L. A., Tomlinson, T. G., and Norton, I. L. (1944). The effect of treatment in percolating filters on bacterial counts. *Journal and Proceedings of the Institute Sewage Purification*, 115–32.

Allen, M. J. and Geldreich, E. E. (1975). Bacterial criteria for ground-water quality. *Ground Water*, **13**, 45–51.

Allen, S. E. (1974). *Chemical analysis of ecological materials*. Blackwell, London.

Allen, T. S. and Kingsbury, R. P. (1973). *The physical design of biological towers.* Proceedings of the 28th Industrial Waste Conference, Purdue University, pp. 462–82.

Al-Mufti, M. M., Sydes, C. L., Furness, S. B., Grime, J. P., and Band, S. R. (1977). A quantitative analysis of shoot phenology and dominance in herbaceous vegetation. *Journal of Ecology*, **65**, 759–91.

American Public Health Association, American Water Works Association and Water Pollution Control Federation (1983). *Standard methods for the examination of water and wastewater*, (15th edn). American Public Health Association Inc, New York.

American Public Health Association, American Water Works Association and Water Pollution Control Federation (1985). *Standard Methods for the Examination of Water and Wastewater*, (16th edn). American Public Health Association Inc., New York.

Anderson, G. K. (1977). Farm and food wates. In *Treatment of industrial effluents*, (ed. A. G. Callely, C. F. Forster, and D. A. Stafford), pp. 245–57. Hodder and Stoughton, London.

Anderson, G. K., Donnelly, T., and McKeown, K. (1984). The application of anaerobic packed bed reactors to industrial wastewater treatment. *Water Pollution Control*, **83**, 491–8.

Anderson, J. G. and Smith, J. E. (1987). Composting. In *Biotechnology of waste treatment and exploitation*, (ed. J. M. Sidwick and R. S. Holdom), pp. 301–25. Ellis-Horwood, Chichester.

Anderson, J. W. (1980). Bioenergetics of autotrophs and heterotrophs. In *Studies in biology*, **126**. Edward Arnold, London.

Anderson, R., Heden, C.-G., and Williams, L. (1980). The potential of algae in decentralized bioconversion systems. In *Bioresources for development*, (ed. A. King, H. Cleveland, and G. Streatfield), pp. 177–83. Pergamon, New York.

Andersson, Y. and Stenstrom, T. A. (1986). Waterborne outbreaks in Sweden—causes and etiology. *Water Science and Technology*, **18** (10), 185–90.

Andrews, J. F. (1971). Kinetic models of biological waste treatment processes. In *Biological waste treatment*, (ed. R. P. Canale), Biotechnology and Bioengineering Symposium No. 2, pp. 5–33. John Wiley, New York.

Andrews, J. F. (1983). Kinetics and mathematical modelling. In *Ecological aspects of used water treatment:* Vol. 3. *The processes and their ecology*, (ed. C. R. Curds and H. A. Hawkes), pp. 113–72. Academic Press, London.

Andrews, J. F. and Graef, S. P. (1971). Dynamic modelling and simulation of the anaerobic digestion process. In *Anaerobic biological treatment*, (ed. R. F. Gould), pp. 126–62. American Chemical Society.

Angelbeck, D. I. and Kirsch, E. J. (1969). The influence of pH and metal cations on aggregative growth of non-slime forming strains of *Zoogloea ramigera Applied Microbiology*, **17**, 435–40.

Anglian Water Authority (1982). *Methane gas as a vehicle fuel.* Anglian Water Authority, Huntingdon.

Anon. (1979*a*). River rescue. *Water and Waste Treatment Journal*, **22**, 38–42.

Anon. (1979*b*). Advanced wastewater treatment: a low-risk proving-ground for experiment and invention. *World Water*, **6**, 29–41.

Anon. (1979*c*). Filamentous micro-organisms in activated sludge. *New Quarterly of the Sanitary Engineering Research Laboratory*, **29**(2), 1–2. University of California, Berkeley.

Anon. (1979*d*). Land cultivation of industrial sludges. *Sludge*, (May/June), 26.

Anon. (1982*a*). Microbial health risks from sludge applied to land. *Water Research*, **16**, 501–2.

Anon. (1982*b*). *Biotechnology Bulletin*, **1**(4).

Anon. (1982*c*). *Biotechnology Bulletin*, **1**(6).

Anon. (1983*a*). *Biotechnology Bulletin*, **2**(7).

Anon. (1983*b*). Biomechanics International plc. *Biotechnology Bulletin Report*, **35**. Biotechnology Bulletin.

Anon. (1984*a*). British Earthworm Technology Ltd. *Biotechnology Bulletin Report*, **44**. Biotechnology Bulletin.

Anon. (1984*b*). *Biotechnology Bulletin*, **3**(2).

Anon. (1984*c*). Bioisolates (Holdings)Ltd. *Biotechnology Bulletin Report*, **45**. Biotechnology Bulletin.

Anon. (1984*d*). Biotechnology Bulletin, **3**(5).

Anon. (1984*e*). Gas makes the wheels go round. *Surveyor*, 8 March, 11–12.

Anon. (1985*a*). Ministers call for survey of beach sewage. *New Scientist*, **107**(1466), 21.

Anon. (1985*b*). Comments on the state of marine science and technology in the UK. *Journal of the Institution of Water Engineers and Scientists*, **39**, 363–70.

Anon. (1986). Sewage to make the desert bloom. *New Scientist*, **112**, 31.

Antonie, R. L. (1976). *Fixed biological surfaces – wastewater treatment. The rotating biological contactor*. CRC Press Inc, Cleveland, Ohio.

Arceivala, S. J., Lakshminarayana, J. S. S., Alagarsamy, S. R., and Sastry, C. A. (1970). *Waste stabilization ponds: design, construction and opertion in India*. Central Public Health Engineering Research Institute, Nagpur, India.

Ardern, E. and Lockett, W. T. (1923). The activated sludge process, Withington Works. *Chemistry and Industry*, 225T-230T.

Argent, V. A., Bell, J., and Emslie-Smith, M. (1977). Animal disease hazards of sludge disposal to land – occurrence of pathogenic organisms. *Water Pollution Control*, **76**, 511–6.

Armitage, P. D., Furse, M. T., and Wright, J. F. (1979). A bibliography of works for the identification of freshwater invertebrates in the British Isles. *Occasional Publication*, **5**. Freshwater Biological Association, Cumbria.

Armstrong, W. (1964). Oxygen diffusion from the roots of some British bog plants. *Nature (London)*, **204**, 801–2.

Armstrong, W. (1982). Water-logged soils. In *Environment and plant ecology*, (ed. J. R. Etherington), pp. 290–330. John Wiley, Chichester.

Athie, D. and Cerri, C. C. (ed.) (1987) The use of macrophytes in water pollution control. *Water Science and Technology*, **19** (10), 1–177.

Atkinson, B., Busch, A. W., and Dawkins, G. S. (1963). Recirculation reaction kinetics and effluent quality in a trickling filter flow model. *Journal of the Water Pollution Control Federation*, **35**, 1307–17.

Atkinson, B. and Fowler, H. W. (1974). The significance of microbial film in fermenters. *Advances in Biochemical Engineering*, **3**, 221–77.

Atkinson, B. and Mavitune, F. (1983). *Biochemical Engineering and Biochemical Handbook*. The Nature Press, London.

Audsley, E. and Knowles, D. (1984). *Feasibility study – straw/sewage sludge compost*. NIAE Report R. 45. National Institute of Agricultural Engineering, Silsoe, England.

Ayres, P. A. (1977). The use of faecal bacteria as a tracer for sewage sludge disposal in the sea. *Marine Pollution Bulletin*, **8**, 283–5.

Aziz, J. A. and Tebbutt, T. H. Y. (1980). Significance of COD, BOD and TOC correlations in kinetic models of biological oxidation. *Water Research*, **14**, 319–24.

Baalsrud, K. (1967). Polluting material and polluting effect. *Water Pollution Control*, **66**, 97–103.

Baalsrud, K. (1975). The case for treatment. In *Discharge of sewage from sea outfalls*, (ed. A. L. H. Gameson), pp. 165–72. Pergamon, London.

Babbitt, H. A. (1947). *Sewerage and sewage treatment*. John Wiley, New York.

Bagnall, L. L., Furman, T. S., Hentges, J. F., Nolan, W. J., and Shirley, R. L. (1974). *Feed and fiber from effluent-grown water hyacinth*. US Environmental Protection Agency, Report 660/2-74-041. Ohio.

Bahr, T. G., King, D. L., Johnson, H. E., and Kerns, C. L. (1977). Municipal wastewater recycling: production of algae and macrophytes for animal food and other uses. *Developments in Industrial Microbiology*. **18**, 121–40.

Bailey, D. A. and Thomas, E. V. (1975). The removal of inorganic nitrogen from sewage effluents by biological denitrification. *Water Pollution Control*, **74**, 497–515.

Baillod, C. R. and Boyle, W. C. (1970). Mass transfer limitations in substrate removal. *Journal of the Sanitary Engineering Division, American Society of Civil Engineers*, **96**, 525–45.

Baines, S., Hawkes, H. A., Hewitt, C. H., and Jenkins, S. H. (1953). Protozoa as indicators in activated sludge treatment. *Sewage and Industrial Wastes*, **25**, 1023–33.

Baines, S., Svoboda, I. F., and Evans, M. R. (1985). Heat from aerobic treatment of liquid animal wastes. In *Composting of agricultural and other wastes*, (ed. J. K. R. Gasser), pp. 147–66. Elsevier, London.

Baker, C. R. and Carlson, R. V. (1978). Oxygen safety considerations. In *The use of high purity oxygen in the activated sludge process*, (ed. J. R. McWhirter), pp. 191–207. CRC Press, Florida.

Baker, M. N. (1949). *The quest for pure water*. American Water Works Association, New York.

Baker, R. A. (1961). A preliminary survey of the mite fauna of sewage percolating filters. M.Sc. Thesis, University of London.

Baker, R. A. (1975). Arachnida. In *Ecological aspects of used-water treatment*, (ed. C. R. Curds and H. A. Hakes), pp. 375–92. Academic Press, London.

Balluz, S. A., Jones, H. A., and Butler, M. (1977). The persistence of poliovirus in activated sludge treatment. *Journal of Hygiene, Cambridge*, **78**, 165–73.

Balluz, S. A., Butler, M., and Jones, H. H. (1978). The behaviour of f2 coliphage in activated sludge treatment. *Journal of Hygiene, Cambridge*, **80**, 237–42.

Balmer, P., Berglund, D. T., and Enebo, L. (1967). Step sludge: a new approach to wastewater treatment. *Journal of the Water Pollution Control Federation*, **29**, 1021–5.

Banks, C. J., Davies, M., Walker, I., and Ward, R. D. (1976). Biological and physical characterization of activated sludge: a comparative experimental study at ten treatment plants. *Water Pollution Control*, **75**, 492–508.

Barber, J. B. and Veenstra, J. N. (1986). Evaluation of biological sludge properties influencing volume reduction. *Journal of the Water Pollution Control Federation*, **58**, 149–56.

Bardtke, D. and Thomanetz, E. (1976). Untersuchung zur Erfassurg der Biomasse von Belebtschlammen durch quantitative Analyse der Desoxyribonukleinsaure (DNS). *GWF Wasser Abwasser*, **117**, 451–3.

Barker, A. N. (1942). The seasonal incidence, occurrence and distribution of protozoa in the bacteria bed process of sewage disposal. *Annals of Applied Biology*, **29**, 23–33.

Barker, A. N. (1946). The ecology and function of protozoa in sewage purification. *Annals of Applied Biology*, **33**, 314–25.

Barker, A. N. (1949). Some microbiological aspects of sewage purification. *Journal Proceedings of the Institute Sewage and Purification* 7–22.

Barker, T. W., Quinn, J. P., and Marchant, R. (1982). The use of a mixed culture of *Geotrichum candidum, Candida krusei* and *Hansennula anomala* for microbial protein production from whiskey distillery sepent wash. *European Journal of Applied Microbiology and Biotechnology*, **14**, 247–53.

Barnard, J. L. and Meiring, P. G. J. (1988) Dissolved oxygen control in the activated sludge process. *Water Science and Technology*. **20** (4/5), 93–100.

Barnes, D. and Bliss, P. J. (1983). *Biological control of nitrogen in wastewater treatment*. Spon, London.

Barnes, D. and Wilson F. (1976). *The design and operation of small sewage works*. Spon, London.

Barnes, D., Forster, C. F., and Johnstone, D. W. M. (1983). *Oxidation ditches in wastewater treatment*. Pitman, London.

Barnes, D., Forster, C. F., and Hrudey, S. E. (ed.) (1984). *Food and allied industries*. Pitman, London.

Barnes, D., Fitzgerald, P. A., McFarland, R., Swan, H., and Schulz, T. (1985). Redox potential – basis of measurement and application. *Effluent and Water Treatment Journal*, **25**, 232–6.

Barraclough, D. H. (1954). Biological filtration of sewage and vacuum filtration of sludge: experimental work at Reading 1949–1954. *Journal and Proceedings of the Institute Sewage Purification* 361–76.

Barrett, K. A. (1972). An investigation into the effects of chlorinated solvents on sludge digestion. *Water Pollution Control*, **71**, 389–403.

Barritt, N. W. (1940). The ecology of activated sludge in relation to its properties and the isolation of a specific substance from the purified effluent. *Annals of Applied Biology*, **27**, 151–6.

Barrow, G. I. (1981). Microbial pollution of costs and estuaries: the public health implications. *Water Pollution Control*, **80**, 221–35.

Barth, E. F., Moore, W. A., and McDermott, G. N. (1965). *Interaction of heavy metals on biological sewage treatment processes*. US Department of Health, Education and Welfare.

Bartlett, R. E. (1981). *Surface water sewerage*. Applied Science Publishers, London.

Bates, M., Zolton, T., and Reynolds, P. (1984). Camplylobacteriosis associated with the consumption of raw milk in British Columbia. *Canada Diseases Weekly Report*, **10**(40), 156–7.

Bates, T. E. (1972). *Land application of sewage sludge*. Project Report No. 1, M.O.E. Project 71-4-1, Ministry of the Environment, Toronto.

Bausum, H. T. *et al.* (1982). Comparison of coliphage and bacterial aerosols at a wastewater spray irrigation site. *Applied Environmental Microbiology*, **43**, 28.

Bayley, R. W. and Downing, A. L. (1963). Temperature relationships in percolating filters. *Institution of Public Health Engineers Journal* **62**, 303–2.

Becker, E. W. (1983). Limitations of heavy metal removal from wastewater by means of algae. *Water Research*, **17**, 459–66.

Becker, J. H. and Shaw, C. G. (1955). Fungi in domestic sewage treatment plants. *Applied Microbiology*, **3**, 173–80.

Beckett, P. H. T. and Davis, R. D. (1982). Heavy metals in sludge – are their toxic effects additive? *Water Pollution Control*, **81**, 112–9.

Bedogni, S., Bregoli, M., and Biglia, A. (1983). Full-scale anaerobic filter treating sugar-mill anionic effluents. In *Proceedings of the Anaerobic Wastewater Treatment Symposium*, pp. 313–4. TNO Corporate Communication Department, The Hague.

Belcher, H. and Swale, E. (1976). *A beginners guide to freshwater algae*. Institute of Terrestrial Ecology. H.M.S.O., London.

Bell, E. (1988). Irish sewage sludges: an evaluation of suitability for disposal to agricultural land. M.Sc. thesis, University of Dublin, Trinity College.

Bell, J. C., Argent, V. A., and Edgar, D. (1978). *The survival of* Brucella abortus *in sewage sludge*, **29**. Proceedings of the Conference in Utilization of sewage sludge on land, Oxford. Water Research Centre, Stevenage.

Bell, J. P. (1973). *Neutron probe practice*. Report No. 19. Institute of Hydrology, National Environment Research Council.

Bell, R. G. (1976). The limitation of the ratio of faecal coliforms to total coliphage as a water pollution index. *Water Research*, **10**, 745–8.

Bellamy, W. D. (1975). Conversion of insoluble agricultural wastes to SCP by thermophylic micro-organisms. In *Single cell protein II*, (ed. S. R. Tannenbaum and D. I. C. Wang), pp. 2163–72. MIT Press, Mass.

Bellan, G. (1967). Pollution et peuplements benthiques sur la substrat meuble dans la region de Marseille. *Revue Internationale d'Oceanographie Medicole*, **8**, 51–95.

Bellinger, E. G. (1980). *A key to common British algae*. Institution of Water Engineers and Scientists, London.

Belser, L. W. (1979). Population ecology of nitrifying bacteria. *Annual Reviews of Microbiology*, **33**, 309–33.

Benarde, M. A. (1973). Land disposal and sewage effluent: appraisal of health effects of pathogenic organisms. *Journal of the American Water Works Association*, **65**, 432–40.

Benedict and Carlson, D. A. (1971). Aerobic heterotrophic bacteria in activated sludge. *Water Research*, **5**, 1023–30.

Benefield, L. D. and Randall, C. W. (1977). Evaluation of a comprehensive kinetic model for the activated sludge process. *Journal of the Water Pollution Control Federation*, **49**, 1636–41.

Benefield, L. D. and Randall, C. W. (1980). *Biological process design for wastewater treatment*. Prentice-Hall, New York.

Benemann, J. R. and Weissman, J. C. (1977). Biophotolysis: problems and prospects. In *Microbial energy conversion*, (ed. H. G. Schlegel and J. Barnea), pp. 413–26. Pergamon, Oxford.

Benemann, J. R., Weissmann, J., Koopman, B., and Oswald, W. J. (1977a). Energy production by microbial photosynthesis. *Nature (London)*, **268**, 19–23.

Benemann, J. R., Koopman, B., Weissmann, J., and Oswald, W. J. (1977b). Biomass production and waste recycling with blue-green algae. In *Microbial energy conversion*, (ed. H. G. Schlegel and J. Barnea), pp. 399–412. Pergamon, Oxford.

Benemann, J. R., Weissman, J. C., and Oswald, W. J. (1979). Algal biomass. In *Microbial biomass*, (ed. A. H. Rose), pp. 177–206. Academic Press, London.

Benjes, H. M. (1980). *Handbook of biological wastewater treatment*. Garland STPM Press, New York.

Bennett, S. (1981). The waste-watchers. *Farmers' Weekly*, **95**(8), v–vii.

Benoit, R. J. (1971). Self-purification in natural waters. In *Water and water pollution handbook*, Vol. I, (ed. L. L. Ciaccio). Marcel Dekker, New York.

Benson-Evans, K. and Williams, P. F. (1975). Algae and bryophytes. In *Ecological aspects of used-water treatment*. Vol. 1. *The organisms and their ecology*, (ed. C. R. Curds and H. A. Hawkes), pp. 153–202. Academic Press, London.

Benton, C., Khan, F., Monaghan, P., Richards, W. N., and Shedden, C. B. (1983). The contamination of a major water supply by gulls (*Larus* sp.): a study of the problem and remedial action taken. *Water Research*, **17**, 789–98.

Beresford, S. A. A. (1983). Cancer incidence and reuse of drinking water. *American Journal of Epidemiology*, **117**, 258–68.

Beresford, S. A. A. (1980). *Water reuse and health in the London area*. Technical Report TR138. Water Research Centre, Stevenage.

Beresford, S. A. A., Carpenter, L. M., and Powell, P. (1984). *Epidemiological studies of water reuse and type of water supply*. Technical Report TR216. Water Research Centre, Stevenage.

Berg, G. (1966). Virus transmission by the water vehicle. II: virus removal by sewage treatment procedures. *Health Laboratory Science*, **3**, 90–100.

Berg, G. (1973) Re-assessment of the virus problem in sewage and in surface and renovated waters. *Progress in Water Technology*, **3**, 87–101.

Berg, L. van den and Lentz, C. P. (1979). *Comparison between up- and down-flow anaerobic fixed film reactors of varying surface-to-volume ratios for the treatment of bean blanching waste*, pp. 319–25. Proceedings of the 34th Purdue Industrial Waste Conference.

Berg, L. van den, and Lentz, C. P. (1980) *Effects of film area-to-volume ratio, film support, height and direction of flow on performance of methanogenic fixed film reactors*. Proceeding of the US Department of the Environment Workshop Seminar on Anaerobic Filters. Howey-in-the-Hill, Florida.

Berkun, M. (1982). Effects of inorganic metal toxicity on BOD, I. Methods for the estimation of BOD parameters, II. *Water Research*, **16**, 559–64.

Bernhard, M. and Zattera, A. (1975). *Major pollutants in the marine environment*, (ed. E. A. Peason), pp. 195–300. Proceedings of the 2nd International Congress on Marine Pollution and Marine Waste Disposal. Pergamon, Oxford.

Berthouex, P. M. and Rudd, D. F. (1977). *Strategy of Pollution Control*. Wiley, New York.

Besselievre, E. B. and Schwartz, M. (1976). *The treatment of industrial wastes*, (2nd edn). McGraw-Hill.

Best, R. (1980). Want not, waste not! Sensible sludge recycling. *Water Pollution Control*, **79**, 307–21.

Beszechits, S. and Lugowski, A. (1983). Waste activated sludge: a potential new livestock feed ingredient. *Process Biochemistry*, **18**, 35–37.

Beveridge, T. J. and Murray, R. G. E. (1976). Uptake and retention of metals by cell walls of *Bucillus subtilus*. *Journal of Bacteriology*, **127**, 1502.

Bhatla, M. N. and Gaudy, A. F. (1965). Role of protozoa in the diphasic exertion of BOD. *Journal of the Sanitary Engineering Division, American Society of Civil Engineers* **91**(SA3), 63–87.

Bick, H. (1972). *Ciliated protozoa: an illustrated guide to the species used as biological indicators in freshwater biology*. World Health Organization, Geneva.

Bicknell, S. R. (1972). *Salmonella aberdeen* infection in cattle associated with human sewage. *Journal of Hygiene, Cambridge*, **70**, 121–6.

Billings, C. H. and Smallhurst, D. F. (1971). *Manual of wastewater operations*, pp. 174–89. Texas State Department of Health, Austin, Texas.

Bissett, K. A. and Brown, D. (1969). Some electron microscope observations on the morphology of *Sphaerotilus natans*. *Giornale di Microbiologia*, **17**, 97–9.

Blackbeard, J. R. and Ekama, G. A. (1984). A survey of activated sludge bulking and foaming in Southern Africa. *IMIESA*, **9**(3), 20–5.

Blackbeard, J. R., Ekama, G. A., and Marais, G. R. (1985). *An investigation into filamentous bulking and foaming in activated sludge plants in Southern Africa.* Research Report W53, Department of Civil Engineering, University of Cape Town.

Blackburn, A. M. (1978). Management strategies dealing with drought. *Journal of the American Water Works Association*, **70**, 51–9.

Blamire, R. V., Goodhand, R. H., and Taylor, K. C. (1980). A review of some animal diseases encountered at meat inspections in England and Wales, 1969–1978. *Vetinary Record*, **106**, 195–9.

Blaser, M., Powers, B. W., Cravens, J., and Wang, W. L. (1978). *Campylobacter enteritus associated with canine infection. Lancet*, **ii**, 979–81.

Bode, H. and Seyfried, C. F. (1984). Mixing and detention time distribution in activated sludge tanks. *Water Science and Technology*, **17** (4/5), 197–208.

Boer, E. de., Hartog, B. J., and Borst, G. H. A. (1984). Milk as a source of *Campylobacter jejuni. Netherlands Milk and Dairy Journal*, **38**, 183–94.

Boll, R. and Kayser, R. (1986). Reuse of wastewaters in agriculture by means of sprinkler irrigation on a farmland area of 300 ha: large scale experience in Germany. *Water Science and Technology*, **18** (9), 163–73.

Bollen, G. J. (1985). The fate of plant pathogen during composting of crop residues. In *Composing of agricultural and other wastes*, (ed. J. K. R. Gasser), pp. 282–90. Elsevier, London.

Bolton, D. H. and Ousby, J. C. (1977). The ICI Deep Shaft effluent treatment process and its potential for large sewage works. *Progress in Water Technology*, **86**, 265–73.

Bolton, F. J., Hinchliffe, P. M., Coates, D., and Robertson, L. (1982). A most probable number method for estimating small numbers of *Campylobacter* in water. *Journal of Hygiene Cambridge*, **89**, 185–190.

Bolton, F. J., Coates, D., and Hutchinson, D. N. (1985). Thirteen month survey of campylobacters in a river system subject to sewage effluent discharge. In *Campylobacter*, Vol. III, (ed. A. D. Pearson, M. B. Skirrow, H. Lior, and R. Rowe), pp. 278–279. Public Health Laboratory Service, London.

Bonde, G. J. (1977). Bacterial indication of water pollution. In *Advances in aquatic microbiology*, (ed. M. R. Droop and H. W. Jannasch), pp. 273–364. Academic Press, London.

Boon, A. G. (1976). Oxygen activated sludge systems. I. Technical review of the use of oxygen in the treatment of wastewater. *Water Pollution Control*, **75**, 206–13.

Boon, A. G. (1978). Oxygen transfer in the activated sludge process. In *New processes of wastewater treatment and recovery*, (ed. G. Mattock), pp. 17–33. Ellis-Horwood, Chichester.

Boon, A. G. and Burgess, D. R. (1972). Effects of diurnal variations in flow of settled sewage on the performance of high-rate activated sludge plants. *Journal of the Water Pollution Control Federation*, **71**, 493–514.

Bopardikar, M. V. (1969). Microbiology of a waste stabilization pond. In *Advances in Water Pollution Research*, (ed. S. H. Jenkins), pp. 595–601. Pergamon, Oxford.

Botha, G. R. (1984). Research related to water pollution control in South Africa. *Water Pollution Control*, **83**, 184–90.

Botto, B. A. and Pawlowski, L. (1983). Reclamation of wastewater constituents by ion exchange. *Effluent and Water Treatment Journal*, **23**, 233–9.

Bouma, J., Baker, F. G., and Veneman, P. L. M. (1974). *Measurement of waste movement in soil pedons above the water table*. Geological and Natural History Survey Information Circular, **27**. University of Wisconsin.

Bow, M. R. and Brown, J. H. (1946). Tularaemia: a report on 40 cases in Alberta, Canada (1931–1944). *American Journal of Public Health*, **36**, 494–500.

Bowmer, K. H. (1985). Detoxification of effluents in a macrophyte treatment system. *Water Research*, **19**, 57–62.

Bowmer, K. H. (1987). Nutrient removal from effluents by an artificial wetland: influence of rhizophere aeration and preferential flow studied using bromide and dye tracers. *Water Research*, **21**, 591–9.

Boyle, J. D. (1984). Effects of biological film on BOD decay rates in rivers. *Water Science Technology*, **16**, 643–51.

Boyle, J. D. and Scott, J. A. (1984). The role of benthic films in the oxygen balance in an east Devon river. *Water Research*, **18**, 1089–99.

Boyles, D. (1984). *Bio-energy: technology, thermodynamics and costs*. Ellis-Horwood, Chichester.

Boyt, F. L., Bayley, S. E., and Zoltec , J. (1976). Removal of nutrients from treated municipal wastewater by wetland vegetation. *Journal of the Water Pollution Control Federation*, **48**, 789–99.

Bradley, D. J. (1974). *Human rights in health*. Ciba Foundation Symposium, **23**. Elsevier, Amsterdam.

Bradley, R. M. (1973). Chlorination of effluents and the Italian concept. *Effluent and Water Treatment Journal*, **13**, 683–9.

Bradshaw, A. D. and Chadwick, M. J. (1980). *The restoration of land*. Blackwell, Oxford.

Braun, R. (1982). *Biogas–Methangärung organischer Abfallstoffe*. Springer, Wien.

Braun, R., Meyrath, J., Stuparek, W., and Zarlauth, G. (1979). Feed yeast production from citric acid waste. *Process Biochemistry*, **14**, 16–20.

Bremond, J. C., Gramet, P. H., Brough, T., and Wright, E. N. (1968). A comparison of some broadcasting equipment and recorded distress calls for scaring birds. *Journal of Applied Ecology*, **5**, 521–9.

Brierley, J. A., Brandvold, D. K., and Popp, C. J. (1975). Waterfowl refuge effect on water quality. I. Bacterial populations. *Journal of the Water Pollution Control Federation*, **47**, 1892–1900.

Brindle, A. (1962). Taxonomic notes on the larvae of British Diptera: 2. Trichoceridae and Anisopodidae. *Entomologist*, **95**, 285–8.

Brink, N. (1967). Ecological studies in biological filters. *Internationale Revue der Gesamten Hydrobiologie*, **52**(1), 51–122.

Brinkhurst, R. O. (1971). *A guide for the identication of British aquatic Oligochaeta*. Scientific Publication, **22**, Freshwater Biological Association, Cumbria.

Briscou, J. (1975). Yeasts and fungi in marine environments. *Société Française Mycologie Medicale Bulletin*, **4**, 159–62.

BSI (British Standards Institution) (1948). *Specification for media for biological filters*, BS 1438. London.

BSI (British Standards Institution) (1971). *Specification for media for percolating filters*, BS 1438. London.

BSI (British Standards Institution) (1972). *Small sewage treatment works*. Code of Practice, CP 302. London.

BSI (British Standards Institution) (1979). *Precision of test methods, part I*. Guide to the determination of repeatability and reproducability of a standard test method. BS 5497. London.

BSI (British Standards Institution) (1983). *Design and installation of small sewage treatment works and cesspools*. Code of Practice, CP 6297. London

BSI (British Standards Institution) (1987) *Precision of test methods, part I*. Guide to the determination of repeatability and reproducability of a standard test method. BS 5497. London.

Brix, H. (1987). Treatment of wastewater in the rhizosphere of wetland plants—the root-zoned method. *Water Science and Technology*, **19** (10), 107–18.

Brock, T. D. (1971*a*). *Biology of microorganisms*. Prentice-Hall, New Jersey.

Brock, T. D. (1971*b*). Microbial growth rates in nature. *Bacteriological Review*, **35**, 39–58.

Brooker, R. and Farnell, G. W. (1979). Kirklees takes the spoil out of colliery waste. *Surveyor*, **153**, 13–15.

Brouwer, R., Mertens, M. J. A., Siem, T. H., and Katchak, J. (1979). An explosive outbreak of *Campylobacter* enteritis in soldiers. *Antonie van Leenwenhoek*, **45**, 517–19.

Brown, J. M., Campbell, E. A., Rickards, A. D. , and Wheeler, D. (1987). *The public health implication of sewage pollution of bathing water*. The Robens Institute of Industrial and Environmental Health and Safety, University of Surrey.

Brown, K. W. *et al.* (1983). The movement of metals applied to soils in sewage effluent. *Water, Air and Soil Pollution*, **19**, 43–54.

Brown, M. J. and Lester, J. N. (1979). Metal removal in activated sludge. The role of bacterial extracellular polymers. *Water Research*, **13**, 817–37.

Brown, M. W. R. and Foster, J. H. S. (1970). A simple diagnostic milk medium for *Pseudomonas aeruginosa*. *Journal of Clinical Pathology*, **23**, 172–6.

Brown, T. J. (1965). The study of protozoa in a difussed-air activated sludge plant. *Water Pollution Control*, **64**, 375–8.

Browning, J. R. and Ives, D. G. (1987). Environmental health and the water distribution system: a case history of an outbreak of Giardiasis. *Water and Environmental Management*, **1**, 55–60.

Bruce, A. M. (1968). The significance of particle shape in relation to percolating filter media. *Journal of the British Granite Whinstone Federation*, **8**(2), 1–15.

Bruce, A. M. (1969). Percolating filters. *Process Biochemistry*, **4**(4), 19–23.

Bruce, A. M. (ed.) (1984). *Sewage sludge stabilization and disinfection*. Ellis-Horwood, Chichester.

Bruce, A. M. and Boon, A. G. (1971). Aspects of high-rate biological treatment of domestic and industrial wastewaters. *Water Pollution Control*, **70**, 487–513.

Bruce, A. M. and Fisher, W. J. (1984). Sludge stabilization – methods and measurement. In *Sewage sludge stabilization and disinfection*, (ed. A. M. Bruce), pp. 23–47. Ellis-Horwood, Chichester.

Bruce, A. M. and Hawkes, H. A. (1983). Biological filters. In *Ecological aspects of used-water treatment*, Vol. 3. *The processes and their ecology*, (ed. C. R. Curds and H. A. Hawkes), pp. 1–111. Academic Press, London.

Bruce, A. M. and Lockyear, C. E. (1982). Uprating sludge treatment processes. *Water Pollution Control*, **81**, 425–43.

Bruce, A. M. and Merkens, J. C. (1970). Recent studies of high-rate biological filtration. *Water Pollution Control*, **69**, 113–48.

Bruce, A. M. and Merkens, J. C. (1973). Further studies of partial treatment of sewage by high-rate biological filtration. *Water Pollution Control*, **72**, 499–523.

Bruce, A. M. and Merkens, J. C. (1975). *Developments in sewage treatment for small communities*, pp. 65–96. Proceedings of the 8th Public Health Engineers Conference, Loughborough University.

Bruce, A. M., Swanwick, J. D., and Ownsworth, R. A. (1966). Synthetic detergents and sludge digestion: some recent observations. *Water Pollution Control*, **65**, 427–47.

Bruce, A. M., Merkens, J. C., and MacMillan, S. C. (1970). Research development in high-rate biological filtration. *Institution of Public Health Engineers Journal*, **69**, 178–207.

Bruce, A. M., Merkens, J. C., and Haynes, B.A.O. (1975). Pilot-scale studies on the treatment of domestic sewage by two-stage biological filtration with special reference to nitrification. *Water Pollution Control*, **74**, 80–100.

Bruce, A. M., Truesdale, G. A., and Mann, H. T. (1967). The comparative behaviour of replicate pilot-scale percolating filters. *Institution of Public Health Engineers Journal* **66**, 151–75.

Bruce, A. M., Truesdale, G. A., and Mann, H. T. (1967). The comparative behaviour of replicate pilot-scale percolating filters. *Journal of the Institution of Public Health Engineers Journal*, **66**, 151–75.

Bruce, D., Zochowski, W., and Fleming, G. A. (1980). Campylobacter infections in cats and dogs. *Veterinary Record* **107**, 200–1.

Bryant, M. P., Barel, V. H., Frobish, R. A., and Isaacson, H. R. (1977). Biological potential of thermophilic methanogenesis from cattle wastes. In *Microbial energy conversion*, (ed. H. G. Schlegel and J. Barnea), pp. 347–59. Pergamon, Oxford.

Bryce, D. (1960). Studies on the larvae of the British Chironomidae (Diptera), with keys to the Chironomidae and Tanypodire. *Transactions of the Society for British Entomology*, **14**, (II), 19–61.

Bryce, D. and Hobart, A. (1972) The biology and identification of the larvae of the Chironomidae (Diptera). *Entomologists Gazette*, **23**, 175–217.

Buchanan, R. E. and Gibbons, N. E. (1975). *Bergey's manual of determinative bacteriology*, (8th edn). Bailliere Tindall and Cox, London.

Buck, J. D. (1977). *Candida albicans*. In *Bacterial indicators/health hazards associated with water*, (ed. A. W. Hoadley and B. J. Dutka), pp. 139–47. ASTM Special Technical Publication, **635**, ASTM, Philadelphia.

Buckley, H. R (1971). Fungi pathogenic for man and animals. 2. The subcutaneous and deep seated mycoses. In *Methods in microbiology*, Vol. 4, (ed. C. Booth), pp. 461–78. Academic Press, London.

Building Research Establishment (1987). *Low-water-use washdown WC's*. BRE Report, Garston, Watford.

Bull, A. T., Holt, G., and Lilly, M. D. (1982). *Biotechnology: international trends and perspectives*. OECD, Paris.

Bu'lock, J. and Kristiansen, B. (eds.) (1987). *Basic biotechnology*. Academic Press, London.

Bungay, H. R. and Bungay, M. L. (1968). Microbial interactions in continuous culture.

In *Advances in Applied Microbiology*, Vol. 10, (ed. W. W. Umbreit and D. Perlman), pp. 269–90. Academic Press, London.

Burman, N. P., Oliver, C. W., and Stevens, J. J. (1969). Membrane filtration techniques for the isolation from water of coli-aerogenes, *Escherichia coli*, faecal streptococci, *Clostridium perfringens*, actinomycetes and microfungi. In *Isolation methods for microbiologists*, (ed. D. A. Shapton and G. W. Gould), pp. 127–45. Academic Press, London.

Burnett, R. C. S., MacLeod, A. F., and Tweedie, J. (1980). An outbreak of salmonellosis in West Lothian. *Communicable Disease Scotland*, Weekly Report, **14**, vi–viii.

Busch, A. W. (1958). BOD progression in soluble substrates. *Sewage and Industrial Wastes*, **30**, 1136–49.

Buswell, A. M. and Mueller, H. F. (1952). Mechanisms of methane fermentation. *Industrial Engineering Chemistry*, **44**, 550–2.

Buswell, A. M., Shiota, T., Lawrence, N., and Meter, I. V. (1954). Laboratory studies on the kinetics of the growth of *Nitrosomonas* with relation to the nitrification phase of the BOD test. *Applied Microbiology*, **2**, 21–5.

Butcher, R. W. (1932). Contribution to our knowledge of the ecology of sewage fungus. *Transactions of the British Mycological Society*, **17**, 112–24.

Butijin, G. D. and Greiner, R. W. (1985) Artificial wetlands as tertiary treatment systems. *Water Science and Technology*, **17**, 1429–31.

Byron, M. (1987). Comparison of the microbiology of activated sludge with operating parameters in domestic and industrial treatment units. M.Sc. thesis, Trinity College, University of Dublin, Dublin.

Cabelli, V. J. (1979). Evaluation of recreational water quality, the EPA approach. In *Biological indication of water quality*, (ed. A. James and L. Evison), pp. 14.1–14.23. John Wiley, Chichester.

Cabelli, V. J. (1980). *Health effects criteria for marine recreational waters*. EPA-600 /1-80-031. Health Effects Research Laboratory, Environmental Protection Agency, Ohio.

Cain, R. B. (1981). Microbial degradation of surfactants and 'builder' components. In *Microbial degradation of xenobiotics and recalcitrant compounds*, pp. 325–66, (ed. T. Leisinger *et al.*). Academic Press, London.

Calaway, W. T. (1963). Nematodes in wastewater treatment. *Journal of the Water Pollution Control Federation*, **35**(5), 1006–16.

Calaway, W. T. (1968). The metazoa of waste treatment processes: rotifers. *Journal of the Water Pollution Control Federation*, **40**, 412–22.

Calaway, W. T. and Lackey, J. B. (1962). *Waste treatment protozoa – flagellata*. University of Florida Engineering Series 3.

Calcutt, T. and Moss, J. (1983). Effective and economic sewage sludge treatment and disposal techniques. *Water and Waste Treatment*, **26**(9), 36–40.

Calcutt, T. and Moss, J. (1984). Sewage sludge treatment and disposal – the way ahead. *Water Pollution Control*, **83**, 163–71.

Calley, A. G., Forster, C. F., and Stafford, D. A. (1977). *Treatment of industrial effluents*. Hodder and Stoughton, London.

Callihan, C. D. and Clemmer, J. E. (1979). Biomass from cellulosic materials. In *Microbial biomass*, (ed. A. H. Rose), pp. 271–88. Academic Press, London.

Cameron, R. A. D., Jackson, N., and Eversham, B. (1983). A field key to the slugs of the British Isles. *Field Studies*, **5**, 807–24.

Canali, R. P., Kehrberger, P. M., Salo, J. E., and Lustig, T. D. (1973). Experimental and mathematical modelling studies of protozoan predation on bacteria. *Biotechnology and Bioengeering* **15**, 707–41.

Canter, L. W. and Knox, R. C. (1985). *Septic tank system effects on groundwater quality.* Lewis, Chelsea, Michigan.

Canter, L. W., Knox, R. C., and Fairchild, D. M. (1987). *Groundwater quality control.* Wiley, Chichester.

Cantino, E. C. (1966). Morphogenesis in aquatic fungi. In *The fungi*, Vol. II, (ed. G. C. Ainsworth and A. S. Sussman), pp. 283–338. Academic Press, London.

Carlson, R. R., Linstept, K. D., Bennett, E. R., and Hartman, R. B. (1982). Rapid infiltration treatment of primary and secondary effluents. *Journal of the Water Pollution Control Federation*, **54**, 270–80.

Carlton-Smith, C. H. and Coker, E. G. (1982). *The manurial value of septic tank sludge.* Report 339-M. Water Research Centre, Stevenage.

Carlucci, A. F. and Pramer, D. (1962). An evaluation of factors affecting the survival of *Escherichia coli* in sea water. III. Antibiotics. *Applied Microbiology*, **8**, 251–4.

Carrington, E. G. (1978). *The contribution of sewage sludges to the dissemination of pathogenic micro-organisms in the environment.* Technical Report TR71. Water Research Centre, Stevenage.

Carrington, E. G. (1979). *The distribution of antibiotic-resistant coliform bacteria in sewage and the effects of sewage treatment on their numbers.* Technical Report TR117. Water Research Centre, Stevenage.

Carrington, E. G. (1980a). *The fate of pathogenic micro-organisms during wastewater treatment and disposal.* Technical Report TR128. Water Research Centre, Stevenage.

Carrington, E. G. (1980b). *The isolation and identification of* Salmonella *spp. in sewage sludges: a comparison of methods and recommendations for a standard technique.* Technical Report TR129. Water Research Centre, Stevenage.

Carter, J. L. and McKinney, R. E. (1973). Effects of iron on activated sludge treatment. *Journal of the Sanitary Engineering Division, American Society of Civil Engineers*, **99**, 135–52.

Carter, K. B. (1984). 30/30 hindsight. *Journal of the Water Pollution Control Federation*, **56**, 302–5.

Casey, T. J. (1981). Developments in anaerobic digestion. *Transactions of the Institute of Engineers in Ireland*, **105**, 25–32.

Casey, T. J. and O'Connor, P. E. (1980). Sludge processing and disposal. In *Today's and tomorrow's wastes*, (ed. J. Ryan), pp. 67–80. National Board for Science and Technology, Dublin.

Caspers, H. (1976). Ecological effects of sewage sludge on benthic fauna off the German North Sea Coast. *Progress in Water Technology*, **9**, 951–6.

Cawley, W. A. (1958). An effect of biological imbalance in streams. *Sewage Industrial Wastes*, **30**, 1174–82.

Cebula, J. and Kutera, J. (1978). Land treatment systems in Poland. In *State of knowledge of land treatment of wastewater*, (ed. H. L. McKim), pp. 257–64. US Army Corps of Engineers, CRREL.

Cech, J. S., Chudoba, J., and Grau, P. (1985). Determination of kinetic constants of activated sludge micro-organisms. *Water Science and Technology*, **17** (2/3), 259–72.

Chabrzyk, G. and Coulson, J. C. (1976). Survival and recruitment in the Herring gull *Larus argentatus*. *Journal Animal Ecology*, **45**, 187–203.

Chakrabarty, A. M. (1980). The biodegradation of PCB and chlorobenzoates. In *Plasmids and transporons: environmental effects and maintenance mechanisms*, (ed. C. Stuttard and K. R. Rozee), pp. 21–30. Academic Press, London.

Chalk, E. and Wheale, G. (1989). The root-zone process at Holtby sewage treatment works. *Journal of the Institution of Water and Environmental Management*, **3**, 201–7.

Chambers, B. (1982). Effect of longitudinal mixing and anoxic zones on settleability of activated sludge. In *Bulking of activated sludge*, (ed. B. Chambers and E. J. Tomlinson), pp. 166–86. Ellis-Horwood, Chichester.

Chambers, B. and Jones, G. L. (1988). Optimisation and uprating of activated sludge plants by efficient process design. *Water Science and Technology*, **20** (4/5), 121–32.

Chambers, B. and Tomlinson, E. J. (1981). *Control strategies for bulking sludge: operator's manual*. Proceedings of the Regional Meeting of the Water Research Centre, Stevenage.

Chambers, B. and Tomlinson, E. J. (1982*a*). The cost of chemical treatment to control the bulking of activated sludge. In *Bulking of activated sludge*, (ed. B. Chambers and E. J. Tomlinson), pp. 264–70. Ellis-Horwood, Chichester.

Chambers, B. and Tomlinson, E. J. (1982*b*). *Bulking of activated sludge: preventive and remedial methods*. Ellis-Horwood, Chichester.

Chan, D. B. and Pearson, E. A. (1970). *Comparative studies of solid wastes management – hydrolysis rate of cellulose in anaerobic fermantal*. SERL Report 70–3. University of California, Berkeley.

Chan, E., Bursztynsky, T. A., Hantzsche, N., and Litwin, T. J. (1982). *The use of wetlands for water pollution control*. Municipal Environmental Research Laboratory, US EPA 600/2–82–086, Environmental Protection Agency, Washington.

Chatterjee, S. N., Arora, B. K., and Gupta, D. R. (1967). Some observations on the utilization of sewage for fish culture in oxidation ponds. *Environmental Health (India)*, **9**, 156–61.

Chaudhuri, N., Engelbrecht, R. S., and Austin, J. H. (1965). Nematodes in an aerobic waste treatment plant. *Journal of the American Water Works Association*, **57**, 1561.

Chen, C. W. and Orlob, G. T. (1972). The accumulation and significance of sludge near San Diego outfall. *Journal of the Water Pollution Control Federation*, **44**, 1362–71.

Chen, Y. R. and Hashimoto, A. G. (1978). Kinetics of methane fermentation. *Biotechnology and Bioengineering* Symposium, **8**, 269–82.

Chen, Y. R. and Hashimoto, A. G. (1980). Substrate utilization kinetic model for biological treatment processes. *Biotechology and Bioengineering* **22**, 2081–95.

Chick, A. J., Finlayson, C. M., and Swarbrick, J. T. (1983). Typhaceae – the cumbungis or bulrushes. *Australian Weeds*, **23**(2), 24–7.

Chin, K. K. and Goh, T. N. (1978). Bioconversion of solar energy: methane production through water hyacinth. In *Clean fuels from biomass and wastes*, pp. 215–28. Institute of Gas Technology, Washington, D.C.

Chipperfield, P. N. J. (1967). Performance of plastic filter media in industrial and domestic waste treatment. *Journal of the Water Pollution Control Federation*, **39**, 1860–74.

Chipperfield, P. N. J. (1968). The development, use and future of plastics in biological treatment. In *Effluent and water treatment manual*. Thunderbird Enterprises, London.

Christensen, D. R. and McCarty, P. L. (1975). Multi-process biological treatment model. *Journal of the Water Pollution Control Federation*, **47**, 2652–64.

Christopher, J., Franklin, H., Bagdasarian, M., and Timmis, K. N. (1981). Manipulation of degradative genes of soil bacteria. In *Microbial degradation of xenobiotics and recalcitrant compounds*, (ed. J. Leisinger, R. Hulter, A. M. Cook, and J. Nuesch), pp. 109–30. Academic Press, London.

Chudoba, J., Ottova, V., and Madera, V. (1973). Control of activated sludge filamentous bulking. I. Effect of the hydraulic regime or degree of mixing in an aeration tank. *Water Research*, 7, 1163–82.

Chumbley, C. G. (1971). *Permissible levels of toxic metals used on agricultural land.* Agricultural Development and Advisory Service Advisory, Paper 10. Ministry of Agriculture, Fisheries and Food, London.

Church, B. D., Erikson, E. E., and Widmer, C. M. (1973). Fungal digestion of food processing wastes. *Food Technology*, 2, 36–42.

Chynoweth, D. P., Ghosh, S., and Henry, M. P. (1981). Biogasification of blends of water hyacinths and domestic sludge. In *Proceedings of the 1981 International Gas Research Conference*, pp. 742–55. Los Angeles.

Chynoweth, D. P. *et al.* (1982). Kinetics of advanced digester design for anaerobic digestion of water hyacinth and primary sludge. *Biotechnology and Bioengineering* Symposium 12, 381–98. Wiley, London.

Cillie, G. G., Henzen, M. R., Stander, G. J., and Baillie, R. D. (1969). Anaerobic digestion. IV. The application of the process in waste purification. *Water Research*, 3, 623–43.

Clapham, G. S. (1980). Sewage development of by-products. In Proceedings of a symposium on energy use and conservation in the water industry, pp. 4.1–4.19. Institution of Water Engineers and Scientists, London.

Clark, C. S. (1983). Health effects associated with wastewater treatment and disposal. *Journal of the Water Pollution Control Federation*, 55, 679–82.

Clark, C. S. (1984). Health effects associated with wastewater treatment and disposal. *Journal of the Water Pollution Control Federation*, 56, 625–7.

Clark, C. S. (1985). Health effects associated with wastewater treatment and disposal. *Journal of the Water Pollution Control Federation*, 57, 566–9.

Clark, C. S. (1987). Potential and actual biological related health risks of wastewater industry employment. *Journal of the Water Pollution Control Federation*, 59, 999–1008.

Clark, C. S., Bjornson, H. S., Linnemann, C. C., and Gartside, P. S. (1984a). Evaluation of health risks associated with wastewater treatment and sludge composting. Part 1. Report PB85–115639. US National Technical Information Series, Springfield, Virginia.

Clark, C. S., Bjornson, H. S., Linnemann, C. C., and Gartside, P. S. (1984b). Evaluation of health risks associated with wastewater treatment and sludge composting. Part 1. Report PB85–115639. US Nat. Tech. Infor. Ser. Springfield, Virginia.

Clark, C. S., Bjornson, H. S., Schwartz-Fulton, J., Holland, J. W., and Gartside, P. S. (1984c). Biological health risks associated with the composting of wastewater treatment plant sludge. *Journal of the Water Pollution Control Federation*, 56, 1269–76.

Clark, J. W., Viessman, W., and Hammer, M. J. (1977). *Water supply and pollution control.* Harper and Row, New York.

Clay, E. (1964). *The fauna and flora of sewage processes: 2. Species list.* ICI Ltd, Paints Division. Research Department. Memorandum PVM 45//732.

Clements, M. S. (1966). Velocity variations in rectangular sedimentation tanks. *Proceedings of the Institute of Civil Engineers*, **34**, 171–200.

Clements, M. S. and Price, G. A. (1972). A two-float technique for examination of flow characteristics of sedimentation tanks. *Journal of Institution of Municipal Engineers*, **99**(2), 53–8.

Coackley, P. (1985). Activated sludge – fundamental process properties and the bulking phenomenon. In *Topics in wastewater treatment*, (ed. J. M. Sidwick), pp. 1–24. Critical Report in Applied Chemistry, **11**. Blackwell, Oxford.

Coackley, P. and O'Neill, J. (1975). Sludge activity and full-scale plant control. *Water Pollution Control*, **74**, 404–12.

Coe, R. L., Freeman, P., and Mattingley, P. E. (1950). *Diptera: Nematorea*. Identification handbook of British Insects, **9**(2). British Entomological Society.

Coetzee, O. J. and Fourie, N. A. (1965). The efficiency of conventional sewage purification works, stabilization ponds and maturation ponds with respect to the survival of pathogenic bacteria and indicator organisms. *Journal and Proceedings of the Institute Sewage Purification* 3, 210–5.

Coker, E. G. (1982). *The use of sewage sludge in agriculture*. Paper presented to IAWPR Conference, South Africa.

Coker, E. G., Davis, R. D., Hall, J. E., and Carlton-Smith, C. H. (1982). *Field experiments on the use of consolidated sewage sludge for land reclamation: effects on crop yield and composition and soil conditions*. Technical Report TR183. Water Research Centre, Stevenage.

Cole, C. A., Stamberg, J. B., and Bishop, D. F. (1973). Hydrogen peroxide cures filamentous growth in activated sludge. *Journal of the Water Pollution Control Federation*, **45**, 829–36.

Collinge, V. K. and Bruce, A. M. (1981). *Sewage sludge disposal: a strategic review and assessment of research needs*. Technical Report: TR166. Water Research Centre, Stevenage.

Collins, O. C. and Elder, M. D. (1980). Experience in operating the Deep Shaft activated sludge process. *Water Pollution Control*, **79**, 272–85.

Collins, O. C. and Elder, M. D. (1982). Experience in operating the Deep Shaft activated sludge process. *Public Health Engineering*, **10**, 153–8.

Committee for Analytical Quality Control (Harmonised Monitoring) (1984). *The accuracy of determination of biochemical oxygen demand in river waters*. Technical Report TR218. Water Research Centre, Stevenage.

Congressional Office of Technology Assessment (1981). *The impacts of genetics: applications to micro-organisms, animals and plants*. Washington, D.C.

Cook, E. E. and Herning, L. P. (1978). Shock load attenuation trickling filter. *Journal of the Environmental Engineering Division, American Society of Civil Engineers*, **104**(EE3), 461–9.

Cook, E. E. and Katzberger, S. M. (1977). Effect of residence time on fixed film reactor performance. *Journal of the Water Pollution Control Federation*, **49**, 1889–95.

Cooke, A. M., Grossenbacher, H., Hogrefe, W., and Hutter, R. (1983). Biodegradation of xenobiotic industrial wastes: applied microbiology complementing chemical treatments. In *Biotech 83*. Proceedings of the International Conference on the Commercial Applications and Implications of Biotechnology. Online Publications, Northwood.

Cooke, J. (1983). Biotechnology in waste treatment. *Water and Waste Treatment Journal*, **26**(1), 24–9.

Cooke, W. B. (1954). Fungi in polluted water and sewage. II. Isolation technique. *Sewage and Industrial Wastes*, **26**, 661–74.

Cooke, W. B. (1958). *Fungi in polluted water and sewage. IV. The occurrence of fungi in a trickling filter-type sewage treatment plant*, pp. 26–45. Proceedings of the 13th Industries Waste Conference Purdue University.

Cooke, W. B. (1959). Trickling filter ecology. *Ecology*, **40**, 273–91.

Cooke, W. B. (1963). *A laboratory guide to fungi in polluted waters, sewage and sewage treatment works*. US Department of Health, Education and Welfare. Public Health Service Publication No. 999–WP–1.

Cooke, W. B. (1987). On the isolation of fungi from environmental samples. *Environmental Technology Letters*, **8**, 133–40.

Cooke, W. B. and Hirsche, A. (1958). Continuous sampling of trickling filter populations. *Sewage and Industrial Wastes*, **30**, 138–55.

Cooke, W. B. and Ludzack, F. J. (1958). Predacious fungus in activated sludge systems. *Sewage and Industrial Wastes*, **30**, 1490–5.

Cooke, W. B. and Pipes, W. O. (1968). *The occurrence of fungi in activated sludge*, pp. 170–82. Proceedings of the 23rd Industrial Waste Conference, Purdue University.

Cooke, W. B. and Pipes, W. O. (1970). The occurrence of fungi in activated sludge. *Mycopathologia et Mycologia Applicata*, **40**, 249–70.

Cooney, C. L., Wang, D. J. C., Wang, S., Gordon, J., and Timinez, M. (1979). Simultaneous cellulose hydrolysis and ethanol production by a cellulolytic anaerobic bacterium. In *Biotechnology in energy production and conservation*, (ed. C. D. Scott). John Wiley, Chichester.

Cooper, P. F. and Atkinson, B. (1981). *Biological fluidised bed treatment of water and wastewater*. Ellis-Horwood, Chichester.

Cooper, P. F. and Wheeldon, D. H. V. (1980). Fluidised and expanded bed reactors for wastewater treatment. *Water Pollution Control*, **79**, 286–306.

Cooper, P. F. and Wheeldon, D. H. V. (1982). Complete treatment of sewage in a two-stage fluidised-bed system. *Water Pollution Control*, **81**, 447–64.

Cooper, P. F., Drew, E. A., Bailey, D. A., and Thomas, E. V. (1977). Recent advances in sewage effluent denitrification. *Water Pollution Control*, **76**, 287–300.

Cork, D. J. and Cusanovich, M. A. (1978). Sulfate decomposition: a microbiological process. In *Metallurgical applications of bacterial leaching and related microbiological phenomena*, (ed. L. E. Murr, A. E. Torma, and J.A. Brierly), pp. 207–21. Academic Press, London.

Cornwell, D. A., Zoltek, J., Patrinely, C. D., Furman, T. de S., and Kim, J. I. (1977). Nutrient removal by water hyacinths. *Journal of the Water Pollution Control Federation*, **49**(1). 57–65.

Coulson, J. C., Butterfield, J., and Thomas, C. (1983). The herring gull *Larus argentatus* as a likely transmitting agent of *Salmonella motevideo* to sheep and cattle. *Journal of Hygiene, Cambridge*, **91**, 437–43.

Coultas, C. L. and Calhoun, F. G. (1975). A toposequence of soils in and adjoining a cypress dome in North Florida. *Soil and Crop Society of Florida Proceedings*, **35**, 186–92.

Cox, A. P. (1977). The petrochemicals and resins industry. In *Treatment of industrial effluents*, (ed. A. G. Calley, C. F. Forster, and D. A. Stafford), pp. 218–28. Hodder and Stoughton, London.

Cox, G. C. (1971). A study of the heat economy of a sewage treatment works operated in conjunction with a thermal power station. M.Sc. Thesis, Imperial College of Science and Technology, London.

Cox, G. C. *et al.* (1980). Use of deep-shaft process in uprating and extending existing sewage-treatment works. *Water Pollution Control*, **79**, 70–86.

Crabtree, K. and McCoy, E. (1967). *Zoogloea ramigera* (Itzigsohn), identification and description. *International Journal of Systematic Bacteriology*, **17**, 1–10.

Craft, T. F. and Ingols, R. S. (1973). Flow through time in trickling filters. *Water and Sewage Works*, **120**(1), 78–9.

Craft, T. F., Eichholz, G. G., and Millspaugh, S. (1972). *Evaluation of treatment plants by tracer methods.* Report ORO–4156–1, USAEC, Washington, D.C.

Cram, E. B. (1943). The effect of various treatment processes on the survival of helminth ova and protozoal cysts in sewage. *Sewage Works Journal*, **15**, 115–38.

Craun, G. F. (1979). Water-borne outbreaks of Giardiasis. EPA–600–9–79–001. US Environmental Protection Agency, Washington.

Crewe, S. M. (1967). Worm eggs found in gull droppings. *Annals of Tropical Medicine and Parasitology*, **61**, 358.

Crewe, W. and Owen, R. (1978). 750,000 eggs a day – £750,000 a year. *New Scientist*, **80**, 344–6.

Critchley, R. F., Davis, R. D., and Fox, J. E. (1982). *Applications methods for utilizing sewage sludge on agricultural land.* Technical Report 317–M. Water Research Centre, Stevenage.

Croll, B. T., Greene, L. A., Hall, T., Whitford, C. J., and Zabel, T. F. (1985). Biological fluidised bed denitrification for potable water. In *Advances in water engineering*, (ed. T. H. Y. Tebbutt), pp. 180–7. Elsevier, London.

Crowther, R. F. and Harkness, N. (1975). Anaerobic bacteria. In *Ecological aspects of used-water treatment*, Vol. 1. *The organisms and their ecology*, (ed. C. R. Curds and H. A. Hawkes), pp. 65–91. Academic Press, London.

Crueger, W. and Crueger, A. (1984). *Biotechnology: a textbook of industrial microbiology.* Science Technology Inc.

Culley, D. D. and Epps, E. A. (1973). The use of duckweed for waste treatment and animal feed. *Journal of the Water Pollution Control Federation*, **45**, 337–47.

Culp, R. L., Wesner, G. M., and Culp, G. L. (1978). *Handbook of advanced wastewater treatment*, (2nd edn). Van Nostrand Reinhold, New York.

Curds, C. R. (1966). An ecological study of the ciliated protozoa in activated sludge. *Oikos*, **15**, 282–9.

Curds, C. R. (1969). *An illustrated key to the British freshwater ciliated protozoa commonly found in activated sludge.* Water Pollution Research Technical Paper No. 12. Water Pollution Research Laboratory, Ministry of Technology. HMSO, London.

Curds, C. R. (1971*a*). A computer-simulation study of predator-prey relationships in a single stage continuous-culture system. *Water Research*, **5**, 793–812.

Curds, C. R. (1971*b*). Computer simulations of microbial population dynamics in the activated sludge process. *Water Research*, **5**, 1049–66.

Curds, C. R. (1973*a*). A theoretical study of factors influencing the microbial population dynamics of the activated sludge process. I. The effect of diurnal variations of sewage and carnivorous ciliated protozoa. *Water Research*, **7**, 1269–84.

Curds, C. R. (1973*b*). A theoretical study of factors influencing the microbial population dynamics of the activated sludge process. II. A computer simulation study to compare two methods of plant operation. *Water Research*, **7**, 1439–52.

Curds, C. R. (1973c). The role of protoza in the activated sludge process. *American Zoology*, **13**, 161–9.

Curds, C. R. (1975). Protozoa. In *Ecological aspects of used water treatment*, Vol. I. *The organisms and their ecology*, (ed. C. R. Curds and H. A. Hawkes), pp. 203–68. Academic Press, London.

Curds, C. R. (1982). British and other freshwater ciliated protozoa: Part I. *Synopses of the British Fauna*, **22**. Linnean Society of London, Cambridge University Press, Cambridge.

Curds, C. R. and Cockburn, A. (1970a). Protozoa in biological sewage-treatment processes: I. A survey of the protozoan fauna of British percolating filters and activated sludge plants. *Water Research*, **4**, 225–36.

Curds, C. R. and Cockburn, A. (1970b). Protozoa in biological sewage-treatment processes. II. Protozoa as indicators in the activated sludge process. *Water Research*, **4**, 237–49.

Curds, C. R. and Cockburn, A. (1971). Continuous monoeric culture of *Tetrahymena pyriformis*. *Journal of General Microbiology*, **66**, 95–108.

Curds, C. R., Cockburn, A., and Van Dyke, J. M. (1968). An experimental study of the role of ciliated protozoa in the activated sludge process. *Water Pollution Control*, **67**, 312–29.

Curds, C. R. and Fey, G. J. (1969). The effect of ciliated protozoa on the fate of *Escherichia coli* in the activated sludge process. *Water Research*, **3**, 853–67.

Curds, C. R., Gates, M. A., and McL. Roberts, D. (1983). British and other freshwater ciliated protozoa: Part II. *Synopses of the British Fauna*, **23**, Linnean Society of London, Cambridge University Press, Cambridge.

Curds, C. R. and Hawkes, H. A. (ed.) (1975). *Ecological aspects of used water treatment*, Vol. I. *The organisms and their ecology*. Academic Press, London.

Curds, C. R. and Vandyke, J. M. (1966). The feeding habits and growth rates of some freshwater ciliates found in activated sludge plants. *Journal of Applied Ecology*, **3**, 127–37.

Curtis, E. J. C. (1969). Sewage fungus: its nature and effects. *Water Research*, **3**, 289–311.

Curtis, E. J. C. (1972). Sewage fungus in rivers in the United Kingdom. *Water Pollution Control*, **71**, 673–83.

Curtis, E. J. C., Delvis-Broughton, J., and Harrington, D. W. (1971). Sewage fungus: studies of *Sphaerotilus* slime using laboratory recirculating channels. *Water Research*, **5**, 267–79.

Curtis, E. J. C. and Harrington, D. W. (1971). The occurrence of sewage fungus in rivers in the United Kingdom. *Water Research*, **5**, 281–90.

Cutler, D. W., Crump, L. M., and Dixon, A. (1932). Some factors influencing the distribution of certain protozoa in biological filters. *Journal of Animal Ecology*, **1**, 141–51.

Dabney, H. L. (1971). *Imhoff tanks*. In *Manual of wastewater operations*, (ed. C. H. Billings and D. F. Smallhurst), pp. 174–89, Texas State Department of Health, Austin, Texas.

Dagley, S. (1981). New perspectives in aromatic catabolism. In *Microbial degradation of xenobiotics and recalcitrant compounds*, (ed. J. Leisinger, R. Hulter, A. M. Cook, and J. Nuesch), pp. 181–8. Academic Press, London.

Dappert, A. F. (1932). Tracing the travel and changes in composition of underground pollution. *Water Works Sewerage*, **76**, 265.

Darcy, H. (1856). *Les Fontaines publiques de la ville de Dijon; distribution d'eau et filtrage des eaux.* Victor Dalmont, Paris.

Dart, R. K. and Stretton, R. J. (1977). *Microbial aspects of pollution control.* Elsevier, Oxford.

Datta, N. (1969). Drug resistance and R-factors in the bowel of London patients before and after admission to hospital. *British Medical Journal*, **2**, 407.

Davis, E. M. and Gloyna, E. F. (1972). Bacterial die off in ponds. *Journal of the Sanitary Engineering Division, American Society of Civil Engineers*, **98**, 59–69.

Davis, J. A. and Unz, R. F. (1973). The microbiology of an activated-sludge wastewater treatment plant chemically treated for phosphorus removal. *Water Research*, **7**, 325–7.

Davis, J. A. and Unz, R. F. (1975). Microbiology and combined chemical-biological treatment. *Journal of the Water Pollution Control Federation*, **47**, 185–94.

Davis, R. D. (1980). *Control of contamination problems in the treatment and disposal of sewage sludge.* Technical Report TR156. Water Research Centre, Stevenage.

Davis, R. D. (1982). Land reclamation – a major use of sewage sludge? *Water Bulletin*, **4**, 5–7.

Davis, R. D. (1987). Use of sewage sludge on land in the United Kingdom. *Water Science and Technology*, **19**(2), 1–8.

Davis, R. D. and Beckett, P. H. T. (1978). The use of young plants to detect metal accumulations in soils. *Water Pollution Control*, **77**, 193–210.

Davis, R. D. and Coker, E. G. (1980). *Cadmium in agriculture with special reference to the utilization of sewage sludge on land.* Technical Report TR139. Water Research Centre, Stevenage.

Davis, R. D. and Carlton-Smith, C. H. (1984). An investigation into the phytotoxicity of zinc, copper and nickel using sewage sludge of controlled metal content. *Environmental Pollution Series B.* **8**, 163–85.

Davis, R. D., Carlton-Smith, C. H., Johnson, D., and Stark, J. H. (1985). Evaluation of the effects of metals in sewage sludge disposal. *Water Pollution Control*, **84**, 380–93.

Davis, R. M. (1979). Biodegradation of high level oil-in-water emulsions. In *Biotechnology in energy production and conservation*, (ed. C. D. Scott). John Wiley, Chichester.

Davy, C. A. E. (1981). Recovery of fruit and vegetable waste. In *Food industry wastes: disposal and recovery*, (ed. A. Herzka and R. G. Booth), pp. 219–29. Applied Science Publishers, London.

Dawson, R. and Riley, J. P. (1977). Chlorine-containing pesticides and polychlorinated biphenyls in British coastal water. *Estuarine, Coastal and Marine Science* **5**, 55–69.

Dean, R. B. (1978). The sewage farms of Paris. In *State of knowledge of land treatment of wastewater*, (ed. H. L. McKim), pp. 241–51. US Army Corps of Engineers, CRREL.

Dean, R. B. and Lund, E. (1981). *Water reuse: problems and solutions.* Academic Press, London.

de Bertoldi, M., Vallini, G., Pera, A., and Zucconi, F. (1982). Comparison of three windrow compost systems. *Biocycle*, **23**, 45–50.

de Bertoldi, M., Vallini, G., and Pera, A. (1983). The biology of composting: a review. *Waste Management and Research*, **1**, 157–76.

de Bertoldi, M., Vallini, G., Pera, A., and Zucconi, F. (1985). Technological aspects of composting including modelling and microbiology. In *Composting of agricultural and other wastes*, (ed. J. K. R. Gasser), pp. 27–41. Elsevier, London.

de Bertoldi, M., Ferranti, M., Hermite, P. L., and Zucconi, F. (1988). *Compost production, quality and use.* Elsevier, London.

De Boer, J. and Linstedt, K. D. (1985). Advances in water reuse applications. *Water Research,* **19**, 1455–61.

Deering, N. and Gray, N. F. (1986). The polluter pays principle – who is really paying? *Effluent and Water Treatment,* **26**, 5/6, 138–42.

Deering, N. and Gray, N. F. (1987). The polluter pays pinciple: a comparison of charging systems in Europe and the USA. *Technical Report* **3.** Water Technology Research, Dublin.

Deinema, M. H. and Zevenhuizen, L. P. T. M. (1971). Formation of cellulose fibrils by Gram-negative bacteria. *Archives Microbiology,* **78**, 42–57.

De Jong, J. (1976). The purification of wastewater with the aid of rush or reed ponds. In *Biological control of water pollution,* (ed. J. Tourbier, R. W. Pierson, and E. W. Furia), pp. 133–9. University of Pennsylvania Press.

Delaine, J. (1981). The economics of effluent treatment (aerobic versus anaerobic). In *Food industry wastes: disposal and recovery,* (ed. A. Herzka and R. G. Booth), pp. 74–84. Applied Science Publishers, London.

Demuynck, M. G., Naveau, H. P., and Nyns, E. J. (1983). Anaerobic fermentation technology in Europe. In *Environmental biotechnology: future prospects,* (ed. J. M. Sidwick). FAST Report FOB51, CEC, Brussels.

Dennis, N. D. and Jennett, J. C. (1974). *Pharmaceutical waste treatment with an anaerobic filter,* pp. 36–43. Proceedings of the 29th Industrial Waste Conference, Purdue University.

DoE (Department of the Environment) (1971). *Inhibition in the anaerobic digestion process for sewage sludge.* Notes On Water Pollution, 53, Water Pollution Research Laboratory, Stevenage.

DoE (Department of the Environment) (1972). *Analysis of raw, potable and wastewaters.* HMSO, London.

DoE (Department of the Environment) (1973). *Treatment of secondary sewage effluent in lagoons.* Notes on Water Pollution, 63. Water Pollution Research Laboratory, Stevenage.

DoE (Department of the Environment) (1974). *Anaerobic treatment processes and methane production.* Notes on Water Pollution, 64. Water Pollution Research Laboratory, Stevenage.

DoE (Department of the Environment) (1975). *Sewage sludge dewatering by filter belt press.* Project Report, 4. Water Engineering Research and Development Division, Department of the Environment. HMSO, London.

DoE (Department of the Environment) (1976). *Pollution control in Britain and how it works.* Central Unit on Environmental Pollution, Department of the Environment. HMSO, London.

DoE (Department of the Environment) (1978). *Sewage sludge disposal data and review of disposal to sea.* Department of the Environment/National Water Council Standing Committee Report, 8. HMSO, London.

DoE (Department of the Environment) (1980). Methods for the examination of waters and associated materials: Dissolved oxygen in natural and wastewaters 1979, by two methods. HMSO, London.

DoE (Department of the Environment) (1983). Methods for the examination of waters and associated materials: Biochemical Oxygen Demand 1981. HMSO, London.

DoE (Department of the Environment) (1984a). *Mersey clean up initiative: the third stage*. Central Office of Information, NW60. HMSO, London

DoE (Department of the Environment) (1984b). *Digest of environmental protection and water statistics*. Government Statistical Service, Department of the Environment. HMSO, London.

DoE (Department of the Environment) (1987). *Digest of environmental protection and water statistics*, No. 9 (1986). HMSO, London.

Department of Scientific and industrial Research (1956). *Water Pollution Research, 1956*, pp. 53–4. HMSO, London.

Department of Scientific and Industrial Research (1963a). *Water Pollution Research, 1962*. HMSO, London.

Department of Scientific and Industrial Research (1963b). Notes on Water Pollution, 23.

Deppe, K. and Engel, H. (1960). Untersuchungen uber die temperaturabhangigkeit der nitritbildung durch *Nitrobacter winogradskyi* buch. bei ungehemmtem und gehemmtem wachstum. *Zentralblatt fuer Bakteriologie, Parasitenkunde*, **113**, 561–8.

de Renzo, D. J. (ed.) (1978). *Nitrogen control and phosphorus removal in sewage treatment*, based on U.S. E.P.A. reports, *Process design manual for nitrogen control*, EPA/625/1–75/007, NTIS PB–259149, and *Process design manual for phosphorus removal*, EPA/625/1–76/001a, NTIS PB-259150; Pollution Technology Review No. 44. Noyes Data Corporation, Park Ridge, New Jersey.

Deufel, J. (1972). die Bakterien – und keimzahlen im oberlauf der Donau bis Ulm. *Archiv fuer Hydrobiologie*, **44**, 1–9.

Dhaliwal, B. S. (1979). *Nocardia amarae* and activated sludge foaming. *Journal of the Water Pollution Control Federation*, **57**, 344–50.

Dias, F. F. and Bhat, J. V. (1964). Microbial ecology of activated sludge. *Applied Microbiology*, **12**, 412–7.

Dias, F. F. and Dondero, N. C. (1967). *Calcium nutrition of* Sphaerotilus. Bacteriological Proceedings, G–150.

Dias, F. F. and Heukelekian, H. (1967). Utilization of inorganic nitrogen compounds by *Sphaerotilus natans* growing in a continuous-flow apparatus. *Applied Microbiology*, **15**, 1083–6.

Dias, F. F., Dondero, N. C., and Finstein, M. S. (1968). Attached growth of *Sphaerotilus* and mixed populations in a continuous-flow apparatus. *Applied Microbiology*, **16**, 1191–9.

Dick, R. I. (1967). Evaluation of sludge thickening theories. *Proceedings of the American Society of Civil Engineers*, **93**, 9–92.

Dick, R. I. (1976). Folklore in the design of final settling tanks. *Journal of the Water Pollution Control Federation*, **48**(4), 633–44.

Dick, R. I. and Ewing, B. B. (1967). Evaluation of activated sludge thickening theories. *Journal of the Sanitary Engineering Division, American Society of Civil Engineers*, **93**, 29.

Dick, R. I. and Vesilind, P. A. (1969). Sludge volume index – what is it? *Journal of the Water Pollution Control Federation*, **41**, 1285–91.

Dickinson, D. (1974). *Practical waste treatment and disposal*. Applied Science Publishers, London.

Dickson, L. (1983). Resazurin reduction method for activated sludge process control. *Environmental Science Technology*, **17**, 407–11.

Diehl, J. C. and Tromp, W. W. (1953). *First report on the geographic and geological distribution of carcinoma in the Netherlands.* Stichtung fer Bevordering van de Psychische Physica, Vol. 1, Leiden.

Dierberg, F. E. and Brezonik, P. L. (1983). Tertiary treatment of municipal wastewater by cyprus domes. *Water Research*, **17**, 1027–40.

Diez, T. and Weijelt, H. (1980). Zur durgervirkung von Mullkompost und Klarschlamm. *Landwirtsch Forschung*, **33**, 47–66.

Dinges, R. (1978). Upgrading stabilization pond effluent by water hyacinth culture. *Journal of the Water Pollution Control Federation*, **50**(5), 833–45.

Dinges, R. (1982). *Natural systems for water pollution control.* Van Nostrand Reinhold, New York.

Dingles, R. and Doersam, J. (1987). The Hornsby bend hyacinth facility in Austin, Texas. *Water Science and Technology*, **19** (10), 41–499.

Dirasian, H. A. (1968*a*). Electrode potentials – significance in biological systems. 1. Fundamentals of measurement. *Water Sewage Works*, **114**, 420–4.

Dirasian, H. A. (1968*b*). Electrode potentials – significance in biological systems. 2. Experiences in waste treatment. *Water Sewage Works*, **115**, 453–6.

Dodd, V. (1980). Land spreading and associated hazards. In *Today's and tomorrow's wastes*, (ed. J. Ryan), pp. 47–66. National Board for Science and Technology, Dublin.

Dodson, C. E. (1981). Anaerobic treatment of food processing wastes and agricultural effluents. In *Food industry wastes: disposal and recovery*, (ed. A. Herzka and R. G. Booth), pp. 85–91. Applied Science Publishers, London.

Dolan, T. J., Bayley, S. E., Zoltek, J., and Herman, A. J. (1981). Phosphorus dynamics of a Florida freshwater marsh receiving treated wastewater. *Journal of Applied Ecology*, **18**, 205–19.

Doman, J. (1929). Results of operation of experimental contact filter process with partially submerged rotating plates. *Sewage Works Journal*, **1**, 555–60.

Domenowske, R. S. and Matsuda, R. I. (1969). Sludge disposal and the marine environment. *Journal of the Water Pollution Control Federation*, **41**, 1613–24.

Donner, J. (1966). *Rotifers*, (trans. H. G. S. Wright). Frederick Warne, London.

Doohan, M. (1975). Rotifera. In *Ecological aspects of used water treatment*, Vol. 1. *The organisms and their ecology*, (ed. C. R. Curds and H. A. Hawkes), pp. 289–304. Academic Press, London.

Downing, A. L. and Hopwood, A. P. (1964). Some observations on the kinetics of nitrifying activated sludge plants. *Schweizerische Zeitschrift für Hydrologie*, **26**, 271–88.

Downing, A. L. and Wheatland, A. B. (1962). Fundamental considerations in biological treatment. *Transactions of the Institute of Chemical Engineers* 40, 91–103.

Downing, A. L., Painter, H. A., and Knowles, G. (1964). Nitrification in the activated sludge process. *Journal and Proceedings of the Institute of Sewage Purification*, 130–58.

Doxat, J. (1977). *The living Thames. The restoration of a great tidal river.* Hutchinson Benham, London.

Drews, R. J. L. C. (1966). Field studies of large scale maturation ponds with respect to their purification efficiency. *Journal and Proceedings of the Institute of Sewage Purification*, 280–94.

Drift, C. van der, Seggelen, E. van, Stumm, C., Hol, W., and Tuinte, J. (1977). Removal of *Escherichia coli* in wastewater by activated sludge. *Applied Environmental Microbiology*, **34**, 315–9.

Dubinsky, Z., Berner, J., and Aavonson, S. (1979). Potential of large scale algal culture for biomass and lipid production in arid lands. In *Biotechnology in energy production and conservation*, (ed. C. D. Scott), pp. 57–68. John Wiley, London.

Dugan, P. R. and Apel, W. A. (1978). Microbiological desulfurization of coal. In *Metallurigical applications of bacterial leaching and related microbiological phenomen*, (ed. L. E. Murr, A. E. Torma, and J. A. Brierly), pp. 223–49. Academic Press, London.

Dunlop, E. H. (1978). Characteristics of sludge from the deep shaft process. In *New processes of wastewater treatment and recovery*, (ed. G. Mattock), pp. 177–200. Ellis-Horwood, Chichester.

Durant, W. (1954). *Our oriental heritage*. Simon and Schuster, New York.

Dutka, B. J. (1973). Coliforms are an inadequate index of water quality. *Journal of Environmental Health*, **36**, 39–46.

Dutka, B. J. (1979). Microbiological indicators, problems and potential of new microbial indicators of water quality. In *Biological indicators of water quality*, (ed. A. James and L. Evison), pp. 18.1–18.24. John Wiley, Chichester.

Dutka, B. J. and Kwan, K. K. (1977). Confirmation of the single-step membrane filtration procedure for estimating *Psuedomonas aeruginosa* densities in water. *Applied Environmental Microbiology*, **33**, 240–5.

Dutka, B. J., Chan, A. S. Y., and Coburn, J. (1974). Relationship between bacterial indicators of water pollution and faecal sterols. *Water Research*, **8**, 1047–55.

Dyson, J. E. B. and Lloyd, L. (1936). The distribution of the early stages of *Metriocnemus longitarus* Goet, (Chironomidae), in sewage bacteria beds. *Proceedings of the Leeds Philosophical and Literature Society*, **3**, 174–6.

Eagle, R. A., Hardiman, P. A., Norton, M. G., and Nunny, R. S. (1979). *The field assessment of effects of dumping wastes at sea. 4. A survey of the sewage sludge disposal area off Plymouth*. Fisheries Research Technical Report **50**. Ministry of Agriculture, Fisheries and Food, Directorate of Fisheries Research, Lowestoft.

Earle, M. (1986). A preliminary survey of percolating filters in Ireland. M.Sc. Thesis, Trinity College, University of Dublin, Dublin.

Earle, M. and Gray, N. F. (1987) *Check list of Irish wastewater treatment plants: percolating filters*. Technical Report **2**. Water Technology Research, Dublin.

Ebner, H. G. (1978). Metal recovery and environmental protection by bacterial leaching of inorganic waste materials. In *Metallurigical applications of bacterial leaching and related microbiological phenomena*, (ed. L. Murr, A. E. Torma, and J. A. Brierly), pp. 195–206. Academic Press, London.

Eck, H. van and Simpson, D. E. (1966). The anerobic pond system. *Journal and Proceedings of the Institute Sewage Purification*, 251–60.

Eckenfelder, W. W. (1961). Trickling filter design and performance. *Journal of the Sanitary Engineering Division, American Society of Civil Engineers*, **87**(SA4), 33–45.

Eckenfelder, W. W. and Argaman, Y. (1978). Kinetics of nitrogen removal for municipal and industrial applications. In *Advances in water and wastewater treatment, biological nutrient removal*, (ed. M. P. Wanielista and W. W. Eckenfelder), pp. 23–41. Ann Arbor Science.

Eden, G. E. (1964). Biological filtration. *Fluid Handling*, **15**(1), 22–8.

Eden, G. E. and Melbourne, K. V. (1960). Radioactive tracers for measuring the periods of retention in percolating filters. *International Journal of Applied Radiation and Isotopes*, **8**, 172–8.

Eden, G. E. and Truesdale, G. A. (1961). Behaviour of a new synthetic detergent in sewage treatment processes. *Journal and Proceedings of the Institute of Sewage Purification*, 30–42.

Eden, G. E., Truesdale, G. A., and Mann, H. T. (1966). Biological filtration using a plastic filter medium. *Journal and Proceedings of the Institute of Sewage Purification*, **6**, 562–74.

Eden, G. E., Brendish, K., and Harvey, B. R. (1964). Measurement and significance of retention in percolating filters. *Journal and Proceedings of the Institute of Sewage Purification*, **6**, 513–25.

Eden, G. E., Bailey, D. H., and Jones, K. (1977). Water reuse in the United Kingdom. In *Water renovation and reuse*, (ed. H. I.Shuval), pp. 397–428. Academic Press, New York.

Edwards, C. A. (1982). Production of earthworm protein for animal feed from potato waste. In *Upgrading waste for feed and food*, (ed. D. A. Ledmon, A. J. Taylor, and R. A. Lawrie), pp. 153–62. Butterworths, London.

Edwards, C. A. (1983). Earthworms, organic waste and food. *Span*, **26**, 106–8.

Edwards, C. A., Burrows, I., Fletcher, K. E., and Jones, B. A. (1985). The use of earthworms for composting farm wastes. In *Composting of agricultural and other wastes*, (ed. J. K. R. Gasser), pp. 229–42. Elsevier, London.

Edwards, C. A. and Lofty, J. R. (1977). *Biology of earthworms*, (2nd edn). Chapman and Hall, London.

Edwards, R. W. and Owens, M. (1965). The oxygen balance of streams. In *Ecology and the industrial society*, (ed. G. T. Goodman, R. W. Edwards, and J. M. Lambert), pp. 149–72. Blackwell, Oxford.

Edworthy, K. J., Wilkinson, W. B., and Young, C. P. (1978). The effect of the disposal of effluents and sewage sludge in the chalk of the United Kingdom. *Progress in Water Technology*, **10**, 479–93.

Eikelboom, D. H. (1975). Filamentous organisms observed in activated sludge. *Water Research*, **9**, 365–88.

Eikelboom, D. H. (1977). Identification of filamentous organisms in bulking activated sludge. *Progress in Water Technology*, **8**, 153–62.

Eikelboom, D. H. (1982). Microscopic sludge investigation in relation to treatment plant operation. In *Bulking of activated sludge*, (ed. B. Chambers and E. J. Tomlinson), pp. 47–62. Ellis-Horwood, Chichester.

Eikelboom, D. H. and Van Buijsen, H. J. J. (1981). *Microscopic sludge investigation manual*. TNO Research Institute for Environmental Hygiene, Delft.

Eilbeck, W. J. and Mattock, G. (1987). *Chemical processes in wastewater treatment*. Ellis-Horwood, Chichester.

Egglink, H. J. (1975). Experience in centrifugation of sewage sludge – aspects of performance and reliability. *Progress in Water Technology*, **7**, 947–58.

Ekama, G. A. and Marais, G. R. (1979). Dynamic behaviour of the activated sludge process. *Journal of the Water Pollution Control Federation*, **51**, 1009–16 or 534–56.

Ekama, G. A., Marais, G. R. and Blackbeard, J. R. (1985). *Exploratory study on activated sludge bulking and foaming problems in Southern Africa (1983–1984)*. WRC Report No. 114/1/85. Water Research Commission, Pretoria, South Africa.

Eliasson, R. (1967). *Aftergrowth of coliforms in estuarine water receiving chlorinated overflow*. Proceedings of the National Symposium on Estuarine Pollution. American Society of Civil Engineers, Stanford, California.

Ellis, B. G. (1974). *The soil as a biological filter*. EPA 660/2–74–002. US Environmental Protection Agency.

Ellis, K. V. and Banaga, S. E. (1976). A study of rotating disc treatment units operating at different temperatures. *Water Pollution Control*, **75**, 73–91.

Emde, W. (1963). Aspects of the high-rate activated sludge process. In *Advances in biological waste treatment*, (ed. W. W. Eckenfelder and J. McCabe), pp. 299–317. Pergamon, Oxford.

Engelbrecht, R. S. (1978). Microbial hazard associated with the land application of wastewater and sludge. *Public Health Engineers*, **6**, 219–64.

Ensign, J. C. (1977). Biomass production from animal wastes by photosynthetic bacteria. In *Microbial energy conversion*, (ed. H. G. Schlegel and J. Barnea), pp. 455–79. Pergamon, Oxford.

Environmental Protection Agency (1975). *Process design manual for nitrogen control*. US Environmental Protection Agency, Washington, D.C.

Environmental Protection Agency (1976). *Quality criteria for water*. EPA/440/9–76/023. US Environmental Protection Agency, Washington, D.C.

Environmental Protection Agency (1978). Water pollution studies 002, 003, 004. Environmental Monitoring and Support Laboratory, Quality Assurance Branch, Cincinnati, Ohio.

Environmental Protection Agency (1981). *Process design manual for land treatment of municipal wastewater*. EPA/625/1–81/013. US Environmental Protection Agency, Cincinnati, Ohio.

Environmental Protection Agency (1983). *Operational manual for stabilization ponds*. U.S. Environmental Protection Agency, Washington.

Environmental Sanitation Information Center (1982). Septic tank and septic systems. *Environmental Sanitation Review*, **7/8**. Asian Institute of Technology, Bangkok.

Eppley, R. W. *et al.* (1972). Evidence for eutrophication in the sea near Southern California coastal sewage outfalls. *Coli Co-op Oceanic Fish Investigation Report*, **16**, 74.

Escritt, L. B. (1965). *Sewerage and sewage disposal*. C. R. Brooke, London.

Escritt, L. B. and Haworth, W. D. (1984). *Sewerage and sewage treatment*. John Wiley, Chichester.

Etherington, J. R. (1983). Wetland ecology. *Studies in biology*, **154**. Edward Arnold, London.

European Communities (1975). Council directive of 16 June, 1975 concerning the quality required of surface water intended for the abstraction of drinking water in member States. *Official Journal of the European Communities*, L194/26–L194/53 (75/440/EEC).

European Communities (1976a). Council directive of 8 December 1975 concerning the quality of bathing waters. *Official Journal of the European Communities*, **19**, L31/1–L31/7.

European Communities (1976b). Proposal for a Council Directive concerning the dumping of wastes at sea. *Official Journal of the European Communities*, C4/3.

European Communities (1979) Council Directive on the quality required of shellfish waters of 30 October 1979. *Official Journal of the European Communities*, L281 (79/923/EEC).

European Communities (1982). Final draft on agricultural use of sewage sludge-proposal for a Council Directive on the uses of sewage sludge in agriculture. *Official Journal of the European Communities*, C/264, 3–7.

European Communities (1984). Proposal for a Council Directive on the use of sewage sludge in agriculture. *Official Journal of the European Communities*, C/154, 6–15.

European Communities (1985). Proposal for a Council Directive on the dumping of waste at sea. COM (85) 373 final. *Official Journal of the European Communities*, C245.

European Inland Fisheries Advisory Commission (1973). *Water quality criteria for European freshwater fish: report on chlorine and freshwater*. Technical Paper **20**. Food and Agriculture Organization, Rome.

Evans, G. O., Sheals, J. G., and MacFarlane, D. (1961). *The terrestrial Acari of the British Isles*, **1**. British Museum of Natural History, London.

Evans, M. R., Hissett, R., Smith, M. P. W., and Ellam, D. F. (1978). Characteristics of slurry from fattening pigs and comparison with slurry from laying hens. *Agriculture and Environment*, **4**, 77–83.

Evans, M. R., Hissett, R., Smith, M. P. W., Thacker, F. E., and Williams, A. G. (1980). Aerobic treatment of beef cattle and poultry waste compared with the treatment of piggery waste. *Agricultural Wastes*, **2**, 93–101.

Evins, C. (1975). *The toxicity of chlorine to some freshwater organisms*. Technical Report TR8. Water Research Centre, Stevenage.

Evison, L. M. (1979). Microbial parameters of raw water quality. In *Biological indicators of water quality*, (ed. A. James and L. Evison). PP. 16.1–16.9. John Wiley, Chichester.

Evison, L. M. and James, A. (1973). A comparison of the distribution of intestinal bacteria in British and East African water sources. *Journal of Applied Bacteriology*, **36**, 109–18.

Ewel, K. C. (1983). Effects of fire and sewage on understorey vegetation in cypress domes. In *Cypress Swamps*, (ed. K. C. Ewel and H. T. Odum). University Presses of Florida, Gainesville.

Fair, G. M. and Moore, E. W. (1934). Time and rate of digestion and its variation with temperature. *Sewage Works Journal*, **6**, 3–13.

Farrah, S. R. and Unz, R. F. (1976). Isolation of exocellular polymer from *Zooglea* strains MP6 and IO6 and from activated sludge. *Applied Environmental Microbiology*, **32**, 33–43.

Farquhar, G. J. and Boyle, W. C. (1971a). Identification of filamentous micro-organisms in activated sludge. *Journal of the Water Pollution Control Federation*, **43**, 603–20.

Farquhar, G. L. and Boyle, W. C. (1971b). Occurrence of filamentous organisms in activated sludge. *Journal of the Water Pollution Control Federation*, **43**, 779–98.

Fattal, B., Peleg-Olevsky, E., Yoshpe-Purer, Y., and Shuval, H. I. (1986a). The association between morbidity among bathers and microbial quality of seawater. *Water Science and Technology*, **18**(11), 56–69.

Fattal, B., Yekutiel, P., and Shuval, H. I. (1986b). Cholera outbreak in Jerusalem 1970, revisited: the evidence for transmission by wastewater irrigated vegetables. In *Environmental epidemiology: epidemiological investigation of community environmental health problems* (ed. J. R. Goldsmith), pp. 49–59. CRC Press, Boca Raton, Florida.

Fattal, B., Tekutiel, P., Wax, Y., and Shuval, H. I. (1986c). Prospective epidemiological study of health risks associated with wastewater utilization in agriculture. *Water Science and Technology*, **18**(10), 199–209.

Fattal, B., Wax, Y., Davies, M., and Shuval, H. I. (1986d). Health risks associated with wastewater irrigation: and epidemiological study. *American Journal of Public Health*, **76**, 977–79.

Fawell, J. K., Fielding, M., and Ridgway, J. W. (1987). Health risks of chlorination – is there a problem? *Water and Environmental Management*, **1**, 61–6.

Feachem, R. (1977). Infectious disease related to water supply and excreta disposal facilities. *Ambio*, **6**, 55–8.

Fazzalari, F. A. (ed.) (1978). *Compilation of odor and taste threshold values data*, ASTM Data Series DS 48A. American Society for Testing and Materials, Philadelphia.

Feachem, R. and Cairncross, S. (1978). Small excreta disposal systems. *The Ross Institute of Tropical Hygiene*, Bulletin 8.

Fegan, L. (1983). How dirty is Dublin Bay? *Technology Ireland*, November, pp. 30–6.

Feldman, B. M. (1974). The problem of urban dogs. *Science*, **185**, 903.

Fenlon, D. R. (1981). Seagulls (*Larus* sp.) as vectors of Salmonellae: an investigation into the range of serotypes and numbers of salmonellae in gulls' faeces. *Journal of Hygiene, Cambridge*, **86**, 195–202.

Fennell, H. (1975). The reason for disinfection. *Water Officers' Journal*, **11**, 19–26.

Fennell, H., James, D. B., and Morris, J. (1974). Pollution of a storage reservoir by roosting gulls. *Water Treatment and Examination* **23**, 5–24.

Ferraiolo, G. and Del Borghi, M. (1987). Bioleaching. In *Biotechnology of waste treatment and exploitation*, (ed. J. M. Sidwick and R. S. Holdom), pp. 236–55. Ellis-Horwood, Chichester.

Ferrero, G. L., Ferranti, M. P., and Naveau, H. (ed.) (1984). *Anaerobic digestion and carbohydrate hydrolysis of waste*. Elsevier, London.

Fetter, C. W., Sloey, W. E., and Sprangler, F. L. (1978). Use of a natural marsh for wastewater polishing. *Journal of the Water Pollution Control Federation*, **50**, 240–307.

Fewkes, A. and Ferris, F. A. (1982). The recycling of domestic wastewater: factors influencing storage capacity. *Building and Environment* **17**, 209–16.

Fewson, C. E. (1981). Biodegradation of aromatics with industrial relevance. In *Microbial degradation of xenobiotics and recalcitrant compounds*, (ed. J. Leisinger, R. Hutter, A. M. Cook, and J. Nuesch), pp. 141–79. Academic Press, London.

Fieldson, R. S. (1985). The economic feasibility of earthworm culture on animal wastes. In *Composting of agricultural and other wastes*, (ed. J. K. R. Gasser), pp. 243–54. Elsevier, London.

Filip, S. D., Peters, T., Adams, V. D., and Middlebrooks, E. J. (1979). Residual heavy metal removal by an algae–intermittent sand filtration system. *Water Research*, **13**, 305–13.

Finch, J. and Ives, H. (1950). Settleability indexes for activated sludge. *Water Sanitary Engineer*. **1**, 186—9.

Finlayson, C. M. (1981). Aspects of the hydrobiology of the Lake Moondarra–Leichhardt River water supply system, Mount Isa. Ph.D. Thesis, James Cook University, North Queensland.

Finlayson, C. M. and Chick, A. J. (1983). Testing the potential of aquatic plants to treat abbatoir effluent. *Water Research*, **17**, 415–22.

Finlayson, C. M. and Mitchell, D. S. (1982). Treatment of rural wastewaters in Australia with aquatic plants: A summary. *Der Tropenlandwirt*, **83**, 153–65.

Finstein, M. S. and Delwiche, C. C. (1965). Molybdenum as a micronutrient for *Nitrobacter*. *Journal of Bacteriology*, **89**, 123–8.

Finstein, M. S. and Heukelekian, H. (1967). Gross dimensions of activated sludge flocs with reference to bulking. *Journal of the Water Pollution Control Federation*, **39**, 33–40.

Finstein, M. S. and Miller, F. C. (1985). Principles of composting leading to maximization of decomposition rate, odor control, and cost effectiveness. In *Composting of agricultural and other wastes*, (ed. J. K. R. Gasser), pp. 13–26. Elsevier, London.

Finstein, M. S., Miller, F. C., Strom, P. F., MacGregor, S. T., and Psarianos, K. M. (1983). Composting ecosystem management for waste treatment. *Biotechnology*, **1**, 347–53.

Finstein, M. S., Cirello, J., MacGregor, S. T., Miller, F. C., and Psarianos, K. M. (1980). *Sludge composting and utilization: rational approach to process control*. Project No. C–340–678–01–1. Final Report, Cook College, Rutgers University, New Jersey.

Fish, H. (1966). Some investigations of tertiary methods of treatment. *Institute of Public Health Engineers Journal*, **65**, 33–47.

Fish, H. (1983). Sea disposal of sludge – the U.K. experience. *Water Science Technology*, **15**, 77–87.

Fitzmaurice, G. D. (1986). The biochemical oxygen demand (BOD) test – an evaluation of its measurement. M.Sc Thesis, University of Dublin, Trinity College.

Fitzmaurice, G. D. and Gray, N. F. (1987*a*). *Biochemical oxygen demand: interlaboratory precision test*. Technical Report **4**. Water Technology Research, Dublin.

Fitzmaurice, G. D. and Gray, N. F. (1987*b*). *Biochemical oxygen demand: a proposed standard methodology for Irish laboratories*. Technical Report **5**. Water Technology Research, Dublin.

Flegal, T. M. and Shroeder, E. D. (1976). Temperature effects on BOD stoichiometry and oxygen uptake rate. *Journal of the Water Pollution Control Federation*, **48**, 2700–7.

Forage, A. J. and Righelato, R. C. (1979). Biomass from carbohydrates. In *Microbial biomass*, (ed. A. H. Rose), pp. 289–313. Academic Press, London.

Forbes, S. A. and Richardson, R. E. (1913). Studies on the biology of the Upper Illinois River. *Bulletin of the Illinois State Laboratory of Natural History*, **9**, 481–574.

Forster, C. F. (1968). The surface of activated sludge particles in relation to their settling characteristics. *Water Research*, **2**, 767–76.

Forster, C. F. (1971). Activated sludge surfaces in relation to the sludge volume index. *Water Research*, **5**, 861–70.

Forster, C. F. (1976). Bioflocculation in the activated sludge process. *Journal of Applied Chemical and Biotechnology*, **26**, 291.

Forster, C. F. (1977). Bio-oxidation. In *Treatment of industrial effluents* (ed. A. G. Cullely, C. F. Forster, and D. A. Stafford), pp. 65–87. Hodder and Stoughton, London.

Forster, C. F. (1985). Biotechnology and wastewater treatment. Cambridge University Press, Cambridge.

Forster, C. F. and Dallas-Newton, J. (1980). Activated sludge settlement – some suppositions and suggestions. *Water Pollution Control*, **79**, 338–51.

Forster, C. F., Crabtree, R. W., Crockett, C. P., and Cluckie, I. D. (1985). Comparison of two models for predicting water quality. *Effluent and Water Treatment Journal*, **25**(5), 161–4.

Fox, J. C., Fitzgerald, P. R., and Lue-Hing, C. (1981). *Sewage organisms – a color atlas*. The Metropolitan Sanitary District of Greater Chicago, Chicago, Illinois.

Fox, J. G., Ackerman, J. I., and Newcomer, C. E. (1983). Ferret as a potential reservoir for human camplobacteriosis. *American Journal of Veterinary Research*, **44**, 1049–52.

Fox, J. G., Zanotti, S., and Jordon, H. V.(1981). The hamster as a reservoir of *Campylobacter fetus* subspecies *jejuni*. *Journal of Infectous Diseases*, **143**, 856.

Foxworthy, J. E. and Kneeling, H. R. (1969). *Eddy diffusion and bacterial reduction in waste fields in the ocean*. University of Southern California, Los Angeles.

Friedman, B. A. and Dugan, P. R. (1968a). Identification of *Zoogloea* species and the relationship to zoogloeal matrix and floc formation. *Journal of Bacteriology*, **95**, 1903–9.

Friedman, B. A. and Dugan, P. R. (1968b). Concentration and accumulation of metallic ions by the bacterium *Zoogloea*. *Developments in Industrial Microbiology*, **9**, 381–8.

Frostell, B. (1981). *Anamet anaerobic-aerobic treatment of concentrated wastewaters*. Proceedings of the 36th Industrial Waste Conference. Purdue University.

Fry, J. C. and Staples, D. G. (1976). Distribution of *Bdellovibrio bacteriovirus* in sewage works, river water and sediments. *Applied Environmental Microbiology*, **31**, 469–74.

Frye, W. W. and Becker, E. R. (1929). The fauna of an experimental trickling filter. *Sewage Works Journal*, **1**, 286–308.

Fullen, W. J. (1953). Anaerobic digestion of packing plant wastes. *Sewage and Industrial Wastes*, **25**, 576–85.

Fuller, R. J. and Glue, D. E. (1978). Seasonal activity of birds at a Buckinghamshire sewage works. *British Birds*, **71**, 235–44.

Fuller, R. J. and Glue, D. E. (1980). Sewage works as bird habitats in Britain. *Biological Conservation*, **17**, 165—81.

Fuller, R. J. and Glue, D. E. (1981). The impact on bird communities of the modernisation of sewage treatment works. *Effluent and Water Treatment Journal*, **21**, 27–31.

Funderburg, S. W. and Sorber, C. A. (1985). Coliphage as indicators of enteric viruses in activated sludge. *Water Research*, **19**, 547–55.

Gaffney, P. E. (1977). Chlorobiphenyls and PCB's: formation during chlorination. *Journal of the Water Pollution Control Federation*, **49**, 401–4.

Galbraith, N. S. (1980). Infection associated with swimming pools. *Environmental Health*, **88**, 31–3.

Galbraith, N. S., Barrett, N., and Stanwell-Smith, R. (1987). Water and disease after Croydon: A review of water borne and water associated disease in the U.K. 1937–86. *Water and Environmental Management*, **1**, 7–21.

Gale, R. S. (1975). Control of sludge filter operation. *Filtration and Separation*, **12**, 74–83.

Gale, R. S. and Baskerville, R. C. (1970). Polyelectrolytes in the filtration of sewage sludges. *Water Pollution Control*, **69**, 660–71.

Gallo, B. J., Andreotti, R., Roche, C., Ryu, D., and Mandels, M. (1979). Cellulose production by a new mutant strain of *Trichoderma veesei* MCG77. In *Biotechnology in energy production and conservation*, (ed. C. D. Scott). John Wiley, London.

Gameson, A. L. H. (1975a) (ed). Discharge of sewage from sea outfalls. Supplement to *Progress in Water Technology*. Pergamon, Oxford.

Gameson, A. L. H. (1975b). Experience on the British coast. In *Marine pollution and marine waste disposal*, (ed. E. A. Pearson and E. D. F. Frangipane), pp. 387–99. Pergamon, London.

Gameson, A. L. H. (1984). *Investigation of sewage discharges to some British coastal waters*. Ch. 8. *Bacterial mortality*, Part 1. Technical Report TR201. Water Research Centre, Stevenage.

Gameson, A. L. H. (1985a). *Investigations of sewage discharges to some British coastal waters:* Ch. 8. *Bacterial mortality*, Part 2. Technical Report TR222. Water Research Centre, Stevenage.

Gameson, A. L. H. (1985b). *Application of coastal pollution research*, 5. *Microbial mortality*. Technical Report TR228. Water Research Centre, Stevenage.

Gameson, A. L. H. and Gould, D. J. (1974). Effects of solar radiation on the mortality of some terrestrial bacteria in sea water. In *Discharge of sewage from sea outfalls*, (ed. A. L. H. Gameson), pp. 209–19. Pergamon Press, London.

Gameson, A. L. H., Bufton, A. W. J., and Gould, D. J. (1967). Studies of the coastal distribution of coliform bacteria in the vicinity of a sea outfall. *Water Pollution Control*, **66**, 501–23.

Garattini, S., Paglianlunga, S., and Scrimshaw, N. S. (ed.) (1979). *Single cell protein – safety for animal and human feeding*. Pergamon, London.

Gardner, D. and Riley, J. P. (1973). Distribution of dissolved mercury in the Irish Sea. *Nature (London)*, **241**, 526.

Gasner, L. L. (1979). Microorganisms in waste treatment. In *Microbial technology*, vol. II. *Fermentation technology*, (ed. H. J. Reppler and D. Perlman), pp. 211–22. Academic Press, London.

Gasser, J. K. R. (ed.) (1980). *Effluents from livestock*. Applied Science Publishers, London.

Gasser, J. K. R. (ed.) (1985). *Composting of agricultural and other wastes*. Elsevier, London.

Gaudy, E. and Wolfe, R. S. (1962). Factors affecting filamentous growth of *Sphaerotilus natans*. *Applied Microbiology*, **9**, 580–4.

Gauntlett, R. B. and Craft, D. G. (1979). *Biological removal of nitrate from river water*. Technical Report TR98. Water Research Centre, Stevenage.

Gayford, C. G. and Richards, J. P. (1970). Isolation and enumeration of aerobic heterotrophic bacteria in activated sludge. *Journal of Applied Bacteriology*, **33**, 342–50.

Geldreich, E. E. (1972). Water-borne pathogens. In *Water pollution microbiology*, (ed. R. Mitchell), pp. 207–41. Wiley Interscience, New york.

Geldreich, E. E., Clark, H. F., and Kabler, P. W. (1965). Faecal coliform organism medium for the membrane filter technique. *Journal of the American Water Works Association*, **57**, 208–14.

Geldreich, E. E., Nash, H. D., Reasoner, D. J., and Tafor, R. H. (1972) The necessity of controlling bacterial populations in potable waters: community water supply. *Journal of the American Water Association*, **64**, 596–602.

Gemmell, R. P. (1974). Revegetation of derelict land polluted by a chromate smelter. *Environmental Pollution*, **6**, 31–7.

Gentelli, E. J. (1967). DNA and nitrogen relationships in bulking activated sludge. *Journal of the Water Pollution Control Federation*, Research Supplement, 39, 10, R31.

Genung, R. K., Hancher, C. W., Rivers, A. L., and Harris, M. T. (1982). Energy conservation and methane production in munipal wastewater treatment using fixed-film anaerobic bioreactors. *Biotechnology Bioengineering Symposium*, **12**, 365–80. John Wiley, New York.

George, E. A. (1976). A guide to algal keys (excluding seaweeds). *British Phycological Journal*, **11**, 49–55.

Gerard, B. M. (1964). *Synopses of British fauna*, Vol. 6. *Lumbricidae (Annelida)*. The Linnean Society of London.

Gerba, C. P., Wallis, C., and Meznick, J. L. (1975). Fate of wastewater bacteria in soil. *Journal of the Irrigation and Drainage Division, American Society of Civil Engineers*, **101**, IR3, 157–73.

Gerick, J. A. (1984). Land application odour study. *Journal of Water Pollution Control Federation*, **56**, 287–91.

Gersberg, R. M., Elkins, B. V., and Goldmann, C. R. (1983). Nitrogen removal in artificial wetlands. *Water Research*, **17**, 1009–14.

Gersberg, R. M., Elkins, B. V., and Goldmann, C. R. (1984). Wastewater treatment by artificial wetlands. *Water Science and Technology*, **17**, 443–50.

Gersberg, R. M., Elkins, B. V., Lyon, S. R., and Goldmann, C. R. (1986). Role of aquatic plants in wastewater treatment by artificial wetlands. *Water Research*, **20**, 363–8.

Geyer, J. C. and Lentz, J. L. (1964). *An evaluation of the problems of sanitary sewer system design*. Final Report, Johns Hopkins University, Baltimore.

Ghosh, S. (1981). Kinetics of acid-phase fermentation in anaerobic digestion. In *Biotechnology in energy production and conservation*, (ed. C. D. Scott), pp. 225–36. John Wiley, London.

Ghosh, S., Henry, M. P., and Klass, D. L. (1980). Bioconversion of water hyacinth–coastal bermuda grass–MSW–sludge blends to methane. *Biotechnology and Bioengineering Symposium* **10**, 163–187.

Gibbons, J. (1968). Farm waste disposal in relation to cattle. *Water Pollution Control*, **67**, 623–6.

Gibbs, D. J. and Greenhalgh, M. E. (1983). *Biotechnology, chemical feedstocks and energy utilization*. Frances Pinter, London.

Gifford, W. H. (1972). Conjoint composting of sewage sludge and straw. *Soil Association Journal*, **17**(1), 27–32.

Giona, A. R., Annesini, M. C., Toro, L., and Gerardi, W. (1979). Kinetic parameters for municipal wastewater. *Journal of the Water Pollution Control Federation*, **51**, 999–1008.

Girdwood, R. W. A., Frickler, C. R., Munro, D., Shedden, C. B., and Monaghan, P. (1985). The incidence and significance of *Salmonella* carriage by gulls (*Larus* spp.) in Scotland. *Journal of Hygiene, Cambridge*, **95**, 229–41.

Glass, J. S. and O'Brien, R. T. (1980). Enterovirus and coliphage inactivation during activated sludge treatment. *Water Research* **14**, 877–82.

Gledhill, P. (1986). Why investing in effluent treatment makes sense. *Technology Ireland*, July/August, 24–26.

Gloyna, E. F. (1971). *Waste stabilization ponds*. WHO Monograph Series, 60.

Gochnarg, I. (1979). The Brazilian national alcohol programme. In *Biomass for energy*. Conference (C20) at the Royal Society, London. UK Section of the International Solar Energy Society.

Godeanu, S. (1966). Contributions to knowledge on Rotifera found in biological waste-water treatment plants. *Studii de Protectia si epuarea apelor*, **7**, 569–99.

Goff, G. D., Spendlove, J. C., Adams, A. P., and Nicholes, P. S. (1973). Emission of microbial aerosols from sewage treatment plants that use trickling filters. *Health Services Report*, **88**, 640–52.

Goldberg, I. (1985). *Single cell protein*. Springer-Verlag, Berlin.

Golovlena, L. A. and Skvyabin, G. K. (1981). Microbial degradation of DDT. In *Microbial degradation of xenobiotics and recalcitrant compounds*, (ed. J. Leisinger, R. Hulter, A. M. Cook, and J. Nuesch), pp. 257–91. Academic Press, London.

Gomes de Sousa, J. M. (1987). Wastewater stabilization lagoon design criteria for Portugal. *Water Science and Technology*, **19** (12), 7–16.

Gondrosen, B., Melby, K., Gregusson, S., and Dahl. O. P. (1985). A waterborne outbreak of campylobacter enteritis in the subarctic region of Norway. In *Campylobacter*, Vol. 3, (ed. A. D. Pearson, M. B. Skirrow, H. Lior, and R. Rowe), p. 277. Public Health Laboratory Service, London.

Gonin, F. R. (1982). Using composted waste for growing horticultural crops. *Biocycle, Journal of Waste Recycling*, **23**, 45–7.

Good, R. E., Whigham, D. F., and Simpson, L. (ed.) (1978). *Freshwater wetlands, ecological processes and management potential*. Academic Press, New York.

Goodfellow, R. M., Cardoso, J., Eglinton, G., Dawson, J. P., and Best, G. A. (1977). A faecal sterol survey in the Clyde estuary. *Marine Pollution Bulletin*, **8**, 272–6.

Gould, D. J. (1976). *Ecological effects of sewage dischargs to the sea: an assessment of research needs*. Technical Report TR26. Water Research Centre, Stevenage.

Gould, D. J. (1977). *Gull droppings and their effects on water quality*. Technical Report TR37. Water Research Centre, Stevenage.

Gowan, D. (1972). *Slurry and farm waste disposal*. Farming Press, Ipswich.

Gower, A. M. (ed.) (1980). *Water quality in catchment ecosystems*. John Wiley, London.

Goyal, S. M. and Gerba, C. P. (1979). Comparative adsorption of human enteroviruses, Simian rotavirus and selected bacteriophages to soils. *Applied Environmental Microbiology*, **38**, 241–7.

Goyal, S. M., Adams, W. N., O'Malley, M. L., and Lear, D. W. (1984). Human pathogenic viruses at sewage sludge disposal sites in the middle Atlantic region. *Applied Environmental Microbiology*, **48**, 758–63.

Grabow, W. O. K. (ed.) (1986). Health related water microbiology 1986. *Water Science and Technology*, **18** (10), 1–268.

Grabow, W. O. K., Burger, J. S., and Nupen, E. M. (1980). Evaluation of acid fast bacteria, *Candida albicans*, enteric viruses and conventional indicators for monitoring wastewater reclamation systems. *Progress in Water Technology*, **12**, 803–17.

Grabow, W. O. K., Coubrough, P., and Nupen, E. M. (1984). Evaluation of coliphages as indicators of the virological quality of sewage polluted water. *Water South Africa*, **10**, 7–14.

Grace, R. A. (1978). *Marine outfall systems: planning, design and construction*. Prentice-Hall, New Jersey.

Grady, C. P. L. *et al.* (1972). Effects of growth rate and influent substrate concentrations on effluent quality for chemostats containing bacteria in pure and mixed culture. *Biotechnology and Bioengineering*, **14**, 391–410.

Grau, P. (1980). Utilisation of activated sludge as fodder supplement. *Environmental Technology Letters*, **1**, 557–70.

Gray, H. F. (1940). Sewerage in ancient and medieval times. *Sewage Works Journal*, **12**, 939–46.

Gray, N. F. (1979). Communication on the pollution of a water supply catchment by breeding gulls and the potential environmental health implications. *Journal of the Institute of Water Engineers Scientists*, **33**, 474–7.

Gray, N. F. (1980). The comparative ecology of highrate plastic, conventional mineral and mixed plastic/mineral media in the treatment of domestic sewage in percolating filters. Vol. I. Ph.D. thesis, Council for National Academic Awards.

Gray, N. F. (1981). Simple retention time analysis in percolating filters using conductivity. *Effluent and Water Treatment Journal*, **21**, 345–7.

Gray, N. F. (1982*a*). A key to the slime-forming organisms of 'sewage fungus'. *Journal of Life Sciences, Royal Dublin Society*, **4**, 97–102.

Gray, N. F. (1982*b*). The use of percolating filters in teaching ecology. *Journal of Biological Educational* **16**, 183–6.

Gray, N. F. (1983*a*). Micro-organisms in waste treatment. In *Microbiology and pollution: problems and perspectives*, pp. 31–47. Royal Irish Academy, Dublin.

Gray, N. F. (1983*b*). Ponding of a random plastic filter medium due to the fungus *Subbaromyces splendens* Hesseltine in the treatment of sewage. *Water Research*, **17**, 1295–1302.

Gray, N. F. (1983*c*). The significance of the flow pattern on retention analysis in percolating filters. *Irish Journal of Environmental Science* **2**, 293–7.

Gray, N. F. (1984*a*). The effect of fungal parasitism and predation on the population dynamics of nematodes in the activated sludge process. *Annals of Applied Biology*, **104**, 143–9.

Gray, N. F. (1984*b*). Ecological consequences of *Ambystegium riparium* growing on a sewage work's percolating filter. *Bulletin of the Irish Biogeographical Society*, **8**, 26–30.

Gray, N. F. (1984*c*). Biological film control in percolating filters using neutron scattering. *Effluent and Water Treatment Journal*, **24**, 203–5.

Gray, N. F. (1985*a*). Heterotrophic slimes in flowing waters. *Biological Reviews of the Cambridge Philosophical Society*, **60**, 499–548.

Gray, N. F. (1985*b*). Universities and the water industry: a personal perspective. *Effluent and Water Treatment Journal*, **25**, 83.

Gray, N. F. (1986). The bathing water directive: a challenge to Ireland. *Technology Ireland*, **18**(12), 15–20.

Gray, N. F. (1987). *Sewage fungus in Irish rivers: a guide to identification, evaluation and control*. Technical Report **6**, Water Technology Research, Dublin.

Gray, N. F. (1988). *Sugar beet processing in Ireland: a case study*. Water Technology Press, Dublin.

Gray, N. F. (1989*a*). The effect of small changes in incubation temperature on the five day biochemical oxygen demand test. *Environmental Technology Letters*, **10**, 253–8.

Gray, N. F. (1989*b*). BOD incubators and temperature variability. *Environmental Technology Letters*, **10**, 259–68.

Gray, N. F. and Clarke, J. (1984). Heavy metals in heterotrophic slimes in Irish rivers. *Environmental Technology Letters*, **5**, 201–6.

Gray, N. F. and Hunter, C. A. (1985). Heterotrophic slimes in Irish rivers. Evaluation of the problem. *Water Research*, **19**, 685–92.

Gray, N. F. and Learner, M. A. (1983). A pilot-scale percolating filter for use in sewage treatment studies. *Water Research*, **17**, 249–53.

Gray, N. F. and Learner, M. A. (1984*a*). Comparative pilot-scale investigation into uprating the performance of percolating filters by partial medium replacement. *Water Research*, **18**, 409–22.

Gray, N. F. and Learner, M. A. (1984b). Estimation of film accumulation in percolating filters: comparison of methods. *Water Research*, **18**, 1509–14.

Green, G. P. (1982). An attempt to predict bulking in activated sludge by the development of a method to measure filamentous organisms. In *Bulking of activated sludge*, (ed. B. Chambers and E. J. Tomlinson), pp. 63–74. Ellis-Horwood, Chichester.

Green, J. and Watts, E. (1973). Sewage effluent lagoons: Zooplankton in theory and practice. In *Symposium on sewage effluent as a water resource*, pp. 36–44. Institution of Public Health Engineers, London.

Greenberg, A. E., Klein, G., and Kaufman, W. J. (1955). Effect of phosphorus on the activated sludge process. *Sewage and Industrial Wastes*, **27**, 277–82.

Greene, L. A. (1978). Nitrates in water supply abstractions in the Anglian Region: current trends and remedies under investigation. *Water Pollution Control*, **77**, 478–91.

Greene, L. A. (1980). Nitrates in groundwaters in the Anglian Region. *Proceedings of the Institute of Civil Engineers*, **2**, 69–86.

Grieve, A. (1978). Sludge incineration with particular reference to the Coleshill plant. *Water Pollution Control*, **77**, 314–23.

Griffiths, I. M. and Lloyd, P. J. (1985). Mobile oxygenation in the Thames estuary. *Effluent and Water Treatment Journal* **25**, 165–9.

Grime, J. P. (1979). *Plant strategies and vegetation processes*. John Wiley, Chichester.

Grunnet, K. (1975). Salmonella *in sewage and receiving waters*. Fadl's Forlag, Copenhagen.

Grunnet, K. and Nielson, B. B. (1969). *Salmonella* types isolated from the Gulf of Aarhus compared with types from infected human beings, animals, and feed products in Denmark. *Applied Microbiology*, **18**, 985–90.

Guarino, C. F., Nelson, M. D., Lazanof, M., and Wilson, T. E. (1980). Uprating activated -sludge plants using rotary biological contactors. *Water Pollution Control*, **79**, 255–71.

Guerrero, R. D. (1983). The culture and use of *Perionyx excavatus* as a protein resource in the Philippines. In *Earthworm ecology from Darwin to vermiculture*, (ed. J. Satchell), pp. 309–13. Chapman and Hall, London.

Haddock, B. A. and Hamilton, W. A. (1977). *Microbial energetics*. Cambridge University Press.

Hakulinen, R. (1988). The use of enzymes for wastewater treatment in the pulp and paper industry – a new possibility. *Water Science and Technology*, **20** (1), 251–62.

Halcrow, W., Mackay, D. W., and Thornton, I. (1973). The distribution of trace metals and fauna in the Firth of Clyde in relation to the disposal of sewage sludge. *Journal of the Marine Biology Association, UK*, **53**, 721–39.

Haldane, J. B. S. (1930). *Enzymes*. Longman, London.

Hall, G. A. and Jones, P. W. (1978). A study of the susceptibility of cattle to oral infection by salmonellas contained in raw sewage sludge. *Journal of Hygiene, Cambridge*, **80**, 409–14.

Hall, E. R. and Murphy, K. L. (1980). Estimation of nitrifying biomass and kinetics in wastewater. *Water Research*, **14**, 297–304.

Hall, J. E. and Davis, R. D. (1983). *Sludge utilization to farmland*. WRC Regional Seminar, Dublin, February 1983. Water Research Centre, Stevenage.

Haller, W. T. and Sutton, D. L. (1973). Effect of pH and high phosphorus concentrations on growth of water hyacinth. *Hyacinth Control Journal*, **11**, 59–69.

Haller, W. T., Sutton, D. L., and Barlowe, W. C. (1974). Effects of salinity on growth of several aquatic macrophytes. *Ecology*, **55**, 891–901.

Hammer, M. J. (1977). *Water and wastewater technology*. John Wiley, New York.

Hao, O. and Hendricks, G. F. (1975*a*). Rotating biological reactors remove nutrients: Part I. *Water Sewage Works*, **122**(10), 70–73.

Hao, O. and Hendricks, G. F. (1975*b*). Rotating biological reactors remove nutrients: Part II. *Water Sewage Works*, **122**(11), 48–50.

Hao, O. J., Richard, M. G., Jenkins, D., and Blanch, H. (1983). The half saturation coefficient for dissolved oxygen: a dynamic method for its determination and its effect on dual species competition. *Biotechnology and Bioengineering*, **35**, 403.

Harder, W. (1981). Enrichment and characterization of degrading microbes. In *Microbial degradation of xenobiotics and recalcitrant compounds*, (ed. T. Leisinger, R. Hutter, A. M. Cook, and J. Neusch), pp. 77–96. Academic Press, London.

Harding, J. P. and Smith, W. A. (1974). *A key to the British freshwater cyclopid and calanoid copepods*. Scientific Publications: **18**, Freshwater Biological Association, Cumbria.

Harding, J. P. C. and Whitton, B. A. (1981). Accumulation of zinc, cadmium and lead by field populations of *Lemanea*. *Water Research*, **15**, 301–19.

Harkness, N. (1966). Bacteria in sewage treatment processes. *Journal and Proceedings of the Institute of Sewage Purification*, 542–57.

Harris, N. P. and Hansford, G. S. (1976). A study of substrate removal in a microbial film reactor. *Water Research*, **10**, 935–43.

Harris, R. F. and Sommers, L. E. (1968). Plate dilution frequency technique for assay of microbial ecology. *Applied Microbiology*, **16**, 330–4.

Harris, R. H. and Mitchell, R. (1973). Role of polymers in microbial aggregation. *Annual Reviews of Microbiology*, **27**, 27–50.

Hartenstein, R. (1981). Sludge decomposition and stabilization. *Science*, **212**, 743–8.

Hartenstein, R., Neuhauser, E. F., and Kaplan, D. L. (1979). *A process report on the potential use of earthworms in sludge management*, pp. 238–41. Proceedings of the 8th National Sludge Conference, Florida. Information Transfer Inc, Maryland.

Hartley, W. R. and Weiss, C. M. (1970). Light intensity and vertical distribution of algae in tertiary oxidation ponds. *Water Research*, **4**, 751–63.

Harvey, B. R., Eden, G. E., and Mitchell, N. T. (1963). Neutron scattering: a technique for the direct determination of the amount of biological film in a percolating filter. *Journal and Proceedings of the Institute of Sewage Purification*, (5), 495–506.

Harvey, R. M. and Fox, J. L. (1973). Nutrient removal using *Lemma minor*, *Journal of the Water Pollution Control Federation*, **45**, 1928–38.

Hashimoto, A. G. (1981). Methane production and effluent quality from fermentation of beef cattle manure and molasses. In *Biotechnology in energy production and conservation*, (ed. C. D. Scott), pp. 481–92. John Wiley, London.

Hatch, R. T. and Menawat, A. (1979). Biological removal and recovery of trace heavy metals. In *Biotechnology in energy production and conservation*, (ed. C. D. Scott), pp. 191–203. John Wiley, London.

Hatcher, P. G. and Keister, L. E. (1976). Carbohydrates and inorganic carbon in New York Bight sediments as possible indicators of sewage contamination. *American Society of Limnology and Oceanography. Special Symposia*, Vol. 2, (ed. M. G. Gross), pp. 240–59. Allen Press, Lawrence, Kansas.

Hatcher, P. G. and McGillivary, P. A. (1979). Sewage contamination in the New York

Bight, Coprostanol as an indicator. *Environmental Science and Technology*, **13**, 1225–9.

Hattingh, W. H. J. (1963a). The nitrogen and phosphorous requirements of the micro-organisms. 1. *Water and Waste Treatment Journal*, **9**, 380–6.

Hattingh, W. H. J. (1963b). The influence of nutrition on the respiratory rate of micro-organisms. 2. *Water and Waste Treatment Journal*, **9**, 424–6.

Hattingh, W. H. J., Kotze, J. P., Thiel, P. G., Toerien, D. F., and Siebert, M. L. (1967). Biological changes during the adaptation of an anaerobic digester to a synthetic substrate. *Water Research*, **1**, 255–77.

Haug, R. T. (1980). *Compost engineering: principles and practice*. Ann Arbor Science Publishers, Michigan.

Haug, R. T. and McCarty, P. L. (1972). Nitrification with submerged filters. *Journal of the Water Pollution Control Federation*, **44**(11), 2087–102.

Haugaard, P. (1984). Cysticercosis and salmonellosis in cattle caused by sewage sludge on Danish farmland. *Waste Management Research*, **2**, 163.

Hauser, J. R. (1984). Use of water hyacinth aquatic treatment systems for ammonia control and effluent polishing. *Journal of the Water Pollution Control Federation*, **56**(3), 219–25.

Hawkes, F. R. and Hawkes, D. L. (1987). Anaerobic digestion. In *Basic biotechnology* (ed. j. Bu'lock and B. Kristiansen), 337–358. Academic Press, London.

Hawkes, H. A. (1951). A study of the biology and control of *Anisopus fenestralis* (Scopoli, 1763), a fly associated with sewage filters. *Annals of Applied Biology*, **38**, 592–605.

Hawkes, H. A. (1952a). The ecology of *Anisopus fenestralis* Scop. (Diptera) in sewage bacteria beds. *Annals of Applied Biology*, **39**, 181–92.

Hawkes, H. A. (1952b). Factors influencing the egress of the sewage fly *Anisopus fenestralis* from bacteria beds. *Proceedings of the Birmingham Natural History Philosophical Society*, **18**(III), 41–53.

Hawkes, H. A. (1955). The effect of periodicity of dosing on the amount of film and the numbers of insects and worms in alternating filters at Minworth. *Journal and Proceedings of the Institute of Sewage Purification*, 48–58.

Hawkes, H. A. (1957). Film accumulation and grazing activity in the sewage filters at Birmingham. *Journal and Proceedings of the Institute Sewage Purification*, 88–110.

Hawkes, H. A. (1959). The effects of methods of sewage application on the ecology of bacteria beds. *Annals of Applied Biology*, **37**, 339–49.

Hawkes, H. A. (1961a). An ecological approach to some bacteria bed problems. *Journal and Proceedings of the Institute of Sewage Purification*, (2), 105–132.

Hawkes, H. A. (1961b). Fluctuations in the aerial density of *Anisopus fenestralis* Scop. (Diptera) above sewage bacterial beds. *Annals of Applied Biology*, **49**, 66–76.

Hawkes, H. A. (1963). *The ecology of wastewater treatment*. Pergamon, Oxford.

Hawkes, H. A. (1965). Factors influencing the seasonal incidence of fungal growths in sewage bacteria beds. *International Journal of Air and Water Pollution*, **9**, 693–714.

Hawkes, H. A. (1972). Biological aspects of river pollution. In *River pollution*, Vol. 2. *Causes and effects*, (ed. L. Klein), pp. 311–42. Butterworths, London.

Hawkes, H. A. (1983a). Activated sludge. In *Ecological aspects of used water treatment*, Vol. 2. *Biological activities and treatment processes*, (ed. C. R. Curds and H. A. Hawkes), pp. 77–162. Academic Press, London.

Hawkes, H. A. (1983b). The applied significance of ecological studies of aerobic processes. In *Ecological aspects of used water treatment*, Vol. 3. *The processes and their ecology*, (ed. C. R. Curds and H. A. Hawkes), pp. 173–334. Academic Press, London.

Hawkes, H. A. (1983c). Stabilization ponds. In *Ecological aspects of used water treatment*, Vol. 2. *Biological activities and treatment processes*, (ed. C. R. Curds and H. A. Hawkes), pp. 163–217. Academic Press, London.

Hawkes, H. A. and Jenkins, S. H. (1951). Biological principles in sewage purification. *Journal and Proceedings of the Institute of Sewage Purification*, 300–18.

Hawkes, H. A. and Jenkins, S. H. (1955). Comparison of four grades of sewage percolating filter media in relation to purification, film accumulation and fauna. *Journal and Proceedings of the Institute of Sewage Purification*, 352–7.

Hawkes, H. A. and Jenkins, S. H. (1958). Comparison of four grades of filter media in relation to purification, film accumulation and fauna of sewage percolating filters operating on alternating double filtration. *Journal and Proceedings of the Institute Sewage Purification*, 221–5.

Hawkes, H. A. and Shepherd, M. R. N. (1971). *The seasonal accumulation of solids in percolating filters and attempted control at low frequency dosing*. Proceedings of the Fifth International Water Pollution Conference, 1970, (ed. S. H. Jenkins), Vol. 1, 1–8. Pergamon, Oxford.

Hawkes, H. A. and Shepherd, M. R. N. (1972). The effect of dosing frequency on the seasonal fluctuations and vertical distribution of solids and grazing fauna in sewage percolating filters. *Water Research*, **6**, 721–30.

Hayes, T. D. *et al.* (1979). *Anaerobic digestion of cattle manure*. First International Symposium on Anaerobic Digestion, University College, Cardiff, pp. 255–88.

Head, P. C. (1980). The environmental impact of the disposal of sewage sludge in Liverpool Bay. *Progress in Water Technology*, **13**, 27–38.

Heaman, J. (1975). Windrow composting – a commercial possibility for sewage sludge disposal. *Water Pollution Control (Canada)*, **133**(14), 30–1.

Hedges, R. M. and Pepper, D. (1983). Reverse osmosis concentration and ultrafiltration separation. In *Applications in biochemical processes, Biotech, 83. Proceedings of the International Conference* on Applications and Implications of Biotechnology. Online Publications, Northwood, London.

Hegg, R. O. and Turner, A. K. (1983). Overland flow as a method of treatment for animal wastes – a review. *Animal Wastes*, **8**, 167.

Helforth, T., Hunter, J. V., and Rickert, D. A. (1970). Analytic and process classification of effluents. *Journal of the Sanitary Engineering Division, American Society of Civil Engineers*, **96**, 779–803.

Heliotis, F. D. and De Witt, C. B. (1983). A conceptual model of nutrient cycling in wetlands used for wastewater treatment: a literature analysis. *Wetlands*, **3**, 134–152.

Helliwell, P. R. (ed.) (1979). *Urban storm drainage*. Proceedings of the International Conference, University of Southampton, April 1978. Pentech Press, London.

Hemens, J. and Mason, M. H. (1968). Sewage nutrient removal by a shallow algal stream. *Water Research*, **2**, 277–87.

Hemming, M. L. (1979). General biological aspects of wastewater treatment including the deep shaft process. *Water Pollution Control* **78**, 312–25.

Hemming, M. L. and Wheatley, A. D. (1979). Low rate biofiltration systems using random plastic media. *Water Pollution Control*, **78**(1), 54–68.

Hemming, M. L., Ousby, J. C., Plowright, D. R., and Walker, J. (1977). Deep shaft – the latest position. *Water Pollution Control*, **76**, 441–51.

Hendricks, C. W. (1972). Enteric bacterial growth rates in river water. *Applied Microbiology*, **24**, 168–74.

Hendricks, C. W (ed.) (1978). *Evaluation of the microbiology standard for drinking water*. USEPA 570/9–78IK–002. Environmental Protection Agency, Washington, D.C.

Henze, M. and Harremoes, P. (1983). Review paper: anaerobic treatment of wastewater in fixed film reactors. *Water Science and Technology*, **15**, 1–90.

Hepher, B., Sandbank, E., and Shelef, G. (1975). Fish feeding experiments with wastewater grown algae. In *Combined systems for algal wastewater treatment and reclamation and protein production*. 2nd Progress Report, Technion, Haifa, Israel.

HMSO (Her Majesty's Stationery Office) (1969). *Report of the joint committee on the use of antibiotics in animal husbandry and veterinary medicine*. Amnd. 4190, London.

HMSO (Her Majesty's Stationery Office) (1971). *Water pollution research, 1970*. Department of the Environment, London.

HMSO (Her Majesty's Stationery Office) (1972). *Analysis of raw, potable and wastewaters*. Department of the Environment, London.

HMSO (Her Majesty's Stationery Office) (1977). *The bacteriological examination of water supplies*. Reports on Public Health and Medical Subjects, 71. Department of the Environment, London.

HMSO (Her Majesty's Stationery Office) (1983a). *Biochemical oxygen demand, 1981*. Methods for the examination of waters and associated materials. London.

HMSO (Her Majesty's Stationery Office) (1983b). *The bacteriological examination of drinking water supplies 1982*. Reports on Public Health and Medical Subjects No. 71. London.

HMSO (Her Majesty's Stationery Office) (1988). *Examining biological filters, toxocity to aerobic bacteria, effect of SRT and temperature 1985–6*. Methods for the examination of waters and associated materials, London.

Hernandez, J. W. and Bloodgood, D. E. (1960). The effects of alkylbenzene sulphonates on anaerobic digestion. *Journal of the Water Pollution Control Federation*, **32**, 1261—8.

Hesseltine, C. W. (1953). Study of trickling filter fungi. *Bulletin of the Torrey Botanical Club*, **80**, 507–14.

Heukelekian, H. (1945). The relationship between accumulation, biochemical and biological characteristics of film, and purification capacity of a biofilter and a standard filter. 1. Film accumulation. *Sewage Works Journal*, **17**, 23–8.

Heukelekian, H. (1947). Use of direct method of oxygen utilization in waste treatment studies. *Sewage Works Journal*, **19**, 875–82.

Heukelekian, H. (1957). In *Biological treatment of sewage and industrial wastes*, (ed. J. McCabe and W. W. Eckenfelder), pp. 25–43. Reinhold, New York.

Heukelekian, H. and Weisberg, E. (1956). Bound water and activated sludge bulking. *Sewage and Industrial Wastes*, **28**, 558–74.

Heukelekian, H. and Balmat, J. L. (1959). Chemical composition of the particulate fractions of domestic sewage. *Sewage and Industrial Wastes*, **31**, 413–23.

Heuvelen, W. van., Smith, J. K., and Hopkins, G. J. (1960). Waste stabilization lagoons: design, construction and operating practices among Missouri Basin States. *Journal of the Water Pollution Control Federation*, **32**, 909–15.

Hickey, R. F. and Owens, R. W. (1981). Methane generation from high strength industrial wastes with the anaerobic biological fluidized bed. In *Biotechnology in energy production and conservation*, (ed. C. D. Scott), 349–413. John Wiley, London.

Hickling, R. A. O. (1977). Inland wintering of gulls in England and Wales, 1973. *Bird Study*, **24**, 78–88.

Hickling, R. A. O. (1977). Inland wintering of gulls in England and Wales, 1973. *Bird Study*, **24**, 78–88.

Hileman, C. H. (1982). Fortified compost product shows promise as fertilizer. *Biocycle, Journal of Waste Recycling*, **23**, 43–4.

Hill, D. T. (1982). A comprehensive dynamic model for animal waste methanogenesis. *Transaction of the American Society of Agricultural Engineers*, **25**, 1374–80.

Hill, D. T. (1983). Simplified Monod kinetics of methane fermentation of animal wastes. *Agricultural Wastes*, **5**, 1–16.

Hill, D. T. and Barth, C. L. (1977). A dynamic model for simulation of animal waste digestion. *Journal of the Water Pollution Control Federation*, **49**, 2129–43.

Hill, D. T. and Nordstedt, R. A. (1980). Modelling techniques and computer simulation of agricultural waste treatment processes. *Agricultural Wastes*, **2**, 135–56.

Hill, G. A. and Grimes, D. J. (1984). Seasonal study of a freshwater lake and migratory waterfowl for *Campylobacter jejuni*. *Canadian Journal of Microbiology*, **30**, 845–9.

Hill, J. M., Mance, G., and O'Donnell, A. R. (1984). *The quantities of some heavy metals entering the North Sea*. Technical Report TR205. Water Research Centre, Stevenage.

Hillman, W. S. and Culley, D. D. (1978). The uses of duckweed. *American Science*, **66**, 442–51.

Hinesly, T. D., Thomas, R. E., and Stevens, R. G. (1978). *Environmental changes from long-term land applications of municipal effluents*. EPA 430/9–78–003. US Environmental Protection Agency.

Hoadley, A. W. (1977). Potential health hazards associated with *Pseudomonas aeruginosa* in water. In *Bacterial indicators/health hazards associated with water*, (ed. A. W. Hoadley and B. J. Dutka), 80–114. ASTM Special Technical Publication 635, Philadelphia.

Hoadley, A. W. and Dutka, B. J. (1977). *Bacterial indicators/health hazards associated with water*. ASTM Special Technical Publication 635, Philadelphia.

Hobson, P. N. (1984). Anaerobic digestion of agricultural wastes. *Water Pollution Control*, **83**, 507–13.

Hobson, P. N. and Robertson, A. M. (1977). *Waste treatment in agriculture*. Applied Science Publishers, London.

Hobson, P. N. and Shaw, B. G. (1971). In *Microbial aspects of pollution*, (ed. G. Sykes and F. A. Skinner), pp. 103–21. Academic Press, London.

Hobson, P. N., Bonsefield, S., and Summers, R. (1974). Anaerobic digestion of organic matter. In *CRC Critical Reviews in Environmental Control*, **4**(2), 131–91, Chemical Rubber Company, Cleveland, Ohio.

Hobson, P. N., Bonsefield, S., and Summers, R. (1981). *Methane production from agricultural and domestic wastes*. Applied Science Publishers, London.

Holderby, J. M. and Moggio, W. A. (1960). Utilization of spent sulphite liquor. *Journal of the Water Pollution Control Federation*, **32**, 171–81.

Holm, H. W. and Vennes, J. W. (1970). Occurrence of purple sulphur bacteria in sewage treatment lagoons. *Applied Microbiology*, **19**, 988–96.

Holm, L. G., Weldon, L. W., and Blackburn, R. D. (1969). Aquatic weeds. *Science*, **166**, 699–709.

Holt, P. E. (1980). Incidence of campylobacter, salmonella and shigella infections in dogs in an industrial town. *Veterinary Record*, **107**, 254.

Holtje, R. H. (1943). The biology of sewage sprinkling filters. *Sewage Works Journal*, **15**, 14–29.

Honda, Y. and Matsumoto, J. (1983). The effect of temperature on the growth of microbial film in a model trickling filter. *Water Research*, **17**, 375–82.

Hoover, S. R. and Porges, N. (1952). Assimilation of dairy wastes by activated sludge. II: the equation of synthesis and oxygen utilization. *Sewage and Industrial Wastes*, **24**, 306–12.

Hopwood, S. R. and Downing, A. L. (1965). Factors affecting the role of production and properties of activated sludge in plants treating domestic sewage. *Journal and Proceedings of the Institute of Sewage Purification*, **64**, 435–52.

Horosawa, I. (1950). Biological studies on activated sludge in the purification of sewage. *Journal of the Japanese Sewage Works Association*, **148**, 62–67.

Houston, J., Dancer, B. N., and Learner, M. A. (1989a). Control of sewage filter flies using *Bacillus thuringiensis* var. *israelensis*—I. Acute toxicity tests and pilot scale trial. *Water Research*, **23**, 369–78.

Houston, J., Dancer, B. N., and Learner, M. A. (1989b). Control of sewage filter flies using *Bacillus thuringiensis* var. *israelensis*—II. Full scale trials. *Water Research*, **23**, 379–85.

Houte, J. van and Gibbons, R. J. (1966). *Studies of the cultivable flora of normal human faeces*. *Antomie van Leeuwenhoek*, **32**, 212–22.

Hoyland, G. and Robinson, P. J. (1983). Aerobic treatment in 'Oxitron' BFB plant at Coleshill. *Water Pollution Control*, **82**, 479–93.

Hoyland, G. and Roland, D. (1984). *Biological filtration of finely-screened sewage*. Technical Report TR198. Water Research Centre, Stevenage.

Hoyland, G., Day, M., and Baskerville, R. C. (1981). *Getting more out of the filter press*. Technical Report TR173. Water Research Centre, Stevenage.

Huang, J. Y. C. and Mandt, M. G. (1978). Pure oxygen activated sludge system transfer oxygen rapidly. *Water Sewage Works*, **125**, 98–103.

Hubbell, J. W. (1962). Commercial and institutional wastewater loadings. *Journal of the Water Pollution Control Federation*, **34**, 962–8.

Huddleston, R. L. and Allred, R. C. (1963). Microbial oxidation of sulphonated alkyl-benzenes. *Developments in Industrial Microbiology*, **4**, 24–38.

Hudson, J. A., Bruce, A..M., Oliver, B. T., and Auty, D. (1988). Operating experiences of sludge disinfection and stabilization at Colburn sewage treatment works, Yorkshire. *Water and Environmental Management*, **2**, 429–41.

Hudson, J. A. and Fennel, H. (1980). Disposal of sewage sludge to land: chemical and microbiological aspects of sludge to land policy. *Water Pollution Control*, **79**, 370–87.

Hughes, A. M. (1961). *The mites of stored food products*. HMSO, London.

Hughes, D. E. (1980). What is aerobic digestion? an overview. In *Anaerobic digestion*, (ed. D. A. Stafford, B. I. Wheatley, and D. E. Hughes), pp. 1–14. Applied Science Publishers, London.

Hughes, D. E. and Stafford, D. A. (1976). Microbiology of the activated sludge process. *CRC Critical Review Environmental Control*, **6**, 232–50.

Hughes, J. M., Merson, M. H., Craun, G. F., and McCabe, L. J. (1975). Outbreaks of waterborne disease in the United States, 1973. *Journal of Infectious Diseases*, **132**, 336–9.

Hughes, D. E. et al. (ed.) (1981). *Anaerobic digestion*. Elsevier, Amsterdam.

Hukker, G. (1979). *Report of working party No. 5: Environmental effects of disposing of sewage sludge on land*. EEC Symposium on characterization treatment and use of sewage sludge. Caderache, France.

Hultman, B. (1971). Kinetics of biological nitrogen removal. *Institut Vattenforsorjm-ingsoch Avloppsteknik samt Vattenkemi, KTH*, Pub. 71:5, Stockholm.

Hultman, B. (1973). Biological nitrogen reduction studies as a general microbiological engineering process. In *Environmental engineering*, (ed. G. Linder and K. Nyberg). D. Reidel, Holland.

Humphrey, A. E. (1975). Economical factors in the assessment of various cellulosic substances as chemical and energy resources. In *Cellulose as a chemical and energy resource*. Proceeding of the Symposium, Biotechnology and Bioengineering, (ed. C. R. Wilke), pp. 49–65. John Wiley, New York.

Humphries, M. D. (1982). The effect of a change in longitudinal mixing on the settleability of an activated sludge. In *Bulking of activated sludge*, (ed.) B. Chambers and E. J. Tomlinson), pp. 261–4. Ellis-Horwood, Chichester.

Hungate, R. E. (1977). Suitability of methanogenic substrates, health hazards, and terrestrial conversion of plant nutrients. In *Microbial energy conversion*, (ed. H. G. Schlegel and J. Barnea), pp. 339–46. Pergamon, Oxford.

Hunter, J. V., Genetelli, E. J., and Gilwood, M. E. (1966). *Temperature and retention time relationships in the activated sludge process*. Proceedings of the 21st Industrial Waste Conference, Purdue University. Engineering Extension Series No. 121, pp. 953–63.

Hunter, J. V. and Heukelekian, H. (1960). *Separation and materials balance of solids fractions in sewage*. Proceedings of the 15th Industrial Waste Conference, Purdue University, pp. 150–63.

Hunter, J. V. and Heukelekian, H. (1965). The composition of domestic sewage fractions. *Journal of the Water Pollution Control Federation*, **37**, 1142–63.

Hunter, M., Motta Marques, M. L. da., Lester, J., and Perry, R. (1988). A review of the behaviour and utilization of polycarboxylic acids as detergent builders. *Environmental Technology Letters*, **9**, 1–22.

Huss, L. (1977). The Anamet process for food and fermentation industry effluent. *Tribune de CEBEDEAU*, **30**, 390–6.

Hussey, B. R. (1975). Ecological studies on percolating filters and stream riffles associated with the disposal of domestic and industrial wastes. M.Sc. Thesis, University of Aston in Birmingham, England.

Hussey, B. R. (1982). Moss growth in filter beds. *Water Research*, **16**, 391–8.

Hutchinson, M. (1974). Microbial aspects of groundwater pollution. In *Groundwater pollution in Europe*, (ed. J A. Cole), pp. 167–202. Water Information Center, Washington.

Hutton, L. G. (1983). *Field testing of water in developing countries*. Water Research Centre/Water and Waste Engineering for Developing Countries Group. Medmenham, England.

Hutton, M. (1982). *Cadmium in the European Community*. Monitoring and Assessment Research Centre Technical Report 26. MARC, Chelsea College, University of London.

Hutzinger, O. and Veerkamp, W. (1981). Xenobiotic chemicals with pollution potential. In *Microbial degradation of xenobiotic and recalcitrant compounds*, (ed. T. Leisinger, R. Hutter, A. M. Cook, and J. Neusch), pp. 3–45. Academic Press, London.

Hynes, H. B. N. (1971). *The biology of polluted waters*. Liverpool University Press.

Idelovitch, E. (1978). Wastewater reuse by biological-chemical treatment and groundwater recharge. *Journal of the Water Pollution Control Federation*, **50**, 2723–40.

Idelovitch, E. and Michail, M. (1980). Treatment effects and pollution dangers of secondary effluent percolation for groundwater. *Progress in Water Technology*, **12**, 949–66.

Idelovitch, E. and Michail, M. (1984). Soil-aquifer treatment – a new approach to an old method of wastewater reuse. *Journal of the Water Pollution Control Federation*, **56**, 936–43.

Iizumi, H., Hattori, A., and McRoy, C. P. (1980). Nitrate and nitrite in interstitial waters of eelgrass beds in relation to the rhizosphere. *Journal of Experimental Marine Biology and Ecology*, **47**, 191–201.

Imhoff, K. R. (1982). Experiences with purification lakes and polishing lagoons. *Water Science and Technology*, **14**, 189–203.

Imhoff, K. R. (1984). The design and operation of the purification lakes in the Ruhr valley. *Water Pollution Control*, **83**, 243–53.

Ingram, W. T. and Edwards, G. P. (1960). *The behaviour of filter biota under controlled conditions*. Proceedings of the Third Biological Waste Treatment Conference. Manhattan College, New York.

Institute of Industrial Research and Standards (1975). *Recommendations for spetic tank drainage systems*. IIRS, SR6.

Institution of Public Health Engineers (1978). *The public health engineering data book 1978–9*, (ed. R. E. Bartlett). Sterling Professional Publications, London.

Institution of Public Health Engineers (1986). *Construction and maintenance of long sea outfalls*. Institution of Public Health Engineers, London.

Institute of Water Pollution Control (1972). *Directory of municipal wastewater treatment plants*, Vols I–IV. Institute of Water Pollution Control, Maidstone.

Institute of Water Pollution Control (1974). *Tertiary treatment and advanced wastewater treatment*. Institute of Water Pollution Control, Maidstone.

Institute of Water Pollution Control (1975). *Glossary of terms used in water pollution control*. Institute of Water Pollution Control, Maidstone.

Institute of Water Pollution Control (1978). *Sewage sludge. III. Utilization and disposal*. Institute of Water Pollution Control, Maidstone.

Institute of Water Pollution Control (1979). *Sewage sludge. I. Production, preliminary treatment and digestion*. Institute of Water Pollution Control, Maidstone.

Institute of Water Pollution Control (1980). *Primary sedimentation*. Institute of Water Pollution Control, Maidstone.

Institute of Water Pollution Control (1981). *Sewage sludge. II. Conditioning, dewatering and thermal drying*. Institute of Water Pollution Control, Maidstone.

Institute of Water Pollution Control (1984). *Preliminary processes*. Institute of Water Pollution Control, Maidstone.

Institute of Water Pollution Control (1986). *Agricultural use of sewage sludge – is there a future?* Institute of Water Pollution Control, Maidstone.

Institute of Water Pollution Control (1987a). *Tertiary treatment and advanced wastewater treatment*. Institute of Water Pollution Control, Maidstone.

Institute of Water Pollution Control (1987b). *Activated sludge*. Institute of Water Pollution Control, Maidstone.

International Standards Organization (1983). *Water quality determination of biochemical oxygen demand after 'n' days (BOD_n), dilution and seeding method*. International Organization for Standardization, ISO 5815.

Irving, T. E. (1977). *Preliminary investigations on the effects of particulate matter on the fate of sewage bacteria in the sea*. Technical Report TR55. Water Research Centre, Stevenage.

Irving, T. E. (1980). *Sewage chlorination and bacterial regrowth*. Technical Report TR132. Water Research Centre, Stevenage.

Irving, L. G. and Smith, F. A. (1981). One-year survey of enteroviruses, adenoviruses and reoviruses isolated from effluent at an activated-sludge purification plant. *Applied Environmental Microbiology*, **41**, 51–9.

Irving, T. E. and Solbe, J. F. de L. G. (1980). *Chlorination of sewage and effects of disposal of chlorinated sewage: a review of the literature*. Technical Report TR130. Water Research Centre, Stevenage.

Isaac, C. G. and McFiggans, A. (1981). Nature and disposal of effluents from malting, brewing and distilling. In *Food industry wastes: disposal and recovery*, (ed. A. Herzka and R. G. Booth), pp. 144–60. Applied Science Publishers, London.

Jack, E. J. and Hepper, P. T. (1969). An outbreak of *S. typhimurium* infection in cattle associated with the spreading of slurry. *Veterinary Records*, **84**, 196–99.

Jackson, C. J. and Lines, G. T. (1972). Measures against water pollution in the fermentation industries. *Pure and Applied Chemistry*, **29**, 381–93.

Jackson, J. D. and Gould, B. W. (1981). *Sewage treatment with water weeds*. Proceedings of the Australian Water and Waste-water Assocation, 1981 National Conference, pp. 7–15. Perth, Australia.

James, A. (1964). The bacteriology of trickling filters. *Journal of Applied Bacteriology*, **27**(2), 197–207.

James, A. and Addyman, C. L. (1974). *By-product recovery from food wastes by microbial protein recovery*. A symposium on treatment of wastes from the food and drink industry, University of Newcastle-upon-Tyne, 8–10 January 1974. Institute of Water Pollution Control, Maidstone.

James, O. S. and Wilbert, R. M. (1962). Infectious hepatitis traced to consumption of raw oysters. *American Journal of Hygiene*, **75**, 90–111.

Jamieson, W., Madri, P., and Claus, G. (1976). Survival of certain pathogenic micro-organisms in sea water. *Hydrobiologia*, **50**, 117–21.

Jank, B. E. and Bridle, T. R. (1983). Principles of carbonaceous oxidation, nitrification and denitrification in single-sludge activated sludge plants. In *Oxidation ditches in wastewater treatment*, (ed. D. Barnes, C. F. Forster and D. W. M. Johnstone), pp. 16–40. Pitman, London.

Jannasch, H. W. (1968). Competitive elimination of Enterobacteriaceae from seawater. *Applied Microbiology*, **16**, 1616–18.

Janus, H. (1965). *The young specialist looks at land and freshwater molluscs*. Burke, London.

Jeger, L. M. (1970). *Taken for granted*. Report of the working party on sewage disposal. Ministry of Housing and Local Government. HMSO, London.

Jenkins, D., Neethling, J. B., Bode, H., and Richard, M. G. (1982). The use of chlorination for control of activated sludge bulking. In *Bulking of activated sludge*, (ed. B. Chambers and E. J. Tomlinson)), pp. 187–206. Ellis-Horwood, Chichester.

Jenkins, D., Richard, M. G., and Neethling, J. B. (1983). Causes and control of activated sludge bulking. *Water Pollution Control*.

Jenkins, D., Richards, M. G., and Daigger, G. T. (1984). *Manual on the causes and control of activated sludge bulking and foaming*. Water Research Commission, Pretoria, South Africa.

Jenkinson, I. R. (1972). Sludge dumping and benthic communities. *Marine Pollution Bulletin*, **3**, 102–5.

Jephcote, A. E., Begg, N., and Baker, I. (1986). An outbreak of Giardiasis. *Lancet*, 730–2.

Jeris, J. S., Owens, R. W., Hickey, R., and Flood, F. (1977). Biological fluidized-bed treatment for BOD and nitrogen removal. *Journal of the Water Pollution Control Federation*, **49**, 816–31.

Jewell, W. J. (1971). Aquatic weed decay: dissolved oxygen utilization and nitrogen and phosphorus regeneration. *Journal of the Water Pollution Control Federation*, **43**, 1457–67.

Jewell, W. J. (1982). *Anaerobic attached film expanded bed fundamentals*. Proceedings of the 1st International Conference on Fixed Film Biological Processes, pp. 17–42. Ohio.

Jin You, X. *et al.* (1982). *Journal of South China Normal College*, **1**, 88–94.

Jofre, J., Bosch, A., Lucena, F., Girones, R., and Tartera, C. (1986). Evaluation of *Bacteroides fragilis* bacteriophages as indicators of the virological quality of water. *Water Science and Technology*, **18** (10), 167–73.

Johnson, M. S., Bradshaw, A. D., and Handley, J. F. (1976). Revegetation of metalliferous fluorspar mine tailings. *Institution of Mining and Metallurgy, Section A*, **85**, A32.

Johnson, R. D. (1973). Land treatment of wastewater. *The Military Engineer*, **65**, 375.

Johnstone, D. W. M. and Carmichael, W. F. (1982). Cirencester carrousel plant: some process considerations. *Water Pollution Control*, **81**, 587–600.

Johnstone, D. W. M., Rachwal, A. J., and Hanbury, M. J. (1979). Settlement characteristics and settlement tank performance in the Carrousel system. *Water Pollution Control*, **78**, 337–56.

Johnstone, D. W. M., Rachwal, A. J., Hanbury, M. J., and Critchard, D. J. (1980). *Design and operation of final settlement tanks: use of stirred specific volume index and mass flux theory*. Paper presented to 33rd International Conference CEBEDEAU, Liege, Belgium, May 1980.

Jolley, R. L. (1975). Chlorine-containing organic constituents in chlorinated effluents. *Journal of the Water Pollution Control Federation,*, **47**, 601–18.

Jones, F., Smith, J. F., and Watson, D. C. (1978). Pollution of a water supply catchment by breeding gulls and the potential environmental health implications. *Journal of the Institution of Water Engineers and Scientists*, **32**, 469–82.

Jones, F. and White, W. R. (1984). Health and amenity aspects of surface waters. *Water Pollution Control*, **83**, 215–25.

Jones, G. E. (1964). Effect of chelating agents on the growth of *Escherichia coli* in seawater. *Journal of Bacteriology*, **87**, 483–99.

Jones, M. B. (1986). Wetlands. In *Photosynthesis in contrasting environments*, (ed. N. R. Baker and S. P. Long), p. 103–38. Elsevier, Amsterdam.

Jones, P. H. and Prasad, D. (1969). The use of tetrazolium salts as a measure of sludge activity. *Journal of the Water Pollution Control Federation*, **41**, R441–R449.

Jones, P. H. and Sabra, H. M. (1980). Effect of systems solids retention time (SSRT or sludge age) on nitrogen removal from activated-sludge systems. *Water Pollution Control*, **79**, 106–16.

Jones, R. E. (1978). *Heavy metals in the estuarine environment*. Technical Report 78. Water Research Centre, Stevenage.

Jones, S. N. (1978). Caroussel plant assessment project No. PA1302RX. Report No. 3. Welsh Water Authority and Water Research Centre, Stevenage.

Josephson, J. (1975). Green systems for wastewater treatment. *Environmental Science and Technology*, **9**(5), 408–9.

Joshi, S. R., Parhad, N. M., and Rao, N. U. (1973). Elimination of *Salmonella* in stabilized ponds. *Water Research*, **7**, 1357–65.

Joslin, J. R., Sidwick, J. M., Greene, C., and Shearer, J. R. (1971). High rate biological filtration: a comparative assessment. *Water Pollution Control* **70**, 383–99.

Kato, A., Izaki, K., and Takahashi, H. (1971). Floc forming bacteria isolated from activated sludge. *Journal of General Applied Microbiology*, **17**, 439.

Kato, K. and Sekikawa Y. (1967). *Fixed activated sludge process for industrial waste treatment*. Proceedings of the 22nd Industrial Waste Conference, Purdue University, pp. 926–49.

Katzenelson, E. and Biederman, N. (1976). Disinfection of viruses in sewage by ozone. *Water Research*, **10**, 629–31.

Keefer, C. E. (1963). Relationship of sludge density index to the activated sludge process. *Journal of the Water Pollution Control Federation*, **35**, 1166–73.

Keefer, C. E. and Meisel, J. T. (1953). Activated sludge studies. IV. Sludge age and its effect on the activated sludge process. *Sewage and Industrial Wastes*, **25**, 898–908.

Keenan, T. (1982). A review of current practices relating to the design and operation of septic tank treatment systems in Ireland. M.Sc. thesis, Trinity College, University of Dublin.

Kehr, D. and Emde, W. (1960). Experiments on the high-rate activated sludge process. *Journal of the Water Pollution Control Federation*, **32**, 1066–80.

Keller, A. J. and Cole, C. A. (1973). Hydrogen peroxide cures bulking. *Water Wastes Engineering*, **10**, E4, E6–E7.

Kelly, D. P., Norris, P. R., and Brierly, C. L. (1979). Microbiological methods for the extraction and recovery of metals. In *Microbial technology: current state, future prospects*, (ed. A. T. Bull, D. C. Ellwood, and C. Rutledge), pp. 263–308. Cambridge University Press.

Kenard, R. P. and Valentine, R. S. (1974). Rapid determination of the presence of enteric bacteria in water. *Applied Microbiology*, **27**, 484–7.

Kennedy, K. J. and Berg, L. van den. (1982). Anaerobic digestion of piggery waste using a stationary film reactor. *Agricultural Wastes*, **4**, 151–8.

Kent-Kirk, J. (1975). Lignin-degrading enzyme system. In *Cellulose as a chemical and energy resource*. Proceedings Symposium. 5. Biotechnology and Bioengineering, (ed. C. R. Wilke), pp. 139–50. John Wiley, New York.

Khaleel, R., Reddy, K. R., and Overcash, M. R. (1981). Changes in soil physical properties due to organic waste applications: a review. *Journal of Environmental Quality*, **10**, 133–41.

Khalid, R. A., Patrick, W. H., and Mixon, M. N. (1982). Phosphorus removal processes from overland flow treatment of simulated wastewater. *Journal of the Water Pollution Control Federation*, **54**, 61–9.

Kiff, R. J. (1972). The ecology of nitrification. I. Denitrification systems in activated sludge. *Water Pollution Control*, **71**, 475–84.

Kiff, R. J. and Brown, S. (1981). *The development of the oxidative acid hydrolysis process for sewage sludge detoxification*. International Conference on Heavy Metals in the Environment. Amsterdam, September, 1981.

Kiff, R. J. and Lewis-Jones, R. (1984). Factors that govern the survival of selected parasites in sewage sludges. In *Sewage sludge stabilization and disinfection*, (ed. A. M. Bruce), pp. 452–61. Ellis-Horwood, Chichester.

Kimerle, R. A. and Anderson, N. H. (1971). Production and bioenergetic role of the midge Glyptotendipes barbipes (Staeger) in a waste stabilization lagoon. *Limnology and Oceanography*, **16**, 646–59.

Kimerle, R. A. and Enns, W. R. (1968). Aquatic insects associated with mid-western stabilization lagoons. *Journal of the Water Pollution Control Federation*, **40**(2), R31–R41.

Kincannon, D. F. and Gaudy, A. F. (1966). Some effects of high salt concentrations on activated sludge. *Journal of the Water Pollution Control Federation*, **38**, 1148–59.

King, E. D., Mill, R. A., and Lawrence, J. (1973). Airborne bacteria from an activated sludge plant. *Journal of Environmental Health*, **36**, 50.

Kinman, R. M. (1972). Ozone in water disinfection. In *Ozone in water and wastewater treatment*, (ed. F. L. Evans), pp. 123–43. Ann Arbor Science Publishing, Michigan.

Kirsch, E. J. (1968). Studies on the enumeration and isolation of obligate anaerobic bacteria from digesting sewage sludge. *Developments in Industrial Microbiology*, **10**, 170–6.

Kirsop, B. H. (1981). Research into utilization of food industry wastes. In *Food industry wastes: disposal and recovery*, (ed. A. Herzka and R. G. Booth), pp. 210–8. Applied Science Publishers, London.

Kist, M. (1985). The historical background to campylobacter infection: new aspects. In *Campylobacter III* (ed. A. D. Pearson, M. B. Skirrow, H. Lior and R. Rowe), 23–28. Public Health Laboratory Service, London.

Kitamori, R. (1971). Changes in the biotic community due to the variation of the polluted water along the coast. *Chiyuku Kaihato*, February, 34–40.

Klass, D. L. and Ghosh, S. (1980). *Methane production by anaerobic digestion of water hyacinth* (Eichornia crssipes). Paper presented at Fuels from biomass and wastes. American Chemical Society, San Francisco.

Kleeck, L. W. van. (1956). Operation of septic tank plants. *Wastes Engineering (New York)*, 23–6.

Klein, L. (1972a). *River pollution. 1. Chemical analysis*. Butterworths, London.

Klein, L. (1972b). *River pollution. 2. Causes and effects*. Butterworths, London.

Klein, S. A. and McGauhey, P. H. (1965). Biodegradation of biologically soft detergents by wastewater treatment processes. *Journal of the Water Pollution Control Federation*, **37**, 857–66.

Kleins, J. A. and Lee, D. D. (1979). Biological treatment of aqueous wastes from coal conversion processes. In *Biotechnology in energy production and conservation*, (ed. C. D. Scott), pp. 379–90. John Wiley, London.

Knackmuss, H. J. (1981). Degradation of halogenated and sulphonated hydrocarbons. In *Microbial degradation of xenobiotics and recalcitrant compounds*, (ed. J. Leisinger, R. Hutter, A. M. Cook, and J. Nuesch), pp. 189–212. Academic Press, London.

Knill, M. J., Suckling, W. G., and Pearson, A. D. (1982). Campylobacters from water. In *Campylobacter: epidemiology, pathogensis and biochemistry*, (ed. D. G. Newell), pp. 281–4. MTP Press, Lancaster.

Knop, E. (1966). Design studies for the Emscher Mouth Treatment Plant. *Journal of the Water Pollution Control Federation*, **38**, 1194–1207.

Kobayashi, M. (1977). Utilization and disposal of wastes by photosynthetic bacteria.

In *Microbial energy conversion*, (ed. H. G. Schlegel and J. Barnea), pp. 443–53. Pergamon, Oxford.

Koerner, E. L. and Haws, D. A. (1979) *Long-term effects of land application of domestic wastewater, Roswell, New Mexico Slow-Rate irrigation site.* EPA 600.2–79–047, US Environmental Protection Agency.

Kokholm, G. (1981). *Redox measurement – their theory and technique.* Radiometer Press, Copenhagen.

Kolkwitz, R. and Marsson, M. (1908). Ökologie der Pflanzlichen saprobien. *Bericht der Deutschen Botanischen Gesellschaft*, **261**, 505–19.

Kollins, S. A. (1966). The presence of human enteric viruses in sewage and their removal by conventional sewage treatment methods. *Advances in Applied Microbiology*, **8**, 145–93.

Komolrit, K., Goel, K. C., and Gaudy, A. F. (1967). Regulation of exogenous nitrogen supply and its possible applications to the activated sludge process. *Journal of the Water Pollution Control Federation*, **39**, 251–66.

Konstandt, H. G. (1976). *Microbial energy conversion*, (ed. H. G. Schlegel and J. Barnea). Erich Goltze, Gotingen.

Kothandaraman, V., Thergaonkar, V. P., Koshy, T., and Ganapati, S. V. (1963). Physio-chemical and biological aspects of Ahmedabad sewage. *Environmental Health (India)*, **5**, 356–63.

Kott, Y., Roze, N., Sperber, S., and Beter, N. (1974). Bacteriophage as viral pollution indicators. *Water Research*, **8**, 165–71.

Kool, H. J. (1979). Treatment processes applied in public water supply for the removal of micro-organisms. In *Biological indicators of water quality*, (ed. A. James and L. Evison), pp. 17.1–17.31. John Wiley, Chichester.

Kool, H. J. and Kranen, H. J. van (1977). Het inaktiversen van diverse micro-organismen in water met behulp van een chloorbehendeling. *RID Rapport CBA*, 78–01.

Krishnaswami, S. K. (1971). Health aspects of land disposal of municipal wastewater effluents. *Canadian Journal of Public Health*, **62**, 36–42.

Kudo, R. P. (1932). *Protozoology.* C. C. Thomas, Illinois.

Kuzma, R. J., Kuzma, C. M., and Buncher, C. R. (1977). Ohio drinking water source and cancer rates. *American Journal of Public Health*, **67**, 725–9.

Laak, R. (1986). *Wastewater engineering design for unsewered areas.* (2nd edn). Technomic Publishing, Basel.

Lack, T. J. and Johnson, D. (1985). Assessment of the biological effects of sewage sludge at a licensed site off Plymouth. *Marine Pollution Bulletin*, **16**, 147–52.

Lackey, J. B. (1924). Studies of the fauna of Imhoff tanks and sprinkling beds. *Bulletin of the New Jersey Agricultural Experimental Station*, **403**, 40–60.

Lackey, J. B. (1925). The fauna of Imhoff tanks. I. Ecology of Imhoff tanks. *Bulletin of the New Jersey Agricultural Experimental Station*, **417**, 1–39.

Lakshminarayana, J. S. S. (1965). Discussion of paper presented by Loedolff. In *Advances in Water Pollution Research* (ed. O. Jaag), Vol. 1. pp. 320–4. Pergamon, Oxford.

Landine, R. C. (1971). Second order and first order kinetics of BOD data. *Water Sewage Works*, **118**, R45–R53.

Larkin, E. P., Tierney, J. T., and Sullivan, R. (1976). Persistence of virus on sewage irrigated vegetables. *Journal of the Environmental Engineering Division, American Society of Civil Engineers*, **102**(EE1), 29–55.

Larkin, S. B. C. (1982). The distribution of agricultural wastes and residues in the United Kingdom. *The Digest*, **6**. Newsletter of the British Anaerobic and Biomass Association Ltd.

Lau, A. O., Strom, P. F., and Jenkins, D. (1984*a*). Growth kinetics of *Sphaerotilus natans* and a floc former in pure and dual continuous culture. *Journal of the Water Pollution Control Federation*, **56**, 41–52.

Lau, A. O., Strom, P. F., and Jenkins, D. (1984*b*). The competitive growth of floc-forming and filamentous bacteria: a model for activated sludge bulking. *Journal of the Water Pollution Control Federation*, **56**, 52–64.

Laudelout, H. and Tichelen, L. V. (1960). Kinetics of the nitrate oxidation by *Nitrobacter winogradskyi*. *Journal of Bacteriology*, **79**, 39–42.

Lavagno, E. *et al.* (1983). An analytic model to study the performance of an anaerobic digester. *Agricultural Wastes*, **5**, 37–50.

Lavender, P. A. (1970). *Percolating filter media properties*. Report to Imperial College, London, in partial fulfilment of requirements for Diploma of Membership of Imperial College.

Lawford, G. R., Kligerman, A., and Williams, T. (1979). Production of high-quality edible protein from *Candida* yeast grown in continuous culture. *Biotechnology and Bioengineering*, **21**, 1163–74.

Lawrence, A. W. and McCarty, P. L. (1970). A unified basis for biological treatment design and operation. *Journal of the Sanitary Engineering Division, American Society of Civil Engineers*, **96**(SA3), 757–78.

Lawrence, P. N. (1970). Collembola (Spring tails) of sewage filters. *Water and Waste Treatment*, **13**, 106–9.

Learner, M. A. (1972). Laboratory studies on the life-histories of four enchytraeid worms (Oligochaeta) which inhabit sewage percolating filters. *Annals of applied Biology*, **70**, 251–66.

Learner, M. A. (1975*a*). The ecology and distribution of invertebrates which inhabit the percolating filters of sewage-works. Ph.D. Thesis, University of London.

Learner, M. A. (1975*b*). Insecta. In *Ecological aspects of used-water treatment*, (ed. C. R. Curds and H. A. Hawkes), pp. 337–74. Academic Press, London.

Le Blanc, P. J. (1974). Review of rapid BOD test methods. *Journal of the Water Pollution Control Federation*, **46**, 2202–8.

Lechevalier, H. A. and Lechevalier, M. P. (1975). *Actinomycetes of sewage treatment plants*. EPA Report No. 600/2–75/031, US Environmental Protection Agency, Ohio.

Lechevalier, H. A., Lechevalier, M. P., and Wyszkowski, P. E. (1977). *Actinomycetes of sewage treatment plants*. EPA Report No. 600/2–77/145, US Environmental Protection Agency, Ohio.

Lechevalier, M. P. and Lechevalier, H. A. (1974). *Nocardia amaerae* sp. nov., an Actinomycete common in foaming activated sludge. *International Journal of Systematic Bacteriology*, **24**, 278–88.

Leclerc, H., Mossel, D. A. A., Trinel, P. A., and Gavini, F. (1976). A new test for faecal contamination. In *Bacterials indicators/health hazards associated with water*, (ed. A. W. Hoadley and B. J. Dutka). ASTM Publication, 635.

Lee, J. A. (1974). Recent trends in human salmonellosis in England and Wales: the epidemiology of prevalent serotypes other than *Salmonella typhimurium*. *Journal of Hygiene, Cambridge*, **72**, 185–95.

Lee, J. V., Bashford, D. J., Donovan, T. J., Turniss, A. L., and West, P. A. (1982). The incidence of *Vibrio cholerae* in water, animals and birds in Kent, England. *Journal of Applied Bacteriology*, **52**, 281–8.

Lee, S.-E., Koopman, B. L., Jenkins, D., and Lewis, R. F. (1982). The effect of aeration basin configuration on activated sludge bulking at low organic loading. *Water Science, Technology*, **14**, 407.

Leeper, G. W. (1978). *Managing the heavy metals on the land*. Marcel Dekker, New York.

Leisenger, T. (1983). Microbial degradation of recalcitrant compounds. In *Environmental biotechnology: future prospects*, (ed. R. S. Holdom, J. M. Sidwick, and A. Wheatley). The European Federation of Biotechnology, CEC, Brussels.

Leisinger, T., Hutter, R., Cook, A. M., and Nuesch, J. (1981) (ed.). *Microbial degradation of xenobiotic and recalcitrant compounds*. FEMS Symposium, 12. Academic Press, London.

Lemmel, S. A., Heimsch, R. C., and Edwards, L. L. (1979). Optimising the continuous culture of *Candida utilis* and *Saccharomycopsis fibuliger* on potato processing wastewater. *Applied Environmental Microbiology*, **37**, 227–32.

Lemmer, H. and Kroppenstedt, R. M. (1984). Chemotaxonomy and physiology of some Actinomycetes isolated from scumming activated sludge. *Systematic and Applied Microbiology*, **5**, 124–35.

Lentz, J. J. (1963). *Special report No. 4 of the residential sewerage research project to the Federal Housing Administration*. John Hopkins University, Baltimore.

Le Roux, N. W. (1982). Odour control at Turiff STW. *Water Research News*, **12**, 6–7.

Le Roux, N. W. and Mehta, K. B. (1980). The biological oxidation of odours. In *Odour control – a concise guide* (ed. F. H. H. Valentine and A. A. North). Department of Industry, Warren Spring Laboratory.

Lettinga, G., Velsen, A. F. M. van, Hobma, S. W., Zeeuw, W. de, and Klapwijk, A. (1980). Use of the upflow sludge blanket (USB) reactor concept for biological waste treatment, especially for anaerobic treatment. *Biotechnology and Bioengineering*, **22**, 699–734.

Lettinga, G. *et al.* (1983). *Upflow sludge blank processes*. In: Proceedings 3rd International Symposium on Anaerobic Digestion, 139–58. Massachusetts.

Levine, M., Luebbers, R., Galligan, W. E., and Vaughan, R. (1936). Observations on ceramic filter media and high rates of filtration. *Sewage Works Journal*, **8**, 701–27.

Lewandowski, T. P. (1974). The use of high-purity oxygen in sewage treatment. *Water Pollution Control*, **73**, 647–55.

Lewin, V. H. and Henley, J. R. (1978). Diffused air supersedes mechanical surface aeration at Oxford. *Effluent Water Treatment Journal*, **18**, 163–5.

Lewis, C. W. (1983). *Biological fuels*. Arnold, London.

Lewis, C. W. (1988). Energy from biomass: present reality and future prospects. *Science Progress, Oxford*, **72**, 511–30.

Liebmann, H. (1949). The biology of percolating filters. *Vom Wasser*, **17**, 62–82.

Liebman, H. (1951). *Handbuch der Frishwasser- und Abwasser-biologie*. Gustav Fischer, Jena.

Liebmann, H. (1965). Parasites in sewage and the possibilities of their extinction. In *Advances in Water Pollution Research*. Proceedings of the 2nd International Conference on Water Pollution Research, Tokyo, 1964, (ed. J. K. Baars), Vol. 2, pp. 269–76. Pergamon, Oxford.

Lighthart, B. and Oglesby, R. T. (1969). Bacteriology of an activated sludge wastewater treatment plant – a guide to methodology. *Journal of the Water Pollution Control Federation*, **41**, R267–R281.

Ligman, K., Mutzler, N., and Boyle, W. C. (1974). Household wastewater characterization. *Journal of the Environmental Engineering, American Society of Civil Engineers*, **100**, 201.

Lin, K. C. and Heinke, G. W. (1977a). Plant data analysis of temperature significance in the activated sludge process. *Journal of the Water Pollution Control Federation*, **49**, 286–95.

Lin, K. C. and Heinke, G. W. (1977b). Variability of temperature and other process parameters: a time series analysis of activated plant data. *Progress in Water Technology*, **9**, 347–63.

Lin, S. D. (1985). *Gardia lamblia* and water supply. *Journal of the American Water Works Association*, **77**, 40–7.

Linden, A. C. van der and Thijsse, G. (1965). The mechanisms of microbial oxidations of petroleum hydrocarbons. *Advances in Enzymology*, **27**, 469.

Linfield, R. (1977). Potato cyst eelworm studies in the Anglian Water Authority. In *Research seminar on pathogens in sewage sludge*. Research and Development Division Technical Note no. 7. Department of the Environment, London.

Linklater, K. A., Graham, M. M., and Sharp, J. C. M. (1985). Salmonellae in sewage sludge and abattoir effluent in south-east Scotland. *Journal of Hygiene, Cambridge*, **94**, 301–7.

Linton, K. B. *et al.* (1972). Antibiotic resistance and transmissible R-factors in the intestinal coliform flora of healthy adults and children in an urban and rural community. *Journal of Hygiene*, **70**, 99–104.

Lister, A. R. and Boon, A. G. (1973). Aeration in deep tanks: an evaluation of a fine-bubble diffused-air system. *Water Pollution Control*, **72**, 590–605.

Little, M. and Williams, P. A. (1971). A bacterial halldo-hydrolase, its purification, some properties and its modification by specific amino acid reagents. *European Journal of Biochemistry*, **21**, 99–108.

Lloyd, L. (1945). Animal life in sewage purification processes. *Journal and Proceedings of the Institute of Sewage Purification*, 119–39.

Lloyd, L., Graham, J. F., and Reynoldson, T. B. (1940). Materials for a study in animal competition. The fauna of the sewage bacteria beds. *Annals of Applied Biology*, **27**, 122–50.

Lo, S., Gilbert, J., and Hetrick, F. (1976). Stability of human enteroviruses in estuarine and marine waters. *Applied Environmental Microbiology*, **32**, 245–9.

Loedolff, C. J. (1965). The function of cladorea in oxidation ponds. In *Advances in Water Pollution Research*, (ed. O. Jaag), Vol. 1, pp. 307–17. Pergamon, Oxford.

Loehr, R. C. (1977). *Land as a waste management alternative*. Ann Arbor Science, Ann Arbor, Michigan.

Loll, U. (1977). Engineering, operation and economics of biodigesters. In *Microbial energy conversion*, (ed. H. G. Schlegel and J. Barnea), pp. 361–78. Pergamon, Oxford.

Lopez-Real, J. and Foster, M. (1985). Plant pathogen survival during the composting of agricultural organic wastes. In *Composting of agricultural and other wastes*, (ed. J. K. R. Gasser), pp. 291–300. Elsevier, London.

Loveless, J. E. and Painter, H. A. (1968). The influence of metal concentration and pH value on the growth of a *Nitrosomonas* strain isolated from activated sludge. *Journal of General Microbiology*, **52**, 1–14.

Lowe, P. (1988). Incineration of sewage sludge: a reappraisal. *Water and Environmental Management*, **2**, 416–22.

Lumb, B. C. and Eastwood, P. K. (1958). The recirculation principle in filtration of settled sewage – some comments on its application. *Journal and Proceedings of the Institute Sewage Purification*, 380–98.

Lumb, C., Brown, D., and Bottomley, M. K. V. (1977). A chloroform problem at Great Watford and its solution. *Water Pollution Control*, **76** 459–67.

Lumbers, J. P. (1983). Rotating biological contractors: current problems and potential developments in design and control. *Public Health Engineering*, **11**, 41–5.

Lumley, D. J. (1985). *Settling of activated sludge*, Publication 6:85. Chalmers University of Technology, Goteborg, Sweden.

Lund, E. and Rønne, V. (1973). On the isolation of virus from sewage treatment plant sludges. *Water Research*, **7**, 863–71.

Lundgren, D. G. and Malouf, E. E. (1983). Microbial extraction of metals. In *Advances in biotechnological processes*, (ed. A. Mizrahi and A. L. van Wezel), pp. 223–49. A. L. Liss, New York.

Lundgren, D. L., Clapper, W. E., and Sanchez, A. (1968). Isolation of human enteroviruses from beagle dogs. *Proceedings of the Society of Experimental Biology Medicine*, **128**, 463–7.

Lundholm, M. and Rylander, R. (1983). Work-related symptoms among sewage workers. *British Journal of Industrial Medicine*, **40**, 325.

Lynch, J. M. and Poole, N. J. (ed.) (1979). *Microbial ecology: a conceptual approach*. Blackwell, Oxford.

Macan, T. T. (1959). *A guide to freshwater invertebrate animals*. Longmans, London.

Macaskie, L. E. and Dean, A. C. R. (1984). Cadmium accumulation by immobilized cells of a *Citrobacter* sp. *Environmental Technology Letters*, **5**, 177–86.

MacGregor, S. T., Miller, F. C., Psarianos, K. M., and Finstein, M. S. (1981). Composting process control based on interaction between microbial heat output and temperature. *Applied Environmental Microbiology*, **41**, 1321–30.

McCalley, D. V. (1980). Sterols as faecal pollution indicators. Ph.D. Thesis, University of Bristol.

McCalley, D. V., Cooke, M., and Nichless, G. (1981). Effect of sewage treatment on faecal sterols. *Water Research*, **15**, 1019–25.

McCarty, P. L. (1964). Anaerobic waste treatment fundamentals. I. Chemistry and microbiology; II. Environmental requirements and control; III. Toxic materials and their control. *Public. Works*, **95**, 91–4; 107–12; 123–6.

McCarty, P. L. (1971). Energetics and bacterial growth. In *Organic compounds in aquatic environments*, (ed. S. D. Faust and J. V. Hunter), pp. 495–504. Marcel Dekker, New York.

McCarty, P. L. (1972). *Stoichiometry of biological reactions*. In. Proceedings of the International Conference Toward a unified Concept of Biological Waste Treatment Design. Atlanta, Georgia.

McCarty, P. L. (1975). Stoichiometry of biological reactions. *Progress in Water Technology*, 7.

McCarty, P. L. and McKinney, R. E. (1961). Salt toxicity in anaerobic digestion. *Journal of the Water Pollution Control Federation*, **33**, 399–415.

McCarty, P. L. and Reinhard, M. (1980). Trace organics removal by advanced wastewater treatment. *Journal of the Water Pollution Control Federation*, **52**, 1907–22.

McCarty, P. L., Beck, L., and St. Amant, P. (1969). *Biological denitrification of wastewaters by addition of organic materials.* Proceedings of the 24th Industrial Waste Conference, Purdue University, pp. 1271–85.

McCornville, J. and Maier, W. J. (1979). Use of powdered activated carbon to enhance methane production and conservation. In *Energy production and conservation*, (ed. C. D. Scott). John Wiley, London.

McFarren, E. F. (1970). Criterion for judging acceptability of analytical methods. *Analytical Chemistry*, **42**, 358–65.

McFeters, G. A., Bissonnette, G. K., Joneski, J. J., Thomson, C. A., and Stuart, D. G. (1974). Comparative survival of indicator bacteria and enteric pathogens in well water. *Applied Microbiology*, **27**, 823–9.

McGarrigle, M. L. (1984). Experiments with a river model. *Irish Journal of Environmental Science*, **3**(1), 40–7.

McGlashan, J. E. (1983) (ed.). Modern trends in sludge management. *Water Science and Technology*, **15**(1).

McInerney, M. L., Bryant, M. P., and Stafford, D. A. (1980). Metabolic stages and energetics of microbial anaerobic digestion. In *Anaerobic digestion*, (ed. D. A. Stafford, B. I. Wheatley, and D. E. Hughes), pp. 91–8, Applied Science Publishers, London.

McIntyre, A. D. (1975). Discussion of 'heavy metals in the marine environment' by A. Jernelov. In *Discharge of sewage from sea outfalls*, (ed. A. L. H. Gameson), pp. 121–2. Pergamon, Oxford.

McIntyre, A. D. and Johnston, R. (1975). Effect of nutrient enrichment from sewage in the sea. In *Discharge of sewage from sea outfalls*, (ed. A. L. H. Gameson), pp. 131–41. Pergamon, Oxford.

McKenna, E. J. and Kallio, R. E. (1965). The biology of hydrocarbons. *Annual Reviews of Microbiology*, **19**, 183–208.

McKinney, R. E. (1957). Activity of micro-organisms in organic waste disposal. II. Aerobic processes. *Applied Microbiology (Baltimore)*, **5**, 167–87.

McKinney, R. E. and Gram, A. (1956). Protozoa and activated sludge. *Sewage and Industrial Wastes*, **28**, 1219–31.

McKinney, R. E. and Weichlein, R. G. (1953). Isolation of floc producing bacteria from activated sludge. *Applied Microbiology*, **1**, 259–61.

McNabb, C. D. (1976). The potential of submersed vascular plants for reclamation of wastewater in temperate zone ponds. In *Biological control of water pollution* (ed. J. Tourbier and R. W. Pierson), 123–32. University of Pennsylvania Press, Pennsylvania.

McNeil, K. E. (1984). The treatment of wastes from the sugar industry. In *Food and allied industries*, (ed. D. Barnes, C. F. Forster, and S. E. Hrudey), pp. 1–68. Pitman, London.

McWhirter, J. R. (ed.) (1978a). *The use of high-purity oxygen in the activated sludge process.* CRC Press, Florida.

McWhirter, J. R. (1978b). Oxygen and the activated sludge process. In *The use of high-purity oxygen in the activated sludge process*, (ed. J. R. McWhirter), pp. 25–62. CRC Press, Florida.

McWhirter, J. R. (1978c). Introduction. In *The use of high-purity oxygen in the activated sludge process*, (ed. J. R. McWhirter), pp. 3–14. CRC Press, Florida.

Madoni, O. and Ghetti, P. F. (1981). The structure of ciliated protozoan communities in biological sewage treatment plants. *Hydrobiology*, **83**, 207–15.

Malherbe, H. H. and Strickland-Cholmley, M. (1967). In *Transmission of viruses by the water route*, (ed. G. Berg), pp. 379–387, Interscience, New York.

Malina, J. F., Ranganathan, K. R., Sagik, B. P., and Moore, B. E. (1975). Poliovirus inactivation by activated sludge. *Journal of the Water Pollution Control Federation*, **47**, 2178–83.

Mandels, M. (1975). Microbial sources of cellulose. In *Cellulose as a chemical and energy resource*. Biotechnology and Bioengineering Symposium, Series 5 (ed. C. R. Wilke), pp. 81–105. John Wiley, New York.

Mandelstam, J. and McQuillen, K. (1973). *Biochemistry of bacterial growth* (2nd edn). Blackwell, Oxford.

Mandt, M. G. and Bell, B. A. (1982). *Oxidation ditches in wastewater treatment*. Ann Arbor Science.

Mann, H. T. (1974). Sewage treatment for small communities. *Environmental Conservation*, **1**, 145–52.

Mann, H. T. (1979). *Septic tanks and small sewage treatment plants*. Technical Report TR107. Water Research Centre, Stevenage.

Mara, D. D. (1974). *Bacteriology for sanitary engineers*. Churchill Livingstone, Edinburgh.

Mara, D. D. and Marecos Do Monte, M. H. (ed.) (1987). Waste stabilization ponds. *Water Science and Technology*, **19** (12), 1–401.

Marais, G. V. R. (1970). Second International Symposium on Waste Stabilization Lagoons, Kansas City.

Maris, G. V. R. (1973). *The activated sludge process at long sludge ages*. Research Report W3. Department of Civil Engineering, Universty of Cape Town, South Africa.

Marcola, B., Watkins, J., and Riley, A. (1981). The isolation and identification of thermotolerant campylobacter spp. from sewage and river waters. *Journal of Applied Bacteriology*, **51**, xii–xiv.

Marcus, J. H., Sutcliffe, D. W., and Willoughby, L. G. (1978). Feeding and growth of *Asellus aquaticus* (Isopoda) on food items from the littoral of Windermere, including green leaves of *Elodea canadensis*. *Freshwater Biology*, **8**, 505–19.

Marine Conservation Society (1987). *The golden list of British beaches*. The coastal Anti-Pollution League, Marine Conservation Society, England.

Martin, D. (1968). *Microfauna of biological filters*. Civil Engineering Bulletin **39**. University of Newcastle upon Tyne, Oriel Press.

Mason, C. F. (1981). *Biology of freshwater pollution*. Longman, London.

Mason, W. T. (1968). *An introduction to the identification of chironomid larvae*. Division of Pollution Surveillance, Federal Water Pollution Control Administration, U.S. Department of the Interior.

Mathur, R. P., Bhargava, R. and Gupta, A. (1986). Anaerobic filtration of a distillery waste. *Effluent and Water Treatment Journal*, **26**, 89–91.

Matsche, N. F. (1977). Blahschlamm-Versuchen und Berkaufung-Weiner Mittelunger. *Wasser-Abwasser-Gewasser*, **22**, 1.

Matson, E. A., Horner, S. G., and Buck, J. D. (1978). Pollution indicators and other micro-organisms in river sediment. *Journal of the Water Pollution Control Federation*, **50**(1), 13–9.

Matsui, K. and Kimata, T. (1986). Performance evaluation of the oxidation ditch process. *Water Science and Technology*, **18**, 297–306.

Matthews, P. J. (1980). *Sewage treatment – its effect on land management and reclamation.* Paper presented to The Reclamation and Development of Contaminated Land. 8–9 December 1980. Scientific and Technical Studies, London.

Matthews, P. J. (1983). The proposal for an EEC Directive on utilization of sewage sludge in agriculture – an introductory review. *Effluent and Water Treatment Journal*, **23**, 94–9.

Matthews, P. J. (1986). EEC proposal for a council directive on the dumping of waste at sea. *Effluent and Water Treatment Journal*, **26**, 96–9.

Maurer, E. W. *et al.* (1965). The effect of tallow based detergents on anaerobic digestion. *Journal of the American Oil Chemistry Society*, **42**, 189–92.

Mehta, D. S., Davis, H. H., and Kingsbury, R. P. (1972). Oxygen theory in biological treatment plant design. *Journal of Sanitary Engineering Division, American Society of Civil Engineers*, **98**(SA3), 471–89.

Melmed, L. N. (1976). Disinfection of municipal wastewater with gamma radiation. *Water, South Africa*, **2**, 131–5.

Mentzing, L. O. (1981). Waterborne outbreaks of *Campylobacter* enteritis in central Sweden. *Lancet* **ii**, 352–4.

Metcalf and Eddy Inc. (1984). *Wastewater engineering: treatment, disposal and reuse.* McGraw Hill, New York.

Meynell, J. P. (1976). *Methane: planning a digester.* Prism Press, Dorset.

Meyrath, J. and Bayer, K. (1979). Biomass from whey. In *Microbial biomass*, (ed. A. H. Rose), pp. 207–69. Academic Press, London.

Middlebrooks, E. J. *et al.* (1979). *Lagoon information source book.* Ann Arbor Science Publishers.

Middlebrooks, E. J. *et al.* (1982). *Wastewater stabilization lagoon design, performance and uprating.* Macmillan, New York.

Miller, R. H. (1974). *The soil as a biological filter.* EPA REport 660/2–74–003. US Environmental Protection Agency.

Milner, H. W. (1953). Algae as food. *Scientific American*, **31**, (October).

Millner, P. D., Marsh, P. B., Snowden, R. B., and Parr, J. F. (1977). Occurence of *Aspergillus fumigatus* during composting of sewage sludge. *Applied Environmental Microbiology*, **34**, 765–72.

Mills, E. V. (1945). The treatment of settled sewage in percolating filters in series with periodic change in the order of the filters: results of operation of the experimental plant at Minworth, Birmingham, 1940–1944. *Journal and Proceedings of the Institute of Sewage Purification*, 35–55.

Ministry of Agriculture, Fisheries and Food. (1973). *Survey of mercury in food.* Working party on the monitoring of mercury and other heavy metals. Supplementary Report. HMSO, London.

Ministry of Health (1936). *Methods of chemical analysis as applied to sewage and sewage effluents.* HMSO, London.

Ministry of Housing and Local Government (1970). *Technical committee on storm overflows and the disposal of storm sewage.* Final Report, 1970. HMSO, London.

Ministry of Technology (1968). *The use of plastic filter media for biological filtration.* Notes on Water Pollution No. 40. HMSO, London.

Ministry of Technology (1970). *Water pollution research, 1969.* HMSO, London.

Mitchell, D. S. (1978). The potential for wastewater treatment by aquatic plants in Australia. *Water*, **5**(4), 15–7.

Mitchell, D. S. (1979). *The potential for using aquatic plants to treat effluent*. Research conference on piggery effluent utilization and disposal. Australian Pig Industry Research Committee, Melbourne.

Mitchell, N. T. and Eden, G. E. (1963). Some measurements of rates of ventilation in a percolating filter. *Water and Waste Treatment*, **9**, 366–70.

Mitchell, R. (1971). Destruction of bacteria and viruses in sea-water. *Journal of the Sanitary Engineering Division, American Society of Civil Engineers*, **97**(SA4), 425–32.

Mitchell, R. (1974). *Introduction to environmental microbiology*. Prentice-Hall, New York.

Mitchell, R. and Chamberlin, C. (1975). Factors influencing the survival of enteric microorganisms in the sea – an overview. In *Discharge of sewages from sea outfalls*, (ed. A. L. H. Gameson), pp. 237–51. Pergamon, London.

Mitchell, R., Yankofsky, S., and Jannasch, H. W. (1967). Lysis of *Escherichia coli* by marine micro-organisms. *Nature (London)*, **215**, 891–3.

Moncrieff, D. S. (1953). The effect of grading and shape on the bulk density of concrete aggregates. *Magazine of Concrete Research*, **5**, 67–70.

Monk, C. D. and Brown, T. W. (1965). Ecological consideration of cypress heads in north central Florida. *American Midland Naturalist*, **74**, 125–40.

Monod, J. (1942). *Recherches sur la croissance des cultures bacteriennes*. Hermann, Paris.

Monod, J. (1949). The growth of bacterial cultures. *Annual Reviews of Microbiology*, **3**, 371–94.

Monod, J. (1950). La technique de culture continui: theorie et applications. *Annals Institute Pasteur*, **79**, 390–410.

Montgomery, H. A. C. and Bourne, B. J. (1966). Modifications to the BOD test. *Journal and Proceedings of the Institute of Sewage Purification*, **65**, 357–69.

Moo-Young, M. (1980). A new source of food and feed proteins: the Waterloo-SCP process. In *Bioresources for development*, (ed. A. King, H. Cleveland, and G. Streatfield), pp. 155–8, Pergamon, New York.

Moo-Young, M., Moreira, A., Daugulis, A. J., and Robinson, C. W. (1979). Bioconversion of agricultural wastes into animal feed and fuel gas. *Biotechnology and Bioengineering Symposium*, Series **8**, 19–28.

Moore, B. (1977). The EEC bathing water directive. *Marine Pollution Bulletin*, **8**, 269–72.

Moore, E. W., Thomas, H. A., and Snow, S. W. (1950). Simplified method for analysis of BOD data. *Sewage Industrial Wastes*, **22**(10), 343–55.

Moore, J. G., Leatherland, T. M., and Henry, K. I. M. (1987). Design and construction of the Kirkcaldy long sea outfall. *Water and Environmental Management*, **1**(2), 185–97.

Moore, M. E. and Todd, J. J. (1968). Sludge production in the contact stabilization process. *Effluent and Water Treatment Journal*, **8**, 551–60.

Moore, M. N. (1980). Cytochemical determination of cellular responses to environmental stressors in marine organisms. *Rapport et Procés-Verbaux des Réunions. Conseil Permanent International pour l'exporation de la mer)*, **170**, 7–15.

Moore, M. N. (1985). Cellular responses to pollutants. *Marine Pollution Bulletin*, **16**(4), 134–9.

Moore, S. F. and Schroeder, E. D. (1970). An investigation of the effects of residence time on anaerobic bacterial denitrification. *Water Research*, **4**, 685–94.

Morgan, H. (1981). The development of an anaerobic process for the treatment of wheat starch factory effluent. In *Food industry wastes: disposal and recovery*, (ed. A. Herzka and R. G. Booth). Elsevier, Barking, Essex.

Morgan, S. A. (1985). A comparison of the survival of indicator bacteria and pathogens in sea water. Ph.D. Thesis, University of Newcastle upon Tyne.

Morris, G. G. and Burgess, S. (1984). Two phase anaerobic wastewater treatment. *Water Pollution Control*, **83**, 514–20.

Morris, J. G. (1975). The physiology of obligate anaerobiosis. *Advances in Microbiology and Physiology*, **12**, 169246.

Morrison, S. M., Martin, K. L., and Humble, D. E. (1973). *Lime disinfection of sewage bacteria at low temperature*. Environmental Protection Technology Series EPA–660/2–73–017. US Environmental Protection Agency, Washington, D.C.

Morrissette, D. G. and Mavinic, D. S. (1978). BOD test variables. *Journal of the Sanitary Engineering Division, American Society of Civil Engineers*, **104**, 1213–22.

Moser, H. (1958). *The dynamics of bacterial population maintained in the chemostat*. Carnegie Institute, Publ. 615, 4. Washington, D.C.

Mosey, F. E. (1974). *Anaerobic biological treatment*. In: Symposium on treatment of wastes from the food and drink industry, pp. 113–28. Institute of Water Pollution Control, University of Newcastle upon Tyne.

Mosey, F. E. (1976). Assessment of the maximum concentration of heavy metals in crude sewage which will not inhibit the anaerobic digestion of sludge. *Water Pollution Control*, **75**, 10–20.

Mosey, F. E. (1983). Anaerobic processes. In *Ecological aspects of wastewater treatment*, Vol. 2. *Biological activities and treatment processes*, (ed. C. R. Curds and H. A. Hawkes), pp. 219–60. Academic Press, London.

Moulton, E. Q. and Shumate, K. S. (1963). *The physical and biological effects of copper on aerobic biological waste treatment process*. Proceedings of the Eighteenth Industrial Waste Conference, Purdue University Engineering. Extension Series No. 115, 602–15.

Mudrack, K. and Kunst, S. (1987). *Biology of sewage treatment and water pollution control*. Ellis-Horwood, Chichester.

Mueller, J. A., Boyle, W. C., and Lightfoot, E. N. (1968). Oxygen diffusion through zoogloeal flocs. *Biotechnology and Bioengineering*, **10**, 331–58.

Mukadi, S. and Berk, Z. (1975). *Feeding experiments of chicken with sewage grown algae*. Second Progress Report, Techmon, Haifa, Israel.

Mulder, E. G. and van Veen, W. L. (1962). The *Sphaerotilus-Leptothrix* group. *Antonie van Leeuwenhoek*, **28**, 236–7.

Mulder, E. G. and van Veen, W. L. (1963). Investigations on the *Sphaerotilus-Leptothrix* group. *Antonie van Leeuwenhoek*, **29**, 121–53.

Munden, J. E. and King, R. W. (1973). High-rate biofiltration system for food pilot plants. *Effluent and Water Treatment Journal*, **13**, 159–65.

Murad, J. L. and Bazer, G. T. (1970). Diplogasterid and rhabditid nematodes in a waste-water treatment plant and factors related to their disposal. *Journal of the Water Pollution Control Federation*, **42**(1), 105–14.

Murphy, M. J., Hall, J. F., Tunney, H., Dodd, V. A., and Fleming, G. A. (1979). An assessment of some effects of the land application of sewage sludge in Ireland. EEC Symposium on characterization, treatment and use of sewage sludge. Caderache, France.

Murr, L. E., Torma, A. E., and Brierly, J. A. (ed.) (1978). *Metallurgical applications of bacterial leaching and related microbial phenomena*. Academic Press, London.

Murtaugh, J. J. and Bunch, R. L. (1967). Sterols as a measure of fecal pollution. *Journal of the Water Pollution Control Federation*, **39**, 404–9.

Nakajima, A., Horikoshi, T., and Saskaguchi, T. (1981). Studies on the accumulation of heavy metal elements in biological systems. XVII. Selective accumulation of heavy metal ions by *Chlorella regularis*. *European Journal of applied Microbiology and Biotechnology*, **12**, 76–83.

National Water Council (1975). Questions in parliament: typhoid. *National Water Council Bulletin*, **42**, 7.

National Water Council (1977). Report of the working party on the disposal of sewage sludge to land. *Department of the Environment and National Water Council Technical Committee Report 5*. National Water Council, London.

National Water Council (1978). *First biennial report: March 1975–January 1977*. Standing Technical Advisory Committee on Water Quality. Department of the Environment and the National Water Council, London.

National Water Council (1981). Report of the sub-committee on the disposal of sewage sludge to land. *Standing Technical Committee Report 20*. Department of the Environment and the National Water Council, London.

National Water Council (1982). Analysing household water demand. *Water Bulletin*, **9**, 6.

Nelson, J. K. and Tavery, M. A. H. (1978). Chemical conditioning alternatives and operational control for vacuum filtration. *Journal of the Water Pollution Control Federation*, **50**, 507–17.

Nemerow, N. L. (1978). *Industrial water pollution*. Addison-Wesley, Massachusetts.

Nethercott, J. R. (1981). Airborne irritant contact dermatitis due to sewage sludge. *Journal of Occupational Medicine*, **23**, 771.

Neuhauser, E. F., Hartenstein, R., and Kaplan, D. L. (1980). *Second progress report on the potential use of earthworms in sludge management*. In Proceedings of the 9th National Sludge Conference, Florida, pp. 175–82. Information Transfer Inc, Maryland.

Newell, P. J. (1981). In *Energy and the uses of solar and other renewable energies*, (ed. F. Voght). Pergamon, London.

Newell, P. J. (1982). Anaerobic digestion of organic effluents. In *Biotechnology and the environment – a seminar*. National Board for Science and Technology, Dublin.

Nichols, D. S. (1983). Capacity of natural wetlands to remove nutrients from wastewater. *Journal of the Water Pollution Control Federation*, **55**(5), 495–505.

Nielsen, C. O. and Christensen, B. (1959). The Enchytraeidae: critical revision and taxonomy of European species. *Natura Jutlandica*. 8/9, 1–160.

Nielsen, C. O. and Christensen, B. (1961). The Enchytraeidae: critical revision and taxonomy of European species. Suppl. I. *Natura Jutlandica*, **10**, 1–23.

Nielsen, C. O. and Christensen, B. (1963). The Enchytraeidae: critical revision and taxonomy of European species. Suppl. II. *Natura Jutlandica*, **3**, 1–20.

Niku, S., Schroeder, D., and Samaniego, J. (1979). Performance of activated sludge processes and reliability-based design. *Journal of the Water Pollution Control Federation*, **51**, 2841–57.

Norouzian, M., Herroz-Zamorano, A., and Perea-Mejia, C. (1987). A technique for the enumeration of protozoa in wastewater. *Environmental Technology Letters*, **8**, 221–4.

Norris, J. R. (1981). Single-cell protein production. In *Essays in applied microbiology*, (ed. J. R. Norris and M. H. Richmond). John Wiley, London.

Norris, P. R. and Kelly, D. P. (1980). Accumulation of metals by bacteria and yeast. *Developments in Industrial Microbiology*, **20**, 299.

Norton, A. A. (1979). *Odours at a sewage treatment works*, Technical Report TR126. Water Research Centre, Stevenage.

Norton, M. G. (1978). The control of monitoring of sewage sludge dumping at sea. *Water Pollution Control*, **77**, 402–7.

Norton, M. G. (1980). Marine disposal by dumping. In *Today's and tomorrow's wastes*, (ed. J. Ryan), pp. 91–9. National Board for Science and Technology, Dublin.

Norton, R. L. (1982). *Assessment of pollution loads to the North Sea*. Technical Report TR182. Water Research Centre, Stevenage.

O'Connor, J. D. and Dobbins, W. E. (1958). Mechanism of reaeration in natural streams. *Transactions of the American Society of Civil Engineers*, **123**, 641–84.

Ohara, H., Naruto, H., Watanabe, W., and Ebisawa, I. (1983). An outbreak of hepatitis A caused by consumption of raw oysters. *Journal of Hygiene, Cambridge*, **91**, 163–5.

Okrend, M. and Dondero, N. C. (1964). Requirement of *Sphaerotilus* for cyanocobalamin. *Journal of Bacteriology*, **87**, 286–92.

Olah, J. and Princz, P. (1986). A new rapid method for determining sludge activity. *Water Research*, **20**, 1529–34.

Old, R. W. and Primrose, S. B. (1980). *Principles of gene manipulation*. Blackwell, Oxford.

Oleszkiewicz, J. A. and Olthof, M. (1982). Anaerobic treatment of food industry wastewaters. *Food Technology*, 78–82.

Oliveira, P. R. C. and Almeida, S. A. S. (ed.) (1987). *The use of soil for treatment and final disposal of effluents and sludge*. *Water Science and Technology*, **19** (8), 1–214.

Oliver, J. H. and Walker, J. F. (1961). The disposal of malting and brewery effluents. *Brewer's Guild Journal*, (February), 81–95.

O'Mahony, M. C. et al. (1986). An outbreak of gastroenteritis on a passenger cruise ship. *Journal of Hygiene, Cambridge*, **97**, 229–35.

Open University (1975). *Environmental control and public health*. Unit 2, Environmental Health PT 272. The Open University Press, Milton Keynes.

Orford, H. E. and Mutusky, F. E. (1959). Comparison of short term with 5-day BOD determination of raw sewage. *Sewage and Industrial Wastes*, **31**, 259–67.

Ormerod, J. G., Grynne, B., and Ormerod, K. E. (1966). Chemical and physical factors involved in the heterotrophic growth response to organic pollution. *Verhandlungen der Internationalen Vereingung fur Theoretische und Angewandte Limnologie*, **16**, 906–10.

Oron, G., Wildschut, L. R., and Porath, D. (1985). Wastewater recycling by duckweed for protein production and effluent renovation. *Water Science and Technology*, **17** (4/5), 803–17.

Ortiz-Canavate, J., Vigil, S. A., Goss, J. R., and Tchobanoglous, G. (1981). Comparison of operating characteristics of a 34 KW diesel engine fueled with low-energy gas, biogas and diesel fuel. In *Biotechnology in energy production and conservation*, (ed. C. D. Scott), pp. 225–36. John Wiley, London.

Oslo and Paris Commissions (1984). *The first decade*. Oslo and Paris Commissions, London.

Oswald, W. J. (1963). In *Advances in biological waste treatment*, (ed. W. W. Eckenfelder and J. McCabe), pp. 357–93, Pergamon, Oxford.

Oswald, W. J. (1972). *Complete waste treatment in ponds.* Proceedings of the Sixth International Water Pollution Research Conference. Pergamon, London.

Oswald, W. J., Golueke, C. G., Cooper, R. C., Glee, H. K., and Bronson, J. C. (1964). Water reclamation of algal production and methane fermentation in waste ponds. In *Advances in Water Pollution Research.* (ed. W. W. Eckenfelder), pp. 141–58. Pergamon, Oxford.

Otter, C. S., den (1966). A physical method for permanent control of *Psychoda* pests at wastewater treatment plants. *Journal of the Water Pollution Control Federation*, **38**, 156–64.

Ottolenghi, A. C. and Hamparian, V. V. (1982). Bacteriology of sewage sludge; examination for the presence of *Salmonella* and *Campylobacter.* In *Abstracts of the Annual Meeting of the ASM*, 214. American Society of Microbiology, Washington.

Ouano, E. A. R. and Mariano, E. E. (1979). BOD/COD ratio and activated sludge design parameters. *Asian Environment*, **11**, 15–21.

Pacham, R. F. (1983). Water quality and health. In *Pollution: causes, effects and control.* (ed. R. M. Harrison). Royal Society of Chemistry, London.

Page, F. C. (1976). An illustrated key to freshwater and soil amoebae. Scientific Publication **34**. The Freshwater Biological Association, Cumbria.

Page, T., Harris, R. H., and Epstein, S. S. (1976). Drinking water and cancer mortality in Louisiana. *Science*, **193**, 55–7.

Painter, H. A. (1954). Factors affecting the growth of some fungi associated with sewage purification. *Journal of General Microbiology*, **10**, 177–90.

Painter, H. A. (1958). Some characteristics of domestic sewage. *Water and Waste Treatment Journal*, **6**, 496–8.

Painter, H. A. (1970). A review of the literature on inorganic nitrogen metabolism. *Water Research*, **4**, 393–450.

Painter, H. A. (1971). Chemical, physical and biological characteristics of wastes and waste effluents. In *Water and water pollution handbook*, Vol. I, (ed. L. L. Ciaccio), pp. 329–63. Marcel Dekker, New York.

Painter, H. A. (1977). Microbial transformation of inorganic nitrogen. *Progress in Water Technology*, **8**, 3–29.

Painter, H. A. (1978). Biotechnology of wastewater treatment. In *The oil industry and microbial ecosystems*, (ed. K. W. A. Chater and H. J. Somerville), pp. 178–98. Heyden, London.

Painter, H. A. (1980). *A survey of filter fly nuisances and their remedies.* Technical Report TR155. Water Research Centre, Stevenage.

Painter, H. A. (1983). Metabolism and physiology of aerobic bacteria and fungi. In *Ecological aspects of used water treatment*, Vol. 2, (ed. C. R. Curds and H. A. Hawkes), pp. 11–75. Academic, London.

Palm, J. H., Jenkins, D., and Parker, D. S. (1980). Relationship between organic loading, dissolved oxygen concentration and sludge settleability in the completely-mixed activated sludge process. *Journal of the Water Pollution Control Federation*, **52**, 2484–2506.

Papaevangelou, G., Kyriakidou, A., Vissoulis, G., and Trichopoulos, D. (1976). Geo-epidemiological study of HBV infections in Athens, Greece. *Journal of Hygiene, Cambridge*, **76**, 229–234.

Parikh, J. K. and Parikh, K. S. (1977). Potential of biogas plants and how to realize it. In *Microbial energy conversion*, (ed. H. G. Schlegel and J. Barnea), pp. 555–91. Pergamon, Oxford.

Parker, C. D., Jones, H. L., and Taylor, W. S. (1950). Purification of sewage in lagoons. *Sewage and Industrial Waste*, **22**, 760–75.

Parker, C. D., Jones, H. L., and Greene, N. C. (1959). Performance of large sewage lagoons at Melbourne, Australia. *Sewage and Industrial Waste*, **31**, 133–52.

Parker, D. S., Jenkins, D., and Kaufman, W. J. (1971). Physical conditioning of the activated sludge floc. *Journal of the Water Pollution Control Federation*, **43**, 1817–24.

Parker, D. S., Jenkins, D., and Kaufman, W. J. (1972). Floc breakup in turbulent flocculation processes. *Journal of the Sanitary Engineering Division, American Society of Civil Engineers*, **98**, 79–99.

Parr, J. F., Epstein, E., and Wilson, G. B. (1978). Composting sewage sludge for land application. *Agriculture and Environment Journal* **14**, 123–37.

Parsons, D. J. (1984). *A survey of literature relevant to the economics of anaerobic digestion of farm animal waste*. Division Note. DN 1225. National Institute of Agricultural Engineering, Silsoe.

Parsons, D. J. (1985). *The economics of anaerobic treatment of dairy cow waste in the UK*. Divisional Note. DN 1276. National Institute of Agricultural Engineering, Silsoe.

Pasveer, A. (1959). A contribution of the development of activated sludge treatment. *Journal and Proceedings of the Institute of Sewage Purification*, **4**, 436–65.

Patrick, F. M. and Loutit, M. W. (1976). Passage of metals in effluents through bacteria to higher organisms. *Water Research*, **10**, 333–5.

Patterson, D. C. (1980). Effluent in the gully – a pipeline to pig profit. *Farmers Weekly*, **13**(93), xxvii–xxxix.

Patterson, D. C. (1981). Silage effluent – new source of pig feed. *Agriculture in Northern Ireland*, **56**, 68–74.

Paulsrud, B. and Eikum, A. S. (1984). Experiences with lime stabilization and composting of sewage sludge. In *Sewage sludge stabilization and disinfection*, (ed. A. M. Bruce), pp. 261–77. Ellis-Horwood, Chichester.

Pay, R. and Gibson, C. R. (1979). The "Carrousel' at Ash Vale and its first year of operation. *Water Pollution Control*, **78**, 326–36.

Peacock, B. D. (1977). Dairy wastes. In *Treatment of industrial effluents*, (ed. A. C. Callely, C. F. Forster, and D. A. Stafford), pp. 204–17. Hodder and Stoughton, London.

Pearce, F. (1981). The unspeakable beaches of Britain. *New Scientist*, **91**, 139–43.

Pearce, F. (1987) Inspired flush in the pan. *New Scientist*, **116**(1589), 38.

Pearce, J. B. (1969). Investigations of the effects of sewage sludge and acid wastes on offshore marine environments. *Marine Pollution Bulletin*, **7**, 5–7.

Pearson, A. D., Lior, H., Hood, A. M., and Hawtin, P. (1985). Longitudinal study of the occurrence, biotype and serogroup of *Campylobacetr jejuni* and *C. coli* isolations from environmental sources in southern England between 1977–1985. In *Campylobacter*, Vol. III, (ed. A. D. Pearson, M. B. Skirrow, H. Lior, and R. Rowe), pp. 279–80, Public Health Laboratory Service, London.

Pearson, A. D., Suckling, W. G., Ricciardi, I. D., Krill, M., and Ware, E. (1977). *Campylobacter* associated diarrhoea in Southampton. *British Medical Journal*, **ii**, 955–6.

Pearson, C. R. (1965). The use of synthetic media in biological treatment of industrial wastes. *Journal and Proceedings of the Institute of Sewage Purification*, 519–24.

Pearson, T. H. and Rosenberg, R. (1978). Macrobenthic succession in relationa to organic enrichment and pollution of the marine environment. *Oceanography and Marine Biology Annual Review*, **16**, 229–311.

Pedersen, T. T. (1982). *Heat energy from animal waste by combined drying, combustion and heat recovery*. Energy from Biomass. Proceedings of the EEC contractors meeting, Brussels, May 1982.

Pelczar, M. J. and Reid, R. D. (1972). *Microbiology*. McGraw-Hill, New York.

Penfold, W. T. and Earle, T. T. (1958). The biology of the water hyacinth. *Ecology monographs*, **18**, 447–72.

Peters, B. G. (1930). Some nematodes met with in a biological investigation of sewage. *Journal of Helminthology*, **8**, 165–84.

Pette, K. C. and Versprille, A. I. (1982). Application of the UASB-concept for wastewater treatment. In *Anaerobic digestion, 1981*, (ed. D. E. Hughes *et al.*), pp. 121–36. Elsevier, Amsterdam.

Pettygrove, G. and Asano, T. (1984). *Irrigation with reclaimed wastewater: a guidance manual*. Third Water Reuse Symposium, August 1984, San Diego, California. AWWA, Denver.

Pfeffer, J. T. (1974). Temperature effects on anaerobic fermentation of domestic refuse. *Bioengineering and Biotechnology*, **16**, 771–87.

Pfeffer, J. T. (1979). *Anaerobic digestion processes*. In First International Symposium on Anaerobic Digestion, University College, Cardiff, pp. 15–36.

Pfeffer, J. T. (1980). Anaerobic digestion processes. In *Anaerobic digestion*, (ed. D. A. Stafford *et al.*), pp. 15–36. Applied Science Publishers, London.

Pfeffer, J. T., Samra, A. A., and Schwegler, D. T. (1965). Trace metals and filamentous micro-organism growth. Proceedings of the 20th Industrial Waste Conference, Purdue University, *Engineering Extension Series*, 118, 608–17.

Phaup, J. D. (1968). The biology of *Sphaerotilus* species. *Water Research*, **2**, 597–614.

Phaup, J. D. and Gannon, J. J. (1967). Ecology of *Sphaerotilus* in an experimental flow channel. *Water Research*, **1**, 523–41.

Phelps, E. B. (1944). *Stream sanitation*. John Wiley, New York.

Phillip, R. *et al.* (1985). Health risks of snorkel swimming in untreated water. *International Journal of Epidemiology*, **14**, 624–7.

Phillips, R. (1986). *Commercial scale worm composting*. Agricultural Wastes Research Unit. Animal and Grassland Research Institute/National Intitute of Agricultural Engineering, England.

Pickford, T. G. (1984). *Thermophilic anaerobic treatment of a high-strength effluent from a pectin extraction plant*. Pollution Abatement Technology Award 1983. Royal Society of Arts, London.

Piecuch, P. J. (1983) The value of safety data. *Journal of the Water Pollution Control Federation*, **55**, 115.

Pieterse, A. H. (1978). The water hyacinth (*Eichhornia crassipes*) – a review. *Abstracts on Tropical Agriculture* (Royal Tropical Institute, Amsterdam), **4**, 9.

Pike, E. B. (1975). Aerobic bacteria. In *Ecological aspects of used-water treatment*, Vol. 1. *The organisms and their ecology*, (ed. C. R. Curds and H. A. Hawkes), pp. 1–64. Academic Press, London.

Pike, E. B. (1978). *The design of percolating filters and rotary biological contractors, including details of international practice*. Technical Report TR93. Water Research Centre, Stevenage.

Pike, E. B. (1981). The control of Salmonellosis in the use of sewage sludge on agricultural land. In *Characterization, treatment and use of sewage sludge*, (ed. P. L'Hermite and H. Ott), pp. 315–29. D. Reidel, Holland.

Pike, E. B. and Carrington, E. G. (1972). Recent developments in the study of bacteria in the activated-sludge process. *Water Pollution Control* **71**, 583–605.

Pike, E. B. and Carrington, E. G. (1979). The fate of enteric bacteria and pathogens during sewage treatment. In *Biological indicators of water quality*, (ed. A. James and L. Evison), pp. 2001–32. John Wiley, London.

Pike, E. B. and Curds, C. R. (1971). Microbial ecology of the activated sludge process. In *Microbial aspects of pollution*, (ed. G. Sykes and F. A. Skinner), pp. 123–47. Academic Press, New York.

Pike, E. B. and Davis, R. D. (1984). Stabilization and disinfection – their relevance to agricultural utilization of sludge. In *Sewage sludge stabilization and disinfection*, (ed. A. M. Bruce), pp. 61–84. Ellis-Horwood, Chichester.

Pike, E. B. and Gameson, A. L. H. (1970). Effects of marine sewage disposal. *Water Pollution Control*, **69**, 355–82.

Pike, E. B., Carrington, E. G., and Ashburner, P. A. (1972). An evaluation of procedures for enumerating bacteria in activated sludge. *Journal of Applied Bacteriology*, **35**, 309–21.

Pillai, J. K. and Taylor, D. P. (1968). *Butlerlius micans* N. sp. (Nematoda: Diplogasterinae) from Illinois, with observations on its feeding habits and a key to the species of *Butlerius* Goodey, 1929. *Nematologica*, **14**, 89–93.

Pinder, L. C. B. (1978). A key to the adult males of the British Chironomidae (Diptera); the non-biting midges. Scientific Publication, **37**, Freshwater Biological Association, Cumbria.

Pipes, W. O. (1965). *Carnivorous plants in activated sludge*. Proceedings of the 20th Industrial Waste Conference, Purdue University, 647–56.

Pipes, W. O. (1966). The ecological approach to the study of activated sludge. *Advances in Applied Microbiology*, **8**, 77–103.

Pipes, W. O. (1967). Bulking of activated sludge. *Advances in Applied Microbiology*, **9**, 185–234.

Pipes, W. O. (1974). Control of bulking with chemicals. *Water and Wastes Engineering*, **11**, 30–4; 70–1.

Pipes, W. O. (1978a). Microbiology of activated sludge bulking. *Advances in Applied Microbiology*, **24**, 85–127.

Pipes, W. O. (1978b) Actinomycetes scum formation in activated sludge processes. *Journal of the Water Pollution Control Federation*, **5**, 628–34.

Pipes, W. O. (1979). Bulking, deflocculation and pin-point floc. *Journal of the Water Pollution Control Federation*, **51**, 62–70.

Pipes, W. O. and Jones, P. H. (1963). Decomposition of organic wastes by *Sphaerotilus*. *Biotechnology and Bioengineering*, **5**, 287–307.

Pirt, S. J. (1975). *Principles of microbe and cell cultivation*. Blackwell, Oxford.

Poduska, R. A. and Andrews, J. F. (1975). Dynamics of nitrification in the activated sludge process. *Journal of the Water Pollution Control Federation* **47**(11), 2599–619.

Pokvovsky, A. (1975). Some results of SCP medico-biological investigations. In *Single cell protein*, Vol. II, (ed. S. R. Tannenbaum and D. I. C. Wang, pp. 475–83. MIT Press, Cambridge, Massachusetts.

Pollitzer, R. (1959). *Cholera*. WHO Monograph Series **43**. World Health Organization, Geneva.

Polprasert, C. and Rajput, V. (1982). *Septic tank and septic systems.* Environmental Sanitation Information Centre, Bangkok.

Pomoell, B. (1976). The use of bulrushes for livestock feed. In *Biological control of water pollution,* (ed. J. Tourbier and R. W. Pierson), pp. 187–9. University of Pennsylvania Press.

Pontin, R. M. (1978). A key to the freshwater planktonic and semi-planktonic Rotifera of the British Isles. Scientific Publications, **38**, Freshwater Biological Association, Cumbria.

Poole, J. E. P. (1984). A study of the relationship between the mixed liquor fauna and plant performance for a variety of activated sludge sewage treatment works. *Water Research,* **18**, 281–7.

Pooley, F. D. (1982). Bacteria accumulate silver during leaching of sulphide ore minerals. *Nature (London),* **296**, 642–3.

Pope, P. R. (1981). *Wastewater treatment by rooted aquatic plants in sand and gravel trenches,* EPA 600/2–81–091, Environmental Protection Agency, Cincinnati.

Pope, W. (1980). Impact of man in catchments: (ii) Roads and urbanization. In *Water quality in catchment ecosystems,* (ed. A. M. Gower), pp. 73–112. John Wiley, Chichester.

Porges, R. (1960). Newer aspects of waste treatment. *Advances in Applied Microbiology,* **2**, 1–30.

Porges, R. and Brackney, D. H. (1962). Industrial waste disposal by stabilization ponds. *Engineering Experimental Station Bulletin,* **68**, 131–40, Louisiana State University.

Porges, R. and Mackenthun, K. M. (1963). Waste stabilization ponds: use, function and biota. *Biotechnology and Bioengineering,* **5**, 255–73.

Porter, K. E. and Smith, E. (1979). Plastic-media biological filters. *Water Pollution Control,* **78**(3), 371–81.

Post, D. C. van der and Engelbrecht, R. J. (1973). The maturation process in the reclamation of potable water from sewage effluent. *Water Pollution Control,* **72**, 457–62.

Postgate, J. R. (1965). Recent advances in the study of sulphate reducing bacteria. *Bacteriological Review,* **29**, 425–41.

Potten, A. H. (1972). Maturation ponds – experiences in their operation in the United Kingdom as a tertiary treatment process for a high quality sewage effluent. *Water Research,* **6**, 375–91.

Pound, C. *et al.* (1977). *Process design manual for land treatment of municipal wastewater.* EPA 625/1–77–008. US Environmental Protection Agency.

Powell, E. O. (1967). In *Microbial physiology and continuous culture: third symposium,* (ed. E. O. Powell *et al.*), pp. 35–55. HMSO, London.

Powledge, T. M. (1983). Prospects for pollution control with microbes. *Biotechnology,* November, 743–55.

Prakasam, T. B. S. and Dondero, N. C. (1967). Aerobic heterotrophic bacterial populations of sewage and activated sludge. 1. Enumeration. *Applied Microbiology,* **15**, 461–7.

Prentis, S. (1984). *Biotechnology: a new industrial revolution.* Orbis Publishing, London.

Price, G. J. (1982). Use of an anoxic zone to improve activated sludge settleability. In *Bulking of activated sludge,* (ed. B. Chambers and E. J. Tomlinson), pp. 259–60. Ellis-Horwood, Chichester.

Priem, R. and Maton, A. (1980). The influence of the content of trace elements in the feed on the composition of liquid manure of pigs. In *Effluents from livestock*, (ed. J. K. R. Gasser), pp. 9–22. Applied Science Publishers, London.

Pringsheim, E. G. (1949). The filamentous bacteria *Sphaerotilus, Leptothrix, Cladothrix* and their relation to iron and manganese. *Philosophical Transactions of the Royal Society, Series B*, **233**, 453–82.

Prior, B. A. (1984). Continuous growth kinetics of *Candida utilus* in pineapple cannery effluent. *Biotechnology and Bioengineering*, **26**, 748–52.

Prost, E. and Riemann, H. (1967). Food-borne salmonellosis. *Annual Reviews in Microbiology*, **21**, 495–528.

Public Health Laboratory Service (1978). Waterborne infectious disease in Britain. *Journal of Hygiene, Cambridge*, **81**, 139.

Public Health Service (1969). *Manual of septic tank practise*. Public Health Service, US Department of Health, Education and Welfare, Ohio.

Pullen, C. J. (1981). Investigations into sludge dewatering using polyelectrolyte conditioners at Bybrook sewage-treatment works. *Water Pollution Control*, **80**, 95–101.

Punchiraman, C. and Hassan, R. S. (1986). Anaerobic biodegradation of a recalcitrant detergent wastewater. *Effluent and Water Treatment Journal*, **26**, 85–8.

Quinn, J. P. and Marchant, R. (1979). The growth of *Geotrichium candidum* on whiskey distillery spent wash. *European Journal of Applied Microbiology and Biotechnology*, **6**, 251–61.

Quinn, J. P. and Marchant, R. (1980). The treatment of malt whiskey distillery waste using the fungus *Geotrichium candidum*. *Water Research*, **14**, 545–51.

Quinn, J. P., Barker, T. W., and Marchant, R. (1981). Effects of the copper content of distillery spent wash on its utilization for single-cell protein production by *Geotrichium candidum, Hansenula anomala* and *Candida knizei*. *Journal of Applied Bacteriology*, **51**, 149–57.

Rachwal, A. J., Johnstone, D. W. M., Hanbury, M. J., and Critchard, D. J. (1982). The application of settleability tests for the control of activated sludge plants. In *Bulking of activated sludge*, (ed. B. Chambers and E. J. Tomlinson), pp. 224–44. Ellis-Horwood, Chichester.

Rachwal, A. J., Johnstone, D. W. M., Hanbury, M. J., and Carmichael, W. F. (1983). An intensive evaluation of the Carrousel system. In *Oxidation ditches in wastewater treatment*, (ed. D. Barnes *et al.*), pp. 132–72. Pitman, London.

Raebel, M. and Schlierf, H. (1980). Ermittlung der aktiven Biomass im Belebtschlamm durch Bestimmung der Desoxyribonucleinsauve (DNS). *Vom Wasser*, **54**, 293–305.

Ramsden, I. (1972). East Keveston RDC sewage treatment programme. *Surveyor*, **140**, 30–1.

Ranchett, J., Prescheux, F., and Menissier, F. (1981). Influence of the method and period of storage of water samples on the determination of BOD, COD and suspended solids. *Techniques et Sciences Municipales*, **76**, 547–51.

Randall, C. W., Benefield, C. D., and Buth, D. (1982). The effects of temperature on the biochemical reaction rates of the activated sludge process. *Water Science and Technology*, **14**, 413–30.

Rau, W. (1967). Untersuchungen über die Lichtabhangige Carotinoidsyntheses. 1. Das Wirkungsspektrum von *Fusarium aquaeductuum*. *Planta*, **72**, 14–28.

Raygor, S. C. and Mackay, K. P. (1975). Bacterial air pollution from an activated sludge tank. *Water, Air and Soil Pollution*, **5**, 47–52.

Razumor, A. S. (1961). Microbial indications of organic pollution of water by industrial effluents, physiology and ecology of the genus Cladothrix. *Microbiology*, **30**, 764–8.

Reddy, K. R. and DeBusk, T. A. (1987). State of the art utilization of aquatic plants in water pollution control. *Water Science and Technology*, **19** (10), 61–79.

Reed, S., Bastian, R., Black, S., and Khettry, R. (1984). *Wetlands for wastewater treatment in cold climates.* Proceedings of the Water Reuse Symposium, III, San Diego, California. AWWA Research Foundation, Denver, Colorado.

Rees, J. T. (1978). Holdenhurt: oxygen boost to seaside works. *Surveyor*, **16**, February, 18–21.

Reilly, W. J., Collier, P. W., and Forbes, G. I. (1981). Cysticercus bovis *surveillance – an interim report.* Communicable Disease Scotland, CDS 81/22, vii.

Reish, D. J. (1973). The use of benthic animals in monitoring the marine environment. *Journal of Environmental Planning and Pollution Control*, **1**, 32–8.

Reisch, D. J. (1974). Induction of abnormal polychaete larvae by heavy metals. *Marine Pollution Bulletin*, **5**, 125–6.

Relbhun, M. and Manka, J. (1971). Classification of organisms in secondary effluents. *Environmental Science and Technology*, **5**, 606.

Reneau, R. B. (1978). Influence of artificial drainage in penetration of coliform bacteria from septic tank effluents into wet tile drainage soils. *Journal of Environmental Quality*, **7**, 23–30.

Reneau, R. B., Elder, J. H., and Weston, C. W. (1975). Influence of soils on bacterial contamination of watershed from septic sources. *Journal of Environmental Quality*, **4**, 249–52.

Rensink, J. H. (1974). New approach to preventing bulking sludge. *Journal of the Water Pollution Control Federation*, **46**, 1888–94.

Rensink, J. H., Leentvaar, J., and Donker, H. J. (1979). Control of bulking combined with phosphate removal by ferrous sulphate dosing. H$_2$O, **12**, 150–3.

Reynoldson, T. B. (1939). The role of macro-organisms in bacteria beds. *Journal and Proceedings of the Institute Sewage Purification*, 158–72.

Reynoldson, T. B. (1941). The biology of the macrofauna of a high rate double filtration plant at Huddersfield. *Journal and Proceedings of the Institute of Sewage Purification*, 109–28.

Reynoldson, T. B. (1947). An ecological study of the enchytraeid worm population of sewage bacteria beds: field investigations. *Journal of Animal Ecology*, **16**, 26–37.

Reynoldson, T. B. (1948). An ecological study of the enchytraeid worm population of sewage bacteria beds: synthesis of field and laboratory data. *Journal of Animal Ecology*, **17**, 27–38.

Richard, M., Jenkins, D., Hao, O., and Shimuzu, G. (1981). *The isolation and characterization of filamentous microorganisms from activated sludge bulking.* Progress Report. 2. Sanitary Engineering Research Laboratory, University of California, Berkeley.

Richard, M. G., Shimizu, G. P., and Jenkins, D. (1984). *The growth physiology of the filamentous organism type 021N and its significance to activated sludge bulking.* Presented at the 57th Annual Conference of the Water Pollution Control Federation, New Orleans, Louisiana.

Richardson, A. (1975). Outbreaks of bovine salmonellosis caused by serotypes other than *S. dublin* and *S. typhimurum. Journal of Hygiene, Cambridge*, **74**, 195–203.

Richardson, M. (1985). *Nitrification inhibition in the treatment of sewage*. Royal Society of Chemistry, London.

Rickert, D. A. and Hunter, J. D. (1967). Rapid fractionation and materials balance of solids fractions in wastewater and wastewater effluents. *Journal of the Water Pollution Control Federation*, **39**, 1475–86.

Rickert, D. A. and Hunter, J. D. (1971). General nature of soluble and particulate organics in sewage and secondary effluents. *Water Research*, **5**, 421–36.

Rickert, D. A. and Hunter, J. D. (1972). Colloidal matter in wastewater and secondary effluents. *Journal of the Water Pollution Control Federation*, **44**, 134–9.

Riemer, M. and Harremoes, P. (1978). Multi-component diffusion in denitrifying biofilms. *Progress in Water Technology*, **10**(5/6), 149–65.

Riley, W. J., Forbes, G. I., Paterson, G. M., and Sharp, J. C. M. (1981). Human and animal Salmonellosis in Scotland associated with environmental contamination 1973–1979. *Veterinary Record*, **37**, 553–5.

Rittmann, B. E., Strubler, C. E., and Tuzicka, T. (1982). Anaerobic filter pretreatment kinetics. *Journal of the Sanitary Engineering Division, American Society of Civil Engineers*, **108**, 900–11.

Rivera, M. A., Lynn, W. R., and Revelle, C. S. (1965). Bio-oxidation kinetics and a second-order equation describing the BOD reaction. *Journal of the Water Pollution Control Federation*, **37**, 1679–92.

Robbins, M. H. (1961). *Short term substitute for BOD test*. Technical Bulletin No. **146**. National Council for Stream Improvement, Inc.

Roberts, J. C. (1973). Towards a better understanding of high-rate biological film flow reactor theory. *Water Research*, **7**, 1561–88.

Roberts, J. C. (1977). Sewage fungus growth in rivers receiving paper mill effluent. *Water Research*, **11**, 603–10.

Roberts, J. C. (1978). Sewage fungus growth in rivers below paper mill discharges. In *New processes of wastewater treatment and recovery*, (ed. G. Mattock), pp. 140–58. Ellis-Horwood, Chichester.

Roberts, P. V. and McCarty, P. L. (1978). Direct injection of reclaimed water into the aquifer. *Journal of the Environmental Engineering Division, American Society of Civil Engineers*, **104**, 933–949.

Robinson, D. A. and Jones, D. M. (1981). Milk borne campylobacter infection. *British Medical Journal*, **282**, 1374–6.

Roe, P. C. and Bhagat, S. K. (1982). Adenosine triphosphate as a control parameter for activated sludge process. *Journal of the Water Pollution Control Federation*, **54**, 244–54.

Rogol, M. *et al.* (1983). Water borne outbreak of *Campylobacter* enteritis. *European Journal of Clinical Microbiology*, **2**, 588–90.

Rogovskaya, C., Lazareva, M., and Kostina, L. (1969). *The influence of increasing temperatures (30–39°C) on the biocenosis of activated sludge and the intensity of decomposition of organic compounds*. Proceedings of the 4th International Conference on Water Pollution Research, pp. 465–76. Academic Press, London.

Rollins, D. M., Roszak, D., and Colwell, R. R. (1985). Dormancy of *Campylobacter jejuni* in natural aquatic systems. In *Campylobacter*, Vol. III, (ed. A. D. Pearson, M. B. Skirrow, H. Lior, and R. Rowe) pp. 283–4. Public Health Laboratory Service, London.

Romantschuk, H. (1975). The Pekilo process: protein from spent sulphite liquor. In

Single cell protein, Vol. II, (ed. S. R. Tannenbaum, and D. I. C. Wang), pp. 344–56. MIT Press, Massachusetts.

Rose, A. H. (ed.) (1979a). Microbial biomass. *Economic microbiology*, Vol. 4. Academic Press, London.

Rose, A. H. (1979b). History and scientific basis of large-scale production of microbial biomass. In *Microbial biomass*, (ed. A. H. Rose), pp. 1–29. Academic Press, London.

Rothman, H. and Barlett, D. (1977). Sewage – pollutant or food? An assessment of alternative ways of disposing of sewage. In *Technology assessment and the oceans*, (ed. P. D. Wilmot and A. Slingerland), pp. 161–7. Westview Press, Colorado.

Rothman, H., Stanley, R., Thompson, S., and Towalski, Z. (1981). *Biotechnology: a review and annotated bibliography*. Frances Pinter, London.

Rowlands, C. L. (1979). Ecological and comparative performance studies on high-rate filter media. Ph.D. Thesis, University of Aston in Birmingham, UK.

Royal Commission on Sewage Disposal (1908). *5th Report*. HMSO, London.

Royal Society (1983). *The nitrogen cycle of the United Kingdom*. Royal Society, London.

Ruttner-Kolisko, A. von (1972). *Rotatoria*. In: Die Binnengewässer, **26**, Das Zooplankton der Binnengewässer. 1, Teil., pp. 99–225. E. Schweizerbart'sche Verlagsbuch-handlung, Stuttgart.

Sabine, J. R. (1978). The nutritive value of earthworm meal. In *Utilization of soil organisms in sludge management*, (ed. R. Hartenstein), pp. 122–30. State University of New York, Syracuse.

Sachs, E. F., Jennett, J. C., and Red, M. C. (1982). Pharmaceutical waste treatment by anaerobic filter. *Journal of the Sanitary Engineering Division, American Society of Civil Engineers*, **108**, 297–313.

Safferman, R. S. and Morris, M. E. (1976). Assessment of virus removal of a multi-stage activated sludge process. *Water Research*, **10**, 413–20.

Salkinoja, Salonen, M. S., Nyns, E-J., Sutton, P. M., Berg, L. van den, and Wheatley, A. D. (1983). Starting up of an anaerobic fixed-film reactor. *Water Science Technology*, **15**, 305–8.

Sambridge, N. E. W. (1972). Commissioning of a primary heated digester. *Water Pollution Control*, **71**, 105–6.

Samuelov, N. S. (1983). Single-cell protein production: review of alternatives. In *Advances in biotechnical processes*, Vol. I, (ed. A. Mizrahi and A. L. van Wezel), pp. 293–335. Liss, New York.

Sanders, P. F. (1982). *River Don 'sewage fungus' study: 1979–1981*. Scottish Development Agency.

Satchell, G. H. (1947). The larvae of the British species of *Psychoda* (Diptea: Psychodidae). *Parasitology*, **38**, 57–69.

Satchell, G. H. (1949). The respiratory horns of *Psychoda* pupae (Diptera: Psychodidae). *Parasitology*, **39**, 43–52.

Sato, T. and Ose, Y. (1980). Floc-forming substrates extracted from activated sludge by sodium hydroxide solution. *Water Research*, **14**, 333–8.

Sattar, S. A., Ramia, S., and Westwood, J. C. N. (1976). Calcium hydroxide (lime) and the elimination of human pathogenic viruses from sewages: studies with experimentally-contaminated (Poliovirus type I, Sabin) and pilot plant samples. *Canadian Journal of Public Health*, **67**, 221–6.

Scarpino, P. V. (1975). Human enteric viruses and bacteriophages as indicators of sewage pollution. In *Discharge of sewage from sea outfalls*, (ed. A. L. H. Gameson), pp. 49–61. Pergamon, Oxford.

Schade, A. L. (1940). The nutrition of *Leptomitus lacteus*. *American Journal of Botany*, **27**, 376–84.

Scherb, K. (1968). Nematoda. In *Troptkorper und Belebungsbecken*, (ed. H. Liebmann), pp. 158–206. R. Oldenbourg, Munich.

Scheuring, L. and Hohnl, G. (1956). *Sphaerotilus natans*, Seine Okologie und Physiologie. *Schriften des Vereins der Sellstoff und Paper-Chemiker und Ingenieure*, Vol. 26.

Schiemer, F. (1975). Nematoda. In *Ecological aspects of used-water treatment*, Vol. 1, (ed. C. R. Curds and H. A. Hawkes), pp. 269–88. Academic Press, London.

Schlegel, S. (1988). Use of submerged biological filters for ntirification. *Water Science and Technology*, **20** (4/5), 177–87.

Schmidt, B. (1948). Der einbruch der tularamie in Europe. *Zeitschrift fuer Hygiene und Infektionskrankhankheiten*, **127**, 139–50.

Schofield, T. (1971). Some biological aspects of the activated sludge plant at Leicester. *Water Pollution Control*, **70**, 32–47.

Schroeder, E. D. (1968). Importance of the BOD plateau. *Water Research*, **2**, 803–9.

Schroepfer, G. J. (1951). Effect of particle shape on porosoity and surface area of trickling filter medium. *Sewage and Industrial Wastes*, **23**, 1356–66.

Schugerl, K. (1987). *Bioreaction engineerings*, Vol. 1. Wiley, Chichester.

Schultze, K. L. (1957). Experimental vertical screen trickling filter. *Sewage and Industrial Wastes*, **29**, 458–67.

Schumate, S. E., Stranberg, G. W., and Parrott, J. R. (1979). Biological removal of metal ions from aqueous process streams. In *Biotechnology in energy production and conservation*, (ed. C. D. Scott), pp. 13–20. John Wiley, London.

Schwab, D., Armstrong, J. H., and Harp, S. (1975). *Septic tank maintenance*. OSU Extension Facts No. 1657. Oklahoma State University, Stillwater.

Schwaartzbrod, J. *et al.* (1987). Wastewater sludge: parasitological and virological contamination. *Water Science and Technology*, **19** (8), 33–40.

Scottish Development Department (1977). *Pilot plant study of the oxygen activated sludge process*. Applied Research and Development Report ARD5. Engineering Division, Scottish Development Department, Edinburgh.

Scottish Development Department (1980). *Settling velocities of particulate matter in sewage*. Applied Research and Development Report ARD 6. Engineering Division, Scottish Development Department, Edinburgh.

Scrimshaw, N. S. (1975). Single cell protein for human consumption – an overview. In *Single-cell protein*, Vol. II, (ed. S. R. Tannenbaum and D. I. C. Wang), pp. 24–5. MIT Press, Massachusetts.

Segar, D. A., Stamman, E., and Davis, P. G. (1985). Beneficial use of municipal sludge in the ocean. *Marine Pollution Bulletin*, **16**, 186–91.

Seidel, K. (1976). Macrophytes and water purification. In *Biological control of water pollution*, (ed. J. Tourbier, R. W. Pierson, and E. W. Furia), pp. 109–21. University of Pennsylvania Press.

Seppanen, H. and Wihuri, H. (ed.) (1988). *Groundwater microbiology: problems and biological treatment. Water Science and Technology*, **20** (3), 1–266).

Severn and Trent Water Authority (1982). *Sludge – a valuable resource*. Severn and Trent Water Authority, Birmingham.

Sezgin, M., Jenkins, D., and Parker, D. S. (1978). A unified theory of filamentous activated sludge bulking. *Journal of the Pollution Control Federation*, **50**, 352–81.

Sharma, B. and Ahler, R. C. (1977). Nitrification and nitrogen removal. *Water Research*, **11**, 897–925.

Shaub, S. A., Bausum, H. T., and Taylor, G. W. (1982). Fate of viruses in wastewater applied to slow-infiltration land treatment systems. *Applied Environmental Microbiology*, **44**, 383–94.

Sheehan, G. J. and P. F. Greenfield (1980). Utilization, treatment and disposal of distillery wastewater. *Water Research*, **14**, 257–77.

Sheehy, J. P. (1960). Rapid method for solving first order equations. *Journal of the Water Pollution Control Federation*, **32**, 646–52.

Sheikh, B., Jacques, R. S., and Cooper, R. C. (1984). *Wastewater effluent reuse for irrigation of raw-eaten food crops: a five year study*. Third Water Reuse Symposium, August 1984. San Diego, California, AWWA, Denver.

Shelef, G., Moraine, R., Meydan, A., and Sandbank, E. (1977). Combined algae production – wastewater treatment and reclamation systems. In *Microbial energy conversion*, (ed. H. G. Schlegel and J. Barnea), pp. 427–41, Pergamon, Oxford.

Shelef, G., Moraine, R., and Oron, G. (1978). Photosynthetic biomass production from sewage. *Ergebnisse der Limnologie*, **11**, 3–14.

Shelton, R. G. J. (1973). Some effects of dumped solid wastes on marine life and fisheries. In *North Sea Sciences*, (ed. E. D. Goldberg), pp. 416–45, MIT Press, Massachusetts.

Shephard, M. R. N. (1967). Factors influencing the seasonal accumulation of solids in bacteria beds. M.Sc. Thesis, University of Aston in Birmingham, England.

Shephard, M. R. N. and Hawkes, H. A. (1976). Laboratory studies on the effects of temperature on the accumulation of solids in biological filters. *Water Pollution Control*, **75**(1), 58–72.

Sherman, V. R., Kawata, K., Olivieri, V. P., and Naparstek, J. D. (1975). Virus removals in trickling filter plants. *Water Sewage Works*, R36–R44.

Sherr, B. F. and Payne, W. J. (1978). Effect of *Spartina alterniflora* root-rhizome system on salt marsh soil denitrifying bacteria. *Applied environmental Microbiology*, **35**, 724–9.

Shetty, M. S. (1971). *Septic tank design, construction and maintenance practices*. Proceedings of the Seminar on Water Supply and Sanitation Problems of Urban Areas, **2**, 21–45. Public Health Engineering Division, Institute of Engineers, India.

Shimzia, T., Wawkimura, K., Tanemara, K., and Ichikawa, K. (1980). Factors affecting sludge bulking in the activated sludge process. *Journal of Fermentation Technology*, **53**, 275–81.

Shore, M., Broughton, N. W., and Bumstead, N. (1984). Anaerobic treatment of wastewaters in the sugar beet industry. *Water Pollution Control*, **83**, 499–506.

Shore, M., Dutton, J. V., Broughton, N. W., and Bumstead, N. (1979). *The effect of pH on the odours of transport and storage waters*. British Sugar Corporation Research Laboratories Report 79/36. Annual Works Managers Meeting, British Sugar Corporation Ltd, Norwich.

Short, C. S. (1988). The Bramham incident, 1980 – an outbreak of water-borne infection. *Water and Environmental Management*, **2**, 383–90.

Shriver, L. E. and Bowers, D. M. (1975). Operational practices to upgrade trickling filter plant performances. *Journal of the Water Pollution Control Federation*, **47**(11), 2640–51.

Shtarkas, E. M. and Krasil'Shchikov, D. G. (1970). *Abst. Hyg. and San.* **35**, 330.

Shuval, H. I. (ed.) (1977). *Water renovation and reuse.* Academic Press, New York.

Shuval, H. I. and Fattal, B. (1980). Epidemiologic study of wastewater irrigation. In *Wastewater aerosols and disease.* Proceedings of the Symposium. Cincinnati, Ohio, 1979, (ed. H. Pahren and W. Jakubowski), pp. 228–38. EPA 600/9–80–028. Health Effects Research Laboratory. US EPA, Cincinnati.

Shuval, H. I., Fattal, B., and Yekutiel, P. (1986*a*). The state of the art review: an epidemilogical approach to the health effects of water reuse. *Water Science and Technology,* **18** (9), 147–62.

Shuval, I., Yekuteil, P., and Fattal, B. (1985). Epidemiological evidence for helminth and cholera transmission by vegetables irrigated with wastewater: Jerusalem—a case study. *Water Science and Technology,* **17**, 433–42.

Shuval, H. I., Yekutiel, P., and Fattal, B. (1986*b*). An epidemiological model of the potential health risk associated with various pathogens in wastewater irrigation. *Water Science and Technology,* **18**(10), 191–8.

Sidwick, J. M. and Holdom, R. S. (1987). *Biotechnology of waste treatment and exploitation.* Ellis-Horwood, Chichester.

Sidwick, J. M. and Lewandowski, T. P. (1975). An economic study of the Unox and conventional aeration systems. *Water Pollution Control,* **74**, 645–54.

Sikyta, B. (1983). *Methods in industrial microbiology.* Ellis-Horwood, Chichester.

Silverman, P. H. and Griffiths, R. B. (1955). A review of methods of sewage disposal in Great Britain with special reference to the epizootiology of *Cysticerous bovis. Annals of Tropical Hygiene and Parasitology,* **49**, 436–50.

Sinclair, C. G. (1987). Microbial process kinetics. In *Basic Biotechnology* (ed. J. Bu'lock and B. Kristiansen), pp. 74–132. Academic Press, London.

Sinskey, A. J. and Tannenbaum, S. R. (1975). Removal of nucleic acids in SCP. In *Single cell protein,* Vol. II, (ed. S. R. Tannenbaum and D. I. C. Wang). MIT Press, Cambridge, Massachusetts.

Skirrow, M. B. (1982). *Campylobacter* enteritis: the first five years. *Journal of Hygiene, Cambridge,* **89**, 175–84.

Skogman, H. (1976). Production of symba-yeast from potato wastes. In *Food from waste,* (ed. G. C. Birch, K. J. Parker, and J. T. Worgan), pp. 167–79. Applied Science Publishers, London.

Skovgaard, N. and Nielsen, B. B. (1972). Salmonellas in pigs and animal feeding stuffs in England and Wales and in Denmark. *Journal of Hygiene, Cambridge,* **70**, 127–40.

Sladka, A. and Ottova, V. (1968). The most common fungi in biological treatment plants. *Hydrobiologia,* **31**, 350–62.

Slater, J. H. and Somerville, H. J. (1979). Microbial aspects of waste treatment with particular attention to the degradation of organic compounds. In *Microbial technology: current state, future prospects,* (ed. A. J. Bull, D. C. Ellwood, and C. Rutledge), pp. 221–61. Cambridge University Press.

Slijkhuis, H. (1983). *Microthrix parvicella,* a filamentous bacterium isolated from activated sludge: cultivation in a chemically defined medium. *Applied Environmental Microbiology,* **46**, 832–9.

Slijkhuis, H. and Deinema, M. H. (1982). The physiology of *Microthrix parvicella* a filamentous bacterium isolated from activated sludge. In *Bulking of activated sludge,* (ed. B. Chambers and E. J. Tomlinson), pp. 75–85. Ellis-Horwood, Chichester.

Sloey, W. E., Sprangler, F. L., and Fetter, C. W. (1978). Management of freshwater wetland for nutrient assimilation. In *Freshwater wetlands: Ecological processes and management potential*, (ed. R. E. Good, D. F. Whigham, and R. L. Simpson), pp. 321–40. Academic Press, New York.

Small, M. (1976). Marsh/pond sewage treatment plants. In *Freshwater wetlands and sewage effluent disposal*, (ed. D. L. Tilton, R. H. Kadlec, and C. J. Richardson), pp. 197–213. University of Michigan, Ann Arbor.

Smith, J. E., Young, K. W., and Dean, R. B. (1975). Biological oxidation and disinfection of sludge. *Water Research*, **9**, 17–24.

Smith, J. L. (ed.) (1986). *Municipal wastewater sludge combustion technology*. Technomic Publishing, Basel, Switzerland.

Smith, M. E. and Bull, A. T. (1976). Studies on the utilization of coconut wastewater for the production of the food yeast *Saccharomyces fragilis*. *Journal of Applied Bacteriology*, **41**, 81—95.

Smith, P. G. and Coackley, P. (1983). A method for determining specific surface area of activated sludge by dye adsorption. *Water Research*, **17**, 595–8.

Smith, R. G. and Schroeder, E. D. (1983). Physical design of overland flow systems. *Journal of the Water Pollution Control Federation*, **55**, 255–9.

Smith, R. J., Twedt, R. M., and Flanigan, L. K. (1973). Relationships of indicator and pathogenic bacteria in stream waters. *Journal of the Water Pollution Control Federation*, **45**, 1736–45, 1736–45.

Smith, R. S. and Purdy, W. C. (1936). Use of chlorination for correction of sludge bulking in activated sludge processes. *Sewage Works Journal*, **8**, 223–30.

Sobsey, M. D. and Cooper, R. C. (1973). Enteric virus survival in algal–bacterial wastewater treatment systems. I. Laboratory studies. *Water Research*, **7**, 669–85.

Solbe, J. F. de L. G. (1971). Aspects of the biology of the lumbricids *Eiseniella tetraedra* (Savigny) and *Dendrobaena rubida* (Savigny) *F. subarubicunda* (Eisen) in a percolating filter. *Journal of Applied Ecology*, **8**, 845–67.

Solbe, J. F. de L. G. (1975). Annelida. In *Ecological aspects of used-water treatment*, (ed. C. R. Curds and H. A. Hawkes), pp. 305–55. Academic Press, London.

Solbe, J. F. de L. G. and Tozer, J. S. (1971). Aspects of the biology of *Psychoda alternata* (Say.) and *P. severini parthenogenetica* Tonn. (Diptera) in a percolating filter. *Journal of Applied Ecology*, **8**, 835–44.

Solbe, J. F. de L. G., Williams, N. V., and Roberts, H. (1967). The colonization of a percolating filter by invertebrates, and their effect on settlement of humus solids. *Water Pollution Control*, **66**(5), 423–48.

Solbe, J. F. de L. G., Ripley, P. G., and Tomlinson, T. G. (1974). The effect of temperature on performance of experimental percolating filters with and without mixed macro-invertebrate populations. *Water Research*, **8**, 557–73.

Sorber, C. A. and Guter, K. J. (1975). Health and hygiene aspects of spray irrigation. *American Journal of Public Health*, **65**, 47–52.

Sprangler, F., Sloey, W., and Fetter, C. W. (1976). Experimental use of emergent vegetation for the biological treatment of municipal wastewater in Wisconsin. In *Biological control of water pollution* (ed. J. Tourbier, R. W. Pierson, and E. W. Furia), pp. 161–71. University of Pennsylvania Press.

Speece, R. E. and McCarty, P. L. (1964). Nutrient requirements and biological solids accumulation in anaerobic digestion. In *Advances in water pollution research*. Proceedings of the First International Conference on Water Pollution Research, 2, (ed. W. W. Ekenfelder), pp. 305–33. Pergamon, Oxford.

Sperber, C. (1950). A guide for the determination of European Naididae. *Zoologiste Bidrag Fran Uppsala*, **29**, 45–78.

Sproul, O. J., Pfister, R. M., and Kim, C. K. (1982). The mechanism of ozone inactivation of water borne viruses. *Water Science and Technology*, **14**, 303–14.

Sridhar, M. K. C. and Oyemade, A. (1987). Potential health risks at sewage treatment plants in Ibadan, Nigeria. *Water and Environmental Management*, **1**, 129–35.

Sridhar, M. K. C. and Phillai, S. C. (1973) Protease activity in sewage sludges and effluents. *Water and Waste Treatment*, August, 35–42.

Stafford, D. A., Wheatley, B. I., and Hughes, D. E. (1980). *Anaerobic digestion*. Applied Science Publishers, London.

Stanbridge, H. H. (1972). Historical account of rotating and travelling distributors for biological filters. *Water Pollution Control*, **71**, 573–7.

Stanbridge, H. H. (1976). *History of sewage treatment in Britain. Biological filtration, 1*. Institute of Water Pollution Control, Maidstone.

Stander, G. J. (1967). *Treatment of wine distillery wastes by anaerobic digestion*. Proceedings of the 22nd Industrial Waste Conference, Purdue University, pp. 892–907.

Stander, G. J. and Meiring, P. G. J. (19665). Employing oxidation ponds for low-cost sanitation. *Journal of the Water Pollution Control Federation*, **37**, 1025–33.

Stanfield, G., Irving, T. E. and Robinson, J. A. (1978). *Isolation of faecal streptococci from sewage*. Technical Report TR83. Water Research Centre, Stevenage.

Stankewich, M. J. and Gyger, R. F. (1978). Nitrification in oxygen-activated sludge systems. In *The use of high-purity oxygen in the activated sludge process*, (ed. J. R. McWhirter) 2, pp. 139–72. CRC Press, Florida.

Steadman, J. R., Maier, C. R., Schwartz, H. F., and Kerr, E. D. (1975). Pollution of surface irrigation waters by plant pathogenic organisms. *Water Research Bulletin*, **11**, 796–804.

Steels, I. H. (1974). Design basis for the rotating disc process. *Effluent and Water Treatment, Journal*, **14**, 431–45.

Steiner, A. E., McLaren, D. A., and Forster, C. F. (1976). Nature of activated sludge flocs. *Water Research*, **10**, 25–30.

Steffen, A. J. and Bedher, M. (1961). *Operation of full-scale anaerobic contact treatment plants for meat packing wastes*. Proceedings of the 16th Industrial Waste Conference, Purdue University, pp. 423–37.

Stensel, H. D., Refling, D. R., and Scott, J. H. (1978). 'Carrousel' activated sludge for biological nitrogen removal. In *Advances in water and wastewater treatment, biological nutrient removal*, (ed. M. P. Wanielista and W. W. Eckenfelder), pp. 43–64. Ann Arbor Science, Michigan.

Stenstrom, M. K. and Poduska, R. A. (1980). The effect of dissolved oxygen concentration on nitrification. *Water Research*, **14**, 643–9.

Stentiford, E. I., Mara, D. D., and Taylor, P. L. (1985). Forced aeration co-composting of domestic refuse and sewage sludge in static piles. In *Composting of agricultural and other wastes*, (ed. J. K. R. Gasser), pp. 42–55. Elsevier, London.

Stephenson, M. *et al.* (1980). *The environmental requirements of aquatic plants*. Publication No. 65, Appendix A. The State Water Resources Control Board, Sacramento, California.

Sticht-Groh, V. (1982). *Campylobacter* in healthy slaughter of pigs: a possible source of infection of man. *Veterinary Record*, **110**, 104–6.

Stickley, D. P. (1970). The effect of chloroform in sewage on the production of gas from laboratory digesters. *Water Pollution Control*, **69**, 585–92.

Stocks, P. (1973). Mortality from cancer and cardiovascular disease in the county boroughs of England and Wales classified according to the sources and hardness of their water supplies, 1958–1967. *Journal of Hygiene, Cambridge,* **71**, 237–52.

Stokes, J. L. (1954). Studies on the filamentous sheathed iron bacterium *Sphaerotilus natans. Journal of Bacteriology,* **67**, 278–91.

Stokes, J. L. and Powers, M. T. (1967). Glucose repression of oxidation of organic compounds by *Sphaerotilus. Canadian Journal of Microbiology,* **13**, 557–63.

Stones, T. (1979). A critical examination of the uses of the BOD test. *Effluent and Water Treatment Journal,* **19**, 250–4.

Stones, T. (1981). A resume of the kinetics of the BOD test. *Water Pollution Control,* **80**, 513–20.

Stones, T. (1982). Re-examining the kinetics of the biochemical oxidation of sewage. *Effluent and Water Treatment Journal,* **22**, 75.

Stones, T. (1985). A review of the biochemical oxidation of sewage. *Effluent and Water Treatment Journal,* **25**, 365–70.

Strauch, D., Havelaar, A. H., and L'Hermite, P. (ed.) (1985). *Inactivation of micro-organisms in sewage sludge by stabilization processes.* Elsevier, London.

Streeter, H. and Phelps, E. (1925). *A study of the purification of the Ohio River.* US Public Health Service Bulletin No. 146. Washington, D.C.

Strom, P. F. and Jenkins, D. (1984). Identification and significance of filamentous micro-organisms in activated sludge. *Journal of the Water Pollution Control Federation,* **56**, 449–59.

Strover, E. L. and McCartney, E. L. (1984). BOD results that are believable. *Water Engineering Management,* **131**, 37–40; 62–6.

Stuckey, D. C. and McCarty, P. L. (1979). Thermochemical pretreatment of nitrogenous materials to increase methane yield. In *Energy production and conservation,* (ed. C. D. Scott). John Wiley, London.

Styles, P. D. (1979). Control of chironomid flies breeding in sewage filters. Ph.D. Thesis, University of Aston in Birmingham.

Suler, D. J. and Finstein, M. S. (1977). Effect of temperature, aeration and moisture on CO_2 formation in bench-scale continuously thermophilic composting of solid waste. *Applied Environmental Microbiology,* **33**, 345–50.

Suominen, A. (1980). Puhtaan hapen käyttö aktiivilietemenetelmässä. *Vesitalous,* **21**, 14–15.

Sutton, D. L. (1974). Utilization of Hydrilla by white amur. *Hyacinth Control Journal,* **12**, 66.

Sutton, P. M., Murphy, K. L., and Jank, B. E. (1977). Nitrogen control: a basis for design with activated sludge systems. *Progress in Water Technology,* **8**, 467–81.

Sutton, P. M., Shleh, W. W., Kos, P., and Dunning, P. R. (1981). Dorr-Oliver's 'Oxitron' fluidized bed water and wastewater treatment process. In *Biological fluidised bed treatment of water and wastewater,* (ed. P. F. Cooper and B. Atkinson), pp. 285–304. Ellis-Horwood, Chichester.

Svedhem, A. and Norkrans, G. (1980). *Campylobacter jejuni* enteritis tranmitted from cat to man. *Lancet.* (i) 713–14.

Svore, J. H. (1968). Waste stabilization and practices in the United States. In *Advances in water quality improvement,* (ed. E. F. Gloyna and W. W. Eckenfelder), pp. 427–34. Water Resources Symposium 1. University of Texas, Austin.

Swanwick, J. D. and Foulkes, M. (1971). Inhibition of anaerobic digestion of sewage sludge by chlorinated hydrocarbons. *Water Pollution Control,* **70**, 58–70.

Swanwick, J. D. and Shurben, D. G. (1969). Effective chemical treatment for inhibition of anaerobic sewage sludge digestion due to anionic detergents. *Water Pollution Control*, **68**, 190–202.

Swanwick, J. D., Shurben, D. G., and Jackson, S. (1969). A survey of the performance of sewage sludge digesters in Great Britain. *Water Pollution Control*, **68**, 639–61.

Swanwick, K. H. (1973). Control of filter pressing at Sheffield. *Water Pollution Control*, **72**, 78–86.

Sydenham, D. H. J. (1968). The ecology of protozoa and other organisms in activated sludge. Ph.D. Thesis, University of London.

Sydenham, D. H. J. (1971). A re-assessment of the relative importance of ciliates, rhizopods, and rotalorians in the ecology of activated sludge. *Hydrobiologia*, **38**, 553–63.

Sykes, J. B. (1981). An illustrated guide to the Diatoms of British coastal plankton. *Field Studies*, **5**, 425–68.

Sykes, R. M., Rozich, A. F., and Tiefert, T. A. (1979). Algal and bacterial filamentous bulking of activated sludge. *Journal of the Water Pollution Control Federation*, **51**, 2829–40.

Tacon, A. G. J., Stafford, E. A., and Edwards, C. A. (1983). A preliminary investigation of the nutritive value of three terrestrial lumbricid worms for rainbow trout. *Aquaculture*, **35**, 187–99.

Tago, Y. and Aiba, K. (1977). Exocellular mucoupolysaccharide closely related to bacterial floc formation. *Applied Environmental Microbiology*, **34**, 308–14.

Taiganides, E. P. (ed.) (1977). *Animal wastes*. Applied Science Publishers, London.

Taiganides, E. P., Chou, K. C., and Lee, B. Y. (1979). Animal waste management and utilization in Singapore. *Agricultural Wastes*, **1**, 129–41.

Takii, S. (1977). Accumulation of reserve polysaccharide in activated sludge treating carbohydrates wastes. *Water Research*, **11**, 79–83.

Talmadge, W. P. and Fitch, E. B. (1955). Determining thickener unit areas. *Industrial Engineering Chemistry*, **47**(1), 38–41.

Tamblyn, T. A. and Sword, B. R. (1969). *The anaerobic filter for the denitrification of agricultural subsurface drainage*. Proceedings of the 24th Industrial Waste Conference, Purdue University, pp. 1135–50.

Tapelshay, J. A. (1945). Control of sludge index by chlorination of return sludge. *Sewage Works Journal*, **17**, 1210–26.

Tare, V.a dn Bokil, S. D. (1982). Wastewater treatment by soils: role of particle size distribution. *Journal of Environmental Quality*, **11**, 596–602.

Tarjan, A. C., Esser, R. P., and Chang, S. C. (1977). An illustrated key to nematodes found in freshwater. *Journal of the Water Pollution Control Federation*, **49**, 2318–37.

Tariq, M. N. (1975). Retention time in trickling filters. *Progress in Water Technology*, **7**(2), 225–34.

Tauk, S. M. (1982). Culture of *Candida* in vinasse and molasses: effect of acid and salt addition on biomass and raw protein production. *European Journal of Applied Microbiology Biotechnology*, **16**, 223–7.

Taylor, D. N., McDermott, K. T., Little, J. R., Wells, J. G., and Blaser, M. J. (1983). *Camplylobacter* enteritis from untreated water in the Rocky Mountains. *American Internal Medicine*, **99**, 38–40.

Taylor, D. W. and Burm, R. J. (1973). Full scale anaerobic filter treatment of wheat starch plant wastes. *American Institute of Chemical Engineers Symposium*, Ser. 69, 30–7.

Taylor, L. R. (1951). An improved suction trap for insects. *Annals of Applied Biology*, **38**, 582–91.

Taylor, L. R. (1955). The standardization of air-flow in insect suction traps. *Annals of Applied Biology*, **43**, 390–408.

Tebbutt, T. H. Y. (1983). *Principles of water quality control*, (3rd edn). Pergamon, Oxford.

Tench, H. B. (1979). Discussion on oxygen transport into bacterial flocs and biochemical oxygen consumption by N. W. F. Kossen. *Progress in Water Technology*, **11**, 21–2.

Terry, R. J. (1951). The behaviour and distribution of the larger worms in trickling filters. *Journal and Proceedings of the Institute of Sewage Purification*, 16–25.

Terry, R. J. (1956). The relations between bed medium and sewage filters and the flies breeding in them. *Journal of Animal Ecology*, **25**, 6–14.

Thauer, R. K., Jungerman, K., and Decker, K. (1977). Energy conservation in chemotrophic anaerobic bacteria. *Bacteriological Review*, **41**, 100–80.

Thiel, P. G. (1969). The effect of methane analogues on methanogenesis in anaerobic digestion. *Water Research*, **3**, 215–23.

Thiel, P. G. and Hattingh, W. H. (1966). Determination of hydrolitic enzyme activities in anaerobic digesting sludge. *Water Research*, **1**, 191–6.

Theis, J. H. and Storm, D. R. (1978). Helminth ova in soil and sludge from twelve U.S. urban areas. *Journal of the Water Pollution Control Federation*, **50**, 2485–93.

Thirumurthi, D. (1974). Design criteria for waste stabilization ponds. *Journal of the Water Pollution Control Federation*, **46**, 2094–2106.

Thoman, J. R. and Jenkins, K. H. (1958). Statistical summary of sewage chlorination practice in the United States. *Sewage and Industrial Wastes*, **30**, 1461–8.

Thomanetz, E. and Bardtke, D. (1977). Studies on the possibility for control of bulking sludge with synthetic flocculating agents. *Korrespondenz Abwasser*, **1824**, 15–7.

Thomas, H. A. (1950). Graphical determination of BOD curve constants. *Water Sewage Works*, **97**, 123.

Thompson, T. J. (1925). Percolating bacteria beds. *Proceedings of the Association of Managers Sewage Disposal Works*, 52–6.

Thostrup, P. (1985). Heat recovery from composting solid manure. In *Composting of agricultural and other wastes*, (ed. J. K. R. Gasser), pp. 167–83. Elsevier, London.

Thunegard, E. (1975). On the persistence of bacteria in manure. *Acta Veterinvia Scandinavica*, **56**, 1.

Tierney, J. T., Sullivan, R., and Larkin, E. P. (1977). Persistence of poliovirus I in soil and on vegetables grown in soil previously flooded with inoculated sewage sludge or effluent. *Applied Environmental Microbiology*, **33**, 109–13.

Tilton, D. L. and Kadlec, R. H. (1979). The utilization of a freshwater wetland for nutrient removal from secondary treated wastewater effluent. *Journal of Environmental Quality*, **8**, 328–34.

Tilton, D. L., Kadlec, R. H., and Richardson, C. J. (ed.) (1976). *Freshwater wetlands and sewage effluent disposal*. University of Michigan, Ann Arbor.

Tobin, R. S. and Dutka, B. J. (1977). Comparison of the surface structure, metal binding and faecal coliform recoveries of nine membrane filters. *Applied Environmental Microbiology*, **34**, 69–79.

Tobin, R. S., Lomax, P., and Kushner, D. J. (1980). Comparison of nine brands of membrane filter and the most probable number methods for total coliform enumeration in sewage contaminated drinking water. *Applied Environmental Microbiology*, **40**(2), 186–90.

Toerien, D. F. (1970). Population description of the non-methanogenic phase of anaerobic digestion. 1. Isolation, characterization and identification of numerically important bacteria. *Water Research*, **4**, 129–48.

Toerien, D. F. and Hattingh, W. H. J. (1969). Anaerobic digestion: 1. The microbiology of anaerobic digestion. *Water Research*, **3**, 385–16.

Toerien, D. F. and Siebert, M. L. (1967). A method for the enumeration and cultivation of anaerobic 'acid-forming' bacteria present in digesting sludge. *Water Research*, **1**, 397–404.

Tomlin, A. D. (1983). The earthworm bait market in North America. In *Earthworm ecology: from Darwin to vermiculture*, (ed. J. Satchell), pp. 331–8. Chapman and Hall, London.

Tomlinson, E. J. (1976a) The production of single-cell protein from strong organic wastewaters from the food and drink processing industries. 1. Laboratory cultures. *Water Research*, **10**, 367–71.

Tomlinson, E. J. (1976b). The production of single-cell protein from strong organic wastewaters from the food and drink processing industries. 2. The practical and economic feasibility of a non-aseptic batch culture. *Water Research*, **10**, 372–6.

Tomlinson, E. J. and Chambers, B. (1979). Methods for prevention of bulking in activated sludge. *Water Pollution Control*, **78**, 524–38.

Tomlinson, T. G. (1941). The purification of settled sewage in percolating filters in series, with periodic change in the order of the filters: biological investigations 1938–1941. *Journal and Proceedings of the Institute of Sewage Purification*, 39–57.

Tomlinson, T. G. (1942). Some aspects of microbiology in the treatment of sewage. *Journal of the Society Chemistry and Industry, London*, **61**, 53–58.

Tomlinson, T. G. (1942). The nitrifying activity of biological film from different depths in percolating filters treating sewage by the processes of alternating double filtration and single filtration. *Proceedings of the Society of Agricultural Bacteriologists*, 5–7.

Tomlinson, T. G. (1946a). *Animal life in percolating filters. Identification of flies, worms and some other common organisms*. Water Pollution Research Technical Paper 9. Department of Scientific and Industrial Research. HMSO, London.

Tomlinson, T. G. (1946b). The growth and distribution of film in percolating filters treating sewage by single and alternating double filtration. *Journal and Proceedings of the Institute of Sewage Purification*, 168–83.

Tomlinson, T. G. and Hall, H. (1950). Some factors in the treatment of sewage in percolating filters. *Journal and Proceedings of the Institute of Sewage Purification*, 338–60.

Tomlinson, T. G. and Hall, H. (1951). Treatment of settled sewage in percolating filters: experiments at Minworth, Birmingham, 1947–1950. *Journal and Proceedings of the Institute of Sewage Purification*, 110–16.

Tomlinson, T. G. and Hall, H. (1953). Treatment of sewage by alternating double filtration with and without settlement of the primary effluent. *The Water and Sanitary Engineer*, **4**, 86–8.

Tomlinson, T. G. and Snaddon, D. H. M. (1966). Biological oxidation of sewage by films of micro-organisms. *International Journal of Air and Water Pollution*, **10**, 865–81.

Tomlinson, T. G. and Stride, G. O. (1945). Investigations into the fly populations of percolating filters. *Journal and Proceedings of the Institute of Sewage Purification*, 140–8.

Tomlinson, T. G. and Williams, I. L. (1975). Fungi. In *Ecological aspects of used water treatment*, Vol. 1, *The organisms and their ecology*, (ed. C. R. Curds and H. A. Hawkes), pp. 93–152. Academic Press, London.

Tomlinson, T. G., Loveless, J. E., and Sear, L. G. (1962). Effect of treatment in percolating filters on the numbers of bacteria in sewage in relation to the composition and size of the filling medium. *Journal of Hygiene, Cambridge*, **60**, 365–77.

Toms, I. P., Owens, M., Hall, J. A., and Mindenhall, M. J. (1975). Observations on the performance of polishing lagoons at a large regional works. *Water Pollution Control*, **74**, 383–401.

Tomson, M. B. *et al.* (1981). Groundwater contamination by trace organics from a rapid infiltration site. *Water Research*, **15**, 1109–16.

Toner, P. F., Clabby, K. J., Bowman, J. J., and McGarrigle, M. L. (1986). *Water quality in Ireland: the current position. I. General assessment.* AFF Report WR/G15. An Foras Forbartha, Dublin.

Topping, G. and McIntyre, A. D. (1972). *Benthic observations on a sewage sludge dumping ground.* ICES CM 72/E30. International Council for the Exploration of the Sea, Charlottenlund.

Torma, A. E., Walden, C. and Branion, R. M. R. (1970). Microbiological leaching of a zinc sulphide concentrate. *Biotechnology and Bioengineering*, **12**, 501–17.

Torno, H. C., Marsalek, J., and Desbordes, M. (1986). *Urban runoff pollution.* NATO ASI Series G10. Springer-Verlag, New York.

Torten, M., Birnbaum, S., Kingberg, M. A., and Shenberg, E. (1970). Epidemiological investigation of an outbreak of leptospirosis in Upper Galilee, Israel. *American Journal of Epidemiology*, **91**, 52–8.

Tourbier, J., Pierson, R. W., and Furia, E. W. (ed.) (1976). *Biological control of water pollution.* University of Pennsylvania Press.

Townsend, C. B. (1937). The elimination of the detritus dump. *Journal and Proceedings of the Institute of Sewage Purification*, (2) 58.

Townshend, A. R. and Knoll, H. (1987). *Cold climate sewage lagoons.* Report EPS 3/NR/1. Environment Canada, Ottawa.

Tromp, S. W. (1955). Possible effects of geophysical and geochemical factors aon development and geographic distribution of cancer. *Schweizensche Zeitschrft für allgemine Pathologie*, **18**, 929.

Truesdale, G. A., Wilkinston, R., and Jones, K. (1962). A comparison of the behaviour of various media in percolating filters. *Journal and Proceedings of the Institute of Sewage Purification*, (4), 325–40.

Tschörtner, U. S. (1967). The determination of chlorophyll *a* in algae and its application in South African oxidation ponds. I. *Water Research*, **1**, 785–93.

Tynen, M. J. (1966). A new species of *Lumbricillus* with a revised check list of the British Enchytraeidae (Oligochaeta). *Journal of the Marine Biological Association (UK)*, **46**, 89–95.

Uhlmann, D. (1969). Discussion on paper presented by Bopardikar, M.V. In *Advances in Water Pollution Research* (ed. S. H. Jenkins), 607–10. Pergamon Oxford.

Unwin, D. M. (1981). A key to the families of British Diptera. *Field Studies*, **5**, 513–53.

Unz, R. F. and Dondero, N. C. (1967*a*). The predominant bacteria in natural zoogloeal colonies. I. Isolation and identification. *Canadian Journal of Microbiology*, **13**, 1671–82.

Unz, R. F. and Dondero, N. C. (1967*b*). The predominant bacteria in natural zoogloeal colonies. II. Physiology and nutrition. *Canadian Journal of Microbiology*, **13**, 1683–94.

Unz, R. F. and Farrah, S. R. (1976). Exopolymer production and flocculation by zoogloea MP6. *Applied Environmental Microbiology*, **31**, 623–6.

Vacker, D., Connell, C. H., and Walls, W. N. (1967). Phosphate removal through municipal wastewater treatment at San Antonio, Texas. *Journal of the Water Pollution Control Federation*, **39**, 750–71.

Valiela, I. and Vince, S. (1976). Green borders of the sea. *Oceanus*, **19**, 10.

Vankova, S., Kasafirek, E., Kupec, J., and Mladek, M. (1980). Beurteilung der Hydrolyseaktivitat von Belbtschlamm mittels spezifischer chromogener Substrate. *Acta Hydrochimica et Hydrobiologica*, **10**, 15–22.

Varma, M. M., Christian, B. A., and McKinstry, D. W. (19740. Inactivation of Sabin oral poliomyelitis type I virus. *Journal of the Water Pollution Control Federation*, **46**, 987–92.

Vavilin, V. A. (1982). Models and design of aerobic biological treatment processes. *Acta Hydrochimica et Hydrobiologica*, **10**, 211–42.

Veen, W. L. van, Mulder, E. G., and Deinema, M. H. (1978). The *Sphaerotilus–Leptothrix* group of bacteria. *Microbiological Reviews*, **42**, 329–56.

Veerannan, K. M. (1977). Effect of sewage treatment by the stabilization pond method on the survival of intestinal parasites. *Indian Journal of Environmental Health*, **19**, 100–6.

Vega-Rodriguez, B. A. (1983). Quantitative evaluation of *Nocardia* spp. presence in activated sludge. M.Sc. Thesis, Department of Civil Engineering, University of California, Berkeley.

Velsen, A. F. M. van, and Lettinga, G. (1979). *Effect of feed composition on digester performance*. In First International Symposium on Anaerobic Digestion, pp. 113–30. University College, Cardiff.

Venosa, A. D. (1972). Ozone as a water and wastewater disinfectant: a literature review. In *Ozone in water and wastewater treatment* (ed. F. L. Evans), pp. 83–100. Ann Arbor Science Publishing, Michigan.

Venosa, A. D. (1983). Current state of the art of wastewater disinfection. *Journal of the Water Pollution Control Federation*, **55**, 457–66.

Viessman, W. and Hammer, M. J. (1985). *Water supply and pollution control*, (4th edn). Harper and Row, New York.

Vincent, A. J. and Critchley, R. F. (1983). *A review of sewage sludge treatment in Europe*. WRC Report 442M. Water Research Centre, Stevenage.

Vincken, W. and Roels, P. (1984). Hypersensitivity pneumonitis due to *Aspergillus fumigatus* in compost. *Thorax*, **39**, 74–5.

Vogel. (1978). *Textbook of quantitative inorganic analysis*, (4th edn). Longman, London.

Vogt, R. L. *et al.* (1982). *Campylobacter* enteritis associated with contaminated water. *Annals of Internal Medicine*, **96**, 292–96.

Vogtmann, H. and Besson, J. M. (1978). European composting methods treatment and use of farmyard manure and slurry. *Journal of Waste Recycling*, **19**, 15–19.

Vuillot, M. and Boutin, C. (1987). Waste stabilization ponds in Europe: a state of the art review. *Water Science and Technology*, **19** (12), 1–6.

Wagner, F. (1982). Study of the causes and prevention of sludge bulking in Germany. In *Bulking of activated sludge*, (ed. B. Chambers and E. J. Tomlinson), pp. 29–46. Ellis-Horwood, Chichester.

Waite, W. M. (1985). A critical appraisal of the coliform test. *Journal of the Institute of Water Engineers and Scientists*, **39**, 341–57.

Waitkins, S. A. (1985). From the PHLS: Update on leptospirosis. *British Medical Journal (Clinical Research)*, **290** (6480), 1502–3.

Waitz, S. and Lackey, J. B. (1959). Morphological and biochemical studies on the organisms *Sphaerotilus natans*. *Quarterly Journal of the Florida Academy of Science*, **21**, 336–60.

Waldichuk, M. (1985). Sewage for ocean disposal – to treat or not to treat? *Marine Pollution Bulletin*, **16**, 41–3.

Walker, A. P. (1982). Quantitative filament counting – a quick, simple method for prediction and monitoring of filamentous bulking. In *Bulking of activated sludge*, (ed. B. Chambers and E. J. Tomlinson), pp. 245–51. Ellis-Horwood, Chichester.

Walker, I. and Austin, E. P. (1981). The use of plastic porous biomass supports in a pseudo-fluidised bed for effluent treatment. In *Biological fluidised bed treatment of water and wastewater*, (ed. P. F. Cooper and B. Atkinson), pp. 272–84. Ellis-Horwood, Chichester.

Waller, C. B. and Hurley, B. J. E. (1982). Some experiences in the control of bulking of activated sludge. In *Bulking of activated sludge*, (ed. B. Chambers and E. J. Tomlinson), pp. 211–21. Ellis-Horwood, Chichester.

Wallis, P. M. and Lehmann, D. L. (ed.) (1983). *Biological health risks of sewage disposal in cold climates*. University of Calgary Press.

Walshe, J. (1988). Biological odour control: the solution to acute odour problems. In *Environmental Health Officers Association Year-book 1988* (ed. C. Smyth), pp. 95–7. Environmental Health Officers Association, Dublin.

Warn, A. E. (1982a). *The calculations of consent conditions for discharges using methods of combining distributions: users' handbook*. Anglian Water Authority, Huntingdon.

Warn, A. E. (1982b). Calculating consent conditions to achieve river quality objectives. *Effluent and Water Treatment Journal*, **22**, 152–5.

Warn, A. E. and Brew, J. S. (1980). Mass balance. *Water Research*, **14**, 1427–34.

Warren, C. F. (1971). *Biology and water pollution control*. W. B. Saunders, Philadelphia.

Waslien, C. I. (1975). Unusual sources of proteins for man. *Critical Review of Food Science and Nutrition*, **6**, 77–151.

Waslien, C. I., Calloway, D. H., and Margen, S. (1968). Uric acid production of men fed graded amounts of egg protein and yeast nucleic acid. *American Journal of Clinical Nutrition*, **21**, 892–7.

Waslien, C. I., Calloway, D. H., and Margen, S. (1969). Human intolerance to bacteria as food. *Nature (London)*, **221**, 84–5.

Wasserman, A. E., Hampson, J. W., and Alvare, N. F. (1961). Large-scale production of yeast in whey. *Journal of the Water Pollution Control Federation*, **33**, 1090–4.

Water Authorities Association (1984). Use and protection of the North Sea and Channel by English Water Authorities. Submitted to the International Conference on the Protection of the North Sea, August 1984, Bonn. *Effluent and Water Treatment Journal*, **24**, 331–4.

Water Data Unit (1979). *Water data, 1978*. Water Data Unit, Reading.

Water Pollution Control Federation (1961). *Operation of wastewater treatment plants. Manual of practice.* 11. Water Pollution Control Federation, Washington.

Water Pollution Research Laboratory (1955). *Report of the Director.* HMSO, London.

Water Pollution Research Laboratory (1967). *Storage of samples before BOD analysis.* WPRL Report 1158. Water Pollution Research Laboratory, Stevenage.

Water Research Centre (1971). *Nitrification in the BOD tests.* Notes on Water Pollution, Water Research Centre, Stevenage.

Water Research Centre (1977). *Bacterial pollution of bathing beaches.* Notes on Water Research 10. Water Research Centre, Stevenage.

Water Research Centre (1987). *Soil injection of sewage sludge: a code of practice.* WRC, Stevenage.

Watling, L., Leatham, W., Kinner, P., Wethe, C., and Maurer, D. (1974). Evaluation of sewage sludge dumping off Delaware Bay. *Marine Pollution Bulletin,* **5**, 39–42.

Watson, D. C. (1980). The survival of *Salmonella* in sewage sludge applied to arable land. *Water Pollution Control,* **79**, 11–18.

Watson, D. C., Satchwell, M., and Jones, C. E. (1983). A study of the prevalence of parasitic helminth eggs and cysts in sewage sludge disposed of to agricultural land. *Water Pollution Control,* **82**, 285–9.

Watson, J. L. A. (1962). Oxidation ponds and use of effluent in Israel. *Proceedings of the Institute of Civil Engineers,* **22**, 21–40.

Watson, W. and Fishburn, F. (1964). Insecticide resistance of filter flies. *Journal and Proceedings of the Institute of Sewage Purification,* 1964, 464–5.

Watson, W., Hutton, D. B., and Smith, W. S. (1955). Some aspects of gas liquor treatment on percolating filter beds. *Journal and Proceedings of the Institute of Sewage Purification,* 73–85.

Weaver, R. W., Dronen, N. O., Foster, B. G., Heck, F. C., and Ferhmann, R. C. (1978). *Sewage disposal on agricultural soils: chemical and microbiological implications. II. Microbiological implications.* EPA 600/2–78–1316. US Environmental Protection Agency.

Webber, M. D., Kloke, A., and Tjell, C. R. (1983). A review of current sludge use guidelines for the control of heavy metal contamination in soils. In *Processing and use of sewage sludge* (ed. P. L'Hermite and H. Ott). Reidal, The Netherlands.

Weddle, C. L. and Jenkins, D. (1971). The viability and activity of activated sludge. *Water Research,* **5**, 621–40.

Weissman, J. R., Eisenberg, D. M., and Benemann, J. R. (1979). Cultivation of nitrogen-fixing blue-green algae on ammonia depleted effluents from sewage oxidation ponds. In *Biotechnology in energy production and conservation,* (ed. C. D. Scott). John Wiley, London.

Welch, E. B. (1980). *Ecological effects of wastewater.* Cambridge University Press.

Weller, J. B. and Willetts, S. L. (1977). *Farm wastes management.* Crosby Lockwood Staples, London.

Wellings, F. M., Lewis, A. L., and Mountain, C. W. (1975). Pathogenic viruses may thwart land disposal. *Water and Wastes Engineering,* **12**, 70–4.

Weninger, G. (1964). Jahreszyklus der Biozonose einser modernen Brockentropf Korperanlage. *Wasser und Abwasser, Beitrage zur Gerwasserf.* **IV**, 96–167.

Weninger, G. (1971).. Das auflereten kleiner Metazoen bei Abbauprozessen in vergleichender Siecht. *Sitz.-Ber. Ost. Akad. d. Wiss.* **179**, 129–58.

Wentsel, R. S., O'Neill, P. E., and Kitchens, J. F. (1982). Evaluation of coliphage detection as a rapid indicator of water quality. *Applied Environmental Microbiology*, **43**, 430–4.

Wert, F. S. and Henderson, U. B. (1978). Feed fish effluents and reel in savings. *Water and Wastes Engineering*, **15**, 38–44.

Westlake, D. F. (1963). Comparisons of plant productivity. *Biological Reviews of the Cambridge Philosophical Society*, **38**, 385–425.

Wheatland, A. B. and Boon, A. G. (1979). Aeration and oxygenation in sewage treatment – some recent developments. *Progress in Water Technology*, **11**, 171–9.

Wheatley, A. D. (1976). The ecology of percolating filters containing a plastic filter medium in relation to their efficiency in the treatment of domestic sewage. Ph.D. Thesis, University of Aston in Birmingham, UK.

Wheatley, A. D. (1981). Investigations into the ecology of biofilms in waste treatment using scanning electron microscopy. *Environmental Technology Letters*, **2**, 419–24.

Wheatley, A. D. (1985). Wastewater treatment and by-product recovery. In *Topics in wastewater treatment. Critical reports on applied chemistry*, Vol. 11, (ed. J. M. Sidwick), pp. 68–106. Blackwell, Oxford.

Wheatley, A. D. (1987). Recovery of by-product and raw materials and wastewater conversion. In *Biotechnology of waste treatment and exploitation*, (ed. J. M. Sidwick and R. S. Holdom), pp. 173–208, Ellis-Horwood, Chichester.

Wheatley, A. D. and Williams, I. L. (1976). Pilot-scale investigations into the use of random-pack plastics filter media in the complete treatment of sewage. *Water Pollution Control*, **75**(4), 468–86.

Wheatley, A. D., Mitre, R. I., and Hawkes, H. A. (1982). Protein recovery from dairy industry wastes with aerobic biofiltration. *Journal of Chemical Technology and Biotechnology*, **32**, 203–12.

Wheatley, A. D., Winstanley, C. I., and Cassel, L. (1983). Biotechnology and effluent treatment. *Effluent and Water Treatment Journal*, **23**, 307–16.

Wheeler, B. D. and Giller, K. E. (1982). Species richness of herbaceous fen vegetation in Broadland, Norfolk, in relation to the quantity of above ground plant material. *Journal of Ecology*, **70**, 179–200.

Wheeler, D. R. and Rule, A. M. (1980). *The role of* Nocardia *in the foaming of activated sludge: laboratory studies*. Presented at Georgia Water and Pollution Control Association, Annual Meeting, Savannah, Georgia.

Whigham, D. F. and Simpson, R. L. (1976). The potential use of tidal marshes in the management of water quality in the Delaware River. In *Biological control of water pollution*, (ed. J. Tourbier and R. R. Pierson), pp. 173–86. University of Pennsylvania Press.

Whipple, G. C., Fair, G. M., and Whipple, M. C. (1927). *The microscopy of drinking water*. Wiley, New York.

Whipple, W. *et al.* (1983). *Stormwater management in urbanizing areas*. Prentice-Hall, New Jersey. 234 pp.

White, G. C. (1978a). *Disinfection of wastewater and water for reuse*. Van Nostrand Reinhold, New York.

White, J. B. (1978b). *Wastewater engineering*. Edward Arnold, London.

White, M. J. D. (1975). *Settling of activated sludge*. Technical Report 11. Water Research Centre, Stevenage.

White, M. J. D. (1976). Design and control of secondary settlement tanks. *Water Pollution Control*, **75**, 459–67.

White, M. J. D. (1982). Discussion on the application of settleability tests for the control of activated sludge plants. In *Bulking of activated sludge: preventative and remedial methods*, (ed. B. Chambers and E. J. Tomlinson), pp. 243. Ellis-Horwood, Chichester.

White, J. B. and Allos, M. R. (1976). Experiments on wastewater sedimentation. *Journal of the Water Pollution Control Federation*, **48**, 1741–52.

White, R. W. G. and Williams, W. P. (1978). Studies on the ecology of fish populations in the Rye Meads sewage effluent lagoons. *Journal of Fish Biology*, 13, 379–400.

Whitelaw, K. and Andrews, M. J. (1988). The effects of sewage sludge disposal to sea-the outer Thames estuary, U.K. *Water Science and Technology*, **20** (6/7), 183–91.

Widdows, J. (1985). Physiological responses to pollution. *Marine Pollution Bulletin*, **16**, 129–34.

Wightman, D., George, P. B., Zirschky, J. H., and Filip, P. S. (1983). High-rate overland flow. *Water Research*, **17**, 1679–90.

Wild, H. E., Sawyer, C. N., and McMahan, T. C. (1971). Factors affecting nitrification kinetics. *Journal of the Water Pollution Control Federation*, **43**, 1845–54.

Wile, I., Miller, G., and Black, S. (1982). *Design and use of articial wetlands*. Proceedings of the Workshop on Ecological Considerations in Wetlands Treatment of Municipal Wastewaters. University of Massachusetts, Amherst.

Wilkie, A., Reynolds, P. J., and Colleran, E. (1983). Media effects in anaerobic filters. In Proceedings of Anaerobic Treatment Symposium, pp. 242–58. TNO Corporate Communication Department, The Hague.

Wilkinson, R. (1958). Media for percolating filters. *Surveyor*, **117**(3433), 131.

Williams, A. (1983). A study of the nutrition of two sewage fungi: *Leptomitus lacteus* and *Fusarium aquaeductuum*. B.A. (Mod.) Thesis, Department of Botany, Trinity College, University of Dublin.

Williams, B. M. (1975). Environmental considerations in Salmonellosis. *Veterinary Record*, **96**, 318–21.

Williams, B. M. (1978). The animal health risk from the use of sewage sludge on pastures. Proceedings of the Conference on the Utilization of Sewage Sludge on Land, Oxford, p. 11. Water Research Centre, Stevenage.

Williams, B. M., Richards, D. W., and Lewis, J. (1976). *Salmonella* infection in the herring gull (*Larus argentatus*). *Veterinary Record*, **98**, 57.

Williams, J. H. (1979). *Utilization of sewage sludge and other organic manures on agricultural land*. EEC Symposium on Characterization, Treatment and Use of Sewage Sludge. Caderache, France.

Williams, J. I. and Pugh, G. J. F. (1975). Resistance of *Chrysosporium pannorum* to an organomercury fungicide. *Transactions of the British Mycology Society*, **64**, 255–62.

Williams, J. P. G. (1982). Introduction. In *Biotechnology: the financial implications*, (ed. J. P. G. Williams), pp. 1–10. City of London Polytechnic.

Williams, I. L. (1971). A study of factors affecting the incidence and growth rate fungi in sewage bacteria beds. M.Sc. Thesis, University of Aston in Birmingham, UK.

Williams, N. V. and Taylor, H. M. (1968). The effect of *Psychoda alternata* (Say.) (Diptera) and *Lumbricillus rivalis* (Levinsen) (Enchytraeidae) on the efficiency of sewage treatment in percolating filters. *Water Research*, **2**(2), 139–50.

Williams, N. V., Solbe, J. F. de L. G., and Edwards, R. W. (1969). Aspects of the distribution, life history and metabolism of the Enchytraeid worms *Lubricillus rivalis* (Levinsen) and *Enchtraeus corornatus* (N. & V.) in a percolating filter. *Journal of Applied Ecology*, **6**, 171–83.

Williams, P. A. (1981). The genetics of biodegradation. In *Microbial degradation of xenobiotic and recalcitrant compounds*, (ed. T. Laeisinger, R. Hutter, A. M. Cook, and J. Nuesch), pp. 3–45. Academic Press, London.

Williams, P. J. (1975). Investigations into the nitrogen cycle in colliery spoil. In *The ecology of resource degradation and renewal*. British Ecological Society Symposium. Blackwell, Oxford.

Williams, W. P., White, R. W. G., and Noble, R. P. (1973) In *Symposium on sewage effluent as a water resource*, pp. 45–53. Institution of Public Health Engineers, London.

Williams-Smith, H. (1970). Incidence in river waters of *Echerichia coli* containing R-factors. *Nature (London)*, **228**, 1286–8.

Williams-Smith, H. (1971). Incidence of R$^+$ *Escherichia coli* in coastal bathing waters in Britain. *Nature (London)*, **234**, 155–6.

Williamson, K. and McCarty, P. L. (1976). A model of substrate utilization by bacterial films. *Journal of the Water Pollution Control Federation*, **48**, 9–24.

Willson, G. B. and Dalmat, D. (1983). Sewage sludge composting in the U.S.A. *Biocycle*, **24**, 20–3.

Willson, G. B. *et al.* (1980). *Manual for composting sewage sludge by the Bettsville aerated-pile method*. US Environmental Protection Agency Report No. EPA–600/880–022.

Wilson, B. (1982). *The development of furnace/heat exchanger systems in which chopped cereal straw is the fuel*. Energy from biomass. Proceedings of the EEC Contractors Meeting, Brussels, May 1982.

Wilson, F. (1981). *Design calculations in wastewater treatment*. Spon, London.

Wilson, I. S. (1960). *Waste treatment*, Pergamon, London.

Wilson, J. G. (1988). *The Biology of Estuarine Management*. Croom Helm, London.

Wilson, T. E., Ambrose, W. A., and Buhr, H. L. (1982). *Operating experiences at low SRT*. Proc. Workshop on design and operation of large wastewater treatment plants. International Association of Water Pollution Reserval and Control.

Windle Taylor, E. (1966). *41st Report on the bacteriological, chemical and biological examination of the London Waters for the years 1963–64*. Metropolitan Water Board, London.

Winfield, B. A. and Bruce, D. A. (1981). The sewage farm reborn. *Public Health Engineer*, **9**, 185–6; 198.

Winkler, M. A. (1981). *Biological treatment of wastewater*. Ellis-Horwood, Chichester.

Winkler, M. (1984). Biological control of ntirogenous pollution in wastewater. In *Topics in enzyme and fermentation biotechnology*, Vol. 8 (ed. A. Wiseman), pp. 31–124. Ellis Horwood, Chichester.

Winkler, M. A. and Cox, E. J. (1980). Stalked bacteria in activated sludge. *European Journal of Applied Microbiology and Biotechnology*, **9**, 235–42.

Winkler, M. A. and Thomas, A. (1978). Biological treatment of aqueous wastes. In *Topics in enzyme and fermentation biotechnology*, (ed. A. Wiseman), pp. 200–79. Ellis-Horwood, Chichester.

Wishart, J. M., Wilkinson, R., and Tomlinson, T. G. (1941). The treatment of sewage in percolating filters in series with periodic change in the order of filters. 1. Results of operation of the experimental plant at Minworth, Birmingham, 1938 to 1940. 2. Biological investigations. *Journal and Proceedings of the Institute of Sewage Purification*, 15–76.

Witherow, J. L. and Bledsoe, B. E. (1983). Algae removal by overland flow process. *Journal of the Water Pollution Control Federation*, **55**, 1256.

Witt, E. R., Humphrey, W. J., and Roberts, T. E. (1979). *Full scale anaerobic filter treats high strength wastes*. Proceedings of the 34th Industrial Waste Conference, Purdue University, pp. 229–34.

Wolfe, R. S. (1971). Microbial formation of methane. *Advances in Microbial Physiology*, **6**, 107–46.

Wolverton, B. C. and McDonald, R. C. (1979). Upgrading facultative wastewater lagoons with vascular aquatic plants. *Journal of the Water Pollution Control Federation*, **51**, 305–13.

Wolverton, B. C., Barlow, R. M.,, and McDonald, R. C. (1976). Application of vascular aquatic plants for pollution removal, energy, and food production in a biological system. In *Biological control of water pollution*, (ed. J. Tourbier and R. W. Pierson), pp. 141–9. University of Pennsylvania Press.

Won, W. D. and Ross, H. (1973). Persistence of virus and bacteria in seawater. *Journal of the Environmental Engineering Division, American Society of Civil Engineers*, **99**, 205–11.

Wood, J. M. and Wang, H. K. (1983). Microbial resistance to heavy metals. *Environmental Science and Technology*, **17**(12), 582A–590A.

Wood, P. C. (1979). Public health aspects of shellfish from polluted waters. In *Biological indicators of water quality*, (ed. A. James and L. Evison), pp. 13.1–13.8. John Wiley, Chichester.

Woods, D. R., Green, M. B., and Parish, R. C. (1984). Lea Marston Purification lake: operational and river quality aspects. *Water Pollution Control*, **83**, 226–42.

Wooten, J. W. and Dodd, J. D. (1976). Growth of water hyacinths in treated sewage effluent. *Economic Botany*, **30**, 29–37.

WHO (World Health Organization) (1971). *International standards for drinking water* (3rd edn).

WHO (World Health Organization) (1973). *Re-use effluents: methods of waste water treatments and health safeguards*. Technical Report Series 517, WHO, Geneva.

WHO (World Health Organization) (1975). *Guides and criteria for recreational quality of beaches and coastal waters*. Regional Office for Europe, WHO, Copenhagen.

WHO (World Health Organization) (1981). *The risk to health of microbes in sewage sludge applied to land*. Euro Reports and Studies 54. WHO, Copenhagen.

WHO (World Health Organization) (1984). *Guidelines for drinking-water quality: 1. Recommendations*. WHO, Geneva.

Wright, C., Kominos, S. D., and Yee, R. B. (1976). Enterobacteriaceae and *Psuedomonas aeruginosa* recovered from vegetable salads. *Applied Environmental Microbiology*, **31**, 453–4.

Wright, E. P. (1982). The occurence of *Campylobacter jejuni* in dog faeces from a public park. *Journal of Hygiene, Cambridge*, **89** 191–4.

Wright, E. P. *et al.* (1983). Milk borne *Campylobacter* enteritis in a rural area. *Journal of Hygiene, Cambridge*, **91**, 227–33.

Wuhrmann, K. (1956). In *Biological treatment of sewage and industrial wastes*, (ed. J. McCabe and W. W. Eckenfelder), pp. 49–65. Reinhold, New York.

Wuhrmann, K. (1963). In *Advances in biological waste treatment*, (ed. W. W. Eckenfelder and J. McCabe), pp. 27–8, Pergamon, New York.

Wyatt, K. L., Brown, P., and Shabi, F. A. (1975). Oxidation processes in the activated sludge process at high dissolved oxygen levels in the application of chemical engineering to the treatment of sewage and industrial liquid effluents. *Institute Chemical Engineers Symposium Series 41*, E.

Yaziz, M. I. and Lloyd, B. J. (1979). The removal of salmonellas in conventional sewage treatment processes. *Journal of Applied Bacteriology*, **46**, 131–42.

Yen, C. H. (1965). *Cholera in Asia*. Proceedings of the Cholera Research Symposium. PHS Pub. 1328, 346.

Young, J. C. (1973). Chemical methods for nitrification control. *Journal of the Water Pollution Control Federation*, **45**, 637–46.

Young, J. C. and Clark, J. W. (1965). Growth of mixed bacterial populations at 20°C and 35°C. *Water and Sewage Works*, **112**, 251–5.

Young, J. C. and Dahab, M. F. (1983). Effects of media design on the performance of fixed-bed anaerobic filters. *Water Science and Technology*, **15**, 321–35.

Young, J. C. and McCarty, P. L. (1969). The anaerobic filter for waste treatment. *Journal of the Water Pollution Control Federation*, **41**, 160–71.

Young, J. C., McDermott, G. N., and Jenkins, D. (1981). Alterations in the BOD procedure for the 15th Edition of 'Standard Methods for the Examination of Water and Wastewaters'. *Journal of the Water Pollution Control Federation*, **53**, 1253–9.

Young, J. C., Thompson, L. O., and Curtis, D. R. (1979). Control strategy for biological nitrification systems. *Journal of the Water Pollution Control Federation*, **51**, 1824–40.

Zachvatkin, A. A. (1941). *Inst. Zool. Acad. Sci. Moscow, N.S.* **28**, 1–465.

Zehlender, C. and Boek, A. (1964). Wachstums und Ernahrungs bedingungen der Abwasserpitzes *Leptomitus lacteus* Ag. *Zentralblatt für Bakteriologie, Parasitenkunde, Infektionskrankheiten und Hygiene*, **117**, 399–411.

Zeikus, J. G. (1977). The biology of methanogenic bacteria. *Bacteriology Review*, **41**, 514–41.

Zeikus, J. G. (1979). *Microbial populations in digesters*. In 1st International Symposium on Anaerobic Digestion, pp. 61–90. University College, Cardiff.

Zeikus, J. G. (1980). Microbial populations in digesters. In *Anaerobic digestion* (ed. D. A. Stafford *et al.*), pp. 61–89. Applied Science Publishers, London.

Zeper, J. and De Man, A. (1970). *New developments in the design of activated sludge tanks with low BOD loadings*. IAWPR Conference, July 1970, San Francisco.

Zerinque, S. P., Rusoff, L. L., and Wolverton, B. C. (1979). *Water-hyacinth – a source of roughage for lactating cows*. February meeting of the American Dairy Science Association (Southern Division). New Orleans, Louisiana.

Zuckerman, M. H. and Molof, A. H. (1970). High quality reuse water by chemical-physical wastewater treatment. *Journal of the Water Pollution Control Federation*, **42**, 437–56.

Zvirbulis, E. and Hatt, H. D, (1967). Status of the generic name *Zoogloea* and its species. *International Journal of Systematic Bacteriology*, **17**, 11–21.

INDEX